IFMBE Proceedings Volume 36

Series Editor: R. Magjarevic

The International Federation for Medical and Biological Engineering, IFMBE, is a federation of national and transnational organizations representing internationally the interests of medical and biological engineering and sciences. The IFMBE is a non-profit organization fostering the creation, dissemination and application of medical and biological engineering knowledge and the management of technology for improved health and quality of life. Its activities include participation in the formulation of public policy and the dissemination of information through publications and forums. Within the field of medical, clinical, and biological engineering, IFMBE's aims are to encourage research and the application of knowledge, and to disseminate information and promote collaboration. The objectives of the IFMBE are scientific, technological, literary, and educational.

The IFMBE is a WHO accredited NGO covering the full range of biomedical and clinical engineering, healthcare, healthcare technology and management. It is representing through its 60 member societies some 120.000 professionals involved in the various issues of improved health and health care delivery.

IFMBE Officers
President: Herbert Voigt, Vice-President: Ratko Magjarevic, Past-President: Makoto Kikuchi
Treasurer: Shankar M. Krishnan, Secretary-General: James Goh
http://www.ifmbe.org

Previous Editions:

IFMBE Proceedings MEDITECH 2011, "International Conference on Advancements of Medicine and Health Care through Technology", Vol. 36, 2011, Cluj-Napoca, Romania, CD

IFMBE Proceedings BIOMED 2011, "5th Kuala Lumpur International Conference on Biomedical Engineering 2011"
Vol. 35, 2011, Kuala Lumpur, Malaysia, CD

IFMBE Proceedings NBC 2011, "15th Nordic-Baltic Conference on Biomedical Engineering and Medical Physics"
Vol. 34, 2011, Aalborg, Denmark, CD

IFMBE Proceedings CLAIB 2011, "V Latin American Congress on Biomedical Engineering CLAIB 2011"
Vol. 33, 2011, Habana, Cuba, CD

IFMBE Proceedings SBEC 2010, "26th Southern Biomedical Engineering Conference SBEC 2010 April 30 – May 2, 2010 College Park, Maryland, USA",
Vol. 32, 2010, Maryland, USA, CD

IFMBE Proceedings WCB 2010, "6th World Congress of Biomechanics (WCB 2010)",
Vol. 31, 2010, Singapore, CD

IFMBE Proceedings BIOMAG2010, "17th International Conference on Biomagnetism Advances in Biomagnetism – Biomag2010",
Vol. 28, 2010, Dubrovnik, Croatia, CD

IFMBE Proceedings ICDBME 2010, "The Third International Conference on the Development of Biomedical Engineering in Vietnam", Vol. 27, 2010, Ho Chi Minh City, Vietnam, CD

IFMBE Proceedings MEDITECH 2009, "International Conference on Advancements of Medicine and Health Care through Technology", Vol. 26, 2009, Cluj-Napoca, Romania, CD

IFMBE Proceedings WC 2009, "World Congress on Medical Physics and Biomedical Engineering",
Vol. 25, 2009, Munich, Germany, CD

IFMBE Proceedings SBEC 2009, "25th Southern Biomedical Engineering Conference 2009",
Vol. 24, 2009, Miami, FL, USA, CD

IFMBE Proceedings ICBME 2008, "13th International Conference on Biomedical Engineering"
Vol. 23, 2008, Singapore, CD

IFMBE Proceedings ECIFMBE 2008 "4th European Conference of the International Federation for Medical and Biological Engineering", Vol. 22, 2008, Antwerp, Belgium, CD

IFMBE Proceedings BIOMED 2008 "4th Kuala Lumpur International Conference on Biomedical Engineering",
Vol. 21, 2008, Kuala Lumpur, Malaysia, CD

IFMBE Proceedings NBC 2008 "14th Nordic-Baltic Conference on Biomedical Engineering and Medical Physics",
Vol. 20, 2008, Riga, Latvia, CD

IFMBE Proceedings APCMBE 2008 "7th Asian-Pacific Conference on Medical and Biological Engineering",
Vol. 19, 2008, Beijing, China, CD

IFMBE Proceedings Vol. 36

Simona Vlad · Radu V. Ciupa (Eds.)

International Conference on Advancements of Medicine and Health Care through Technology

29th August – 2nd September 2011, Cluj-Napoca, Romania

Springer

Editors

Simona Vlad
Electrotechnics Department
Technical University of Cluj-Napoca
Memorandumului 28
400114 Cluj-Napoca,
Romania
E-mail: Simona.Vlad@et.utcluj.ro

Radu V. Ciupa
Electrotechnics Department
Technical University of Cluj-Napoca
Memorandumului 28
400114 Cluj-Napoca,
Romania
E-mail: Radu.Ciupa@et.utcluj.ro

ISSN 1680-0737
ISBN 978-3-642-22585-7 e-ISBN 978-3-642-22586-4

DOI 10.1007/ 978-3-642-22586-4

Library of Congress Control Number: 2011931875

The IFMBE Proceedings is an Official Publication of the International Federation for Medical and Biological Engineering (IFMBE)

Typesetting & Cover Design: Scientific Publishing Services Pvt. Ltd., Chennai, India.

Printed on acid-free paper

9 8 7 6 5 4 3 2 1

springer.com

Foreword

Health systems are increasingly dependent on technologies. Today, appropriate infrastructure and medical equipment are indispensable for the delivery of effective preventive and curative health services.

In many exciting areas of biomedical research, our advancing knowledge promises to revolutionise the way disease is detected, treated and prevented. Interdisciplinary research is vital for fields such as structural biology, bioinformatics, post-genomics, nanotechnology and imaging to proceed.

Frequently however, the implications of technologies in terms of increasing recurrent costs, additional required support services, change in medical practice and training needs are underestimated. As a result, the widespread irrational use of technologies leads to a wastage of scarce resources and weakens health systems performance. To avoid such problems, a systematic and effective Health Technology System must be developed and introduced, requiring the support and commitment of decision makers of all levels of the health system.

The 3rd conference on Advancements of Medicine and Health Care through Technology - MediTech2011 aims to provide opportunities for the Romanian professionals involved in basic research, R&D, industry and medical applications to exchange their know-how and build up collaboration in one of the most human field of science and techniques. The conference is intended to be an international forum for researchers and practitioners interested in the advance in, and applications of biomedical engineering to exchange the latest research results and ideas in the areas covered by the topics.

Another objective is to improve communication and encourage collaboration among all health care professionals through the presentation and discussion of the new research and current challenges.

We believe the reader will find the proceedings an impressive document of progress to date in this rapidly changing field. Those who participated in this conference represent some of the leading authorities in the field. Their presentations and discussions have pinpointed current thinking and new leads for future work.

All papers submitted for presentation went through a review process and were evaluated by two reviewers from 12 countries (whose effort and hard work reflect their commitment and dedication to the profession). Those individuals whose papers were chosen for presentation at the conference submitted manuscripts to be published in these *Proceedings*.

Sincere appreciation and gratitude are expressed to members of the Scientific Advisory Committee for their diligent efforts in effecting a most timely and provocative conference. To document the current thinking in such a publication is most important and should serve as a useful guide for others working in the field.

Professor Radu V. Ciupa
MediTech2011 Conference Chair

Organization

Organizers

Romanian National Society for Medical Engineering and Biological Technology
Technical University of Cluj-Napoca

Partners

Medical University of Vienna
The University of Sheffield

Conference Chair

Radu V. Ciupa - *Technical University of Cluj-Napoca, Romania*

Honorary Committee

Radu Munteanu	Technical University of Cluj-Napoca, Romania, Rector, President of the Romanian Academy of Technical Sciences, Cluj-Napoca subsidiary.
Helmut Hutten	University of Technology Graz, Austria

Scientific Advisory Committee

Salem Abdel-Badeeh (Egypt)
Mariana Arghir (Romania)
Radu Badea (Romania)
Corina Botoca (Romania)
Cornel Brisan (Romania)
Alfonso Bueno (Spain)
Anca Buzoianu (Romania)
Alexandru Gh. Catana (Romania)
Radu Ciorap (Romania)
Radu V. Ciupa (Romania)
Hariton Costin (Romania)
Eugen Culea (Romania)
Giuseppe D'Avenio (Italy)
Vanessa Diaz-Zuccarini (UK)
Winfried Mayr (Austria)
Petru Mircea (Romania)
Ioana Moisil (Romania)
Alexandru Morega (Romania)
Mihaela Morega (Romania)
Calin Munteanu (Romania)
Andrew Narracott (UK)
Radu Negoescu (Romania)
Ramon Pallas-Areny (Spain)
Sever Pasca (Romania)
Dan Pitica (Romania)
Catalin Popa (Romania)
Dan V. Rafiroiu (Romania)
Marius N. Roman (Romania)

Gabriele Dubini (Italy)
John Fenner (UK)
Horia Hedesiu (Romania)
Rod Hose (UK)
Adrian Iancu (Romania)
Beriliu Ilie (Romania)
Ioan Jivet (Romania)
Josipa Kern (Croatia)
Matthias Krenn (Austria)
Mircea Leabu (Romania)
Patricia Lawford (UK)
Eugen Lupu (Romania)
Dan Mandru (Romania)
Janez Rozman (Slovenia)
Corneliu Rusu (Romania)
Marcel Rutten (The Netherlands)
Anne-Virginie Salsac (France)
Dan I. Stoia (Romania)
Mihai Tarata (Romania)
Daniela Tarnita (Romania)
Ioan Tarnovan (Romania)
Vasile Topa (Romania)
Mirela Toth-Tascau (Romania)
Mircea Vaida (Romania)
Kari Vehmaskoski (Finland)
Uludag Yildiz (Turkey)

Local Organizing Committee

Radu Badea
Catalin Curta
Laura Darabant
Victoria Man
Petru Mircea
Mihai S. Munteanu
Anca I. Nicu
Dan V. Rafiroiu
Marius N. Roman
Simona Vlad

Table of Contents

Health Care Technology

Medical Devices, Measurement and Instrumentation

Medical Imaging, Image and Signal Processing

Modelling and Simulation

Molecular Bioengineering

Biomechanics

AWARD – An Innovative Training Tool for Accessible Built Environment

C. Aciu, N. Cobirzan, and M. Brumaru

Technical University of Cluj-Napoca, Faculty of Civil Engineering, Cluj-Napoca, Romania

Abstract— **The goal of this paper is to present some of the results of the Leonardo da Vinci Project no. 07/0227-L/LLP-LdV-TOI-2007 - HU_001 "Accessible World for All Respecting Differences" - AWARD. The main objective of the project was to develop a teaching material in the field of barrier-free built environment by applying the concepts of universal design, in order to include the necessary knowledge in the curricula of vocational and higher technical institutions, thus contributing to changing mentality and attitude towards this important part of population. The teaching material is learner-oriented and may be used by a large range of vocations, also facilitating the Lifelong Learning Programme of practicing professionals and craftsmen.**

Keywords— **rehabilitation engineering, assistive technology, training tool, universal design, built environment.**

I. Introduction

We are all physically disabled at some time in our lives. Children, a person with a broken leg, a parent with a pram, an elderly person, are all disabled in one way or another. Those who remain healthy and able-bodied all their lives are few. It is important for the environment to be barrier-free and be adapted to fulfill the needs of all people equally. The needs of the disabled coincide with the needs of the majority. Planning for the majority implies planning for people with varying abilities and disabilities [1].

As defined by the Rehabilitation Act of 1973, Amended 1998 Rehabilitation engineering is the systematic application of engineering sciences to design, develop, adapt, test, evaluate, apply, and distribute technological solutions to problems confronted by individuals with disabilities in functional areas, such as mobility, communications, hearing, vision, and cognition, and in activities associated with employment, independent living, education, and integration into the community [2].

Rehabilitation engineers are key to the development and delivery of 'assistive technology', a term which refers to technologies and principles that meet the needs of and address the barriers confronted by individuals with disabilities in a range of life areas. Assistive technology is often associated with education, rehabilitation, employment, transportation, independent living, and recreation [2].

II. Accessible world for all respecting differences

Accessible World for All Respecting Differences (AWARD) was a 2 years Leonardo da Vinci Program (Project no. 07/0227-L/LLP-LdV-TOI-2007 –HU_001) aiming at the development of a teaching material in the field of barrier-free built environment [3].

The **project goal** was to produce electronic teaching material demonstrating the concepts of universal design in the built environment. This has been structured in a 'triple matrix system' (Environment/Element-Trade-Disability), covering the complexity of the related vocations and professions. These illustrate barriers to accessibility which are not recognized by the majority (92-94%) but prove insurmountable to people with disabilities and could easily be solved by responsible thinking professional and crafts people conversant with the principles of universal design. Examples of such illustrations will include accessibility requirements for wheelchair users and simulation of impaired vision.

The **objective** is to include the necessary knowledge in the curricula of vocational and higher technical institutes by providing learner-oriented teaching material for a large range of vocations and facilitating the LLL of practicing professionals and craftsmen. The consortium includes institutes of higher education of varying expertise in the built environment, an organization working in the field of sociological research and occupational therapy and an organization representing people with disabilities completed by five subcontracted vocational schools, evaluating the teaching material in practice. Some of the members have departments of nursing, health care, one also runs vocational training, an other one deliver short courses in vocational schools.

The work plan encompasses a complete suite of electronic teaching materials covering all related vocations as completely as possible. To raise empathic approach (in particular that of architects and decision makers) active participation of people with disabilities has been recorded in different environments.

Examples show maneuvers with wheelchair from different directions and from the perspective of wheelchair users as well as distorting optics to illustrate the problems of visually impaired people etc [4].

Apart from the traditional forms of teaching, the material can be used in distance learning and, due to the popular medium, it may also be used for individual, informal LLL. The teaching material is complemented by a handbook for teachers/tutors. The basic material in English is translated into the languages of the consortium members, resulting in national versions by adjusting the details to the national legislation in force and craftsmanship practice. The modules were tested in vocational institutes and the outcomes were disseminated at national and international seminars.

The innovative content and form of the electronic teaching material aims at the integration of the considerable minority of disabled and elderly people.

The main target groups of the AWARD software are:

- teachers/trainers of vocational schools;
- teachers in higher education institutions who want to get the basic information concerning the accessible built environment and attached design rules;
- professionals (designers, practicians) in any domain of the built environment;
- decision makers, any other actor involved in the adaptation and administration of the built environment.

Both, regular training and/or retraining of practicing craftspeople and professionals are considered.

Other partners bringing the own contribution to the project were:

- Vocational Secondary and High schools, Universities – concerned with the education in the field of the built environment in order to:
    - initiate and develop training (pilot) courses in the field of the accessible built environment, thus ensuring the complementarity of the studies in vocational education at all levels;
    - contributing to changing mentality towards disabled people and also increasing awareness concerning their right to a dignified, independent life;
    - contributing to the improvement of the teaching modules;
- Associations / Foundations concerned with the support and promotion in any form of the disabled persons rights;
- Voluntary partners: experts with experience in the field, for revising the elaborated materials, establishing contacts and promoting the implementation of the project results.

III. The program structure

There are many languages to create hypertext (Macromedia FLASH, Microsoft PowerPoint, Adobe PDF, Microsoft Compiled HTML Help ...).

The most popular way to share information is:

- creating hypertext;
- handling text, images, videos, ...;
- managing external documents (PDF, DOC, XLS, ...);
- having internal or external links (from one file to another);
- software-independent is using HTML (Hypertext Markup Language); it is the language of the Web sites, but is not designed to handle data-bases.

To elaborate different queries an interface was created using Visual Basic language under Windows O.S.

According with the Leonardo da Vinci project outline the electronic teaching material has to:

- be structured in a triple matrix system (Environment/Element, Trade, Disability) (Fig. 1);
- cover the complexity of the related trades;
- facilitate the search of information either on the basis of a defined part of the environment and/or on the basis of a relevant trade and/or on the basis of disabilities.

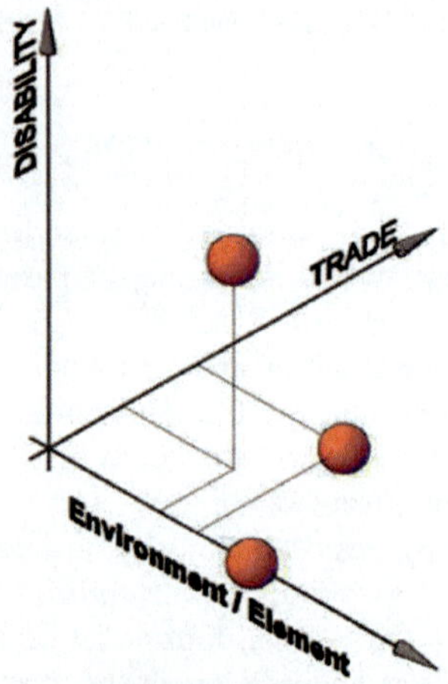

Fig. 1 The triple matrix system [3]

In the triple matrix system there are 3 possible options:

1. For a given **disability**, may be found different interrelated **environments** and **professions** that are involved in making those specific environments **accessible**.
2. Given the **profession/trades**, different **disabilities** and **environments** can be found, if practically correlated.
3. Given the **environment**, can be found the **professions** involved for solving the accessibility problems imposed by different **disabilities.**

For example, a possible combination might be: what should be done by a plumber in a bathroom for a blind user.

This modularized structure should facilitate the complex overview of the environment as well as (decomposing the structure by trades and disabilities) providing separate comprehensive teaching material for vocational institutes.

The allocation of modules is based on three macro categories:

- **Environment / Element** ($n_1 = 26$), represents the spaces where people with disabilities live, work, etc. or the equipment used for their daily activities (Table 1);
- **Trade** ($n_2 = 14$), the vocations to build the Environment/Element (bricklayer, building locksmith, upholsterer, tiler, carpenter, electrician, gardener, joiner, cabinet maker, painter, paver, plasterer, plumber, sign writer, system integrator);
- **Disability** ($n_3 = 4$), the different types of impairments suffered by people with disabilities: motor (mobility, strength, dexterity), visual (blind, low vision), hearing, cognitive (Alzheimer, Down,…).

Table 1 Environment / element [3]

Environment	Element	
Generic elements (which do not depend on the use of the space or are applicable in many space)	1.	Assistive technology (hoists, wheelchairs, etc.)
	2.	Automation, smart buildings
	3.	Communication, signage
	4.	Doors, corridors and windows
	5.	Elevators, lifts
	6.	Handrails
	7.	Lighting and visual impairments
	8.	Ramps
	9.	Safety in use, fire protection, evacuation
	10.	Stairs
	11.	Tactile and contrasting floor covering
Outdoor environment	12.	Sidewalks, trails and pathways
	13.	Pedestrian crossings
	14.	Kerb ramps
	15.	Parking
	16.	Public places, green areas
	17.	Public transport facilities
Indoor spaces	18.	Bath and WC
	19.	Bed, living and hotel
	20.	Classrooms (interactive communication), cinema, theatre, auditorium
	21.	Dining, restaurants
	22.	Historical buildings incl. archaeological areas
	23.	Kitchen
	24.	Lobby, reception, public service (e.g. post office, bank)
	25.	Offices, work places
	26.	Shops

IV. AWARD SOFTWARE APPLICATION

The program is an **interface** to give the user a quick access to the information request.

The software interface automatically supports different national languages (English, Italian, Hungarian, Romanian and Slovenian).

The first window shows general information regarding the project with the list of the European partners and their local coordinators (Fig. 2).

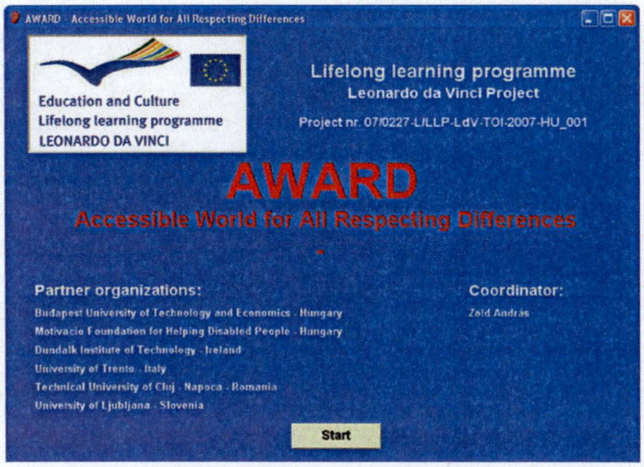

Fig. 2 AWARD programs interface [3]

The 'triple matrix system' (Fig. 3) enable one or more selections from the lists (ENVIRONMENT / ELEMENT; TRADE; DISABILITY).

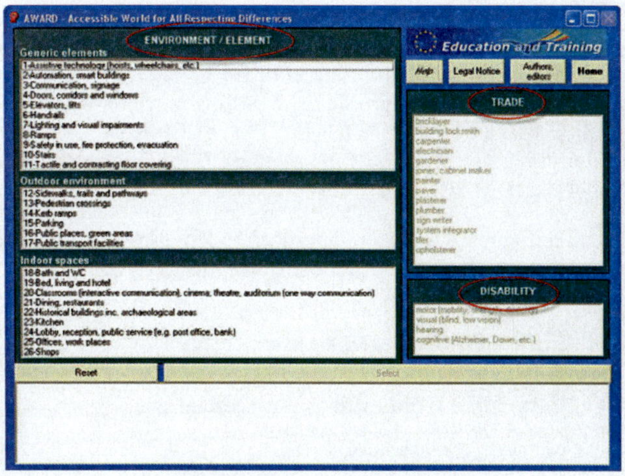

Fig. 3 The triple matrix system [3]

Possible choices:

1. User selects only one item from the 'ENVIRONMENT / ELEMENT' list;
2. User selects one item from the 'ENVIRONMENT / ELEMENT' list and one item from the 'TRADE' list;
3. User selects one item from the 'ENVIRONMENT / ELEMENT' list and one item from 'DISABILITY' list;
4. User selects one item from the 'ENVIRONMENT / ELEMENT' list, one item from the 'TRADE' list and one item from the 'DISABILITY' list (Fig. 4).

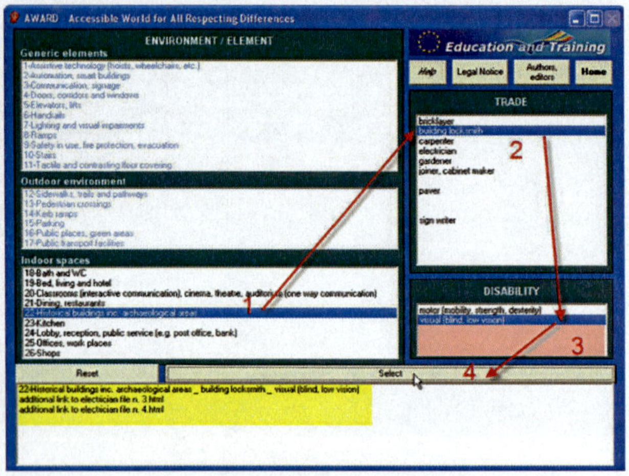

Fig. 4 Selection of the items [3]

Note 1: the software automatically shows only DISABILITIES having information for selected TRADE.
Note 2: it may be that there are multiple links to html files.

Pressing the graphical button 'Select' in the lower window will result in one or more titles appearing; these are hyperlinks to the appropriate html files containing information on how to solve problems regarding barriers to accessibility for the 'ENVIRONMENT / ELEMENT' selected from the point of view of the selected 'TRADE' and for the specific 'DISABILITY'.

To facilitate the empathic approach video clips illustrate the problems which are not recognizable for „average people". Clips illustrate the:

- maneuvers of wheelchair users;
- the use of tactile floor covering;
- the visual environment as it is seen in case of:
  - low visual acuity;
  - gun barrel vision;
  - macular degeneration.

V. Conclusions

Accessibility needs to put in practice a series of indicators which would be a true social integration of the differences resulting in a strategy to change attitudes. This would be the responsibility of the national and local public authorities and of the civil society as well.

The AWARD program is meant to be used by people involved in activities, of any kind and at any level, related to Accessibility and Universal Design.

In the program, best practice examples, standard solutions as well as unwanted (disappointing) examples of lack of concern and sloppiness are shown.

The main advantage of the Award program consists in the vast material, which covers (with solutions) the most important parts of the BUILT ENVIRONMENT / DISABILITY / TRADE; mainly, given the environment, can be found the professions involved for solving the accessibility problems imposed by different disabilities.

Acknowledgment

The AWARD software has been elaborated within the Leonardo da Vinci Program (Project no. 07/0227-L/LLP-LdV-TOI-2007 –HU_001).

References

1. Accessibility for the Disabled - A Design Manual for a Barrier Free Environment at www.un.org/esa/socdev/enable/designm/intro.htm
2. Rehabilitation engineering at www.engenhariadereabilitacao.net/ arquivo/Brochura_CNERAustraliano.pdf
3. Accessible World for All Respecting Differences (AWARD) at http://constructii.utcluj.ro/award/
4. Wheelchair Homes Design Guidelines at www.selondonhousing.org/Documents/ 080530%20WC%20guide%20May%2008.pdf
5. Building for Everyone, NDA – National Disability Authority at http://www.nda.ie/cntmgmtnew.nsf/0/EBD4FB92816E8BB480256C83 0060F761

Author: Aciu Claudiu
Institute: Technical University of Cluj-Napoca, Faculty of Civil Engineering
Street: 28 Memorandumului Street
City: Cluj-Napoca
Country: Romania
Email: claudiu.aciu@cif.utcluj.ro

The Development of the ALPHA Rollator

A. Abrudean

Drive Medical, Bistrita, Romania

Abstract— **After a short presentation of the most important mobility aids, details about rollators are given and few representative examples are described. The original design of ALPHA rollator is presented and the constructive and technological aspects are emphasized. The paper presents the developed prototypes and their advantages as well as the results of the experimental tests.**

Keywords— **Rollator, design, prototype.**

I. INTRODUCTION

People are unique amongst the members of the animal kingdom, succeeding in manipulating the environment to an extent never proven possible by any other living organism. Furthermore, we are the only species to ever develop two-legged walking. Bipedalism, meaning two-legged walking, is one of the characteristics of the human species, freeing the arms, opening 'the door' for being able to handle with the surrounding world [1].

For humans, maintaining the bipedal posture is thus essential for carrying out the everyday tasks in their homes, in the communities that they attend. Any deviation from normal walking and from the maintaining of the bipedal posture, which reduces the individual's skills of interacting with the surrounding world is called mobility deficiency.

Walking rehabilitation is an important step in the process of recovering and assisting mobility-deficient people. Both rehabilitation and assistance are performed by using the mobility equipment, which are devices designed to support walking or to improve people's mobility [2].

There are various mobility equipment, starting from the mere walking stick to crutches, crutches combined with a forearm support, walking frames, manual or electrical trolleys, elevators and other similar devices [3].

In their turn, the walking frames can be of several types : mere fixed metal frames, which are the most stable for walking and which can be endowed with 3 or 4 supporting points, walking metal frames, combined metal frames with wheels upfront, and metal frames with wheels in all their 3 or 4 supporting points, which are termed rollators, as well.

Figure 1 presents a fixed metal frame with 4 supporting points. The 4-point walking frames are similar, except that they have knuckles, which enable the movement of the side frames.

Fig. 1 Walking frame

The rollators enable the most fluid walking, as the walking movement is replaced by the movement of rotation by means of the wheels that touch the rolling surface. Yet, the person who uses the rollator has to possess the ability of controlling both himself/herself and the walking rate. In addition, the using people are provided the possibility of coordinating the action of the rollator brakes to the walking movement, if need arises. Generally speaking, a rollator can be beneficial for a balance problem person [4]. Figure 2 shows a few examples of rollators.

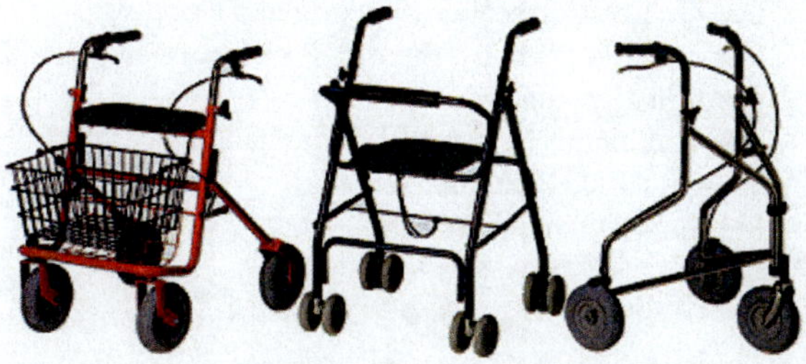

Fig. 2 Various rollators

This work here further focuses on this type of devices for mobility purposes.

The engineering part that tackles with the study, the development and the manufacture of these devices is named rehabilitation engineering.

II. THE ANALYSIS OF SIMILAR PRODUCTS

We will further review a few representative kinds of rollators now existing on the market [5]:

S. Vlad and R.V. Ciupa (Eds.): MEDITECH 2011, IFMBE Proceedings 36, pp. 5–8, 2011.
www.springerlink.com

A. Migo Rollator (Fig. 3)

- A steel-framed standard rollator
- Both indoor and outdoor utilizations
- Double function brakes : braking and stopping
- A very simple folding mechanism, with the handle within the seat
- The continuous adjustment of the handles height
- Weight (basket included) : 9.9 kg
- Maximum loading : 130 kg

Fig. 3 Migo Rollator

B. Let's Go Rollator (Fig. 4)

- An aluminum alloy-framed rollator
- Indoor utilizations
- Double function brakes : braking and stopping
- Very simple folding ; the rollator remains in a standing position after folding
- The continuous adjustment of the handles height
- Weight (basket included) : 6.9 kg
- Maximum loading : 100 kg

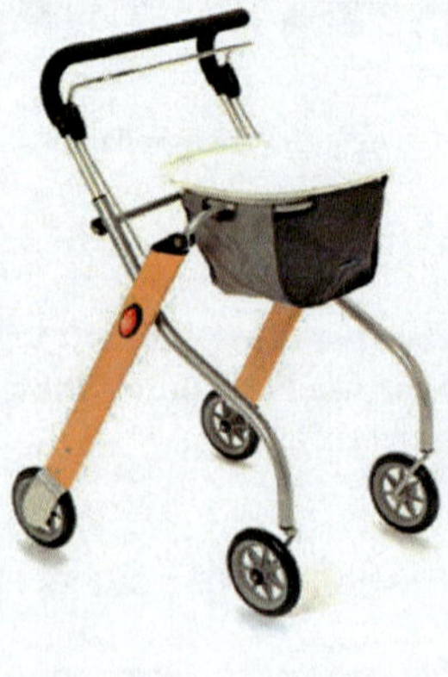

Fig. 4 Let's Go Rollator

C. Diamond Deluxe Rollator (Fig. 5)

- Easy to change from a rollator into a transporting trolley
- An aluminum alloy-framed rollator
- Both indoor and outdoor utilizations
- Double function brakes : braking and stopping
- Very simple folding, by bringing together the side frames
- The incremental adjustment of the handles height
- Weight (feet support included) : 10.3 kg
- Maximum loading : 130 kg

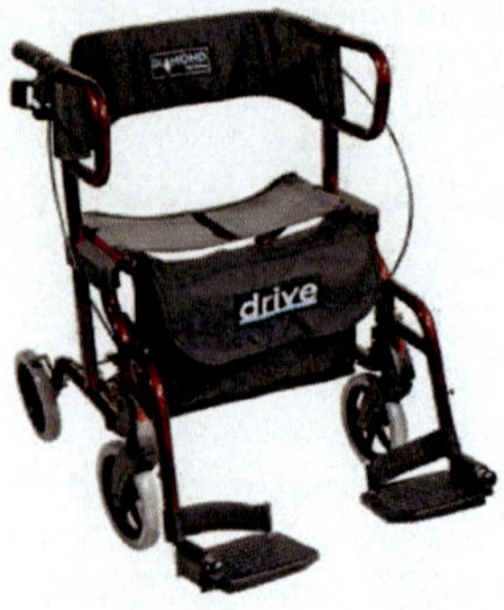

Fig. 5 Diamond Deluxe Rollator

III. DESCRIPTION OF THE CONSTRUCTIVE VARIANT

A. The Selection of the Constructive Variant

In selecting the constructive solution we took into account the requirements below, which we deemed to be prioritized by the potential customers when choosing a rollator:

- An aluminum alloy-framed rollator
- Both indoor and outdoor utilizations
- Braking cables hidden inside the frame
- The braking force – set up at the level of the wheel
- A large support of the folding system for a better stability
- A folding system with inter-blocking options in the open and closed positions
- A handle incremental height adjustment by using a string system for the free play elimination
- With an aim to have the least space occupied : the possibility of disassembling the front wheels and of folding the rear wheels jointly with part of the frame
- The possibility of overcoming big obstacles when used outdoors
- Weight : around 7 kg
- Maximum loading : 130 kg

B. The Design

In defining the design I tried to approximate the new designing line of the Scandinavian rollators (Fig. 6 and Fig. 7)), while taking into account the aforesaid requirements. For complying with the requirement related to the obstacle overcoming, I decided to use two sizes for the wheels, namely big wheels upfront (240 mm) and small wheels rearwards (200 mm), inspired from the construction of the Enduro motorcycles.

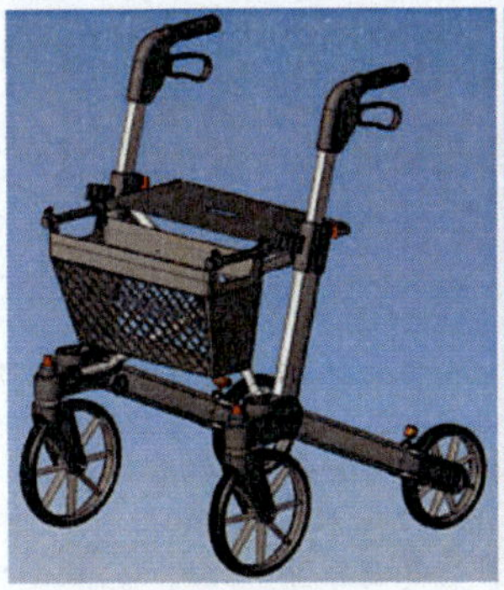

Fig. 6 Alpha Rollator – view 1

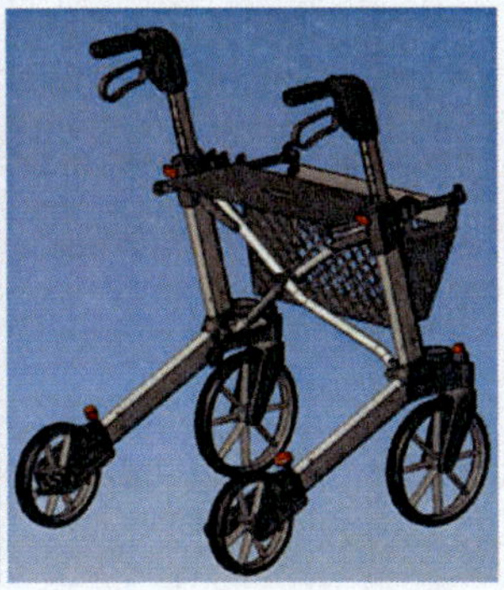

Fig. 7 Alpha Rollator – view 2

The product is prepared in such a way as by changing some sizes of the metal elements to obtain a family of products, that is to say an equal wheeled rollator or a small sized rollator meant to children or to short people, for whom the lower limit of adjusting the handles position is too high.

C. Technological Issues

Ever since the design phase I have taken into consideration the technological issues regarding the product manufacture, so as to have the lowest possible manufacturing costs . Taking into account these aspects, the materials selected for manufacturing do not call for special requirements and the processing operations needed for the elements involve low technological costs.

For the resistance frames I chose the aluminum alloy pipe, for the section thickness and form to be easy to manufacture and individualized at the same time. As for the plastic parts, the material was generally chosen after several simulations to various strength tests and material injection simulations, for any possible resistance problems that might subsequently come up to be removed ever since the design phase. Figure 8 sets out an example thereof.

My idea was to extensively use the plastic parts, for processing and manufacturing costs purposes, however aiming at the fact that the injection moulds needed for their creation should have a most simple form, wherever possible, and also in order to join as many parts as possible for the injection moulds needed. For creating the seat I used non water-retaining textile materials with a reduced specific elongation. As for the shopping basket, I used a combination of full material and net for an appearance that be both pleasant and functional, in order to be folded more easily.

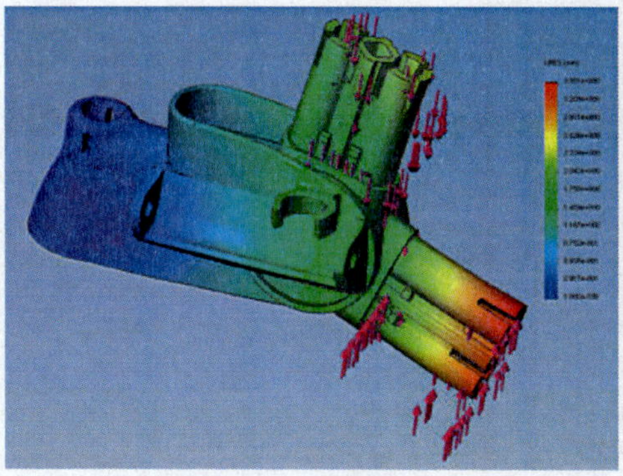

Fig. 8 Simulation of a plastic element deformation

D. The Prototypes Creation

An important step for the manufacture preparation is to make prototypes, in order to have a most accurate assessment of the future product. In this respect, for creating the plastic parts I used the rapid prototype method (Fig.9). The material used had properties similar to the material selected for the production, so that an 80 % testing could be performed in the end. Little form modifications were made as well, in order to be able to use aluminum alloy profiles that can be currently found on the market.

The prototypes were of an utmost importance, in order to have a first market feed-back concerning the new product.

Fig. 9 Prototype part

With an aim of having an as ample assessment as possible I made two versions of rollators : the basic version with unequal wheels and the derived version with equal wheels (Fig. 10 and Fig. 11).

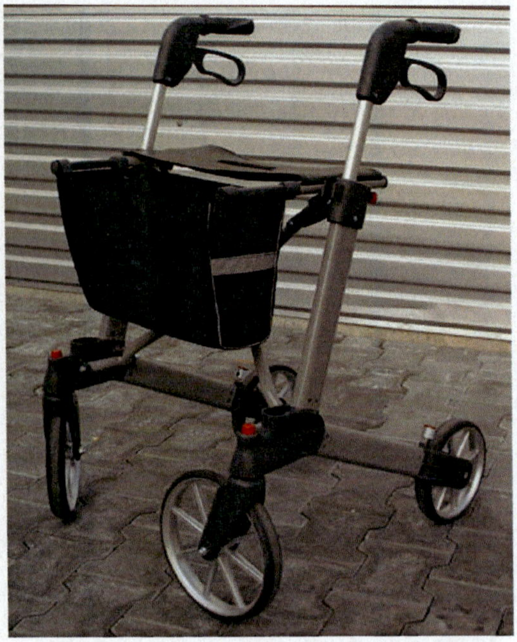

Fig. 10 The ALPHA prototype – the basic variant

Fig. 11 The ALPHA prototype – the two versions

E. The Testing Operations

For market placing purposes, the rollator has to comply with certain requirements and specific tests need to be carried out – the ones described in the ISO 11199-2:2005 standard [6]. These ones include, amongst others, the dynamic test, which consists in applying a force equaling 80 % of the maximum capacity on the handles. The number of cycles is 200,000 and the application frequency is 1 Hz. The latter test that I wish to specify is static and it consists in applying, for 5 seconds, a force equaling 120 % of the maximum capacity onto the handles raised in the maximum position, after which the plastic deformation should not exceed 1 % of the height with the handles raised in the maximum position.

F. Advantages

- A new constructive solution that meets all the requirements set out at section A.
- Simple manufacturing technologies, which lead to a low manufacturing cost.
- The creation of three different models, with no need to alter the technological flow.
- Easy mounting and demounting for storing or transporting purposes.
- Given that the braking cables are hidden on the inside, the possibility of their being damaged or hooked during walking is eliminated, or else accidents may take place.

G. Originality Elements

The product contains several original elements, the most important ones being the central plastic part, which comprises all the elements that link the other components, the solution for joining the frames with the plastic elements (the central part, the rear fork), with no visible exterior elements, a constructively very simple guiding system of the small bar, the braking force adjustment system and so forth.

IV. Conclusions

The release into production of a new product supposes a detailed study both from the standpoint of the new constructive solution and from the viewpoint of the market response to that respective product.

The market feedback after the prototypes display is of an utmost importance.

The simulations made during the product designing process bring forth the reduction of the remedial steps after the products creation for homologation testing purposes.

References

1. Hong, Y., Bartlett, R. (2008) Handbook of Biomechanics and Human Movement Science, Routlege, London.
2. Kommu, S. (2007) Rehabilitation Robotics, I-Tech Education and Publishing, Viena.
3. Zajicek, M., Brewster, S. (2004) Design principles to support older adults, Univ Access Inf Soc, 3:111-113
4. Mandru, D. et al (2009)., User centered design of rehabilitation engineering systems, Acta Technica Napocensis, series: Applied Mathematics and Mechanics, 52:. 475-480.
5. www.drivemedical.de
6. ISO 11199-2:2005

Ethical Issues in Nanomedicine

F. Graur[1,2], R. Elisei[1], A. Szasz[1], H.C. Neagos[1,2], A. Muresan[1], L. Furcea[1,2], I. Neagoe[2], C. Braicu[2], G. Katona[3], and M. Diudea[3]

[1] Surgical Clinic III, Cluj-Napoca, Cluj, Romania
[2] "Iuliu Hatieganu" University of Medicine and Pharmacy, Cluj-Napoca, Cluj, Romania
[3] Babes-Bolyai University, Cluj-Napoca, Cluj, Romania

Abstract— **Nanomedicine is a top developing domain. However, there are neither sound rules, nor a proper legislation that might regulate the ethics in this domain. Here we discuss the main topics pertaining to the ethics in nanomedicine at this time. From issues such as equity and discrimination, to profound modifications of the human body, far beyond the limit of human and non-human, there are many ethical aspects that can give rise to conflicting opinions.**

Keywords— **ethics, nanomedicine, nanotechnology.**

I. Introduction

Developing technologies that combine several areas have been constantly growing in recent years. Technologies that use nanostructures, nanotechnologies, are about to revolutionize the healthcare system in the near future.

Nanostructures will be able to act at molecular levels unattainable until now, revolutionizing medical therapies and therapeutic protocols. Nanotechnology and research in this area have experienced an extremely quick development in recent years, enjoying a very generous grant. Nanotechnology was able to break through the existing barriers across multiple disciplines and to become a multidisciplinary science which combines not only physical sciences, chemistry and engineering, but also molecular biology research and other healthcare segments.

II. Background

In the next decades nanotechnologies will revolutionize the healthcare system and the practice of medicine. There are drug delivery nanoparticles already in use, testing or development, and they will make it possible for us to reach targets such as chemotherapy, nanowiring for brain activity monitoring, nanoscaffolds for neural tissue repair and nanofiber brain implants. [1]

Nanomedicine refers to that side of the healthcare system that uses nanodrugs, which constitute a heterogeneous group of drugs that display, in general, unique properties, due to their nanoscale dimensions (nanometer to micron), many of these properties being fundamentally different from those of their macroscopic analogues.[2]

We will examine the development of nanotechnology and debate on the ethical issues to be considered when developing prudent policies regarding nanotechnologies.[1] The specific ethical regulations regarding nanomedicine should take into account all the ethical issues pertaining to human dignity, such as privacy, non-discrimination, informed consent, equity, etc. [3]

III. Equity and nondiscrimination

Ever since the idea came out, nanotechnology research has been enjoying an astounding development in many areas. The greatest progress and development was achieved by the pharmaceutical companies in drug delivery, domain which is already yielding significant results. Today, drug delivery covers over 78% of global sales in nanomedicine. At the moment, a new level of nanomedicine is in research, thanks to advances in pharmacogenetics and pharmacogenomics: the drug delivery system research was specifically targeted, this thrusting into limelight the concept of personalized medicine. [2] All this research is being done by the greatest pharmaceutical companies worldwide. In 2006 an amount of over 12.4 billion USD was spent, globally, by corporations, governments and venture capitalists, with 13% more than the one spent in 2005 and the amounts are increasing every year.[4] Specialists forecast that the nanotechnology pharmaceutical applications market will be around 18 billion dollars per year.[5] Even a cursory glance at the 2007 report will lead us to the conclusion that the US demand for nanomedical products will rise to over 53 billion US dollars in 2011 and to 110 billion in 2016.[6] Due to the fact that nanomedical products are extremely expensive, as the investors have to cover their investments, the nanomedical products market is expanding only in developed countries like the USA or Western European Countries. In developing countries, like those from Eastern Europe, the governments and the people cannot afford the nanomedical products. Thus, the following question arises: how ethical is it to implement very expensive nanomedical

S. Vlad and R.V. Ciupa (Eds.): MEDITECH 2011, IFMBE Proceedings 36, pp. 9–12, 2011.
www.springerlink.com

therapies for patients and medical systems? This situation is creating a gap between the state of the health care system in developed countries and that of the health care system in developing countries. Moreover, an increasing number of specialists are emigrating to Western European Countries to have access to new nanotechnologies in the medical care system. [7] This will ultimately lead to a lack of specialists in developing countries and to poor medical care and assistance for the people who live there.

In the short term, these new medical technologies will be out of reach for many people of lower socioeconomic status or for those in developing countries, because the price of the new products will be really high until the original patents on them expire and the respective generic products arrive on the market. However, in the future there is a possible scenario where only the rich have access to novel treatments, while the poor are denied even the knowledge of their diseases.[2] On the other hand, issues such as biopiracy, intellectual property theft and greed of the multinationals have been proffered as reasons for the unavailability of essential drugs to the poorest and neediest people in the world. Moreover, the Western companies make efforts to keep the prices high by continually citing the need to reward the innovation as a justification for stronger patent laws; that makes them spend more on reformulating preexisting drugs than on innovations.[8]

There is a fear regarding the future "nano-device" society, the divide between the rich and the poor being likely to expand because of expensive nanomedical drug treatments and huge implants prices.[9]

The massive investments in nanomedicine areas and the research and development in nanotechnology, as well as the lack of policy related to research funding and to educational programs have caused nanoethics to lag behind, and we should not have allowed this to happen.[10]

IV. Treatment versus Enhancement

The boundary between enhancement and therapy is rather relative, because both are based on the concept of the "normal", which is also very relative. Consequently, the following question arises: when should the novel medical procedure or treatment be considered therapy and when should it be viewed as enhancement? An answer to this question is needed because the novel medical procedures, therapy and diagnosis methods can be used to enhance the human body or mind.[2] It is difficult even for the scientists to draw a clear-cut line between therapy and enhancement.[11] On the other hand, Norm Daniels argues that any intervention designed to restore and preserve a species-typical level of functioning for an individual should count as therapy [12] and the rest as enhancement, while Eric Juengst states that therapy is health related, aimed at pathologies which compromise health, whilst enhancement aims at improvements which are not health related.[13]

All the science fiction scenarios of today are getting closer and closer to reality due to the novel technology which is creating nanoscale devices that can be inserted into the human body for diagnosis, treatment or enhancement. Here comes the ethical problem regarding the distinction between therapy and enhancement. In the case of the invention of an artificial oxygen boost red blood cell – called respirocyte – which holds a reservoir of oxygen, the question above will definitely arise.[14] These cells could help heart attack victims to survive and would allow their heart to continue breathing oxygen from these artificial cells until the emergency team can offer them proper medical treatment. But, at the same time, these cells could be used by an athlete to enhance his body and boost his performance.[15] How ethical and moral is it to use these cells for enhancement?

As regards the nano-chip implants, there may also appear some ethical issues, such as hybrid humans, which will be on the interface of human and machine. Nano-chip brain implants may lead to the "Lego set mentality" [16] of "Nanomed-human beings". There is also the ethical issue of affecting the human brain with these nanochip brain implants which may go beyond the healing purpose and impact on the preservation of human identity [9], this being one of the ethical principles pertaining to human dignity that we mentioned above.

V. Toxicity and Biocompatibility of Nanostructures

The most highly developed sector of nanomedicine is the nanodrug delivery and therapy system, which uses a wide range of materials to deliver active agents to different parts of the human body. The behavior of nanomaterials is often unpredictable, because they may behave differently in vivo as compared to in vitro: nanoparticles can disintegrate into smaller particles that are toxic to the human body, or they may aggregate into larger particles as well.[17] Because of that, it is ethically desirable to design short- and long-term studies to determine whether nanomedicines really are more effective and safer for humans than conventional drugs. Nevertheless, when trying to run those trials, there may appear some other difficulties, such as problems with comprehension and with understanding as regards the informed consent, given the complexity of nanotechnologies. Moreover, the long-term effects of using nanomedicines still are, to this day, largely unknown.[2]

To ensure that the new drugs can be used safely by the people, nanomedicine companies should be required to conduct long-term studies on nanomedical products following their introduction on the market: post-marketing surveillance or Phase IV studies. But this type of studies is poorly practiced because they are not required by current laws. Although it is clear that new regulations should be introduced, this is to be done with care, so as to prevent overregulating tendencies, because those would have a chilling effect on research, development, commercialization efforts and fair access of the public to nanomedicines.[18]

VI. Informed consent

Nanotechnology risks communication problems between the promoters of the study and the subjects participating in the research; therefore, conflicts with the public opinion are often a difficult challenge. Applicable laws and ethical rules require investigators to inform potential subjects or their legal representatives of the study's objectives, of the procedures to be performed, of the benefits, of the risks the participants may be subject to during the study, of the alternatives that they have, of the privacy protection that they are offered and of any additional information that the subject needs in order to decide whether or not to take part in the study.[19] The participants most often underestimate the risks to which they subject themselves by taking part in such a study, and are likely to overestimate the benefits.[20] During the study of nanomaterials, investigators must inform subjects of the possible existence of risks that may not be obvious at that particular time.[21]

VII. Boundary between living and non-living

When we start using nanomaterials in medicine, some fundamental questions about human nature and human enhancement are bound to arise.[22] One such question refers to how many implantable nanodevices it would take for a person to no longer be considered a human being.[2] Such questions spring to mind when one imagines futuristic scenarios about our nanomedical future: the "nanofuture". The conventional boundary between living and non-living is blurred in these scenarios, because, due to the socio-economic inequity, unfair competition and discriminations might raise some other ethical issues. We can consider, for instance, exactly who is eligible for enhancement and who is not. Should military people and athletes be considered morally eligible to have some parts of their bodies enhanced to increase their body performance, whilst other people should not?[2]

VIII. Hyperdiagnosis and hyper therapy

In a medical future where a single chip will be able to check every other second if there are some abnormalities in the body cells or fluids and in the blood flow, and also to detect any mutation on the gene level, there arises the question of what a disease actually implies. Is it enough to detect an individual defective cell or a minor abnormal change in the blood chemistry to set up a diagnostic? In this context, we have to reconsider what the syntagm "healthy person" means and what the phrase "a person who has a disease" refers to, as well as how far, how deep to molecular level, we should go when establishing a diagnosis. This new kind of nanodevices, or nanochips (lab on a chip) which can detect almost any abnormality in the human body, from the blood chemistry to the ability to detect a single cancerous cell, cannot only diagnose current illnesses or diseases, but can also analyze the DNA for diseases that may appear in the future.[23] These nanochips can generate panic, increased anxiety and fear about illness, being thus likely to cause psychosocial harm. What is more, these nanodevices, which are real repositories of medical information about the health status of any one of us, must not be accessible to insurance companies, because, if they are, the harm such devices may cause will outweigh their potential benefits.[2] Consequently, inevitable ethical issues arise, issues concerning the patient's right to know, right not to know and duty to know, which are inherent to the principle of autonomy.[24] On the other hand, nanochip implants which monitor a person's health raise serious privacy concerns, for instance in what regards those who monitor the information from the chip, those who have access to all this information, and whether or not these chips may be used for other purposes than the medical ones, purposes such as enhancement, for example.[25] Nanomedicine aims to provide all the relevant medical information regarding the patient, including the prevision of malignant tumors for example, by means of this "lab-on-a-chip", which is able not only to read the human genetic code, but also to issue forecasts about the medical future of the patient.[16]

IX. Responsibility

In the nanomedical era, fast diagnoses and speedy treatment possibilities will be common. What if this leads to mistaken diagnoses and inappropriate treatment for diseases that, maybe, do not actually exist in a certain patient's case? How far should we trust the new discoveries in nanomedicine? What if there appear on the market instruments for a nanomedical entire body scan that may provide diagnostics and treatment, but which could be bought by people as easily as the blood pressure measure instruments used today? The problem is that people will avoid going to the doctor if they are able to buy "their personal nanomedical doctor", due to the fear and high costs of medical assistance and treatment, and of medical insurance. [2]

There is also the question regarding the doctors' responsibility, which might be shifting away from the doctors to the patients, as the medicine of the future will be minimally invasive and will increase the patients' autonomy by reducing the amount of personal interaction.[25]

X. Conclusion

As Ferrari et. al. conclude, we should not take our concerns to the extreme by over-analyzing nanomedicine ethical issues: "the greatest risk in nanomedicine may well be in letting our concerns paralyze our action and not taking advantage of the full, revolutionary potential that nanotechnology in medicine can offer humankind" [26].

The future innovations in nanomedicine should be analyzed from an ethical, political and social point of view. The most important thing for a proactive analysis of the new nanotechnologies is to make sure that the proper authorities discover and debate, from as early as the project phase, the ethical issues that can appear. In order for this to become reality, a very tight collaboration is required between those who are developing the novel nanotechnologies and the ethics specialists. This is the only way to ensure that the nanomedical technologies will get the ethical and social acceptance of the people. [27]

References

1. Resnik D, Tinkle S. (2007) Ethics in Nanomedicine. Nanomedicine (Lond). June; 2(3): 345–350.
2. Bawa R and Johnson S. (2008). Emerging issues in nanomedicine and ethics. In: Allhoff F and Lin P (editors.), Nanoethics: Emerging Debates, Springer, Dordrecht (in press)
3. Reisch MS: (2007) Nano goes big time. Chemical & Engineering News 85(4), 22-25.
4. Reflections on ethical aspects of nanomedicine. http://www.capurro.de/nanoethics.html
5. Hunt W. (2004) Nanomaterials: nomenclature, novelty and necessity. Journal of Materials. Vol 56 no. 10: 13-18.
6. Report. (2007) Nanotechnology in Healthcare. The freedonia Group, Inc. Cleveland, Ohio
7. Chen, L.C. and Boufford, J.I. (2005) Fatal flows-doctors on the move. N Engl J Med 353, 1850-1852
8. Saini A (2007) Making the poor pay. New Scientist 193(2597), 20
9. Capurro R: Reflections on Ethical Aspects of Nanomedicine http://www.capurro.de/nanoethics.html
10. Cameron NM de S (2006).The NELSI Imperative: nano ethics, legal and social issues and federal policy development. Nanotechnology Law & Business 3(2), 159-166
11. Naam (2005), More Than Human. Broadway Books, New York. p.5 www.morethanhuman.org.
12. Daniels Norm (1992) Growth Hormone Therapy for Short Stature: Can We Support the Treatment/Enhancement Distinction? Growth: Genetics & Hormones 8.S1: 46-8.
13. Juengst Eric (1997) Can Enhancement Be Distinguished from Prevention in Genetic Medicine?, Journal of Medicine and Philosophy 22 125-42.
14. Freitas Robert A. Jr. (1998) Exploratory Design in Medical Nanotechnology: A Mechanical Artificial Red Cell Artificial Cells, Blood Substitutes, and Immobil. Biotech. 26411-430.
15. Lin P, Allhoff F (2006) Nanoethics and Human Enhancement:A Critical Evaluation of Recent Arguments. Nanotechnology Perceptions 2 (2006) 47–52
16. Bruce D: Ethical Issues in Nano-Medicine and Enhancement : an Overview. www.edinethics.co.uk/nano/nanoeth9a-med.doc
17. Oberdorster G, Oberdorster E, Oberdorster J (2005). Nanotoxicity: an emerging discipline evolving from studies of ultrafine particles. Environ. Health Persp. 113, 823-839
18. Strom B (2006) How the U.S. drug safety system should be changed. JAMA 295, 2072-2075
19. Donaldson K. (2006) Resolving the nanoparticles paradox. Nanomedicine. ;1:229–34.
20. Ethics and Nanomedicine – Analysis Of The Issues and Principles To Be Faced By The Medical Application of Nanotechnology. http://www.azonano.com/Details.asp?ArticleID=1705
21. Resnik D, Tinkle S. (2006) Ethical issues in clinical trials involving nanomedicine. Contemp Clin Trials. Nov 17; Considers ethical issues in nanomedicine clinical trials.
22. Beyond Therapy (2003) Biotechnology and the Pursuit of Happiness. President's Council on Bioethics http://www.bioethics.gov/reports/beyoundtherapy/
23. Kuiken T (2011) Nanomedicine and ethics: is there anything new or unique? Wiley Interdisciplinary Reviews: Nanomedicine and Nanobiotechnology. Volume 3, Issue 2, pages 111–118, March/April
24. Evers J, Aerts S, De Tavernier J. (2008) An ethical argument in favor of nano-enabled diagnostics in livestock disease control. Nanoethics, 2:163–178.
25. Spagnolo AG, Daloiso V. (2008) Outlining ethical issues in nanotechnologies. Bioethics, 23:394–402.
26. Ferrari M, Philibert MA, Sanhai WR. (2009) Nanomedicine and society. Clin Pharmacol Ther, 85:466–467.
27. The Institute of Medicine: The Future of Drug Safety: Promoting and Protecting the Health of the Public. www.nap.edu/catalog.php?record_id=11750#toc

Author: Graur Florin
Institute: University of Medicine and Pharmacy Cluj-Napoca
Street: Croitorilor 19-21
City: Cluj
Country: Romania
Email: graurf@yahoo.com

Innovative Technologies in Vitreo-Retinal Surgery for Rhegmatogenous Retinal Detachment

S.D. Talu[1], S. Rus[2], I. Tamasoi[2], and C. Dragos[2]

[1] "Iuliu Hatieganu" University of Medicine and Pharmacy/Ophthalmology, Cluj-Napoca, Romania
[2] Emergency County Hospital/Ophthalmology, Cluj-Napoca, Romania

Abstract— **Vitreo-retinal surgery techniques have significantly improved over time and increased the success rates in the surgery of Rhegmatogenous Retinal Detachment (RRD). The purpose of this study is to evaluate the results of the primary vitrectomy for RRD. We have conducted a retrospective study on all the RRD cases that we have operated on by pars plana vitrectomy between October 2009 – October 2010 (70 cases). The description of the surgical techniques concerns the vitrectomy itself and the associated maneuvers: drainage of the subretinal fluid, retinopexy (endolaser/exocryo), internal tamponade. The peroperative complications are presented and the results are evaluated 3 months after surgery as: success, simple recurrence, Proliferative Vitreo-Retinopathy (PVR), Cystoid Macular Edema (CME). All the posterior vitrectomies have been performed with the Accurus machine (Alcon). We used the 20 gauge system and a high speed vitrectomy probe (2500 cuts/minute). The vitrectomy has been as complete as possible in all situations, up to the vitreous base. The subretinal fluid has been drained by PerFluoroCarbon Liquids (PFCL) injection in 65 cases (92.85%) or by fluid/air exchange in the remaining 5 cases (7.14%). The retinopexy has been performed with a cryoprobe in 46 cases (65.71%) and with the endolaser fiber in 24 cases (34.28%). The endolaser cerclage has been associated in 26 cases (37.14%) and we used the silicone oil for the internal tamponade in all the situations. We have accidentally injured the retina in 5 cases (7.14%) and touched the lens in 4 cases (5.71%). Success has been achieved in 58 cases (82.85%), simple recurrence has been identified in 8 cases (11.42%), PVR in 3 cases (4.28%), CME in one case (1.42%). The technical advances are emphasized (high-speed cutting rate, 3D system), with implications in the surgery of RRD. The modern vitrectomy techniques have proven their efficacy in the treatment of RRD.**

Keywords— **vitreo-retinal surgery, rhegmatogenous retinal detachment.**

I. Introduction

About 100 years ago, Rhegmatogenous Retinal Detachment (RRD) was essentially untreatable, with an estimated success rate of 1 in 1000 [1].

The purpose of the treatment in RRD is to reattach the retina and maintain it attached. The means to achieve this goal vary according to the type of detachment and the experience of the surgeon. As the surgeons became more familiar with vitrectomies in resolving cases of vitreous pathology and complex retinal detachments, the advantages of an internal approach (vitreous cavity) could also be useful for simpler cases [2].

Vitreo-retinal surgery techniques have significantly improved over time and increased the success rates in the surgery of RRD, as compared to the traditional methods. Lately, the vitrectomy techniques have became the indication of choice in RRD for most surgeons, as the cause of the disease is addressed directly: in the vitreous cavity [3].

II. Purpose

The purpose of this study is to emphasize the major impact of the innovative technologies in vitreo-retinal surgery for RRD, as revealed by our personal experience.

III. Method

We have conducted a retrospective study on all the RRD cases that we have operated on by pars plana vitrectomy between October 2009 – October 2010.

The following parameters are analyzed: the lens status (phakic – the natural lens present, pseudophakic – artificial lens, aphakic – the absence of any lens), the retinal detachment extent (number of detached quadrants), number of retinal tears, size of the worst tear, aggravating conditions (choroidal detachment, hypotony, vitreous hemorrhage), PVR stage (0, A, B, C1). Grade A is limited to the presence of vitreous cells or haze. Grade B is defined by the presence of rolled or irregular edges of a tear or inner retinal surface wrinkling, denoting subclinical contraction. Grade C is recognized by the presence of preretinal or subretinal membranes. Grade C is further delineated as being anterior to the equator (grade C1) or posterior to the equator (grade C2) [4].

The description of the surgical techniques concerns the vitrectomy itself and the associated maneuvers: drainage of the subretinal fluid, retinopexy (endolaser/cryopexy), choice of the internal tamponade.

S. Vlad and R.V. Ciupa (Eds.): MEDITECH 2011, IFMBE Proceedings 36, pp. 13–18, 2011.
www.springerlink.com

The peroperative complications are presented and the results are evaluated 3 months after surgery, as follows: success, simple recurrence, Proliferative Vitreo-Retinopathy (PVR), Cystoid Macular Edema (CME).

IV. RESULTS

Between October 2009 – October 2010, we have performed the primary vitrectomies in 70 RRD cases.

In 42 cases (60%) the patients were phakic, in 25 cases (35.71%) they were pseudophakic and in the remaining 3 cases (4.28%) they were aphakic.

The extent of the RRD is illustrated in table 1.

Table 1 Extent of RRD

RRD extent (quadrants)	Number of cases	%
1	10	14.28
2	22	31.42
3	25	35.71
4	13	18.57

The retinal detachment has been caused by a single tear in 55 cases (78.57%), while multiple retinal tears have been identified in 15 cases (21.42%). The data regarding the size of the worse retinal tear are presented in table 2 and the PVR classification – in table 3.

Table 2 Size of the worse retinal tear

Size of the worse retinal tear	Number of cases	%
Hole	10	14.28
Normal	45	64.28
Large	12	17.14
Giant	3	4.28

A hole is a round defect in all the retinal layers, with a diameter of no more than 1- 2 mm. A normal retinal tear is larger than a hole, but smaller than 90 degrees. The dimension of a large retinal tear is comprised between 90 – 180 degrees and a giant retinal tear is defined by a more than 180 degrees size.

Table 3 PVR classification

PVR stage	Number of cases	%
0	0	0
A	44	62.85
B	18	25.71
C1	8	11.42

The PVR develops as a complication of the RRD and is the biggest obstacle to successful retinal reattachment surgery. The name is derived from *proliferation* (of the pigment epithelial and glial cells) and *vitreo-retinopathy* (to include the involved tissues: the vitreous humour and the retina). The factor that causes the PVR is the retinal tear. The pigment epithelial cells get into contact with the vitreous cavity through the tear and initiate a fibrous proliferation that makes the retina rigid, immobile and less compliant [4]. Therefore, the RRD has to be operated as soon as possible, before the PVR becomes advanced (stage C).

Vitreous hemorrhage has been associated in 16 cases (22.85%).

The posterior vitrectomy has been performed in all cases with the Accurus machine (Alcon), which works with a Venturi pump. We used the 20 gauge system and a high speed vitrectomy probe (2500 cuts/minute). The vitrectomy has been as complete as possible in all situations, up to the vitreous base. We did not associated the Internal Limiting Membrane (ILM) peeling and we did not perform retinotomies for the subretinal fluid drainage. It has been drained by the injection of PerFluoroCarbon Liquids (PFCL) in 65 cases (92.85%) or by fluid/air exchange in the remaining 5 cases (7.14%).

The retinopexy has been performed externally (with a cryoprobe) in 46 cases (65.71%) and internally (with the endolaser fiber) in 24 cases (34.28%). The endolaser cerclage has been associated in 26 cases (37.14%) and we used the silicone oil for the internal tamponade in all the situations.

Regarding the per-operative complications, we have accidentally injured the retina in 5 cases (7.14%) and touched the lens in 4 cases (5.71%).

The 3 months results are illustrated in table 4.

Table 4 The 3 months results

The 3 months result	Number of cases	%
Success	58	82.85
Simple recurrence	8	11.42
PVR	3	4.28
CME	1	1.42

V. DISCUSSION

Over the past 10 years, vitrectomy techniques have undergone a revolution, as a result of a rather intuitive than science-based approach [5]. High-speed cut rates and the 3D technology (dual dynamic) have created the optimal conditions for the surgeons to perform with maximum efficiency [5]. These progresses have been validated particularly in the

RRD surgery, where the vitrectomy needs to be as complete as possible, in an eye with a detached, mobile retina.

Posterior vitrectomy has obvious advantages in rhegmatogenous retinal detachment surgery over the ab externo techniques: it allows the complete evaluation and removal of the vitreous base (fig. 1) and the identification of all the retinal breaks; it removes the media opacities (fig. 2); the subretinal fluid is drained under direct control; endolaser retinopexy is made possible and the eye refraction is not modified after surgery [2, 6].

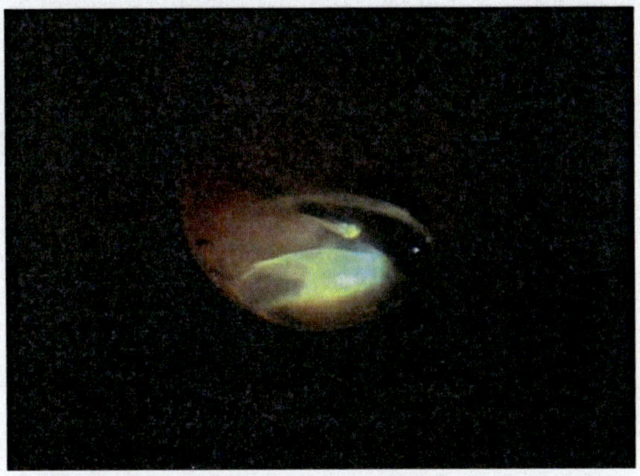

Fig. 1 Evaluation and removal of the vitreous base

The base is the most peripheral part of the vitreous and plays an important part in the pathogenesis of the RRD. Fig. 1 demonstrates how the vitrectomy probe removes the vitreous base from the surface of the retina, while the assistant indents the eye wall.

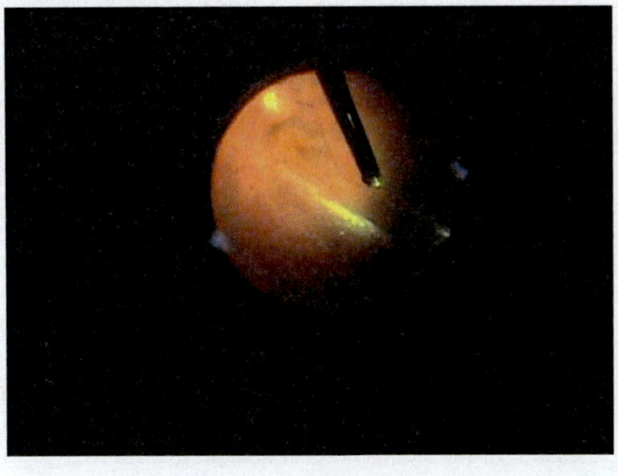

Fig. 2 Removal of the vitreous opacities

Fig. 2 illustrates the intraoperative aspect of the vitreous cavity and of the retina. The vitrectomy probe "cleans" the vitreous cavity, allowing the retina to reattach during the subsequent phases of the surgery.

During vitrectomy, we work in a complex environment, containing substances with varying viscosity and density: vitreous, blood, saline, attached and detached retina [1, 5].

The flow is dependent on the density of the aspirated material and the cutting rate. The peristaltic pump with constant pedal depression decreases the gradient of pressure when aspirating less viscous material with the same flow. The Venturi pump of the Accurus machine does not allow the direct control of flow. This is the time when the high-speed cutting rate intervenes: it reduces the difference in aspiration flow due to fluid viscosity, thus enhancing the safety of the Venturi pump. Using the high-speed cutters, these tissues can be managed without worrying about the sudden increase in the aspirating flow that occurs when aspirating saline after a dense material. High-speed cutting reduces the trans-orifice pressure variation that occurs with each port open/close cycle. As a result, surgically induced retinal motion decreases, which is particularly important in retinal detachment surgery, as it significantly reduces the risk of iatrogenic retinal tears. High cutting rates result in a decreased flow per port-opening cycle, less fluctuation in pressure, and a greater fluidic stability. At the same flow rate, high-speed cutting reduces iatrogenic tractions and increases vitreous chamber stability, allowing the surgeon to approach the retina safely and closely, which is crucial in retinal detachment surgery [1, 2, 5].

Before high-speed cutting was available, surgeons were forced to use a lot of heavier- than- water liquids (perfluorocarbon liquids - PFCL) in many cases of primary vitrectomy for retinal detachment to overcome the motion of the peripheral retina while trying to remove the vitreous. In our series, we have used PFCL to stabilize the retina during vitrectomy in 7 of the 70 cases (10%) , where the retina was very bullous and totally detached. Taking into account the high cost of these substances, reducing the expenses of surgery is another important advantage of the high-speed cutting rate.

The 3D technology enables the surgeon to control all parameters by using the foot pedal. The surgeon is also more independent with this technology, because it is no longer necessary to ask a nurse to change the parameters. Cutting and vacuum are interchangeable parameters that control flow which drives vitreous movement, and our ability to control that flow gives us precision. Apparently, the third dimension of 3D control is flow. It is safer to start with high-speed cutting and lower vacuum. With this parameter setting, we have less flow during surgery. This is safer in two conditions: when we first enter the eye and when we start working close to the retina. When doing high-speed cutting and the vitreous flow in the port is low, or the vitrectomy is taking too long, stepping on the pedal increases vacuum while decreases the cut rate. This adjustment will give us more flow, so that the removal of the vitreous is faster. Two major advantages emerge from this interplay between high- and low-speed cutting: faster and safer surgery [7].

In vitrectomies for RRD, we used high cutting rates and low vacuum when working close to the retina, while when in the mid vitreous (away from the retina), we increased the vacuum and simultaneously decreased the cutting rate, by stepping on the pedal.

Why is the flow low with high cutting rates? The explanation is given by the so-called aperture limited flow concept: with a high cutting rate, the port is open for a shorter period of time. Basically, this means that at higher cutting rates, the length of time the port remains open is shorter than the time it is at lower cutting rates. This provides a second control of the flow. We can adjust the vacuum, which obviously affects flow, and we can also control the length of time that the cutting port is actually open. Thanks to the 3D technology, we were able to concentrate on the patient's eye rather than on gauges and settings [5].

As postulated by Jules Gonin, the retinopexy is mandatory for retinal reattachment. During vitrectomy, this can be achieved either by exocryo (fig. 3) or by endolaser (fig. 4).

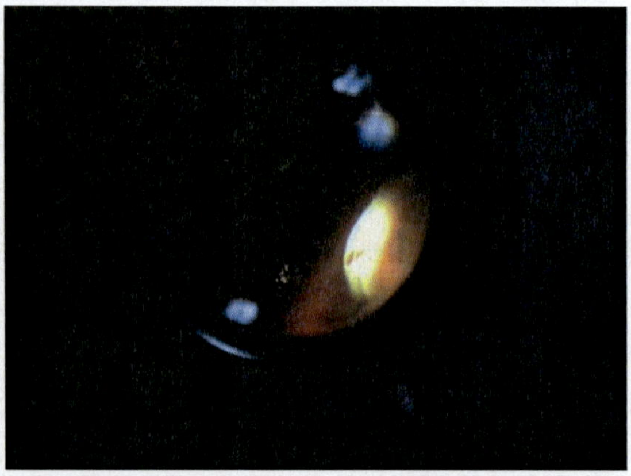

Fig. 3 Retinopexy of a retinal tear with exocryo

The application of the cryoprobe on the eye wall at the level of the retinal tear is followed by the retinal whitening at its site. The effect will be the creation of an adherent chorio-retinal scar that guarantees the healing of the retinal tear (fig. 3)

The goal of the endolaser retinopexy is to surround the retinal tear with 2 – 3 rows of endolaser impacts. The effect will be the same as after cryopexy: the healing of the retinal tear (fig. 4).

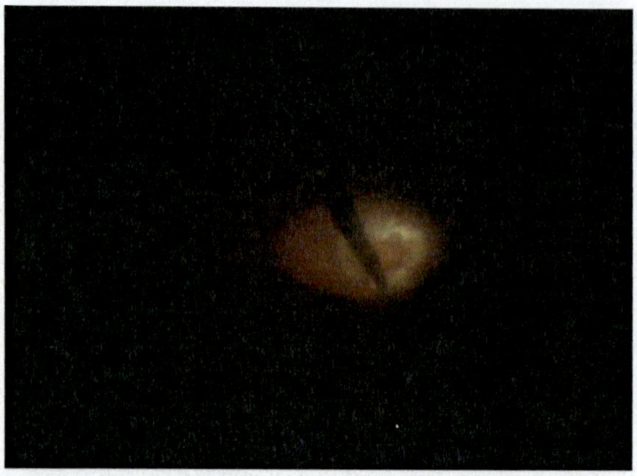

Fig. 4 Retinopexy of a retinal tear with endolaser

Debate continues about the advantages and disadvantages of each technique. On our series, cryo has been used more often: 46 cases (65.71%), because of the peripheral location of the retinal tear and easier access to it. Cryotherapy causes dispersion of the retinal pigment epithelial cells (promoters for PVR) and breakdown of the blood-retinal barrier. Therefore, after cryo is applied, the pigmented epithelial cells have to be removed from the vitreous cavity by vitrectomy. Cases of CME have been reported before the "vitrectomy era" after cryotherapy [8], but this association is not necessarily causative, given the fact that macular pathology also occurs following posterior vitreous detachment without tears [2]. We report one case of CME among the 46 patients treated by cryo.

Many studies have examined the risk factors for PVR development with contradictory results [9, 10, 11]. The 3 cases of postoperative PVR on our series were identified in the cryo patients. Our small number of cases cannot sustain the idea that cryo is the cause of all the above mentioned conditions. In the literature there is no strong evidence that the use of cryo produces inferior results to laser photocoagulation, their selection being a matter of surgeon preference [1,2]. However, the effect of endolaser retinopexy is faster than the one of cryo: it creates an almost instant adhesion, although this is not up to full strength for about 10 days [1]. Lately, we have started to use more frequently the endolaser retinopexy, as we became more familiar with the indentation techniques (required for the endolaser retinopexy), making the identification and visualization of the tear easier.

Basically, two types of silicon oil have been injected, according to the viscosity: the low viscosity silicon oil (1000 cs or 1300 cs) and the high viscosity one (5000 cs or 5500 cs). The tamponade efficacy is similar for the two types of silicon oil, but the high viscosity oil has a lower tendency to emulsification, which is the reason why we selected it when we anticipated the need for a longer tamponade.

The elevated viscosity of the silicon oil requires higher infusion pressures as compared to other liquids. According to the Poiseuille's law:

$$\Delta P = \frac{8\mu LQ}{\pi r^4} \quad (1)$$

where: ΔP= pressure drop, μ=viscosity, L=length of the tube, Q=rate of the volumetric flow, r = radius of the tube. In other words: Q (the flow in a tube) is directly proportional with the 4th power of the radius and inversly proportional with the length of the tube. Therefore, the silicon oil needs to be injected through short, high diameter and relatively rigid tubes [1].

There are three modalities to inject the silicon oil in the eye: exchange with fluid (saline or BSS), exchange with air, exchange with heavier- than- water liquid.

In the first variant, simultaneously with silicon oil infusion, the retina is flattened by the internal aspiration of the subretinal fluid. As the specific gravity of the oil is subunitary, it settles above the saline. The extrusion needle is placed in the fluid phase (vitreous cavity and through the retinal break) which is passively/actively evacuated. If the drainage is active, the negative pressure can occur at the tip of the extrusion needle (silicon oil infusion rate laggs), with the risk of collapse and retinal incarceration in the extrusion needle [1].

The heavier than water fluid/silicon exchange is similar with the water/silicon exchange, except that the retina is already flattened (it is no need to remove the subretinal fluid) [1].

In the third variant, initially, the retina is flattened by fluid/air exchange and by the internal drainage of the subretinal fluid. The next step is the laser retinopexy (in the air-filled eye), followed by the silicon oil injection, whereas the air exits through the sclerotomy. The silicon oil continues to be injected up to the sclerotomies (in the phakic and pseudophakic eyes) or iris plane (in the aphakic eyes) [1].

We preferred in all situations the second variant, as it proved safer in our hands, even if it takes a supplementary step: the exchange of fluid with air.

A so called „sandwich" technique can be used if the tears are located in the middle part of the retina: the heavier than water fluid is injected up to the posterior margin of the tear, then the fluid/air or fluid/silicon oil exchange is performed, concomitently with the subretinal fluid drainage through the tear, until the retina is flattened and then the exchange is continued [1]. We have used this technique in 5 cases, with a very good intra- and postoperative outcome. This method decreases the risk for subretinal migration of the heavier than water fluid (giant retinal tears) and the volume of fluid needed.

As the silicon oil has a subunitary specific gravity, it floats and in the aphakic eyes it blocks the pupil and occupies the anterior chamber. Therefore, in all the silicon oil filled aphakic eyes, we performed a periferal iridectomy. It needs to be infero-nasal and relatively large. In the saline filled eye, we used the vitreo-cutter which has been set on aspiration only, placed it behind the iris and then set a very low cutting rate (100 – 300 cuts/minute).

The superficial tension of the silicon oil relatively to water is much lower than the one of gas relatively to water. As consequence, the silicon oil passes much more easily through retinal breaks under traction as compared with gas. Therefore, before injecting the silicon oil, we must be sure that there are no longer tractions at the level of the retinal break (s).

The simple recurrence situations (8 cases) have been reoperated by the same technique and all of them had a good anatomical outcome: the retina reattached. The causes of the recurrences have been represented by the incomplete removel of the vitreous base in phakic eyes. Given the fact that the lens prevented us from a complete vitreous base dissection, we have performed the lens extraction followed by the completion of vitrectomy.

The main factor that prevented the final retinal reattachment has been the PVR, which is directly related to the duration of the retinal detachment and the size of the retinal tear on our series. All the 3 cases that developed PVR postoperatively had a longer than 2 months duration and big retinal tears.

Ultimately, the therapeutical option in RRD is up to the surgeon, according to the ideology, personal experience, intuition and equipment.

The surgeon must have experience in vitreo-retinal surgery, understand the physical and optical properties of the substances that are used as tamponade during and after surgery and must be capable to control the intra- and postoperative complications.

VI. Conclusions

1. Posterior vitrectomy has proven its efficiency in the treatment of the Rhegmatogenous Retinal Detachment on our series: after one surgery, the attachment rate has been 82.85%. After the second surgery, it has increased to 92.84%.

2. Proliferative Vitreo-Retinopathy has been the factor that prevented the final retinal reattachment, being related to the duration of the retinal detachment and the big size of the retinal tear: 3 cases – 1.42%.

3. The impressive progress in vitreo-retinal surgery techniques over the past 10 years has allowed us a much safer, faster and more physiological approach in the surgery for rhegmatogenous retinal detachment.

ACKNOWLEDGMENT

This work has been financially supported by the Romanian Ministry of Education, Research and Youth, through The National University Research Council, Grant PN–II–ID–PCE–2007–1, code ID_459, 2007 - 2010.

REFERENCES

1. Aylward G W (2006), Optimal Procedures for Retinal Detachments, in Retina. Elsevier, Philadelphia
2. Heimann H, Kirchhof B (2006), Primary Vitrectomy in Rhegmatogenous Retinal Detachment, in Retina. Elsevier, Philadelphia
3. Mitry D, Fleck B W, Wright A F, Campbell H, Charteris D G (2010), Pathogenesis of Rhegmatogenous Retinal Detachment. Predisposing Anatomy and Cell Biology, Retina 30 (10): 1561-1572 DOI: 10.1097/IAE. 0b013e3181f669e6
4. Spirn M J, Regillo C (2008), Proliferative Vitreoretinopathy, Retinal Physician, at http://www.retinalphysician.com
5. Talu S D, Lazan N, Lupascu S, Sebestyen E, Toader L (2009), Indications of the Modern Vitreo-Retinal Surgery Techniques, IFMBE Proc, vol. 26, International Conference on Advancements of Medicine and Health Care through Technology, Cluj-Napoca, Romania, 2009, pp 15-18
6. Kloti R (1983), Amotio-chirurgie ohne sklaerindellung. Primäere vitrektomie, Klin. Mbl Augenheilk 182: 474 - 478
7. Charles S, Calzada J, Wood B (2007) Diabetic Retinopathy, in Vitreous Microsurgery. Lippincott Williams & Wilkins, Philadelphia
8. Kimball R W, Morse P H, Benson W E (1978), Cystoid macular edema after cryotherapy, Am J Ophthalmol 86: 572-573
9. Cowley M, Conway B P, Campochiaro P A et al (1989), Clinical risk factors for proliferative vitreoretinopathy, Arch Ophthalmol, 107: 1147-1151
10. Bonnet M, Fleury J, Guenoun S et al (1996), Cryopexy in primary rhegmatogenous retinal detachment: a risk factor for post-operative proliferative vitreoretinopathy, Graefe's Arch Exp Ophthalmol, 234: 739-743
11. Kon C H, Asaria R H Y, Occleston N L et al (2000), Risk factors for proliferative vitreoretinopathy after primary vitrectomy: a prospective study, Br J Ophthalmol, 84: 506-511.

Author: Simona –Delia Talu
Institute: "Iuliu Hatieganu" University of Medicine and Pharmacy
Street: 3 – 5, Clinicilor
City: Cluj-Napoca
Country: Romania
Email: simonatalu@yahoo.com

PerFluoroCarbon Liquids in Vitreo-Retinal Surgery – Personal Experience

S.D. Talu[1], I. Tamasoi[2], C. Dragos[2], and S. Rus[2]

[1] "Iuliu Hatieganu" University of Medicine and Pharmacy/Ophthalmology, Cluj-Napoca, Romania
[2] Emergency County Hospital/Ophthalmology, Cluj-Napoca, Romania

Abstract— **The combination of the physical and chemical properties of the PerFluoroCarbon Liquids (PFCL) has propeled their clinical use not only for the repair of the complex retinal detachments, but also for a range of intraocular surgical maneuvres. The purpose of this study is to outline the status of the use of PFCL in vitreo-retinal surgery, as it emerges from our personal clinical experience. We have used the PFCL in 82 vitrectomies over a one-year time period (October 2009 – October 2010). The PFCL were injected into the vitreous cavity in patients with the following conditions: rhegmatogenous retinal detachment (RRD), tractional retinal detachment (TRD), combined retinal detachment (rhegmatogenous and tractional) and severe proliferative diabetic retinopathy (PDR). In the RRD group (65 cases), success has been achieved in 55 cases (84.61%), simple recurrence has been identified in 6 cases (9.23%), Proliferative VitreoRetinopathy – in 3 cases (4.61%) and Cystoid Macular Edema – in 1 case (1.53%). The remaining 17 cases in which PFCL have been used (20.71%) were all caused by diabetes. In the severe PDR subgroup, the disease has been stabilized in all the 3 cases. In the TRD subgroup, success has been obtained in all the 4 cases (retinal reattachment), whereas in the combined retinal detachment subgroup (RRD+TRD) the retina has reattached in 8 cases (80%) and maintained detached in 2 cases (20%) after surgery. PFCL have proved their efficacy in vitreo-retinal surgery.**

Keywords— **vitreo-retinal surgery, PFCL.**

I. Introduction

The PerFluoroCarbon Liquids (PFCL) have initially been developed as blood substitutes, given to their high ability to transport oxygen and their biological inaction [1]. The first experimental studies with PFCL in ophthalmology investigated their value as vitreous substitutes in healing the inferior retinal breaks induced in rabbits [2,3]. Chang et al. have studied, for the first time, the PFCL in humans [1]. The combination of the physical and chemical properties (transparency, high specific gravity, imiscibility with water) has propeled the clinical use of these liquids not only for the repair of the complex retinal detachments, but also for a range of intraocular surgical maneuvres [4].

II. Purpose

The purpose of this study is to outline the status of the PFCL in vitreo-retinal surgery, as it emerges from our personal clinical experience.

III. Method

We have retrospectively included in this study all the cases in which PFCL have been used during the posterior vitrectomy: Rhegmatogenous Retinal Detachment (RRD), Tractional Retinal Detachment (TRD), combined rhegmatogenous and tractional retinal detachment (RRG+TRD) and Proliferative Diabetic Retinopathy (PDR).

The following parameters are analyzed: the clinical condition that required the PFCL use, the technical peculiarities imposed by each type of disease, the intra- and postoperative complications related to PFCL.

In the retinal detachment groups (rhegmatogenous, tractional or combined), the 3 months results are defined as: success, simple recurrence, Proliferative Vitreo-Retinopathy (PVR) and Cystoid Macular Edema (CME).

In the PDR group, the 3 months results are defined as the stabilization or progression of the disease.

The significant progress related to the PFCL use is emphasized.

IV. Results

We have used the PFCL in 82 vitrectomies for a one-year period of time (October 2009 – October 2010). The PFCL have been injected in the vitreous cavity in the following conditions: rhegmatogenous retinal detachment (RRD), tractional retinal detachment (TRD), combined retinal detachment (rhegmatogenous and tractional) and severe proliferative diabetic retinopathy (PDR).

Case distribution according to the clinical condition that required the PFCL use is illustrated in table 1.

All the posterior vitrectomies have been performed with the Accurus machine (Alcon), we used the 20 gauge system and a high speed vitrectomy probe (2500 cuts/minute).

S. Vlad and R.V. Ciupa (Eds.): MEDITECH 2011, IFMBE Proceedings 36, pp. 19–22, 2011.
www.springerlink.com

Table 1 Clinical conditions that required PFCL use

The clinical condition	Number of the cases	%
RRD	65	79.26
TRD	4	4.87
RRD + TRD	10	12.19
Severe PDR	3	3.65

The *rhegmatogenous retinal detachment* has been caused by a single tear in 50 cases (76.92%), while multiple retinal tears have been identified in 15 cases (23.07%). The data regarding the size of the worse retinal tear are presented in table 2.

A hole is a round defect in all the retinal layers, with a diameter of no more than 1- 2 mm. A normal retinal tear is larger than a hole, but smaller than 90 degrees. The dimension of a large retinal tear is comprised between 90 – 180 degrees and a giant retinal tear is defined by a more than 180 degrees size.

Table 2 Size of the worse retinal tear

Size of the worse retinal tear	Number of the cases	%
Hole	10	15.38
Normal	42	64.61
Large	10	15.38
Giant	3	4.61

The vitrectomy has been as complete as possible in all situations and we did not associate the Internal Limiting Membrane (ILM) peeling.

The retinopexy has been performed externally (with a cryoprobe) in 45 cases (69.23%) and internally (with the endolaser fiber) in 20 cases (30.76%). The endolaser cerclage has been associated in 26 cases (40%) and we used the silicone oil for the internal tamponade in all the situations.

In one case the PFCL has migrated under the retina, but it has been extracted easily. We recorded no postoperative complication related to PFCL.

The anatomical 3 months results are illustrated in table 3.

Table 3 The 3 months results in the RRD cases

The 3 months result	Number of cases	%
Success	55	84.61
Simple recurrence	6	9.23
PVR	3	4.61
CME	1	1.53

The remaining 17 cases in which PFCL have been used (20.71%) were all caused by diabetes. In the severe PDR subgroup, the disease has been stabilized in all the 3 cases. In the TRD subgroup, success has been obtained in all the 4 cases (retinal reattachment), in the combined retinal detachment subgroup (RRD+TRD) the retina has reattached in 8 cases (80%) and maintained detached in 2 cases (20%) after surgery.

V. Discussion

PFCL are completely fluorinated synthetic analogs of the carbohydrates, containing carbon-fluor bonds. In the saturated PFCL, the stability of the carbon-fluor bond makes the liquid biologically inactive and stable at temperatures of 400 – 500 °C. They are colourless, odorless, have high density (specific gravity 1.6 – 2.1) and low viscosity (2 – 3 cs at 25°C) [1].

The most common indication for the use of the PFCL on our series has been represented by the rhegmatogenous retinal detachment (RRD) without proliferative vitreoretinopathy (PVR): 65 cases (79.26%). The purpose was to drain the subretinal fluid, allowing the further retinopexy and internal tamponade.The high specific gravity as compared to water allows the hydrokinetic manipulation of the detached retina, by displacing the subretinal fluid anteriorly, through the periferal retinal breaks. Thus, the posterior retinotomies for drainage become useless.This property also permits them to flatten the detached retina delicately and uniformly. The PFCL have been injected in the vitreous cavity after the complete vitrectomy in 58 cases (89.23%). In the remaining 7 cases (10.76%), the injection of PFCL has been performed earlier during surgery, in order to stabilize a very mobile and bullous retina. In this latter situation, the vitrectomy has been continued toward the perifery and then the PFCL injection has been completed. The rationale of PFCL injection in RRD is to push the subretinal fluid toward the periferal break, thus eliminating it through it and flattening the retina. Besides this effect, the PFCL helps in identifying occult breaks: the flattening posterior force displaces the subretinal fluid through the periferal break and a line appears at the mixture of two different liquids [1].

In the RRD provoqued by big tears, located in the middle part of the retina, we used the so called „sandwich" technique: the PFCL is injected up to the posterior margin of the tear, then the fluid/air exchange is performed, concomitently with the subretinal fluid drainage through the tear, until the retina is flattened and then the exchange is continued [1]. We"ve used this technique in 5 cases, with a very good intra- and postoperative outcome. This method decreases the risk for subretinal migration of the heavier than water fluid and the volume of PFCL needed.

A subtype of retinal detachment that has benefited substantially from the PFCL use is the one provoqed by a giant retinal tear (180 degrees or more). Before the PFCL era, the repair of giant retinal tears with vitrectomy required turning the patient in the prone position during surgery, allowing to a gas bubble, as it was progressively injected into the eye, to unfold the inverted retinal flap of the giant tear. The PFCL bring a major improvement, permitting the retinal flap to be inverted with the patient in the normal supine position. The significant advantages of this approach include the gentle manipulaiton of the tear and the possibility to treat it (retinopexy) with the retina in the proper position, reducing the risk of retinal pigment epithelial dispersion. After the PFCL have been introduced in the clinical practice, the success rate in retinal detachment with giant retinal tear has improved considerably (more than 90% reattachments) [1]. On our series, we had only three cases of retinal detachment with superior giant retinal tear (180 degrees) and they all had a favorable outcome after posterior vitrectomy with PFCL use. It was no need for lens removal and scleral buckle.

Before the PFCL/air exchange, care must be taken in order to remove all the anterior subretinal fluid, as it may result in the slippage of the tear. If this occurs, the PFCL is reinjected after the air is replaced with saline. Then the PFCL/silicone oil exchange is performed directly via an automated infusion system. Slippage of the retina is prevented by the mechanical advantage of oil over gas. As a relatively incompressible liquid, the incoming oil meniscus engages the edge of the tear as the PFCL is removed. None of the two giant retinal tear cases on our series have complicated with retinal slippage.

The significant tamponade force of PFCL is approximately 6 times higher than that of the fluorosilicon oil and therefore, it has been attributed the role of a „third hand" stabilizing the retina during peeling and membrane delamination [1,5]. This property has been validated on our cases that implied membrane dissection: TRD, tractional and rhegmatogenous retinal detachment and PDR (20.74%). All these situations were the consequence of advanced diabetes, which explains the less number of cases as compared to the RRD category, given the fact that the diabetic retinopathy screening has improved substantially over the last years.

The most severe situation has been represented by the combined retinal detachment (rhegmatogenous and tractional). The complexity of the surgical act in these circumstances is given by the necessity to dissect very thick and adherent membranes from the surface of a mobile retina, with the risk of enlarging the already present retinal tear (s) and producing additional ones. This is the circumstance where without the use of the PFCL, the accomplishment of the surgical goals had not been possible. The PFCL injection has maintained the retina attached during the delicate maneuvers of epiretinal tissue dissection.

The optical clarity of these liquids and their very similar to water refractive index guarantee the optimal visualization during surgery and allow the intraoperative delivery of the laser energy on the attached retina. The properties of PFCL create ideal conditions for laser delivery during surgery: their boiling point being higher than the one of water, there is no risk of intraocular vaporization; they do not absorb the laser radiations (wavelengths between 488 – 810 nm), allowing the complete penetration of the laser energy through the liquid bubble [6]. Endolaser photocoagulation has been necessary in all the cases: for the treatment of the retinal tear (retinopexy – in the rhegmatogenous and combined retinal detachments), cerclage (in the RRD cases) or panretinal photocoagulation (in the diabetes cases).

Their imiscibility prevents the penetration of blood or of the silicone oil in the PFCL bubble, thus maintaining a very good level of visualization during surgery [1,5]. This has been particularly important in the diabetes cases that bleed frequently, even if we have administered preoperatively anti-VEGF (Vascular Endothelial Growth Factor).

The low superficial tension of PFCL (and subsequently, the high interfacial tension) decreases the risk of their subretinal migration. These substances are cohesive: they stay in a one, big bubble (fig.1). We have had only one case of subretinal PFCL migration, but we,ve extracted it easily, thanks to its cohesiveness and low viscosity.

The low viscosity is another desirable quality of these liquids (0.8-8.03 cs at 25°), permitting their easy injection and aspiration, through very thin instruments (fig.1).

Figure 1 shows the intravitreal injection of a PFCL bubble, through a very thin and long cannula. The retina is attached under the bubble and still detached in the periphery. As the PFCL is injected, the subretinal fluid is pushed through the tear and the retina is progressively attaching.

This is also very useful in two circumstances: as a diagnostic tool (to evaluate the areas of residual retinal traction) and during the fluid/air exchanges [1,4].

The aspiration of the PFCL from the eye at the end of surgery is mandatory, because of their toxicity and dispertion. The dispertion (fragmentation of the large bubble in smaller ones) occurs at 2-3 days after surgery, at the interface PFCL-vitreous fluid. The consequences are: the loss of the optical clarity (interfering with the visualisation of the retina), the passage of the bubbles through the retinal breaks (impending the reattachment of

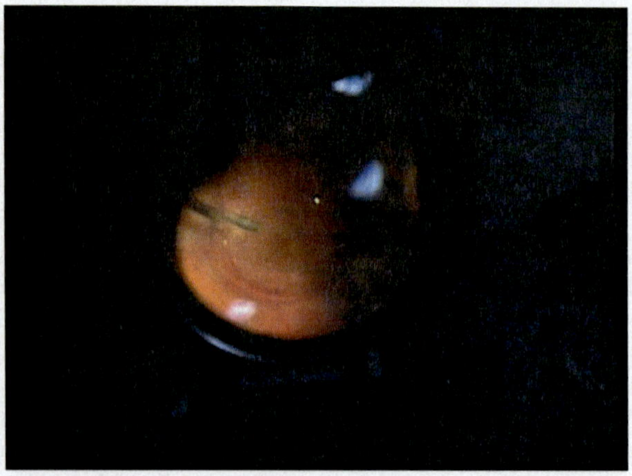

Fig. 1 Injection of PFCL liquid

the retina) and the blockage of the trabeculum by macrophages that have ingested the small bubbles (generating the secondary glaucoma) [1,7]. We,ve never left deliberately PFCL in the eye, but in one case we,ve discovered a few bubbles under the retina in the next day (they reached that space through inferior retinal breaks). We removed them, because they did not allow the retina to flatten inferiorly. Some authors report the good tolerability of very small volumes of PFCL, but in this particular case, the location of the substance under the retina forced us to re-opeate.

The toxicity of the PFCLs is generated by several facts: due to the high specific gravity, they compress the retina, leading to the loss of the external plexiform layer; later, the displacement of the photoreceptors nuclei in the external segment, the distortion of the external segment and the pigment epithelium atrophy have been identified; their concentration in polar impurities (H2) allows the absorbtion of lipoproteins and proteins, inducing a fibroblastic reaction and the formation of the preretinal membranes [1,7].

The litterature reports situations of PFCL identification in an eye after removing the silicon oil. When placed together in water, between the two liquids (PFCL and silicon oil) there is a natural attraction, because the interfacial energy between them is low. Therefore, when the silicon oil is infused in the eye, any drops of PFCL will adhere to the surface of the oil. When the PFCL volume increases and mixes with the silicon oil, the combination will have a specific gravity higher than the one of water [1].

The possible complications related to the PFCLs use in vitreo-retinal surgery are: the subretinal migration (0.9%), if the fish egg phenomenon occurs or if there are tractions at the margins of the retinal tear; the small rezidual bubbles (1 - 11%), which are well tolerated; if large amounts of PFCL remain in the eye, they induce a macrophagic response, translated by cell deposits on the lens, the cilliary body and the peripheral retina; in the aphakic eyes, the contact of the PFCL with the cornea generates the loss of the endothelial cells with its subsequent opacification [8].

VI. Conclusions

1. The PFCL have proven their physical and chemical qualities as adjuvant tools in vitreo-retinal surgery according to our experience.

2. The most common indication for the use of the PFCL on our series has been represented by the rhegmatogenous retinal detachment (RRD): 65 cases (79.26%).

3. The PFCL have brought a major improvement in the management of the RRD with giant retinal tear: all the 3 cases with superior giant retinal tears had a good outcome.

4. The combined rhegmatogenous and tractional retinal detachment represents a circumstance where without the use of the PFCL, the accomplishment of the surgical goals had been impossible. The PFCL injection has maintained the retina attached during the delicate maneuvers of epiretinal tissue dissection in all the situations.

5. We recorded no postoperative complication related to the PFCL use on our series.

Acknowledgment

This work has been financially supported by the Romanian Ministry of Education, Research and Youth, through The National University Research Council, Grant PN–II–ID–PCE–2007–1, code ID_459, 2007 - 2010.ID_459.

References

1. Chang S, Choi Kwun R (2006), Perfluorocarbon Liquids in Vitreoretinal Surgery, in Retina. Elsevier, Philadelphia
2. Stolba U, Krepler K, Pflug R et al (1997) Experimental vitreous and aqueous replacement wih perfluorophenanthrene: clinical, histologic, and electrophysiologic results. Retina 17: 146 – 153
3. Zeana D, Becker J, Kuckelkorn R et al (1999) Perfluorohexyloctane as a long term vitreous tamponade in the experimental animal. Experimental perfluorohexyloctane substitution. Int Ophthalmol, 23:17-24
4. Meinert H, Roy T (2000) Semifluorinated alkanes – a new class of compounds with outstanding properties for use in ophthalmology. Eur J Ophthalmol, 10: 189 – 197
5. Charles S, Calzada J, Wood B (2007) Diabetic Retinopathy, in Vitre ous Microsurgery. Lippincott Williams & Wilkins, Philadelphia
6. Azzolini C, Brancato R., Trabucchi G et al (1994) Endophotocoagulation through perfluorodecalin in rabbit eyes. Int Ophthalmol, 18:33
7. Sparrow J R, Matthews G P, Iwamoto T et al (1993) Retinal tolerance to intravitreal perfluoroethylcyclohexane liquid in the rabbit. Retina 13: 56 – 62
8. Vote B, Wheen L, Cluroe A et al (2003) Further evidence for proinflammatory nature of perfluorohexyloctane in the eye. Clin Experiment Ophthalmol 10: 408 – 414

E-NOTES Transumbilical Cholecystectomy

F. Graur[1,2], A. Szasz[1], R. Negru[1], H.C. Neagos[1,2], R. Elisei[1], A. Muresan[1], and L. Furcea[1,2]

[1] Surgical Clinic III, Cluj-Napoca, Romania
[2] University of Medicine and Pharmacy "Iuliu Hatieganu", Cluj-Napoca, Romania

Abstract— **Laparoscopic cholecystectomy has become the gold standard for the symptomatic gallbladder lithiasis. Usually, the procedure is performed using 4 trocars. Reducing the number of trocars is one of the ways of minimizing the invasiveness of this procedure. Single port laparoscopy through the umbilical scar, also known as "embryonic natural orifice transumbilical endoscopic surgery" (E-NOTES) was developed as an attempt to improve the aesthetic aspect after surgical procedures and to reduce the morbidity of minimally invasive surgery.**

We report a case of transumbilical laparoscopic cholecystectomy without the use of clips, for symptomatic gallbladder lithiasis, the sealing of the cystic duct and cystic artery being obtained by using a LigaSure 5 mm. A 35 year old young male with history of pain in the upper abdomen and right upper quadrant, nausea and vomiting was diagnosed with gallbladder lithiasis after an ultrasound examination. The gallbladder was of normal size, with multiple gallstones. Under general anesthesia, a pneumoperitoneum was created. Three trocars were inserted into the peritoneal cavity for this intervention through the SILS port, through a single incision at the umbilical scar. The cholecystectomy was performed in a retrograde manner. One subhepatic drainage tube was inserted. The operative time was 50 min. There was no intra- or postoperative complications. The patient was discharged in the second postoperative day after a normal course. The 1 month follow-up was normal.

Keywords— **laparoscopic cholecystectomy, SILS, LigaSure, E-NOTES.**

I. Introduction

The first transumbilical cholecystectomy was performed by Navarra in 1997, followed shortly by Piskun in 1999 [1,2].

The fundamental idea of this new and revolutionary approach is having all „entry points" in the same place, the umbilicus, resulting in only one postoperative wound, almost invisible. Although this method is safe and feasible, it cannot be applied to all patients, being extremely difficult in obese patients. Other NOTES techniques are using different natural orifices in order to approach the peritoneal cavity, such as vagina, rectum or mouth, but these techniques require a total different set of instrument, extremely expensive and inaccessible for us at the moment and there is limited experience with these approaches.

II. Case report

We report the case of a 35 years old male who has been experiencing pain in the right upper quadrant of the abdomen for the last few months, nausea and vomiting especially after meals. An ultrasound examination was performed, showing gall bladder lithiasis. The patient was admitted one day before surgery. At admission, all laboratory tests proved within limits. No preoperative treatment was necessary.

The procedure was performed under general anesthesia with orotracheal intubation. The patient was placed in dorsal decubitus in anti-Trendelenburg position. A 2,5 cm vertical incision was performed inside the umbilicus penetrating all the layers of the abdominal wall in order to access the peritoneal cavity. Using small retractors we inserted the SILS device. In order to have enough room to manipulate the instruments inside the abdomen there is the need of inflating the peritoneal cavity with carbon dioxide (procedure called pneumoperitoneum insufflation), this being a standard procedure in all laparoscopic interventions. The SILS device has a special port for insufflating and maintaining the pneumoperitoneum. Intraabdominal pressure was kept at 12 mmHg throughout the intervention. The SILS device allows the insertion of 3 trocars, one of them being used for the optic system and the other two for surgical instruments. The optic system consists in a 5 mm diameter rigid video camera with a 30° direction of view connected to a monitor. This system is used in all laparoscopic procedures. The rest of the surgical instruments we used are common laparoscopic instruments (graspers, electrocautery, aspiration system, LigaSure sealer) (Fig.1). This system provides all necessary instruments for operation inserted using the same entry point, but in order to perform the cholecystectomy there is still need of holding the gall bladder in a still and elevated position.

In order to achieve that without inserting an additional trocar we performed the suspension of the gall bladder using a straight needle with a 2-0 silk stitch which was passed through the abdominal wall, through the fundus of the

S. Vlad and R.V. Ciupa (Eds.): MEDITECH 2011, IFMBE Proceedings 36, pp. 23–25, 2011.
www.springerlink.com

Fig. 1 The SILS device inserted through the 2,5 cm umbilical incision with the optic system, insufflation tube and surgical graspers attached

gall-bladder and again through the abdominal wall right under the 12th right rib, providing the necessary suspension of the gallbladder. Perforating the gall bladder means minor biliary leak, with no prejudice at all for the patient, all the bile being aspirated at the end of the procedure. A minor inconvenient is the lack of tension in the gall bladder wall after being emptied of bile, making the dissection of the surrounding peritoneum of the gallbladder a little more difficult. The next step of the operation is the isolation of the cystic duct and cystic artery (dissection of the Calot triangle). This was done in an usual manner using a grasper and the electrocautery. After isolating the cystic duct and artery we used the 5mm LigaSure sealer to seal and cut the cystic duct and artery. Ordinary laparoscopic procedures make use of a clip applier to seal these elements with clips before being cut, but the dimension of the applier (10 mm in diameter) made it impossible to use in our case (Fig.2, Fig.3).

The LigaSure sealer proves to be reliable and safe in preventing hemorrhage and bile leakage, literature reporting the safe use of LigaSure on much larger vessels or biliary structures [3,4,5]. Dissecting the gallbladder from its liver site was done in a usual manner, using the hook electrocautery. The extraction of the gall bladder was performed through the umbilical incision along with the extraction of the SILS device.

A drain tube was inserted under the liver using the same umbilical entry point and the umbilical wound was sutured in a usual manner using separate fascial and skin stitches. The total duration of the operation was 50 minutes. The patient had a simple evolution being discharged on the 2nd day after surgery. No complications occurred during admittance and during the first month follow-up. The aestethic result was perfect, practically no scar shows at the site of umbilical incision (Fig.4).

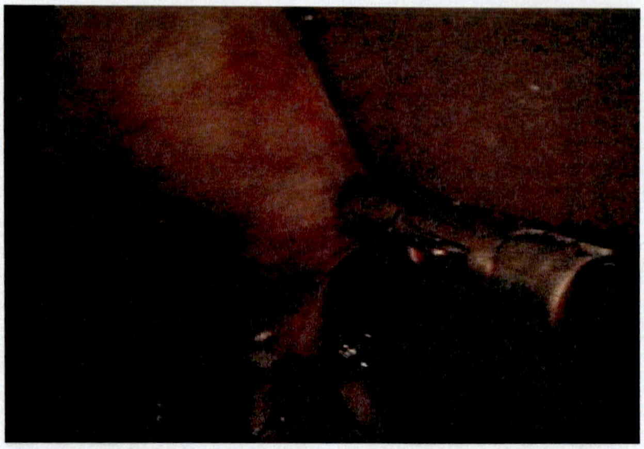

Fig. 2 The dissection of the Calot triangle using a grasper and a hook-tip electrocautery

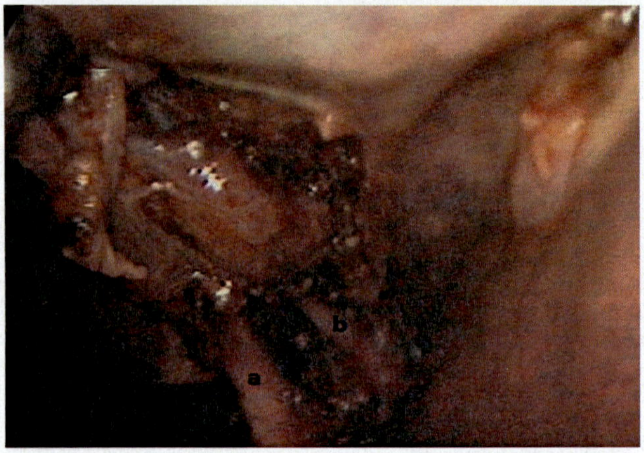

Fig. 3 The Calot triangle after dissection - notice the cystic duct (a) and the cystic artery (b) before being cut

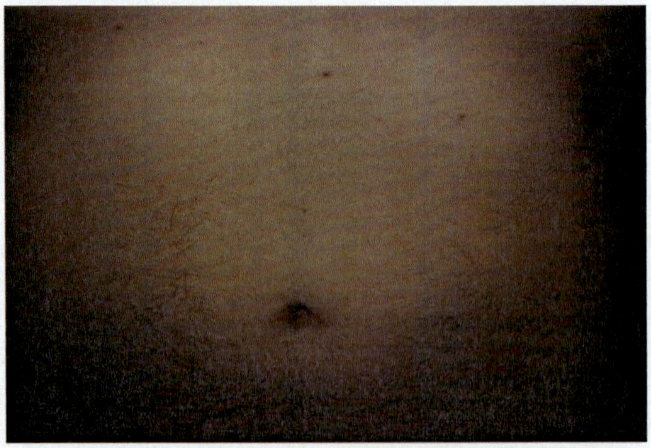

Fig. 4 The aspect of the abdominal wall one month after surgery (notice a previous appendectomy scar on the left inferior corner of the picture)

III. Conclusions

Laparoscopic cholecystectomy became a „gold standard" in gallbladder lithiasis in less than 5 years since this method was developed [6], but the aesthetics requirements went a step forward and NOTES (natural orifice transluminal endoscopic surgery) occurred in 2004. E-NOTES makes use of embryonic natural orifice (umbilicus) as path of access to the peritoneal cavity. Along with technical developments, NOTES became safer and easier to use, but there is still need of improvement in equipments. E-NOTES uses various types of umbilical ports with multiple working channels. One of the basic principles of laparoscopy is assuring enough space between instruments, in order to have enough room so that the instruments won't touch each other.

With NOTES came the need of a different type of instruments, angulated graspers, flexible optic systems, instruments which unfortunately are unavailable to us due to the high cost. Recently, many companies have developed special designed NOTES instruments. Angulated graspers may be found in different shapes, all of them trying to achieve the same goal of providing enough room between instruments inserted through the same entry point. Angulated graspers may come with a predefinite shape or they can be adjusted by the surgeon. Another kind of innovative graspers allow the tip of the grasper to move according to the movement of the handle.

In our case, using standard laparoscopic instruments lead to technical difficulties and increased the operating time, almost doubling it. The SILS port draws its name from „single incision laparoscopic surgery" and it's meant to provide a single access point to the abdominal cavity for all necessary instruments. It has an insufflation port and three trocars for optical system and working instruments (Fig. 5).

As a difference from standard laparoscopic cholecystectomy we used the LigaSure device for sealing the cystic artery and bile duct. Usually we apply titan clips but we do not have a 5mm clip-applier. The LigaSure sealer is a bipolar electrothermal device meant for sealing blood vessels no larger than 7mm in diameter [7].

It was proved that the LigaSure is safe for organ dissection, including bile duct sealing [8]. The thermal injury caused by this device is confined to 2mm around the surgical site, which means it's a very accurate instrument [4,5].

E-NOTES is part of a larger concept: NOTES, which makes use of all natural orifices (mouth, vagina, rectum) as access point to the abdominal cavity. Accessing the abdominal cavity only through mouth or vagina involves the use of cutting-edge technology and instruments, available only in a few surgical sites around the world. No matter the access point, the principles of surgery are the same.

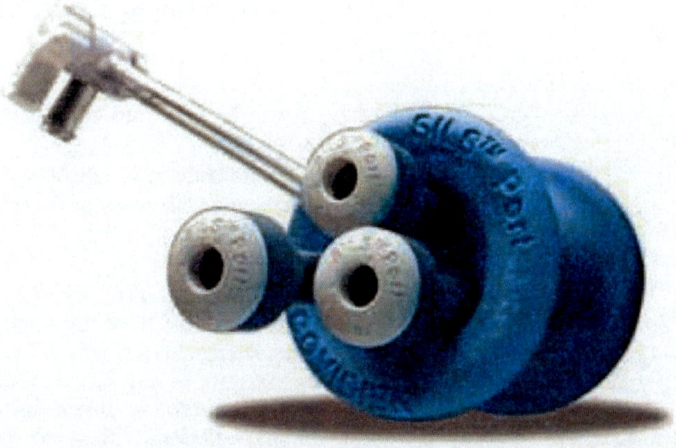

Fig. 5 SILS port

E-NOTES operations are currently gaining a lot of ground due to the growing interest and demand in aesthetics and because they are considered a safe alternative to laparoscopic or classic surgery.

Acknowledgment

The authors gratefully acknowledge the financial support provided by the research grants awarded by the Romanian Ministry of Education, Research and Innovation.

References

1. Navarra G, Pozza E, Occhionorelli S, Carcoforo P, Donini I. One-wound laparoscopic cholecystectomy. Br J Surg. 1997;84:695
2. Piskun, G. and Rajpal, S., Transumbilical laparoscopic cholecystectomy utilizes no incisions outside the umbilicus. *J Laparoendosc Adv Surg Tech*, 9:361–364, 1999
3. Campbell PA, Cresswell AB, Frank TG, Cuschieri A. Real-time thermography during energized vessel sealing and dissection. Surg Endosc. 2003;17(10):1640–1645
4. Diamantis T, Kontos M, Arvelakis A, et al. Comparison of monopolar electrocoagulation, bipolar electrocoagulation, Ultra-Cision, and LigaSure. Surg Today. 2006;36:908–913
5. Nii A, Shimida M, Ikegami T, et al. Efficacy of vessel sealing system for major Glisson bundles and major bile ducts. J Hepatobiliary Pancreat Surg. 2008;15:522–527
6. Gadacz TR, Talamini MA, Lillemoe KD, Yeo CJ (1990) Laparos copic cholecystectomy. Surg Clin North Am 70:1249–1262
7. Fried GM. Hemostatic tools for the gastrointestinal surgery: ultrasonic coagulator vs. bipolar ligation. J Gastrointest Surg. 2001;5:216-218
8. Schulze S, Krisitiansen VB, Fischer Hansen B, et al (2002) Sealing of cystic duct with bipolar electrocoagulation. Surg Endosc 16:342–344

Author: Dr. Florin Graur
Institute: Surgical Clinic III, Cluj-Napoca
Street: Croitorilor 19 - 21
City: Cluj-Napoca
Country: Romania
Email: graurf@yahoo.com

Multiagent System for Monitoring Chronic Diseases

D. Floroian[1], L. Floroian[2], and F. Moldoveanu[1]

[1] Transilvania University of Brasov/Automatics Department, Brasov, Romania
[2] Transilvania University of Brasov/Physics Department, Brasov, Romania

Abstract— **The aim of this paper is to present a society of intelligent agents linked into a multiagent system with the scope of monitoring a large group of human beings in order to detect possibilities of estimating catching chronic diseases. For this purpose a heterogeneous group of patients was taken into observation into approximately 15 years period. The group is provided by a family doctor and is formed by many social categories, many age increments and some different cities. The paper will prove that such monitoring will detect in the first stage or will detect the possibilities of being ill based only on historical dates of the patients. Obviously the names of the patients were not taken into multiagent system, but in real cases the family doctor could prevent a real patient about the danger of his life style or medical historical dates.**

Keywords— **intelligent agents, chronic diseases, multiagent system, health care.**

I. Introduction

Nowadays the chronic diseases are increasing and the associated problems (cost of medications, cost of medical examinations, social exclusions, etc.) are more and more expensive. In the same time the number of persons with chronic diseases increases year by year. It is obviously interesting to find a way to prevent such things to happen in the case that we could determine what type of behavior or what type of previous disease leads to a chronic disease.

Furthermore, if will be possible to have an automatic equipment (or software program in this case) to detect the formation of chronic diseases, will be possible (at least theoretical) to reduce the number of patients with chronic diseases.

At Brasov County official dates reveal the fact that about 20% of population is in a prediabetes stage, according to national statistics. Another 10% of population has an immediate risk of heart attack. This situation implies a sustained effort to prevent and detect the chronic diseases.

This paper presents a study made on 1874 patients from different locations, which evolve on a 15 year period. In this purpose each patient was agentified using a program computer. For each person in the group there is an intelligent agent who takes the basic information about patient and in the same time, the medical information about that patient. In the real life these information are connected with the name of the real patient. In our study the name was replaced by numbers for medical confidence because the medical dates are real.

Furthermore each agent enrolls it to a multiagent system. The multiagent systems provide possibility of the agents' collaborations and the interaction with the user interface. Each agent will provide needed information on demand.

Our program will display based on agents the evolution of different diseases along these years for this group of patients. In the future we want to develop the program in order to automatically alert the family doctor about the chronic diseases which will potentially develop for a specific patient. This is possible because the multiagent system learns some basic behaviors about information that it have.

The paper is organized as follow: in second section we present the supporting concepts as "multiagent system" and "intelligent agents"; in section III we present an example with 1874 group of patients and connection between chronic diseases and them. Finally in conclusion section we present the results.

II. Supporting concepts

A. *Multiagent System*

In the context of this paper, an agent is a computer program associated either to some hardware component (a sensor connected to a person) or to some specific information about a person and which can run continuously and autonomously. The multiagent system (MAS) provides a set of individual agents who can communicate between them for achieving a purpose and with a central station for monitoring process. The agents within the system model interact with the agents who represent the components or operations to find the needed information with minimal resources and in optimal time.

The agents' interaction is done under the control of the coordination agent (which also provides the graphic user interface) and looking for key information. When a satisfactory solution cannot be found in the current structure it will try to looking for the alternative options with gradually relaxed constraints, allowing subsystems to regroup. This

S. Vlad and R.V. Ciupa (Eds.): MEDITECH 2011, IFMBE Proceedings 36, pp. 26–31, 2011.
www.springerlink.com

reflects the Belief Desire Intention – BDI architecture (together with the feedback for system states updates). The basic requirements for the MAS architecture are that the system model must supply enough information about the patient to allow the planning, scheduling, and reconfiguration of the dates to be carried out under gradually relaxed constraints. The model also needs to avoid centralized control which does not characterize the multiagent technology, in order to allow subsystems to autonomously negotiate each other and to collaborate across system boundaries to form new groupings [1].

Also the subsystems must be able to carry out collectively simulation of discrete events, in any grouping, in order to evaluate restructured decisions. Based on the analysis, each subsystem on the system hierarchy needs to be modeled as an agent. An agent should be able to send and receive messages to or from other agents and the environment, and to make decisions to take part in bidding and negotiation. There should also be a mechanism to support dynamic simulation of discrete events [2, 3].

In order to implement the described MAS, the BDI architecture is very convenient because the environment is not fully known and it is changes over time. Epistemic Deontological Axiological (EDA) architecture is not so useful because the norms which are included will not help in this situation. Also Layered architectures or Logic Based Agent architectures are too close to a centralized control in described situations [4, 5].

In this paper we consider that the communications between agents are FIPA compliant (Foundation for Intelligent Physical Agents) like described in [6]. Also for MAS implementation we use JADE (Java Agent DEvelopment Framework) which is a free tool available online [7]. Furthermore the ontology was integrated in JADE with Protégé 2000 [8, 9, 10, 11].

For success implementing the BDI architecture a belief revised (BR) function is necessary. This function is intended to revise the knowledge database (the beliefs of the agents) in order to maintain the consistency of the database. In our special case this function is necessary because the beliefs of the agents could be wrong in case of an unpredictable situation is involved in FMS flow. In these cases the BR function will correct the beliefs of the agents. The feedback flow of the BR function is represented in figure 1 (adapted from [12]).

As is seen in the figure 1, the agents of the MAS can perceive only a part of the variables of the patients. Using the beliefs gathered initially from the patient's model and the new perceived information, the BR function rebuilds the unperceived information. Together with the update rebuilt information these knowledge represent the new set of beliefs of the agents.

For this approach we focus for a MAS where the environment is a dynamic system with a fixed structure where the value of some system variables (necessary for BR function implementation) is unknown. In this paper we consider a practical case study for implementing such concepts.

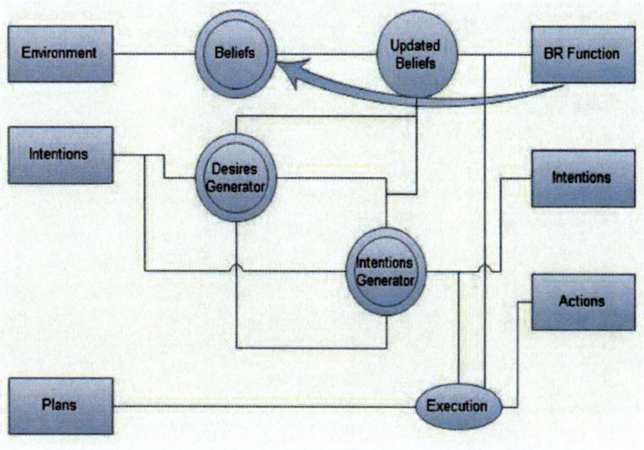

Fig. 1 The BR function generates an update belief

B. Patients Agentifying

For integrating of the information in MAS we need a procedure for connecting of the information with MAS's coordination agents. This procedure is called patient agentifying and implies that for each patient in the group an agent to be created. That agent must retain both medical and personal information. The personal information is needed for infrastructure purpose and also for statistical dates. The medical information is needed for experiment. As we mentioned before the name are replaced by number in this study case for confidential purpose. In the real cases the names must be accessible only to the medical or authorized staff.

This procedure is difficult and takes a long time. But this procedure is needed only one time. After this all will happen automatically. The user must complete or update the information only when it is necessary [13].

After agentifying process the MAS could access all information that it needs. The access to the information is made only by agent's way. In this stage we have a collection of intelligent agents that cooperate in order to achieve their purpose [4].

In this process the user has a graphic user interface (GUI) that helps the input of the information. This GUI is represented in figure 2. The GUI will provide information that is needed for the agentifying process. The action for the submit button implies that an agentifying agent is solicited to complete the agentifying process. By this action all the information is used to setup the patient's agent. Thus a specific agent associated to a specific patient is obtained. This process is called agentifying of patients [14, 15, 16, and 17].

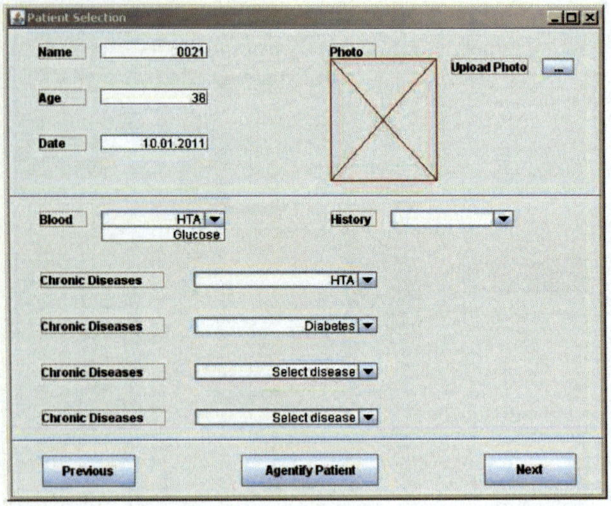

Fig. 2 The GUI interface

GUI is configurable in order to permit many kinds of diseases to be monitored.

In our MAS the cooperation and coordination between PTAxs agents is important because they share the same goals. In the same time is important to notice that the mechanism that governs coordination is the interaction protocols.

Cooperation (to achieve a certain goal) is based on two phases. First decompose the main task into several small tasks (i.e. finding diabetes mellitus is decomposing in: search for already declared patients, search for inducing diabetes diseases, search for other emerging factors, reduce the redundancies, compose the outputs). Several decomposition plans can be generated according to the available resources and the capabilities of the cooperating entities. In the second phase each subtask is executed and after all the *generateOutput* task will contain the results. These entire subtasks are running concurrently. This implies a competition between agents because each of them needs to obtain their task and also report information to others. The competition is regulated by a coordination agent in order to keep the accuracy of dates.

III. STUDY CASE

In order to prove the above concepts we use a specific group of 1874 patients. These patients are from different four locations (near Brasov).

They are monitored over a 15 years period of time for chronic diseases. We want to obtain some results which will indicate the chronic diseases rates and to obtain a link that we could use in the future to predict the occurrence of chronic diseases in a specific population.

In our study case we choose to monitor: diabetes mellitus, dyslipidemia, hypertension, obesity, ischemic cardiomyopathy (see fig. 3).

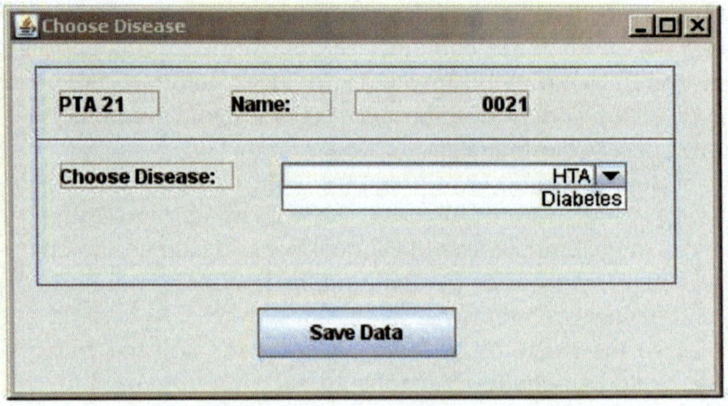

Fig. 3 Chosen the chronic disease

In our case a patient's agent will retain only information about age, chronic diseases and historical dates. However it could be programmed to retain much more information.

This study case will only take care about a few chronic diseases because its role is to prove the procedure. In the future we want to extend this example to permit prediction of all chronic diseases occurrence.

In the first stage we agentified all 1874 patients, using the GUI presented in figure 2. After this process we obtained 1874 agents. These agents are coordinated by a coordination agent (CA). This agent has an interface role between patient's agents and MAS. Also this agent preserves the accuracy of dates by rules (these rules modify the behavior of the agents). There is not a subordinating relationship but a cooperative relation. In this way the user could obtain the results, the agents could communicate and the MAS principle is not broken [1].

The MAS is structured like in figure 4. There are 1874 patient agents (PTAx) and 1 CA. These agents communicate each other's in order to cooperate. Also PTAxs communicate with CA. For interaction with people there is a GUI (see fig. 2).

For the study of the MAS's agents we use JADE's GUI. This interface reveals the agents that constitute the MAS. Also the JADE has a core that runs silent in background. Only at specific request the GUI is displayed (see fig. 5).

For our purposes, we have adopted the description of an agent as a software program with the capabilities of searching and cooperate. This implementation is made in JADE because this development tools are very versatile and could be very well integrated with others development tools (like Protégé-2000 and Java). Also JADE is an open source FIPA compliant Java based software framework for the implementation of multiagent systems. It simplifies the implementation of agent communities by offering runtime and agent programming libraries, as well as tools to manage platform execution and monitoring and debugging activities. These supporting tools are themselves FIPA agents. JADE offers simultaneously middleware for FIPA compliant multiagent systems, supporting application agents whenever they need to exploit some feature covered by the FIPA standard (message passing, agent life cycle, etc.), and a Java framework for developing FIPA compliant agent applications, making FIPA standard assets available to the programmer through Java object-oriented abstractions.

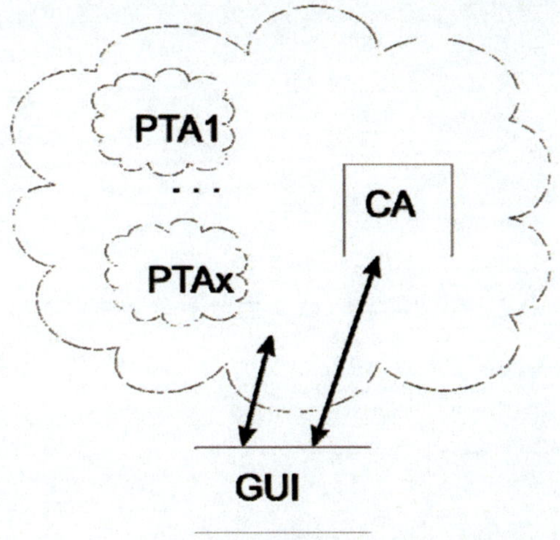

Fig. 4 MAS architecture

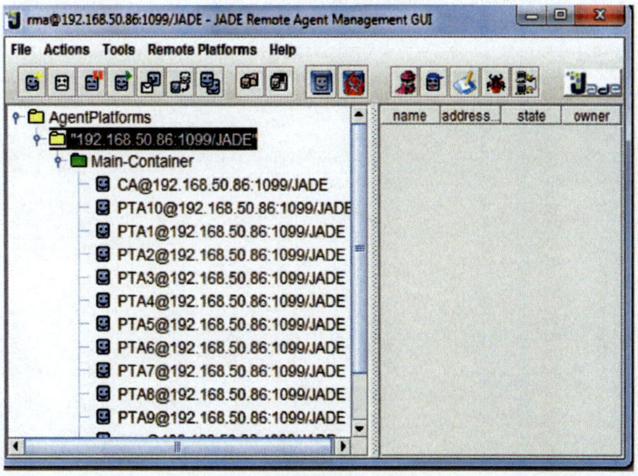

Fig. 5 MAS as seen in JADE

The general management console for a JADE agent platform (RMA – Remote Monitoring Agent), like in figure 5, acquires the information about the platform and executes the GUI commands to modify the status of the platform (creating new agents, shutting down containers, etc.) through the AMS (Agent Management System).

The agent platform can be split between several hosts (provided that there is not firewall between them). The agents are implemented as one Java thread and Java events are used for effective and lightweight communication between agents on the same host.

The parallel tasks can be still executed by one agent, and JADE schedules these tasks in a more efficient (and even simpler for the skilled programmer) way than the Java Virtual Machine (VM) does for threads.

Several Java VMs, called containers in JADE, can coexist in the same agent platform even though they are not running in the same host as the RMA agent. This means that a RMA can be used to manage a set of VMs distributed across various hosts.

Each container provides a complete run time environment for agent execution and allows several agents to concurrently execute on the same host.

The DF (Directory Facilitator), AMS, and RMA agents coexist under the same container (main-container) together with the patient's agents. To facilitate message reply, which, according to FIPA, must be formed taking into account a set of well-formed rules such as setting the appropriate value for the attributes in-reply-to, using the same conversation-id, etc., the method createReply() is defined in the class that defines the ACL (Agent Communication Language) message.

Different types of primitives are also included to facilitate the implementation of content languages other than SL (Standard Languages), which is the default content language defined by FIPA for ACL messages. This facility is made with Protégé 2000 as depicted in figure 6.

Protégé presents also a graphical view of ontology classes which facilitate understanding the fact that from the point of view of the programmer, a JADE agent is simply a Java class that extends the base agent class. It allows inheriting a basic hidden behaviour (such as registration, configuration, remote management, etc.), and a basic set of methods that can be called to implement the application tasks of the agent (i.e. send/receive ACL messages).

Moreover, user agents inherit from their Agent superclass some methods to manage agent behaviours. Also this diagram can be represented in UML (Unified Modelling Language) because behaviours are implemented as hierarchy of classes. The Protégé 2000 connects to PTAxs JADE Agents by including a Protégé configuration file in the Java compiler. The ontology is created in Protégé 2000 and MAS's classes are presented in figure 6.

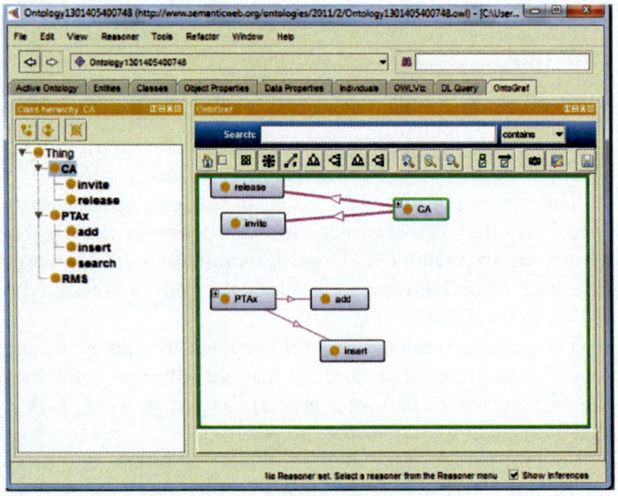

Fig. 6 MAS's ontology realized in Protégé 2000

The outputs are revealed in figures 7, 8, 9, 10 and 11. In these figures are presented the evolutions for diabetes mellitus, dyslipidemia, hypertension, obesity, ischemic cardiomyopathy for PTAxs included in MAS. These figures don't represent simple histograms of the respective illness in time; they are the outputs of the specific PTAx. These outputs reflect the searches and also the incidence based on the other illness declared in patient's clinical dates. The time for searching, cross-over searching and data-link is minimized by the MAS algorithm and is made in one step. This is the advantage of using this MAS. This is possible because the PTAxs are not simple searching tools.

In these figures could be seen that a spectacular growth has occurred in recent years. This problem is a big issue and preventing and detecting the chronic diseases in the first stages are imperative.

In the future we intend to extend the MAS on a very large number of patients for early detecting these diseases based on historical medical dates.

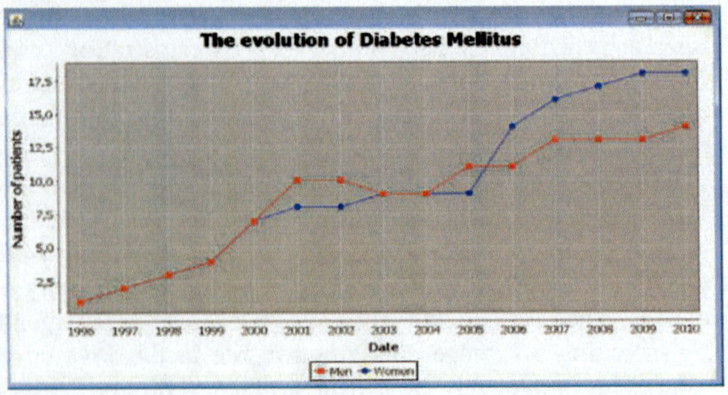

Fig. 7 Diabetic evolution

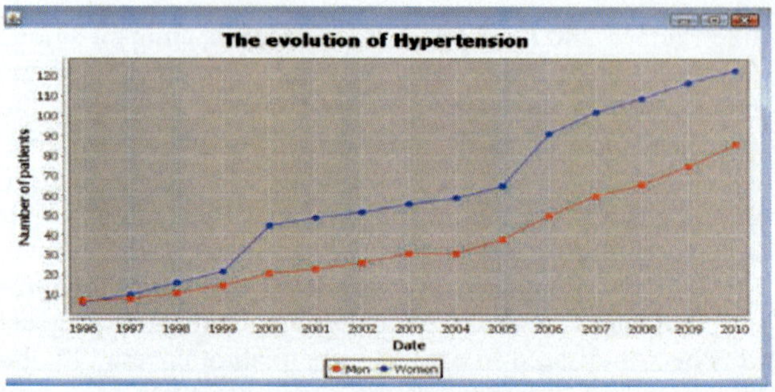

Fig. 8 Arterial hypertension evolution

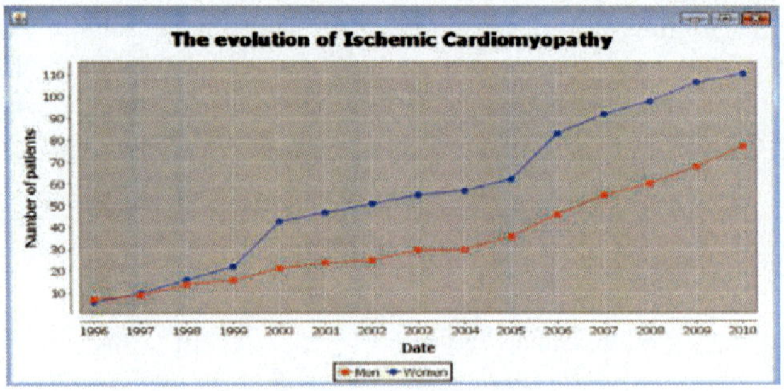

Fig. 9 Ischemic cardiomyopathy evolution

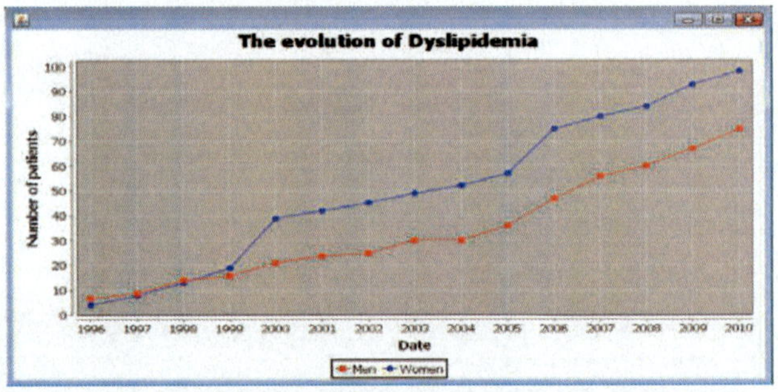

Fig. 10 Dyslipidemia evolution

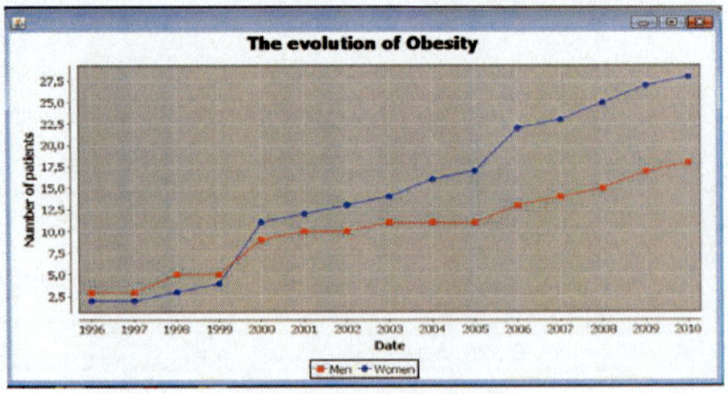

Fig. 11 Obesity evolution

IV. Conclusions

The chronic diseases problem is very serious and the ways to prevent and to detect earlier is also a big issues. This application starts to resolve the second part of the problem. Agent-based computing has great potential in addressing this problem by using distributed resources and reducing the programming effort. In this paper, we present a new approach based on agent-oriented programming and the concept of collaborative intelligent agents.

We develop a multiagent system that consists of patient representative agents. To obtain satisfaction for patients receive, the patient representative agents are designed to be as informative and timely as communicating with people by GUI. In our paper we used Java in the initial pseudo-natural language parser, XML as the knowledge base and XQL as the query language. This paper represents a quite new area and has not been developed to its full capability. Future work includes enhancing the capabilities of agents and the communication, collaboration, and coordination mechanisms in the multi-agent system, improving the pseudo-natural language parser to make it understand the user's input and synonyms better, supporting a more advanced query setup to provide accurate replies in the circumstance of multiple inexact matching, and developing mechanisms for the agents to learn automatically from the users, history profiles, and information on the Web. The ultimate goal is to detect in the first stage the chronic diseases develop.

Also the use of MAS is imperative because of its flexibility. Thus the system can be very easily switched for other diseases and the time for obtaining results remain very short even in the case of millions of patients.

References

1. Floroian D (2009) Multiagent Systems, Albastra, Cluj-Napoca, Romania (in Romanian)
2. Barata J, Camarinha-Matos L, Onori M (2005) A Multiagent Based Control Approach for Evolvable Assembly Systems. Industrial Informatics, INDIN '05. Proc. of the 3rd IEEE Int. Conf., 2005, 478-483
3. Zhang D Z, Anosike A, Lim M K (2007) Dynamically Integrated Manufacturing System (DIMS) – A Multiagent Approach. IEEE Trans. Syst., Man, & Cybern. – Part A, Syst., Humans, 37(5), 824-850
4. Wooldridge M J, Jennings N R. (1995) Agent Theories Architecture and Languages: A Survey, Intelligent Agents. In Proc. Workshop on Agents Theories, Architecture and Languages (ECAI-94), Amsterdam, Holland, 1995, pp. 1–39
5. Jennings N R (2000) On Agent-based Software Engineering. In Artificial Intelligence, no. 117, pp. 277-296
6. FIPA Communicative Act Library Specification (XC00037J). FIPA - Foundation For Intelligent Physical Agents. At http://www.fipa.org/specs/fipa00023/index.html
7. JADE at http://jade.tilab.com/
8. Protégé-2000 at http://protege.stanford.edu
9. Ontoprise: Semantics for the Web at http://www.ontoprise.de
10. The Semantic Web Community Portal at http://www.semanticweb
11. WebOnto. Knowledge Media Institute (KMI), The Open University at http://eldora.open.ac.uk: 3000/webonto/
12. Barata J, Camarinha-Matos L (2003) Coalitions of Manufacturing Components for Shop Floor Agility - the CoBASA Architecture. Int. J. of Networking and Virtual Organisations, 2(1), 50-77
13. Smirnov A, Pashkin M, Chilov N, et al. (2003) Knowledge Source Network Configuration Approach to Knowledge Logistics. Int J General Systems, Vol. 32(3), pp. 251-269
14. Walczak S (2003) A Multiagent Architecture for Developing Medical Information Retrieval Agents. J. Med. Syst., Vol. 27, No. 5, 479-498
15. Shang Y, Shi H (1999) A Web-based multi-agent system for interpreting medical images World Wide Web 2 (1999) pp.209–218
16. Bousquet F, Le Page C. (2004) Multi-agent simulations and ecosystem management: a review. Ecol. Model. 176, 313–332
17. Linard C, Ponc on N, Fontenille D, Lambin EF (2008) A multiagent simulation to assess the risk of malaria reemergence in southern France. Ecolog. Mod. 220:160–174. DOI:10.1016/j.ecolmodel.2008.09.001

Author: Dan Floroian
Institute: Transilvania University of Brasov
Street: 29, Eroilor Blvd.
City: Brasov
Country: Romania
Email: dan.floroian@unitbv.ro

Real Time Biostatistics Software: Application in Acute Myeloid Leukemia Assessment

A. Bacarea[1], B.A. Haifa[2], M. Marusteri[2], M. Muji[3], A. Schiopu[2], D. Ghiga[4], M. Petrisor[2], and V. Bacarea[4]

[1] University of Medicine and Pharmacy/Pathophysiology Department, Tirgu Mures, Romania
[2] University of Medicine and Pharmacy/ Medical informatics and biostatistics Department, Tirgu Mures, Romania
[3] Petru Maior University/ Engineering Department, Tirgu Mures, Romania
[4] University of Medicine and Pharmacy/ Medical Research Methodology Department, Tirgu Mures, Romania

Abstract— **the aim of this paper is to present an useful software in medical research. The new concept proposed is "real time biostatistics" and its application in Acute Myeloid Leukemia. For this purpose open source software (wxWidgets, R and SQLLite3) are used. The cases were patients with AML from Hematological Department, County Emergency Clinical Hospital Tirgu Mures. We created a friendly interfaced software that allows appropriate data collection and real time update of statistical parameters as each case is introduced.**

The medical importance derives from the possibility to have valid study and to see in each moment a change in the evolution of patients.

Collaboration between specialists (hematologist, PC programmer, biostatistician and methodologist) was really important for accomplishing our goal.

Keywords— **real time biostatistics, acute myeloid leukemia, open source.**

I. Introduction

Acute myeloid leukemia (AML) is a malignant disease with very standardized treatment and diagnosis techniques with a complex patient management. The incidence of AML is low, but it is increasing. Given the fact that the disease is rare one of the major concerns of the medical researcher is to be certain that each patient is accurately registered. Any data loss can reduce the level of significance of the study results.

The authors experience says that a data collection using a database management system doubled by statistic software in front of it could be a successful tool in order to obtain early warnings concerning data quality or, worse, treatment misconduct.

We have already developed a software application for data collection and analysis in AML. [1] Studying the results from previous similar studies we have found that an important tool for this application is missing: the data error warning in the very moment of data collection. Though the fact that the patient management is very well standardized, the evolution of the disease has a high variability and this reflects in the recorded parameters. Our application performs real time biostatistics and offers a snapshot of the results in every moment of the patient data entry allowing the formulation of hypotheses even if the number of cases is rather small.

Our aim is to extend the existing data collection software in order to incorporate the entire statistical protocol that will allow us to have a better real time view on the patient's status.

II. Material and methods

The software mentioned above has a user friendly design aiming to help the investigator to achieve a proper and complete data collection. In order to accomplished these tasks we used wxWidgets [2] library creating the user interface, and R functions [3] in order to estimate the level of statistical significance (p value) in the same time with the data entry. SQLite3 [4] was used as Database Management System. We have also use an open source software in developing these applications, in order to avoid legal issues (software license) and to provide to a potential investigator with low or no founding a more useful PC assisted tool for his research.

In this context we aim to propose a real time biostatistics software with practical application in AML patient management, concerning the survival analysis curves (Kaplan Meyer) and log rank test. The factors implied in stratifying the survival can be selected in the same time.

Each processed patient can modify the results in real time concerning the statistical outcomes (p value and Kaplan Meyer curves).

The patients with AML selected and included in this study were hospitalized in the Hematological Department, County Emergency Clinical Hospital Tirgu Mures.

III. Results

We created a GUI in wxWidgets to collect the desired medical data. We collected information regarding: age, date

S. Vlad and R.V. Ciupa (Eds.): MEDITECH 2011, IFMBE Proceedings 36, pp. 32–35, 2011.
www.springerlink.com

of diagnosis, date of death, blood count values, the CD leukocyte markers detected by flow cytometry. Survival period is automatically calculated in months, years or days of investigator's free will. You can enter and edit in real-time database cases, changes that are made are saved in real time and statistical results are automatically updated. Standardizing the entry modes of data does not allow incomplete data entry and the interference of data with subsequent processing of data. For binary data, it will also enable the introduction of binary values and also a variant with lack of value.

We have created a very friendly interface in order to asses in real time Kaplan Meyer curves (right window). Also we can evaluate a significant difference between the survival curves applying log rank test (left window). (Fig. 1) Each patient recorded can produce effects in drawn curves at once he/she is introduced in the database.

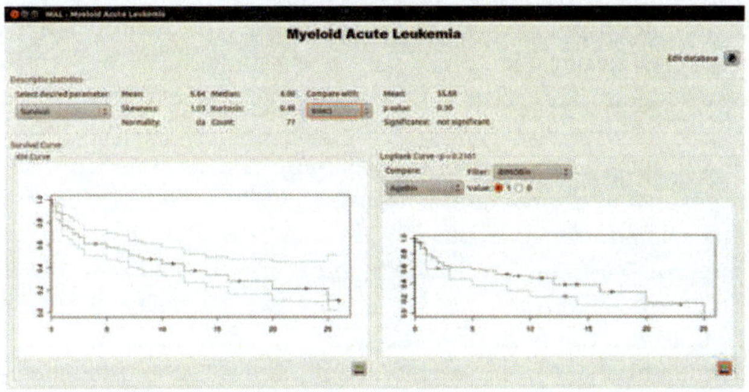

Fig. 1 Real time analysis interface with automatic survival curve (Kaplan Mayer) with 95% Confidence Interval (right window) and log rank test (left window)

For assessment of and testing for survival according to blood count values at diagnosis is needed binarization of quantitative results. This operation is specifically permitted in the program presented by changing the threshold value of binarization. It is known that there is no threshold value, for which the survival prognosis can be said to be good or bad (with some exceptions). In order to clarify these challenges the program provides the solution (Fig. 2).

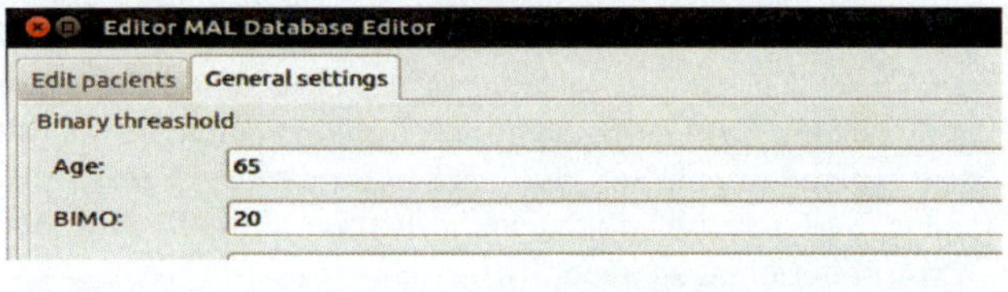

Fig. 2 Menu "General Settings" for establishing threshold values for laboratory measurements

The variable of interest which could generate differences in survival can be selected using a "combo box" (Fig. 3).

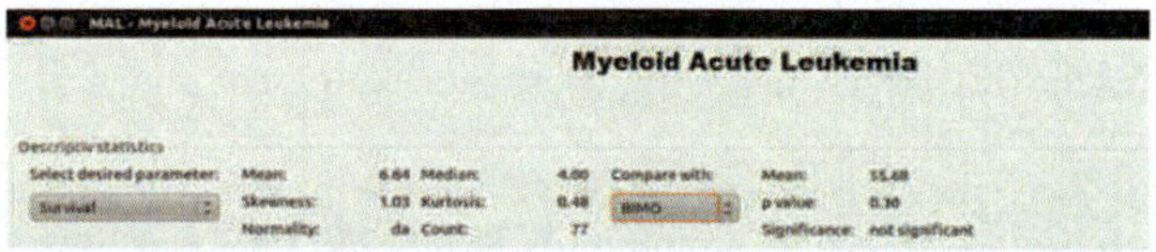

Fig. 3 Selecting the interest variable

Descriptive statistical estimators are calculated instantly and the results of log rank test as inferential analysis are listed above. (Fig. 4)

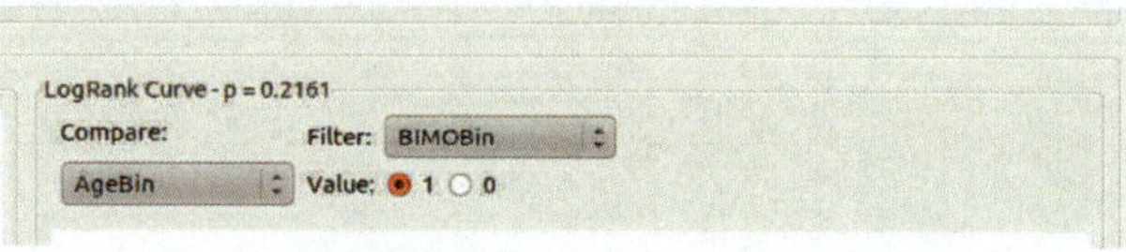

Fig. 4 Real time p value for log rank test

The time factor. In AML the incidence is low, so each new case is important to be carefully managed. By introducing records (patients with full data) the outcome is changing. In real time we can evidence of changes that can be really important for the treatment and the prognosis of the disease.

According to "real time biostatistics" the concept that we proposed, our software can provide a step by step update of information.

When introducing real cases in the AML database in order to compare survival periods, the program will draw a Kaplan Meyer curve for the studied variables including each new value recorded in. Also, using the "combo box" "Compare:" for selection of the binary variable which is suspected to modify the survival, a log rank test is performed and the outcome is displayed above.

To achieve the goal we have tested differences issued by adding new records in two different points in time. The results concerning descriptive statistics are shown below (Fig. 5).

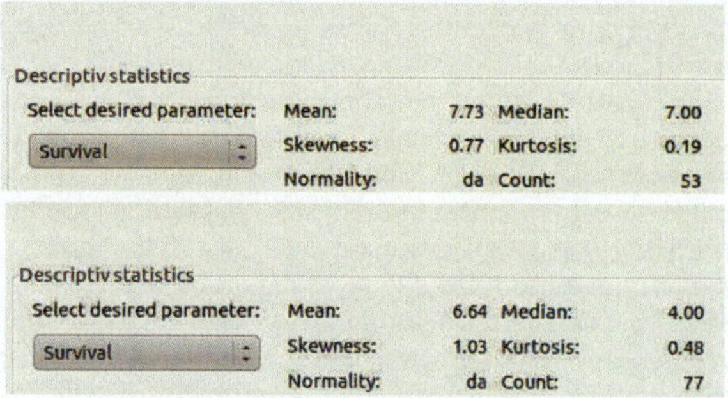

Fig. 5 Real time descriptive statistics

The program calculates descriptive statistics parameters as shown above (mean, skewness, normality test, mean median, kurtosis).

The software we propose gives visual information to the physician, at two or more points in time, the Kaplan Meyer curves, and the log rank test results for the variables tested. (Fig. 6)

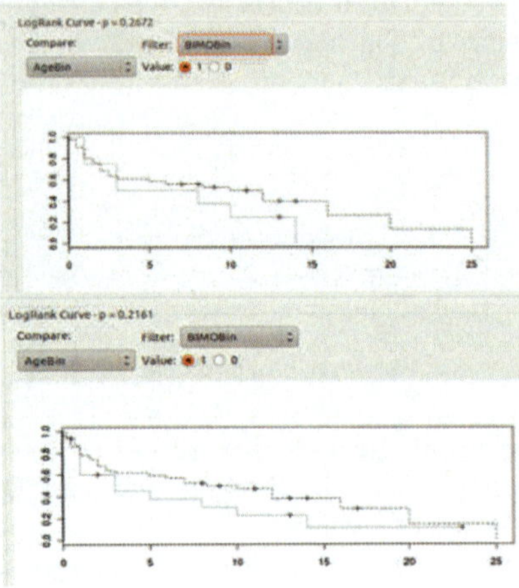

Fig. 6 Real time Survival curves (log rank test)

IV. DISSCUSIONS

For writing this program we used an open source GUI library called wxWidgets. This library is an open source construction kit for writing complex C++ interfaces and applications for a variety of platforms (Windows, Linux, Mac OS X and even Pocket PC). This library has been around for a long time now, contains mature code with a lot of functionality that helps a programmer in its work for building quality software. Its cross platform intrinsic nature allows us to create a software that behaves identically in all major operating systems and that is a big plus in the research activity, allowing for all involved researcher to use the operating system that suites them best without fear that it will create incompatibilities in the collected data.

In order to store the information for the patients we needed a database management system. We wanted our software to be portable and easily maintainable so we decided to use an embedded DBMS, SQLite3. It supports a big part of the SQL standard, it is fast and requires low memory overhead. But the most important facility is that it keeps the data in a single file that can be moved to another computer without affecting data integrity. We avoided in this way the bloating of the software that a regular DBMS would have added, the necessity of keeping a server for the data and to have a network to access it, in the same time being able to use the flexibility that the SQL language offers in data processing.

This wxWidgets library unfortunately doesn't contain any specialized section for statistical/biostatistical processing so we had to use something else for that. We decided to go with "R", a free software environment and language for statistical computing and graphics. It is a well known statistical software in the academic community. The preference for this software is given by the fact that it is also an open source software, it is supported on most platforms and has a modular design and has a shell that allows for scripting and bulk processing. Its facilities can be extended using packages downloaded from different mirrors. Many of its statistical extensions have been written by third party developers and researchers from different universities. So, it offers solutions for a variety of different problems making it a logical choice for inclusion in our program. The main problem was how to interface the R software with the GUI written in wxWidgets. The solution has been given by two extensions of "R", Rcpp and RInside. The first one, Rcpp transforms the R specific objects in objects more similar with C++ design and implementation, allowing C/C++ code to be run from inside the R shell. The former, RInside, as its name suggests contains the headers and libraries needed to interface with Rcpp objects. Using this two packages we were able to use the R statistical abilities in order to see how our results evolved in time as we could add more and more patients to the database.

Another facility that we considered useful in the program was the way information is saved during data entry. Any modification to the patient data is saved instantly without the need for any special action from the user. This helps a lot when there is a lot of data to be entered and also prevents loose of data if the user forgets pressing a "Save" button.

Any modification to the data also triggers a recalculation of the statistical parameters involved in the study so that the researcher can always have a clean and clear picture of the results.

Medically it is important to asses a difference between the mean of "Survival" variable. This finding should alert the physician if the mean decrease or increase. He/she must reevaluate the patients in order to detect which is the cause of this event. Evaluating a cohort of patients only after a full data collection can not give such information in real time, and a perturbation given can influence in both ways the AML patient management.[5]

Our precedent studies of prognosis in AML showed the necessity of such real time software. AML is an malignant disease with poor outcome concerning the survival period and the improvement of treatment is expected to produce real time changing of the survival curve. AML is the disease where there is no rule regarding some parameters (cell blood count); there is no typical pattern of the disease that makes the software more valuable concerning an uniform data entry and data quality control.[6]

The existence of such of software is useful from the methodological point of view because a more accurate study design is possible; possibility of bias regarding data collection is diminished and that certify the validity of the study.

We have started the development of this software since 2008, relying on the lab hematologist, methodologist and biostatician's experience and the possibility of real time interpretation of data came out like a natural necessity.

So it is a team where everybody's contribution it is considered of equal importance.

V. Conclusions

The presence of open source software helps the researcher to develop useful tools in each medical field of interest.

Team work is a very important in the development of such software that will be used in particularly medical research.

Acknowledgment

This paper is partially supported by the Sectorial Operational Programme Human Resources Development, financed from the European Social Fund and by the Romanian Government under the contract number POSDRU/89/1.5/S/60782.

References

1. Anca Bacârea, Bogdan Adnan Haifa, Marius Muji, Alexandru Şchiopu (2011) Software Application For Data Collection And Analysis In Acute Myeloid Leukemia. Applied Medical Informatics Vol. 28, No. 1/2011, Pp: 16-22.
2. ***Wxwidgets Cross Platform Gui Library [Online][Cited Decemcer 2010]. Available From : Url: Http://Www.Wxwidgets.Org
3. *** R Project For Statistical Computing [Online][Cited Decemcer 2010]. Available From : Url: Http://Www.R-Project.Org
4. *** Sqlite [Online] [Cited 2010 December]. Available From : Url: Www.Sqlite.Org
5. Davis, Bh; Holden, Jt; Bene, Mc, Et Al. (2006). 2006 Bethesda International Consensus Recommendations On The Flow Cytometric Immunophenotypic Analysis Of Hematolymphoid Neoplasia: Medical Indications Cytometry Part B-Clinical Cytometry 72: S5-S13
6. Vardiman J W, Thiele J, Arber D A , Et Al. (2009) The 2008 Revision Of The World Health Organization (Who) Classification Of Myeloid Neoplasms And Acute Leukemia: Rationale And Important Changes Blood 114: 937-951

Author: V. Bacarea
Institute: University of Medicine and Pharmacy
Street: 38 Gh. Marinescu
City: Tirgu Mures
Country: Romania
Email: bacarea@yahoo.com

Database External Level Architecture for Use in Healthcare Information Systems

P. Olah[1], D. Dobru[1], R.V. Ciupa[2], M. Marusteri[1], V. Bacarea[1], and M. Muji[2]

[1] University of Medicine and Pharmacy, Tirgu-Mures, Romania
[2] Technical University of Cluj-Napoca, Cluj-Napoca, Romania

Abstract— **Healthcare Information Systems rely very often on complex database structures and require complex processing of the data in order to offer relevant information to the user. In an attempt to extend the benefits deriving from the structural approach present at the internal level of such a database to its external level, this paper proposes a database external level architecture based mostly on server side views, supported by client side views and a minimal number of server side stored procedures.**

Keywords— **healthcare information systems, database, user interface.**

I. Introduction

The external level of the databases used in Healthcare Information Systems is in general designed and built by the application development team. This approach is probably enforced by the fact that Healthcare Information Systems require highly specialized solutions from an Information Technology perspective [1]. We argue in this paper that although this level is usually application oriented, it should also benefit from the advantages of structure orientation provided by the design of the underlying database.

The denormalisation of the data, via views and stored procedures, is done usually by the application developers or at their request. This approach presents three drawbacks:

- low time-performances of the resulting system, especially when there are complex application requirements
- an inconsistent behavior of the different parts of the user interface built by different application designers but based on the same data collection
- difficulties in system maintenance and further development deriving from lack of control over the external level of the database

In this paper we will propose a database external level architecture that is more data oriented [2], based mostly on server side views, supported by client side views and a minimal number of server side stored procedures. This architecture offers fairly complex functionality while greatly reducing the above mentioned drawbacks of the traditional approach.

II. Remote views versus internal views

The two basic functions of the user interface are to retrieve data from the database and to offer the user the possibility to modify that data.

Queries are, in most cases, more than a simple data access tool. They often embed a significant part of the systems "intelligence". Their construction consumes a significant amount of man-hours, the structural information of the query level being of vital importance for a large system.

There are several ways to store the structure of a query. The three most usual storage forms are:

- *Client procedures* - Often programmers use SQL syntax in their client code to directly access the data in the database. This technique may work fine in case of small systems, but when things get bigger and most of all complex, the lack of centralized control over this information will inevitably lead to a dead end. Maintenance and consistency assurance will be almost impossible.
- *Views* - Building a layer of views to access data is a better way to start. Views are actually labeled snippets of SQL code, thus providing a structured way to store query architecture information. Views can be stored either directly in the database or in the data environment of the client application (if the development software of the client supports it). A notable advantage of views is the possibility to refer another view, thus providing a way to build hierarchical architectures.
- *Server-side Stored Procedures* - Stored procedures are usually used to implement complex procedural business logic but they also provide a good way to access the data from the interface. Stored procedures can retrieve and update data using parameters passed by the caller (in this case by the user interface of client application)

To develop our HIS (Healthcare Information System), we have used views to implement the external database level. Most of our views are external (i.e. stored in the client's data environment). This approach offered us the advantage of *having automated data updating*. The client development

S. Vlad and R.V. Ciupa (Eds.): MEDITECH 2011, IFMBE Proceedings 36, pp. 36–39, 2011.
www.springerlink.com

environment we used features it's own relational machine, offering a remote view mechanism. This means that the structure of the query is stored in the client's database, but the actual execution is done on the database server. The interface works with a local cursor, the client environment granting an automatic dynamic generation of *select, update, delete* commands on the server.

Another major advantage in using external views is that they can be *parameterized*. This offers a very easy and natural way to implement dynamic filtering. The back-draw of this approach is the fact that the system becomes *dependent on the client application,* part of the data source being stored at client level.

The independence of the data source against the application can be achieved either by using internal views or by implementing a layer of stored-procedures to provide data access. Any of these approaches would store the structural information of the query level directly in the database.

Internal views do not support parameters, thus to provide dynamic filtering they need to be queried either trough stored procedures or trough client code. An alternative would be to migrate the parameters into the database using a special table, which will be included in the views as needed. This would replace every filter clause with a join. Overall time performance of the system may drop.

Internal views also have some severe limitations regarding update and deletion of rows from the base tables. The updating mechanism must be provided at the client level. Depending on the client environment, this may be transparent to the programmer or it may need some extra code to be implemented.

Stored procedures represent a sound way to provide data retrieval and modification services. They are able to retrieve record-sets to the client application and they accept parameters.

Some maintenance problem may although arise. The structure of these procedures is not standardized as is the SQL syntax, thus being programmer-dependent. Conventions may be enforced, but at a large scale 100% adherence to a set of conventions is difficult to achieve. It is also difficult to build hierarchical query structures using stored-procedures.

Building a hybrid architecture, which uses internal views to provide a basic query level and stored procedures to provide dynamic filtering and data modification services, would probably be a good solution.

III. Solutions for the External Level Design and Administration in Medical Information Systems

We will present in this chapter some of the most important issues related to the external database level that we have encountered during the design and development of our HIS. The suggested solutions have been successfully implemented on a real system.

A. *The Filter Processing*

Working with large data sets yields the need for a dynamic filtering mechanism. One of the ways to implement such an instrument is to use parameterized client-side views. Specialized interface objects provide the values of the parameters at run-time, the view being re-queried every time a filtering criteria gets changed by the user.

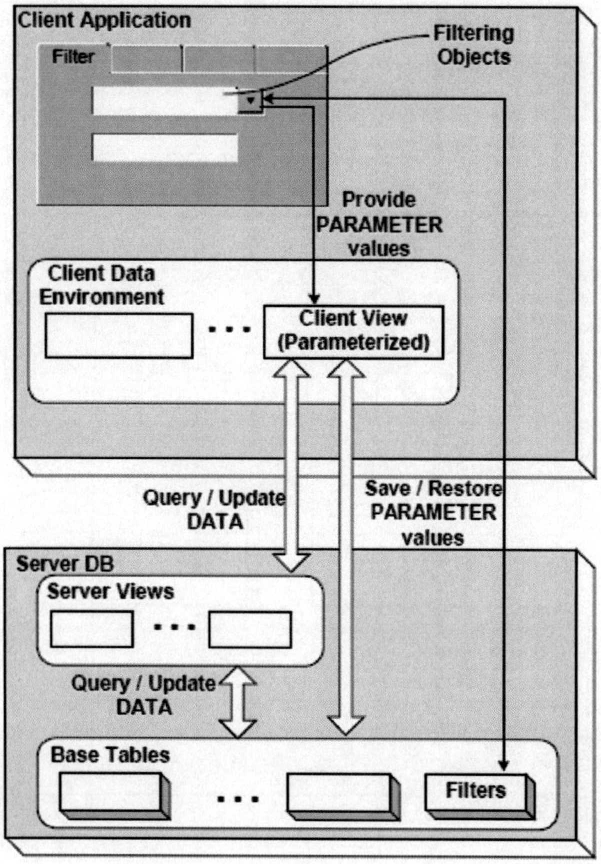

Fig. 1

Remote views are executed on the server machine, thus using them does not affect the overall performance of the system. Although they have a disadvantage regarding the system architecture: part of the business logic becomes embedded in the client application. An alternative would be the use of parameterized stored procedures to retrieve and update data. This technique would grant the independence of the data source against the application level by embedding all the business logic into the database.

B. The Order Processing

Server –side views do not allow an order clause to be specified. The ordering of the returned rows must be specified in the queries that refer these views. Client views, on the other hand, support an ordering clause. However, the ordering cannot be specified using parameters, so the solution for dynamic ordering is to build more than one remote-views which are identical except for the ordering expression.

On application level a specialized interface object selects the current record-set depending on the ordering specified by the user. All the other interface objects will refer to the data provided by this current record-set.

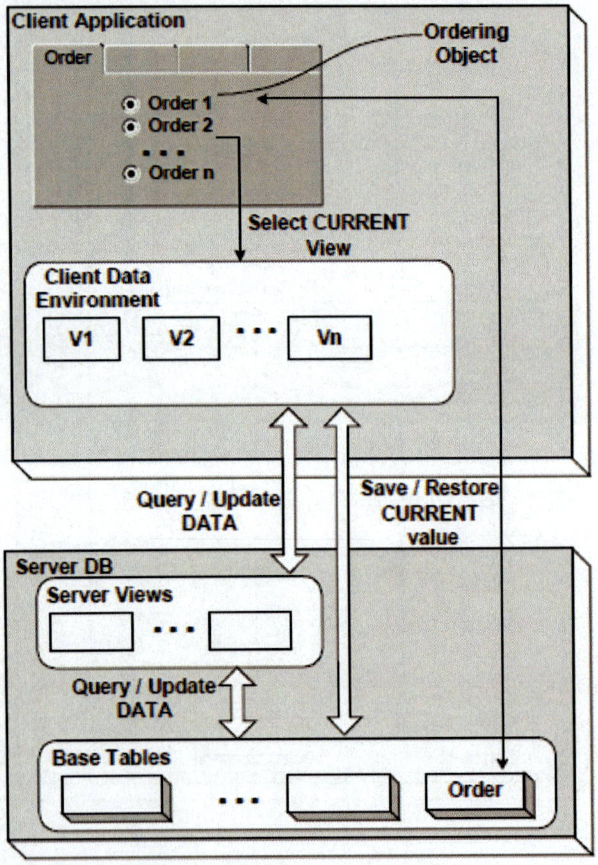

Fig. 2

C. The Synchronization Process

One of the main concepts, related to the interface model we propose for a HIS, is synchronization. We consider this to be a vital mechanism for the dynamic behavior of the user interface. There are three major types of synchronization present in our interface model, each of them being implemented in a different way, regarding the external level of the database.

a) Synchronization between Primitive Objects

One of the best ways to reduce the error susceptibility of an information system is to provide the user with limited and valid choices. A set of choices often depends on another choice made earlier during the interaction with the system.

A common mechanism to implement such a dependency is filtering the future options based on a current selection. To accomplish this, we have used at the external level parameterized views.

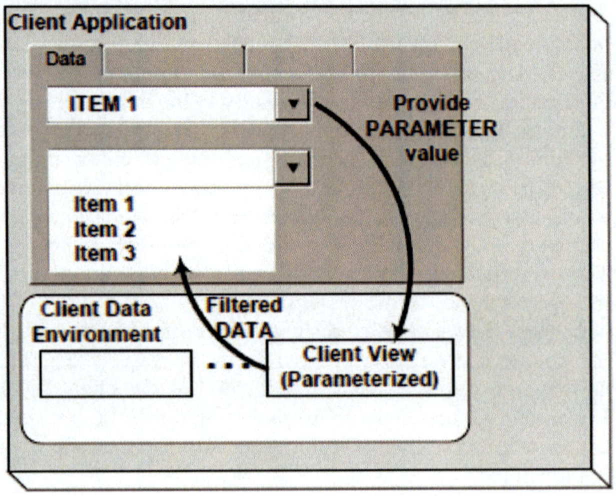

Fig. 3

Every time the user makes a selection one or more parameters are declared and one or more views, which will provide the data for future choices, are re-queried. The process is executed by interface objects instantiated from specialized object classes.

b) Synchronization between Pages

Our interface model uses page-frames to grant the user access to the entities from the data source. Each page represents a different instrument (i.e. detail view, list view, filtering, ordering, links to other page-frames). Referring the same underlying data, the pages must implement some synchronization mechanism to offer a consistent view to the end-user. A simple technique is used in order to accomplish this goal: all the interface objects refer the same record-set in the client data environment. This provides an implicit synchronization.

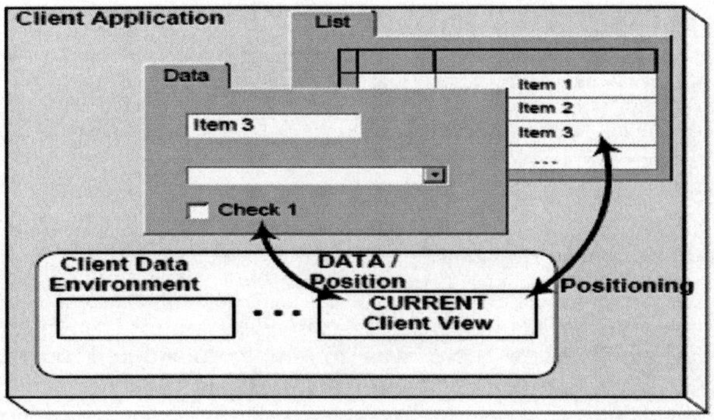

Fig. 4

For instance the grid on the list-view page positions the current record indicator of the record-set. Upon changing the inspection mode to the detailed-view, the controls on this page will read and modify the data from the current record. The filtering objects will re-query only the current record-set whilst the order object will set one record-set as the default cursor for all the other objects.

c) Synchronization between Page Frames

The interface model we have developed suggests that every entity should be represented by a page-frame. But, what about related entities? Typical one-to-many (1-n) relationships are usually represented in master-slave type screens. This technique represents more than one entity in the same interface object (form) denying the independent representation for one or more of these entities. Instead of this approach we chose to represent each entity separately providing a synchronization / de-synchronization mechanism.

Regarding the external database level this basically means to filter one record-set (child cursor) according to data from the current record in another record-set (parent cursor).

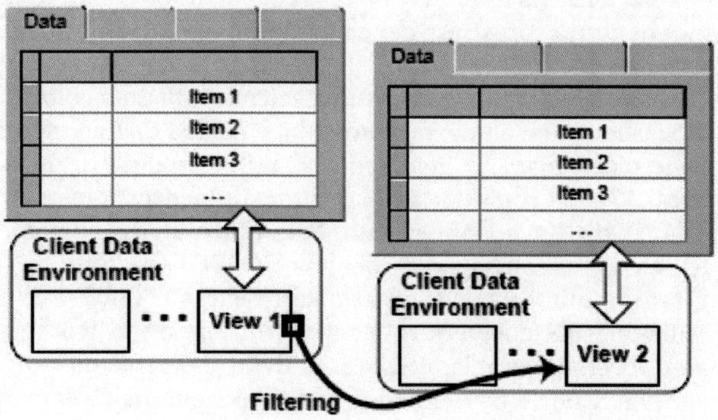

Fig. 5

The relevant data from the current record of the parent is extracted using some information stored in interface object properties. This data is then loaded into variables, which in turn are passed to the parameterized view representing the child in the relationship.

De-synchronization is done by passing neutral values to the child view, so that it will not be filtered according to these specific criteria.

IV. Conclusions

The combined use of server side and client side views (provided that the client programming environment supports them) offers sufficient flexibility to build user interface elements of basic to medium complexity. This approach yields the benefits of keeping the development of the external level of the database, which hosts the user views according to the ANSI/SPARC three levels architecture [3], under control by the database designers themselves. If more complex functionality is needed for the user interface, a structured system of stored procedures is needed to complete the interface of the database towards the application level of the system.

Acknowledgment

This paper is partly supported by the Sectorial Operational Programme Human Resources Development (SOP HRD), financed from the European Social Fund and by the Romanian Government under the contract number POSDRU 64331.

References

1. Muji M, Ciupa R V et al (2009) Database Design Patterns for Healthcare Information Systems, IFMBE Proceedings 26, pp. 63–66, MEDITECH 2009
2. Lewis, Bill. (2007) Data-Oriented Application Engineering: An Idea Whose Time Has Returned. The Data Administration Newsletter - TDAN.com. January 1, 2007
3. Date, C. J. (2003) An Introduction to Database Systems (8th edition).s.l. : Addison-Wesley

Author: Peter Olah
Institute: Univ Med & Pharm Tg-Mures
Street: 38 Gheorghe Marinescu
City: Tirgu-Mures
Country: ROMANIA
Email: olah_peter@yahoo.com

Beat-by-Beat Variability of QT ECG-Interval Holds a Well-Founded Promise for Clinical Cardiology: A Review

R. Negoescu

National Institute of Public Health, Bucharest, Romania

Abstract— **The advent of reliable software access to beat-by-beat QT interval series opened a novel insight to sympathetic control of ventricular myocardium.**

Experimental measurement of time constant in the RR-QT system (τ_{QT}) using pharmacological or electrophysiological manipulations of heart rate (HR) has documented that τ_{QT} size may introduce in borderline subjects in normal life settings episodes of functional long-QT syndrome.

In type I young diabetics in the earliest phase of autonomic dysfunction, spectral QT and RR analyses suggest preferential vagal down-regulation but preservation of sympathetic responsiveness of ventricles to life challenges.

Both findings signal an early risk worth to be dealt with by specific preventive steps.

Keywords— **Beat-by-beat QT and RR time series, power spectra of variability, long-QT syndrome, young type I diabetics, minimal autonomic dysfunction.**

I. Introduction

The beat-by-beat variability of the ECG's RR interval - currently referred to by heart rate (HR) variability (HRV) has been introduced to the cardiology clinic in '70s [1], and its measurement instrumentation and software techniques are now stabilized and largely accepted by the clinical user [2].

QT interval is the sum of ventricular depolarization and repolarization durations with beat-by-beat variability residing by far in repolarization changes. First efficient algorithms for automatic, software measurements of beat-by-beat QT intervals in high temporal resolution ECG facsimiles were published in 1992 [3]. Subsequently, experimental work in open chest, anesthetized dog conducted by Huang, Negoescu et al [4] put face to face direct intramyocardial pressure and QT under various mechanical and pharmacological interventions; closely parallel patterns have documented the indexation value of QT interval for the force of ventricular contraction on a beat-by-beat basis.

RR and QT are related by the classical Bazett's relationship that predicts mean QT corresponding to a standard mean RR of 1 second, in the form of corrected QT: $QT_c = QT/[RR]^{1/2}$. This is a "static" equation, because implies averaging over epochs of several minutes, and has been challenged when there was minutely considered beat-by beat instead of mean QT and RR intervals [5].

Beat by beat QT variability amounts aproximately 1/10 of that of RR interval; this time domain relation corresponds to a contrast of about 1/100 in the spectral domain.

QT Fourier spectral power derived from high resolution ECG, clusters the same way as in the better known RR spectrum, that is within 0.1 Hz, low frequency (LF) or "sympathetic" band (standardized now at 0.04-0.15 Hz) and "respiratory" band (standardized 0.15-0.4 Hz) around the breathing rhythm [2].

Since QT traces partly reproduces variations seen in RR, question arose whether QT-LF power relates to RR interval's LF modulation, mostly brought by right cardiac vagus, or, differently, whether QT-LF mainly reflects sympathetic control of ventricles.

In this vein, a study suggested QT-LF spectral power has a double subordination: during relaxed rest it follows mostly RR-LF fluctuation; during mental stress a RR-independent contribution is added, which presumably reflects the sympathetic status of ventricles [6].

While sympathetical *message* became clear enough within the QT-LF band, the neural systems involved remain unclear since both α_1 and β_1 - (sympathetic) receptors and, still, muscarinic (vagal) stimulation can modulate repolarization in myocardial fibers in the 0.1 Hz rhythm present in every autonomic efferent traffic to the heart.

Sympathetic (exciting) and/or vagal (stabilizing) vehicle of such a modulation may be consequential for QT-LF reading in terms of stress-dependent arrhythmia and sudden death risk.

Subsequently, a study using selective pharmacological blockade of the above receptors documented that an adrenergic mechanism is involved in QT-LF response to mental stress [7]. To filter out atrial control influences transmitted to QT by its coupling with RR, other study computed RR-by-QT mean squared coherence (MSC) spectrum to get QT-LF's idio(proper to)ventricular fraction (IV QT-LF) by removing RR-coherent influences [8]. Putting to work this new spectral variable, it appeared that QT-LF features ordinarily a better sensitivity to mental stress but its IV fraction is robust vis-à-vis of confusing influences from either RR-LF or up and downs in mean RR [9].

S. Vlad and R.V. Ciupa (Eds.): MEDITECH 2011, IFMBE Proceedings 36, pp. 40–43, 2011.
www.springerlink.com

II. Long time-constant in the RR - QT coupling

The long-QT syndrome (LQTS) is a rare (1/20,000) congenital or acquired (medicinal drugs -related) heart condition in which a delayed repolarization follows a quite normal depolarization (excitation) of the heart, so that interval QT is abnormally prolonged; this way Bazett corrected QT is usually over 440 ms in males and over 460 ms in females [10,11]. LQTS is associated with syncope (fainting) due to ventricular arrhythmias, possibly of the type known as torsade de pointes, that can then deteriorate into ventricular fibrillation and ultimately into sudden cardiac death (SCD). In terms of measurement, a QT_c above 440 milliseconds is considered longer than normal, suggesting either a borderline setting or a clear cut LQTS. Subjects with borderline QT_c - i.e. 450 to 470 ms in males and 460 to 480 in females - have a prolongation of QT_c, not enough to clearly make a diagnosis. The average QT_c for someone who has long-QT syndrome is 490 ms [12]. Sudden death in untreated long-QT patients is ten times higher than that recorded in general population and underlain primarily by more or less abrupt vagal withdrawal, and secondarily by sympathetic going-up [13].

Observations in normals collected after the advent of software measurement of QT interval suggested a slow response of QT when RR briskly changes 100 ms or more as under rapid tilting or under speedy standing up, i.e. a large time constant (τ_{QT}) in the RR - QT dynamic system.

A large τ_{QT} may keep QT a significant lapse of time after a brisk decrease in RR fairly over the value a prescribed by Bazzet, inducing for that time a prolonged QT_c setting, similar in appearance (because it is functional) with the pathologic long-QT syndrome. The more, borderline (at the limit of LQTS) but otherwise healthy subjects may experience during trivial life episodes a risk comparable with that of the patent LQTS cases when a new atrial depolarization (P-wave) is initiated well over the tail of an abnormally prolonged T-wave and may trigger arrhythmic events (like torsade de pointes).

First study dealing with clinical assessment of true τ_{QT} [14] used 15 observations (obs.) in normal 10 young males, using the exponential-like decrease of beat-by-beat QT in response to a step decrease in RR got by right atrial pacing. A positive RR step was then created by ceasing the pacing and allowing HR to recover towards sinus rhythm. Given the exponential-like dynamics of QT response, the time constant τ resulted directly as the interval from beginning of change to intersect of the tangent to the exponential curve with the final level. Other day investigators manipulated HR using vagal or sympathetic & vagal pharmacological blockade using standard protocols (propranolol 0.2 mg/kgbody over 15 min and/or atropine 0.04 mg/kgbody over 2 min). Assuming exponential-shaped response curves both in RR and QT, effectual (or true) τ_{QT} results as difference between experimental τ_{RR} and τ_{QT} (21 obs.).

All interventions were done at relaxed rest while recording a thoracic ECG lead with best-visible T-wave, digitized at 1ms. Beat-by-beat RR and QT time series (ms) were derived using a published detection software resulting in QTs within +/- 1.12 ms of the values got by the tangent method used in the clinic. Finally RR and QT time series were evenly re-sampled at 400 ms for time plot analysis.

Under moderate level RR step decreases (mid-signal excitation), RR - QT system behaved quite linearly: various τ_{QT} were fairly close to 48 s. With larger negative RR steps there was a larger τ_{QT} dispersion. Conversely, with larger magnitude RR step increases from pacing to sinus rhythm, τ_{QT} was about 55 s. τ_{QT} dependence on step's direction and magnitude points to the non-linearity of the system.

During blockade, the true τ_{QT} (about 40 s) was lower than τ_{QT} found under various pacing interventions, suggesting a certain speeding-up of the RR - QT system when vagal control gets away through atropinization.

Subsequently investigators simulated using Matlab the QT_c overshoots during/after leading edges of sustained (5 min) vagal withdrawal that mimics current life stresses in active healthy people.

Figure 1 presents live recordings of RR, QT and QT_c plots in a quasi-denervated heart, diabetic female patient featuring borderline basal QT_c (469 ms). During a moderate HR acceleration from supine to standing (1st event from left to right) QT_c remains 45 s around 510 ms, just over-passing what investigators called standard long-QT syndrome *arrhythmia risk surface,* circumvented by the actual QT_c curve and the long-QT threshold taken here at QT_c=500 ms.

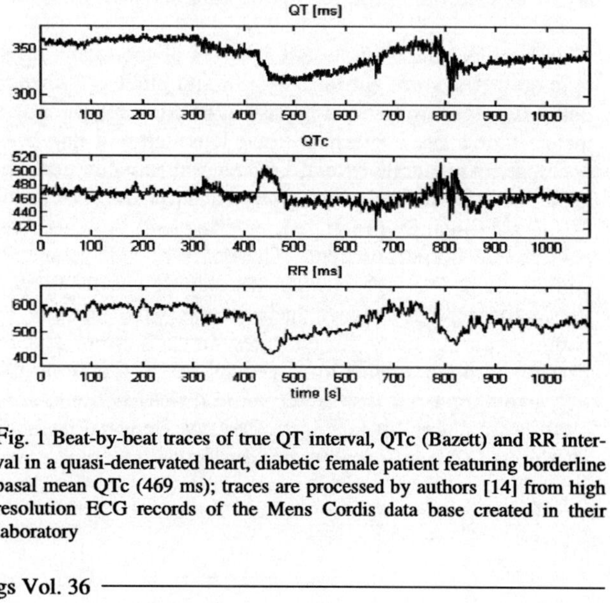

Fig. 1 Beat-by-beat traces of true QT interval, QTc (Bazett) and RR interval in a quasi-denervated heart, diabetic female patient featuring borderline basal mean QTc (469 ms); traces are processed by authors [14] from high resolution ECG records of the Mens Cordis data base created in their laboratory

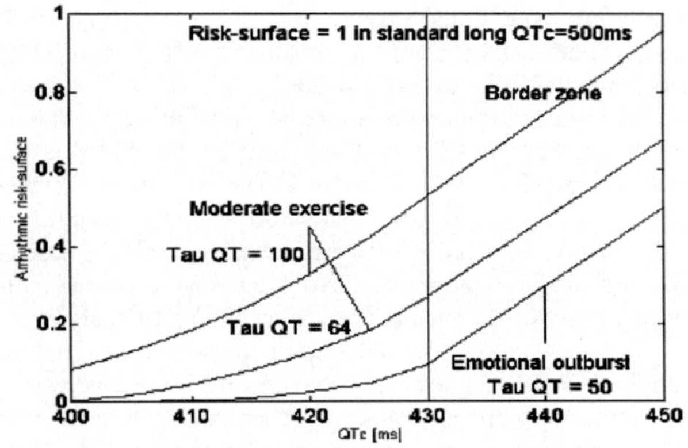

Fig. 2 Risk surfaces got by computer simulation [14].

Simulation dealt with QT_c overshoots occurring on HR outbursts that could possibly take place in sustained, short-term (5 minutes) life challenges in healthy. Figure 2 shows dependence on basal QT_c of the arrhythmia risk surface in man under:

a) an "emotional outburst" molded on pacing to pacing data (τ_{RR} =2 s, τ_{QT} =50 s), and

b) a "moderate exercise induction" with an atropine-like vagal withdrawal pattern (τ_{RR}=24 s, with either normal τ_{QT} =64 s or longer τ_{QT} =100 s).

Ordinate value 1 corresponded to the risk surface for a diagnosed long-QT syndrome patient with basic QT_c = 500 ms over the same 5 minute epoch.

Risk surfaces for borderline QT_c subjects approach quite worrisome the arrhythmic risk of the long-QT patient; during such a functional long-QT episode in borderline normal subjects, a sympathetical extra stimulus may encounter an arrhythmic terrain close to that of patients with high SCD risk.

In healthy young adults RR changes of moderate amplitude are always present in daily life and elicit QT changes delayed according to a time constant of 40 to 60 seconds. It means that a not-neglectible time lapse after a quick RR decrease of moderate extent, QT interval remains over that prescribed for normals by Bazett equation, i.e. correected QT is significantly prolonged, entering and perhaps overpassing the borderline zone. The delay of QT versus RR could possibly be higher when RR step-downs are bigger - i.e. in the case of sudden tachycardias - that are not uncommon during daily life episodes. In such circumstances, passing through minute-long epochs of patent long-QT-like arrhythmia risk distributed more or less frequent along daily normal activities of normal or quasi-normal individuals (in the QT length sense) appears as unavoidable.

III. The earliest autonomic deregulation in type I diabetes

Diabetes mellitus is an important risk factor for cardiovascular diseases: 75 percent of all diabetes deaths occurs due to cardiac or cerebro-vascular causes. Risk of myocardial infarction in diabetics is 3-5 times higher than in nondiabetics; whereas risk of stroke is double. Cardiac impairment by sino-atrial and ventricular dysinnervation, including sometimes inter-ventricular sympathetic unbalance and long QT syndrome, is apriori the most important risk factor for SCD in old diabetic patients.

On this background, auto- or cross- spectral analysis of RR and QT beat-by-beat variability, as derived from high temporal resolution (1-2ms) telemetric or Holter ECG recordings, became attractive for early detection of the cardiac autonomic lesion in young insulin-dependent diabetic patients, whose central mechanisms modulating efferent traffic to the heart are active.

Such a study [15] focused on 27 consecutive insulin-dependent patients, free of significant clinical cardiovascular or neurological involvements, who laid 10 min in relaxed rest (baseline), then experienced a stress- interview - 10 min, and finally stood up (standing) - 7 min. A thoracic EGG lead was digitized at 1 ms, RR & QT intervals were detected and Fourier-transformed over 3 minute steady state RR epochs to get low frequency spectral power (LF: 0.04-0.15 Hz). RR-by-QT mean square coherence (cross) spectrum was used to unveil the RR-independent (proper to ventricles or idio-ventricular, IV) fraction from the QT (auto) spectrum.

Investigators detected mild autonomic impairments in 16 patients while standing in form of either RR shortening > 200 ms; or normal RR-shortening but no RR variance; or yet stiff RR around 600 ms & no variance; these patients were excluded from what follows. The other 11 patients (5 males and 6 females, 34.9 ± 6.7 years SD, diabetes history of 7.5 y ± 5.4 years) were compared with 11 healthy gender and age-paired controls from lab's data base.

Table 1 compares averages of patients/controls groups under the same intervention (baseline, stress or standing), or of pairs of interventions within the same group. Inter-groups mean RR differed highly significantly in standing, on a background of mild tachycardia in patients. RR-LF is significantly lower in diabetics versus normals in all interventions. This held true for QT-LF (except standing) but did not for IV QT-LF. Intra-groups, there were similar responses to interventions in all variables, except patients' mean RR that did not shorten under stress.

In patients versus controls, significant decreases of RR-LF in every intervention paralleled by mild to moderate decreases of mean RR suggest withdrawal of vagal drive, known as vehicle of every RR modulation.

Table 1 Effects of the interventions and health status on the main RR and QT spectral variables. There were 11 patients in every group; entries are mean group averages +/- standard deviations.

Mann-Whitney U - Wilcoxon: ⟷ $p < 0.01$, ⟷ $p < 0.05$, ⟷ p=0.06

Variable	mean RR [ms]		RR-LF [ms²]		QT-LF [ms²]		IV QT-LF [ms²]	
Intervent	norms	diabs	norms	diabs	norms	diabs	norms	diabs
Stress	**785** ± 54	**758** ± 51	**1662** ± 932	**555** ± 320	**4.80** ± 2.2	**2.35** ± 1.6	**2.24** ± 1.4	**1.30** ± 0.9
Baseline	***849*** *±75*	***775*** *±90*	***606*** *±329*	***246*** *±186*	***2.19*** *±2.9*	***0.81*** *±0.5*	***1.22*** *±2.1*	***0.46*** *±0.3*
Standing	**735** ± 61	**641** ± 70	**613** ± 369	**231** ± 132	**3.50** ± 2.0	**4.10** ± 4.2	**1.70** ± 1.6	**1.53** ± 1.4

Differently, similarity between controls and patients with respect to IV QT-LF response suggests preservation of sympathetic drive to ventricles under mental stress.

Investigators concluded that in young insulin-dependent diabetic patients in the earliest stage of disease a sympathetic ventricular response to mental stress as found in normal subjects coexists with a vagal withdrawal at sino-atrial level. Taken together these findings point to an early arrhythmic risk that should not be neglected.

IV. Conclusion

Beat-by-beat QT interval time series and its variability in spectral domain prove useful in detecting subclinical regulatory disturbances in subjects or patients, worth to be dealth with primary or secondary prevention.

With the first illustration, direct experimental estimations of time constant in the RR-QT system (τ_{QT}) using pharmacological or electrophysiological manipulations of heart rate (HR) has documented through computer simulation that τ_{QT} size may introduce in borderline subjects under current life challenges episodes of functional long-QT syndrome featuring "risk surfaces" comparable with those of patent long-QT patients associating a sudden death risk 10 times higher than general population. Such a functional long-QT phenomenon suggests that anti-arrhythmic protection (beta blocking drugs) ought be preventively extended over borderline long-QT subjects as well.

In the same vein, spectral RR-by-QT analyses conducted in type I young diabetics in the earliest phase of autonomic dysfunction have suggested preferential parasympathetic down-regulation while preservation of sympathetic drive to ventricles under mental stress introduces an early arrhythmic risk to be contained by behavioral and pharmacologic steps.

References

1. Eckberg DL, Dempsey JA, Pack AI, (1995) Respiratory Sinus Arrhythmia and Other Human Cardiovascular Neural Periodicities Regulation of breathing, eds.
2. Camm J, Malik M, (1996) Heart rate variability: standards of measurements, physiological interpretation, and clinical use, Eur. Heart J. , 17, 354-381.
3. Negoescu R, Porges SW, Goddart P, Richardson DC, Carlton B, (1992) Short term oscillations in systolic time intervals, Proceed. IEEE-EMBS Ann. Int. Conf. Paris, France, V. 14, pp. 506-507.
4. Huang MH, Negoescu R, Horackova M, Wolf S, Armour JA, (1996) Polysensory response characteristics of dorsal root ganglion neurones that may serve sensory functions during myocardial ischaemia, Cardiovascular Research, 32, p. 503-515.
5. Huang MH, Ebey J, Wolf S, (1991) Heart rate-QT interval Relationship During Postural Change and Exercise. A possible connection to cardiac contractility, Integr. Physiol. Behav. Science, 26, pp. 5-17.
6. Negoescu R, Dincă-Panaitescu S, Filcescu V, Ionescu DD, Wolf S (1997), Mental stress enhances the sympathetic fraction of QT variability in an RR-independent way, Integr. Physiol. Behav. Science, 32, no. 3, p. 220-226.
7. Negoescu R, Dincă-Panaitescu S, Ionescu DD, Filcescu V (1997) , Mechanism of 0.1 Hz modulation of QT interval by mental stress is adrenergic, Eur. Heart J.– XIX Congr. Eur. Soc. Cardiol., Stockholm.
8. Dincă-Panaitescu S, Ionescu DD, Filcescu V, Negoescu R, (1996) Spectral RR*QT coherence can spot the sympathetic-drive-to-ventricles fraction of QT power spectrum, Eur. Heart J.– XVIII Congr. Eur. Soc. Cardiol., 17, p.102, Birmingham..
9. Dincă-Panaitescu S, Dincă-Panaitescu M, Achim A, Negoescu R, (1999) Idioventricular low frequency oscillation in QT interval responds univocally to RR confusing kinds of mental stress, Integr. Physiol. Behav. Science, 34, no. 2, p. 10-18.
10. Moss AJ, Robinson JL, (2002) The Long-QT Syndrom, Circulation, 105:784.
11. Rossenbacker T, Priori SG, (2007) Clinical diagnosis of long-QT syndrome: back to the calliper, European Heart Journal 28(5):527-528.
12. www. qtsyndrome.ch.
13. Schwartz PJ, (2005) Management of long-QT syndrome, Nat Clin Pract Cardiovasc Med. V 2(7), pp. 346-51.
14. Negoescu R., Dinca-Panaitescu S, Ionescu DD, Achim A, (2009) Long time-constant of the RR - QT coupling may cause episodes of functional long-QT in borderline subjects, 2nd Int Conf on e-Health and Bioengineering - EHB, Iasi-Constanta.
15. Negoescu R., Istratescu O, Dinca-Panaitescu S (2009), The earliest autonomic dysfunction in type I young diabetics: preferential vagal down-regulation but preservation of sympathetic responsiveness of ventricles to life challenges, Int Conf Eur. Assoc. Cardiovasc. Prev. Rehab. EUROPREVENT, Stockholm.

Author: R. Negoescu
Institute: National Inst. Public Health
Street: 1-3, Dr Leonte Str.
City: Bucharest
Country: Romania
Email: radu.negoescu@insp.gov.ro

Monitoring System for a Medical Facility Using the OPC Platform

V.D. Zaharia and F. Dragan

Department of Electric Measurements, Faculty of Electric Engineering, Technical University of Cluj-Napoca, Cluj-Napoca, Romania

Abstract— **Efficient parameter monitoring and real-time medical data acquisition and presentation can prove to be crucial, especially in a medical unit treating patients with severe affections and highly unstable conditions. This paper proposes a monitoring system for the vital parameters of patients hospitalized in a Intensive Care Unit based on the OPC architecture.**

Keywords— **parameter, monitoring, real-time, vital, OPC.**

I. INTRODUCTION

Real-time patient monitoring in a medical unit is a big concern of today's medical system, especially in the case of Intensive Care Units (ICU). The complexity of the patient monitoring issue grows as the number of patients and parameters to be monitored is getting higher, while the medical staff has a constant number of members, often required to perform different tasks in various areas of the medical unit. Therefore, a centralized monitoring system may be a key factor in the improvement of the medical care activities.

A viable solution for such a centralized system is provided by the OPC (OLE for Process Control) platform, which has been designed to eliminate the communication incompatibilities between distributed devices using different communication interfaces and protocols.

II. DATA ACQUISITION OF PHISIOLOGYCAL PARAMETERS

Intensive care unit equipment includes patient monitoring, life support and emergency resuscitation devices and diagnostic devices. The concern of this present paper, however, is to propose an interconnection method for the monitoring devices in the ICU, and an interfacing method for real-time data presentation to the medical staff assigned to the monitored unit.

Depending on the affections or the state of each patient, various physiologic parameters are monitored by using different devices for each of them. Some of these parameters and the monitoring devices are presented below:

- Acute care physiologic monitoring system—comprehensive patient monitoring systems that can be configured to continuously measure and display a number of parameters via electrodes and sensors that are connected to the patient. These may include the electrical activity of the heart via an EKG, respiration rate (breathing), blood pressure, body temperature, cardiac output, and amount of oxygen and carbon dioxide in the blood. Each patient bed in an ICU has a physiologic monitor that measure these body activities. All monitors are networked to a central nurses' station.
- Pulse oximeter—monitors the arterial hemoglobin oxygen saturation (oxygen level) of the patient's blood with a sensor clipped over the index finger.
- Intracranial pressure monitor—measures the pressure of fluid in the brain in patients with head trauma or other conditions affecting the brain (such as tumors, edema, or hemorrhage). These devices warn of elevated pressure and record or display pressure trends. Intracranial pressure monitoring may be a capability included in a physiologic monitor.
- Apnea monitor—continuously monitors breathing via electrodes or sensors placed on the patient. An apnea monitor detects cessation of breathing in infants and adults with high risk of respiratory failure, displays respiration parameters, and triggers an alarm if a certain amount of time passes without a patient's breath being detected. Apnea monitoring may be a capability included in a physiologic monitor. [1]

This paper proposes an interconnection and interfacing method for the medical equipment enumerated above based on the OPC Client/Server architecture.

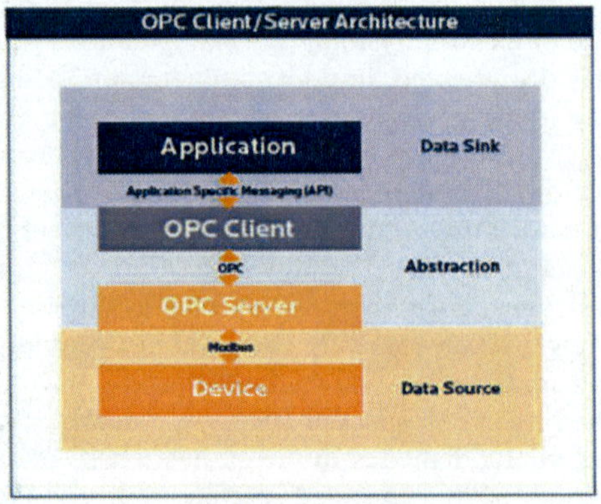

Fig. 1 The OPC Client/Server Architecture [2]

S. Vlad and R.V. Ciupa (Eds.): MEDITECH 2011, IFMBE Proceedings 36, pp. 44–47, 2011.
www.springerlink.com

The OPC architecture consists of two main elements: the OPC Server and the OPC Client applications. These two elements represent an intermediate layer in the communication process between a data source and a data sink. [2]

An OPC Server is a software application, a "standardized" driver, specifically written to comply with one or more OPC specifications. OPC Servers are connectors that may be thought of as translators between the OPC world and a Data Source's native communication protocol or interface. The OPC Client and Server applications relationship being bi-directional, this means OPC Servers can both read-from and write-to a Data Source. The OPC Client/OPC Server relationship is also a Master/Slave one which means one OPC Server will only transfer data to/from a Data Source if an OPC Client commands it to. [3]

An OPC Client software is written to communicate with OPC "connectors". It uses messaging defined by an appropriate OPC Foundation specification. OPC Clients represent a data-sink. They initiate and control communications with OPC Servers based on the request of the embedding application. OPC Clients translate a given application's communication requests into an OPC equivalent request and send it to the appropriate OPC Server for processing. In turn, when OPC data returns from the OPC Server, the OPC Client translates it back into the application's native format, so the application can properly work with the data. [3]

Given the OPC platform's ability to overcome connection and communication issues which may occur between monitoring devices in a medical facility, and due to its components structure, this paper proposes a concept of integrated monitoring system, based on an OPC Client/Server backbone.

III. PROPOSED APPLICATION EXAMPLE

In order to connect the data acquisition system to a remote terminal for data presentation and logging purposes, an OPC Client/Server platform has been configured.

The OPC Server allows the user to define communication channels with the distributed devices. Each channel is configured according to the remote devices' communication interfaces and protocols. The next step of the OPC Server's configuration is adding devices in their respective communication channels, which have been defined according to their specifications.

Finally, tags are defined for each device connected to the channel. The tags connect the OPC Server to memory registers in the remote devices. Each tag corresponds to a parameter monitored by the remote device. In the OPC Architecture tags have three attributes: **value, quality** and **timestamp**.

Each device defined in the OPC Server application can parent one or more tag groups, which can be organized by the user in order to meet the particular needs generated by the structure of the system being supervised.

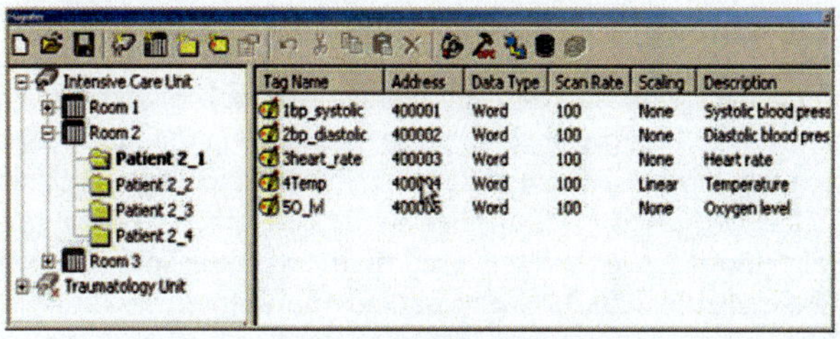

Fig. 2 The OPC Server Application

Given the present paper's purpose to propose a interfacing model for a medical unit, the OPC Server application will be configured according to the following table:

Table 1 OPC Server configuration

OPC Server item	Monitored entity item
Channel	Medical unit/floor
Device	Room
Tag group	Patient identifier/bed
Tags	Physiological parameters: - heart activity – EKG - respiration rate - blood pressure - body temperature - O and CO_2 levels in blood

During the OPC Client configuration sequence, a connection to the desired server must be established, and afterwards, items are added to the respective server connection. Each item defined in the OPC Client is bound with a tag defined in the OPC Server application. Each tag's value, quality and timestamp are displayed in the OPC Client application, after the connection to the server is established.

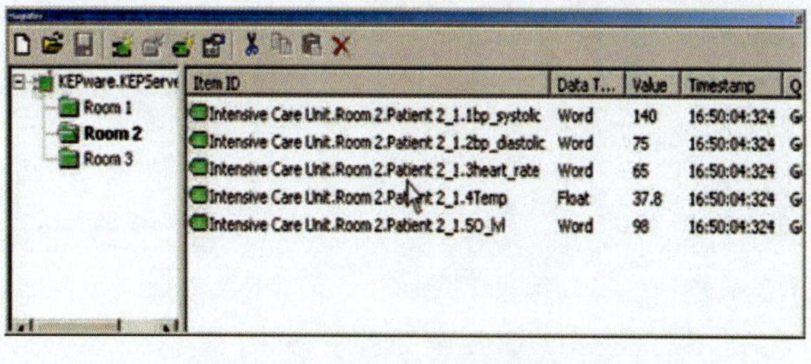

Fig. 3 The OPC Client Application

However, in order for the data to be easily read and interpreted by the medical personnel, the paper proposes the implementation of a customized Human Machine Interface (HMI) application, for explicit information display. Furthermore, the HMI application can be programmed to visually and acoustically warn the authorized personnel in case one or more parameters' values are higher or lower than the predefined values. The limit values for each parameter can be modified by the specialized personnel via the same interface, according to each patient's current state and affections.

The proposed HMI application is conceived to have multiple interfacing layers. The main window presents the whole medical unit. Every patient represented in this window has a warning indicator attached, which is activated when a monitored parameter falls out of the preset interval, considered to be safe.

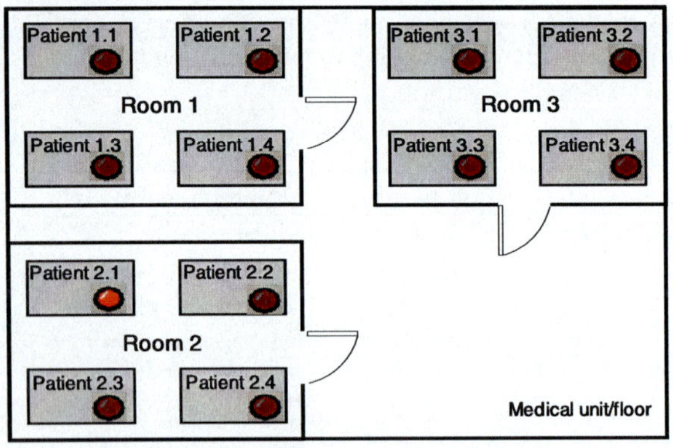

Fig. 4 The main window of the HMI application

The supervisory personnel can choose to view each patient's detailed information by clicking on its respective icon in the main window.

The "Detailed information" window displays real-time values for the monitored physiological parameters. For each parameter, a warning light is attached to indicate weather or not the parameters' values are in the predefined interval. [4]

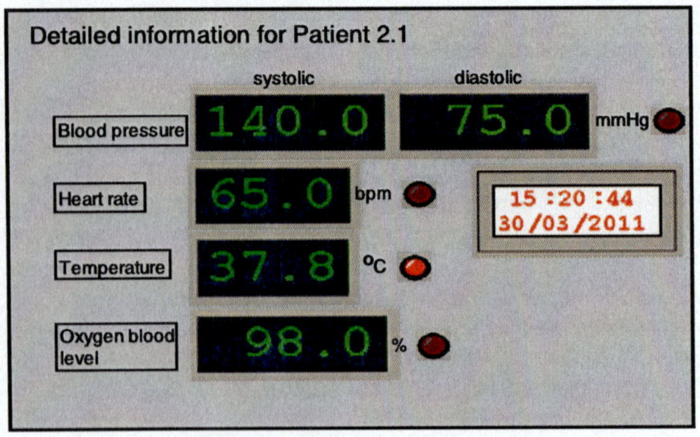

Fig. 5 The "Detailed information" window

Besides showing the real-time values of the monitored parameters, the HMI application can be configured to display the time evolution of the monitored parameters (blood pressure, heart rate, body temperature, level of Oxygen in blood etc.) for each patient. By clicking on each parameter's tag, a new window is opened, displaying the time variation of the respective parameter.

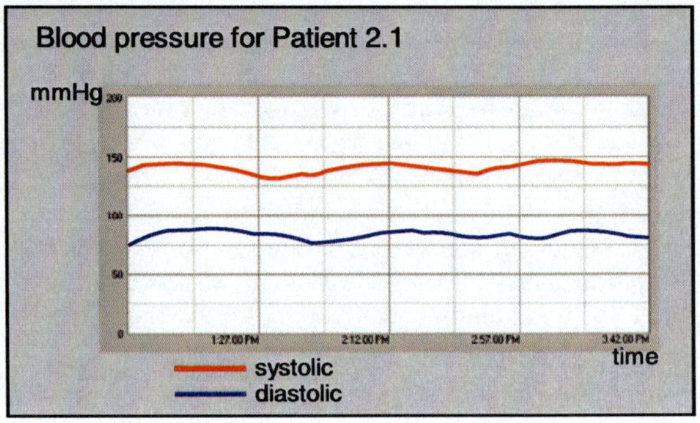

Fig. 6 Evolution in time of blood pressure for a certain patient

The OPC Client application can also be configured to send an SMS alert to one or more members of the medical staff, containing the respective patient's coordinates (room, name) and the abnormal parameter values.

Another feature presented by the OPC Client application that can be used in a system for medical parameters monitoring is the "DataLogger" plug-in. This plug-in facilitates the connection of an online OPC Client application to a database management application, such as Microsoft Access. [5]

The data collected from the monitored medical unit is stored in a customized database which is bound to the OPC Client application via the DataLogger module. The information stored in the database can be presented in the form of printed daily reports, or it can be exported to other applications, such as Microsoft Excel, for further calculations and interpretation.

The data logging module allows the configuration of an alarms and events logging database [6], which is populated with data only in case that the monitored parameters' values do not fit the predefined intervals, considered as safe values for each patient.

By using the data logging method described above, statistic reports regarding specific parameters, patient evolution and unit efficiency can easily be generated.

The information logged in the database associated to a certain medical unit can be linked to information contained in databases created for other medical units from the same hospital, providing real-time information on all the patients being hospitalized at a certain moment. Statistic reports concerning similar medical units in a certain hospital or from numerous hospitals can be easily generated.

id	BP_S	BP_D	HR	%_O	Temperature	QUALITY	TIMESTAMP
1	140	67	65	97	37.1	192	3/2/2011 3:19:43 PM
2	141	69	65	98	37.1	192	3/2/2011 3:20:43 PM
3	141	71	67	99	37.2	192	3/2/2011 3:21:06 PM
4	141	74	70	97	36.9	192	3/2/2011 3:22:07 PM
5	142	75	74	98	36.8	192	3/2/2011 3:23:07 PM
6	144	75	76	96	36.9	192	3/2/2011 3:24:07 PM
7	145	77	76	95	37	192	3/2/2011 3:25:07 PM
8	144	81	74	94	37.1	192	3/2/2011 3:26:07 PM
9	146	83	70	96	37.2	192	3/2/2011 3:27:17 PM
10	149	72	69	98	37.3	192	3/2/2011 3:28:19 PM
11	152	77	73	97	37.3	192	3/2/2011 3:29:21 PM
12	151	76	75	98	37.3	192	3/2/2011 3:30:22 PM
13	147	78	69	99	37.4	192	3/2/2011 3:31:22 PM
14	147	81	77	97	37.4	192	3/2/2011 3:32:22 PM
15	149	78	73	98	37.4	192	3/2/2011 3:33:22 PM
16	144	79	65	96	37.5	192	3/2/2011 3:34:29 PM
17	144	67	63	95	37.5	192	3/2/2011 3:35:30 PM
18	143	69	65	94	37.5	192	3/2/2011 3:36:33 PM
19	141	65	65	96	37.5	192	3/2/2011 3:37:34 PM
20	141	64	67	98	37.5	192	3/2/2011 3:38:35 PM
21	142	69	70	98	37.6	192	3/2/2011 3:39:35 PM
22	145	66	74	96	37.6	192	3/2/2011 3:40:37 PM
23	148	65	76	95	37.6	192	3/2/2011 3:41:37 PM
24	149	68	76	94	37.7	192	3/2/2011 3:42:52 PM
25	150	71	74	96	37.7	192	3/2/2011 3:43:52 PM
26	148	71	70	98	37.8	192	3/2/2011 3:44:52 PM
27	150	71	69	97	37.9	192	3/2/2011 3:45:52 PM
28	147	69	73	98	37.9	192	3/2/2011 3:46:52 PM
29	145	69	75	99	37.9	192	3/2/2011 3:47:52 PM
30	143	68	69	97	37.8	192	3/2/2011 3:48:52 PM
31	142	66	77	98	37.7	192	3/2/2011 3:49:52 PM
32	141	65	73	97	37.7	192	3/2/2011 3:50:26 PM

Fig. 7 The database generated for a certain patient

IV. Conclusions

The system concept introduced in this paper represents an easy to use and scalable application for a monitoring system in a medical facility.

Such a system presents numerous advantages for the medical facility it is implemented in, as well as for the entire medical system, in case it would be generally adopted by hospitals and clinics.

The patient's status can be observed in real-time from a remote monitoring central unit, saving the medical personnel the time necessary to perform routine checks on patients, time which can prove to be crucial, especially in intensive care units.

Another important feature of such a distributed system for monitoring a medical unit is the capability to warn the specialized personnel by various means (visually, acoustically, SMS) in case an emergency occurs. Therefore, qualified medical assistance can be offered to the needing patient as fast as possible.

The ability to log the data collected from every patient is also a major advantage of such a system, therefore obtaining statistical reports on individual patients or on medical units from a hospital becomes an easy and less time-consuming task.

Although the presented system is only in a conceptual phase at this time, due to its multiple predicted advantages, the authors of this paper wish to realize an actual experimental system for data collection from medical equipment. The connection between the medical devices and the central monitoring station should be done via the OPC Server/Client platform. Web based remote monitoring interfaces can also be easily developed.

Furthermore, it is the authors' intention to integrate such a monitoring system on a multi-touch cooperative interface, suitable for medical unit management, as described in [7].

Acknowledgment

This paper was supported by the project "Doctoral studies in engineering sciences for developing the knowledge based society-SIDOC" contract no. POSDRU/88/1.5/S/60078, project co-funded from European Social Fund through Sectorial Operational Program Human Resources 2007-2013.

References

1. Savino J. S., William Hanson III C., Gardner T. J., (2000) Cardiothoracic Intensive Care: Operation and Administration, Seminars in Thoracic and Cardiovascular Surgery
2. Kominek D., (2009) OPC: The Ins and Outs to What It's About, Matrikon Inc.
3. OPC Foundation (2003) OPC Data Access Custom Interface Standard Version 3.00,
4. Davies B., Morris T., (1993) Physiological Parameters in Laboratory Animals and Humans, Pharmaceutical Research, Volume 10, Number 7, pp 1093-1095
5. OPC Foundation (2003) OPC Historical Data Access Specification Version 1.20
6. OPC Foundation (2002) Alarms and Events Custom Interface Standard Version 1.10
7. Crişan S., Zaharia V.D. et al. (2009) A Multitouch Collaborative Solution for Measurement Data Visualization, XIX IMEKO World Congress: Fundamental and Applied Metrology, Proceedings, Lisbon, Portugal. 2009, pp 421-424

The Use of Medical Devices in Self Monitoring of Chronic Diseases

S. Mirel[1], S. Pop[2], E. Onaca[2], S. Domnita[2], and V. Mirel[3]

[1] University of Medicine and Pharmacy "Iuliu Hatieganu", Faculty of Pharmacy, Cluj-Napoca, Romania
[2] University of Medicine and Pharmacy "Iuliu Hatieganu", Faculty of Medicine, Cluj-Napoca, Romania
[3] National Institute for Research and Development of Isotopic and Molecular Technologies, Cluj-Napoca

Abstract— **The range and complexity of medical devices used in homecare are increasing. Home medical devices must be safer, easier to use and more accessible for all users. The safety and the efficiency of the medical device used depend on the device's design, users' ability to operate with devices and the used environments for devices. In order to investigate the proper operation of the medical device and to identify sources and the nature of difficulties in their use, there were investigated three self monitoring medical devices which are most frequently used by the patients of Diagnostic and Treatment Center Cluj-Napoca for their chronic disease management. The compliance problems, focused on difficulties which may occur during the use of medical device, suggested that the proper functioning of the medical device depend, first of all, on user' competency.**

Keywords— **medical device, self monitoring, safety, usability.**

I. Introduction

A modern healthcare system uses complex medical technologies, which refers not only to sophisticated medical procedures and new drugs, but also to medical devices. Efforts to improve patient safety and the quality of healthcare delivery must take into account the omnipresence of medical technology. Diagnosis, monitoring, treatment and rehabilitation require the use of some medical devices specific for hospitals, and for home environments. In this context, over the past decade, the use of home healthcare medical device is in a continuous increase.

Home-care is more convenient for monitoring chronic diseases. Self-monitoring of characteristic parameters is recognized as an integral adjunctive tool for specific management of some chronic diseases such as heart disease, diabetes and respiratory problems. These specific parameters (blood pressure, blood glucose, peak expiratory flow rate etc) could give indications about the current status of chronic illnesses and are necessary for optimizing the patient's treatment [1-5]. In order to monitor the medical condition of patients it is important to make the medical devices used more accessible in an application, but also to use the most adequate and the most compatible ones [6, 7].

The transfer of healthcare from hospital to home environment means that these home devices are managed by the user. Due to the fact that the technological developments become more complicated and the delivery of high-tech home care is likely to grow, it is compulsory to deal with three important aspects: the effectiveness, the safety of the medical devices and compliance regarding their utilization [8-10]. The term "user-friendly" is used, generally, to characterize a device as easy to operate or not needing special training. There are considered two components which interact with each other: *user-acceptance* (the extent to which the user agrees using the technology) and *user-competence* (the abilities required to use the technology effectively).

The safety and effective use of medical devices depends on design issues and on cognitive and physical capabilities of the users. The technical dimensions regarding device's physical properties (appearance, size, noise, weight etc.), functionality (fit with the task, usefulness, complexity), mechanical or electrical system (power sources, supplies etc.) and safety, will affect user-acceptance. In this context, the principles of universal design applied to medical devices are: equitable use, ease of use, flexibility in use, perceptible information, tolerance to error, low physical effort and adequate size and space for approach [11]. These self monitoring medical devices vary in their technical and clinical features, and the differences regarding the accuracy of results obtained were reported [12-15]. The human dimensions, which include skills and knowledge, autonomy (self-efficacy, independence, mobility), sociability (social rules, patients' and relatives' attitudes) will influence user-competence. Furthermore, there must be taken into account the fact that people with varying levels of education and technical skills become users of health technology [16 -17].

II. Experimental

A. *Materials and Method*

We selected 3 self monitoring medical devices (SMMD) which are more frequently used by the patients of Diagnostic and Treatment Center Cluj-Napoca for their chronic disease management:

- blood pressure monitors (BPM) -for essential arterial hypertension management

- blood glucose meters (BGM) - for diabet management),
- peak-flow-meters (PFM) - for asthma management.

Our study relies on individual interviews (conducted face to face and through questionnaires) and by directly observing the patients. The interview techniques are used to evaluate the importance of the use of medical devices in monitoring the chronic disease and the consequences associated with the use of SMMD. The patients recruited were requested to complete a questionnaire investigating their SMMD practice. Conversation analysis provided detailed data on the impact of new technologies upon self-monitoring system, and the interactions between patients and health professionals regarding medical device used.

In order to have a complex variety of viewpoints, we included participants of varying age, gender and socio-economic status who lived in the area of Cluj-Napoca city, Romania. The study included 75 patients, with known essential arterial hypertension, diabetes or asthma (under the adequate treatment for the specific diagnosis). The demographic characteristics of the patients included in the study are presented in Tables 1.

The study includes two aspects.

The first questionnaire focused on the reason of choice and use of the medical device (duration of use, the frequency for performed measurements with the SMMD, the control of correct results, the training available or necessary and the importance of self-monitoring (Tables 2).

Table 1 Characteristics of the subjects (medical devices users)

Characteristics of the subjects		Medical device used		
		Blood pressure monitors	blood glucose meters	peak-flow-meters
Number of patients		35	28	13
Mean age (years)		55.8	50, 2	32,5
Age group distribution (years)	10-19	0	3	7
	18-39	7	6	3
	40-59	11	9	2
	60-80	17	10	1
Sex distribution	M	24	13	5
	F	11	15	8
Educational level	Gymnasium	13	7	5
	High school	10	13	2
	Faculty	12	8	6
Disabilities (limitation by other illnesses)	Yes	7	13	1
	No	28	12	12

The second part of the study regarded the usability problems, focusing on the difficulties or errors which may occur during the use of medical device.

Statistical analysis

The results were statistically analyzed using the *Microsoft Office 2003* software. We considered statistical significance $p<0.05$.

Table 2 Characteristics of the use of medical devices

Parameters	Characteristics	BPM (n= 35)	BGM (n=28)	PFM (n= 13)
Duration of use (years)	total duration of use	4	5	1,4
Frequency of use	1–3 times/month	1	0	2
	1–5 times/week	7	8	5
	daily	10	13	4
	more times/day	17	7	2
Person recommending a SMMD	Healthcare person	21	25	11
	Non-healthcare	6	1	1
	Self initiative	8	2	1
Influence in choosing a SMMD	Physician (specialist, GPs)	11	13	11
	Pharmacist	5	8	1
	Sales person (non-healthcare)	14	3	1
	Commercial interests	5	4	0
Training for using the SMMD	Physician. (specialist, GPs)	8	21	12
	Pharmacist	5	3	0
	Sales person	7	2	0
	Self-educated	14	3	1
	Other	6	1	0
Difficulties using SSMD	Yes	27	19	8
	No	8	9	5
Control of SMMD	No	29	24	13
	Yes	5	3	0
Considerents regarding the ability required to operate SMMD	Experience	21	7	1
	Dexterity	4	8	3
	Learning ability	4	4	1
	Medical training	6	9	8
The importance of SMMD in monitoring chronic condition	Very important	28	25	11
	Important	7	3	1
	Insignificant	0	0	1

B. *Results and Discussion*

The data showed that there is not a greater tendency to self-monitoring for active subjects (between 40-60 years) and more educated patients' group than to those ageing (over 60 years) and with medium level of education (44 vs. 42 %, respectively 51 vs. 49 %). The children's user number was lower than the other groups mentioned above, in concordance with the low number patients included for these diagnostics. It must be mentioned that only the costs of glucometers and reagent strips for glucose are covered by the national health system; the others monitoring devices investigated (and their accessories) are supported by the patients.

The majority of patients are influenced by their health-care personnel (especially GPs) when initiating self-monitoring. When an SMMD is recommended it is compulsory to ensure that users are able to safely operate the device. This is why, it is highly suggested that this should be done by the health-care personnel who obtain the relevant medical history of the patient. Moreover, the use of the devices and a correct monitoring on patients must be checked by keeping regular contact with their health-care providers.

Regarding the choice of a certain type of medical device, most of patients are influenced by other non-healthcare personnel or by commercial interests (in the case of blood pressure monitors), or respect the physicians recommendation (for peak-flow-meters). The specific education of the patients is often provided by the personnel working in the specialized clinics as well as by GPs. But many patients do not receive structured education regarding the performed measurements using a medical device. Only 10% of the subjects are used to control their SMMD, in order to verify if the medical devices used shows the correct result.

The abilities of patients, considered necessary for a correct use of a medical device were: experience (38%), medical training (30%), dexterity (20 %), and learning ability (12%). 70% of the patients reported a variety of difficulties when using SSMD. Regarding the role of the SMMD used, most of the subjects considered that it is very important to facilitate the monitoring of their chronic condition.

In the second part of the study we investigated the difficulties which may occur during the use of medical device. The patients used different devices commercially available the most frequently ones used by our patients were:

- blood pressure monitors: by *Omron, Microlife, Sensacare, Braun, A&D, Hartmann., etc*
- blood glucose meters Accu-Check products (by *Roche Diagnostics*), One Touch products (by *Lifescan*), PalmLab (by *Sand Biotechnology Inc*), etc.
- peak-flow-meters: Mini-Wright Peak Flow Meter (by *Clement Clarke*), Asmaplan (by *Vitalograph*), Piko (*Ferraris Respiratory Europe Ltd*)

and their technical characteristics are presented in Tables 3a-3c [12-15,18-20];

Table 3a Characteristics of the Blood pressure meters used

	Device	Memory	PC connection	Irregular Pulse Detection
Upper arm device	MIT Elite *Omron*	90	No	Yes
	MM10 IT *Omron*	84	Yes	Yes
	M1 Plus *Omron*	90	No	Yes
	BP A100 *Microlife*	40	No	Yes
	SAA-102 *Sensacare*	30	No	Yes
	Tensoval duo control *Hartmann*	2x 30	No	Yes
	UA-631 *A&D*	30	No	Yes
Wrist device	R7 *Omron*	90	optional	No
	BP W100 *Microlife*	200	No	Yes
	SAW-102 *Sensacare*	30	No	Yes
	BPM2510 *Braun*	30	No	Yes

Table 3b Characteristics of the Blood glucose meters used

product	Sample size (µl)	Test time (s)	Memory (readings)	PC conection	Alternative site testing
One Touch Select	5	5	350	Yes	Yes
One Touch Ultra	1	5	150	Yes	Yes
One Touch Ultra Easy	1	5	500	No	Yes
Accu-chek Compact	1,5	8	100	Yes	Yes
Accu-chek Advantage	4	40	480	Yes	No
Accu-chek Active	2	5	200	No	No
MultiCare	3	10	250	Yes	No
EasyGluco	1.5	10	200	Yes	No
GlucoSmart	2	10	100	No	No
GlucoS-martPlus	2	10	250	No	No
Palmlab	0.3	10	200	No	No
CleverChek TD4227A	0.7	7	450	Yes	Yes

The automatic devices for self-measurement of blood pressure (which use, generally, the oscillometric technique) are: upper arm and wrist devices. Their technical characteristics could affect the accuracy and reproducibility of the results.

The glucometers differ in several technical aspects such as: amount of blood needed for each test, testing speed, ability to store test results in memory, overall size, cost of the monitoring (device and test strips used).

Table 3c Characteristics of the peak-flow-meters used

Device	Characteristics
Mini-Wright Peak Flow Meter	Range: 50-800 L/min Accuracy: ± 10% Memory: 200 tests
Peak-Flow-Meter ASMAPLAN	Range: 50-800 L/min Accuracy: ±10 % Memory: 96 tests
Peak-Flow-Meter PIKO	Range 15 - 999 L/min Accuracy: ±5 Memory: 96 tests

The performance characteristics of peak flow meters are dependent on the type of transduction system, but the measurement depends significantly on patient's effort and technique.

The most frequent and common sources of errors are: the improper calibration of the device, the battery discharged, disobeying the conditions of measuring recommended by the doctors (Table 4).

Table 4 Common sources of errors

Medical Device	Sources of errors
Blood pressure monitors	- the arm is not kept at heart level during measurement (or incorrect position of the wrist)) - the measures performed don't respect the condition recommended - cuff size not corresponding with the size of patient arm)
Blood glucose meters	- the quantity of blood applied to the reagent strip is insufficient; - the blood is not left on the reagent strip for an adequate period of time - the reagent strips are expired or are stored improperly; - the test is performed under wrong conditions of temperature and humidity; - patients are dehydrated.
Peak-flow-meters	- the voluntary effort of the patient is insufficient - the muscular strength of the patient is poor - the technique used is inadequate - the measures are performed before the use of an inhaled treatment - the cleaning of PEF devices is inadequate - disassembling of the device

Difficulties using certain devices can be caused by a variety of factors and could lead to errors: "use error" and "user error". There are numerous sources of errors associated with the use of medical technology. Problems which could appear depend on the medical devices used, but is often assumed to be caused by the failure of the patient to perform the measurements

The most important factors which can lead to errors are:

- *Medical device design problems*

Every interface between a human and a device contains opportunities for error. Any change in device's design (the replacement of the SMMD type) can lead to extra errors. In order to be safely used by patient, generally, the medical devices are designed by the manufacturer following the idea of simplicity or standardization. The patients' preferences are for devices which having no (or few) external function buttons and work automatically (with little or no human intervention). The effort required by the user to apply a sample correctly was reported.

- *Medical device labeling and information package*

A medical device requires an adequate design, with adequate labeling for the user and a complete information package. The major factor contributing to the user errors is the difficulty that patients have understanding written instructions provided with devices. The confusions caused by new terminology were observed. 60% of our subjects report dissatisfaction with the quality of instructional materials accompanying medical devices, regarding the understanding of labeled instructions. This is the reason why, the information packaging (graphics and text) is preferred. Some of the subjects required an alternative guide such as video or audio.

- *"Transferred" device.*

Some medical devices are handed down from one patient to another, without the intervention / recommendation of physicians or pharmacist. In this case it is possible that the "transferred" device is not appropriate for the intended user, or it is not accompanied by complete, accurate and adequate information.

- *Age and Disability of the User*

The greatest users of home-health services are older people (due to the fact that diabetes, heart and respiratory problems are the top chronic diseases of this age group). The first problem is regarding the understanding of instructions (written or training). In addition to these, there are mentioned the difficulties regarding memory, vision, hearing and language barriers.

We can conclude that patients with physical disabilities or incapacitated by other illnesses are disadvantaged (physically and psychologically) in the use of the new technologies. Their general health, mental and emotional state or the effects of their medication on their abilities must be taken into consideration. For now, the technology is not designed for these special categories.

Other factors which affect the compliance of the patients are:

- *Failing to follow maintenance and calibration procedures*

Medical devices with a measuring function require calibration and special procedures. Checking or setting the calibration of a SMMD guarantees the accuracy of the measurement results.

- *Failing to communicate regularly with health-care professionals*

We could not establish a correlation between the medical product used and the success of monitoring, but we managed to observe a relation between the correct use of device and the results of self monitoring.

It is important that the subjects performing the measurement using SMMD are clearly instructed about the procedure. This implies that the specialized personnel (health-care personnel and sales personnel) must explain in detail the procedure to the patient. A demonstration of the full procedure by the operator to the subject is also required.

A patient needs to know: what SMMD measures; what are the types of SMMD available; how to prepare themselves for performing measurements; what are the common problems encountered in performing the measurement; how to achieve the results; how to interpret the results obtained. 30% of the subjects don't understand how the devices are used, and over 60% of the subjects were not informed about the situations that are likely to cause incidents. The results of our study are presented in figure 1.

An educational effort coming from the medical personnel (physicians, nurses, pharmacists) could improve patient's SMMD performance. An adequate training and support from physicians and nurses focused on skills and routines, increasing the user-friendliness of technology. In this context, we considered necessary to develop educational programs for health-care professionals, physicians, pharmacist and nurses in order to improve their skills and their ability to participate in efforts to enhance patient safety regarding medical devices. The role of pharmacist in recommending, choosing and training regarding must be increased.

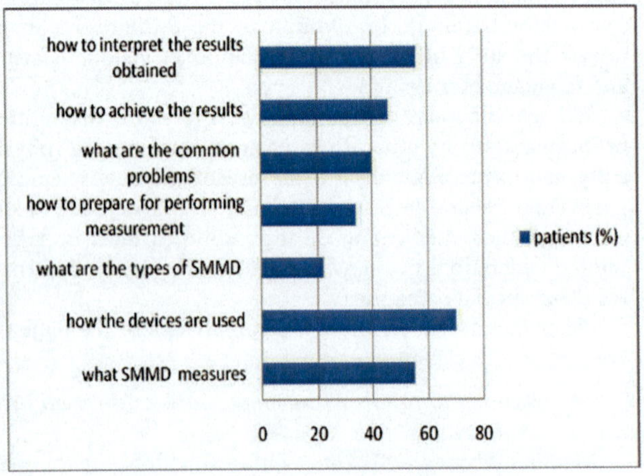

Fig. 1 The results of training regarding SMMD

The specific parameters for certain diseases can be measured by the patient himself and could give indications about the current status of chronic illnesses, which are necessary for optimizing the patient's treatment. However, the optimal interpretation of these measurements and the recommendation of the adequate treatment require physicians' knowledge.

III. Conclusions

The proper operation of the medical device depends on the device's design and on the competency of its user. This is why, when a SMMD is recommended, we have to ensure that users are able to safely operate the device. Medical devices used are of various types available, with different technical characteristics, but we could not establish a direct correlation between the product used and the success of monitoring of chronic disease. The study suggested a strong relation between the correct use of device and the results of self monitoring. The information about the correct use of medical devices must be accurately communicated and recorded in order to obtain the compliance of the patient. The self monitoring using specific medical devices must be checked by keeping regular contact with their health-care providers.

References

1. Gutknecht D R (2005) Review: home or self blood pressure monitoring improves clinic blood pressure in essential hypertension. Evid Based Med10: 40.
2. McManus R J , Ryan A, Greenfield S et all (2007) Self measurement of blood pressure: a community survey Journal of Human Hypertension 21:741–743
3. Bell P M, Walshe K (1983) Benefits of self monitoring of blood glucose. Br Med J (Clin Res Ed) 286 (6373): 1230–1231
4. Franciosi M , Pellegrini F, De Berardis G et all (2001) The Impact of Blood Glucose Self- Monitoring on Metabolic Control and Quality of Life in Type 2 Diabetic Patients. Diabetes Care 24 (11):1870–1877
5. Tierney W M, Roesner J F, Seshadri R et all (2004) Assessing symptoms and peak expiratory flow rate as predictors of asthma exacerbations. J Gen Intern Med 19:237
6. Kaye L W and Davit J (1995) Importation of high technology services into the home in: New developments in home care services for the elderly: Innovations in policy, program, and practice. The Haworth Press, New York
7. J. Zhang, Patel V L, Johnsonet T D et all (2009) Evaluating and Predicting Patient Safety for Medical Devices with Integral Information Technology. Advances in Patient Safety 2: 323-336
8. Lun K C (1995) New user interfaces. International Journal of Biomedical Computing 39: 147–150
9. Lehoux P (2004) Patients' perspectives on high-tech home care: a qualitative inquiry into the user-friendliness of four technologies. BMC Health Serv Res 4: 28
10. Rogers W A, Essa I A, Fisk A D (2007)Designing a technology coach. Ergonomicsin design 3:17-23

11. Follette Story M (2007) Applying The Principles Of Universal Design To Medical Devices in Medical instrumentation: accessibility and usability consideration, CRC Press, 83-91
12. Johnson RN, Baker JR (1999) Analytical error of home glucose monitors: a comparison of 18 systems. Ann Clin Biochem 36:72–79
13. Johnson R N, Baker JR (1998) Accuracy of devices used for self-monitoring of blood glucose. Ann Clin Biochem 35: 68-74
14. Koyama H, Nishimura K, Ikeda A et al (1998) Comparison of four types of portable peak flow meters (Mini-Wright, Assess, Pulmograph and Wright Pocket meters)". Respir Med 2(3): 505-511
15. Stewart MJ, Gough K, Padfield PL (1995) The accuracy of automated blood pressure measuring devices in patients with controlled atrial fibrillation. Journal of Hypertension 13 (3): 297-300
16. Hickman J M, Rogers WA, Fisk A D (2007) Training Older Adults To Use New Technology. The Journals of Gerontology Series B: Psychological Sciences and Social Sciences 62: 77-84, valabil la http://psychsoc.gerontologyjournals.org/cgi/content/
17. Mykityshyn A L, Fisk AD, Rogers W A (2002) Learning to Use a Home Medical Device: Mediating Age-Related Differences with Training Human Factors: The Journal of the Human Factors and Ergonomics Society 44: 354-364
18. Accu-Chek Blood Glucose Monitoring Systems at http://www.accu-chek.com
19. Home products at http://www.omronhealthcare.com/products
20. Manuel devices at http://www.dableducational.org/spygmomanometers/devices

A Wireless System for Monitoring the Progressive Loading of Lower Limb in Post-Traumatic Rehabilitation

F. Neaga[1], D. Moga[2], D. Petreus[3], M. Munteanu[4], and N. Stroia[2]

[1] Salaj County Hospital, Zalau, Romania
[2] Technical University of Cluj-Napoca, Faculty of Automation and Computer Science, Cluj-Napoca, Romania
[3] Technical University of Cluj-Napoca, Faculty of Electronics, Telecommunications and Information Technology, Cluj-Napoca, Romania
[4] Technical University of Cluj-Napoca, Faculty of Electrical Engineering, Cluj-Napoca, Romania

Abstract— **The paper presents a low cost system for wireless monitoring of the progressive loading of lower limb. This system contains a smart device able to measure the plantar pressure, to compare it with predefined levels and to signal the patient in order to avoid excessive loading of the limb in case of patients following post-traumatic rehabilitation.**

Keywords— **Wireless sensor, post-traumatic rehabilitation, plantar pressure.**

I. Introduction

Plantar pressure pattern has been extensively used for diagnosis of foot disorder[14]. Pressure plantar distribution, in conjunction with finite element analysis was also used as a clinical tool to explore the effects of various insoles models[7].

Kinetic studies also require measuring plantar pressure. In this field studies have been conducted with shoes with built-in sensors, such as F-Scan System, which measure force from output with dozens to hundreds of elements, allowing to detect both plantar force and distribution. As indicated in [9], such devices obtain a variety of information but are too costly and complex for practical use such as in rehabilitation, being limited mostly to research.

It is proven in many studies that the instrumentation of inner soles or shoes is essential in characterizing the behavior of the force distribution on the feet [6]. In Rehabilitation Engineering the measurement of the efforts exerted by the lower limbs is necessary for providing the knowledge of the plantar surface forces distribution as well as for giving to the patients, with spinal cord injury or diabetics, the contact sensation with the ground [6].

In the last four decades, a considerable number of attempts were made to use transducers and load cells mounted in or on soles of shoes.

Several examples are enumerated in [5]: strain gauge, spring element under the heel and the fore foot of each shoe to measure vertical reaction and shear forces; circular strain gauges strap under the sole of the shoes for measuring the force in vertical direction, piezoelectric sensor to assess ground reaction force.

Pressure sensors are based on a variety of technologies. The authors of [8] consider that as plantar pressure measurement is considered, the choice is reduced to only two fundamental types: resistive and capacitive.

Usually the physical mechanism involved is that of contact resistance, as in devices commercially known as FSR (Force Sensing Resistor).These are roughly linear in a given range of pressure and care should be taken in considering the border effects and geometry of sensor. Such devices offer the advantage of organization in large resistive arrays with a low impedance that promises good noise immunity of measurements.

The capacitive sensors (generally based on the modification of the thickness of an elastic material present between the plates of a capacitor) are also widely used for pressure measurement. The geometry of the sensor and the border effects should also be carefully considered. It is important to point out that the measurement with capacitive sensor is not as fast as the measurement with resistive ones, as only time-varying test signals may be used. The noise and interference issues are of critical importance, since the small capacitance leads to high impedance devices. An attempt to avoid such problems, by using a PZT ceramic type of sensor (lead zirconate-titanate) is suggested in [13].

The diversity of the approaches in terms of sensor technologies([18], [23], [24], [25]), measurement and simulation techniques ([7], [20], [21]) or measurement data aggregation methods is motivated by the clinical value of such devices ([1], [2], [3], [14],[26], [27]), by the search for feasible and affordable implementations and by their usage in the design process of other devices ([4],[17]). From an implementation point of view, in-shoe plantar pressure measurement provides a challenging measurement environment for the sensor, leading to a associated challenges like crosstalk between elements, errors due to the bending forces and difficulty of calibration [12].

S. Vlad and R.V. Ciupa (Eds.): MEDITECH 2011, IFMBE Proceedings 36, pp. 54–59, 2011.
www.springerlink.com

As reported in [11], the most commonly reported pressure value is the single maximum pressure recorded at critical areas of body, but the average pressure is a more stable measure and gives a better overall picture of interface pressure in disabled people than maximum pressure. A correlation based study of the plantar pressure maps [11] shows that the maximum and average pressure maps reveal almost the same information in the case of static pressure measurement, while for the data measured at normal and fast walking speeds, the correlation coefficients between the cumulative, maximum or average pressure maps had similar magnitudes.

When examining the pressure measurement devices (PMDs), [8] points out that differences in sensor technology, matrix spatial resolution, pressure range, sampling rate, calibration procedures, raw data post-processing, ageing, lead to significant differences in PMD overall accuracy, but it is worth to mention that the performance of a PMD is in fact the result of many factors (mechanical assembling, scanning electronics, and data transfer protocol).A classification of the measurement method is given in [10] (see Figure 1).

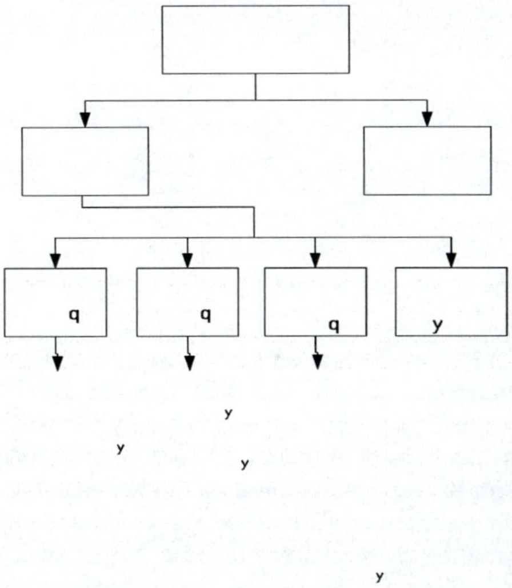

Fig. 1 Classification of pressure measurement methods after [10]

The system described in this paper is intended to the measurement of the force appearing at the foot-sole interface in order to prevent an exaggerate load of then lower limb in the case of the patients going through a post-traumatic rehabilitation process. The measurement of the force appearing at the foot-sole interface is done with a single pressure transducer. The conversion of the pressure into force value is done trough an algorithm implemented on a microcontroller. The measured values can be transferred through a wireless interface. The patient is warned upon the reaching of a physician prescribed force level threshold by acoustic and optical signaling. Such feedback allows the patient to correct the forces developed in walking or in stationary positions to levels appropriate for the post-traumatic rehabilitation process.

II. Instrumented soles

The instrumented soles are the solution for embedding the measurement devices (force or pressure transducers), and the associated hardware for processing and communication of the measurement data.

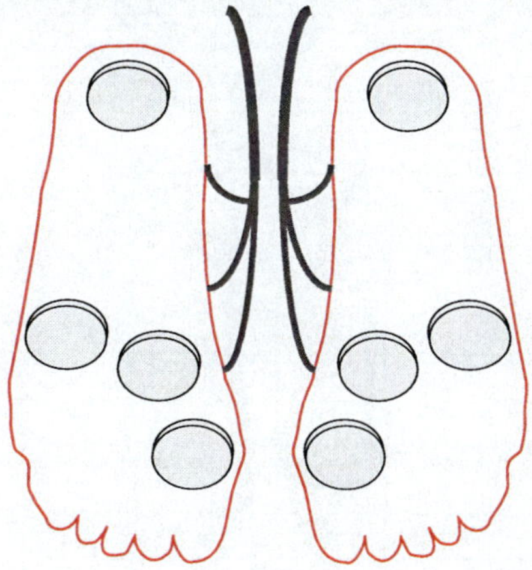

Fig. 2 Instrumented inner sole as presented in [6]

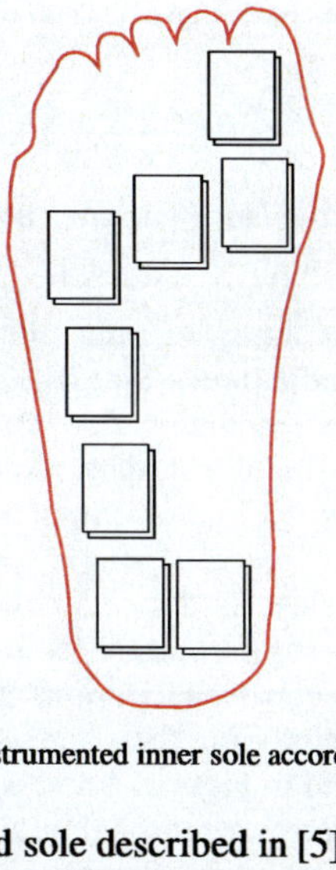

Fig. 3 Instrumented inner sole according to [5]

The instrumented sole described in [5] has total thickness of 15mm and is divided in eight areas defined in relation to the dynamic roll of the foot during the stance phase of gait: two transducers corresponding to the heel (internal and external), two corresponding to the external side of the sole (behind and fore foot), three under the five metatarsal heads and one under the big toe, adding up to eight areas (and transducers) for one shoe (see Figure 3).

Another implementation, described in [6], mentions a reduced number of measurement areas (see Figure 2).

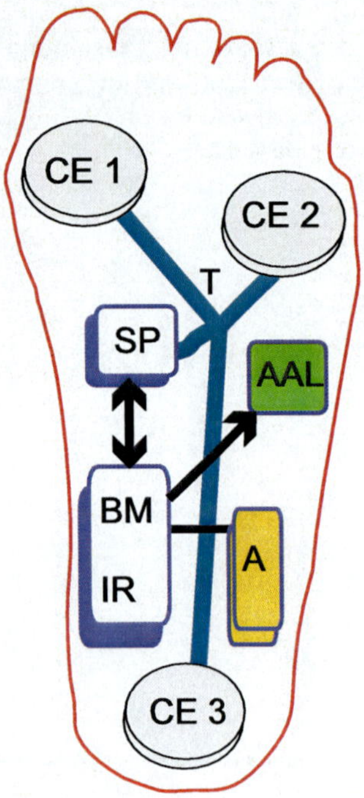

Fig. 4 Proposed setup of the instrumented sole

III. METHOD AND RESULTS

A. *Functionality and Hardware Architecture*

The idea of using a microprocessor-based data-acquisition system for measuring plantar pressures is reported in early works like [28], [29], but without the integration of wireless communication, its implementation are in the manner of umbilical data-acquisition systems. An example of such wired implementation is given in Figure 5 after [10].

A compact wireless hardware architecture for a pressure measurement system appropriate for implementation in the form of an instrumented sole without any wires is presented in this section starting from the setup of [19] (see Figure 4). The setup described in Figures 4 and 6 offers the advantage of measuring the pressure through a single transducer connected to a network of three flexible chambers placed under three plantar areas of maximal pressure. The in-sole device is built around a measurement and control block **BM** having a radio interface **IR,** a nonvolatile memory for calibration data and force thresholds as well as an optical and acoustic warning block **(AAL)**. All subsystems are powered through the power supply **A.**

For the application described in this paper, the value of interest is that of maximal pressing force at foot-sole interface.

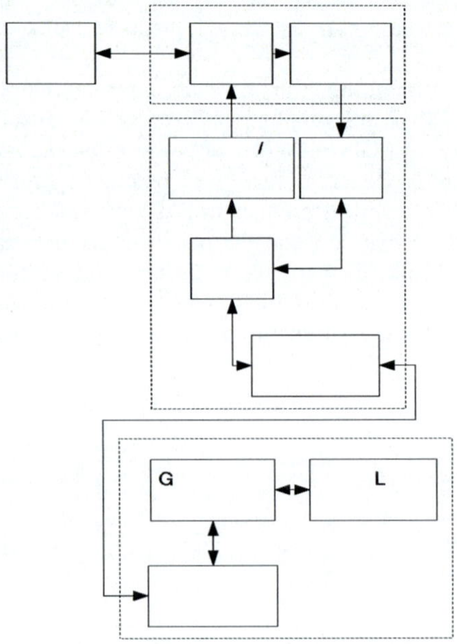

Fig. 5 Wired data-acquisition system for plantar pressure

The output of the pressure transducer **SP** is converted into a numerical value by the **A/D Conversion** block, and further transformed into a force value by the software running on the **Microcontroller.** This value is further compared with the threshold stored on the non-volatile memory. When the maximum of the force values **Fmax**, measured in an programmable measurement time **Tmas,** at a programmable sampling frequency **Fs,** is greater than the current prescribed threshold FP, the measurement and control block generates the appropriate driving signals for the optical and acoustic warning block **(AAL)** (Figure 7) .

The advantage of using the elastic chambers **CE1**, **CE2** and **CE3** connected together through the tube **T** to the pressure transducer **SP**, is that the value sensed by **SP** is always the maximum one due to the balancing of the pressure between the chambers **CE1**, **CE2** and **CE3**.

The operation of the system is controlled by a list of parameters **(T, Fs, FP, D)** stored as a table in the on-volatile memory of the **Microcontroller**.

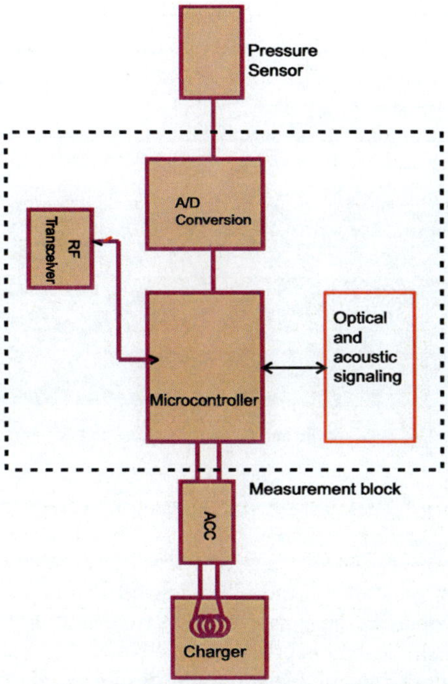

Fig. 6 In-sole wireless device

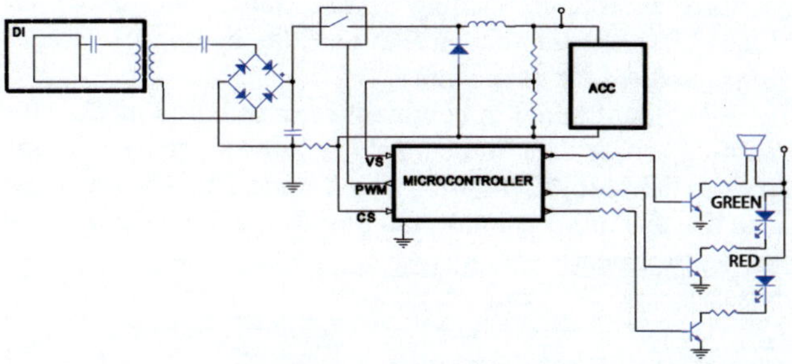

Fig. 7 Schematic of the in-sole device

The radio interface **IR** is implemented with an **RF Transceiver** block that offers bidirectional communication. It allows the physician to read or modify the current values of these parameters in a wireless manner. For gait studies, plantar studies or calibration activities, the measured values can be continuously transmitted toward other systems with compatible transceivers at a rate **Fs/D** (where **D** is the parameter defining a decimation factor). Since a continuous transmission would lead to a high current consumption from **ACC,** the measurement and control block invalidates this function when in the parameters table the current value of **D** is 0.The power block **A** consist from a battery **ACC** and a charger block. The charging control, monitoring of the current consumption, as well as contactless transfer of the energy through inductive coupling from an external charging device **DI**, are functionalities implemented based on the electronics of the charger block (see Figure 7) and on the microcontroller software control routines.

Since the situation of battery exhausting should be avoided, the **Microcontroller** processes the current monitoring data of **ACC** read from the charger block and generates a specific warning sequence for the **AAL,** in order to warn the patient in time.

A prototype device was developed containing an insole device according to the described architecture (Figure 8). Its implementation is based on low-cost silicon pressure transducer, 8-bit microcontroller, serial A/D converter, and WirellesUSB transceivers. The device communicates with a wireless hub that connects to an USB port of a computer running the applications for configuration, calibration and data-acquisition.

Fig. 8 Prototype shoe with in-sole device and warning device

B. *Wireless Communication with the In-Sole Device*

The wireless communication with the in-sole device using the WirelessUSB N:1 Protocol is presented in what follows in accordance with [15], and[16]. This protocol is optimized for low-power Sensors that require battery life to be measured in years, not days or months. In order to accommodate extremely power-sensitive Sensors the Hub is typically assumed to be powered constantly by an external power supply. The WirelessUSB N:1 Hub (Figure 9) interfaces with a local host in order to form a N:1 Star Network. The WirelessUSB N:1 protocol utilizes the unlicensed 2.4 GHz Industrial, Scientific, and Medical (ISM) band for wireless connectivity. Each WirelessUSB N:1 network uses a subset of channels spread across the 2.4 GHz frequency band in order to minimize the probability of interference from other WirelessUSB networks, while reducing the number of possible channels each Sensor must search in order to find the current channel being used by the Hub.To assure the quality of the radio link, Pseudo-Noise Codes (PN Codes) are used. These codes are used in order to achieve the special matched filter characteristics of DSSS communication. The length of the PN Code results in different communication characteristics. All devices in a network must use the same PN Code and channel in order to communicate. In order to have an optimum data transfer over the radio link several methods of error correction are used in the WirelessUSB N:1 network: Chip error correction., Bit error correction, Cyclic Redundancy Check (CRC), and ACK/re-transmission. There are a number of parameters that are used to define and identify a WirelessUSB N:1 network which will be shortly explained in what follows.

Each WirelessUSB radio contains a 4-byte *Manufacturing ID* (MID). During Bind Mode the Hub notifies the Sensor of the current channel being used for Data Mode (*Network Channel)*. All packets transmitted between bound devices use a single PN Code, which is determined by the Hub(*Network PN Code*). During the bind procedure the Hub assigns each Sensor an 8-bit or 16-bit device ID (*Device ID*), which is used to uniquely identify each Sensor. *The Network Checksum Seed* is XORed with each byte in the header and payload when determining the checksum value for all packets sent between bound devices (Data and ACK packets). *The Network CRC Seed* is used to seed the CRC calculation for all packets sent between bound devices (Data, ACK, and Network Ping packets) in order to reduce the possibility of packets from neighbouring systems being accidentally received as valid packets. The WirelessUSB N:1 Protocol operates in a series of specific modes. The *Automatic Bind Mode* allows the Sensor to retrieve the network parameters from the Hub. All Sensors must be bound before they can exchange data with the Hub. The *Seeded Bind Mode* will be described in more detail in what follows, for both Sensor and Hub operation.

The *Channel Selection Mode* is used by the Hub to find an available channel; channels are unavailable if they are being used by another network with the same PN code, or if there is excessive noise on the channel.

The *Channel Search Mode* is used by the Sensor to discover the current channel used by the Hub. Upon entering Channel Search Mode the Sensor selects a channel from the Network Channel Subset. The Sensor alternately transmits Data Packets containing its Device ID and listens for an ACK Packet from the Hub. If the Sensor does not receive an ACK, it selects the next channel in the Network Channel Subset and repeats the procedure. If an ACK is received the Sensor exits Channel Search Mode and enters Data Mode.

The *Data Mode* allows application data to be transmitted between the Sensor and the Hub. When the Sensor application has data to send to the Hub the Sensor creates a Data packet and listens for a response (either an ACK or Data packet). The Sensor may also send an empty Data packet to the Hub in order to poll the Hub for data. If no response is received, the Sensor retransmits the packet. If the received response is a Data packet, then the Sensor may respond immediately with an ACK or wait until the next transmitted Data packet to acknowledge the received Data packet from the Hub. In Data Mode the Hub listens for Data packets from the Sensors. When a valid Data packet is received the Hub responds with an ACK packet or a Data packet, if there is application data to be sent back to the Sensor.

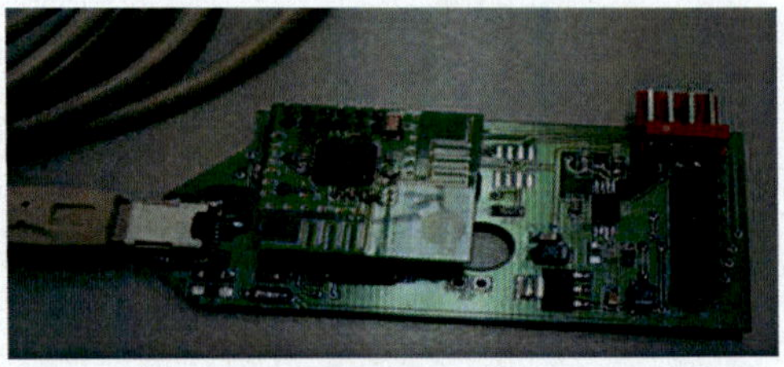

Fig. 9 Prototype hub device connected to USB port

IV. Comments and further developments

A complex, yet low-cost, wireless system that can allow a patient to start a controlled rehabilitation treatment in hospital or at home is proposed. Visual and audible feedback for the patient is provided by a sole integrated device.

The guidelines of the proposed design are compactness, low power and usage of low cost electronic components for making the system affordable. Once the physician's prescribed acceptable loading levels (force thresholds) are loaded into the microcontroller unit, the appropriate signaling is assured for the patient.

When configured in continuous transmission mode, after reading sensor data, the embedded software sends the samples to the host PC for archiving (Figure 10). The physician can use this information to monitor the gait and the progress of the treatment.

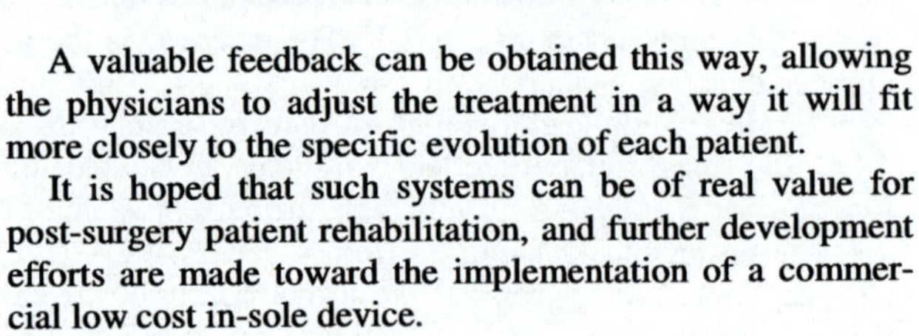

Fig. 10 Sequence acquired in Matlab during walking

A valuable feedback can be obtained this way, allowing the physicians to adjust the treatment in a way it will fit more closely to the specific evolution of each patient.

It is hoped that such systems can be of real value for post-surgery patient rehabilitation, and further development efforts are made toward the implementation of a commercial low cost in-sole device.

ACKNOWLEDGMENT

The work presented in this paper was partly supported through the research project PN II IDEI – "Arhitectura complexa de monitorizare si transfer a datelor medicale".

REFERENCES

1. Abu-Faraj, Z. O. (1995). *A Holter type system for recording plantar pressures: Development and clinical applications.* PhD thesis, Marquette University, Milwaukee, Wisconsin. Dissertations (1962 - 2010) Access via Proquest Digital Dissertations. Paper AAI9600843. http://epublications.marquette.edu/dissertations/AAI9600843.
2. Abu-Faraj, Z., Harris, G., Chang, A.-H., and Shereff, M. (1996). Evaluation of a rehabilitative pedorthic: plantar pressure alterations with scaphoid pad application. *IEEE Transactions on Rehabilitation Engineering*, 4(4):328 - 336.
3. D'Août, K., Lescrenier, K., Van Gheluwe, B., and De Clercq, D., editors (2008). *Advances in Plantar Pressure Measurements in Clinical and Scientific Research.* Shaker Publishing BV.
4. Eneslow Pedorthic Institute. Eneslow, The Foot Comfort Center. A Pedorthic Guide to Shoe Modifications. www.eneslow.com.
5. Faivre, A., Dahan, M., Parratte, B., and Monnier, G. (2004). Instrumented shoes for pathological gait assessment. *Mechanics Research Communications*, 31(5):627-632.
6. Faria, U. and A. de Carvalho, A. (2002). Implementation of strain gage transducers for monitoring forces exerted by upper and lower limbs of patients. *Brazilian Journal of Biomedical Enginnering*, 18(3):163-172.
7. Fasolo, S. G. P., Azevedo, D. F. O., and Fumagalli, M. A. (2008). Análise da distribuicão de Pressão Plantar para Diferentes Formatos de Palmilhas de Silicone através de Lementos Finitos. In Muller-Karger, C., Wong, S., and Cruz, A., editors, *IV Latin American Congress on Biomedical Engineering 2007, Bioengineering Solutions for Latin America Health*, volume 18, pages 705-709. Springer Berlin Heidelberg.
8. Giacomozzi, C., editor (2010). *Assessment of pressure measurement devices (PMDs) for their appropriate use in biomechanical research and in the clinical practice.* Istituto Superiore di Sanitá. Abstract book.
9. Hirasawa, M., Okada, H., and Shimojo, M. (2008). The Development of the Plantar Pressure Sensor Shoes for Gait Analysis. *Journal of Robotics and Mechatronics*, 20(3):289-295.
10. Kalamdani, A. A. (2006). Development and characterization of a high-spatial-temporal-resolution foot-sole-pressure measurement system. Master's thesis, The Robotics Institute Carnegie Mellon University Pittsburgh.
11. Karki, S., Lekkala, J., Kaistila, T., Laine, H.-J., Maenpaa, H., and Kuokkanen, H. (2009). Plantar pressure distribution measurements: An approach to different methods to compute a pressure map. In *XIX IMEKO World Congress. Fundamental and Applied Metrology*. September 6-11, Lisbon, Portugal.
12. Karki, S. (2009). *Film-type Sensor Materials in Measurement of Physiological Force and Pressure Variables.* PhD thesis, Tampere University of Technology.
13. Lee, S. S., Shin, K.-Y., and Mun, J. H. (2006). Development of an algorithm for a PZT ceramic foot pressure sensor. *Key Engineering Materials*, 321-323:1111-1114. www.scientific.net.
14. Lung, C. W. and Yang, S. (2005). Does hallux deformity affect the plantar pressure distribution? In *Proceedings of the 7th Symposium on Footwear Biomechanics*.
15. Moga, D., Dumitrean, M., Stroia, N., and Petreus, D. (2010). Remote Monitoring of Industrial Systems Health. In *2nd IFAC Symposium on Telematics Applications*, volume Telematic Applications.
16. Moga, D., Munteanu, R. A., Dumitrean, M., Dobra, M., and Moga, R. (2008). Wireless System for Remote Tilt Measurement in Monitoring and Control Applications. *Advances in Electrical and Computer Engineering*, 8(2):32-36.
17. Mueller, M. (1999). Application of Plantar Pressure Assessment in Footwear and Insert Design. *Journal of Orthopaedic and Sports Physical Therapy*, 29(12):747-755.
18. Munk-Stander, J. (2006). Evaluation of Piezoelectric Film Sensors for In-Shoe Pressure Measurement. Technical Report 06/04, Department of Computer Science, University of Copenhagen. (Joint technical report with University of Cambridge).
19. Munteanu, R. I., Moga, D., Neaga, F. C., Petreus, D., Dumitrean, R. M., Munteanu, M. S., and Vladareanu, L. Sistem de monitorizare a incarcarii progresive a membrului inferior in recuperarea posttraumatica. Brevet OSIM.
20. Nicolopoulos, C. and Barnett, S. Plantar pressure review using the FSCAN system. www.figroup.com/PDF/PLANTAR_.PDF.
21. Nicolopoulos, C., Giannoudis, P., and Stergiopoulos, K. (2001). History and literature review of plantar pressure measurement studies and techniques (pelmatographise). *Acta Orthopaedica et Traumatologica . Journal of Hellenic Association of Orthopaedic Surgery and Traumatology*, 52(4). www.acta-ortho.gr/v52t4_7.html.
22. Orlin, M. N. and McPoil, T. G. (2000). Plantar Pressure Assessment. *Physical Therapy*, 80(4):399-409.
23. Patil, S. L., Thatte, M. A., and Chaskar, U. M. (2009). Development of Planter Foot Pressure Distribution System Using Flexi Force Sensors. *Sensors Transducers Journal*, 108(9):73-79.
24. Wahab, Y. and Abu Bakar, N. (2011). MEMS Biomedical Sensor for Gait Analysis. In Laskovski, A. N., editor, *Biomedical Engineering Trends in Electronics, Communications and Software*. InTech.
25. Wang, W.-C., Ledoux, W. R., Sangeorzan, B. J., and Reinhall, P. G. (2005). A shear and plantar pressure sensor based on fiber-optic bend loss. *Journal of Rehabilitation Research and Development*, 42(3):315-326.
26. Wertsch, J. J., Webster, J., and Tompkins, W. J. (1992). A portable insole plantar pressure measurement syskm. *Journal of Rehabilitation Research and Development*, 29(1):13-18.
27. Wervey, R., Harris, G., and Wertsch, J. (1997). Plantar pressure characteristics during stair climbing and descent. In *Proceedings of the 19th Annual International Conference of the IEEE Engineering in Medicine and Biology Society, 1997.*, volume 4, pages 1746 - 1748.
28. Zhu, H., Harris, G. F., Wertsch, J. J., Tompkins, W. J., and Webster, J. G. (1991). A Microprocessor-Based Data-Acquisition System for Measuring Plantar Pressures from Ambulatory Subjects. *IEEE Transactions On Biomedical Engineering*, 38(7):710-714.
29. Zhu, H., Maalej, N., Webster, J. G., Tompkins, W. J., Bach-Y-Rita, P., and Wertsch, J. J. (1990). An Umbilical Data-Acquisition System for Measuring Pressures Between the Foot and Shoe. *IEEE Transactions On Biomedical Engineering*, 37(9):908-911.

Author: Florian NEAGA
Institute: Salaj County Hospital, Zalau, Romania
Street: Simion Barnutiu 67
City: Zalau
Country: ROMANIA
Email: neagaf2000@yahoo.co.uk

Telemonitoring of Vital Signs – An Effective Tool for Ambient Assisted Living

H. Costin[1,2], C. Rotariu[1], F. Adochiei[3], R. Ciobotariu[3], G. Andruseac[1], and F. Corciova[4]

[1] 'Gr.T. Popa' University of Medicine and Pharmacy, Faculty of Medical Bioengineering, Iasi, Romania
[2] The Institute of Computer Science, Romanian Academy – Iasi Branch, Romania
[3] 'Gh. Asachi' Technical University, Iasi, Romania
[4] Institute of Cardiovascular Diseases, Iasi, Romania

Abstract— **Telemedicine and e-health solutions provide (chronic) patients and elderly people services that enhance their quality of life. Advances in wireless sensor network technology and the overall miniaturization of their associated hardware low-power integrated circuits have enabled the design of low-cost, miniature and precise physiological sensor modules. These modules are capable of measuring, processing, communicating one or more physiological parameters, and can be integrated into a wireless personal area network (WPAN). This paper is dedicated to the most complex Romanian telemedical pilot project, TELEMON, that implemented a system for automatic and complex telemonitoring, everywhere and every time, in real time, of the vital signs of persons with chronic illnesses, of elderly people, of those having high medical risk and of those living in isolated regions. The main objective of this pilot project is to enable personalized medical teleservices delivery, and to act as a basis for a public service for telemedical procedures in Romania and abroad.**

Keywords— **telemedicine, telemonitoring, wireless technology, embedded systems, ambient assisted living.**

I. Introduction

The demographic change and ageing in Europe, seen as a major time aggravated thread to the European societies, implies not only challenges but also opportunities for the citizens, the social and healthcare systems as well as industry and the European market.

Emerging information and communications technologies have considerable potential for enhancing the lives of many older people and effectively approach the concept of persons' wellbeing, by using advances in biomedical wearable sensors, wireless sensor networks and technologies for remote monitoring. In this respect, the paradigm of Ambient Assisted Living (AAL) fosters the provision of equipment and services for the independent or more autonomous living of elderly people, via the seamless integration of info-communication technologies within homes and residences, thus increasing their quality of life and autonomy and reducing the need for being institutionalized or aiding them when it happens [15].

In the UK, for example, health officials believe ICT enabled self-care could potentially reduce general practitioners visits by 30% and hospital admissions by 40%. Moreover, length of hospital stays and days off work could also be reduced by 40%. Successful aging means maintain physical, cognitive, and social activities, live an independent life and maintain a good quality of life. Senior citizens wish to remain living at home for as long as possible, despite the appearance of motor and/or cognitive impairment.

Telemonitoring is one of the alternatives that provide users and their families with confidence and satisfaction, since it allows elderly patients with chronic diseases or very frail to live independently in their own home with direct contact to the professionals, relatives and friends [14].

Our project enables to design a secure multimedia transmission system (medical records, digital images, video, and text) in order to enhance the telemedical consultancy services. The main objective of this project is to enable personalized teleservices delivery and patient safety enhancement based on an earlier diagnosis with medical telemetry using biosignals, images, text transmissions, and also applying the suitable treatment according to the remote medical experts' recommendations [1], [8]. In this way the medical risks and accidents are diminished. The TELEMON system acts as a pilot project destined to the implementation of a public e-health service, "everywhere and every time", in real time, for people being in different hospitals, at home, at work, during the holidays, on the street, etc.

II. Materials and methods

A. *Medical Monitoring*

The main objective of this project is the achievement of an integrated system, mainly composed by the following components in a certain area: a personal network of wireless medical sensors on the ill person (Figure 1), a personal server on the same patient (a Personal Digital Assistant - PDA), and a personal computer (PC). After local signal processing, according to the specific monitored feature, the salient data are transmitted via internet or GSM/GPRS to

S. Vlad and R.V. Ciupa (Eds.): MEDITECH 2011, IFMBE Proceedings 36, pp. 60–65, 2011.
www.springerlink.com

the database server of the Regional Telemonitoring Centre. The personal network of sensors includes at least one medical device for vital signs acquisition (ECG, heart rate, arterial pressure, oxygen saturation, body temperature, respiration), or a fall detection module, all these components having radio micro-transmitters and allowing an autonomic movement of the subject.

The project also approaches the situation of *the mobile patient*. The data processing is done by a PDA with GPS localization, and data transmission is obtained by using the GSM / 3G module of the PDA.

Concerning the application programs, they act and correlate on two levels: a local data processing, near the patient, as well as another processing on database central server. So, the general software architecture is a client-server one, and the project develops a *SOA – Service Oriented Architecture* - which is a standards-based approach to manage services made available by different software packages for reuse and reconfiguration [2]. Services are written to protocols that are interchangeable, unlike many interfaces that are unique to each software application and typically vary by hardware platform, software language, and operating system.

The results of data processing include different locally generated alarms, transmitted to the central server, to the family or specialist doctor, to the ambulance or to a hospital. Other results of on server data processing are different medical statistics, necessary for the evaluation of health status of the subject, for the therapeutic plan and for the healthcare entities. The TELEMON system is built around a database server that receives data from local subsystems and also from mobile subsystems. The transferred information to database server are represented by those data above the limits and by medical recordings. The database server stores the recordings and is capable to send alarms to the ambulance service and patient's doctor. Also, the database server can be connected to another database server, for example a hospital server, in order to send the patient's medical data. The subsystems are connected to the database server through an Internet (if it is available) or a GSM connection.

So, the *Local Subsystem* for telemonitoring of the patient (Figure 1) is built around a PDA, which wirelessly receives data from the patient, process them and send them to the central database.

The following medical devices were built by us to allow a good monitoring of the vital parameters [7], [8], [9], [12], [13]:

a) A *3-leads ECG module* to record and transmit data through a radio transceiver interface. This system allows detection of various abnormalities of electric heart activity, focusing only on those that can be life threatening and thus a medical emergency [10][11]:

- rhythm modifications: severe bradychardia (< 45/min) or tachycardia (> 140/min or even asystola – the heart rate equals 0 for at least 3 sec.);
- recently installed AV blocks;
- signs of myocardial ischemia: new, significant pathological Q wave, elevation of the ST segment > 200 µV or depression of the ST segment < –150 µV, negative T wave;
- enlargement of the QRS complex > 0, 13 sec;
- prolonged QT interval > 0, 45 sec.

The module is recommended to patients prone to heart complications or at risk for myocardial/vascular problems, which represents more than 80% of the elderly population.

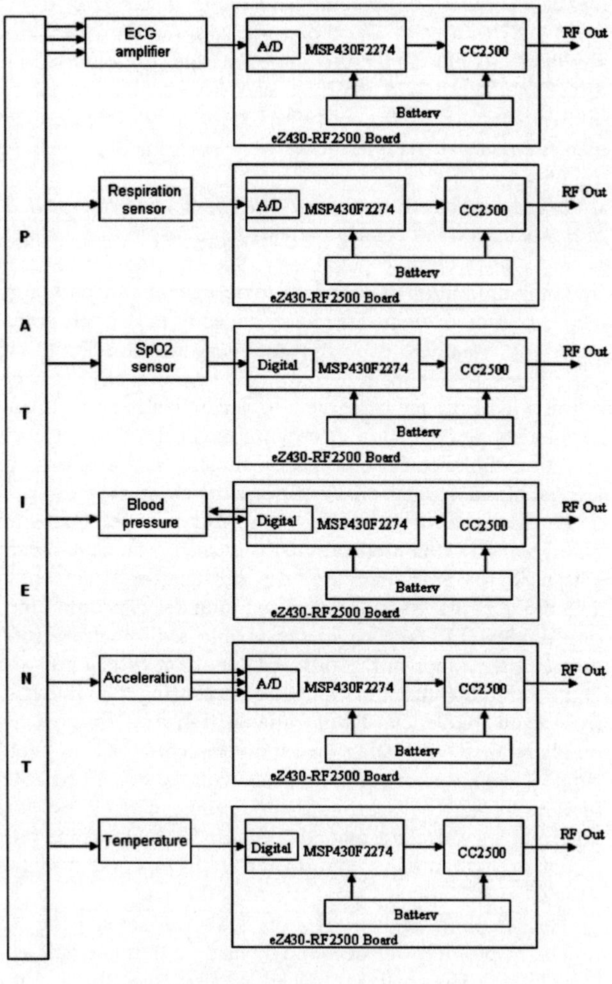

Fig. 1 The personal subsystem for telemonitoring

The 3-leads ECG amplifier is a custom device made by us by using instrumentation amplifiers [17]. It has for each channel a gain of 500, is AC coupled and has a band limited to 35 Hz. The high common mode rejection (>90 dB), high input impedance (>10 MΩ), the fully floating patient inputs are other features of the ECG amplifier. The sampling frequency for the ECG signal was set at 200 Hz.

b) The *arterial pressure module*, with serial interface. This module identifies significant variation of blood pressure such as hypotension (BP < 90 mmHg) or hypertension (BP > 160 mmHg) and is extremely important for elderly persons, who are very prone to this kind of oscillations. Postural hypotension is one of the most frequent situations a senior person must deal with; its complications are severe, impairing the quality of life and becoming life threatening (e.g.: falls, syncope, and stroke). For the blood pressure measurement, a commercially available A&D UA-767PC BPM [18] was used. The blood pressure monitor measures simultaneous blood pressure and pulse rate. It includes a bi-directional serial port interface used to transfer the measurement results to the eZ430-RF2500 at 9600 bps. The measurements of BP and pulse were performed at each 30 minutes.

c) The *oxygen saturation module* (*SpO_2*) and *the respiration module*. Many elderly patients have respiratory insufficiency due to chronic pulmonary diseases, mostly induced by smoking and/or exposure to toxic agents and pollution. The evolution of their respiratory capacity is strongly influenced by weather, exposure to allergens, humidity and compliance to treatment therefore detection of oxygen saturation will allow the caregivers to decide between adjusting oxygen supplementation or seeking medical advice or even call an ambulance for emergency situations. A decrease of arterial blood oxygen < 90% triggers the alarm system.

For the measurement of SpO_2 a Micro Power Oximeter board from Smiths Medical [19] was used. The same board was used for SpO_2 measurement and heart-rate detection. The probe is placed on a finger and includes two light emitting diodes (LEDs), one in the visible red spectrum (660 nm) and the other in the infrared spectrum (905 nm). The Oximeter computes the SpO_2 by measuring the intensity from each frequency of light after it transmits through the body and then calculating the ratio between these two intensities. The Oximeter communicates with the eZ430-RF2500 module through asynchronous serial channel at CMOS low level voltages. Data provided includes % SpO_2, pulse rate, signal strength, plethysmogram and status bits and is sent to the eZ430-RF2500 at a baud rate of 4800 bps.

The sampling frequency for the SpO_2 was set at 1 Hz.

The respiration was detected by using a thoracic belt as a transducer. The belt is placed around the thorax and generates a high-level, linear signal in response to changes in thoracic circumference associated with respiration. The sampling frequency for the respiration signal was set at 10 Hz.

d) The *body temperature module* gives important information about occurrence of fever, especially for persons with mild cognitive impairment who cannot sense temperature modifications (> 38 °C or < 35°C).

For the body temperature measurement a TMP275 temperature sensor was used [20]. The TMP275 is a 0.5°C accurate, two-wire, serial output temperature sensor. The TMP275 is capable of reading temperatures with a resolution of 0.06°C and is connected to the eZ430-RF2500 by using the I^2C bus. The accuracy of the sensor for the 35-45°C interval is below 0.2°C and the conversion time for 12 data bits is 220 ms typically. The sampling frequency for the temperature was set at 1 Hz.

e) The *fall detection module* should be recommended to all senior persons who live alone. Elderly are exposed to often falls due to several causes: postural hypotension (induced by inadequate hydration, cervical spondilosys with vertebrobasilar circulatory problems or even inappropriate medication for hypertension); inappropriate house conditions such as poor lighting conditions, narrow halls or staircases, slippery surfaces which predispose losing balance and fall; sensory disturbances (visual, postural) that induce imbalance and fall; inappropriate shoeing and/or clothing.

An elderly who falls is a medical emergency and recovering his/her health and mental state is extremely important.

The module for fall detection is based on tri-axial ADXL330 accelerometer [21]. Linear accelerations were measured to determine whether motion transitions were intentional. The sampling frequency for the acceleration was set at 25 Hz.

These modules transmit data to a PDA through radio transceivers (eZ430-RF2500 boards), can operate in the 2.4 GHz band, and have 5m /10m range indoors /outdoors.

The eZ430-RF2500 is a complete MSP430 [19] wireless development tool providing all the hardware and software for the MSP430F2274 microcontroller and CC2500 2.4 GHz wireless transceiver [4].

Operating on the 2.4 GHz unlicensed industrial, scientific and medical (ISM) bands, the CC2500 provides extensive hardware support for packet handling, data buffering, burst transmissions, authentication, clear channel assessment and link quality. The radio transceiver is also interfaced to the MSP430 microcontroller using the serial interface.

The USB interface enables eZ430-RF2500 to remotely send and receive data through USB connection using the MSP430 Application UART.

B. *Patient Localization*

Mobile patients are localized on different maps by using the GPS module of the PDA device worn by those patients and a special application named *TelemonMap*.

The link between the medical data base and the geographic data base is made by means of a specialized tool, the *Project file TelemonMap*.

The geographic data base consists of vector maps (built from different layers) covering the area of possible displacements of patients. We have used such a map for Iasi County, in which the details for Iasi city are at streets and buildings level and is composed by 16 layers. In vector maps the layers are visible or hidden, according to the detail level (map scale).

TelemonMap is a MDI-type application developed in C++ language, by using the menu Microsoft Visual Studio 2008 and displaying different windows.

The synchronization between the medical and geographic data bases is made every 10 seconds by default.

Every generated alarm for the medical status of the patient is automatically transmitted to the geographic data base, from where the position of the patient given by GPS coordinates is transpose on the corresponding vector map.

Fig. 2 The patient localization and alarm generating

III. Results

Figure 3 shows *the personal server*, which was implemented by means of a PDA (HTC X7500). The PDA has the following technical specifications: CPU Intel XScale PXA270 at 624MHz, 128MB RAM, 256MB ROM, 8GB HDD, a large TFT display with resolution 640 x 480 pixels, WiFi, GSM/GPRS and Bluetooth (client/host) interfaces and runs Windows Mobile 5 as operating system.

This medical monitor provides transparent interfaces to the wireless medical sensors, to the patient, and to the central server.

Its USB interface is realized by using a serial-to-USB transceiver (FT232BL) from FTDI [9] and enables eZ430-RF2500 radio module to remotely send and receive data through USB connection using the MSP430 Application UART. All data bytes transmitted are handled by the FT232BL chip.

The software on the Personal Server receives real-time patient data from the sensors and processes them to detect anomalies. The software working on the Personal Server (Figure 4) was written by using C# from Visual Studio.NET, version 8. The software displays temporal waveforms, computes and displays the vital parameters and the status of each sensor (the battery voltage and distance from the Personal Server). The distance is represented in percent, based on RSSI (received signal strength indication), measured on the radio power present in a received radio signal).

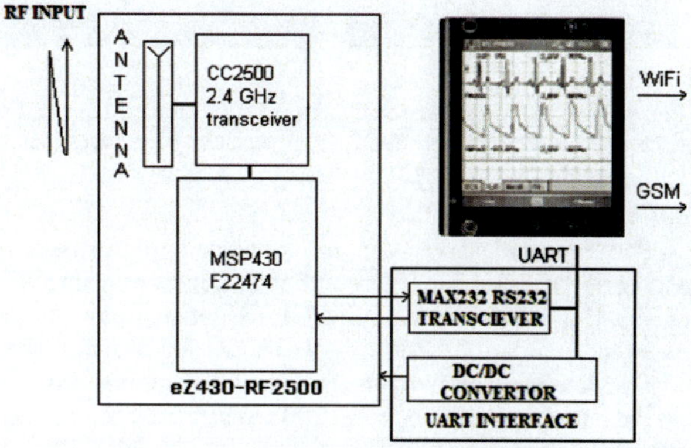

Fig. 3 The Personal server (block diagram)

If the patient has a medical record that has been previously entered, information from the medical record (limits above the alarm) become active and is used in the alert detection algorithm. So, when an anomaly is detected in the measured patient vital signs, the Personal server software application generates an alert in the user interface and transmits the information to the TELEMON Server.

For instance, the following physiological conditions cause important alerts:

- low SpO2, if SpO2 < 90%;
- bradycardia, if HR < 40 bpm;
- tachycardia, if HR > 150bpm;
- HR change, if ΔHR / 5 min > 20%;
- HR stability, if max HR variability from past 4 readings > 10% ;
- BP change if systolic or diastolic change is > ±10% .

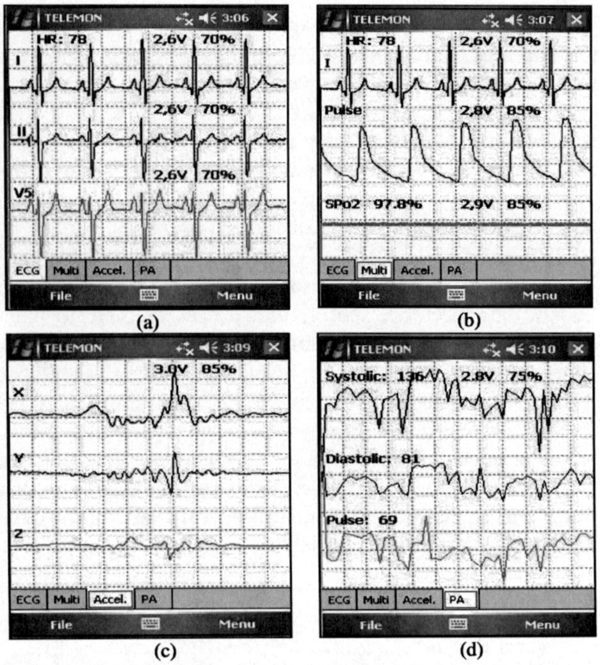

Fig. 4 The Personal server interface: (a) 3 ECG traces, (b) one ECG trace, pulse waveform and SpO2, (c) 3 accelerometer traces, (d) systolic, diastolic pressure and pulse diagrams, measured each 30 minutes

The Personal server can be connected to the database server of the Regional Telemonitoring Centre by using WiFi (default, if available) or GSM/GPRS connections. In both cases the standard TCP/IP protocol is used to transfer data.

The database server of the Regional Telemonitoring Centre is based on SQL Server 2008 platform running on a Windows platform.

The programs are written in Visual Studio .NET 2008 and SQL LITE (for data bases). Basically, the server is composed of a console application and a relational database. The database stores the following patient data: a list of patients and personal information, numerical values of vital signals and alarms, and other medical records related to the patient. The medical records summaries the medical history diagnostics including medications.

The client module comprises the software working on the patient, physician or medical expert computer. The software was written in C# with Microsoft Visual .NET running on a Windows platform and uses the standard TCP/IP protocol to download data from server.

The client module has the following facilities: GUI (Graphic User Interface) for vital signs waveforms, displays the patient's parameters and alarms received from each sensor, management of the patient information and view of clinical records.

While the server is running, the client can start a session from anywhere in the Internet by accessing the server's connection port and providing a proper login and password.

IV. Conclusions

In this paper it is presented a project that is the base for developing a secure multimedia, scalable system, designed for medical teleconsultation and telemonitoring services. The main goal is to build a complete pilot system that will connect several local telecenters into a regional telemedicine network. This network enables the implementation of teleconsultation, telemonitoring, homecare, urgency medicine, etc., for a broader range of patients and medical professionals, including elderly people and those people living in rural or isolated regions [1][3][6].

The Regional Telecenter in Iasi/Romania, situated within the Faculty of Medical Bioengineering, allows local connection of hospitals, diagnostic and treatment centers, as well as a local network of family doctors, patients, paramedics and even educational entities. As communications infrastructure, we have developed a combined mobile-internet (broadband) links.

The proposed system is also used as a warning tool for monitoring during normal activity or physical exercise.

Such a regional telecenter acts as a support for the developing of a regional medical database, which should serve for a complex range of teleservices such as teleradiology, telepathology, telediagnosis, and telemonitoring of different medical parameters. It should also be a center for continuous training and e-learning tasks, both for medical personal and for patients.

In this context, the telemonitoring of elderly people who want to preserve as long as possible their independence but are of medical risk (cardiovascular, respiratory, proneness to falls) represents one of the most important solutions to an ageing Europe.

Besides the medical and technical objectives, TELEMON project also proposes important economic objectives:

- the decrease of budgetary and personal expenses dealing with the unjustified transport of patients to the hospital;
- „zero costs" for the hospitalization in the case of patients who may be treated at home.

Overall, the low cost of the entire system, including its maintenance and operating costs, is entirely justified by the great benefits for health monitoring and care.

Acknowledgment

This work is financed by a grant from the Romanian Ministry of Education, Research and Youth, within PNCDI_II program, contract No. 11-067/2007. Its web address is at www.bioinginerie.ro/telemon.

References

1. Shnayder V, et al. (2005) Sensor Networks for Medical Care. Tech. Rep. TR-08-05, Division of Eng. and Applied Sciences, Harvard Univ.
2. Shorey R. (2006) Mobile, Wireless and Sensor Network: Technology, Applications and Future Directions, Wiley
3. Costin H, Rotariu C, et al. (2008) Integrated system for real time monitoring of patients and elderly people. Ukrainian Journal of Telemedicine and Medical Telematics 6(1): 71-75
4. MSP430 datasheet at http://www.ti.com/MSP430
5. CC2500 datasheet at http://www.ti.com/CC2500
6. Rotariu C, Costin H, et al. (2009) A Low Power Wireless Personal Area Network for Telemedicine, Proc. of the 4th European Conf of the Int. Federation for Medical and Biological Engineering 22: 982-985
7. http://www.smiths-medical.com/Userfiles/oem/OEM.31392B1.pdf
8. http://www.lifesourceonline.com/and_med.nsf/html/UA-767PC
9. FT232 datasheet at http://www.ftdichip.com/FT232
10. Pan J, Tompkins WJ (1985) A real-time QRS detection algorithm, IEEE Trans. Biomed. Eng. 1985; BME-32: 230-236
11. Tompkins WJ (ed.) (1993) Biomedical Digital Signal Processing: C-Language Examples and Laboratory Experiments for the IBM PC., Englewood Cliffs, NJ: PTR Prentice Hall
12. Urrusti JL, Tompkins WJ (1993) Performance evaluation of an ECG QRS complex detection algorithm, Proc. Annual Int. Conference of the IEEE Engineering in Medicine and Biology Society, pp 800–801
13. Hamilton PS (2002) Open Source ECG Analysis Software. E. P. Limited, Somerville, Mass, USA
14. Carone G, Costello D (2006) Can Europe Afford to Grow Old ? Finance and Development 43(3)
15. Vincent C, Talbot L, Deaudeline I, Garceau M (2003) Telemonitoring for frail elderly: is it relevant? In G. Craddock et al. (eds.) Assistive Technology-Shaping the Future, IOS Press
16. AAL Program at http://www.aal-europe.eu
17. Rotariu C, Costin H, Arotăriţei D and Dionisie B. (2008) A Wireless ECG Module for Personal Area Network, Buletinul Institutului Politehnic din Iaşi, Tome LIV (LVIII), Fasc. 1, AUTOMATIC CONTROL and COMPUTER SCIENCE Section: 45-54
18. UA767PC Blood Pressure Monitor, http://www.lifesourceonline.com/and_med.nsf/html/UA-767PC
19. Micro Power Oximeter Board, http://www.smiths-medical.com/upload/products/PDF/OEM/196002_Micro-Power-Oximeter.pdf
20. TMP275 Temperature Sensor, 0.5°C Digital Out Temperature Sensor, http://focus.tij.co.jp/jp/lit/ds/ symlink/tmp275.pdf
21. http://www.analog.com/en/mems-sensors/inertial-sensors/adxl330/products/product.html

Author: Hariton Costin
Institute: University of Medicine and Pharmacy, Iasi, Romania
Street: 700454, M. Kogalniceanu street, 9-13
City: Iasi
Country: Romania
Email: hcostin@gmail.com

eHealth – Unified Healthcare Logical Space through Applied Interoperability

M. Rusu[1], C. Lelutiu[2], N. Todor[3], and G. Saplacan[1]

[1] Company for Applied Informatics, Cluj-Napoca, Romania
[2] Technical University of Cluj-Napoca, Cluj-Napoca, Romania
[3] Department of Biostatistics and Informatics, Cancer Institute "Ion Chiricuţa", Cluj-Napoca, Romania

Abstract— **The current paper presents a patient-centric approach that implements a unified healthcare logical space model for medical information environment. We focused our work on interoperability features with extensive usage of standards in order to develop a distributed information framework: a pilot system called CardioNET. This eHealth system was designed to integrate medical services from various healthcare providers and improve the quality of services, through some of latest medical and IT&C technologies. Modern healthcare activities require a patient-centric vision, where patients must receive medical attention or treatment anytime, regardless of their physical location. CardioNET embodies this approach where hardware, software and medical activities become "services" of a "logical cardio-health care domain-space". This distributed environment also offers tools for remote interactions between patients, doctors, medical entities (e.g. hospitals, labs) and authorities. Based on international domain standards (IDC10, LOINC, HL7), the system creates an infrastructure for interoperability and data exchange using widely accepted formats (HL7 messages, or XML records). High level protocols (UDDI, SOAP/HL7 and HTTP) provide the presented framework the means to exchange of HL7 or XML compliant messages between the systems' main healthcare actors. CardioNet subsystems have specialized metadata registries and shared data repositories, which altogether create a distributed healthcare pilot environment for medical decisions support, research and educational activities.**

Keywords— **e-health, interoperability, tele-medicine.**

I. INTRODUCTION

Healthcare data flows and associated data structures are very complex and are formalized with a lot of methods, by different institutions. Joined-up, these „objects" are more difficult to handle when patients with multiple problems are treated by several specialists in jurisdictionally different locations. Solutions? Domain standards utilization to achieve the full health systems interoperability. All aspects of the medical processes: data flows, the practice of Evidence-Based Medicine (EBM - one of the most important developments in the clinical practices over the last years) require the integration of clinical expertise with the external facts and with patient's life parameters and circumstances, to create a realist Electronic Health Record (EHR) as support for medical decisions. Recent developments in healthcare data standardization processes create large interoperability opportunities between different healthcare information systems. Nowadays IT&C, offers the possibility to quickly develop, access, change and share meaningful information, about patients and their health, beyond organization boundaries.

The domain's standards utilization must be extended beyond their primary definitions, in order to achieve interoperability and bring together disparate systems. For healthcare entities the goal is cross-borders access to observations, reports and results of medical procedures (trials, claims, infectious disease reports, patient summaries), access that will eventually extend across jurisdictional (national or regional) borders. Interoperability is a prerequisite for the process of the old Health Information Systems reengineering, that will reduce the costs, errors, delays, and development repetition efforts.

The transformation of healthcare services depends critically on interoperability, enabling computers to share data and deliver information from where it originates to where they are needed. When interoperability will be a commonplace, patients, clinicians, managers, and researchers will enjoy secure access to the right information at the right time and at the right place, leading to better patient outcomes and fewer mistakes.

The current paper presents the results of an interdisciplinary research effort to develop a framework for medical data exchange. We present here, an enterprises-cross border service-oriented approach, CardioNet distributed information system, HL7&IHE based.

II. BACKGROUND: STANDARDS, ORGANISATIONS, TRENDS

Worldwide the HIS (Healthcare Information Systems) interoperability is one of the core themes. The US Federal Health Information Technology Strategic Plan, states: "to effectively exchange health information, health IT systems and products must use consistent, specific data and technical standards", [1]. The main goal of NESSI(Networked

S. Vlad and R.V. Ciupa (Eds.): MEDITECH 2011, IFMBE Proceedings 36, pp. 66–71, 2011.
www.springerlink.com

European Software and Services Initiative) is to develop a visionary unified European Strategy for Software and Services[2]. The National Health System (NHS) Informatics Review 2008, set out a vision to support patient-centered care in a way that empowers patients to be more involved in their care and staff to improve Great Britain's NHS performance [3].

For the lack of space we present only a small number of examples, still there are a lot of other countries who make large efforts towards this coming globalized market, not only in the healthcare field.

Besides international organizations for standards development: ISO, CEN, BSI, ANSI, IEEE, we remind here several specialized standards and organizations in healthcare area: SNOMED, IHE, HL7, ICD10, DICOM, LOINC, and some IT&C standards: SOA, SOAP, ISO/OSI, SaaS, Web Services and UDDI. All of them are deeply implied in the interoperability implementation processes between healthcare information systems.

SNOMED CT - Systematized Nomenclature of Medicine Clinical Terms, provides a comprehensive clinical terminology, analogous to a dictionary [4].

ICD-10 - International Classification of Diseases (ICD), was endorsed by the World Health Assembly in May 1990 and came into use in WHO Member States as from 1994[5].

LOINC - Laboratory Object Identifier and Numerical Code, LOINC® and RELMA® are trademarks of U.S. Regenstrief Institute, Inc. Codes and other information from the LOINC are used in electronic messages building for labs test results and clinical observations [6].

DICOM - Digital Imaging and Communications in Medicine standard is required by all EHR systems that include images' information, as a part of the patient records [7].

HL7 - Health Level 7, a non-profit organization, developing standards for the exchange of clinical and administrative data. HL7 provides a grammar as standardized structures for healthcare communications using messages [8].

IHE - Integrating the Healthcare Enterprise was established in 1999 by the Healthcare Information Systems and Management Society (HIMSS) and the radiological Society of North America (RSNA) to improve the way healthcare computer systems share information. IHE has defined an integration profile called Cross-enterprise Document Sharing (XDS) [9]. IHE – XDS allows healthcare documents to be shared over a wide area network, between hospitals, primary care providers, and social services. Documents are discovered using UDDI-Universal Description Discovery and Integration and exchanged using SOAP and HTTP protocols, largely based on HL7 messages standard, while SQL is used for information retrieval. The model developing within the U. S. suggests that medical data sharing will happen first at a local level, as part of Regional Health Information Organizations (RHIO) and then between RHIOs.

Usually healthcare information systems have been organized hierarchically, with the government at the top, then healthcare-provider organizations (hospitals), followed by departments, clinicians, and eventually the patient, at the end of chain. This hierarchy reflects the flow of power, authority and money, but has little in common with the natural flow of healthcare data, resulted from the actual care of patients. In reality „patient care looks like a social network, where each individual patient is in the centre of a healthcare net"[10]. Nowadays, the interoperability between healthcare systems is a challenge and a corner-stone. The documentation for all the previously presented aspects runs to thousands of pages and creates a steep learning curve and barrier for starting point.

III. UNIFIED HEALTHCARE LOGICAL SPACE - CARDIONET PROJECT

The new patient-centric vision is quite different: the center of systems is the patient with his data and episodic or long term problems, and not the healthcare organizations. This model is based on continuous healing relationships, customized according to individual patient needs and values, with the patient as the ultimate source of control. Knowledge is shared, information flows freely, and decision-making is evidence-based. Transparency and collaboration are common behaviors, patient needs are anticipated, and effort is devoted toward reducing any activity that delivers no benefit to the patient. An electronic health record (EHR) is not necessarily stored as a single physical entity in a centralized system.

The idea is to build patient records on the fly from a variety of clinical documents created by different healthcare organizations. The record required (EHR-Electronic Health Record) can be aggregated into a single coherent record, at request, from data stored at various geographical locations. We take all these new ideas to design and implement a „Unified Healthcare Logical Space"- UHLS pilot for our healthcare information distributed system.

The CardioNet project is an interdisciplinary applicative research project in concordance with the Romanian Ministry of Health public health program. It was designed to comply with national and international standards and trends in cardiology. The UHLS distributed environment includes several local (Cluj-Napoca) healthcare organizations and aimed:

- to create a surveillance and medical data acquisition unified infrastructure;
- to create a set of medical applications deployed at different medical entities and institutions;
- to create a distributed medical data repository built upon a domain ontology (cardiology);

- to share a set of software services necessary for real-time medical data acquisition, classification, decision support, statistical analysis, long term storage and controlled access.

The system was designed as SOA model where hardware, software and medical activities become services. CardioNet connects a group of healthcare entities that have agreed to work together using a common set of policies and to share a common infrastructure. Figure 1 shows an overview of the CardioNET's „medical logical space", emphasizing the actors of the system, their roles and interactions. During the implementation stages of the CardioNet distributed environment the following main integrated subsystems were identified:

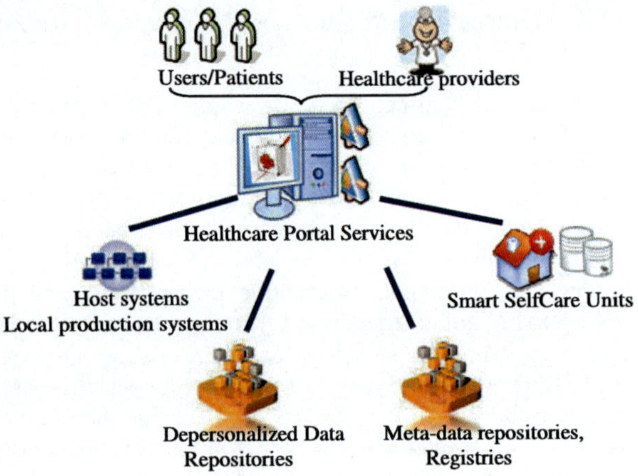

Fig. 1 UHLS – CardioNET Overview

- Host systems, as local production systems (with local operational databases, h-DB) [11]:
    a. General Practitioner systems [Figure 2];
    b. Analysis Laboratory systems [Figure 3];
    c. Hospital systems [Figure 4].
- Portal services: education, resources registries and repositories, data exchanges [11] [Figure 5];
- Smart Self Care Units - SSCU as home systems with personal databases, p-DB [12].

At the portal level, "patient-centric" medical services are discovered using specialized registries that among other functions, allow controlled access to data stored in the shared repositories.

In this boundary-less environment, the proper identification (patients, practitioners, and healthcare facilities) is a key feature for the platform subsystems. Currently four types of centralized registries were defined: Persons, Providers, Facilities and Locations. For each registry a master index was built – set of software tools that assure better identification and access to the resource. EHR-Electronic Health Record will be dynamically built at run-time, as Virtual Patient Records, from a variety of clinical events and documents managed by different healthcare organizations and providers, who agreed to operate in this way.

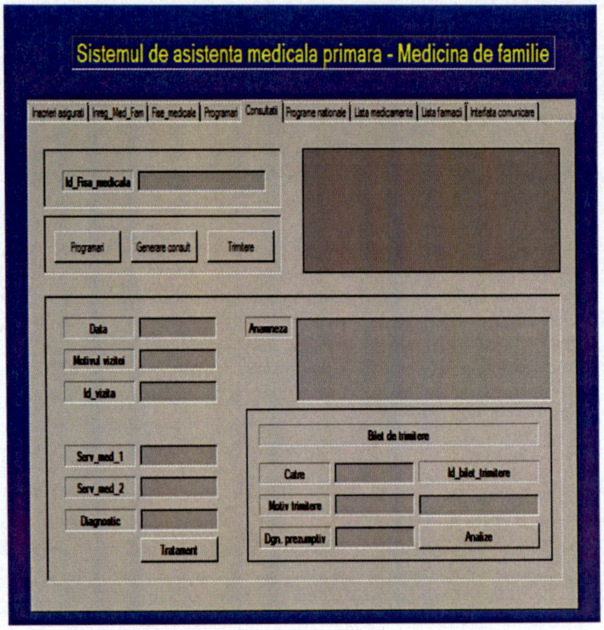

Fig. 2 UHLS- General Practitioner – Observations UI

The innovation is the logical and physical separation of the indexed metadata built from current information registrations, used then to retrieve documents through this unified virtual space. CardioNet distributed subsystems can operate in two modes:

Local Mode as an enterprise-internal mode.

Network Mode as integrated components of a logical unified virtual space.

In the local mode CardioNet enables actors to operate intranet (enterprise-internal mode), with local administrative and clinical data. In the network operating mode, CardioNet system provides secured access to local databases and offers a range of services for: health insurers, clinical decision-making entities, management authorities or other health services providers/consumers (depersonalized databases analysis, resources monitoring, etc.).

A. *Local Hub Centers (L-HC) or Portals*

L-HCs' repositories and services assure controlled access to electronic medical record containing patients' medication histories, lab results, allergies and other vital health information. The access is available, not only to the doctors, but to pharmacists too, as well as other members of the patients' team of care providers. The CardioNet portal (located at SC CIA SA project partner) enables users (consumers) to retrieve different types of documents (letters, results, images) contained in one or more repositories in a quick and reliable manner. We also created the way to integrate „in house – standalone" systems: for instance we integrated in our system a Radiotherapy Information System [Figure 3], running Filemaker 9 server located at Cancer Institute "Ion Chiricuţa" [13].

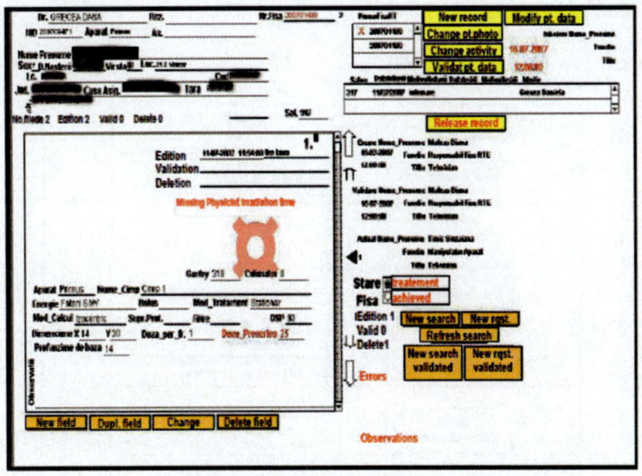

Fig. 3 Radiotherapy system – Observations UI, [13]

B. Host Systems

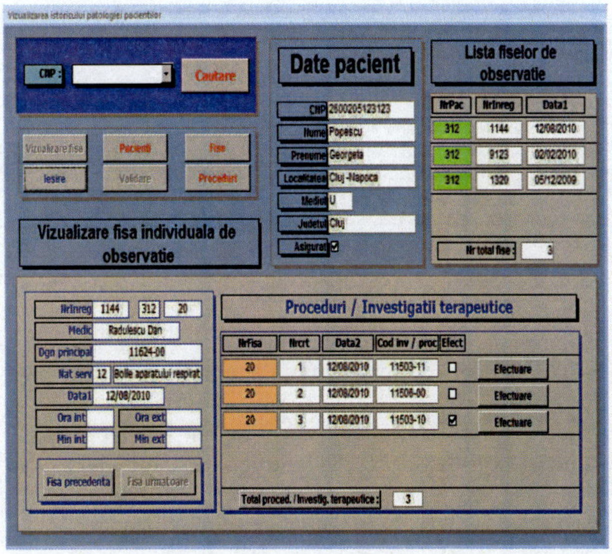

Fig. 4 Clinical host system – Observations UI

Host Systems are built of information systems and database servers [Figure 4], access points and fixed/mobile interconnected medical devices. The clinical information system was tested at clinical hospital UMF Medicala V (project partner). The CardioNet portal provides the infrastructure required to collect information from the mentioned local systems into shared repositories. At the portal level, "patient-centric" services [Figure 5] was tested and will be assured using data stored into the local repositories.

C. UHLS through Registries, Web Services, Master Indexes and HL7 Messages

UHLS critical functions are provided through portal registries and web services collections. [Figure 5] presents the UHLS portal – unifying B2B&B2C services, grouped for: User authentication, Demographic matching and Patient registration, Resources for unique identification, Identifiers mapping, Records and documents discovering services, Records and documents exchange services, etc.

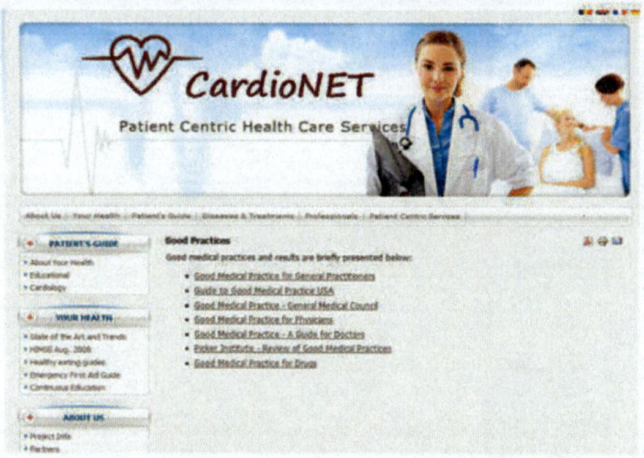

Fig. 5 UHLS CardioNet portal – Educational section

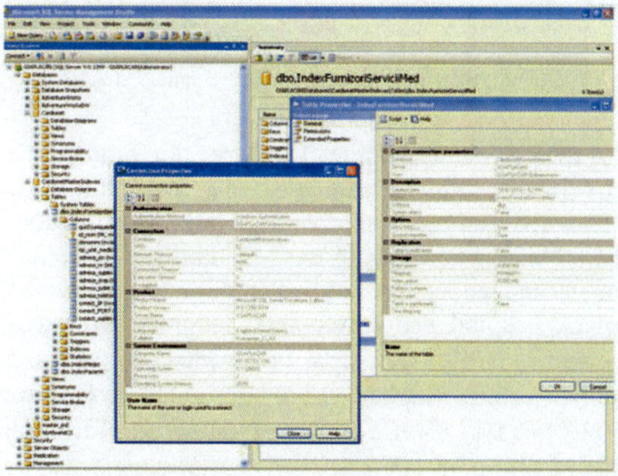

Fig. 6 UHLS-Portal providers' master index tables

***Portal resources:** master indexes, registries and web services.* CardioNet portal implements master indexes and registries for: Persons, Providers [Figure 6], Facilities and Locations. Master indexes by GUIDs (global unique identifiers) and dedicated methods provide the means to uniquely identify any networked and shared resource. Using UDDI services(Universal Description Discovery and Integration) on the CardioNet portal the resources are discovered, identified and eventually used to call web services and collect data from production systems (using HL7 messages). All the methods of the web services [Figure 7] use SOAP messages in order to expose functionality: both the requests and the replies are embedded in soap envelopes that respect the format of each particular web method. Invocation of the web methods returns data objects available in the system's metadata repository. Cross platform interoperability is obtained by serializing data objects in widely accepted standards (such as HL7 and XML).

Fig. 7 UHLS – Portal web service example: PatientByID

HL7 Messages. Cardionet-IUI is a pilot project that provides interoperability through HL7 User Interface. The system was designed based on HL7&IHE and LOINC standards. This approach proposes standardized HL7 messages formats and codes for automatic data exchanges between clinical, hospital and commercial laboratories to complete Electronic Health Records of production systems. A sample HL7 message instance shown in the following:

MSH|^~\&||^IOCN^Labs|||200808141530||ORU ^R01|123456789|P|2.4
PID|||123456^^^SMH^PI||POPESCU^VASILE||19620114|M|||4 Republicii^ClujNapoca^^^MM1 9DL
PV1|||5N|||||G123456^DR POPESCU ION
OBR|||54321|666777^CULTURE^LN|||20080802|||||||SW^^^FOOT^RT|C987654
OBX||CE|500152^AMP|01||||R|||F
OBX||CE|500155^SXT|01||||S|||F
OBX||CE|500162^CIP|01||||S|||F

Note : The OBX segment repeats information about the susceptibilities of detected organism (linked by using the Observation Sub-ID field).

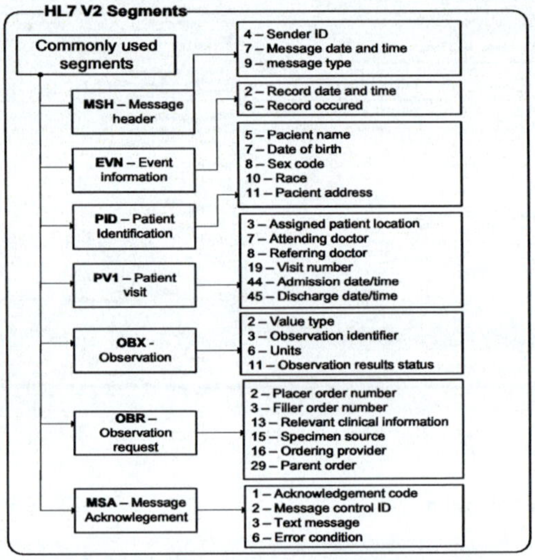

Fig. 8 List of the main HL7 v2.x used segments

This approach has selected a subset of HL7 v2.x [Figure 8] messages for observations and patient registration activities. [Figure 9] - UI sample to present data exchanging between Clinical and Labs information systems [14].

SSCU – Smart Self Care Units. A typical SSCU's is composed of several medical sensors for patients' vital sign monitoring. The data acquisition layer in the SSCUs has two versions: medical sensors are integrated in custom hardware platforms or the sensors are embedded in commercial solutions available at different vendors [15][16]. Data collected at the SSCU is stored locally on the patient's PDA and computer and eventually transferred in the CardioNET system for persistent long-term storage.

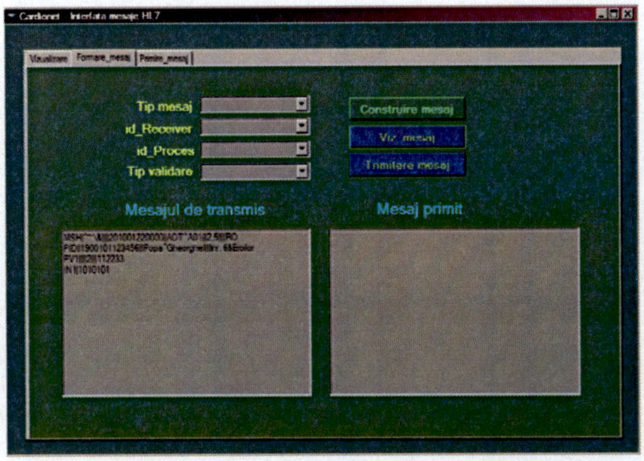

Fig. 9 Cardionet – HL7 exchanging message UI sample

IV. CONCLUSIONS AND FUTURE WORK

The implementation of a distributed eHealth system is a complex task that involves: remote data acquisition and monitoring, data logging and information exchange between medical entities, applications and users. This paper presents a model for an eHealth system providing solutions to both medical and IT related problems including the remote monitoring of patients' medical status. The information gathered through web-services is preserved into specialized databases built as a domain metadata repository. This approach reviles complex relations between different concepts involved in a medical act and assures both interoperability and transparent exchange of data between different medical applications while providing support for better medical diagnoses and treatment. The proposed pilot solution was implemented for monitoring and treating patients with cardio-vascular diseases. This approach reduces significantly the time spent by patients in hospitals, allows remote monitoring of patients with chronic diseases and facilitates a flexible patient-doctor interaction over the Internet. The project partners interoperability tests show that HL7 standard is a reliable solution and can be applied for intercomunication at application level.

As future work, the authors intend to add more elements of intelligence to the system, such as: data-mining procedures for specialized clinical trials, statistical evaluation facilities and alternative decision support services.

ACKNOWLEDGMENT

This work was supported by the National Agency for Scientific Research - Romanian Ministry of Education and Research (grant number CNMP-11001).

REFERENCES

1. The ONC-Coordinated Federal Health Information Technology Strategic Plan: 2008–2012: Using the Power of Information Technology to Transform Health and Care. Department of Health and Human Sciences, 3 June, 2008.
2. NESSI - Networked European Software and Services Initiative – www.nessi-europe.eu
3. Health Informatics Review. DH July 2008, Ref 10104.
4. SNOMED, - http://www.ihtsdo.org/snomed-ct/snomed-ct-publications/.
5. ICD-10, - "International Classification of Diseases (ICD)" .
6. LOINC – Logical Observation Identifiers Names and Codes (LOINC)
7. DICOM, - Digital Imaging and Communications in Medicine, http://medical.nema.org/.
8. HL7 , - http://www.hl7.org/.
9. IHE-XDS, - IT Infrastructure Technical Framework Volume 1 (ITI TF-1), http://www.ihe.net/Technical_Framework/upload/ihe_iti_tf_2.0_vol1_FT_2005-08-15.pdf .
10. T. Benson, Principles of Health Interoperability HL7 and SNOMED, HI,DOI 10.1007/978-1-84882-803-2_1, © Springer-Verlag London Limited 2010.
11. M. Rusu, G. Saplacan, Gh. Sebestyen, C. Cenan, L. Krucz, T. Nicoara, N. Todor - Distributed e-Health System with Smart Self-care Units, ICCP Aug. 2009 Cluj-Napoca, ROU.
12. M. Rusu, T. Nicoara, C. Cenan, Gh. Sebestyen, V. Olaru, G. Saplacan, D. Radulescu – „Smart Home Healthcare Unit in Healthcare Systems", International School on Cyber-Physical and Sensor Networks, "SensorNets 2009", Monastir 17-21 dec 09, Tunisia.
13. Todor N., Cernea I.V., Chis A., From Passive to Active in the Design of External Radiotherapy Database at Oncology Institute "Prof. Dr. Ion Chiricuță" from Cluj-Napoca, Applied Medical Informatics, 25 (2009), 7-15, ROU.
14. C. Lelutiu, M. Rusu, T. Nicoara, N. Todor, Gh.Sebestyen, G. Saplacan, eHealth systems interoperability through HL7 interfaces, AQTR 28-30 May 2010, p. 228-233, Cluj-Napoca, RO
15. Easy EKG Pocket, http://www.atesdevice.it/site/ecg/ecgpocket.html
16. Corscience NiBP 2010, http://www.corscience.de/en/medical-engineering/oemodm-solutions/oem-modules/nibp-2010-blood-pressure.html

Medical Services Optimization Using Differential Evolution

F.-C. Pop[1], M. Cremene[1], M.-F. Vaida[1], and A. Serbanescu[2]

[1] Faculty of Electronics, Telecommunications and IT, The Technical University of Cluj-Napoca, Romania
[2] The Faculty of Dental Medicine, "Iuliu Hatieganu" University of Medicine and Pharmacy, Cluj-Napoca, Romania

Abstract— **This paper proposes a method to compose and optimize medical services as business workflows. Such a workflow consists in a set of abstract services, and for each abstract service there are several concrete services. Since each medical service has different QoS (Quality of Service) parameters such as *response time*, *rating*, *distance* and *cost*, determining the optimal combination of concrete services that realize the abstract services of the business workflow is an NP hard problem. Recent proposals for solving NP optimization problems indicate the Genetic Algorithms (GA) as the best approach for complex workflows. But this problem usually needs to be solved at runtime, a task for which GA may be too slow. We propose a new approach, based on Differential Evolution (DE), that converges faster and it is more scalable and robust than the existing solutions based on Genetic Algorithms.**

Keywords— **Services, Composition, QoS, Optimization, Selection, Genetic Algorithms, Differential Evolution.**

I. Introduction

A. Background

The Service Oriented Architecture (SOA) model has become very popular in enterprise environments, where the complicated business logic is implemented by combining the functionality of various services. First, the business functions are defined. These functions represent the set of activities used to manage the assets of the organization in their various states. Then, the business functions are further decomposed into services, which implement the logic required to realize defined functions.

In software engineering, SOA defines how to discover and integrate disparate applications from different platforms into web-based applications. For example, one image processing application can be composed of several independent software components, each of them realizing a different function: enhancements, rotation, segmentation etc. and each of these components can be offered by a different *service provider*. Such a process that combines the functionality of multiple services is called *service composition*, and the resulted application is called a *composite service*.

In medicine, a service provider could be, for instance, a dental office, which offers various dental treatment services. A composite medical service can then be defined as any medical activity that requires the patient to benefit from two or more different medical services.

The patient (or the user of a service) is called a *service consumer*. A contract (formal or informal) is defined between the service provider and the service consumer to specify the level of service. This contract is called the Service Level Agreement (SLA). For example, the SLA between the dentist and the patient for a dental implant service may include the amount of time the implant is guaranteed to last, the cost of the medical procedure or the average rate of success. Such attributes represent the Quality of Service (QoS) properties of a service.

Two services that provide the same functionality often have different QoS properties. For example, many clinics offer a similar range of medical tests, but promote different prices and require different amounts of time to deliver the results. One may be cheaper, but require longer time to provide the results than a more expensive service.

B. Motivating Example

A composite service can be described as a process that involves the execution of several activities according to a workflow. An example workflow for a series of clinical tests is depicted in Fig. 1. This workflow consists of the following activities:

- *S1.* *Assisted General Diagnosis* for reading the patient's symptoms and classifying them in one of the 3 (example) categories: *Heart Disease Symptoms*, *Digestive System Symptoms* or *Other Symptoms*. According to the assigned category, the patient is then scheduled for specific medical tests.
- *S2.* *Cholesterol Test* for measuring the cholesterol level,
- *S3.* *Cardiac Exam* for investigating signs of any cardiovascular pathology,
- *S4.* *Endoscopy,* where the digestive tract is investigated,
- *S5.* *Physician Consultation* for having a physician examine the patient's symptoms and the results of the scheduled investigations,
- *S6.* *Send Test Results* that ensures the delivery of the patient's investigations result to his home and/or the location of his medical records.

S. Vlad and R.V. Ciupa (Eds.): MEDITECH 2011, IFMBE Proceedings 36, pp. 72–77, 2011.
www.springerlink.com

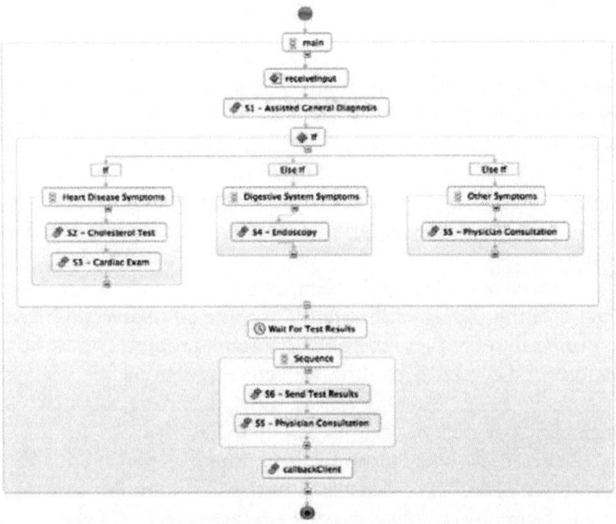

Fig. 1 An abstract process containing several clinical tests

Executing an activity means invoking a service. For each activity, which is assimilated to an *abstract service* (*S1, S2,* ... in Fig. 1), several concrete services exist. Each concrete service has different QoS properties. For describing the QoS we use the following parameters: *response time (t), rating (r), distance (d)* and *cost (c)*.

In software engineering, the *response time* (t) is a measure for the performance of a service. It represents the round-trip time between sending a request and receiving the response. In the medical world, the response time represents the duration for which a patient benefits from a medical service. The user *rating* (r) represents the score the patient uses to reward a medical service after accessing it. The *distance* (d) is a numerical description of how far apart are the patient and the medical service. The *cost* (c) is the price to pay for using each service.

The QoS of the composite service is obtained by aggregating the QoS of the component services. The aggregation rules are described in the section *Proposed approach*.

Given m abstract services and n concrete services for each abstract service, there are n^m possibilities. The search space is a discrete one since for each abstract service we need to choose one concrete service and any combination is possible. We have a combinatorial optimization problem here. An exhaustive search algorithm is very inadequate because the solution should be found at runtime. Finding the solution with the optimal QoS is an NP-hard problem.

Numerous existing proposals for NP optimization problems indicate Genetic Algorithms (GA) as the preferred approach. The particularity of QoS optimization is that it's usually done at runtime, where a fast algorithm is preferred. This fact and also the need to improve the accuracy and the exploration of the solutions space motivated us to propose a new approach, based on Differential Evolution (DE). According to the experimental results this method proved to outperform the GA in terms of convergence time and scalability.

C. *Outline*

The next section presents some of the existing NP optimization solutions. Section three contains some introductory aspects about Differential Evolution. Section four presents the proposed approach based on DE. In section five we show some numerical experiments and we compare the proposed method with the existing approaches. The last section contains the conclusions and future work.

II. RELATED WORK

The NP optimization problem stated previously is well known in domains like Service Oriented Computing (SOC) and Search-based Software Engineering (SBSE). We found it discussed in [2, 3, 7, 14, 18] and other papers. In the literature, various solutions are proposed based on different approaches such as: integer programming (greedy algorithms), genetic algorithms and hill climbing algorithms. In this section we present what we considered the most relevant of these proposals.

Genetic algorithms versus linear programming. G. Canfora et al. [2] have compared a linear integer programming [16] based algorithm with a genetic algorithm. As a case study, they considered a workflow containing 8 distinct abstract services. The number of available concrete services per abstract service was set to: 5, 10, 15, 20 and 25. The comparison was based on the convergence time that was considered proportional to the CPU user time. The authors used an elitist GA where only the best 2 individuals are copied to next generations, a crossover probability of 0.7, a mutation probability of 0.01 and a population of 100 individuals. The selection mechanism adopted was the roulette wheel selection. Their conclusion was that, in contrast with linear integer programming (the widely adopted approach at the moment), GA is able to deal with QoS attributes having non-linear aggregation functions. Also, GA can scale-up when the number of concrete services per abstract service increases. When the workflow size and the number of concrete services per abstract service are limited and there is no need to use non-linear aggregation functions, integer programming is however preferable.

A genetic algorithm for services deployment optimization. Yves Vanrompay et al. [14] also propose to use genetic algorithms for mobile service composition and deployment.

In this case, the problem is formulated slightly different: there is a system consisting of several nodes on which a composite service can be deployed in a distributed manner. The goal is to deploy the composite service onto a set of connected nodes in a way that the allocation meets the given QoS constraints and minimizes the communication cost between the nodes. A set of constraints are added to the problem model for specifying if a certain component can be deployed on a specific node. The authors prove that GAs provide a scalable mechanism which offers improvements over relevant solutions.

Genetic algorithms versus greedy algorithms. Liu Xiangwei et al. [7] also suggest that genetic algorithms are a good approach for semi-automatic service composition. The paper presents an independent global constrains-aware Web service composition method based on extended Color Petri net (eCPN) and a genetic algorithm (GA). The authors compared the genetic algorithm with a greedy algorithm and the conclusion was that GA has higher execution efficiency and success rate.

Weise et al. [15] also compare genetic algorithms with greedy algorithms and conclude that GAs offers a good exploration of the solutions space but they are slower than the greedy algorithm. Other advantages of the genetic algorithms approach are the generality and the extensibility.

The large majority of existing proposals indicate *genetic algorithms* as the best approach for large search spaces: complex composite services with numerous abstract services and numerous concrete services. One of the main advantages of the GA is scalability.

Some existing research, as for instance Tusar and Filipic [13], show that for some general optimization problems, the algorithms based on Differential Evolution (DE) [11] performed significantly better than the corresponding genetic algorithms. This fact motivated us to choose a DE-based approach.

III. Differential Evolution

The DE algorithm was introduced by Storn and Price [11]. DE is a population based, stochastic, and continuous function optimizer [12] where distance and direction information from the current population is used to guide the search process [4]. DE is known to be able to handle non-differentiable, nonlinear, and multimodal objective functions, to be easy to use, and to converge consistently to the global optimum in consecutive, independent trials.

Essentially, for each individual of the population (target vector $x_i(t)$), a mutant vector $m_i(t)$ is first generated by adding the weighted difference (difference vector) between two randomly chosen vectors (parameter vectors $p_{i1}(t)$ and $p_{i2}(t)$) to a third chosen vector (base vector $b_{i3}(t)$) as follows:

$$m_i(t) = b_{i3}(t) + F \cdot (p_{i1}(t) - p_{i2}(t)) \quad (1)$$

where $i \neq i_1 \neq i_2 \neq i_3$; i_1, i_2 are randomly and uniformly chosen between 1 and the population size and $F \in \Re^+$ is the scaling factor, controlling the amplification of the differential variation.

Secondly, one child, called *trial* vector, is obtained by crossover of the mutant vector and the target vector. Finally, the target vector is replaced by the best of either the trial or target vector.

One issue in using Differential Evolution derives from the fact that DE was originally proposed to solve problems defined in a continuous domain and the problem we want to solve is discrete. Since the objective functions we want to optimize are of the form $f : D \rightarrow \Re$, where D is a discrete domain, DE can't be used in its canonical form.

Several methods to apply differential evolution for discrete variables were discussed in the literature [1, 6, 9, 17], two of which are discussed below: *TruncDE* and *XueDE*.

TruncDE was proposed by Lampinen and Zelinka [6] for applying DE to integer-valued problems. They maintain floating-point variables for internal DE computations, and truncate the values when evaluating the fitness function $f(y_i)$, where

$$y_i = \begin{cases} x_i & \textit{for continuous variables} \\ INT(x_i) & \textit{for discrete variables} \end{cases} \quad (2)$$

$x_i \in D$ and INT is a function that converts a floating-point number to an integer by truncation.

For finite discrete domains, the authors propose that instead of attributing the actual discrete values to x_i, this should store the index of the discrete value in the corresponding subset of values. Then, this problem can be handled as an integer problem.

Xue et al. [17] replace the mutation operator of DE with a conditional operator based on three probabilities: greedy probability p_g, mutation probability p_m and crossover probability p_c. A new individual is generated with the following rule:

$$y_i = \begin{cases} x_{best_j} & r \leq p_g \\ rand(\Omega_j) & p_g < r \leq p_g + p_m \\ x_{a_j} & p_g + p_m < r \leq p_g + p_m + p_c \\ x_j & \textit{otherwise} \end{cases} \quad (3)$$

where r is a random number, x_{bestj} is the individual with the highest fitness value from the population, Ω_j contains all the possible values for allele j, x_{a_j} is a randomly selected individual from parent population that is distinct with x_j.

IV. Proposed Approach

A. Services Technologies

Several technologies for creating and executing business workflows (such as the one depicted in Fig. 1.) exist: WS-BPEL (Web Services Business Process Execution Language) [10], WSCI (Web Service Choreography Interface), and others.

The most widely used standard for composing services, WS-BPEL, was chosen as service model. In WS-BPEL, a business process (workflow) consists in a set of activities that are executed according to some control structures. Such control structures include: *flow, sequence, switch* and *while*.

Flow is used to define concurrent activities. A flow completes when all its activities did complete. A *sequence* is a set of activities that are executed one after the other. *Switch* selects between any number of *case* branches based on a condition. *While* is used to create conditional loops.

B. Genotype

Let $S_A=\{S_{A1}, S_{A2}, .., S_{Am}\}$ be the set of abstract services from a business workflow and $S_{Ci}=\{S_{Ci,1}, S_{Ci,2}, .., S_{Ci,n}\}$ the set of concrete services that can realize the abstract service S_{Ai} and $Q_{i,j}=(t, r, a, c)$ the vector of QoS properties (*response time - t, rating - r, distance - d and cost - c*) for $S_{Ci,j}$.

For the problem of services QoS optimization, the genome is usually encoded as a vector of integers: the ordinal value represents the identity of the abstract service and the cardinal value corresponds to the concrete service or to the execution node.

The genome encoding is depicted in Fig. 2 and was initially proposed in [2]. It consists in an array of integer values and has the length equal to the number of abstract services in S_A. Each gene stores the index of the concrete service that realizes the corresponding abstract service.

C. Fitness Assignment

The fitness is assigned to a composite service function of its QoS attributes. But the composite service QoS is not given. Thus, it is necessary to compute the QoS of a composite service starting from the QoS of the concrete services called by that composite service. This operation is called QoS aggregation.

The aggregation operations depend on the composite service architecture. Table 1 shows how the aggregate QoS is computed for each control structure. For *flow* and *sequence* the QoS vector for individual services is sufficient to evaluate the aggregate QoS. For example, since *flow* executes several activities in parallel, the total response time is given by the maximum response time of all executed activities.

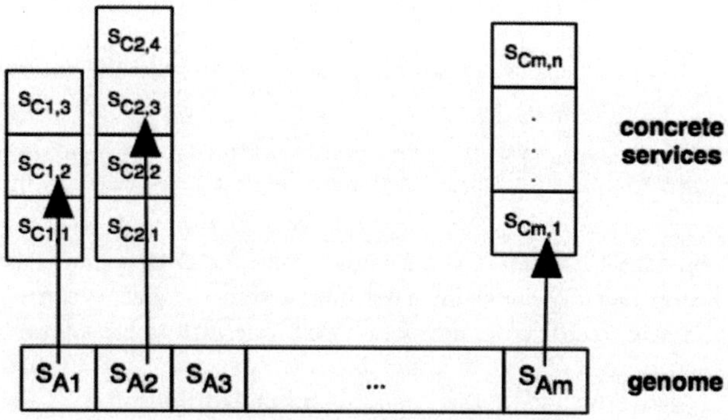

Fig. 2 Genome encoding [2]

Table 1 QoS Aggregation

Control struct. / QoS Property	Flow	Sequence	Switch	While
Response Time (T)	$\max_{i\in 1..m}(t_i)$	$\sum_{i=1}^{m} t_i$	$\sum_{i=1}^{m} p_i \cdot t_i$	$k \cdot t$
Rating (R)	$\prod_{i=1}^{m} r_i$	$\prod_{i=1}^{m} r_i$	$\sum_{i=1}^{m} p_i \cdot r_i$	r^k
Distance (D)	$\sum_{i=1}^{m-1} d_i$	$\sum_{i=1}^{m-1} d_i$	$\sum_{i=1}^{m-1} p_i \cdot d_i$	$k \cdot d$
Cost (C)	$\sum_{i=1}^{m} c_i$	$\sum_{i=1}^{m} c_i$	$\sum_{i=1}^{m} p_i \cdot c_i$	$k \cdot c$

In case of the *switch* construct, the BPEL process needs to be monitored at runtime during multiple executions, to determine the probabilities p_i associated to each case branch, $\sum_{i=1}^{m} p_i = 1$.

p_i represents the probability to select case branch *i*. In case of the *while* loop, the average number of iterations *k* is also determined during monitoring.

To evaluate the quality of each potential solution, we consider an aggregate objective function (AOF) similar to the one proposed by Canfora et al. [2]:

$$F(y) = \frac{w_1 \cdot R}{w_2 \cdot T + w_3 \cdot D + w_4 \cdot C} \tag{4}$$

where w_i are the weights that correspond to the importance of each QoS property to the user and *R, T, D, C* are the aggregate QoS values for the business workflow.

V. Numerical experiments and evaluation

In order to test our solution, we implemented the following algorithms:

1. *TruncDE* - the DE algorithm based on Lampinen and Zelinka's proposal [6] with the parameters: scaling factor $F = 0.95$, jitter $F\_NOISE = 0.001$ and crossover constant $Cr = 0.95$. The strategy used for Differential Evolution is *DE/best/1/bin*. This means that the base vector is the best vector from the population, one difference vector is considered for generating the new vector and uniform crossover is used, based on a binomial distribution.

2. *XueDE* - the DE algorithm proposed by Xue et al. [17] with the following parameters: *DE/best/1/bin* strategy, scaling factor $F = 0.9$, jitter $F\_NOISE = 0.25$ and crossover constant $Cr = 0.95$. The probabilities for the conditional operator in equation (3) are: greedy probability $p_g = 0.1$, mutation probability $p_m = 0.65$ and crossover probability $p_c = 0.2$.

3. *IntGA* - the GA algorithm proposed by Canfora et al. [2] with the parameters: *uniform* crossover where one parent is selected using *tournament selection* and the second parent is selected using *roulette-wheel selection*, the tournament size is *5*. The mutation probability suggested in [2] is $p_m = 0.01$.

For all these algorithms, the population was limited to 100 individuals, which were evolved for 1000 generations. We conducted experiments for 25 scenarios that include all combinations of $m \in \{10, 20, 30, 40, 50\}$ abstract services and $n \in \{10, 20, 30, 40, 50\}$ concrete services. Each scenario ran 100 times and the results were averaged. All algorithms were implemented using ***ECJ*** version 20 [8].

Figures 3 – 7 show most significant numerical results for two of the considered test scenarios.

A case with a business workflow consisting in $m=10$ abstract services, each of them having $n=10$ concrete alternative services was evaluated. The results are depicted in Fig. 3. Within the first 80 generations all algorithms find a very good individual. Then, the best fitness of the population increases at a very slow rate. The fastest algorithm for this scenario is *TruncDE*.

A more complex scenario, involving a business workflow consisting in $m=20$ abstract services, each of them having $n=40$ alternatives is presented in Fig. 4. We notice that when increasing the complexity of the problem, *XueDE* becomes the fastest algorithm to converge, while *TruncDE* is the slowest. *IntGA*'s performance is above average, being comparable to the best *DE* in every scenario.

Since our aggregate fitness function *(4)* is composed of several objectives, some requiring to be maximized, others requiring to be minimized, we present the evolution of the objectives represented by the distance, cost and rating for the second scenario in Fig. 5 – 7.

These results show that the proposed DE approach (*TruncDE* and *XueDE*) outperforms the genetic algorithm proposed by Canfora et al. *IntGA* [2]) for solving the NP-hard problem of QoS-based service optimization.

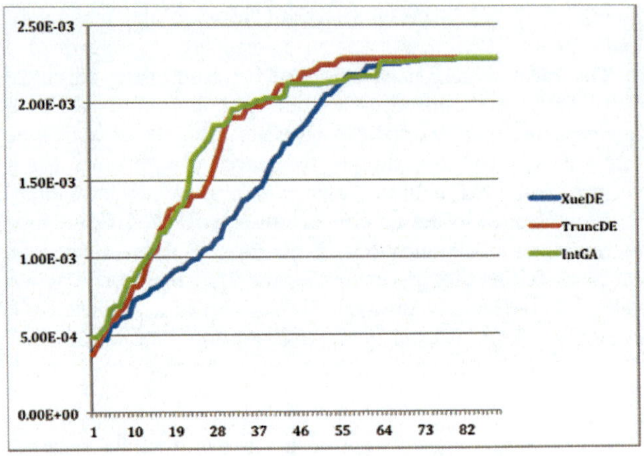

Fig. 3 The evolution of the best fitness over 90 generations for m=10 abstract services and n=10 concrete services

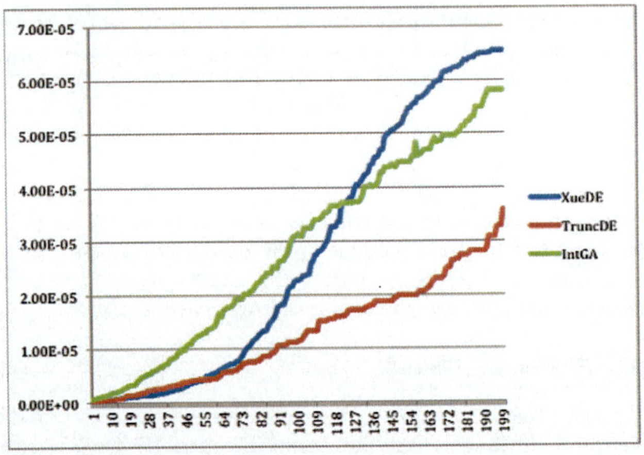

Fig. 4 The evolution of the best fitness over 200 generations for *m=20* abstract services and *n=40* concrete services

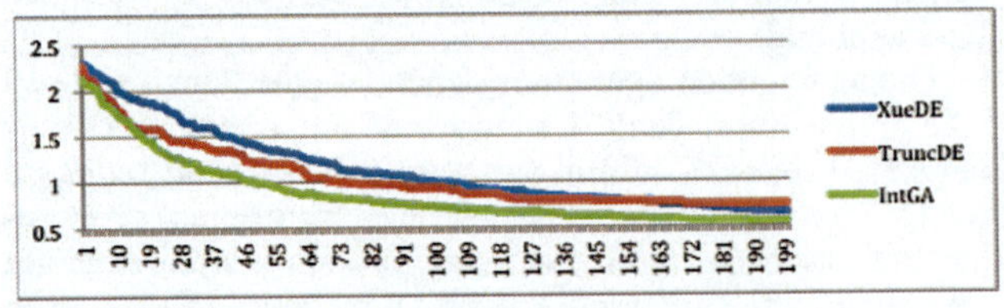

Fig. 5 Distance minimization during 200 generations for *m=20* abstract services and *n=40* concrete services

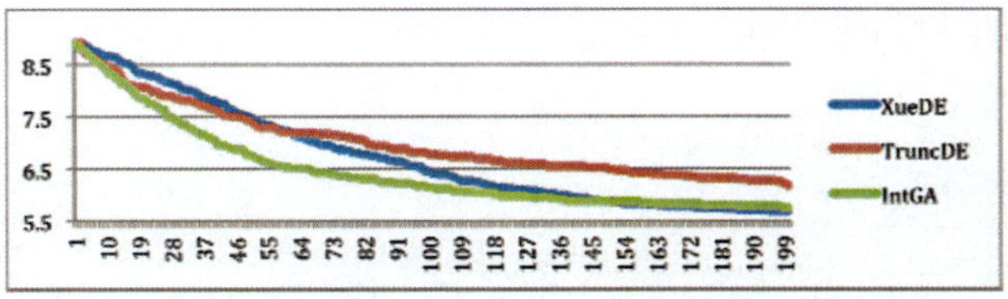

Fig. 6 Cost minimization during 200 generations for *m=20* abstract services and *n=40* concrete services

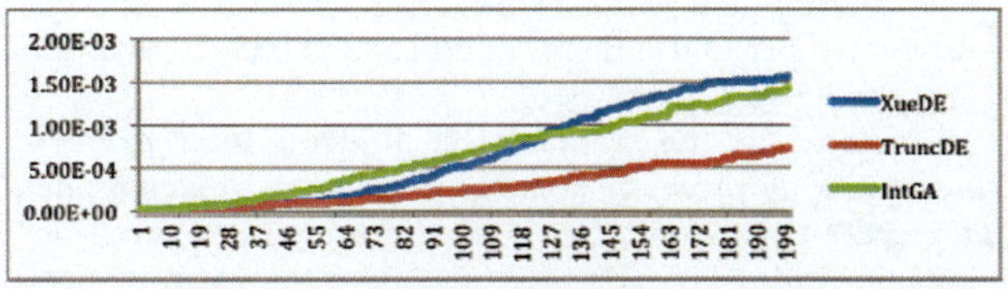

Fig. 7 Rating maximization during 200 generations for *m=20* abstract services and *n=40* concrete services

VI. Conclusion

This paper proposes a method to compose and optimize medical services as business workflows. Such a workflow consists in a set of abstract services, and for each abstract service there are several concrete services. Since each medical service has different QoS parameters such as *response time*, *rating*, *distance* and *cost*, determining the optimal combination of concrete services that realize the abstract services of the business workflow is an NP hard problem.

To solve this problem, we propose a new solution, based on Differential Evolution. We implemented two Discrete DE algorithms from the literature *TruncDE* [6] and *XueDE* [17] which we adapted to solve the services QoS optimization problem. We compare these algorithms with the genetic algorithm proposed by Canfora et al. in [2] – *IntGA*.

The results show that the approach based on DE outperforms the genetic algorithms. *TruncDE* proved to be suited for scenarios of low complexity (up to 15 abstract services, each of them having up to 40 alternatives), while *XueDE* was superior for scenarios of medium and high complexity. The performance of *IntGA* was average, but it was not the slowest to converge in any of the test scenarios.

As future work, we intend do some more comparative experiments with other meta-heuristics such as: hill-climbing, simulated annealing and others. Another future direction is to develop a solution based on multi-objective optimization algorithms.

Acknowledgment

This project was supported by the national project code TE 252 financed by CNCSIS-UEFISCSU.

References

1. M. Zhang, S. Zhao, and X. Wang. Multi-objective evolutionary algorithm based on adaptive discrete differential evolution. In Proceedings of the Eleventh conference on Congress on Evolutionary Computation, CEC'09, pages 614–621, Piscataway, NJ, USA, 2009. IEEE Press.
2. G. Canfora, M. Di Penta, R. Esposito, and M. L. Villani. An approach for qos-aware service composition based on genetic algorithms. In Proceedings of the 2005 conference on Genetic and evolutionary computation, GECCO'05, pages 1069–1075, New York, NY, USA, 2005. ACM.
3. D. Comes, H. Baraki, R. Reichle, M. Zapf, and□K. Geihs. Heuristic approaches for qos-based service selection. In ICSOC 2010, Lecture Notes in Computer Science, 2010.
4. A. P. Engelbrecht. Computational Intelligence: An Introduction. John Wiley and Sons, 2nd edition, 2007.
5. J. Kennedy and R. Eberhart. A discrete binary version of the particle swarm algorithm. In Systems, Man, and Cybernetics, 1997. 'Computational Cybernetics and Simulation'., 1997 IEEE International Conference on, volume 5, pages 4104 –4108 vol.5, Oct. 1997.
6. J. Lampinen and I. Zelinka. Mechanical engineering design optimization by differential evolution, pages 127–146. McGraw-Hill Ltd., UK, Maidenhead, UK, England, 1999.
7. X. Liu, Z. Xu, and L. Yang. Independent global constraints-aware web service composition optimization based on genetic algorithm. Intelligent Information Systems, IASTED International Conference on, 0:52–55, 2009.
8. S. Luke. Ecj - a java-based evolutionary computation research system.
9. G. Onwubolu and D. Davendra. Scheduling flow shops using differential evolution algorithm. European Journal of Operational Research, 171(2):674 – 692, 2006.
10. Organization for the Advancement of Structured Information Standards (OASIS). Web Services Business Process Execution Language (WS-BPEL) Version 2.0, April 2007.
11. R. Storn and K. Price. Differential evolution - a simple and efficient adaptive scheme for global optimization over continuous spaces. Technical Report TR-95-012, March 1995.
12. R. Storn and K. Price. Differential evolution - a simple and efficient heuristic for global optimization over continuous spaces. Journal of Global Optimization, 11:341-359, 1997.
13. T. Tusar and B. Filipic. Differential evolution versus genetic algorithms in multiobjective optimization. In Proceedings of the 4th international conference on Evolutionary multi-criterion optimization, EMO'07, pages 257–271, Berlin, Heidelberg, 2007. Springer-Verlag.
14. Y. Vanrompay, P. Rigole, and Y. Berbers. Genetic algorithm-based optimization of service composition and deployment. In Proceedings of the 3rd international workshop on Services integration in pervasive environments, SIPE'08, pages 13–18, New York, NY, USA, 2008.
15. T. Weise, S. Bleul, D. Comes, and K. Geihs. Different approaches to semantic web service composition. In
16. L. Zeng, B. Benatallah, A. H.H. Ngu, M. Dumas,□J. Kalagnanam, and H. Chang. QoS-aware middleware for web services composition. IEEE Trans. Softw. Eng., 30:311–327, May 2004.
17. F. Xue, A. Sanderson, and R. Graves. Multi-objective differential evolution and its application to enterprise planning. In Robotics and Automation, 2003. Proceedings. ICRA '03. IEEE International Conference on, volume 3, pages 3535 – 3541 vol.3, 2003.
18. L. Alboaie, T. Barbu. Reputation System User Classification Using a Hausdorff-Based Metric. Computational Intelligence for Modelling Control & Automation, 2008 International Conference on, pp. 1035-1040, Dec. 2008

Direct Response Advertising in Romanian Dental Field: A Qualitative Analysis

A. Constantinescu-Dobra

Technical University of Cluj-Napoca/Electrical Engineering Faculty/Department of Electrical Machines, Marketing and Management, lecturer, Cluj-Napoca, Romania

Abstract— ***Romanian dental industry*** **is in its initial stage of development, while *the market* has a huge potential for future growth. Hence, the advertiser from this field should integrate direct response with image based advertising for reaching more efficiency and effectiveness. The paper aims at identifying the degree of direct response advertising vs. image advertising, as a marketing tool for dental services and products. The assessment will be distinctly processed by different types of products. Moreover, the study tries to draw some correlation between dental market development and the above mentioned rate, during the latest five years.**

The performed research is a qualitative one, based on content analysis of 748 advertisements, published in professionals' magazines, between 2006 and 2010.

The research outcomes reflect a balanced advertising for materials, equipments, services and reveal strong image advertising for the dental products end-users. The economic crisis has changed the pattern in dental advertising, mainly by emphasizing rather on emotion than on rational reason.

Keywords— **direct response advertising, imagery advertising, content analysis, small and medium enterprises, dental industry.**

I. Introduction

The marketing challenge facing dentistry today is how dental community continues to build image while creating demand. Direct marketing advertising spending by health care providers has raised from 200 million in 1990 to 800 million in 1998 and 1.2 billion in 2008 USD [1]. Advertising by dentist is relatively a recent phenomenon. The research in this field reveals that advertising and marketing clearly have an important place in the future of dental care services [2].

The increased "mass market" appeal of dental services alongside of growing competition in the market (at a product/manufacturer and service delivery level) and an increase in brand level awareness creates a greater need for marketers to use effectively the direct response advertising.

The present study performs *the first assessment of Romanian direct response and image advertising among dental products and services*, using an adapted methodology. We are going to analyze the differences between direct responses advertising and image advertising (taking into consideration their practical advantages) and, also, the integrative manner to use them, in order to maximize their efficiency in today's dental field.

Findings will aid marketers in planning their strategies and tactics, to promptly react to changes in Romanian dental market.

In the following, we shortly present the main evidence from global dental industry and its advertising, emphasizing on the Romanian market particularities, in order to create connection with direct response- image advertising degree.

II. Dental industry today

The dental industry is one of the most attractive segments of the healthcare industry, with an estimated size of about 18.8 billion USD in 2008. Latest, the market growth reached 4.6% percent annually [3], [4]. The UE markets have experienced also constant growth, in recent years. Some factors that explain the constant expanded demand for dental services are the following [5]:

- Growing acceptance and reduced stigma towards cosmetic dental surgery,
- The strong competition in materials and equipments production catalyzed technical progress in the field,
- Price reduction of cosmetic treatments allows a wider variety of individuals to benefit,
- Media coverage has created a high level of consumer interest in such procedures and services.

The National Institute of Statistics [6] states that in Romanian dental market are acting 8071 private centers (clinics, medical dentist and other societies). The growing rate per year is around 11%. The Romanian dental equipment is imported mainly from Germany, Italy, Czech Republic, Hungary, United Kingdom and recorded a significant growth. All the distributors and manufacturers are small and medium enterprises.

According to recent research reports [7], [8], [9] *Romanian dental industry* is undeveloped. During the last years, the rapid expansion and growth of the dental care industry, due to globalization and increasing awareness of consumers for treatment alternatives, will contribute to the development

S. Vlad and R.V. Ciupa (Eds.): MEDITECH 2011, IFMBE Proceedings 36, pp. 78–83, 2011.
www.springerlink.com

of both dental services and equipments market in Romania. On the other hand, the *Romanian dental market* reaches 300 million Euros [10] and has a huge potential for future growth.

Various factors such as low costs, highly trained surgeon, dentists and specialist, short treatment period, developing the dental tourism in the country and investments in the latest technologies will catalyze the Romanian dental industry in coming years.

III. Direct Advertising and Image Advertising – Literature Review

Direct response advertising aims at closing a sale or a transaction in short time. Image advertising means building brand equity, which establishing a brand's value proposition in the minds of consumers.

In the past, image advertising and direct response advertising have been analyzed relatively isolated from one another [11]. Recent research finds multiple way of increasing communication efficiency by integrating features belonging to both mentioned orientation.

Image marketing is about building an image in the marketplace, while direct response (or direct marketing) is all about producing calls, driving customer traffic and getting immediate revenue. On image marketing side, the return of investment is not going to occur immediately.

On the other hand, *direct-response advertising* is exclusively designed to solicit a direct response which can be tracked and measured. It supposes direct communication between the viewer and the advertiser. Furthermore, direct-response campaigns perform best if the underlying strategies and tactics are highly competitive.

Direct-response advertising is characterized by four primary elements [12]:

- An offer
- Sufficient information for the consumer to make a decision whether to act
- An explicit "call to action"
- Means of response (typically multiple options such as a toll free number, web page, and email)

Referring to the *print advertising*, we will present some findings from psychological visual imagery research, relevant for our study. The outcomes focus on using image advertising benefits for long term instead of direct advertising. We mention the most important ones [13], [14]:

1. The larger visual images produce better learning or determine positive brand attitude. The picture-then–word superiority held (regardless of whether the target response was to recognize the picture or to recall the word).
2. Attention holding is important for evaluative response (the longer the stimulus is attended to beyond two seconds, the better). Learning is facilitated if the order is picture-word rather word-picture.
3. Greater use of variation on a theme for print advertising enhances more attention. Varied but related illustrations are consistent with novel-but – familiar principle.

The studies reveal that is only one situation where illustration size is not important: direct response adverting of the informative "*long to*" variety. Also, the typical technique in direct response advertising is to provide the reader with as much information as possible in order to achieve a "stimulus sufficient" decision.

From a different point of view, recent interest has been directed toward integrating the two disciplines [15], [16]. The investigation of the effective way of melding the two techniques conducted to several benefits of both traditional and direct marketers. The factors that distinguish direct response advertising from image advertising are examined form multiple points of view: percentage of advertise image, main purpose, emphasizing on product or brand, listing several contact possibilities, frequency, segmentation or the quality of advertise image[17].

A very important presumption for this study is the following: for efficient direct response advertising, it's important that a brand's equity have been communicated in advance of the consumer's purchase decision. It requires long-term perspective and understanding of the many ways in which advertising works. In this regard, it is vital to take into account all of the marketing stimuli that affect consumer purchase behavior, not just that which occurred just prior to purchase.

Narrowing the perspective, a study [18] concerning the attitude toward *dental advertisement*, conducted in USA, reveals that more than half of the consumers feel that newspaper and professional magazines are appropriate media for dentist to use in advertising. Another research find that [2] the great majority of persons interested in dental services and products watch dental advertisement in weekly magazines, health or beauty magazines or professionals magazines. We have also to take into consideration another aspect too: the adoption of cosmetic dental procedures in an extended level by mainstream consumer created a shift from "type of procedure" to advertising at brand level. Finally, an important particularity of advertising in dental field (and in healthcare, in general) is the mass-media low influence for the end-users. It is more important the opinion of leaders, prescribers and reference groups [19] in choosing such a product. Hence, *advertiser acting in dental care should emphasize on leaders first, after that on final consumers.*

IV. METHODOLOGY

A. *Research Objectives*

The main goal of this research is to investigate the degree of direct response advertising in Romanian dental magazines. The main objectives derived from the mentioned goal are:

- Assessing the penetration of direct response advertising as a marketing tool for dental services and products (materials, equipments, laboratory services, educational services);
- Evaluating the degree of image vs. direct response orientation of different types of products/services in dental care advertisements;
- Revealing the evolution of this examined rate in the latest five years

B. *Research Methodology*

The performed research is a *qualitative* one, based on *content analysis* of 748 advertisements published in the last 5 years (from 2006 to 2010).

The data was collected from Dental Target magazine. It is a Romanian publication which has national coverage and appears four times per year. We chose this print magazine due to its audience (600000/magazine) and its targets: the medical dentist, the dental technician, the nurse and also the patients. Manufacturers, importers and distributors of dental materials and equipments, dental cabinets and laboratories post information in this magazine.

Content analysis of print advertising is mainly used for examine and measuring the visual and verbal elements of advertisements or to analyze the feelings, beliefs or attitudes of a culture in their environment.

In order to reach our objective, we adapted an existing method propose by Peltier *et all* [11] and those modified by Seiz [17], fitted for cosmetic products. In this study, the mentioned method was adjusted by adding suited criteria to the dental field. From the beginning, we added more constraints in operational direct response advertising in magazines to the mentioned studies determination, like dimension *advertising specific web address* to purchase the product. In the original methodology there are present: advertisements with the specific address, local phone number, a toll free number, a fax number of where to purchase the brand, a mail in coupon.

To assess the degree of direct response versus image based in each advertisement, we uses the key dimensions proposed by Peltier *et al.* and some other new, specific to dental industry.

The scoring system was designed in a way to reflect differences for the following eight dimensions:

- no response device/strong response device,
- awareness attitudes/response,
- uses imagery/ does not use imagery,
- uses emotion/uses reason,
- uses little information/much information,
- attitudes toward brand/attitudes toward product,
- high message frequency/low message frequency and
- high production values/low production values.

As we already underline before, in Romanian dental market are acting mainly the importers and distributors. This is the reason why in the present research, we have added another dimension (criteria) for the study: printed message/advertisement is promoting the manufacturer or the distributor. In our previous assessment we observed in a lot of cases when the distributor emphasize both on the producers brand and on the product. Our presumption was the following: *when the advertisements focus only on distributor, direct response orientation prevails*. In the same time, when the producer is most advertised, the image orientation is present.

For our purpose, an Osgood scale (semantic differential scale) was designed for evaluating the changes taking place in Romanian dental advertising in recent years.

Every dimension was assessed by five intervals where 1 means image-majority, 3 - balanced and 5 direct response-majority). Peltier *et al.* researches state that an *image majority advertisement* focuses on the image of the brand and the response devices are not compulsory. It can imply attitudes or emotion (happiness, love, sadness etc.). Also, an image advertisement has little (technical/specific) information, high production value (the highest quality materials, very good resolution, vivid colors, quality paper etc.). In an effort to build the image of the brand, the frequency of the message will be increased (more than one advertisement per edition and one per each semester's edition of the magazine).

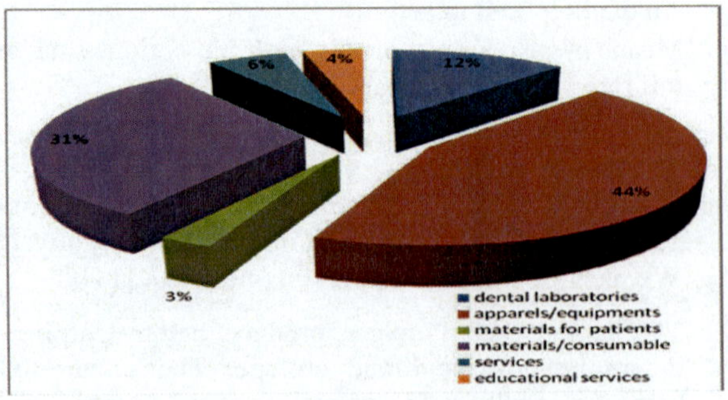

Fig. 1 The study's advertisements by type of products/services

In a *direct response majority advertisement*, the message can be communicated through different other devices than magazines, and the frequency in a year can be lower. The product is more important as opposed to the image of the brand. Moreover, in a direct response-majority advertisement, the focus is toward reasons to buy the product and not to create emotion. The direct response ad can include one or several response devices such as a reply card as well as a free phone number. The quality of production will not be as high as in an image-majority advertisement.

V. FINDINGS AND DISCUTION

The data processed in SPSS 16 are synthesized in table 1, 2 and figure 2. Figure 1 display the analyzed advertisements by type of dental categories of products and services. A total of 748 dental related ads were examined, 93 for dental laboratory services, 329 for equipments, 21 for materials and supplies offered to the final consumer, 231 for materials and supplies offered to the practitioners, 32 for educational trainings, courses and 42 for other related services.

A. *Direct Response Advertising as a Marketing Tool for Dental Field*

The most common direct response vehicle was postal addresses, web-page addresses and phone numbers. In 32% of the cases were also included the toll free number, the majority of them in 2010 editions. Only a few ads included a fax number in the ad. The average values were computed for each advertisement from the sample, displayed in table 1. If the mean score is between 1-2 points, we have considered that is image majority advertising A balanced one has the score is between 2.01-3.99 points, and direct response advertising with a score between 4 and 5 points.

Surprisingly, after clustering all the advertisements based upon the mentioned 9 criteria, we can observe *strong response devices* (4.47 points). This means that almost every ad publishes at least three ways for contacting the provider.

Regarding image vs. direct response majority, although the evaluated ads had some sort of direct response device, *with the exception of materials for end-users, overall, advertisements was primarily balanced*. This evidence was expected given the early stage of development of both Romanian dental market and industry. Many dental equipment producers companies are multinationals and are focusing rather on brand's image, targeting sophisticated market segments. But, as opposed to materials and consumables, the products may differ, delivering a wide range of product benefits. Although image is important in delivering the message to the target audience, reason comes more into play in the description of the product benefits.

Furthermore, the materials for patients are based on image advertising, which is often employed when there is little difference in the brands or prevails on products where the focus is on the personality associated with the brand name as opposed to product benefits (cosmetic dental products/services).

Table 1 Scores by each measured dimensions

Research dimensions	Mean	Std. Dev
response device	**4.47**	1.082
Awareness attitude/response	2.58	1.390
uses imagery/ do not uses imagery	2.75	1.047
uses emotion/reason	3.93	.996
uses information/ do not uses	2.89	1.194
attitudes toward brands/products	2.66	1.147
message frequency	3.54	1.275
production values high / low	**2.09**	1.334
Producer/Distributor	3.98	1.187
Valid N (list wise)		

In dental ads from the sample, the distributors prevail (mean score 3.98 points) and the main argument is rational, not emotional (3.93 points). This result was expected, dental equipments being more technical and high technology products. Ads highlight both the brand and the product and have a very good visual quality. Frequency of the messages is rather low in the same edition of the magazine but, during a year, it reaches a medium rate.

B. *Direct Response Advertising by Specific Category of Dental Products/Services*

Analyzing in depth the degree of image vs. direct response orientation and grouping the different types of products/ services, we obtained mean scores depicted in table 2.

We find interesting the *materials for end-users* (patients) score 1.86 points, indicating that the ads *is image majority*. For the other category of equipments, scores range between 2.93 and 3.49, pointing a *balance ad*.

Educational services (trainings for practitioners, medical doctors and students or long term courses for learning the newest technologies), services provided by dental laboratory and other dental services (like cosmetic procedures) followed the same advertising pattern: at least four possibilities of response, balance between image and information, focus on services benefits and not on brand, very high production value.

Table 2 Direct response advertising scores by specific categories of dental products/services

Types of product/services advertisements	Mean	Std. Dev
dental laboratories	3.49	.822
apparels/equipments	3.39	1.120
materials for patients	1.86	1.047
materials/consumable	3.32	.934
services	3.02	1.043
educational services	2.93	1.211
Valid N (list wise)	748	

On the other hand, ads for *equipments and materials* seem to be similar too, with two exceptions: equipments account the highest response devices (mean score 4.79 points) and material ads are mainly based on reasons than emotions (4.19 points - the highest mean score from all categories).

The surprise comes from materials for patients (for example Lacalut Active). Those ads emphasize exclusively on image (1.24 points), the attitude is toward the brand only (1.98 points), an average of two prints per edition and the best quality from all sample (1 point!). All the request for an image majority are accomplished, even the poor response devices (1.52 points).

According to previous studies [17], cosmetics ads in Europe are fairly balanced. For dental cosmetics we have found an image orientation. This gap may come from the early stage of Romanian hygiene market development, mainly due to the very low level of consumer education in dental prevention and care, in comparison with the same indicator in more developed countries (West European markets) [20]. This means that brand consolidation strategy is recommended to the entry stage on the market.

C. *Direct Response Advertising Evolution in Time*

In order to assessing the changes taking place in Romanian dental market, the research is covering a five years period (between 2006 and 2010). Over the five years period, the frequency of direct response advertisements doesn't show a significant seasonal pattern and a steady decrease in use over time. Most of the direct response advertisements are in the second semester of 2008, probably as a result of the dramatic increase of dental market in analyzed period. Direct response equipments ads were predominant during the last semester of every year. The reason could be the implementation of the predicted annually plan through budget validation. The most investments in equipments are made by the end of the year [6], when the company knows the certain sales indicators. No pattern was apparent for direct response dental laboratory or services ads. We can observe that direct response ads for educational services are more frequent since 2008. Figure 2 depict direct response advertisements average score, during the mentioned period.

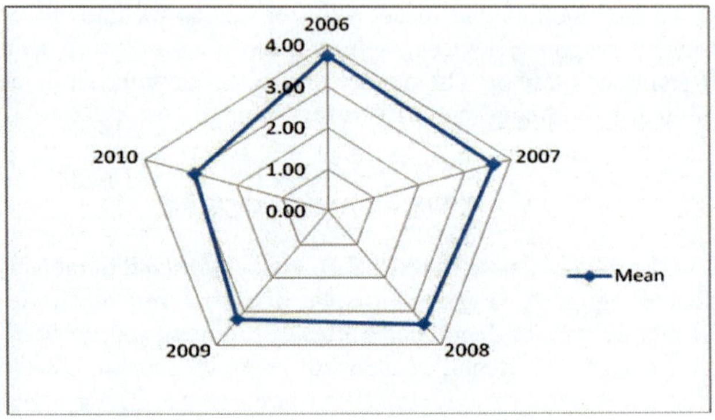

Fig. 2 The study's advertisements mean by years

The findings reveal a constant decrease in direct response ads dimensions every year (from a mean score of 3.89 points in 2006 to 2.03 points in 2010). Evaluating the average values for each analyzed criteria, we observe a *disruptive evolution in dental message frequency and in persuasive ads messages* (by using emotion or reason). If during the years 2006 and 2007 clearly prevails the rational reasons for argumentation (mean score 4.24 points, respectively 4.2 points), since 2009, the average value sharply decreases (to 2.58 points in 2010), meaning the emotions are use in a large extent.

Moreover, the same situation occurs in ads frequency by lowering the prints number in 2009 and 2010 (from a score mean 4.18 points in 2006, to 3.14 points in 2010).

The economic crisis could be the explanation for this phenomenon. In 2009, the dental market growth has diminished and the actors in the field have not invested in additional technologies in the same manner as in previous period. In the last two years, direct response advertising and image advertising become balanced. The providers are focusing more and more on brands consolidation.

This situation is atypical taking into consideration the best practice in advertising. As we have mentioned already, the existing research demonstrate that efficiency for industrial product/services consist in two steps: building the brand first, and after that developing direct-response advertising. Although, the advertising strategies implemented by Romanian dental actors followed the opposite trends. During the expansion period of the entire economy, the producers/distributors obtained unexpected results in terms of sales with no (or little) additional efforts.

But, in 2009, when the crisis effects have affected Romanian markets, their sales begun to stagnate. The crisis sharps the competition battle and the dental providers were constraint to adapt rapidly their strategies. We think this is the reason why they have begun just recently to plan and implement advertising strategies, based on brand image.

VI. CONCLUSIONS

The general conclusion of the study is the following: dental materials, equipments and services use balanced advertising, while image prevails in materials for patients advertisements. Through this integrative approach, employing direct response along with image based strategies, advertisers can achieve a positive brand image as well as increase sales. Lately, the most prevalent direct response vehicles included local phone numbers and web addresses of where to purchase the brand. In the effort of integrating direct response vehicles, dental companies should include customer service strategies to handle the influx of product inquiry calls. For being more effective, dental providers should emphasize first on brad and after that on direct response advertising. The economic crisis has changed the pattern in dental advertising, mainly by focusing rather on emotion than on rational reason and by lowering the frequency rate.

FUTURE RESEARCH

The author future actions will be: 1) to check through specific marketing research means in what degree the distributers and producers from dental field have achieved their objectives through the implemented image/response advertising campaign. 2) Future research might look at the other European Countries and find correlations between the rate of image/response orientation in dental field.

ACKNOWLEDGMENT

This study has been performed in the framework of PN II-RU research grant "Development of an electronic marketing plan for Romanian SMEs in order to increase their competitiveness", PD 58/2010, supported by CNCSIS-UEFISCSU.

REFERENCES

1. Direct Marketing Association (2008) Direct Marketing Facts and Figures in the Health Services Industry, http://www.marketingcharts.com/direct/one-third-of-health-services-advertising-spend-is-for-dm-campaigns-3375
2. Press J, Simms C (2010) Segmenting cosmetic procedures markets using benefit segmentation: A study of the market for tooth whitening services in the United Kingdom, *Journal of Medical Marketing* (2010) 10, 183–198
3. Global Dental Industry: An Analysis (2009) http://www.reportlinker.com/p0167104/Global-Dental-Industry-An-Analysis.html
4. Koncept Analytics, Research and Markets: Global Dental Market Report: The Updated 2010 Edition (2010) http://www.businesswire.com/news/home/20110120005058/en/Research-Markets-Global-Dental-Market-Report-Updated
5. De Jongh A., Oosterink F., van Rood Y. and Aartman I. (2008) Preoccupation with ones appearance: A motivating factor for cosmetic dental treatment? *British Dental Journal* 204 (12): 691
6. Institutul National de Statistica , www.insse.ro
7. FRDCenter market entry services Romania (2010)The Dental Services and Equipment Market in Romania, http://www.slideshare.net/JackieBP/the-dental-services-and-equipment-market-in-romania-demo-april-2010,
8. RNCOS (2010) Romania Dental Market Analysis, http://www.rncos.com/Report/IM276.htm,
9. Euromonitor International (2010) Romania : economy statistics and industry reports, http://www.reportlinker.com/r0792/Romania-industry-reports.html
10. Rosu S (2009) Clinicile stomatologice, in criza de timp: pacinetii la fel de multi ca si inainte de recesiune, Capital, 30 september 2009
11. Peltier, J.W., Mueller, B., Rosen, R.G. (1992), Direct response versus image advertising, Journal of Direct Marketing, Vol. 6 No.1, pp.40-7
12. Stone B, Jacobs R (2008) Succesfull direct marketing Methods, McGrowHill
13. John R. Rossiter (1982), Visual imagery: applications to advertising, in Advances in Consumer Research Volume 09, pp 101-106
14. Ship TA (2010) Advertising, Promotion and other aspects of integrated margeting communications, 8-th edition, SouthWest Cengage Learning
15. Low G S (2000), "Correlates of Integrated Marketing Communications," Journal of Advertising Research, 40 (May/June), 27-39
16. Peltier J W, Schibrowsky J, Schultz D (2002 In Press), "Leveraging Customer Information to Develop Sequential Communication Strategies: A Case Study of Charitable Giving Behavior," Journal of Advertising Research.
17. Seiz V (1998) Direct response advertising in the US and European markets: a content analysis of fashion products, European Business Review, vol 98, no 5, pp268-275
18. Johns H.E; Moser H R (**1989**) How Consumers View Dental Advertising, Health Marketing Quarterly, Volume 6, Issue 4 October 1989 , pages 3 - 16
19. Catana G A (2009) Marketingul serviciilor de ocrotire a sanatatii, Alma Mater, Cluj-Napoca
20. Constantinescu-Dobra A. (2009) The competitive Strategies on Romanian Cosmetic market, Marketing from information to decision, 2-nd edition, Risoprint, pp137-148

Author: Constantinescu-Dobra Anca
Institute: Technical University of Cluj-Napoca
Street: Memorandumului, no. 28
City: Cluj-Napoca
Country: Romania
Email: anca.constantinescu@mae.utcluj.ro

Software System for Medical Device Management and Maintenance

C. Luca and R. Ciorap

"Gr.T.Popa" University of Medicine and Pharmacy, Iasi, Romania

Abstract— **Concerning the inpatient care the present situation is characterized by intense charges of medical technology into the clinical daily routine and an ever stronger integration of special techniques into the clinical workflow. Medical technology is by now an integral part of health care according to consisting general accepted standards. Purchase and operation thereby represent an important economic position and both are subject of everyday optimization attempts. For this purpose by now exists a huge number of tools which conduce more likely to a complexness of the problem by a comprehensive implementation. In this paper the advantages of maintenance system architecture on the medical device management are shown and we are presenting original software that can be implemented in any hospital without being necessary the existence of a clinical engineering department.**

Keywords— **Medical device, maintenance, health technology management, maintenance protocols.**

I. Introduction

This article deals with the present situation of service management as well as facility management of medical technology and demonstrates a new integrative approach to optimize the medical device management using a software system. The concept appeared against the backdrop of foreseeable social and technical developments. As a result of social change acting, hospitals and medical technology (equipment and services) providers have to rise to new challenges at the healthcare market. Exemplary to mention are:

- a change in the demographic structure towards an older society [1],
- a hair-trigger financial situation of hospitals which is determined through a severe regimented service rebate and a perspicuous slowdown in investment [1],
- high expectations within the society regarding the quality offer of health care,
- the problem of the acquisition of specialized professionals, in the regarded case especially in the region of medical technology services [2]
-the consisting obligations of an incessant quality assurance and also the documentation of services rendered [1].

In modern and forward thinking Hospitals, it has been realized that prevention is better than cure, and that proactive, rather than reactive management brings the best results.

This is true for managing human resources, to change policies and (just as importantly) for managing the medical devices that all Hospitals use for diagnostic and clinical care [2].

II. Methodology

Many devices require regular and effective maintenance to operate correctly and meet their design specifications. The consequences of ineffective maintenance can be huge in terms of patient care, personnel morale and management time [2].

Such consequences are often overlooked or miscalculated because device breakdowns are not just a cause of lost time but have a direct effect on patient throughput, efficiency and thus waiting lists. Therefore, the importance of effective maintenance to reduce the occurrence of such incidents cannot be overstated [3].

The management of this maintenance for single or few devices is relatively simple.

However, most Hospitals have thousands of medical devices and to correctly maintain these devices can be difficult or impossible without a formalized computerized scheduling system (database). Medical Physics departments should have a structured approach to planned preventive maintenance projects and should implement a computerized system, or audit, refine and improve the effectiveness of the existing implemented system [4].

Maintenance Planning constitutes a sustainable level of activity (dependent upon adequate resources) commencing with the development of the maintenance plan, implementation and performance review [5].

The difference between planning and scheduling needs to be clear. These are differing areas worthy of differing measurement and improvement initiatives.

Planning can occur at any stage during the life of a works order. An electronic indicator in the work-order system needs to be able to identify the work-order by 'status' i.e. planned, ad-hoc, remedial, etc.

In this manner work orders requiring parts, procedures, documents, skills or specialist test equipment can easily be focused [3].

S. Vlad and R.V. Ciupa (Eds.): MEDITECH 2011, IFMBE Proceedings 36, pp. 84–89, 2011.
www.springerlink.com

A work order cannot be considered planned until all of these have been properly considered and made available. The analysis of workflows for medical equipment maintenance can be done on the basis of analyzing the actors, the working steps and the triggering events for maintenance interventions [4].

First, an actor is the person involved in the maintenance service. The main kinds of actors are the following:

- Clinical staff, medical qualified personnel (e. g. doctors, nurses, medical technical assistants, laboratory personnel, etc.). They are the users of the equipment.[6]
- In-house (maintenance) technicians in charge of medical equipment and other employees of in-house maintenance areas (e. g. building services)
- Technicians from external maintenance service providers (medical technology manufacturer, service provider, etc.)

Second, a working step is any task required for maintenance purposes. Some of them can be the following:

- **Alarm generation** (fault detection), i.e. the action that fires the maintenance procedure because the occurrence of a possible misbehavior has been detected.
- **Fault diagnosis**, i.e. the identification, isolation and location of the fault and possible causes. All the actors could be involved in this task.
- **Service intervention**: this is cut normally into the so-called first line service which is often ensued by captive medical technician and if necessary by the following service of an external service technician (medical technology manufacturers, other service providers or the like) [7].
- **Documentation**: services rendered and expenditures (time, spare parts, etc.) will be documented in conclusion and future tasks will be specified (e.g. a short-term inspection of the result of maintenance).

And third, a trigger event is a condition that fires some of the working steps of the maintenance workflow. The main triggers are the following:

- scheduled service intervention: via an a priori agreed appointment (e. g. for a scheduled maintenance, safety-related inspection or medical engineering inspection),
- spontaneous service intervention: through the occurrence of a device error.[7]

The single components (actors, events and working steps) interact in the course of the service workflow according to the particular device classes in different manner. Thereby, different device classes (Magnetic Resonance Imaging (MRI) vs. syringe pump) make very different demands on a service management. Several criteria can be used to classify medical equipment, as for example:

- Location:
  - ✓ stationary medical technology (e.g. MRI)
  - ✓ mobile medical technology (e.g. syringe pump)
- Cost:
  - ✓ high investment volume (e.g. MRI)
  - ✓ low investment volume (e.g. syringe pump)
- Amount:
  - ✓ high quantity (e. g. syringe pumps)
  - ✓ low quantity (e.g. MRI)
- Diagnosis capabilities:
  - ✓ Small: devices without diagnosis capabilities and non repairable equipment.
  - ✓ Medium: devices with the ability for auto diagnosis
  - ✓ Large: devices without prerequisites for auto diagnosis [3].

As a function of the device class there are different approaches with regard to the operational costs and the technologic effort concerning the service management maintenance.

Risk Based Inspection Planning reduces maintenance costs and improving efficiency.

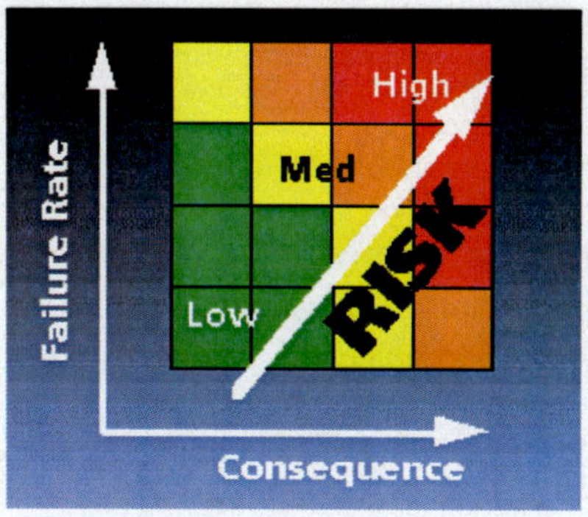

Risk Matrix

Fig. 1 Risk Based Inspection Planning

It is becoming ever more important for medical devices and their component parts to work longer at a minimal increase to costs, improving patient care with higher levels expected in terms of reliability and quality. Risk-based inspection (RBI) can provide an auditable response to by reducing medical device downtime through optimizing inspection and testing strategies, and carrying out those inspections where the devices are normally located.

The major benefits this offers the operator are:

- ✓ Reduction in downtime
- ✓ Reduction in the total inspection needed to justify safe operations

- ✓ Provision of intelligent 'common sense' inspection and testing regimes
- ✓ Quantification of real risk.

Too often, the maintenance of equipment and its associated inspections is regarded by budget holders as an "add-on" cost. Risk Based Inspections enable budget holders not only meet their bottom line targets but also their statutory requirements through schemes based on due diligence.[3]

III. RESULTS

Medical devices are complex repairable systems consisting of a large number of interacting components which perform a system's required functions. A repairable system, upon failure, can be restored to satisfactory performance by any method except replacement of the entire system. Medical devices usually undergo several types of tests/inspections during their life cycles as described here:

• *Acceptance Test*
A series of qualitative and quantitative tasks designed to verify the safety and performance of newly received equipment, as well as conformity to applicable codes, regulations and standards.

• *Operational Check*
Visual and operational check of the equipment's safety and functionality typically performed at the beginning of the day or work period, or just before using equipment on a patient.

• *Safety and Performance Inspection (SPI)*
A set of qualitative and quantitative tasks designed to verify the safety and performance of each piece of equipment by detecting potential and hidden failures and taking appropriate actions.

These tests/inspections are scheduled and it is very important that we remember the periodicity of them. This is why this software system that we developed it is very important for medical devices proper functioning.

After accomplishing the acceptance test for a newly received device, SPIs are scheduled to be performed periodically. The software system can help the biomedical engineer to program the safety and performance inspection. If any problem is found at inspection, corrective actions are taken to restore the device or its defective parts to an acceptable level. This operation it is also written in the maintenance part of the program. By having all the problems that occurred with that medical device a set of failure preventive actions may be taken to prevent future failures and/or restore device function; these include part replacement, calibration, lubrication, etc. to address age or usage related deterioration.

When a device fails while it is in use, the problem is reported by the operator, and again appropriate actions (corrective maintenance) are taken. When the repair of a device is no longer technically feasible or cost effective, replacement becomes the best or the only option. With the help of this software we can calculate the costs in case of repair, because he has a data base with the prices of all spare parts and consumable for each medical device that it is in the hospital. The user has the possibility to update the prices every ear, at the signature of a new contract with the service or supplies company. Figure 2 describes major tests and actions performed during a device's life cycle.

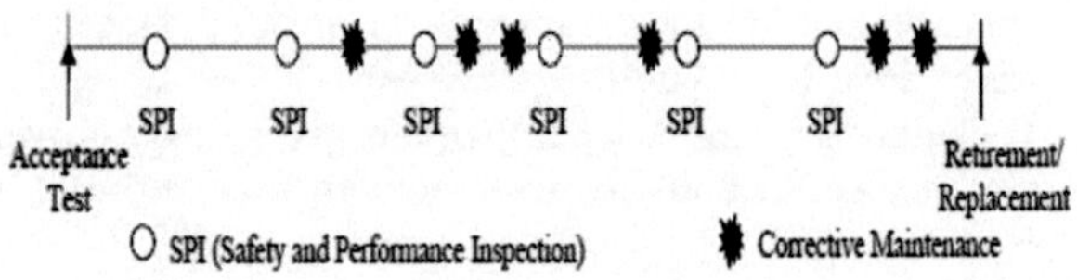

Fig. 2 Major tests and actions performed during a device's life cycle

Currently, most hospitals merely follow manufacturers recommended intervals for periodic SPI of devices. SPI intervals differ from 6 to 12 months depending on the device type and risk level. Class III (high risk) devices such as defibrillators should be inspected every 6 months, and class II (medium risk) devices like ECGs should be inspected annually. However, the optimality and even the necessity of these recommended intervals are questionable. It is essential to establish an evidence-based inspection or maintenance regimen derived from analysis of field data.

Medical Device Management and Maintenance System Architecture

Currently in Romanian hospitals doesn't exist a software tool for medical device management besides accounting software although the preventive maintenance it is very important for the medical device proper functioning. This software:

- will provide a pool of medical devices, all required information about these devices, and the maintenance protocols for the devices.
- will also contain complete repair and maintenance history of a specific device.
- will create optimal maintenance schedule for devices.
- will enable the service technician to carry out and report maintenance/repair processes .
- thus will prevent/minimize possible future failures.
- will provide an inventory of medical devices, spare parts and consumables for each device.

The software system that we developed it is original because we used Visual FoxPro program, not Access or Excel that are simply inventory programs. With this software the user can build his own medical device database, spare parts and consumable database or he can generate the maintenance monthly report. He can also generate a report for spare part or consumables annual requirements or cost of individual devices in terms of maintenance.

The application will have a powerful evaluating facility.

Evaluation of the medical devices, the technicians and the maintenance processes will be made automatically using scoring databases. These databases will contain grading of processes/properties which will be added up to make evaluations. Keeping information of and evaluating the technical staff is especially important as a significantly large proportion of total human errors occur during the maintenance phase.

While planning maintenance scheduling, the number and the availability status of maintenance equipment should be taken into consideration. Similarly, the number, status, and expertise of technicians should be an input data. Availability of medical device to be maintained and its maintenance history are also required.

Maintenance frequency is affected by medical device related factors like:

- ✓ recommendations of the manufacturer and advisory bodies,
- ✓ age of the device,
- ✓ past history of the device,
- ✓ experience and knowledge of the user, and frequency, environment and nature of use.

In case of failure, the action to be taken is to adjust a new maintenance schedule after approximating repair period; meanwhile the technician/user will be directed to a repair request form for the medical device.

These operations will not be done manually; underlying software will run after each newly entered/updated maintenance document and will make maintenance scheduling.

Qualitative tests mainly consist of visual inspection of the main parts/components of a device. For a general infusion pump, these include testing its chassis/housing, line cord, battery/charger, etc. Quantitative tests include measuring parameters of a device to check whether or not the parameters are in control. Grounding resistance and maximum leakage currents are among the quantitative tests for a general infusion pump.

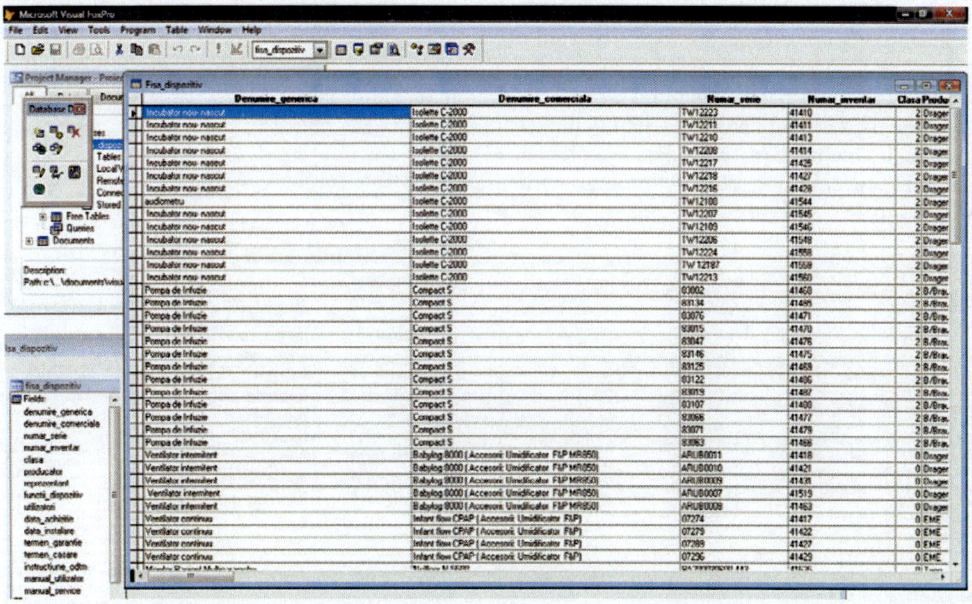

Fig. 3 Database with medical device

A work order presents all PM checks and actions, such as cleaning, lubricating or replacement of a device or its parts. A work order is also created for an acceptance test of a newly received device.

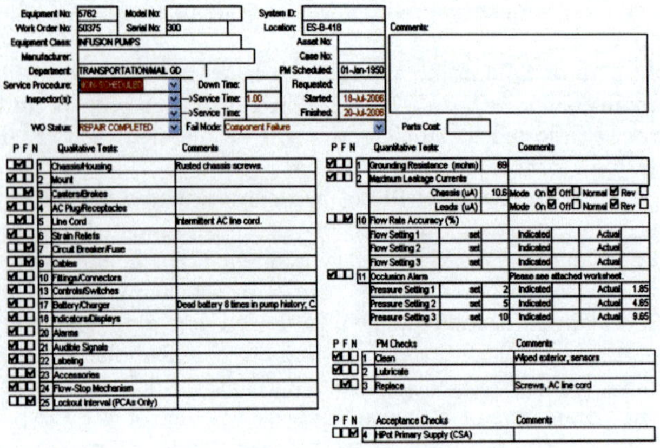

Fig. 4 A typical work order

The access is limited with technician's authorities. The technician is supplied with the maintenance protocol and the maintenance history of the device. Besides, he is informed about the jobs ahead he is responsible for and is alerted for incomplete jobs.

Operation Modes can be further divided into three.

The first process is forming databases of all devices, maintenance protocols, technicians, trainings, maintenance equipment, maintenance calendar, etc. Recording, update and deletion are main processes in terms of data management.

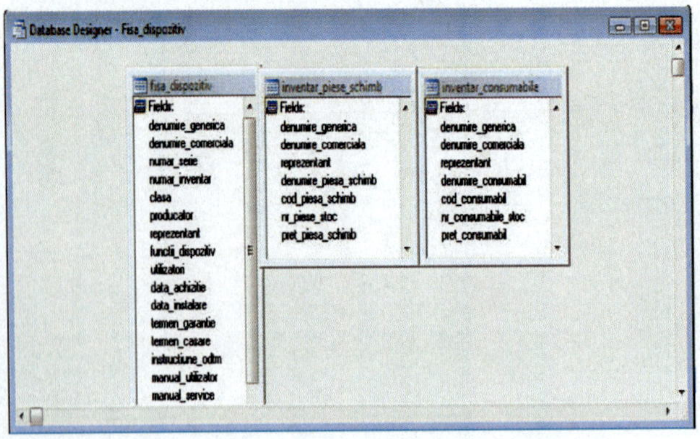

Fig. 5 Examples of principal fields in databases

The second process is the display of the database; it includes construction of many reports.

✓ Preparing proper reports, it will be able to track the devices in the hospital, the status of devices, the most problematic devices, the problems experienced (frequency and reasons either), the cost of individual devices in terms of maintenance, etc.

✓ The reporting process will help the directors in:

- following the medical equipment,
- finding out which devices are most problematic,
- which are most expensive to maintain,
- finding out the most experienced problems and the reasons of problems etc.

The third process is designing and implementing an optimal maintenance schedule.

✓ An underlying program will be run in cases of maintenance entry or update (like case of failure).

✓ A change in technician record or maintenance accessory record (i.e. ammeter is impaired) will also cause the program to run.

✓ The program will make a new schedule, optimizing many parameters like technician, maintenance device, medical device availability, recommended device maintenance period, approximate repair duration, historical data about maintenance activity of the medical device, etc.

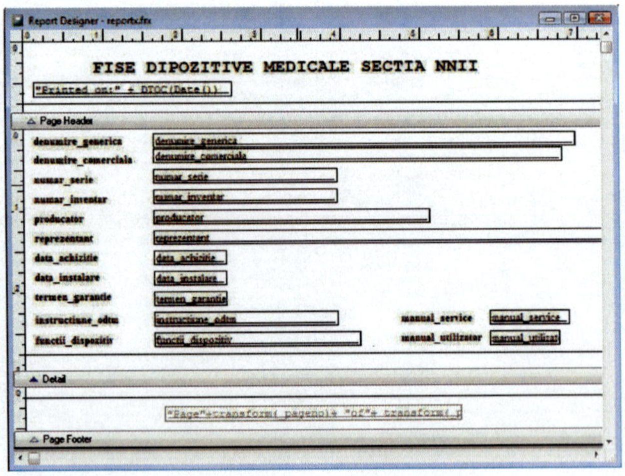

Fig. 6 Example of report designer

✓ The beginning to manage this is creating and maintaining of database files required for optimization. The rest is to implement a well built algorithm for optimizing maintenance schedule.

A new test record / maintenance form will be created for each maintenance process, which will contain:

- ✓ instructions,
- ✓ a to do list which will be checked out by the technician as the process goes on,
- ✓ and some information space to be filled by the technician, like the nature of the problem, the work carried out, the measurement values, the results of the work done i.e. succeeded or not, etc.

These maintenance records will be associated with the device being processed, and will totally constitute the repair and maintenance history of the device.

IV. Conclusions

Hospitals have thousands of medical devices and to correctly maintain these devices can be difficult or impossible without a formalized computerized scheduling system (database). The maintaining process it is a very important part for any medical device. If a preventive maintenance it is correctly done and at time the costs with the medical device will be cheaper.

This Software System for Medical Device Management and Maintenance provide a database of medical devices, all required information about these devices, and the maintenance protocols for the devices. With this software the user can build its own medical device database, spare parts and consumable database or can generate the maintenance monthly report. The user can also generate a report for spare part or consumables annual requirements or cost of individual devices in terms of maintenance.

Because the application software contain complete repair and maintenance history of a specific device, will prevent/minimize possible future failures.

References

1. Atles L.A , *Practicum for Biomedical Tehnology and management issues.* Kendall/ Hunt Publishing Company, Dubuque 2008.
2. Beatriz.l, J.Melendez, O.Grosser, *Towards Medical Device Manitenance Worklow monitoring* , Deutschland, 2007.
3. Ion R.A. Sonnesmans PJM, *Planing maintenance for medical devices,* version B, 2009
4. EN ISO 9001:2000, *Quality management systems - Requirements,* Part4.1: General requirements, 17-18.
5. *DrägerRemote Service,* Dräger Medical AG & Co. KG Lübeck 2007
6. Fries RC. *Reliable Design of medical devices.* Boca Raton:CRC, 2005
7. Bangemann T., Rebeuf X., Reboul D., Schulze A., Szymanski J.Thomesse J.P., Thron M., Zerhouni N. PROTEUS: *Creating distributed maintenance systems through an integration platform.* Computers in Industry 57 (2006), pp. 539-551.

Author: Radu Ciorap
Institute: “Gr.T.Popa” University of Medicine and Pharmacy
Street: Kogalniceanu 9-13
City: Iasi
Country: Romania
Email: radu.ciorap@bioinginerie.ro.

A Novel Concept of a Braille Device Based on Wax Paraffin Actuator

A.M. Alutei

Technical University of Cluj-Napoca / Mechanisms, Precision Mechanics and Mechatronics Department, Cluj-Napoca, Romania

Abstract— **The aim of this paper is to investigate and develop a novel refreshable Braille device actuated by thermal actuators based of phase change transformation material. The paper starts with a short introduction about some representative tactile devices, the importance of those devices into society, and highlighting the author's previous work in the field. Second paragraph presents the proposed design of a Braille cell, third paragraph discuss aspects regarding the thermal analysis of the proposed model along with numerical results. Forth paragraph deals with the presentation of the developed prototype along with results from the conducted test of its performances.**

Keywords— **Braille device, paraffin, actuator, Assistive Technology, design.**

I. Introduction

The recent developments in industry and technology have enhanced people's requirements for information availability. One of the challenges is to satisfy the need of information for the blind and visually impaired people in order to integrate them into society. Tactile devices are used for this purpose [1].

Numerous studies have been performed on tactile sensation of human beings and tactile information transmission methods, and a lot of devices have been proposed. A Braille display is an electronic device, that allows the visually impaired to read the contents of a display one text line at a time in the form of a line of Braille characters. Solutions based on mechanical needles actuated by electromagnetic technologies, piezoelectric crystals, shape memory alloys, pneumatic systems, and others have been proposed. Other methods, such phase change materials, and in particularly wax paraffin are under investigation [2-4]. Additionally, thermally driven actuators (including thermo-mechanical, phase change and shape memory methods) require cooling to reverse their action. This can occur through passive thermal radiation, or via active cooling systems, both electrical and mechanical. In [4], a refreshable Braille cell actuated using paraffin wax micro-actuators has been developed for Braille displays. In addition, this actuator has also been applied to micro-fluidic bulk-micro-machined micro-valves, micropipettes and micro-grippers.

Our previous works were focused on the design and development of an interface for a Braille device [7].

II. Design model of a cell

For the development of active Braille dots, the standard dimensions for an actuator are taken into consideration. This prototype has the overall dimensions doubled and all further results will be reported to this. The design of the developed Braille cell prototype and the corresponding dimensions elements are presented in figure 1 and described as follows.

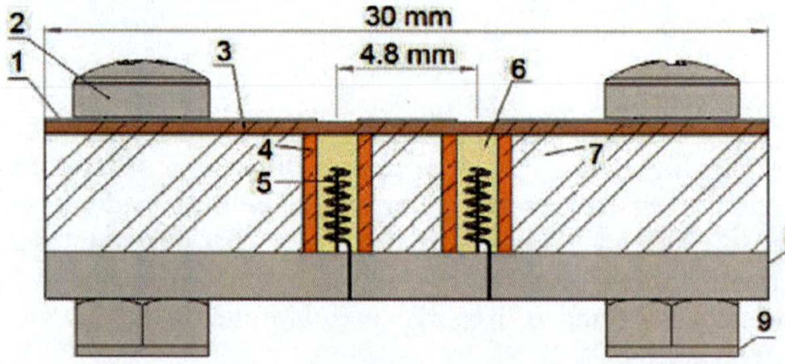

Fig. 1 Section through the Braille cell

On the support plate 8 made of aluminum, plate 5 which has 6 holes Ø3 [mm] in form of the 2x3 matrix is mounted. Thermal insulation between the Braille dots is assured because Plexiglas thermal conductivity is 0.19 [$Wm^{-1}K^{-1}$]. Inside each hole a copper bush 4 is inserted to obtain a good thermal conductivity and improving the actuation result. In this case, generated heat at the resistive mini heater can produce thermal disturbance to the other neighbor actuators. For this reason, the important roles of added case for insulate against the generated heat and to allow interoperability with other modules is fully justified. The wax paraffin 6 is melted by mini heater 5 in form of a spring made from NiCr alloy and mounted in the center of the recipient. The flexible membrane 3 is maintained on the plate with a 0.1 mm thick aluminum plate 1. All this plates are put together and fixed with four M3 screws 2 and nuts 9, to perform sealing to the entire device. The distance between Braille dots is 2.8 [mm] and a pin diameter has Ø2 [mm]. The prototype Braille cell measures 30mm×30mm×8 mm.

The reason for the huge volume expansion in paraffin when melted is that paraffin is crystal in its solid form, i.e. the molecules are packed close together. The more crystal in

S. Vlad and R.V. Ciupa (Eds.): MEDITECH 2011, IFMBE Proceedings 36, pp. 90–93, 2011.
www.springerlink.com

solid phase the larger volume expansion is to be expected at transition.

III. THERMAL ANALYSIS

In this case, generated heat from the resistive mini-heater was imposed to rise above 64 [^{o}C]. Resistive heating wire (mini-heater) is made from a NiCr alloy. Thermal conduction problem take into consideration details regarding properties of the materials.

The actuation response time of the mini actuator is determined by the melting rate of the paraffin wax. Given the thermal properties and the measured melting time of the paraffin wax, the melting process within the actuator can be analyzed. The phase changing process is dependent on the thermal conduction, convection, diffusion and thermal properties of the material.

Thermo physical properties of the wax paraffin [8] are: latent heat capacity 206 [J g^{-1}]; specific heat 2.4 [J /g °C]; density 750 [kg m^{-3}]; thermal conductivity 0.19 [Wm^{-1}K^{-1}]. In accordance with the presented data, temperature and expansion behavior properties will be examined for optimization and improvement.

A. Thermal Actuator Conduction Problem

Conduction refers to that mode of heat transfer ΔQ that occurs when exists a temperature gradient in a medium. The energy is transported from the high-temperature region to the low-temperature region by molecular activities. The steady-state thermal behavior of the element can be modeled using Fourier's law. Heat changed by a system (a body) with the environment, in a basic thermodynamic process, during which the temperature of the system suffers an infinitely small variation is expressed according [9, 10] as follows:

$$\Delta Q = c \cdot m \cdot \Delta T \tag{1}$$

In accordance with it, specific heat c = 2.4 [J /g °C] depends on the nature of the body, and its thermodynamic state. Numerical determination was made to increase the temperature of a 5 grams paraffin rod which is inside the Braille dot container from figure 3 with presented dimensions. The working process is done between ambient (initial) T_a=20 [°C] and melted (final) T_f=68 [°C].

$$\Delta T = (T_f - T_a) \tag{2}$$

The amount of heat needed to raise the temperature is:

$$\Delta Q = 696[J] \tag{3}$$

B. Melting Process

Based on presented thermal properties of paraffin wax, the thermal diffusivity k [m^2/s] represents the heat conductivity of the material λ = 0.19 [Wm^{-1}K^{-1}] reported at specific heat capacity c = 2.4 [J/g °C] and paraffin density depend of its phase. The density values of material in solid and liquid phase are tacking from "*Densities of a microcrystalline paraffin wax in the solid and liquid phase*" table presented in [11].

At temperature of 26.9 [°C] the material is in solid phase whit density ρ_{solid} = 923600 [g m^{-3}] and at temperature of 93.3 [°C] the material is in liquid phase whit density ρ_{liquid} = 793500 [g m^{-3}]. In [10] the equation for the calculus of the thermal diffusivities is mentioned as follows:

$$k = \frac{\lambda}{c\rho} \tag{4}$$

After the equation was applied for the proposed Braille cell model, the thermal diffusivities for the *phase change material* are k_{solid} = 9.97e-8 and k_{liquid} = 1.05e-7. The melting interface moves from the centre to the surface by means of heating (Fig. 2).

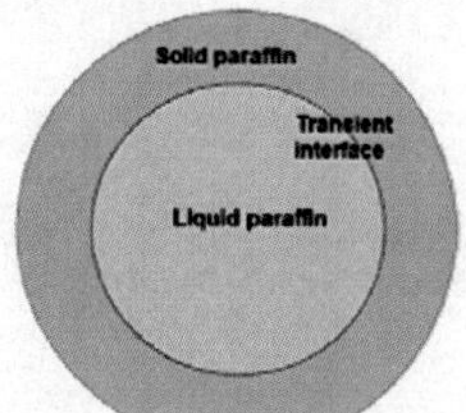

Fig. 2 Melting process of the paraffin rod

For the wax paraffin rod in 1 second, mini heater sizing calculation was made using equation (1) and phase change material data presented in previous paragraph. The power applied for mini heater to be able to melt 5 grams of wax paraffin in 1 second need to be at 580 [W].

C. Steady-State Thermal Analysis Inside of Paraffin Rod

A steady-state thermal analysis calculates the effects of steady thermal loads on a system or component. Engineer/analysts often perform a steady-state analysis before doing a transient thermal analysis, to help establish initial conditions. A steady-state analysis also can be the last step of a transient thermal analysis; performed after all transient effects have diminished.

We use this analysis system to determinate temperature variation and heat flux inside of the paraffin rod. The model was developed in Solid Works software and imported in

ANSYS Workbench module for performing the thermal analysis.

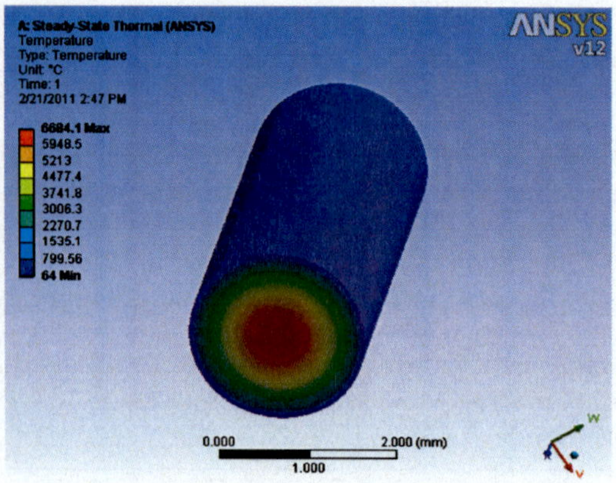

Fig. 3 Steady-State Thermal analysis of the paraffin rod

Figure 3 provide us data about the heat dispersion in the paraffin wax from the interior of the recipe. In 1 second, at an applied heat from the mini-heater of 580 [W] the paraffin wax from the recipe reaches the liquid phase; therefore the pin reaches the height of 1.04 [mm].

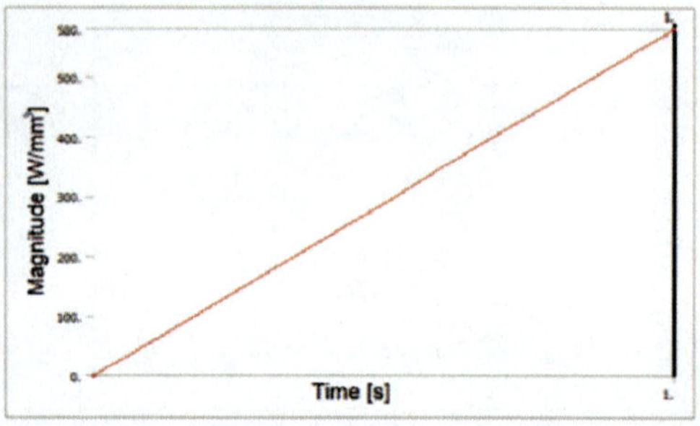

a

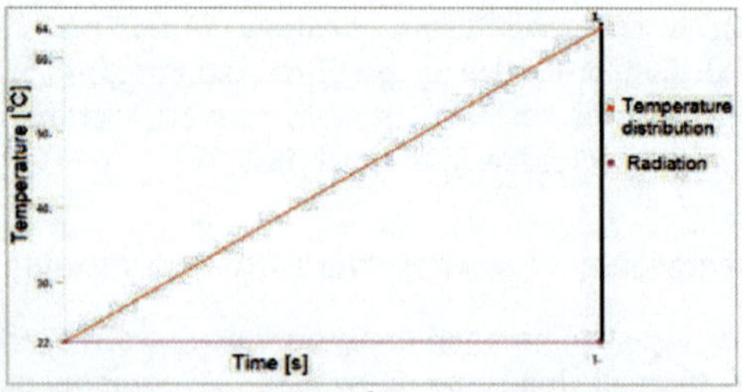

b

Fig. 4 a) Internal heat generation; b) Temperature distribution and radiation from environment

On the graphs from figure 4 the rising magnitude and heating temperature are presented. At 1 second the magnitude reached a value of 580 [W] and the minimum temperature reached a value of 64 [°C].

IV. Experimental results

For testing of the proposed actuator performances an experimental setup was developed. The proposed actuator prototype was designed, and based on the assembly drawings the prototype was obtained as presented in figure 5.

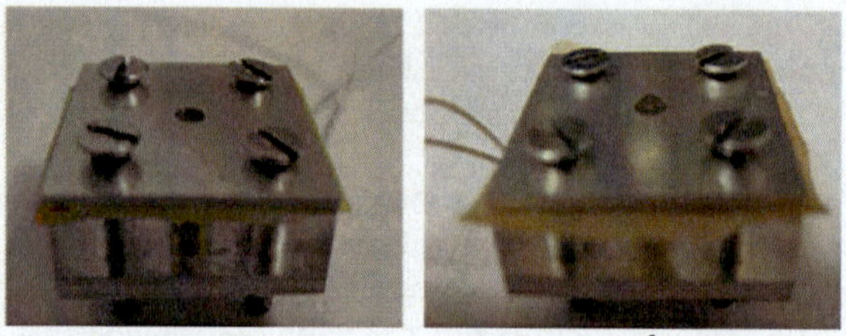

a b

Fig. 5 Picture of the developed actuator: a) actuator in position 0; b) actuator in position 1

Based on the model, the response time of the actuator for the necessary imposed displacement was measured using dial indicator and the time using a chronometer. The measurements of the response time were conducted for the displacement of the actuator from 0 - 0.5 [mm]. In figure 6 the experimental setup for the conducted measurements is presented.

At a supply voltage of 2 [V] applied to the mini-heater and a variation of the current between 1.3 ÷ 2 [A], the time of the actuator's displacement between 0 ÷ 0.5 [mm] was measured along with the time of the free cooling process.

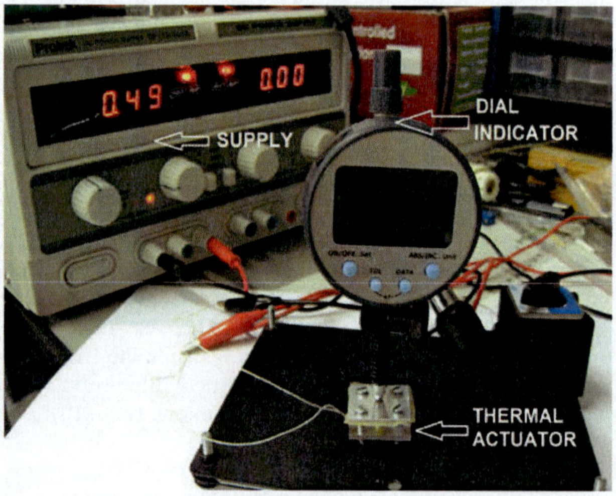

Fig. 6 The experimental stand

The obtained results from the measurements are presented on the graphs in figure 7.

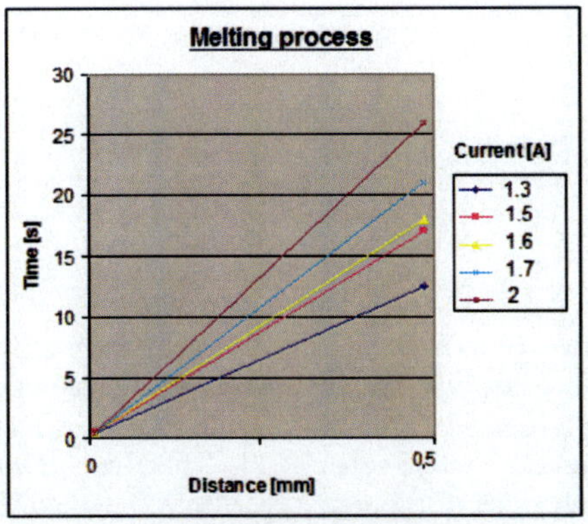

a

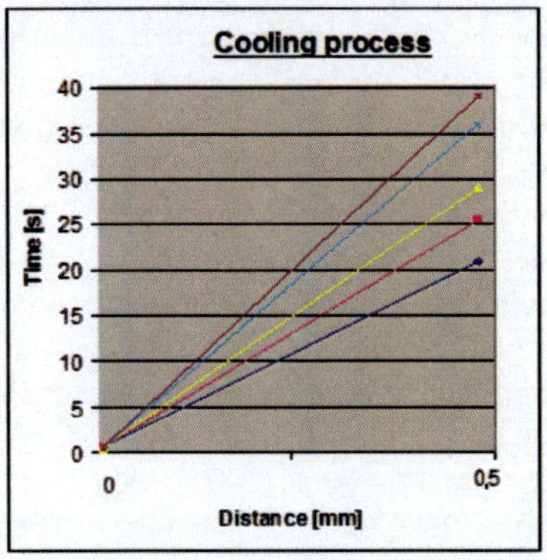

b

Fig. 7 Actuation and return to initial position process depending of some parameters: a) Melting process; b) Free cooling process

The obtained experimental results show that the developed actuator prototype can provide a sufficient displacement caused by the thermal expansion of the active material, in a short time and at good performances, and also proves that the presented theoretical analysis is correct.

V. Conclusions

The results of this paper are included in the steps for development of future Assistive Technologies applications and represent the development and test of thermal actuators integrated in such systems. One of the most relevant applications of thermal actuators in the field of Assistive Technologies is represented by the Braille devices.

This paper represents proof-of-concept demonstration of a novel active based phase-change Braille dots, which exploits changes in volume, pressure, temperature and the different chemical properties of paraffin during its solid-liquid-solid phase transitions, and does not require a large consumption of power.

The experimental prototype was made and tested so far with a single recipient who represents a Braille dot. Overall, the materials and methods used in this work allow for cost-effective and rapid fabrication of the future prototypes. For the developed experimental setup the authors focused on a simple design.

In our future work we intend to optimize de actuator and incorporate him in a miniaturized Braille module attachable to a large variety of visually impaired aid devices. Furthermore, the cooling improvement process will be investigated with a cooler and a Peltier module.

Acknowledgment

This paper was supported by the project "Doctoral studies in engineering sciences for developing the knowledge based society-SIDOC" contract no. POS-DRU/88/1.5/S/60078, project co-funded from European Social Fund through Sectorial Operational Program Human Resources 2007-2013.

References

1. Grigas V, Tolocka RT, Ziliukas P (2007) Dynamic interaction of fingertip skin and pin of tactile device, Journal of Sound and Vibration, vol. 308, pp. 447-457, Elsevier Ltd.
2. Mad Saad S, Razaly F, Md Zain MZ et al. Development of Piezoelectric Braille Cell Control System Using Microcontroller Unit (MCU), WSEAS Transactions on Circuits and Systems Issue 6, Volume 9, pp 379-388, ISSN: 1109-2734, June 2010.
3. Chouvardas VG, Miliou AN, Hatalis MK, Tactile displays: Overview and recent advances, Displays Volume 29 pp 185–194, Elsevier Ltd.2008.
4. Lee JS, Lacyszyn S (2007) Thermal analysis for bulk-micromachined electrothermal hydraulic microactuators using a phase change material, Sensors and Actuators A vol. 135, pp 731–739
5. Aluţei AM, Chetran B, Mândru D (2010) Development of an interface for refreshable Braille devices, International Conference on Mechtronic Systems and Materials MSM2010 Opole, Poland, Abstracts Book, pp 17, ISBN 978-83-60691-78-6.
6. Trp A (2005) An experimental and numerical investigation of heat transfer during technical grade paraffin melting and solidification in a shell-and-tube latent thermal energy storage unit, Solar Energy vol. 79 pp. 648–660, Elsevier Ltd.
7. Moaveni S (2003) Finite Element Analysis Theory and Application-with ANSYS Second Edition, Pearson Education Inc., New Jersey ISBN 0-13-111202-3
8. Benenson W, Harris JW, Stocker H et al. (2000) Handbook of Physics, ISBN 0-387-95269-1
9. Freund M, Csicós R, Keszthelyi S et al. (1982) Paraffin products, Elsevier Inc. Budapest, ISBN 0-444-99712-1.

Author: Alexandra-Maria Alutei
Institute: Technical University of Cluj-Napoca
Street: Memorandumului Street No. 28
City: Cluj-Napoca
Country: Romania
Email: Alexandra.alutei@mmfm.utcluj.ro,
Alexandra.alutei@gmail.com

Considerations on Electromagnetic Compatibility for Medical Devices

M.I. Buzdugan, T.I. Buzdugan, and H. Balan

Technical University from Cluj-Napoca, Romania

Abstract— **The paper deals with the electromagnetic compatible design of medical products, starting from the early stage of the design process (the prototype stage). This way of thinking is presented for the case of one kinetic therapy equipment. In the functional block diagram of the equipment are pointed out the main electromagnetic interference paths. The specific test setup and measurements in accordance with the European conducted immunity standards for electromagnetic compatibility are presented. Two kinds of measurement were performed: inter-system measurements (to the mains supply port) and intra-system measurements (to the printed circuit board and the internal cables). Several solutions for mitigating electromagnetic interference problems which greatly improved the electromagnetic immunity of the equipment are also depicted. The main conclusion which must be drawn from the present paper is that designing for electromagnetic compatibility is the most energy saving and cost effective policy.**

Keywords— **EMC, EMI, PCB, electromagnetic immunity, EFTs, surge.**

I. INTRODUCTION

Improper operation of ECG and EEG monitors, apnea monitors, anesthetic gas monitors, and other medical devices due to electromagnetic interference, determined the regulation bodies to look carefully at these occurrences and establish regulations by which medical equipments must possess sufficient immunity to operate as intended in the presence of interference.

The development of eHealth services and of electronic health records (EHRs) which include computers and computer networks impose low-noise design principles. Considering actual web-based EHR systems, patient-centric and patient moderated approaches will be widely deployed. Besides, there is an emerging market of so called personal health record platforms [1].

All these considerations are not conceivable without a minimum knowledge of electromagnetic compatibility (EMC) techniques on medical devices.

Technical literature is abounding in EMC definitions.

The most well-known (which almost became classic) consider EMC as "the ability of an electronic system to function properly in its intended electromagnetic environment and not be a source of pollution to that electromagnetic environment", but probably the most synthetic and comprehensive one, simply considers EMC "the absence of effects due to electromagnetic interference" [2]. This last definition includes both sides of the term, i.e. electromagnetic emissions and electromagnetic immunity (susceptibility).

The European regulations consist of directives and standards. The directives are very general and define the essential requirements for a product to be marketed in the EU [3]. The standards provide one way (but not the only one) to comply with the directive. Currently, over 50 different standards (product-specific standards, as well as four generic standards) are associated with the EMC directive.

Speaking about medical devices, it must be said that the European Community regulates electromagnetic emissions and immunity through the EN-60601-1-2 standard (Medical Electrical Equipment—Part 1: General Requirements for Safety; Section 2: Collateral Standard: Electromagnetic Compatibility—Requirements and Tests) as well as through the EN-55011 standard (Limits and Methods of Measurement of Radio Disturbance Characteristics of Industrial, Scientific and Medical Radio Frequency Equipment).

Recall that in most situations, assuring compliance with EMC standard prescriptions involves an extensive series of tests.

A system designed with complete disregard for EMC will almost always have problems when testing begins. On a contrary, when EMC regulations are mastered and handled correctly, the designer should be able to produce equipments with most of the potential problems eliminated prior to initial testing.

Probably the most important aspect of becoming effective in EMC design is to begin thinking about the nonideal behavior of electrical components in addition to the ideal behavior that everyone has been taught to keep in mind.

If someone thinks only in terms of ideal behavior of electrical and electronic components, he will not be able to see or anticipate the nonideal electrical paths and hence will not be able to look after other possible causes of conducted or radiated interferences. Therefore he will have inadvertently reduced our possibilities for correcting EMC problems and will not have the ability to see a schematic beyond its appearance, the so called hidden schematic.

S. Vlad and R.V. Ciupa (Eds.): MEDITECH 2011, IFMBE Proceedings 36, pp. 94–99, 2011.
www.springerlink.com

Solutions at the latest stage of design usually involve the addition of extra components that are not integral parts of the circuit. Penalties paid include the added engineering and testing costs, as well as the cost of the mitigation components and their installation [4].

The transfer of electromagnetic energy (with regard to the prevention of interference) is broken into four subgroups: radiated emissions, radiated susceptibility, conducted emissions, and conducted susceptibility.

For ease of measurement and analysis, radiated emissions are assumed to predominate above 30 MHz, while conducted emissions are assumed predominant below 30 MHz.

There is of course no magic changeover at 30 MHz, but typical cable lengths tend to resonate above 30 MHz, leading to anomalous conducted measurements, while measurements radiated fields below 30 MHz will of necessity be made in the near field closer to the source, giving results that do not necessarily correlate with real situations.

At higher frequencies, mains wiring becomes less efficient as a propagation medium, and the dominant propagation mode becomes radiation from the equipment or wiring in its immediate vicinity.

II. Conducted Immunity of a Medical Equipment

As mentioned earlier, the electromagnetic compatibility problem should be solved at the prototype design stage, where the solutions adopted to control interference are easier to implement, less expensive, and above all, more efficient.

In this respect the authors have tried to achieve electromagnetic conducted immunity for a prototype of a medical device, namely a kinetic therapy equipment (actually a treadmill).

The prototype's component parts are: (i) a single phase power converter, SYSDRIVE 3G3EV - AB015M - CUE, 1.5kW (OMRON), pulse width modulation (PWM) controlled, in the range from 0 VDC until 10 VDC, which is driving a 460 W three phase squirrel cage induction motor; (ii) associated transducers which sense the speed of the treadmill's band and the patient's heart rate; (iii) an associated computer using a serial through interface USB which controls the equipment's operation and the medical procedures; (iv) a data acquisition and control board which is provided with an ATmega16 microcontroller (16K Bytes In-System Programmable Flash); (v) a firmware package. The functional blocks of the treadmill equipment are depicted in figure1, along with the main conducted EMI inter-system (1, 2, 4) and intra-system (3, 5) paths.

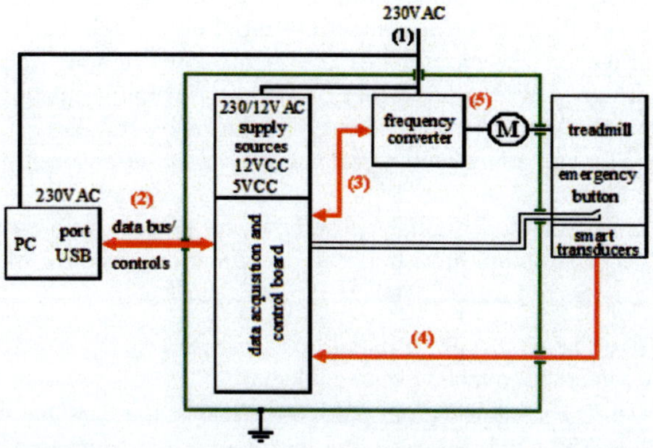

Fig. 1 Block diagram of the treadmill

Obviously there are also secondary conducted and radiated EMI paths, but this approach focus mainly on conducted electromagnetic immunity at the data acquisition and control board level and on the physical link between the board and the frequency converter (interference path 3), suspected to be the essential sources of corruption for the firmware package.

The frequent firmware's corruption was the main problem to be solved, because the medical recovery procedures and the information regarding the patients' evolution were often lost. It was also obvious that the product could not be put on the market in this situation.

Another path of high interest was the interference path 1, namely the product's AC power cord; consequently the attention was also focused on it, because noise currents generally may determine a very damaging effect.

A. *Specific Immunity Design for Medical Devices*

In literature, a non-exhaustive list of rules is proposed to be considered during the *circuit design* and/or *prototyping* step, starting from the partition of the circuit into critical and non-critical sections, to the selected circuit topology that minimizes interference, intrinsic noise and PCB layout. In the same time, at the PCB level, the following steps must be taken into account: proper grounding, then eventually local shielding and finally extra-filtering if necessary [5].

Medical equipments must meet also the requirements included in the generic standard EN 60601-1: 2006, according to a possible risk for electrical shock, which could occur whenever an operator can be exposed to a part at a voltage exceeding 25 V_{RMS} or 60 VDC, while an energy risk is present for circuits with residual voltages above 60 V or residual energy in excess of 2 mJ.

Obviously, the enclosure of the device is the first barrier of protection that can protect the operator or patient from intentional or unintentional contact with these hazards. Beyond the electrical protection supplied by the enclosure, however, the circuitry of the medical instrument must be designed with other safety barriers to maintain leakage currents within the limits allowed by the safety standards.

Since patient and operator safety must be ensured under both normal and single-fault conditions, regulatory agencies have classified the risks posed by various parts of a medical instrument and have imposed specifications on the isolation barriers to be used between different parts [6].

- The first type of part is the *accessible part*, a part that can be touched even accidentally.
- The second type of part is the *live part*, a part that when contacted can cause the leakage current to ground or to an accessible part of the equipment to exceed the limits established by the standard.
- The third type of part comprises *signal-input and signal-output parts*, referring to circuits used to interface a medical instrument to other instruments.
- The fourth and most critical part of a medical instrument is that which deliberately comes into physical contact with the patient. Such a part, called an *applied part*, may include a number of patient connections which provide an electrical pathway between it and the patient.

The level of electrical shock protection provided to patients by the isolation of applied parts classifies them as follows:

- *Type B*: applied parts that provide a direct ground connection to a patient.
- *Type BF* (the F stands for "floating"): indicates that the applied part is isolated from all other parts of the equipment to such a degree that the leakage current flowing through a patient to ground does not exceed the allowable level even when a voltage equal to 110% of the rated power line voltage is applied directly between the applied part and ground.
- *Type CF*: similar to type BF, but refers to applied parts providing a higher degree of protection, to allow direct connection to the heart.
- The use of F-type applied parts is preferable in all cases to type B applied parts. This is because patient environments often involve simultaneous use of multiple electronic instruments connected to the patient.

In any case, type B applied parts are prohibited whenever patient connections provide either low-impedance or direct connections to the patient. In the specific case of the treadmill, the equipment must satisfy to the BF class prescriptions for EMC.

When properly used, grounding is a powerful technique to increase the electromagnetic immunity of a system. In many practical situations, two grounds are not expected to have the same potential, no current is allowed to flow through the interface. This problem is particularly critical in medical equipment, where it may happen that an electrode applied to the human body must be connected to a particular ground whose potential might differ from the earth potential by as much as several hundreds volts!

Evidently, the purpose of the various isolation barriers is to ensure that leakage currents are maintained within safe values even when a single-fault condition occurs. Three types of leakage currents are defined within the standards:

- *Ground leakage current:* current flowing from all mains parts through or across the insulation into the protective ground conductor of the grounded power cord
- *Enclosure leakage current:* total current flowing from the enclosure and all accessible parts (but excluding applied parts) through an external conductive connection other than the protective ground conductor to ground or another part of the enclosure
- *Patient leakage current:* current flowing from the applied part by way of the patient to ground or flowing from the patient via an F-type applied part to ground;

B. PCB Electromagnetic Immunity

Figure 2 depicts the multilayer PCB of the prototype, containing the data acquisition block (in the center of the board), the control (in the left side) and the linear power supply (in the right side).

One can observe that the partition of the circuit into critical and non-critical sections that minimizes both interference and intrinsic noise is respected.

In order to isolate the patient connected transducers from the PCB we have adopted quad channel low input current hermetically sealed ceramic optocouplers (red rectangle in figure 2).

This double conversion of energy is the key to near-perfect isolation between circuits and as a result, electromagnetic susceptibility is greatly improved.

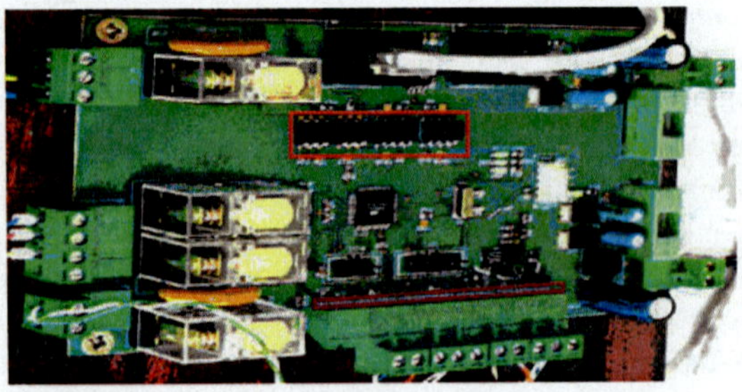

Fig. 2 The main PCB

Typical performance parameters of an optocoupler are: maximum isolation voltage 1.5 kV, isolation resistance 1-TΩ at 500 VDC and residual coupling capacitance of order 1.5 pF at1 MHz (single channel high speed optocouplers 6N136 were used [7]).

In the same time, I/O signal cable was protected with the addition of transient voltage suppression bidirectional diodes (TVS - SR12: working voltage 24 V, capacitance 10 pF and leakage current 2 μA [8]) where the cables enter the product, which provide a simple solution to increase the EMI and ESD immunity level of the circuit, especially for common mode problems (brown rectangle in figure 2).

III. IMMUNITY TESTS

In this prototype stage of the equipment, according to EN-60601-1-2 generic standard, we referred to the immunity standard EN 61000-4 [9] therefore the treadmill was tested for:

- electrical fast transients (EFTs), at 1 kV at power line for plug-connected equipment and at 0.5 kV for signal lines (EN-61000-4-4)
- surge, at 1kV differential mode at power line, 2 kV common mode at power line; signal lines are not tested (EN-61000-4-5)

The electrostatic discharge (ESD) test (EN-61000-4-2) is irrelevant in the prototype stage because it is applied to the enclosure, while most high-voltage transient disturbances are applied to the cables.

The tests were performed using the BEST*emc* (Schaffner Instruments) immunity test equipment [10] and the WinBEST firmware [11].

Deenergizing inductive loads such as relays or contactors will produce short bursts of high-frequency impulses.

Figure 3 shows the wave shape of the European Union's EFT/burst test impulse, which consists of a burst of 75 pulses, repeated every 300 ms for a duration of not less than 1 min. Each individual pulse has a 5 ns rise time and a 50 ns pulse width with a repetition frequency of 5 kHz (see figure 3). The amplitude of the individual pulses is 71 kV on ac power lines and 70.5 kV on dc power as well as signal and control lines and is applied as a common-mode voltage. The test generator has a source impedance of 50 Ω.

The EU's surge requirements are not intended to simulate a direct lightning strike to the ac power line. Rather, they are intended to simulate voltage surges on the power line caused by a nearby lightning strike or the downing of a utility pole resulting from an accident or storm. Voltage surges can also be caused by the inductance of the power line when high current loads are suddenly switched OFF.

The test setup for EMFs is depicted in figure 4.

The surge test generator is designed to produce a 1.25 μs rise time and a 50 μs pulse width (between 50% amplitude points) voltage surge into an open circuit (figure 5), as well as a 8 ms rise, 50 ms pulse width (50% amplitude points) current surge into a short circuit. The effective source impedance of the test generator is 2 Ω.

The voltage surge only has to be applied to ac and dc power lines, both common-and differential-mode, not the signal lines. On the ac power line, the voltage level is 72 kV line to ground and 71 kV line to line. On the dc power line, the voltage level is 70.5 kV line to ground and line to line. A 70.5 kV pulse must also be applied to any and all ground conductors.

The test setup for surge is depicted in figure 6.

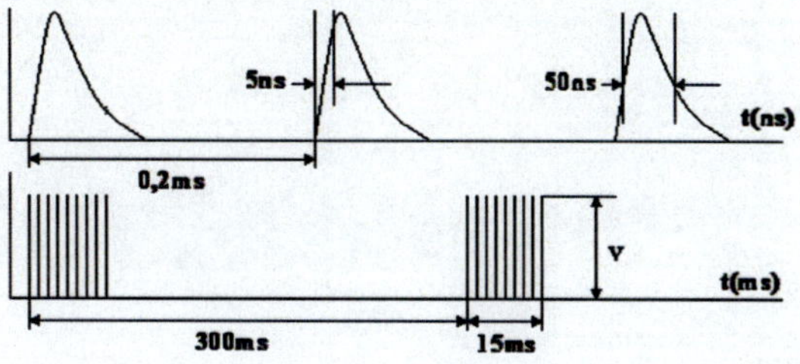

Fig. 3 EFTs test waveform

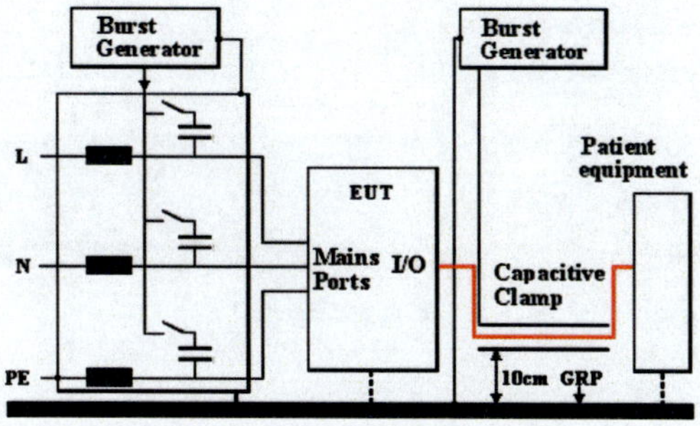

Fig. 4 EFTs test setup

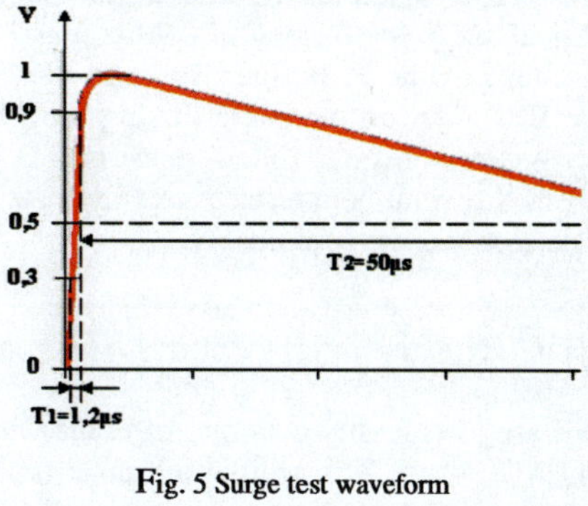

Fig. 5 Surge test waveform

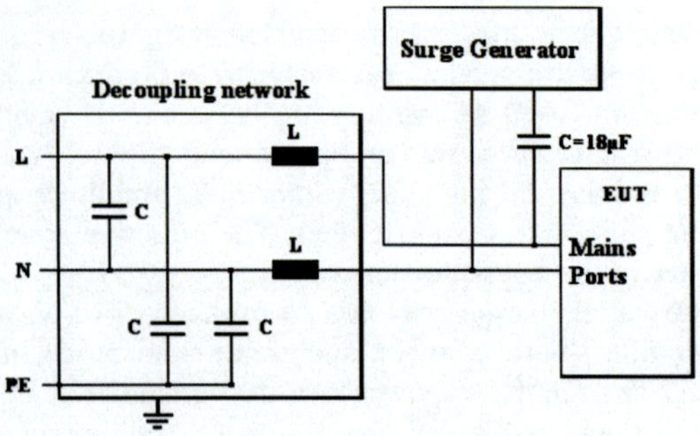

Fig. 6 Surge test setup

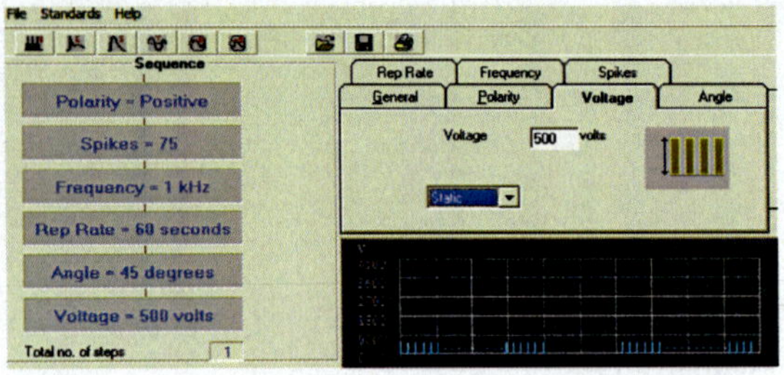

Fig. 7 EFTs test

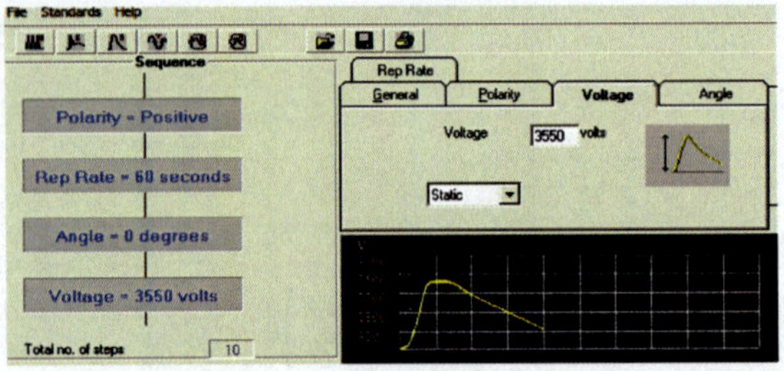

Fig. 8 Surge test

Figures 7 and 8 depict the two tests performed with the BESTemc immunity test equipment, using WinBEST software, namely for EMFs and surge.

The prototype passed all the tests at the mains port both for EFTs and surge (performed in common and differential mode) because of the EMI filter fitted on the input, but it failed the EFTs test on the cable linking the PCB and the frequency power converter (interference path 3 in figure1). Consequently, several countermeasures in order to mitigate the EFTs effect became compulsory.

IV. SUSCEPTIBILITY COUNTERMEASURES

Ferrites are low-density ceramic materials with composition Fe_2O_3XO, where X is a different metal (cobalt, nickel, zinc, manganese, etc.) which has reduced electrical conductivity and therefore very low eddy currents. Ferrites are very useful for high-frequencies.

There are several types of ferrite cores or beads: single or multihole beads, tubular or split, spherical, etc. Beads can be slipped over wires or cables and component leads can be passed through.

The concept of complex magnetic permeability permits the formal separation of the cores into two components: $\mu = \mu' - j \cdot \mu''$. Its real part, denoted by μ', is related to the ability of the material to concentrate magnetic flux, an ideal inductive component (without losses) $X = \omega \cdot L_0 \cdot \mu'$, where L_0 represents the air coil inductance (without magnetic core). The imaginary part, μ'', is related to the dissipation of magnetic energy, as it flows through the material, a resistive component, frequency dependent, which quantifies the losses in the material of the magnetic core $R = \omega \cdot L_0 \cdot \mu''$. The impedance $|\underline{Z}| = \sqrt{R^2 + X^2}$ and the loss tangent angle $(\tan\delta = \mu'' / \mu')$ are both function of frequency.

This formal approach may represent a basic characteristic that allows discriminating efficiently between inductors and ferrites [12].

Figures 9 and 10 depict the reactive, respectively the resistive part of the impedance of ferrites, which show beyond any doubt that only the NiZn material is effective for frequencies higher than 100 MHz, where the common mode cores become dissipative. The other ferrite materials present almost zero impedance beyond 100 MHz.

As a general rule, ferrites intended for use in interference control exhibit large loss tangent at high frequencies. This means that a large amount of the magnetic energy is transformed into heat inside the ferrite bead. Consequently, ferrite beads are among the very few devices able to transform noise energy into heat.

On each extremity of the cable was placed a snap ferrite sleeve with plastic case (catalog code 7427122 [13]), intended to reduce interference problems on data lines and electronic devices, easy to mount on already installed cables.

The plastic case puts high pressure on the ferrite for improved attenuation properties and reduces the risk of ferrite breaking due to impact or vibrations.

Typical impedance of the snap ferrite sleeve for one turn is $Z_1 = 155\ \Omega$ at 25 MHz and $Z_2 = 273\ \Omega$ at 100 MHz, while for two turns is $Z_1 = 650\ \Omega$ at 25 MHz and $Z_2 = 744\ \Omega$ at 100 MHz; Curie temperature is 150°C and the NiZn material is 4W620 according to the manufacture's catalog.

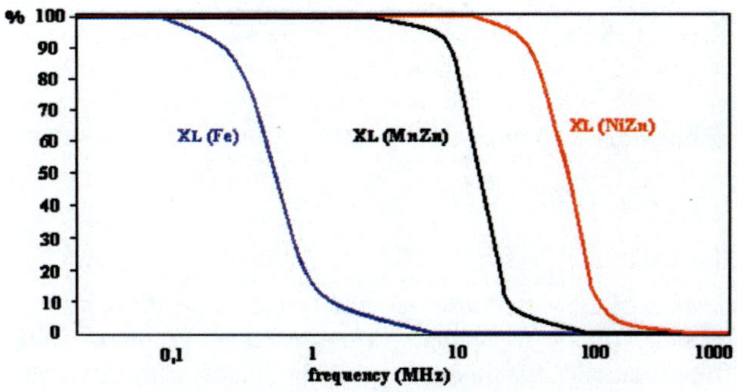

Fig. 9 Reactive component of the ferrite impedance

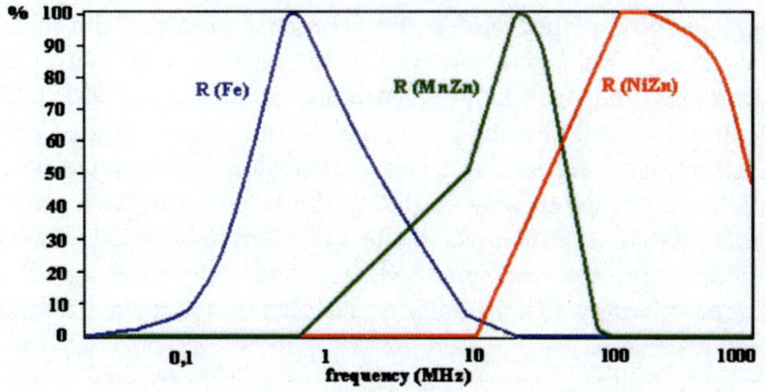

Fig. 10 Resistive component of the ferrite impedance

Two turns from the cable were wound in common mode around the snap ferrite sleeve on each side of the linking cable, the EFTs tests being repeated after. This time, the prototype passed the tests.

V. Conclusions

In conclusion it can be said that everyone involved in the design of an electric supplied product, should not only think about its functionality, but in the same time about its electromagnetic compatibility. This way of thinking about design could save his energy and at least money.

A true designer anticipates EMC problems at the beginning of the design process, finds the remaining problems in prototypes stage and tests the final prototype for EMC as thoroughly as possible. This way, EMC becomes an integral part of the electrical, mechanical, and in some cases, software/firmware design of the product.

In the present case, the analyzed product was in a prototype stage. Consequently, only the appropriate EMC rules intended for this stage of design, along with the specific standardized tests and a few possible actions to take for achieving preliminary electromagnetic compatibility compliance were depicted.

Keep in mind that each stage of the design implies different tests as will be pointed out further.

In the early stage of the breadboard, starting from the block diagram, the designer should focus on PCBs and cables linking several blocks. Simultaneously he must apply proper grounding to every part of the system, separating grounds, if necessary.

Once these precaution measures are taken into account, the prototype should be tested for conducted immunity at the mains ports and on internal cables (intra-system).

In a next stage of design, after carefully considering enclosures (according to EMC shielding prescriptions) and general grounding of the system, a new set of tests must be completed, this time exhaustive, including conducted immunity at the mains ports, radiation immunity, electrostatic discharge and also conducted and radiated emissions.

If the product is powered by the low voltage network we must also verify its compliance with power quality standards (EN 50160:2000 [14]), i.e. slow voltage changes, voltage sags or dips ($\leq$ 1 min), short interruptions ($\leq$ 3 min), temporary and transient over-voltages and harmonic voltages (THD < 8%).

Immunity to magnetic fields (standard EN-61000-4-8) could also be tested (optional).

Finally, if the product still presents EMC compliance problems, shielding reinforcement methods or extra EMI filtering should be taken into consideration.

References

1. Slamanig D, Stingl C, (2010), Electronic Health Records: An Enhanced Security Paradigm to Preserve Patient's Privacy, Biomedical Engineering Systems and Technologies, BIOSTEC 2009, Springer-Verlag Berlin Heidelberg.
2. Williams T, (2007), EMC for Product Designers, Newnes.
3. Directive 2004/108/EC of the European Parliament and of the Council, Official Journal of the European Union 31.12.2004.
4. Ott HW, (2009), Electromagnetic Compatibility Engineering, John Wiley & Sons, Inc.
5. Vasilescu G, (2005), Electronic Noise and Interfering Signals, Springer-Verlag Berlin Heidelberg.
6. Prutchi D, Norris M, (2005), Design and Development of Medical Electronic Instumentation, John Wiley & Sons, Inc.
7. http://www.fairchildsemi.com/pf/6N/6N136.html
8. http://www.bourns.com/data/global/pdfs/Bourns_Eth_ESD_Prot_Port Note.pdf
9. EN 61000-4: 2004 Electromagnetic Compatibility (EMC) Part 4: Testing and measurement techniques.
10. SCHAFFNER INSTRUMENTS: BEST *emc* Operating Instructions.
11. SCHAFFNER INSTRUMENTS: WinBEST Software Operating Instructions.
12. Gerfer A, Ralf B, Zenker H, (2002) Trilogy of Inductors, Swiridoff Verlag.
13. WŰRTH ELEKTRONIK (2007) EMC& Inductive Solutions.
14. EN 50160: 2000 Voltage Characteristics of Electricity Supplied by Public Distribution Systems

Environmental Effects on the Center's Offset of the Kistler Force Plate

I. Serban, I.C. Rosca, B.C. Braun, and C. Druga

Fine Mechanics and Mechatronics / Advanced Mechatronics Systems, Transilvania University, Brasov, Romania

Abstract— **Kistler Force Plate (FP) (model 9286AA) with the corresponding software Bioware (2812A1-3, Version: 3.2.6.104) were used for the research described in the present paper. The software contains a number of equations for the calculation of all the variables obtained in the form of amplified electrical signals from the piezoelectric sensors incorporated in the FP. After a few measurements taken on the plate with the purpose of analyzing the influence of environment on the equilibrium of the human body we've obtained the set of values and graphs that the software offers. Knowing the equations, that the software uses to determine the forces and moments, we made a calculation using both the equations and the values obtained to get a closer look at the errors that can appear. As a result we observed a variance of the constant parameters representing the distance from the center of the plate to the center of each sensor. This article exposes the graphs obtained, the variance of the parameters that should normally be constant. The study was taken at Transilvania University of Brasov.**

Keywords— **Kistler, Bioware, Offset, Center of Pressure, fidelity.**

I. INTRODUCTION

Although the scientific literature for kinematic data has established and consistently reported the estimation and propagation of errors [1], FP data are often taken as error-free. Assuming FP data to an unjustifiably high precision and acceptably accurate is potentially problematic. This is obvious in case the data is used further in other equations to calculate other parameters. In this case the error will propagate affecting the final result.

According to other authors there are some investigators, however, who have tried to assess the accuracy [2,3,4,5] and reliability of their measurements, finding inaccuracies in estimates of the center of pressure position, especially toward the platform edges, and identified poor calibration and differences in the individual characteristics among the load cell amplifiers as possible causes.

The FP contains calibration data that is usually available from manufacturers; researchers should not rely that this values are retained following installation and over time. [1]

According to the software's information some eight-channel plates have Center of Pressure enhancement variable that, when activated, implement an algorithm based on these variables to more accurately calculate the center of pressure of the force plate. The Center of Pressure (COP) Improvement algorithm is valid for plates with integrated amplifiers only. This is not available for the model of plate used in this experiment as it uses an external amplifier.

Accepting that, generally, the true origin of the strain gauge force-plate is not at the geometric center of the plate surface, considering the problems in the manufacturing process. The manufacturers usually go through a series of calibrations and estimate the position of the true origin. [7]

Fidelity (Repeatability) is a metrological characteristic that reflects the quality of getting the same result for a set of different measurements, in the same environmental conditions, using the same method, the same equipment and the same operator. The fidelity and accuracy, together, provide the precision. [8]

According to these references there is necessary to determine more exactly the center of pressure location in order to eliminate the errors that can appear further in the calculation of the other parameters. [10, 11,12]

This article highlights the need to calculate more accurate the center of pressure and the need to evaluate the influence of the environment on sensors during measurement and the way the software uses the algorithm to counteract the errors.

In order to eliminate impedance from such factors as cable interference, electrical inductance, and temperature and humidity variations, the FP should be kept in a controlled environment and tested frequently, as suggested by Dainty and Norman (1987) and Bartlett et al. (1997) [1]. This was not possible because the laboratory is not yet fully equipped to register the modifications from the environment's parameters, but it is envisaged such an equipment.

All the subjects provided informed consent prior to the test. They have been informed about the protocol of the experiment and the devices that were used.

II. EXPERIMENTAL INSTALLATION

A 600 mm x 400 mm piezoelectric platform (Type 9286AA; Kistler Instrumente AG, Winterthur, Switzerland), mounted to the ground floor according to manufacturer's instructions, was used in this experiment. The platform was interfaced to a laptop via an 8-channel charge amplifier

S. Vlad and R.V. Ciupa (Eds.): MEDITECH 2011, IFMBE Proceedings 36, pp. 100–105, 2011.
www.springerlink.com

(Type 5606A; Kistler Instrumente AG) and a 16-bit analog-to-digital converter (ADC). To show how the values vary, determining the center of the plate, there has been made a couple of experiments. Two of them took into consideration the values obtained from two subjects (gender: male and female; mean age: 26 years; mean height 180 cm; male body mass: 77 kg; female body mass: 97 kg) in static equilibrium, anatomical position with eyes open; while five experiments used a rigid body (metrological load of 10kg). The two subjects, with no pathologies, were placed in anatomical position with eyes opened so that their center of pressure would have slight variations. The rigid body was placed in different positions according to a fix pattern made out of paper with the two dimensional geometrical center of the plate drawn on it. The participants provided informed consent before the test. For each measurement the sampling frequency was 100 Hz and the period was of 15 sec resulting 1500 samples.

The Kistler platform used in this experiment is constituted, as shown in figure 1, from four piezoelectric sensors. Their center is equally spaced, depending on the axis of symmetry, from the center of the FP.

Following the action of forces on quartz crystals, piezoelectric crystals that formed sensor gives a very small electric charge value and a high output resistance. For this reason it is necessary to use a signal amplifier. The amplifier is connected further through two acquisition boards (analogue and digital). The main purpose of the amplifier is to transform the outputs of the piezoelectric sensors in standard signals (tension between ±1V and 10V with low impedance), to make compensation or linearization and to limit the frequency in the required interval. Acquisition boards are designed to perform specified sampling and converting of the signal analog / digital.

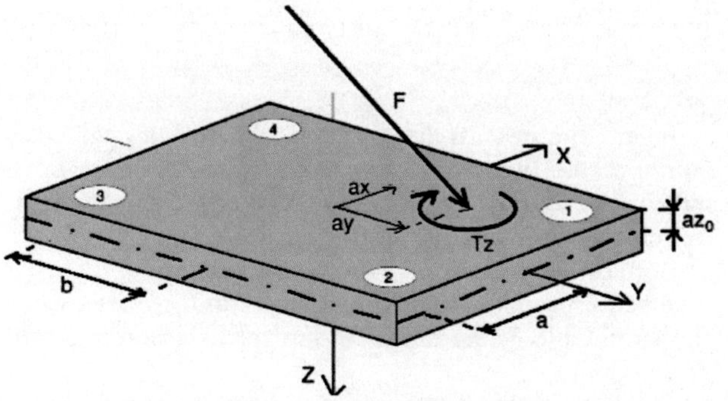

Fig. 1 Kistler platform

The technical data made available by the manufacturer is presented in table 1.

According to figure 1 and to the manual, there are three parameters (a, b, and az0) that should be constant, representing the distances (coordinates), on the three axes, corresponding to the center of the force plate, as shown in figure 1.

Table 1 Datasheet for Kistler FP model 9286AA [6,9]

Technical data	Symbol/ Measuring unit	Value
Linearity	%FSO	<±0,5
Hysteresis	%FSO	<0,5
Length	mm	600
Width	mm	400
Height	mm	35
Weight	kg	17.5
Measuring range	Fx, Fy/ kN	-2,5…2,5
	Fz/ kN	0…10
Operating temperature	t/ ^{0}C	0…60
Threshold	Fx, Fy, Fz/ mN	<10
Natural frequency	f_0(x,y)/ Hz	≅350
	f_0(z)/ Hz	≅200

Table 2 Equations that contain the constants marked in red

Parameter	Calculation	Description
Fx	=fx12+fx34	Medio- lateral force
Fy	=fy14+fy23	Anterior-posterior force
Fz	=fz1+fz2+fz3+fz4	Vertical force
|Ft|	=sqrt(Fx*Fx+Fy*Fy+Fz*Fz)	Resultant force
Mx	=b*(fz1+fz2-fz3-fz4)	Plate moment about x-axis
My	=a*(-fz1+fz2+fz3-fz4)	Plate moment about y-axis
Mx^1	=b*(fz1+fz2-fz3-fz4)+Fy*az0	Plate moment about top plate surface
My^1	=a*(-fz1+fz2+fz3-fz4)-Fx*az0	Plate moment about top plate surface
Mz	=b*(-fx12+fx34) +a*(fy14-fy23)	Plate moment about z-axis
Tz	=Mz-Fy*az+Fx*ay	Free moment, Vertical torque
COFx	=Fx/Fz	Coefficient of Friction in x direction
COFy	=Fy/Fz	Coefficient of Friction in y direction
|COFxy|	=Sqrt(COFx*COFx+COFy*COFy)	Coefficient of Friction resultant
ax	=(Fx*az0-My)/Fz =-My^1/Fz	x-Coordinate of force application point (COP)
ay	=(Fy*az0-Mx)/Fz =-Mx^1/Fz	y-Coordinate of force application point (COP)

Table 2 contains the force plate output signals equations, mentioned in the Bioware software manual. All of them depend on the values of the three parameters (a, b, and az0) and on the values of the vertical forces (fz1, fz2, fz3, fz4) and horizontal forces (fx12, fx34, fy14, fy23).

III. Results

Considering the equations presented in table 2 and the values of the forces (fz1, fz2, fz3, fz4) and plate moment about x-axis and respectively y-axis (Mx, respectively My), obtained using the Kistler FP, we calculated the values of the parameters that determine the center of the FP. The graphics obtained from this experiment are shown below (figure 3, figure 4, figure 5, figure 6, figure 7) as a function of the parameters value depending on time.

The parameters that determine the center of the force plate are, as mentioned in table 2 and figure 1, distance (a) between the center of the plate and the center of the sensor along x axis, distance (b) between the center of the plate and the center of the sensor along y axis, and distance (az0) between the center of the plate and the surface of the plate along z axis.

We will take into consideration only the first two parameters.

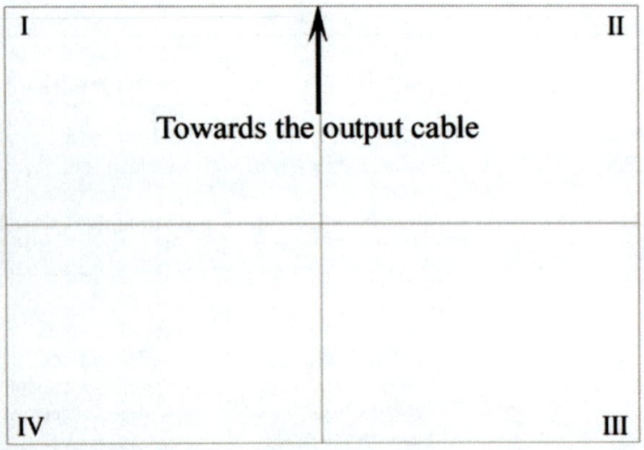

Fig. 2 The pattern, used on the FP, with four quadrants and orientation

In order to know the exact position of the feet for further measurements, there has been established a pattern (figure 2), for each subject and one for the metrological load, that fits the force plate and is divided into four quadrants with a marker, for orientation, towards the cable that communicates with the amplifier of the force plate.

The experiments were taken during a period of 15 seconds at an acquiring rate of 100Hz resulting in a number of 1500 values.

The measurements taken from the two subjects reveal a great variance of the parameters that should normally be constant.

Both of the subjects were asked to maintain their equilibrium as stable as possible in an orthostatic posture with their feet in a normal base of support with a width between 5- 10 cm. [13]

The results from the Bioware software have been saved in a txt file so that the values could be evaluated further on in another software. We used Microsoft Office Excel 2007 to calculate and extract the parameters of the center of the force plate from the equations mentioned in table 2.

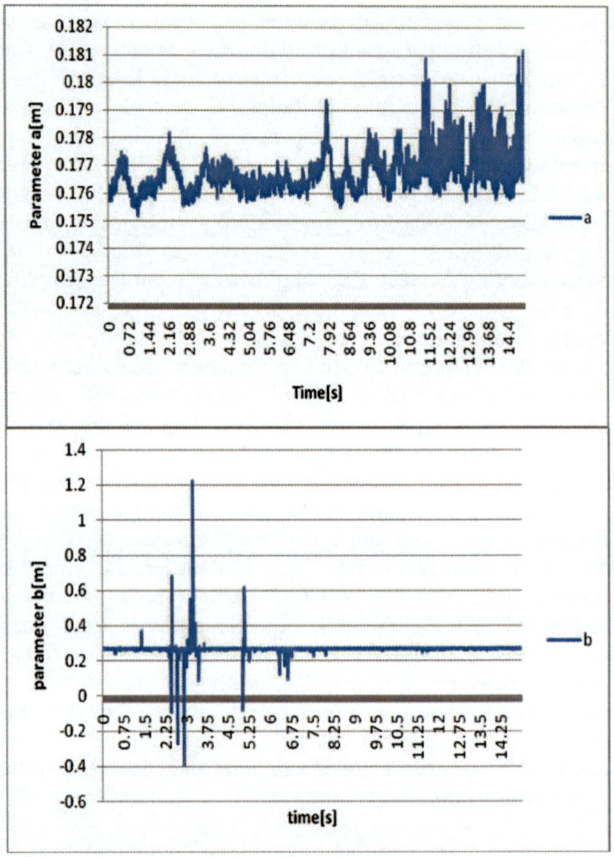

Fig. 3 Variation of the parameters- male case

Figure 3 represents the variation, obtained through calculation, of the two parameters (a and b) in the case of the male while figure 4 is a representation of the same parameters calculated in the case of the female.

Both of them illustrate a great variance during the measurement, especially for the b parameter in the male case, as shown in table 3 (the ranges of the values according to the gender).

Considering this we have made some further experiments with a metrological load placed in six positions, three of them are illustrated (random- figure 5, middle- figure 6, quadrant 1- figure 7), while three of them (quadrant 2, quadrant 3, quadrant 4) are shown as calculated ranges in table 4.

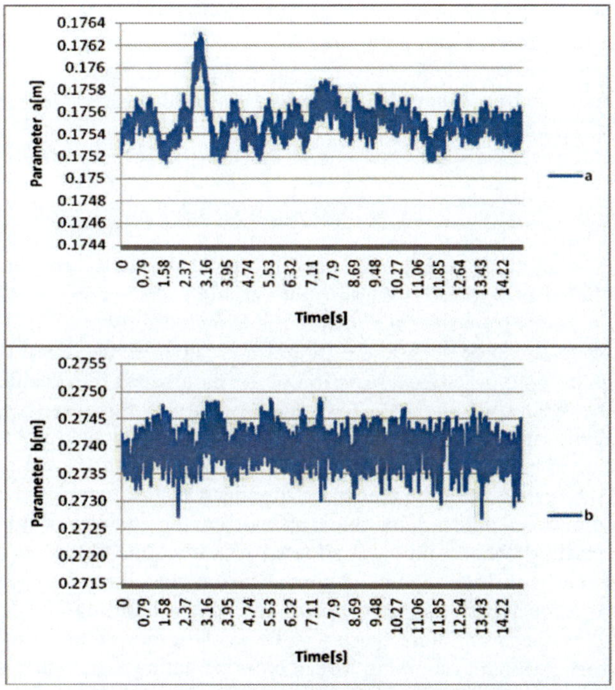

Fig. 4 Variation of the parameters- female case

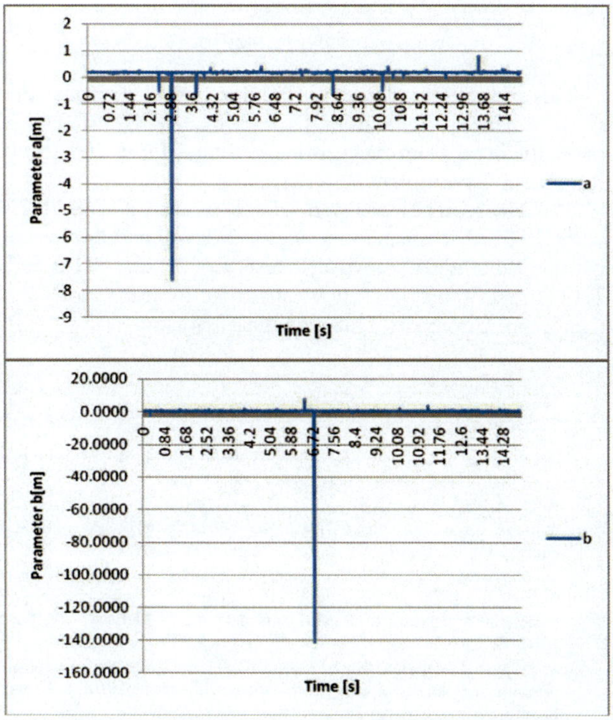

Fig. 5 Variation of the parameters- metrological load placed random

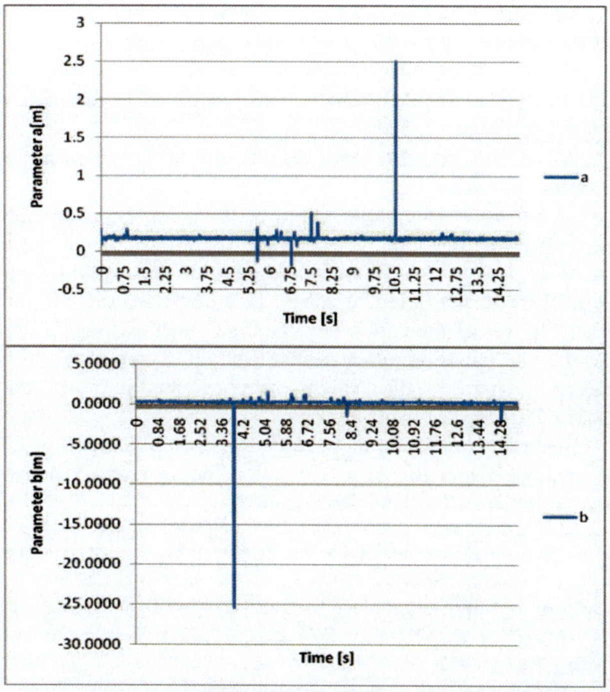

Fig. 6 Variation of the parameters- metrological load placed middle

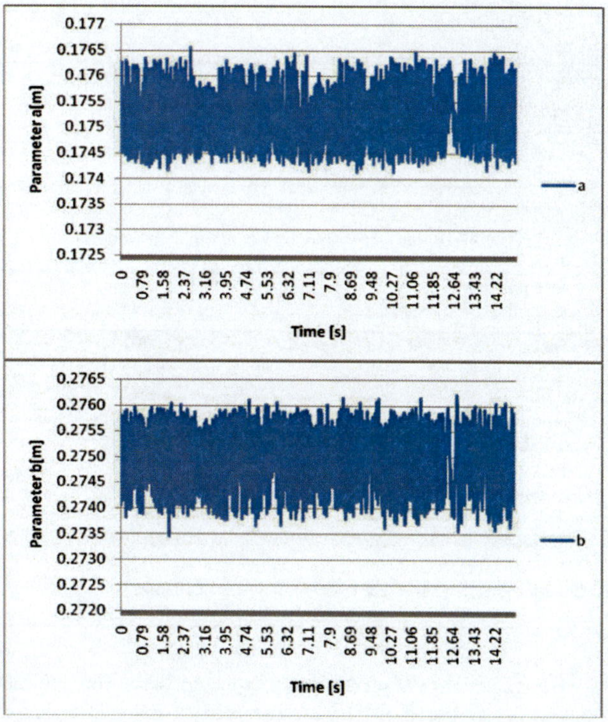

Fig. 7 Variation of the parameters- metrological load placed quadrant 1

Only the first two positions (random and middle) of the load indicate a great range of a and b parameters.

Figure 7 is introduced only as a sample as it is similar to the other placements of the metrological load in the middle of the quadrants 2, 3 and 4.

All of the measurements exhibit a great range in a stochastic manner.

The most considerable range of the samples, calculated according to equation 1, was obtained in the cases of the metrological load for the random position and middle position. Due to limitation of space, in the article, and the fact that the rest of the experiment's range wasn't as significant as the one's in the figures above (figure 3, figure 4, figure 5, figure 6, figure 7), their values were presented in table 3 and table 4, instead of using figures.

The range of a sample, in statistics, is an indication of the statistical dispersion of a population, being greater than or equal to twice the standard deviation.

$$\Delta_x = MAX(x_i) - MIN(x_i) \quad (1)$$

where: x- parameters a respectively b; and i- sample of the parameter a respectively b, i takes values in the interval from 1 to 1500.

Table 3 Range of the samples according to gender

Gender	Symbol	Range Value [m]
Male	Δa	0.005973
	Δb	1.617733
Female	Δa	0.001167
	Δb	0.002189

Table 4 Range of the samples according to the position

Position	Symbol	Range Value [m]
Random	Δa	8.397912
	Δb	148.835317
Middle	Δa	2.680757
	Δb	26.948181
Quadrant 1	Δa	0.002428
	Δb	0.002765
Quadrant 2	Δa	0.002302
	Δb	0.002789
Quadrant 3	Δa	0.001894
	Δb	0.001774
Quadrant 4	Δa	0.002374
	Δb	0.002112

Range, as it is approximately equal to twice the standard deviation, is the most illustrative tool to reveal the amplitude of the parameters that should represent the center of the force plate.

Other statistical tools (mean, standard deviation, variance, and covariance) can be used to evaluate the variance of the values further on.

IV. Conclusions

According to this article and to the all the references, included in it, there is an abstract variation of the considered constants from the equations of the Bioware software.

As it can be seen in the figures and tables above, there is an obvious stochastic variation of the parameters that decide the center of the FP. The most important three ranges, downward, are in the cases of the metrological load positioned randomly, positioned in the middle and of the male. The reason could be the small vibrations that occur from the vibrations induced by the traffic from the outside of the building. The building is situated in a circulated area on a street with high traffic. The experiments were taken during daytime at rush hour at the third level of the building.

In the near future there will be made a more closely investigation determining the factors affecting the center's values of the FP, in a controlled environment, looking for a solution to the stochastic variation of those parameters.

Acknowledgment

This paper is supported by the Sectoral Operational Programme Human Resources Development (SOP HRD), ID 6600 financed from the European Social Fund and by the Romanian Government References.

These researches are part of the Grant PNII-IDEI 722 with CNCSIS Romania and we have developed the investigations with apparatus from Research Project "CAPACITATI" and "IDEI 722" in Mechatronic Researches Department from University Transylvania of Brasov.

References

1. Psycharakis S. G., Miller S. (2006) Estimation of errors in force platform data, Res Q Exerc Sport. 77(4):514-8
2. Bobbert M.F., Schamhardt H.C. (1990) Accuracy of determining the point of force application with piezoelectric force plates, Journal of Biomechanics, Volume 23, Issue 7, Pages 705- 710
3. Schmiedmayer H., Kastner J. (1999) Parameters influencing the accuracy of the point of force application determined with piezoelectric force plates, Journal of Biomechanics, Volume 32,Iissue 11, Pages 1237-1242
4. Middleton J., Sinclair P. and Patton R. (1999) Accuracy of centre of pressure measurement using a piezoelectric force platform, Clinical Biomechanics, Volume 14, Issue 5, Pages 357-360

5. Schmiedmayer H., Kastner J. (2000) Enhancements in the accuracy of the center of pressure (COP) determined with piezoelectric force plates are dependent on the load distribution, Mendeley, Volume 122, Issue: 5, Pages: 523-527
6. http://www.kistler.com/mediaaccess/000-158e-05.08.pdf (22.May.2011)
7. VISOL at http://www.kwon3d.com/theory/grf/cop.html
8. Ciocirlea A. (2005) Metrologie industrială, vol.1, Printech, Bucuresti
9. DIRECT INDUSTRY at http://pdf.directindustry.com/pdf/kistler/biomechanics-measuring- systems-for-performance-diagnostics-and-gait-and-balance-analysis-an-sports-and-medicine/5346-73492.html
10. Rabuffetti M., Ferrarin M. and Benvenuti F. (2001) Spot check of the calibrated force platform location, Medical and Biological Engineering and Computing, Volume 39, Number 6, Pages 638-643
11. Holden J.P., Selbie W.S., Stanhope S.J. (2003) A proposed A proposed test to support the clinical movement analysis laboratory accreditation process, Gait&Posture, Volume 17, Issue 3, Pages 205-213
12. Collins S.H., Adamczyk P.G., Ferris D.P., and Kuo A.D. (2009) A simple method for calibrating force plates and force treadmills using an instrumented pole, Gait&Posture, Volume 29, Issue 1, Pages 59-64
13. Cuccurullo S. (2004) Gait Analysis - Physical Medicine and Rehabilitation Board Review, Demos Medical Publishing, New York, ISBN-10: 1-888799-45-5

Author: Ionel SERBAN
Institute: Transilvania University
Street: Vlad Tepes
City: Brasov
Country: Romania
Email: ionel.serban@unitbv.ro
serban_ionel1984@yahoo.com

Transformation Design – A New Method for Developing Medical Products

H. Waedt[1], M. Popa[1], and P. Manea[2]

[1] Technical University of Cluj-Napoca, Romania
[2] TEMCO, Cluj-Napoca, Romania

Abstract— **„Transformation design" is the generic name of a new scientific method, appeared at the beginning of the 21st century [1], that is about to become a new independent scientific discipline [2]. The new method gains more importance in today's conditions, when the effects of the world economic crisis are not completely eliminated.**

Simplifying things, it can be said that transformation design is a process through which new, but obsolete, systems are brought to the requirements of the current quality standards [2].

Periodic mandatory consultations (chest X-ray) for certain occupational categories were made in Romania up till the EU pre-accession phase (2004 – 2006). For this were used the Micro-Radio-Photography systems (MRF), which were based on the Roentgen rays scanning principle. These systems weren't compliant with new regulations in force and, due to the specific legal and economic situation of Romania, extremely strict conditions were imposed to be satisfied until the accession, respectively till January 2007. Thus, the Romanian government had to regulate and plan the decommissioning of all MRF systems. In this situation were 21 MRF systems, new versions, which were installed just a few years earlier and recently put into operation in various clinics in Romania. These 21 MRF systems were technically new, but their principle of operation, Roentgen ray chest scanning, was no longer compliant with valid legal regulations in July 2006. The article describes the finding of a technical solution to solve the situation of the 21 MRF systems removed from service. This is possible by means of transformation design, which offers the possibility to transform these systems, whose principle of operation is contrary to legal norms in force, into X-ray systems with a new operating principle, compliant to these rules.

Keywords— **transformation design, Roentgen standard, interdisciplinary process, standards, extended SID.**

I. INTRODUCTION

Through the application of the transformation design method was intended that these systems will not be completely removed from service, but as many components and their functions to be integrated into the new systems, compliant with current legal provisions. On the other hand was aimed that, by conversion, the new systems will be able to be installed and used in existing medical spaces. The materialization of these objectives ultimately led to a reduction in the economic losses arising from compulsory process to align Romania with EU legislation.

II. TRANSFORMATION DESIGN

Transformation design is an interdisciplinary scientific method that has as purpose the alteration of existing systems (socio-political or technical) in order to keep them updated in the accordance to the social or legal changes that take place in the medium that they are used. The arguments for which the transformation design method proves to be the only realistic solution to achieve the objectives set forth in this article, in comparison with other known methods, are as follows:

The Standard Design method [3, 4] is applicable for systems or products, already on the market, at which are necessary modifications, either for the creation of new versions or for the elimination of any reliability risks, identified due to users' complaints [5].

The designing of a new system version is an extremely laborious and complex task, which does not change the basic product, which remains on the market. The new version of the system is usually intended for a new category of beneficiaries, without changing the principle of operation.

Standard engineering method [6] is applicable when a product on the market is replaced by a new one. The replacement process unfolds over time and involves a start phase and a finishing one. This means that, till the end of the process of replacement, on the market will exist, at the same time, both new and old products.

It is obvious that none of the methods described above is applicable for a system that is still on the market, although it is out of order, due to the principle of operation that is not meeting yet the new requirements for approval, and should be therefore transformed into a system based on an approvable principle of operation.

Thus, compared to those above, transformation design is the only method that offers concrete ways to perform and finish a process of such complexity with manageable costs.

III. THE IMPORTANCE OF TRANSFORMATION DESIGN

Transformation design broadly defines an interdisciplinary process oriented toward meeting the human needs. Through transformation design is also aimed the achievement of lasting changes in individual behaviour, in form and functioning

S. Vlad and R.V. Ciupa (Eds.): MEDITECH 2011, IFMBE Proceedings 36, pp. 106–109, 2011.
www.springerlink.com

of the systems and of the private or public organizations, in order to accelerate social progress [7].

Transformation design is presented as a phased, iterative process, applied to big and complex problems and many times without being limited only at the social aspects of these. In the application of this method problems are analyzed, chronologically and unitary, in interdependency, the components being grouped according to new criteria. The practice of transformation design method involves the analysis of the problems in a holistic way and to a lesser extent reductive, as single elements, in order to understand both the interdependence and mutual influence [8]. Transformation design involves, among others, applying techniques and methods of conception and design, already established, in new non-traditional areas, which leads to unprecedented solution and results. Transformation design, as emerging field, uses classic design disciplines such as:

- Service design
- User centred design
- Conception design
- Informational design
- Industrial design
- Graphic design
- Systems design
- Interactive design

Although much has been written and extensively discussed in recent years in scientific circles about the economic value and the necessity of introducing transformation design [2], the implementation of these innovative ideas took place barely in 2004, when "The Design Council", the strategic body for design in Great Britain, decided the forming of the RED group, which self-defined as objectives not only the development of new solutions and concepts (think-tank), but also an "active" action mode (do-tank) in order to "transform" the public services [7]. This action was a response to the request of the Prime Minister Tony Blair to create "redesigned" public services, "transformed", centred on the needs of the user (patient, passenger, victim of a crime, etc.) [13].

IV. TRANSFORMATION DESIGN PROCESS

Because of the novelty, the stages and processes embedded in transformation design are still not covered by rules or standards. Based on the gathered experience from the projects where transformation design was already applied and on published scientific papers on this topic [15], the following methodology (order) for process flow was crystallized, methodology presented in the present paper for the application of this method in industrial projects:

- End user's (beneficiary) point of view analysis.
- Market and product scope analysis for the out of use product and for the transformed product. For medical products the analysis will address both the international health sector and the national health sector, as a destination and use domain of these.
- Analysis of political influence and legal regulations for the application domain of the "transformation" process. The analysis will cover both the legal base and medical aspects related to the use of two systems – the one out of use and the "transformed" one.
- The analysis of the economic situation will focus initially on MRF systems with international research on the areas of use and also on justification of use up till the stopping of use.
- Basic "know-how" resources. The analysis will have as objective the basic operating principles for both systems out of operation and for the "transformed" system Development phase, the basic concept.
- Transformation design: Basic concept development phase.
- Transformation design: The design-development phase of the product and the improvement of its quality. The main activities of this phase will be:
    - Specifying of the usage conditions of the system. The design (development) of the product in Engineering.
    - Checking product safety
    - Simulation in meeting the imposed requirements and risk analysis.
- EU Directives and Regulations requirements for product design-development domain.
- EU requirements concerning the approval of transformed medical products.

In conclusion, the processes set out above will be the structure and the approaching way in carrying out the activities for the conversion of those 21 MRF systems through the application of the transformation design method from this paper.

V. ANALYSIS OF MARKET SITUATION

In accordance with the terms from the point above, in this paragraph, the world wide prevalence of different types of X-ray systems will be analyzed in order to make a ranking in descending order, starting with the most used, and to identify the development trends of this economic sector in the future. This analysis is an important criterion for evaluating the economic gain from the usage of MRF systems converted (transformed) in standard X-ray systems, in the terms of discontinuing the regular preventive radiological controls. For the reasons stated above will be addressed, in developing the analysis, the same strategy applied by a company that produces medical equipment, namely the gathering and the selection of data and information at the international speciality trade shows (exhibitions). Thus is presented the situation at the two big international medical exhibitions specialized on Roentgen equipment, were the attending of the companies from this area is substantial. It is about the Radiological Society of North America (RSNA) in Chicago and the European Congress of Radiology (ECR) in Vienna [9, 10]. RSNA, as a scientific organization, was founded in 1915 with the purpose of setting up and imposing high standards both in the field of radiological analysis and other related scientific domains and also in order to support the study and research related to these areas. In close connection with these occupations, the society publishes two specialized publications, the magazines: Radiology [11] and RadioGrafics [12] and awards three gold medals annually to scientific personalities with outstanding merits in the field of radiology. Participating companies exhibit here their most representative products for the radiological equipment market sector. Thus it is observed that the companies participating in this event expose both the basic versions of both systems and derived variants from the first ones, which allows detailed information necessary for a correct analysis of the radiological equipment market sector.

When asked why it is necessary to analyze the market in the content of this paper, the answer is that the transformation of the 21 MRF systems should be looked at and developed in close touch with current trends on the Roentgen devices market sector. Before their removal from service, the 21 MRF type systems were used mainly for radiological consultation in mandatory preventive check for certain groups of population – especially in public services – in accordance with the law at that time in Romania. For this reason, the load of these systems could be determined, respectively planned, with sufficient accuracy, the number of patients that were to be consulted being usually known. For systems that will be converted by transformation design, the situation is completely changed and the process of transformation must take firstly into account the concrete conditions of the new Roentgen equipment on the market, and also of the new EU regulations, valid from the end of 2006 in Romania, which have eliminated the preventive, mandatory radiological controls for certain groups of population. The analysis of Roentgen devices and systems presented in the 2007 edition of RSNA (Table 1) is important for the transformation of the 21 systems and by competitive reasons, because it provides information on the prevalence (ranking) for different products, and also about the trends for future development in this economic sector. Further, samples from this analysis will be presented with explanations and arguments for the placement on a particular rank of the different types of X-ray systems in a ranking type classification list (Table 1).

Table 1 Ranking type classification list

Type of System	unitati	Percent	Marketing Ranking Nr.:
Mammography	29	10,43	1
Magnetic Resonance	27	9,71	2
Toplift	22	7,91	3
Elevator patient table	22	7,91	4
CT; CT scanner	20	7,19	5
C-Arm mobile	18	6,47	6
Patient table for MR	16	5,76	7
Digital-System	16	5,76	8
Patient table for CT	15	5,40	9
Patient table mobile	15	5,40	10
Wallstand tilt	14	5,04	11
Wallstand manual	11	3,96	12
C-Arm	8	2,88	13
U-Arm mobile	8	2,88	14
Urology	6	2,16	15
System with elevator patient table	6	2,16	16
Fluoroscopy	6	2,16	17
System with patient table	5	1,80	18
Mobile Swivel-arm	5	1,80	19
System with patient table and C-Arm	2	0,72	20
Mammographic System with patient table	2	0,72	21
Ceiling sistem with C-Arm	1	0,36	22
Patient table	1	0,36	23
Wallstand mobile	1	0,36	24
Dental	1	0,36	25
Telescope column with C-Arm	1	0,36	26

Thus is crystallized the idea that between the systems currently on the market, the most interesting for the transformation of the 21 MRF systems, are those for chest X-ray analysis with the patient upright or sitting position. From this point of view the vertical (wall) stand systems for chest consults, with the patient upright or in sitting position, shows the greatest similarity with the 21 MRF systems to be transformed. The market analysis based on the attendance degree at the largest international exhibition specialized on radiological equipment - RSNA 2007 in Chicago - showed that 48 systems, respectively 17.2% of total, are intended for thoracic consultations (Thorax). This category of radiological consultations (Thorax) can and should be integrated in the new transformed system. The analysis results confirm the validity of the premise that the 21 MRF systems in Romania that will be converted by transformation design, will end up with a high degree of use (patient charge). Regarding the theme, the main objective of transformation design method, for converting the 21 MRF systems in standard X-ray systems, is the compliance with EU standards relating to medical products (systems) of this type. Along with this main objective, in the technical specification of the transformation of the 21 MRF systems is included the following increase in quality (Fig. 1).

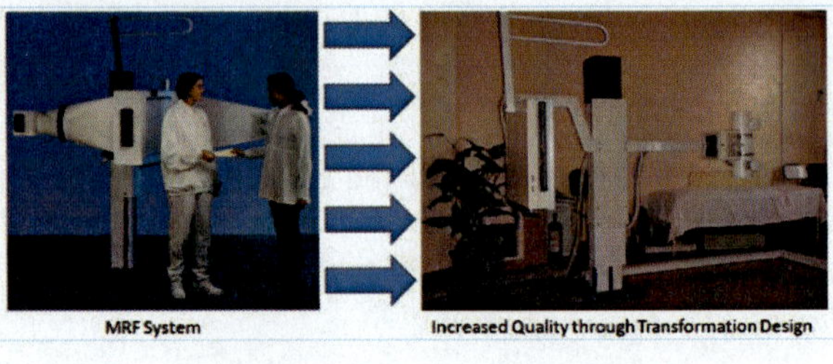

Fig. 1 Transformation design realisation

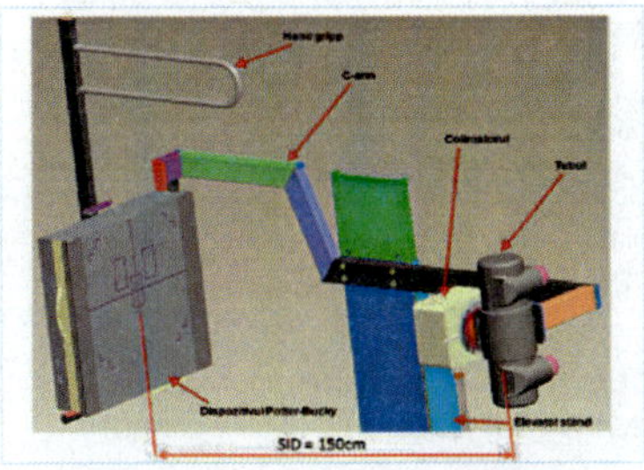

Fig. 2 Extended SID to 150 cm

Extending the source-image-distance (SID) - Distance between the outbreak of the Roentgen tube and film - from 120 cm to 150 cm (Fig. 2).

This extension is necessary to meet the requirements imposed by Thorax examinations of anterior-posterior and lateral type [14]. Thus the following types of radiological examinations on the transformed systems will ensure a high degree of charging patients, with positive consequences for the damping of the costs caused by conversion: General posterior-anterior pulmonary examination (back to front) (Fig. 3a), Lateral pulmonary examination (Fig. 3b).

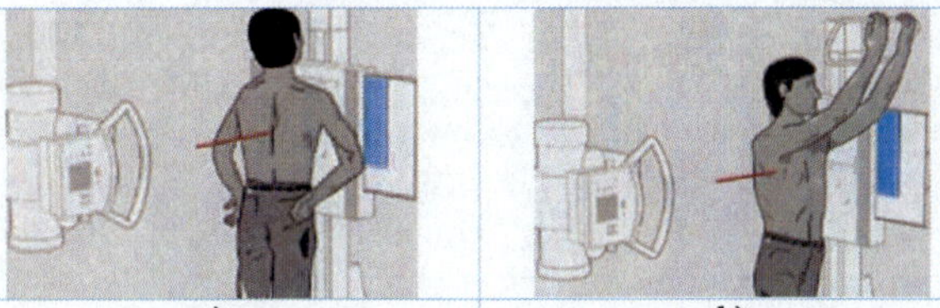

a) b)

Fig. 3 Type of examination: a) general posterior-anterior pulmonary examination, b) lateral pulmonary examination

Abdominal examination from the left side position (with the patient on the mobile table); posterior-anterior abdominal examination (back to front with to the patient standing); Trans-thoracic examination of the shoulder; Side examination of the scapula (with the patient in standing position); Side neck examination; Cervical inclined examination, Thoracic spine area examination (with the patient supine on the mobile table); Side hip joint examination after Sven Johansson (with the patient placed on the table); Axial kneecap examination (with the patient placed on the mobile table); Ankle joint examination (with the patient standing).

VI. Conclusions

Was investigated and found that the only efficient method, both technically and economical, through which the change of the operating principle of those 21 MRF systems removed from service is achievable, is transformation design [15]. Applying transformation design is recommended and justified in situations arising from decisions of authorities (higher courts) with a regulatory role in the respective field, or right in this situation were the MRF systems from Romania, to which, following the decision of the Ministry of Health, the exploitation was prohibited and were removed from service, although they were in perfect technical condition for operation. Following the phases and making the economic and technical analyses imposed by the application of transformation design method finally led to the development of the solution for systems conversion, solution accepted by the Ministry of Health of Romania. In this paper was also analyzed the solution to convert the MRF systems in standard digital X-ray systems, showing that, at first, this solution is not the best, due to economic and time related reasons. The chosen solution does not exclude, however, a later conversion of the transformed MRF systems into digital X-ray systems.

References

1. Zuboff, S., Maxim, J. The Support Economy, Viking Penguin, New York, 2002 – ISBN 0-670-88736-6
2. Pine B. Joseph, Gilmore James H., The Experience Economy. Harvard Business School Press, Boston, 1999.
3. Roloff Matek, Maschinenelemente, Normung, Friedr. Vieweg & Sohn, Braunschweig / Wiesbaden, 2003
4. Engelken Gerhard, Konstruktionsmanagement, Wiesbaden 2001
5. *** EN ISO 13485, Medizinprodukte - Beuth Verlag
6. Wagner W., Konstruktionsverfahren, Wiesbaden 2002
7. *** RED PAPER 02 Transformation Design http://www.designcouncil.info/mt/RED/transformationdesign /TransformationDesignFinalDraft.pdf
8. Goodare, H. and Lockwood, S., Involving patients in clinical research, British Medical Journal, Sep 1999, 319: 724-725
9. *** European Radiology, The official journal of the ESR, Springer Verlag, Vienna 2007
10. *** ECR 2008, The Anual Meeting of the European Society of Radiology, ROBIDRUCK, AT-1200 Vienna, 2008
11. *** Radiological Society of North America, Radiology November 2010 257:586-587
12. *** RadioGraphics, The Journal of continuing medical education in radiology, October 2010 30:1441-1443*** Liaison Committee, Oral Evidence given by the Tony Blair MP, HC (2003–04) 310-ii, Q 17
13. Hoxter Von Erwin A., Röntgenaufnahmetechnik, AG Siemens. Berlin: 1991. - ISBN 3-8009-1566-9
14. Waedt Horst, Popa Marcel und Manea Pompiliu, Quality Management and Regulatory Affairs for Medical Products. 2010 IEEE AQTR 2010 THETA 17, May 28-30 2010, Cluj-Napoca, Romania
15. *** RED PAPER 02 Transformation Design http://www.designcouncil.info

The Effects of Exposure of the Human Body to RADON. Integrated Measurements Performed in Alba County, Romania

L.E. Muntean[1], D.L. Manea[1], and C. Cosma[2]

[1] Faculty of Civil Engineering, Technical University of Cluj-Napoca, Romania
[2] Faculty of Environmental Science, Babes-Bolyai University, Cluj-Napoca, Romania

Abstract— **In many countries, radon is the second most important cause of lung cancer after smoking. The proportion of lung cancers attributable to radon is estimated to range from 3 to 14% . The hygiene requirements, the people's health and environment protection should be observed according to the regulations in force. In such a context, is included the protection against radioactive emissions that appear inside buildings, if one takes into account the consequences they can have for human health, as radon is the main radiation source for the population; it contributes with about 57% to the annual effective dose in general to radiation, though in some areas this percentage can increase to over 95%.**

Keywords— **Radon, lung cancer, exposure dose, CR-39 trace detectors, Alba County.**

I. INTRODUCTION

Radon is a noble gas, having three natural isotopes: $^{219}R_n$ (actinon), $^{220}R_n$ (thorium) and $^{222}R_n$ (radon), descendents of $^{223}R_a$, $^{224}R_a$, $^{226}R_a$, formed in the radioactive series of ^{235}U, ^{232}Th, respectively ^{238}U, 219R_n.

Radon and thorium are produced in the radioactive disintegration of radium ($^{226}R_a$ and $^{224}R_a$) which is to be found in the soil and in building materials. After radium disintegration, the radon atoms formed migrate in the air in the rock capillary vessels and, by diffusion or due to pressure differences, they are transported in the atmosphere[1].

In the dwelling buildings, radon is accumulated and inhaled during breathing, but it can be exhaled in a relatively short period of time, as it does not form chemical bonds.

Radon descendents are solid radioactive isotopes adherent to any particle with which they interact or are deposited on surfaces, leading to the radiation of living bodies. Radon ($^{218}P_o$, $^{214}P_b$, $^{214}B_i$) and thorium ($^{212}P_b$, $^{212}B_i$) descendents penetrate human body by inhaling, ingestion, or though the skin, by contact with contaminated water.

After inhaling, radionuclides deposit on the bronchial mucosa, irradiate the bronchial epithelium transferring to the target cells important amounts of energy.

If inhaled, radon can enter the bronchial epithelium and can expose the sensitive cells to radiation (figure 1).

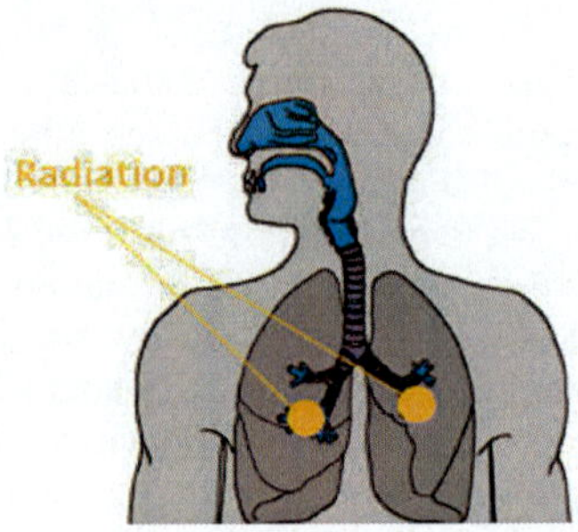

Fig. 1 Inhalation of radon decay products [2]

At international levels, recent epidemiological surveys (USA, Canada, U.K., Sweden, China, Japan, The Czech Republic etc.) as well as a Common European Project based on the joint analysis of 13 studies in epidemiology performed in European countries (France, Belgium, Germany, Luxembourg) have highlighted the connection between the risk of lung cancer and radon concentration, even if the levels of radon concentrations are normal, that is of 40-300 Bq/m³ [3].

It was estimated the risk of lung cancer increases by 8.4% if the radon concentration grows with 100 Bq/m^3, compared to the data found in the North American study which estimated the risk increase of 11% in the same conditions. The relationship between dose and response was linear, no threshold being marked [4]. It was estimated that the risk of lung cancer increases proportionally to the expose to radon. In the absence of other death causes, the absolute risk of contracting lung cancer before age 75 at radon concentrations of 0; 100; 200 and 400 Bq/m^3 for both smoking people and non-smoking people is given in Table1.

Table 1 The risk of contracting lung cancer before age 75

	The risk of cancer (%) at radon concentration of: 0; 100; 200; 400 Bq/m^3			
Non-smoking	0,41%	0,47%	0,55%	0,67%
Smoking	10%	12	13%	16%

S. Vlad and R.V. Ciupa (Eds.): MEDITECH 2011, IFMBE Proceedings 36, pp. 110–113, 2011.
www.springerlink.com

It was found out that the risk for lung cancer in the case of smokers is approximately 25 times greater than the risk to lifelong non-smokers [5, 6].

The World Health Organisation (WHO) has reduced ten times the value of the tolerable level of radon, to 100 Bq/m^3, with the main aim of diminishing the danger for health in dwelling buildings. According to the Ministry of Health Order no. 381/05.04.2004, in Romania, inside a building the design level of radon should be maximum 200 Bq/m^3, for buildings erected after 2005 and 400 Bq/m^3 for buildings built before 2005.

The total exposure dose is distributed according to Figure 2 thus: 80.9% of exposure to natural radiation and 19.1% of exposure to artificial sources. The percentage of 80.9% represents the sum of radiations of radon, thoron and their descendents 46.3%; the terrestrial gamma radiation from the natural nuclides 16.4%; cosmic radiation 10.0%; internal potassium^{40}, radium^{226} and other radionuclides radiation 8.20%. The percentage of 19.1% is given by the sum of the radiations originating in medicine 17.8%; radiation from atmospheric deposits of nuclear tests 0.35%; radioactive discharges of nuclear industry 0.04%; professional exposure0.04%; radiations from other radioactive sources 0.35%; post Chernobyl radiation 0.52%. Figure 1 presents the contribution of the natural irradiation sources for the Romanian population.

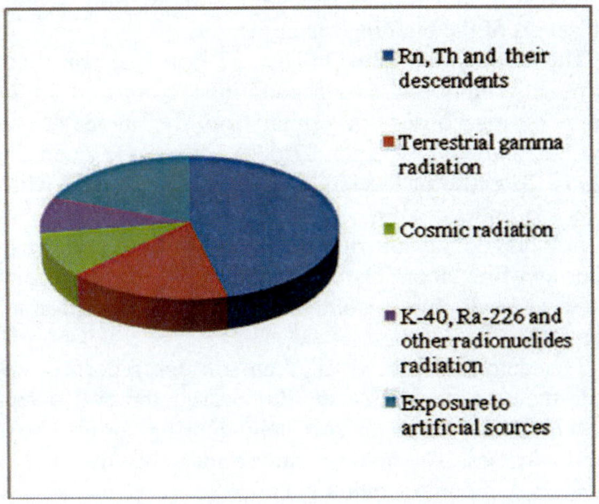

Fig. 2 Distribution of the total exposure dose

II. The radon's effects to the human body

It was found that, if in dwellings, the radon level is over 4 pCi/L (~150kBq/m^3), the body is exposed to a radiation level that is 35 times higher than the level admitted by nuclear medicine.

A pupil or an undergraduate spending 8 hours/day, 180 days of a year in a classroom of level 4 pCi/L is subjected to a radioactivity level that exceeds over 10 times the admitted level for a population living in the vicinity of a nuclear power station. At present, at world level, it is appreciated that at 4 pCi/L (~150kBq/m^3), the death risk is of 1%/year, 1000 times higher than in the case of the majority of carcinogenic products.

The radon-related death causes in USA during an interval of one year compared to other death cases are given in Figure 3.

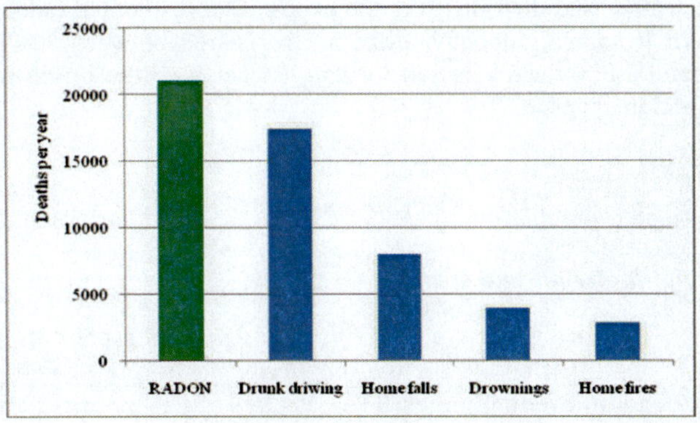

Fig. 3 The Radon Health Risk [7, 8]

The main disease due to extended exposure to radon is lung cancer (lung carcinoma). The effect of the exposure to radon can be expressed in four great syndromes that are typical of nuclear radiation exposures: the cerebral syndrome, the gastro-intestinal syndrome, the cardiovascular syndrome, the haematopoietic syndrome.

Everywhere in the world, lung cancer is the most frequently met form of cancer with males. In the European Union, 21 % of the male cancer of this origin is present; the neoplasm death rate is 29 %. Epidemiological data for females in the European Union present considerably smaller figures, with 5 % less for incidence and 9 % less deaths due to lung cancer.

The risk factors for lung cancer are mainly given by:

- exposure to cigarette smoke;
- occupational and environmental carcinogenic agents, such as the professional exposure to asbestos, exposure to radon, arsenic, hallogenous ethers, aromatic hydrocarbons, nickel;
- previous lung diseases: chronic obstructive bronchopneumopathy, diffuse interstitial fibrosis, benign asbestosis.

Lung cancer is the main "killing" cancer in America, both in males and females, accounting 160,000 deaths per year, a figure that is larger than in the case of breast cancer, prostates cancer and colon-rectum cancer. While in other cancers the death rate has decreased during the last 60 years, the incidence and mortality related to age in lung cancer have increased.

In Romania, approximately 13,000 new cases are diagnosed every year only 5% having some chance of healing. The continuous increase of the incidence of lung cancer in both sexes is evident as a phenomenon typical of the second half of the last century [9].

Recent research shows that at least 10% of the total number of lung cancer, representing about 0.8% of the deaths, find their origin in the exposure to residential radon. In Romania, annually, there are registered between 1,000 and 2,000 deaths caused by lung cancer with the origin in radon exposure [5].

III. Experimental research

A. *Materials and Methods*

The method of type CR-39 trace detectors represents a solution that allows for the long-term monitoring of radon found in dwelling places. The detectors of CR-39 traces are manufactured from plastics, are sensitive to alpha rays, have a small size, reduced costs, can be easily used in the filed and lab as well to detect the radon found inside homes.

The technique used consists of fixing the CR-39 track detector, made of plastic, in a small box called radapot (cylindrical Ø23x40mm), provided with holes that allow radon to enter the box. The top of the box is covered with a polyethylene film, that does not allow radon descendents in the air to get in the box, but it will enter by diffusion, in a percentage of 95%.

To highlight the traces in the CR-39 detector, it should be treated with a solution of NaOH of 6.25 moles, in a thermostat fixed at 90°C temperature. The optimal time of treatment is found in the range of 5 to 6 hours. The number of the traces in the detector shall be proportional to the radon concentration in the location of measurement [3]. The reading of the traces detected by detectors is made with RadoSys-2000 (Figure 4), in fact a set of tools for the integrated measurement of radon activity.

Fig. 4 The RadoSys2000 set of tools [10]

B. *Results and Discusion*

The CR-39 trace detectors were placed in eight homes in the Alba County as follows: six in the city of Alba Iulia, six in the village called Cetea, two in the commune Vintu de Jos and two in the village Pianu de Sus. The detectors were left inside the homes (living rooms and bedrooms) for 90 days, in the autumn-winter season.

The house in Alba Iulia, 7 Basarabiei str. is formed of a partial basement and ground floor, with a natural rock foundation, concrete floor and masonry (bearing walls) for structure. The detectors were placed at the ground floor in the dwelt vicinity of the area without basement; the values found there are quite high: 326 and 292 Bq/m^3, though smaller that the maximum admitted values.

The home in Alba Iulia, 12 Basarabiei str. contains a basement, a ground floor, the first floor, having concrete foundation, concrete flooring and the strength structure (bearing walls) made of masonry. The detectors were placed at the ground floor in the bedroom and living room which were two neighbouring rooms. The recorded values were 74 and 63 Bq/m^3, the explanation resides in a smaller influence of the soil, the values obtained coming only from the emissions of the building materials.

The home in the Cetea village, 37 Principală str. presents extremely high radon concentrations compared to other neighbouring houses. Dissimilar from the homes at no. 34 and no.57, the house at no. 37 does not include a cellar (and this is the cause of the radiation differences found with the detectors under analysis).

In Pianu de Sus, the detectors were situated at the ground floor and first floor. The concentration at ground floor was found to be double as compared to the concentration at the first floor.

The detectors placed in Vintu de Jos recorded values over the admitted value of 400 Bq/m^3, the two detectors placed in this home in two neighbouring rooms showing similar values. The home is not permanently dwelt, and an excessive amount of radon is collected in this house as it is not ventilated.

The values obtained in this experimental study are given in Table 2.

As all the houses where detectors were placed were built before 2005, we used as a comparative standard the value of 400 Bq/m^3. It is found that only in the case of detectors used in the commune of Vintu de Jos the mentioned limit value is exceeded. The detectors placed in one of the houses in the Cetea village (no.37) showed values a little below 400 Bq/m^3.

Table 2 The radon concentration in Alba County

Crt. no.	Place/street	No.	Radon concetration (Bq/m^3)		
			0-100	100-200	>200
1	Alba Iulia (Basarabiei St.)	7	-	-	326 292
2		10	96	101	-
3		12	74 63	-	-
4	Cetea (Principală St.)	37	-	-	368 389
5		132	-	123 104	-
6		211	61	-	222
7	Pianu de Sus (Principală St.)	172	49 80	-	-
8	Vintu de Jos (Telman St.)	19	-	-	514 581

IV. Conclusions

The originality of this study consists in identifying some of the areas from Alba County whose radon concentration levels exceed the maximum admitted limit, regulated by the Ministry of Health Order no. 381/05.04.2004 which specifies 400 Bq/m^3 as the admitted value for the buildings erected before 2005. In this cases, methods for reducing the concentration of radon are provided, based on pressure modifications together with the application of an extraction fan or the implementation of a radon-resistant membrane.

Similar studys were made by other research teams from Romania in Bihor and Cluj Countys, but also in Unirea from Alba County, from which it has resulted a mean value of 87 Bq/m^3 much smaller than the mean value of 215 Bq/m^3, obtained in this study.

It is estimated that about a third of the lung cancers caused by radon could be avoided if the radon concentration were reduced in houses where the concentration of radon usually exceeds 200 Bq/m^3.

Acknowledgment

This paper was supported by the project "Doctoral studies in engineering sciences for developing the knowledge based society-SIDOC" contract no. POSDRU/88/1.5/S/60078, project co-funded from European Social Fund through Sectorial Operational Program Human Resources 2007-2013.

References

1. Cosma C., Jurcut T., Radonul si mediul inconjurator, Ed. Dacia, Cluj-Napoca, 1996
2. http://www.unce.unr.edu/radon/, Accessed at: 24 March 2011
3. S. Forkapic, I. Bikit, Lj. Conkic, M. Veskovic, J. Slivka, M. Krmar, N. Žikic-Todorovic, E. Varga, D. Mrđa, Methods of radon measurement, FACTA UNIVERSITATIS, Series: Physics, Chemistry and Technology Vol. 4, No 1, 2006, pp. 1 – 10, avalable at: http://facta.junis.ni.ac.rs/phat/pcat2006/pcat2006-01.pdf, Accessed: 24 March 2011.
4. http://www.radosys.com/, Accessed: 24 March 2011.
5. Darby S., Hill D., Auvinen A., Barros-Dios J. M., Baysson H. et. al., 2006, Residential radon and lung cancer-details results of a collaborative analysis of individual data on 7148 persons with lung cancer and 14208 persons without lung cancer from 13 epidemiologic studies in Europa. Scandinavian Journal of Work, Environment and Health, 32, suppl. 1:1-84, 2006, Available at: www.sjweh.fi/download.php?abstract_id=982&file_nro=1, Accessed: 24 March 2011.
6. S Darby, D Hill, A Auvinen, J M Barros-Dios, H Baysson, F Bochicchio, et al., Radon in homes and risk of lung cancer: Collaborative analysis of individual data from 13 European case-control studies, British Medical Journal paper, Available at: http://www.radonnews.org/radonrisksstudies/europeanpoolingBMJ-Dec2004.pdf, Accessed: 15 March 2011.
7. A Citizen's Guide To Radon http://www.epa.gov/radon/pdfs/citizensguide.pdf, Accessed: 24 March 2011.
8. http://www.scribd.com/doc/21918034/RADIATIILE-SI-RADIOPROTECTIA, Accessed: 24 March 2011.
9. NCI Analysis of Home Radon Studies Finds Small Increase in Lung Cancer Risk, in Line with Predictions from Miner Studies- National Institute of Health, 1996, Available at: http://www.nih.gov/news/pr/dec96/nci-31.htm, Accessed: 24 March 2011.
10. http://www.ukradon.org/article.php?key=risksradon, Accessed: 24 March 2011.
11. World Health Organization, Radon and cancer, Fact sheet No. 291, updated September 2009, Available at: http://www.who.int/mediacentre/factsheets/fs291/en/index.html, Accessed: 15 March 2011.
12. Miron L., Mihăiescu T., Cancerul bronhopulmonar- Aspecte de practică clinică şi tratament, Iaşi, 2002
13. Scivyer C., Radon protection for new dwellings, Available at: http://www.herefordshire.gov.uk/docs/Radon_in_Dwellings.pdf, Accessed: 24 March 2011.
14. Catelinois O., Rogel A., Laurier D. et coll., Lung cancer attributable to indoor radon exposure in France: impact of the risk models and uncertainty analysis. Environ Health Perspect 114 (9): 1361-1366, Published online 2006 May 20, Available at: http://www.ncbi.nlm.nih.gov/pmc/articles/PMC1570096/, Accessed: 24 March 2011.
15. Cosma C., Dicu T., Dinu A., Begy R., Radonul şi cancerul pulmonar, Editura Quantum, Cluj-Napoca, 2009
16. Jonathan M. Samet, Residential radon and Lung cancer: end of the story?, Journal of toxicology and Environmental Health, Part A, 69: 527-531, 2006, ISSN: 1528-7394, Accessed at: 15 March 2011
17. x x x, Ministry of Health Order no. 381/05.04.2004 on the approval of sanitary norms for the safe conduct of nuclear activities

Author: Muntean Lavinia Elena
Institute: Tehnical University of Cluj-Napoca, Faculty of Civil Engineering
Street: 28 Memorandumului Street
City: Cluj-Napoca
Country: Romania
Email: lavinia.muntean@cif.utcluj.ro

Education of Arms' Users in Dealing with Technological Activities

M. Baritz[1] and D. Cotoros[2]

[1] University Transilvania Brasov, Precision Mechanics and Mechatronics Department, Brasov, Romania
[2] University Transilvania Brasov, Mechanics Department, Brasov, Romania

Abstract— **The methodology is based upon the research conducted by help of two subjects' samples, whose upper limbs skills will be tested and studied using the dedicated equipment in our laboratory. One sample consists of implant users and the other one of non-implant users, in order to be able to set the normal limits and compare them to the performances reached before and after the skills reeducation. Performance is usually expected from people involved in all kinds of activities without taking into consideration the possibility that some of them might be subjected to accidents or diseases, affecting their working capacity. Our goal is not only to improve the skills of healthy persons but also to offer a chance to a normal and productive life to those slightly disabled persons that posses the will and the motivation to pursue it.**

Keywords— **biomechanics, arm, sensorial gloves, range of motion.**

I. INTRODUCTION

A various series of applicative researches highlighted the necessity of conscious and repetitive learning of functional handling during different work stages in order to obtain correct and efficient actions of the superior system (arm-hand-fingers) even if the workers are in a good shape or have implants or other disabilities at the arm level. [1]

The concept of training is related most of the time to the conscious actions that aim at assimilating a skills' system, developing competencies and knowledge interests, conception forming, unlike education, which includes all the conscious actions of the individual with the purpose of transforming the entire personality of the person subjected to the educative action and also to develop and to re-find the lost abilities due the implant action.

Because training deals with the revealing of learning processes in all domains, it has a general character and represents the starting point in the analysis of investigation, understanding and correlation methodologies. [7, 8, 9]

Every human subject acting into a training program meant to establish the handling abilities has to participate with an open mind, with good attention and wishes to collaborate with the trainer. The information perception or processing represents an important part of the complex called cognitive learning environment because it involves the use of senses, in a simple or combined way, signals transmission to the central unit, response understanding, correlation with the adjacent information and finally the reactive response, accomplishing a **structural methodical assembly (SMA).**

The assembly is an interactive process involving an operator (user) and the handled objects, and hence simulation environments must be able to react according to the user's actions in real time and according with their physical problems. Furthermore, the action of the user and the reaction of the environment must be presented in an intuitively and comprehensible way.[2]

The human factor which is the investigated subject represents in this structure the central platform that collects and directs the information, correlates their structure with the "system requirements" and coordinates the activities in an organizational or cooperative way inside the cognitive learning environment to accomplish the training program.

A cognitive learning environment assumes the consideration of several acting dimensions such as: operational dimension, the normal one, innovative level and not last, the psycho-social and ergonomic dimensions. [4]

By the development of these dimensions, the cognitive learning can develop a series of skills and competencies and allows the passing to an adjacent level of communication and rehabilitation of some temporarily diminished functions like hands movements before or after implant, or in different effort conditions.[12]

In order to accomplish a standardized methodology for analyzing the cognitive learning way and the adaptability of the human factor in case of handling abilities we studied the possibility of using a flexible and ergonomic system based on sensorial gloves and also different sort of training computer games specially designed for this sort of movements.

II. EXPERIMENTAL SETUP

The experimental equipment consists of a set of sensory gloves, an assembly of devices for anthropometric measurements, physiological parameters and a computer system to record and process the information obtained from the gloves sensors.

At this integrated system it can be possible to add dedicated computer games – with different difficulties levels

S. Vlad and R.V. Ciupa (Eds.): MEDITECH 2011, IFMBE Proceedings 36, pp. 114–119, 2011.
www.springerlink.com

allowing the human subjects to develop and to maintain the handling abilities.

At the same time we used a video surveillance system for the subjects' reactions and attitudes during the experiment to correlate all data at the final moment; these video investigations being necessary to analyze other aspects of the human behaviour and to take correct decisions.

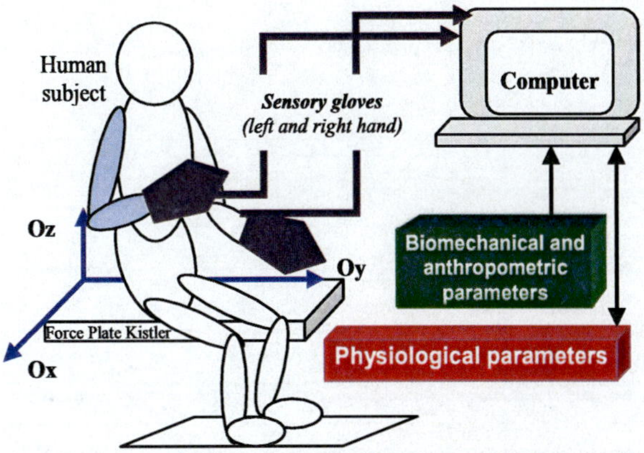

Fig. 1 Experimental setup

The cognitive environment was monitored during an experiment performed upon, first on a human subject without upper implant and after that on a sample of 1subject having an implant in the right hand, aimed at knowing, assessing, determining, modifying and finally improving the handling skills in long term actions. [3]

The first stage of the experiment consisted of completing a uniform physical environment evaluation along the entire duration of the investigations, as far as temperature (temperature in laboratory was 21^{0}C); humidity (air humidity 80%), vibrations, noises and light stimuli are concerned and also atmospheric pressure (755 mmHg).

There also was necessary to acquire anthropometric data (for squeezing, grasping, twisting motions, etc.) for the subject and further adjusting the sensory gloves to these values.

All these parameters are necessary to establish a common modelling base to measure and to evaluate the human body behaviour in cognitive learning process.

In the second stage we measured the physiological parameters of the human subject (weight, height, age, pulse, blood pressure) in relaxed stance, without any other general health problems and with a good metabolism (example: blood pressure 155/82 mmHg, pulse 78, face temperature 36,7^{0}C, height 170 cm, weight 62 kg).

III. Results and conclusions

The used sensory glove (***Meditutor*** type no.4 with dedicated software) allows the subject to learn in "a virtual way" how to accomplish anatomic motions (at the motion limit of joints by flexion/extension both for fingers and hand joint) and also to understand and perform the hand normal physiological motions.

We noticed that when the experiment starts and the subject is suitably trained, the response materialized in "motion range" with the characteristic quantities: motions speed, cycles/sec and active/passive deficits, accurately express the subject behavior along the entire duration of the recording, established for 13 seconds each stage of recording.

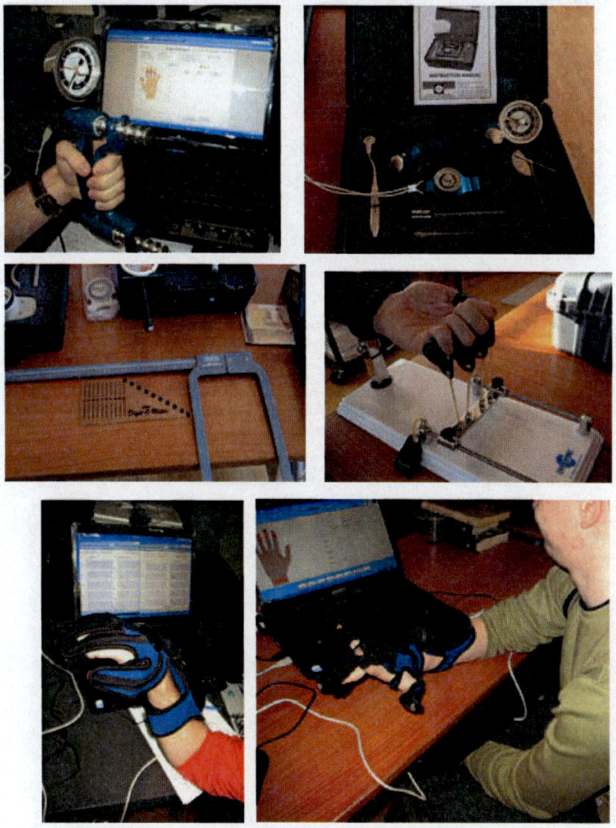

Fig. 2 Recording equipment

The data recording by help of the sensory gloves is preceded also by a calibration for subject measurements and the recordings sequence was established according to the importance of the anatomical elements namely: first we analyze the fingers, then the joint and finally the entire motion of the arm (right and left arms for comparing the movements). Initially we started with the analysis of the data obtained by recording the rhythmic movements of the fingers, with the arm supported by the elbow on the table and respectively the arm in a vertical downward position, no support. The fingers movements were in the same sense, velocity and rhythm and the subjects in a normal effort state (e.g. recording the dynamic response of the fingers movements for a subject, with the hand supported on the table) (see fig 4.)

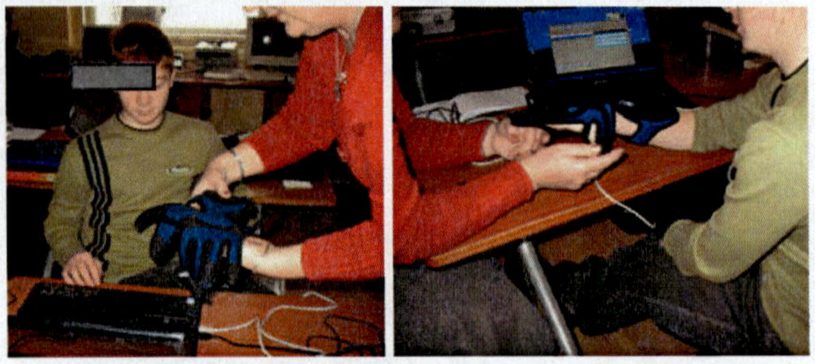

Fig. 3 Training steps to record hand movements

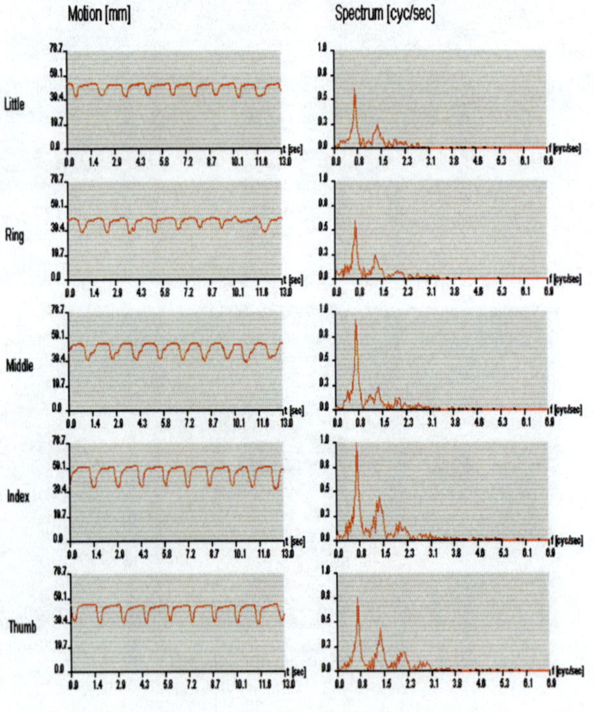

Fig. 4 Recordings of fingers movements

We may observe and analyze that the hand motion on vertical direction with no support allows performing more fingers motion cycles, more rhythmic, with higher velocities and amplitudes.

After subjected to dynamometer controlled effort (20 min with dynamometer in the right hand), we record the same motion succession (fingers rhythmic motion with the arm supported on the elbow or in vertical position) and then we calculate the differences or correlations between motions and stages (first graph represents the motion of the little finger, then the others until the last graph representing the thumb-opposable finger motion) (see fig.4. and 5.)

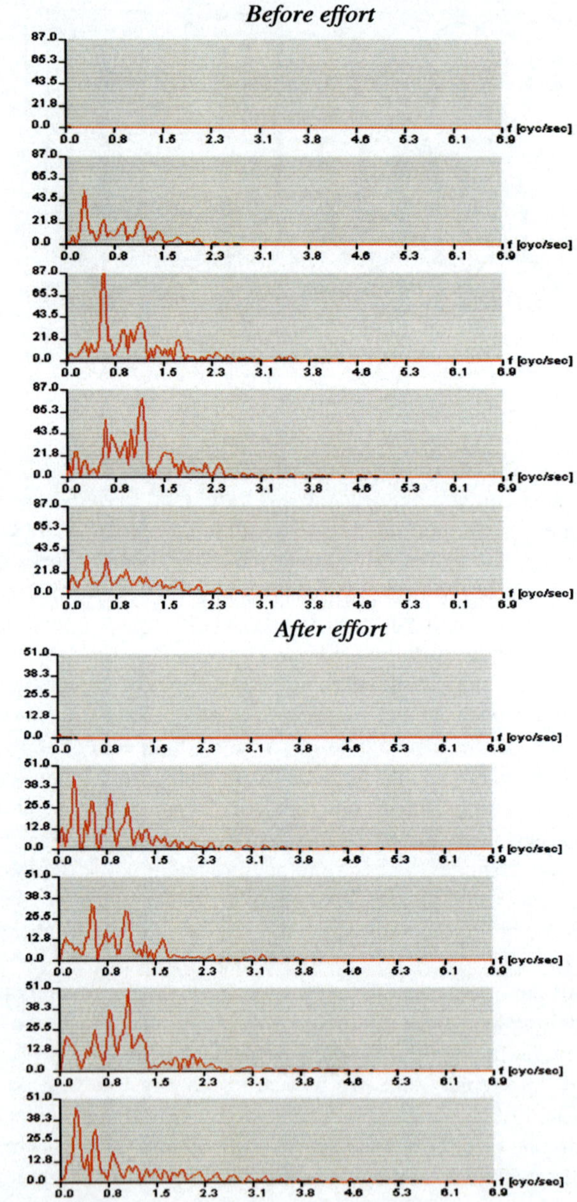

Fig. 5 Recordings of the hand movements before and after induced and controlled effort

In order to entirely assess the fingers motions and respectively the hand and its joint, we go to the next stage of determining the influence of a finger motion upon the others, influence that is determined when the user has to perform an activity involving all the fingers and also the hand joint (musicians, computer users, etc.) (see fig.6).

This analysis is performed in the same conditions as the previous recordings, aiming at the correlation of the obtained values with the skill and performance degree. The fingers motion was thus performed that each finger executed a complete cycle, the others being consciously maintained outside the motion.[12]

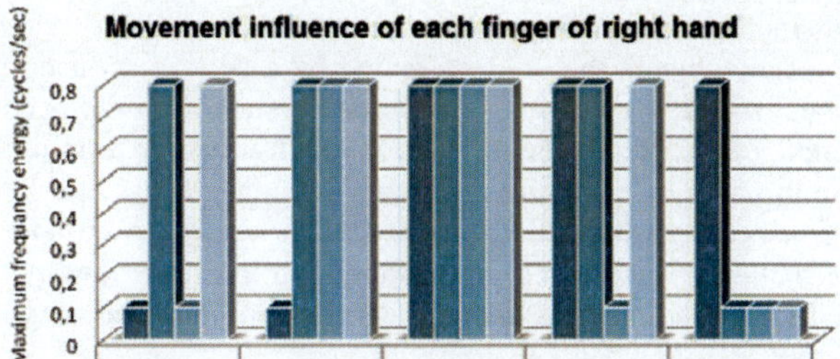

	1	2	3	4	5
Series1	0,1	0,1	0,8	0,8	0,8
Series2	0,8	0,8	0,8	0,8	0,1
Series3	0,1	0,8	0,8	0,1	0,1
Series4	0,8	0,8	0,8	0,8	0,1

Fingers of hand: 1-thumb; 2-index; 3-middle; 4-ring; 5-little.

Fig. 6 Effects of each fingers movements

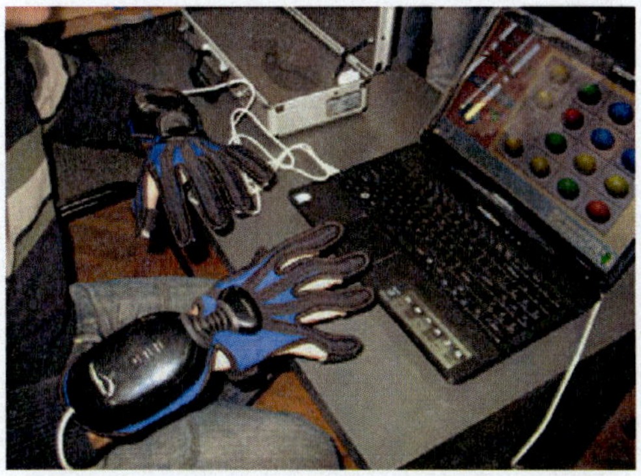

Fig. 7 Training and learning movements for fingers and wrist playing dedicated computer game

Investigations of abilities and developing these abilities in hand movements continue with a training of human subject to "play" on the computer software (see fig.7), by using the fingers and hand during 120 sec, at speed level 10, with vertical movement of a wrist and horizontal movement for the fingers. In this way at the first game the subject makes a training for his abilities and obtains the range of value 25^{+}(ball caught) and 20^{-} (ball missed) and at the second game the new score is 32^{+}(ball caught) and 14^{-} (ball missed).[10]

After that, the human subject made controlled medium effort systematic movements using a dynamometer system, during 15 minutes and then the ranges of motion were again measured (see fig.8). Flexion-extension motion of each finger is transformed in length units because the system initially was calibrated between anatomical and physiological positions and measured these positions.

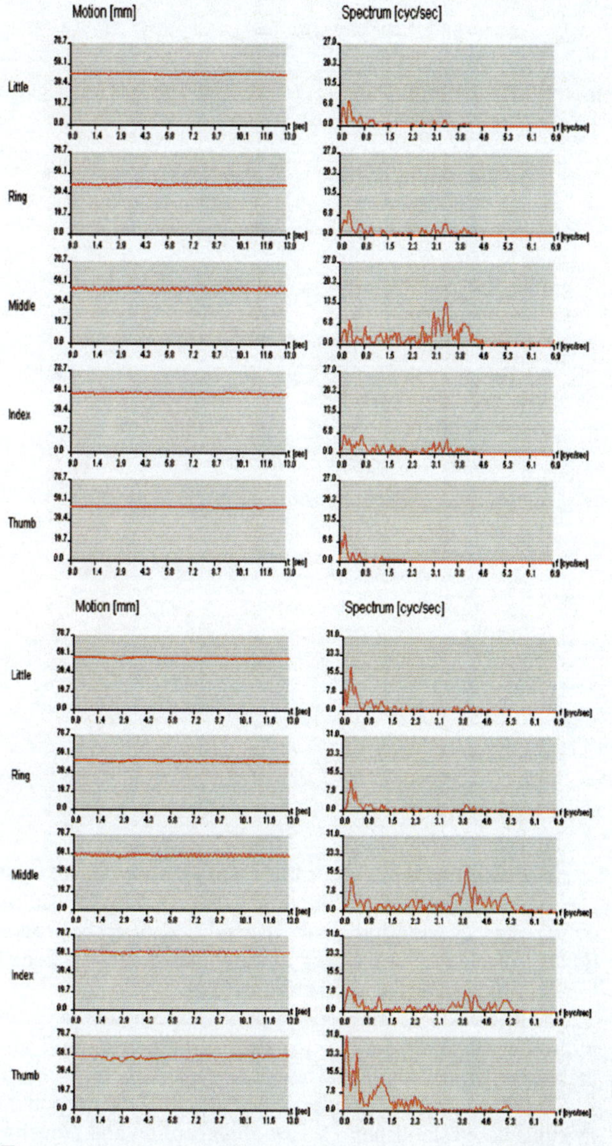

Fig. 8 Recordings of the hand movements before and after induced and controlled effort in second stage of investigation – after the computer game

It is obvious from the recordings that the abilities in range of motion (ROM) (see fig.9) of this subject were improved after training stage even if he makes some controlled effort with his right hand. Spectrum of movements measured in cycles/sec., represents analyze of movements typology (frequencies of ROM for each fingers). [6]

These qualities are shown also in the measurements of the energy deficit in active/passive actions with the fingers and wrist.

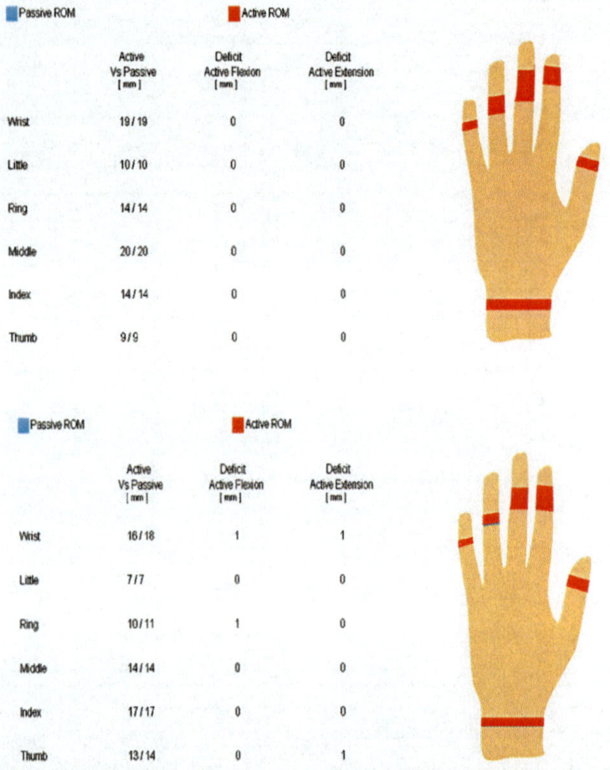

Fig. 9 Recordings of the ROM in active and passive actions after training with computer game

As a result of these recordings and analyses of all samples we can conclude that the movements of fingers also the movements of hand joint will be modified directly proportional to the level of training at computer games dedicated to rehabilitation of the hand abilities.[10]

These modifications consist of increasing the range of motion with small deficits for flexion of fingers and also increasing of measured coefficient at dynamometer.

Cognitive virtual motion applied for the repetitive motions case establishes a series of perception and adjusting mechanisms that are conditioned by the physical environment where the experiment takes place and by the subjects understanding skills.

The learning process is completed by the exercising procedure of a test for increasing manoeuvrability and avoiding muscular blockings.

By these tests the subject learns and adjusts mechanisms in order to increase performances. [12]

From the analysis performed on the subjects sample we found other important aspects:

- the subjects with a high degree of attention along the experiment learned and understood from the first testing the working mechanism so that they responded quickly and the recordings presented uniformity, similar amplitudes, obvious motions for all fingers;
- the subjects that were affected by a certain acoustic or visual stimulus changed the recorded response by the occurrence of some levels or motion or frequency amplitude diminishing; the cognitive virtual learning process is a lasting process, it must be practiced from simple to complex, permanently monitored and adjusted to the new conditions and must be changed when learning reaches the routine limit;
- the subjects responded much better to the testing when their physical and psychical status was within normal limits and there were no kind of stimuli (audio or visual).

These investigations are developed upon a set of subjects, analyzed and finalized using a handling skills improvement procedure in order to correlate with the performed activities and respectively with the effort degree.

The information was processed and analyzed using a software module compatible to the information recorded by the sensor glove.

In future researches we will establish a configuration to analyze the influences of different positions of hand and fingers also for hand joint, from left and right hand at the same time, to evaluate these influences on working abilities and comfort.

Also other future investigation is orientated on the integrated correlated technique investigations (ICTI), using analyze of variance (ANOVA) method to establish the level of correlations between the hand range of motion and influences of upper limb implant or hand disabilities.

Acknowledgment

These researches are part of the Grant PNII-IDEI 722 with CNCSIS Romania and we've developed the investigations with equipments from Research Project PNII-IDEI 722 in *Advanced Mechatronic Researches* Department from University Transilvania Brasov.

REFERENCES

1. M. Lawo, H. Witt, H. Kenn, T. Nicolai, R. Leibrand. "A Glove for Seamless Computer Interaction –Understand the WINSPECT".
2. Thomas A., (2010), Development of a biomechanical hand model for study of hand posture, strength and musculoskeletal disorders, Final Report for Pilot Research Training Projects in Occupational Health and Safety;
3. User guide Hand Rehabilitation System, (2010), operation manual for sensory gloves;
4. Satoshi I., et al., (2009), A hand rehabilitation support system with improvements based on clinical practices", the 9th International Symposium on Robot Control (SYROCO'09) Nagaragawa Convention Center, Gifu, Japan, September 9-12;
5. http://www.inrets.fr/ur/lbmh/ergo/IndexTheme.html accessed in July 2010
6. Baritz M., Cotoros D., Cristea L., Balcu I., (2009), Analyses of noise effects on standing human body stability, 9th WSEAS International Conference on SIGNAL, SPEECH AND IMAGE PROCESSING (SSIP '09), Budapesta, sept.;
7. Tozeren A. (2000), Human Body Dynamics: Classical mechanics and Human Movement, Springer-Verlag New York, Inc. ISBN 0-387-98801-7
8. Mohamad D. et al., (2010), Integration of Comfort into a Driver's Car Seat Design Using Image Analysis, American Journal of Applied Sciences 7 (7): 937-942, ISSN 1546-9239, Science Publications;
9. Pennestrı E. et al, (2007), Virtual musculo-skeletal model for the biomechanical analysis of the upper limb, *Journal of Biomechanics 40, 1350–1361, Elsevier Ltd,*
10. Baritz M. (2010), Computing Techniques to analyze the Human Hand Movements, Proceedings of International Conference on Optimisation of the Robots and Manipulators Calimanesti, Romania, 28-30 May, ISBN 978-981-08-5840-7;
11. Pitarch E., **(2007),** Virtual Human Hand, Grasping Strategy and Simulation, Academic Dissertation PhD thesis Universitat Politehnica de Catalunya (UPC);
12. Baritz M., Cristea L., Cotoros D., Rogozea L., (2008), Human body biomechanical stability evaluation affected by automated movements, 3rd International Conference "Optimization of the Robots and Manipulators" OPTIROB 2008 - PREDEAL Romania, 30 May- 1 June;

Author: Mihaela Baritz
Institute: University Transilvania from Brasov
Street: B-ul Eroilor nr.29
City: Brasov
Country: Romania
Email: mbaritz@unitbv.ro

Strain Measurement in an Elastic Material under Large Deformation Using Optical Reconstruction Methods

I. Zwierzak, J.W. Fenner, and A.J. Narracott

Medical Physic Group, Department of Cardiovascular Science, University of Sheffield, UK

Abstract— **The aim of this study was to explore photogrammetry- optical imaging method used to describe the local variation of strain in an elastic material undergoing large deformation. The material under investigation was marked with reference points (to identify the relative deformation) and imaged with two stereoscopically positioned cameras during a tensile test (strain rate 10 mm/ 60 s). The cameras were calibrated using the Bouguet calibration toolbox in Matlab to characterize their optics and establish their positions. Two dimensional (2D) images from both cameras were investigated to obtain the three dimensional (3D) position of the markers. The deformation of the elastic material was determined by 3D reconstruction of the deformed geometry. Strain results derived from Matlab were in good agreement (99 % ± 0.3 %) with a manual measurement using a caliper. Camera calibration and reconstruction accuracy results indicated an uncertainty of dot separation of the order 30 μm, with inter-point distance of approximately 3 mm.**

The feasibility of application of photogrammetry for strain measurement through 3D reconstruction of an investigated object has been demonstrated in this study. The optical technique shows potential for development to investigate local strain variation in elastic biological materials.

Keywords— **Restenosis, Strain, Photogrammetry, Calibration, Reconstruction.**

I. INTRODUCTION

One third of all natural deaths in the modern world are associated with cardiovascular disease (CVD). Most commonly atherosclerosis, which causes stenosis in arteries that supply blood to the heart [1]. Both non-invasive (angioplasty, angioplasty with stenting) and invasive (vascular bypass) treatments are used to restore normal blood flow in the coronary arteries and prevent heart attack. Unfortunately, restenosis after coronary intervention still remains an unsolved and important clinical problem [1]. The amount of neointimal hyperplasia in response to stent deployment and vessel wall injury has been shown to be related to the magnitude of the stresses and strains caused during implantation [2].

Many numerical and experimental studies have investigated stent expansion and arterial wall stresses caused by the interaction between the stent and the blood vessel. This problem is often examined using computer simulations because of the cost and difficulty of *in vitro* studies.

Finite element (FE) models have been employed by researchers to study the behaviour of different stent designs [3] and strut thickness [2]. Such studies have examined the local stress distribution generated in the vessel wall, which is seen to vary for different stent designs [3]. It has been shown that thicker stent struts result in greater stresses [2], with the hypothesis that increased local stresses lead to greater vascular injury. This hypothesis is supported by clinical evidence from the implantation of stents with differing strut thickness. It was found that thicker struts are correlated with increased neointimal hyperplasia [4]. Numerical models are excellent research tools, especially when physical models are difficult to create. Nevertheless, it is necessary to carry out experimental work to confirm the results of computational studies [5].

A previous study which considered the strain of arterial tissue caused by stent expansion was reported by Squire *et al* [6]. An image acquisition system was developed to capture arterial deformation of *in vitro* bovine coronary artery and *in vivo* rabbit femoral artery marked with contrast ink. A single CCD camera was used to track stent deployment and arterial deformation during its expansion. To determine the strain of the expanded stent in 3D, the location of markers captured in 2D were backprojected onto a 3D cylindrical model. The method of using high-contrast ink to mark the vessel and measure strain during stent expansion was found to be effective reporting of absolute strain with errors of less than 0.1 % (for a field of view 3 mm by 12 mm). Full 3d geometry reconstruction for assessment of soft tissue deformation is reported in a study by Lujan et. al [7]. Stereo digital cameras where used to determine strain, reporting maximum errors of the order 0.1%.

Detailed information describing 3D optical surface reconstruction for strain measurement in an aneurysm model has been presented by Randall [8]. A balloon model with reference points on its surface was captured during expansion and strain was calculated from images obtained using stereoscopically positioned digital cameras. Calibration was performed using the Bouguet Calibration Toolbox [8] and subsequently, the object was reconstructed in 3D to calculate the strain. Common points in the stereo image pair were computed as a location in 3D space (x,y,z coordinates). The

S. Vlad and R.V. Ciupa (Eds.): MEDITECH 2011, IFMBE Proceedings 36, pp. 120–123, 2011.
www.springerlink.com

calibration and reconstruction process was performed in MatLab.

This paper focuses on the development of the stereo-photogrammetric technique, to provide accurate measurements for strain characterization in elastic materials undergoing large deformation. In the future, strain measurement at vascular length scales will be employed to investigate stent expansion in: i) vessel analogues and ii) coronary arteries *in vitro*.

II. MATERIAL AND METHODS

A. Test Rig Setup

Images from two stereoscopically positioned digital Canon Powershot A40 cameras (Fig1) were used to reconstruct the object under investigation in 3D. The technique requires the object to be placed within the field of view of both cameras to form a stereoscopic camera rig. The distance between the cameras and the object was measured manually. The cameras were separated by ~70 ° (Fig.2).

Fig. 1 Experimental setup.

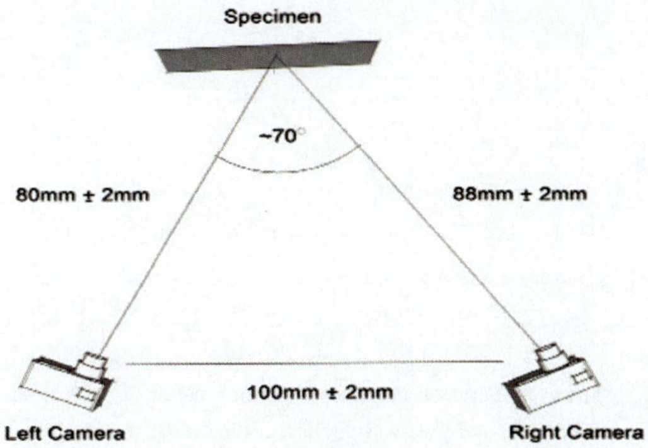

Fig. 2 Distance between cameras and the specimen.

In order to reconstruct 3D geometry, single and stereo calibration of the camera system is required.

B. Camera Calibration and Reconstruction Procedure

A square calibration grid object (with edge length of 12 mm) positioned manually at 15 various orientations was captured by each camera. The process required manual identification of the four extreme grid corners. Calibration (single and stereo) was carried out to characterize cameras optics and establish their relative positions. The physical camera rig is characterized by intrinsic (focal length, principal point, skew coefficient, lens distortion and pixel error) and extrinsic parameters (relative positions of the grids with respect to the camera). The Bouguet Camera Calibration Toolbox for Matlab [9], which has a reputation as an accurate and robust tool [8], was applied.

Reconstruction of arbitrary geometry captured by stereo imaging was achieved by software previously developed in house in Matlab. Highly contrasting points (Fig.4) marked on the object surface (to identify the relative deformation) were employed for effective reconstruction. Reconstruction required selection of the same point in the left and right 2D images. An epipolar line drawn between the focal point of one camera and the chosen point of the first image assisted with point selection on the second image. The centre of each picked point was determined by a centre of gravity algorithm, which places a correspondent point at the centre of the area manually selected. Triangulation of the selected point, in the 3D space, was derived from the intersecting epipolar lines from the two cameras [8]. Two sets of 3D coordinates from right and left camera reference frames were generated.

To ensure accuracy of the calibration, routine assessments were carried out with a fixed camera setup and three different size square calibration grids (with edge lengths of 12, 24 and 36 mm); calibration data was then used to provide reconstruction of a control object.

C. Tensile Test

The methods described above establish the surface geometry of the object of interest. Strain was determined by capturing geometry changes as the material was deformed. An experimental tensile test was conducted using a GOODBRAND GBX testing machine to stretch a flat elastic sheet, with deformed geometry captured simultaneously by the two cameras. The strain rate was 10 mm/60 s. The material was progressively stretched at one end over a 60 s period, the machine was stopped every 5 s and photos were taken (with an initial photo taken before stretching). The captured images were used to carry out 3D reconstruction (*section B*) of the deformed material and calculate the strain.

The measurement of strain required the tracking and reconstruction of the same dots in deforming object. Simple tensile strain (ε) was measured as the extension (*dx*) per unit length (*x*) between reference points:

$$\mathcal{E} = \Delta x / x \tag{1}$$

III. RESULTS

A. Quality of Calibration

Fig.3. depicts the 3D rig geometry determined by the stereo calibration result, providing the extrinsic parameters which describe the grid orientations in relation to both cameras. Accurate calibration is confirmed by accurate representation of the physical stereo system. The stereo rig geometry derived from camera calibration (Fig.3) was shown to be in good agreement (97 % ± 0.5 %) with manual measurements (Fig.2).

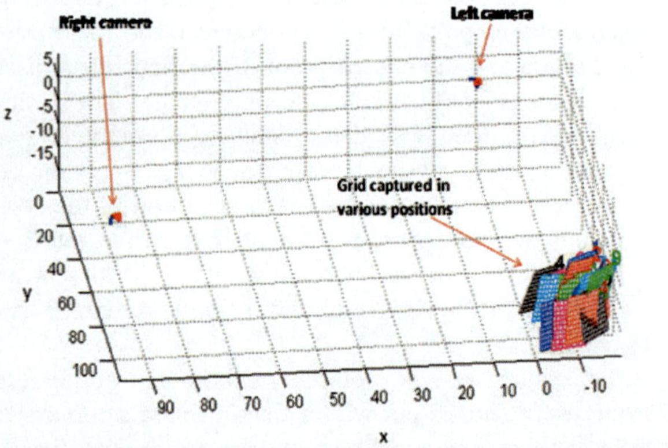

Fig. 3 Extrinsic parameters presented in 3D plot for stereo camera calibration.

The camera calibration results were consistent with accurate and robust techniques for all calibration grid sizes and an uncertainty of dot separation of the order 30 μm, with inter-point distance of approximately 3 mm.

B. 3D Reconstruction and Tensile Test Results

Figure 4 displays the initial and final image of the stretched elastic material. 3D reconstruction of reference point separation was used to calculate measures of strain.

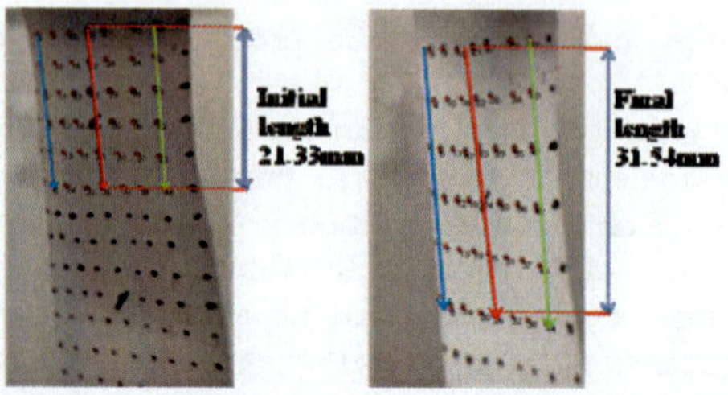

Fig. 4 The position of marker points. Initial and final length (before and after tensile test) of the elastic material.

The values from strain measurement of three regions of the elastic material (blue, red, green) (Fig.4) are presented in graphical form in figure 5. Strain was calculated as a total strain, ε_T, from the first to the last point on a line and as a mean, ε_{av}, of the strain of neighboring points, ε_i, for comparison (Table 1). Measurements were obtained after 60±1.4 s of stretching with a 10 mm/ min ± 0.21 mm/ min strain rate. The derived results agree within 99 % ± 0.3 % of a manual measurement using a caliper. Average values of strain, ε_{av}, calculated for each line differ slightly (0.5 % ± 0.2 %) from the total strain (ε_T) measurement.

Table 1 Strain results: εav - average value of εi's, ε_T - the total strain between the first and last point, εi - the strain between neighboring points.

The strain [%] between marker points			
Investigated line	Measurement with caliper ± 0.6 mm	$\varepsilon_{av.} = \left(\sum_i^n \varepsilon_i\right)/n$	ε_T
Left line	43.7	43.3	43.5
Central line	48.0	48.2	48.3
Right line	51.1	51.4	51.5

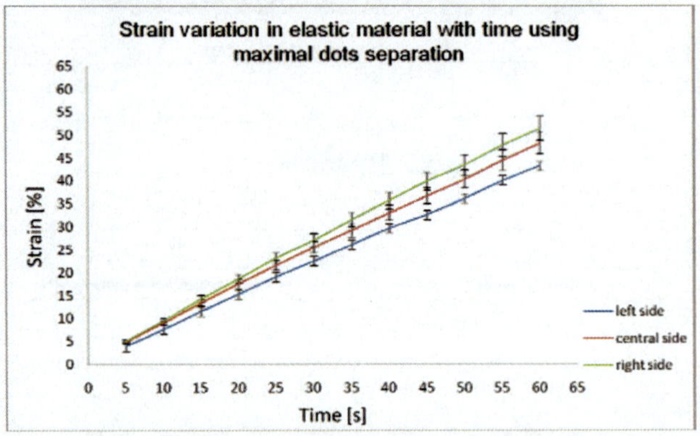

Fig. 5 Strain ε_i, between neighboring points variation with time: blue- left side, red– central and green- right side of the elastic material.

IV. Discussion

This experimental work was carried out to measure the strain in an object undergoing large deformation. It demonstrates the application and assesses the robustness and accuracy of calibration and reconstruction techniques.

From 2D images of a deformed material, obtained from two stereoscopically positioned cameras, 3D reconstruction of markers was carried out for strain measurement. 3D images enable visualization of the object shape and can characterize the object length, width and depth. This approach offers superior determination of the object surface because it provides full 3D geometry, which is important for adequate strain analysis.

Camera calibration and reconstruction accuracy results indicated an uncertainty of dot separation of the order 30 μm, with inter-point distance of approximately 3 mm. This is encouraging, but requires further improvement if the technique is to be applied at smaller length scales.

From the results presented in Figure 5 it is observed that larger deformation occurs on one side of the elastic material. This effect may arise from material non-uniformity or non-uniform boundary conditions and requires further investigation. The error from strain results (less than 1 %) may be associated with a number of factors: performance of the centre gravity algorithm, marker size and separation, quality of the camera optics. This study utilized consumer cameras and results are likely to improve using high resolution equipment and appropriate lenses.

Future experimental effort is required, associated with improvement of the specimen marking method (to identify the relative deformation of elastic material), development of camera calibration and 3D reconstruction techniques in Matlab.

This work aims ultimately to investigate stent deployment in the vessel analogue and real pig coronary artery *in vitro*. Characterization of the vessel wall strain caused by stents may deepen understanding of its variation due to stent design and deployment technique. This may help to improve stent design to minimize wall injury and neointimal hyperplasia causing restenosis.

V. Conclusions

Computational and experimental studies have emerged in recent years to study stresses on the vessel wall caused by stents. Investigations of the stent/vessel wall interaction are important in predicting the stresses and strains induced after stent implantation. This study presents application of the stereo- photogrammetric technique for strain calculation in elastic materials undergoing large deformations, demonstrating accuracy of calibration and reconstruction techniques. The stereo- photogrammetric method produced acceptable errors at the present length scales and will be improved in future work for examination of stent deployment in vessel analogues and pig coronary arteries *in vitro*.

Characterization of the local vessel wall strain will provide a deeper understanding of the strain variation during stent deployment and may help to prevent in- stent restenosis.

Acknowledgment

This research is funded by the European Commission, through the MeDDiCA ITN (www.meddica.eu, Marie Curie Actions, grant agreement PITN-GA-2009-238113).

References

1. Duraiswamy N., Schoephoester R. T. et al (2007) Stented artery flow patterns and their effects on the artery wall, Annu Rev Fluid Mech, 39: p 357- 382.
2. Gijsen F.J., Migliavacca F. et al (2008) Simulation of stent deployment in a realistic human coronary artery, Biomed Eng Online, 7: p.23
3. Lally C., Dolan F. et al (2005) Cardiovascular stent design and vessel stresses: a finite element analysis, J Biomech, 38(8): p. 1574-81.
4. Pache J., Kastrati A. et al (2003) Intracoronary stenting and Angiographic Results: Strut Thickness Effect on Restenosis Outcome (ISAR- STEREO- 2), Am J Cardiol, 41(8):1289-92.
5. Wang Y. X., Yi H.eta al (2005) Experimental Research on Balloon-expandable Endovascular Stent Expansion, Conf Proc IEEE Eng Med Biol Soc, 3: p. 2272-5.
6. Squire, J.C., Rogers C. et al (1999) Measuring arterial strain induced by endovascular stents, Med Biol Eng Comput, 37(6): p. 692-8.
7. Lujan T., Lake S. et al (2005) Simultaneous Measurement of Three-Dimensional Joint Kinematics and Ligament Strains with Optical Methods, Vol. 127/193
8. Randall D., Poster presentation, 17th Conference of the European Society of Biomechanics, 2010 Edinburgh.
9. Bouguet J. (2008), Calibration Toolbox in Matlab, http://www.vision.caltech.edu/bouguetj/calib_doc/

Author: Iwona Zwierzak
Institute: Medical Physics Group,
Department of Cardiovascular Science,
Faculty of Medicine, Dentistry and Health
Street: University of Sheffield, Beech Hill Road
City: Sheffield, S10 2RX
Country: United Kingdom
Email: i.zwierzak@sheffield.ac.uk

Hand Vein Biometric Authentication in Optical Multi-touch Systems

S. Crisan, I.G. Tarnovan, B. Tebrean, and T.E. Crisan

Technical University of Cluj-Napoca, Department of Electric Measurements, Faculty of Electrical Engineering, Cluj-Napoca, Romania

Abstract— **Multi-touch systems and their applications are entering a mainstream phase where fast access to visual data and collaborative environments are key factors for an informational age. However, most systems are envisioned as public workstations where resources are shared equally between the users with few or no restrictions on the content delivered by the applications. While real-time identification of the users of a multi-touch system is a difficult subject to tackle, authentication using biometric parameters could provide a hierarchy based content management system. Based on previous research in both hand vein biometric detection and multi-touch systems, this paper describes a hand vein authentication device usable in a touch/object sensing multi-user collaborative environment.**

Keywords— **Multi-touch, optical systems, hand veins, biometry.**

I. INTRODUCTION

Due to the recent researches in tactile systems, multi-point collaborative environments have become affordable and the suite of applications covers important domains such as engineering, computer assisted design, medical data visualization, asset management, media etc.[1][2][3] Whether in a table or wall form, multi-touch devices offer concurrent access to visual information allowing multiple users to share and visualize parameters pertaining to the chosen field.

One important problem can be identified in the vast majority of the implemented touch systems and it is related to the identity of a given user interacting with the tactile device. Due to the inconsistency of the touches performed by a user during a work session and the different angles of approach for left and right hand, determining a user's identity is one of the most challenging problems in multi-touch sensing.

User identification for multi-touch, multi-user devices has been thoroughly researched in the past years and several methods have been devised. In most cases users are given external devices in order to be recognized by the touch sensing system. In the case of DiamondTouch by Mitsubishi Electric Research Laboratories –one of the most advanced user identification systems- an array of antennas is embedded in the touch surface. Each antenna transmits a unique signal. Each user has a separate receiver, connected to the user capacitively, typically through the user's chair. When a user touches the surface, antennas near the touch point couple an extremely small amount of signal through the user's body and to the receiver. [4]. However, DiamondTouch is a front projected tactile surface with the displayed information suffering from visual occlusions due to the users' hands. Movement around and near the table is also restricted because of the nature of the identification method described earlier in the paper. Dohse et al. [5] suggest a hand-tracking method to differentiate touches from different users. The system uses a top mounted camera to track the users' hands on the tactile surface. In this method, close touches from different hands are harder to distinguish and furthermore, users leaning over the table might obscure the hand positions. In [6] the authors extend this technique by introducing a biometric hand-contour based system. It allows users to identify by laying their hands flat on the surface. While this does not require the users to possess external devices, the system needs them to re-identify every time they moved a hand off the table [7]. In addition, hand contour detection has a low discrimination ratio in comparison to other biometric techniques and might provide an increased number of false acceptances. The research presented in [8] describes a wristband user identification system based on pulsed light that allows the system to perceive the users using individual codes. This method uses the nearest neighbor approach and is prone to failure when users are controlling the touch surface in close vicinity.

Based on previous research in the field of biometrics and touch sensing [9][10] this paper offers a biometric implementation with no external devices in order to address the user authentication problem in tactile systems and suggests further research in the field of hand vein biometrics in order to provide a real-time identification method.

II. HAND VEIN DETECTION IN OPTICAL SYSTEMS

A biometric system is essentially a pattern-recognition system that recognizes a person based on a feature vector derived from specific physiological or behavioral characteristic that the person possesses. Vein pattern detection fully complies with this definition and it provides many important biometric features:

- Uniqueness and permanence of the vein pattern during the lifespan of an individual;

S. Vlad and R.V. Ciupa (Eds.): MEDITECH 2011, IFMBE Proceedings 36, pp. 124–127, 2011.
www.springerlink.com

- Non-contact detection procedure in most sensing applications;
- Almost impossible to forge or copy due to the nature of the scanning;
- The biometric parameter is hidden from general view and an external source of radiation is required for vein scanning;
- The vein pattern is intricate enough to allow sufficient criteria for positively detecting various subjects even identical twins or to distinguish vein models in both hands from the same user.

An optical system can be used to emit near infrared radiation towards the subject's hand. Absorbed radiation can be perceived as a darker color on a filtered CCD camera. This method is able to detect veins but not arteries due to the specific absorption of infrared radiation in blood vessels. Also, superficial veins in the back of the hand are preferred for biometric detection since the model is easier to acquire than in the case of palm veins or fingers.

A sketch of an actual vein detection system is shown in figure 1.

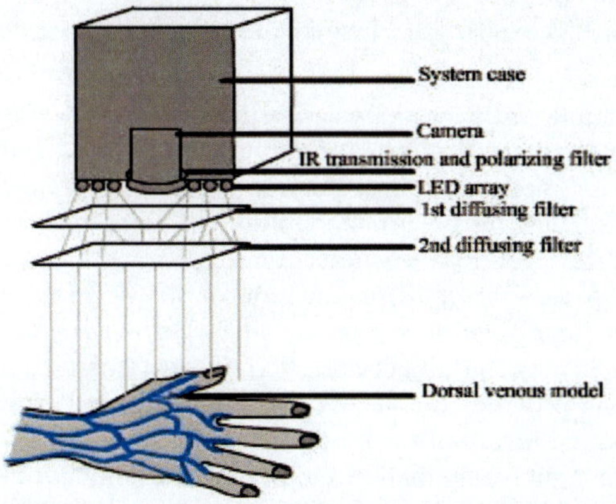

Fig. 1 A basic hardware vein detection system (sketch)

The infrared radiation is absorbed in a different way in water and various types of organic compounds. In order to achieve visual penetration through the respective tissue and to take advantage of the sensitivity of CCD cameras to near infrared radiation, lighting should be performed under a very tight optical window, in the range of 750... 940nm. [11].

The scanning principle used to acquire the pattern of the veins in the back of the hand shares several similarities to the methods used in optical tactile systems. The wavelength range for the emitted radiation is the same since multi-touch devices register touches based on the reflection of near infrared radiation from the coupling points. Light diffusion is also an important requirement of a touch device in order to avoid uneven illumination.

The raw image of the vein pattern detected by the IR-sensitive CCD camera looks similar to the image presented in figure 2.

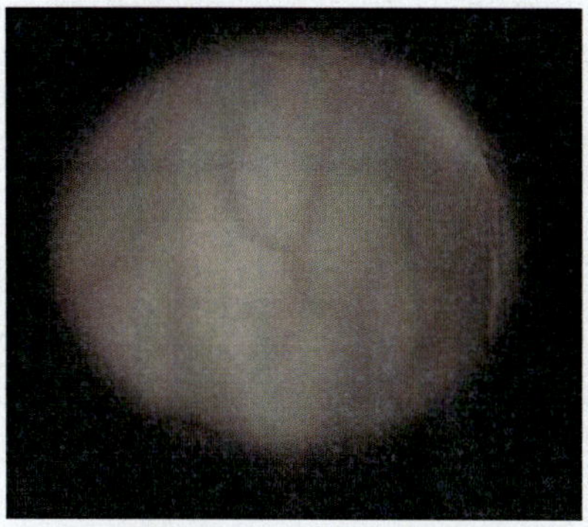

Fig. 2 Raw image of the vein pattern

The acquired image suffers a series of transformations in order to extract the vein pattern from the background information. Previous research from the authors described a sequence of methods and algorithms employed in the recognition of vein structures [10]. Together with a contrast variation identification method that insures liveness detection of the biological probes presented by the users, biometric authentication in multi-touch systems retains the same low error rates calculated in standalone vein recognition applications.

In order to translate the hand veins recognition phase in a tactile system several modifications must be applied to the detection algorithms.

The new set of software algorithms that govern the required transformations to the raw image can be described as follows:

- The image is converted to grayscale with a bit depth equal to 8 and the individual pixel values are fed to a 2 dimensional array. This phase is coupled with the fiducial detection of a touch system in order to determine that a pattern can belong to a vein model and is not treated as a regular touch object.
- The background surrounding the hand is removed from the image array based on analysis of the individual pixel values
- The image is centered and a mean filter with a 5x5 kernel window is apply to reduce acquisition noise from the camera
- The image is binarized and the vein model is extracted using an adaptive threshold operation with a sliding kernel.

- The obtained vein pattern is thinned using a set of 19 conversion rules (15 for general thinning and 4 for diagonal lines –as described in [10])
- The structure is then optimized by pruning the extraneous branches of the pattern and by verifying the connectivity rules specific to a real vein model (veins can not close on each other; veins can not be horizontal in respect to the hand plane and they need to be connected etc.)
- The complete set of vein features is extracted. Important parameters such as total length of the model, distance and angles between vein branches, number of node and terminations of the pattern are counted and saved as a biometric record of the user in a dedicated database.
- On each subsequent scan, the processed data is compared to the saved information in a database in order to ascertain the identity of the user.

Since the camera resolution and speed are not important factors in the retrieval of hand vein models, most infrared sensitive cameras used in optical systems are viable to be used in vein based user identification and authentication.

In addition, most optical touch systems use powerful multi-core computers in order to register and track touch points over sequential frames in real time. In this scenario, the low computational power required for the vein processing algorithms, either in a user authentication or identification phase will not significantly slow down the touch sensing application.

Integrating a vein processing module in an optical multi-touch system does not require significant changes in the layout and design of a preexisting device. Moreover, accurate liveness detection allows the system to check for vein patterns without the need for external fiducials, preprogrammed areas on the tactile surface or additional context gestures.

The following chapter will address the fusion of both technologies in order to create an identity/context aware multi-user touch system.

III. BIOMETRIC INTEGRATION IN TACTILE DEVICES

User identification is perceived in a different manner than user authentication. Authentication allows the user to present his credentials at the beginning of the work session and may require occasional proof. An identification system, however, tracks continuously the identity of the user over the span of the interaction and is inherently more difficult to design and implement.

In most multi-touch devices, content is public and there is no information delivery restriction based on roles or ranks [12]. The usual security concerns related to identity theft, PIN/ID stealing, difficulties in remembering the credentials needed for the system apply even on authentication type touch systems. Biometric verification offers authentication based on physical and behavioral characteristics that are harder to forge or copy than identification numbers or passwords.

By combining the vein pattern recognition with a suitable optical system, authentication for hierarchy-based content delivery can be achieved. However, not all optical multi-touch, multi-user devices are able to allow the integration of the biometric module.

A multi-touch system that can prompt the user for biometric credentials when sensitive information is required has been designed by the authors and is based on the multi-touch table described in [9] with the addition of a vein pattern recognition module with integrated liveness detection[10].

As presented in [12], most optical tactile systems are based on one detection technique or a combination of two of the following methods:

- FTIR (Frustrated Total Internal Reflection)[2]
- DI (Diffused Illumination)[1][12][14]
- DSI (Diffused Surface Illumination)[12]
- LLP (Laser Light Plane)[1][13]

When comparing these methods, the biometric module is compatible with the optical sensing device that allows object and pattern detection on the tactile surface. Both FTIR and LLP methods do not provide native support for object tracking and thus vein recognition is inapplicable. While DSI offers good object detection, experiments have shown that the amount of diffusion reduces the accuracy of vein model identification by up to 40%. Diffused Illumination (DI) allows good object detection, intricate pattern recognition and provides the necessary illumination for penetrating the tissue in front of the blood vessels.

The design schematic of a Diffused Illumination system is shown in figure 3.

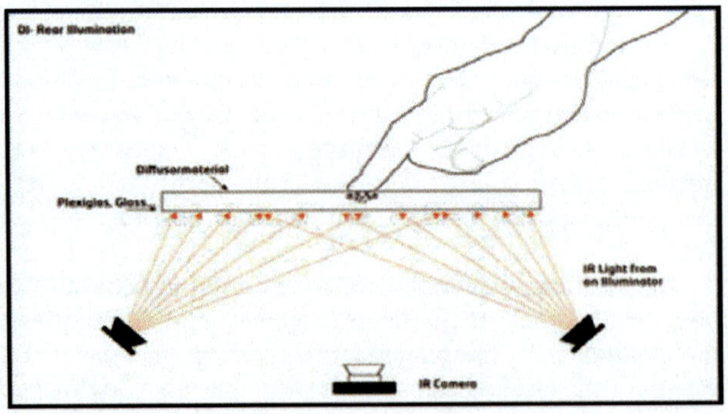

Fig. 3 Design schematic of a Diffused Illumination multi-touch system

In small factor touch tables, a single CCD camera is preferred [12]. Depending on the distance between the camera and the touch surface, the quality of the acquired images may vary, affecting the False Acceptance Rate (FAR) and the False Rejection Rate(FRR) of the biometric module. User authentication is easier to be performed on multiple camera setups where fiducials and special small sized object patterns are clearly recognized by the multi-touch device. Hand vein user recognition can also be integrated without modifications into a popular touch device such as the Microsoft Surface [14]. With a 5 camera setup, the Surface is able to detect handwriting and patterns similar in size and shape with the human vein models.

Experiments were conducted with both the original 612 record database used in [10], and with a new set of biometric records acquired with the touch table camera. The user places the hand anywhere on the tactile surface with the palm facing up. The software detection module identifies the hand as a biometric object using the liveness detection test and starts the scanning process. Since the original images were acquired with a different system under different lighting conditions and with varying distances, the recognition algorithm was able to identify less than 72% of the users. After introducing the optimization algorithm for touch systems the value has improved to 84%. On records scanned and authenticated with the tactile device, the system was able to recognize 97% of the users and the false acceptance rate was below 0.13%.

IV. Conclusions

Without the use of external devices, user identification in tactile systems still represents a challenge for the scientific community. Authentication based systems do exist and the introductory chapter has presented several practical implementations. This paper has tried to offer an alternative to other authentication solutions by introducing a biometric parameter in the process of user detection. As in other applications, biometric detection has several advantages over identification numbers and passwords since the risk of identity theft and other security concerns is greatly diminished.

Identifying vein patterns from the back of the hand or palm can be combined with optical touch tables in order to provide hierarchy based content delivery where access to sensitive information is limited. Authentication checks are applied sequentially for continuous access to sensitive data.

Real time user identification without the use of external devices is also the subject of an ongoing research and true identification/authentication touch systems are the main focus for future studies.

Acknowledgment

This paper was supported by the project "Development and support of multidisciplinary postdoctoral programmes in major technical areas of national strategy of Research - Development - Innovation" 4D-POSTDOC, contract no. POSDRU/89/1.5/S/52603, project co-funded by the European Social Fund through Sectoral Operational Programme Human Resources Development 2007-2013.

This paper was supported - conference fees, research equipment and materials- by CNCSIS–UEFISCSU research grant PNII – IDEI 2008, code 180, contract 616/2009.

References

1. Bill Buxton, "Multi-Touch Systems that I Have Known and Loved" at http://www.billbuxton.com/, 2007
2. Jeff Han. "Low-Cost Multi-Touch Sensing through Frustrated Total Internal Reflection" *Proceedings of the 18th Annual ACM Symposium on User Interface Software and Technology* 2005
3. C Forlines, D Wigdor, C Shen, and R Balakrishnan. Direct-touch vs. mouse input for tabletop displays. Proceedings of the SIGCHI conference on Human factors in computing systems, (2007)
4. P H Dietz and D Leigh. DiamondTouch: A multi-user touch technology. In Proc. UIST, pages 219–226, 2001
5. K C Dohse, T Dohse, J D Still, and D J Parkhurst. Enhancing multi-user interaction with multi-touch tabletop displays using hand tracking. In Proc. ACHI, pages 297–302, 2008.
6. M Ringel, K Ryall, C Shen, C Forlines, and F Vernier. Release, relocate, reorient, resize: fluid techniques for document sharing on multi-user interactive tables. In Proc. CHI, pages 1441–1444, 2004.
7. D Schmidt, M K Chong, and H Gellersen. Hands-Down: Hand-contour-based user identification for interactive surfaces. In *Proc.* NordiCHI, 2010
8. T. Meyer, D. Schmidt, IdWristbands: IRbased User Identification on Multitouch Surfaces in ITS 2010 Poster, 2010
9. S. Crisan, Zaharia V. D., Brender, L. V., Crisan, T.E, A multi-touch collaborative solution for measurement data visualisation, XIX IMEKO World Congress, Lisbon Portugal, 2009
10. S. Crisan, I.G. Tarnovan, T.E. Crisan, Radiation optimization and image processing algorithms in the identification of hand vein patterns, CS&I, Volume 32, Iss 3, Pages 130-140, 2010
11. Nadort, Annemarie, *"The hand vein pattern used as a biometric feature"*, Nederlands Forensisch Instituut, 2007
12. Cetin, G., Bedi, R., Sandler, S. (Eds.) Multi-touch Technologies. 1st edition at http://www.nuigroup.com 2009
13. Hodges S, Izadi S, Butler A, Rrustemi A and Buxton B (2007) Thin-sight: versatile multi-touch sensing for thin form-factor displays. Proceedings of the 20th annual ACM symposium 2007.
14. Microsoft Surface. at http://www.surface.com, 2007

Author: Septimiu Crisan
Institute: Technical University of Cluj-Napoca
Street: Memorandumului 28
City: Cluj-Napoca
Country: Romania
Email: septimiu.crisan@mas.utcluj.ro

Spatial Correlation in Mechanical Heart Valve Leakage Jets

G.L. Wang, G. D'Avenio, C. Daniele, and M. Grigioni

Istituto Superiore di Sanità, Dept. of Technology and Health, Rome, Italy

Abstract— **Turbulent flow can be generated by prosthetic devices at the blood stream level, causing a mechanical stress on the blood particle. This mechanical stress can entail adverse impacts of prostheses' recipients, such as hemolysis.**

The experimental research is on the quantitative investigation of the flow field of mechanical bileaflet valve using Particle Image Velocimetry(PIV). Two major measurements focused on leakage jets study were taken. One is standard nozzle measurement and the other one is mechanical valve measurement. These quantitative data could be utilized as basis for a CFD simulation of relevant phenomena. The self-preservation of leakage jets was evaluated with a new method spatial correlation.

Keywords— **Particle Image Velocimetry, Mechanical Bileaflet Valve, Leakage Jets, Turbulent Shear Stress, Spatial Correlation.**

I. Introduction

Mechanical heart valves, specifically bileaflet valves, are widely applied in the replacement of diseased native valves. Compared to biological heart valves, these prosthetic valve replacements exhibit good long-term mechanical durability but can suffer functionally from increased blood element damage, a strong propensity for thrombus formation, and a need to maintain an anticoagulant drug therapy [1]. Turbulent flow can be generated by prosthetic devices at the blood stream level, causing a mechanical stress on the blood particle. This mechanical stress, can entail adverse impacts of prostheses' recipients, such as hemolysis.

In this report, Particle Image Velocimetry (PIV), which is an optical method of fluid visualization, will be employed to analyze complicated flow flied, such as those generated by mechanical heart valves (which are characterized by high velocity gradients and turbulence). PIV is a direct derivative of a series of PIV measurements made at transonic speeds with sub-micron seeding particles with wide applications for fluid measurements[2]-[5]. PIV techniques was firstly achieved at MIT (Massachusetts Institute of Technology) by P.J.Bryanston-Cross[6], and then extended by C.E.Towers[7] at the ARA(Aircraft Research Association) to achieve a long range application. PIV is the reference technique for evaluation of the potential blood trauma of a given implant, in terms of turbulence-related stresses. Moreover, the occurrence of flow stagnation/separation can be also detected, enabling the assessment of the potential for thrombogenesis of a given design.

In this paper, PIV techniques were used to measure the leakage flow of mechanical bileaflet valve. Our study was aimed at the quantitative investigation of the flow field of mechanical bileaflet valve. The previous works were mainly conducted to investigate the flow field in open and closed state of leaflet at hinge plane[1][8]. We scanned the 2D flow field from the outside of the distal hinge to the inside, while keeping the leaflets closed. The scanned scale was 1 mm for each slice. After analysis of the 2D measurement, a 3D presentation of flow field was submitted to evaluate the leakage of the valve close to the region of two hinges. This study could provide a experimental basis for CFD simulation. Additionally, the properties of flow field of leakage were evaluated as average velocity field, Turbulent Intensity (TI), Turbulent Shear Stress(TSS). We also introduced the method of spatial correlation to characterize the development of leakage jets.

II. Methods

A. *Experiments Setup*

The experiments were conducted in the steady-flow experimental in vitro chamber. During the 2D slices acquisition, the valve was constantly kept in the closed position, in order to study the same flow field throughout the series of measurement. A large chamber with optimized optical access for PIV measurement (Fig. 1) was employed in the steady-flow loop for mimicking inlet and outlet of valve, such as ventricle and atrium for mitral valve. A plastic valve holder was placed between the chambers and securely held the prosthetic heart valve under study (mechanical bileaflet valve, size 23mm, pyrolitic-carbon occluder). The valve was inserted in a silicone rubber ring, which assured easy positioning of the valve in the seating. The circulation fluid was forced by hydraulic pump in a reservoir (reservoir 1) placed above the rest of system and connected to the sealed measurement chamber 1 to supply constant pressure. The incoming flow from ventricle kept close state of the valve and leakage of valve passed into measurement chamber 2. Measurement chamber 2 remained exposed to the atmosphere and the outflow of it, as well as the outflow from the

S. Vlad and R.V. Ciupa (Eds.): MEDITECH 2011, IFMBE Proceedings 36, pp. 128–135, 2011.
www.springerlink.com

constant-head tank, was conducted into a pooling reservoir (reservoir 2). At last, the pooling reservoir was connected to the inlet of the pump to achieve a close hydraulic circuit. Since the tubes connection the reservoirs to chambers had a small hydraulic resistance, a negligible value of the pressure head was lost across those connections. The isolation of the outflow section of the reservoir from the incoming fluid, as provide by the baffles arrangement, guaranteed that the height of the fluid column driving the flow was constant during the measurements, thereby assuring the stationary flow. Finally, a contribution of the flow-generating system to the observed effects of valve was excluded.

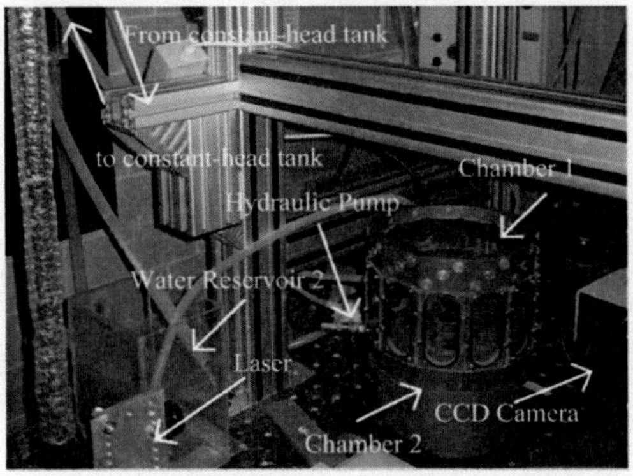

Fig. 1 Hydraulic flow circuit

The axes for the experiments are illustrated in Fig.2. The central plane of valve defined as z=0. The positive y was identified from the edge of valve holder and into the upstream chamber. The positive x axis was defined from centre of valve to right side of valve. In measurement system, the optical access provided flow field 3mm away from valve seating in the atrium side.

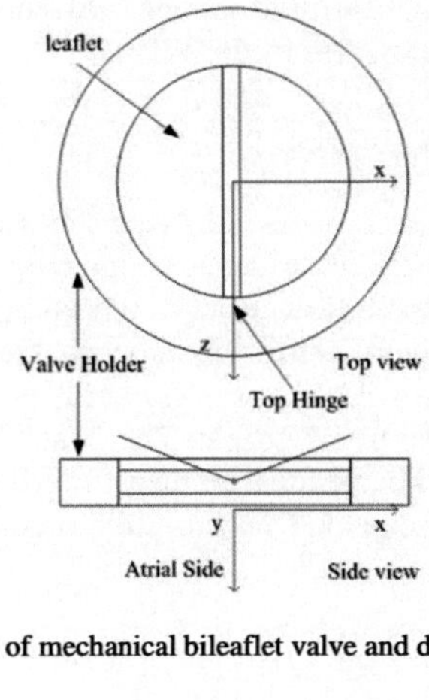

Fig. 2 Sketch of mechanical bileaflet valve and definition of axis

The blood analogue fluid was distilled-water (viscosity of 1cP and a density of 1.0 g/cm^3 at room temperature). To accommodate PIV measurement, the fluid was seeded with neutrally buoyant silver-coated 10 µm diameter ($\rho = 1.1 g/cm^3$) hollow glass spheres. Each flow measurement was captured by a CCD camera in frame-straddle mode. A synchronizer was applied to control the laser pulses and frame acquisition of camera. In this experiment, double-frame images are needed to be recorded for each PIV analysis to determine to displacement of particles.

The laser pulses were obtained with a commercially available dual laser pulsed PIV system (Quantel, France). To illuminate double-frame images, two Nd:YAG lasers were used. Each laser generated a light beam in the infrared at a 1064 nm wavelength. This beam was subsequently passed through a frequency doubler that produced light at 532 nm in the green region of the spectrum. The Insight 3G PIV software controlled the two lasers triggered with the dual-frame image acquisition through the synchronizer. The shutter delay of camera could vary from 10 to 500 μs to provide double-frame images corresponding to each laser pulse. Beam combination optics combined with light sheet optics(cylindrical lens with a spherical focusing lens) were used to convert the two pulsed beams into pulsed light sheets with an average thickness of roughly 1 mm.

Applying a PIVCAM CCD camera, the PIV images were acquired with a resolution of 1,024*1,018 pixels and a sample frequency of 10 frames/s. A region of interest of 16mm*14mm was focused using a Nikon AF Micro-Nikkor 60mm lens.

B. *Analysis of Double-Frame Images*

The software Inside 3G was selected to deal with the analysis of double-frame images. The image acquisition procedure was already finished during the flow image measurements. For one cycle of measurement, 1000 double-frame images were captured. The extracted digital data was input to Insight 3G for next step analysis.

In the process of the image analysis, we used the Cross-Correlation algorithm to indentify the displacement of particles. The two signals analyzed are given by double-frame images, containing the information of the local displacement of particles, and they are not independent. In this project, FFTCorrelator was applied as the PIV cross correlation engine. It computed the correlation using Fast Fourier Transform (FFT). This algorithm required the spots to be square powers of two and have same size.

The grid engines of PIV cross correlation were Nyquist Grid and Recursive Nyquist Grid. Nyquist Grid was selected for fast results and it was the classic PIV grid. We controlled the size of interrogation window, and viewed the highest spatial frequency of velocity vector was limited by the size of interrogation window. Nyquist Grid set vectors with the x spacing equal to half width of interrogation window and the y spacing equal to half height of interrogation window. This gave a vector grid with 50% spot overlap fitting the Nyquist sampling criteria. Recursive Nyquist Grid was used for increased accuracy or higher spatial resolution, compared to the Nyquist Grid. It processed the images in two passes. The first processing pass computed the vector field at the starting spot sizes with Nyquist. The results of the first processing pass are used to optimize the spot offsets for the second processing pass. As the final size of interrogation window decreasing by a factor of two, the resolution of vector field increased by a factor of two.

Validation macro was applied to the vector field after each processing pass. In recursive processing two validation macros were used for each pass. Validation macro had functions of spurious vectors removing and holes filling through interpolation. Spurious vectors could be induced by the results of correlation of random pairing of particle images. Holes on the velocity field was caused by in-plane and out-plane motion or low seeding density with a low correlation signal strength.

In this section, the extracted data was applied to determine the fluid dynamics properties of the measured flow. By averageing of 1000 velocity field measurements, mean value of the velocity along x direction $\overline{U}_x$ and mean value of the velocity along y direction $\overline{U}_y$ were calculated in the first procedure. The root mean square values of the velocity fluctuation were determined as equation (1) along x direction and y direction.

$$u'_x = \sqrt{\overline{u_x^2}}, \quad u'_y = \sqrt{\overline{u_y^2}} \tag{1}$$

The turbulence shear stress (TSS) is defined as the equation (2).

$$TSS = \rho \overline{u'_x u'_y} \tag{2}$$

ρ is density of fluid.

However, TSS was not qualified to estimate the maximum TSS, because it was related to a fixed (x,y) reference system. To solve this problem, it was helpful to refer to a suitable coordinate system (ξ, η, ζ), in which the TSS tensor was cast in a diagonal form:

$$T = \rho \overline{u'_i u'_j} = \begin{pmatrix} \sigma_1 & 0 \\ 0 & \sigma_2 \end{pmatrix} (i, j = \xi\eta\zeta). \tag{3}$$

where σ_1, σ_2 (in the conventional ordering scheme, $\sigma_1 > \sigma_2$) were the two principal in-plane normal stress(PNS). These two PNS were calculated as equation (4).

$$\sigma_1 = \rho \max_{2D}\left\{\overline{u^{*2}}\right\}, \quad \sigma_2 = \rho \min_{2D}\left\{\overline{u^{*2}}\right\} \tag{4}$$

where u^{*2} was a squared velocity fluctuation component in an arbitrary direction in the (x,y) plane and could be determined by the Eigenvalue of the tensor matrix.

As it is known from stress analysis (Malvern, 1969), the maximum TSS is defined simply by the equation (5).

$$TSS_{max} = \frac{\sigma_1 - \sigma_2}{2} = \frac{\rho}{2}\left(\max_{2D}\left\{\overline{u^{*2}}\right\} - \min_{2D}\left\{\overline{u^{*2}}\right\} \right) \tag{5}$$

The maximum TSS could be used to evaluate the possibility of hemolysis occurring in blood flow[9].

III. RESULTS AND DISCUSSION

A study of flow field of nozzle and hinge leakage of mechanical bileaflet valve was conducted applying PIV. The nozzle and valve were seated in the centre of the test chamber. The slices of flow field were scanned by shifting laser sheets as 1mm for nozzle and 0.5mm for valve respectively. The PIV data was recorded on (x,y) plane, and z axis was the shift direction of laser sheet. The average velocity fields of flow, illustrated in Fig 6 and Fig 9, were calculated from 1000 measurements on each scan. During the whole procedure of scanning, the valve was kept in closde state, in order to rule out the possibility of leaflets' different repositioning, after successive closures of the valve.

A. *Nozzle Measurements*

An axysymmetric nozzle has been built according to the standard EN ISO 5840:2006. It is currently been used as a reference for leakage flow testers. According to the results of nozzle measurement, a strong symmetrical jet with a shape of top-hat was shown. The jet was gradually decreased as being away from the centre of nozzle. The quantities of properties of flow field for nozzle measurement was illustrated in Fig.4. The maximum value of TSS was 30Pa according to Fig.4(c).

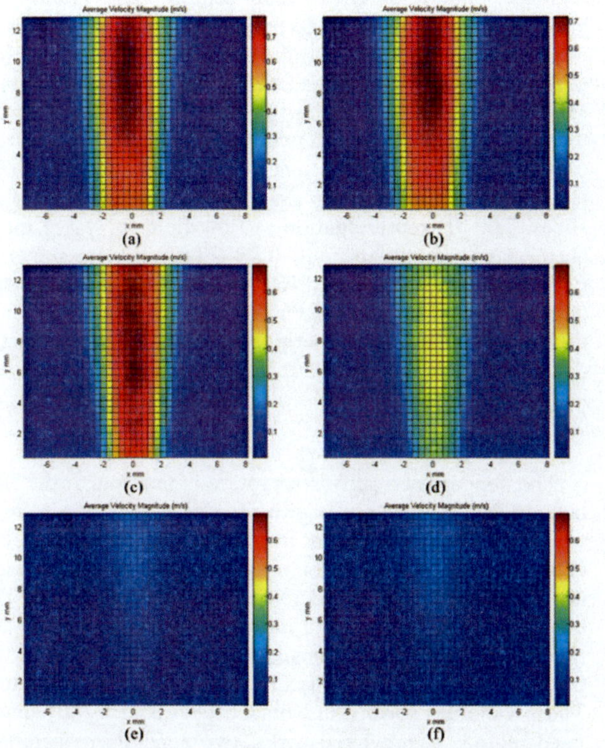

Fig. 3 Average velocity field for nozzle measurement: (a) Z=0mm, (b) Z=1mm, (c) Z=2mm,(d)Z=3mm,(e)Z=4mm,(f)Z=5mm

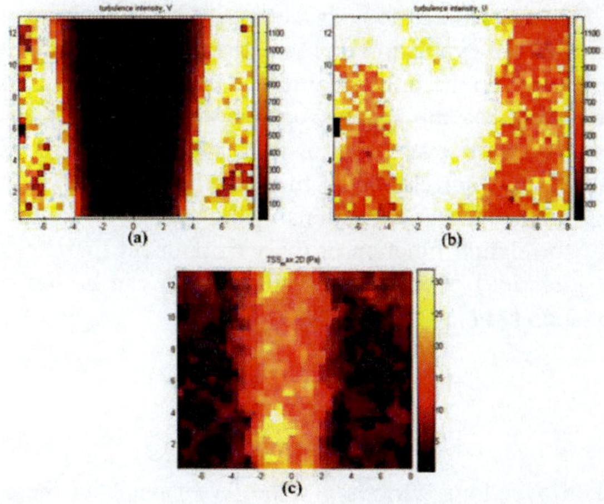

Fig. 4 Quantitative properties of flow field from nozzle measurements on Z=0mm: (a) Turbulent Intensity along Y axis, (b) Turbulent Intensity along X axis, (c) TSS_2D

B. *Valve Measurements*

Fig. 5 shown hinge jets on different slices from 0mm to 7mm as centre z=0mm at hinge plane. From the results of the hinge leakage measurement of valve, two strong jets were clearly observed on each side from the leakage of left and right hinges. The maximum of velocity of hinge jets was 0.17m/s shown in Fig. 5(a) and two slight peripheral leakage jets were observed on each side of flow field.

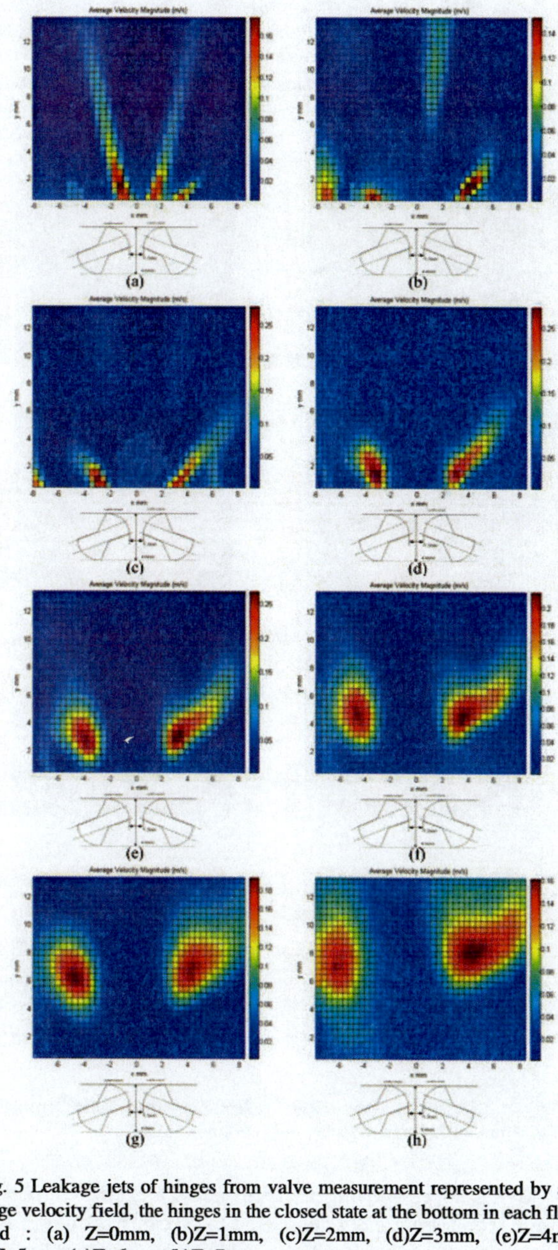

Fig. 5 Leakage jets of hinges from valve measurement represented by average velocity field, the hinges in the closed state at the bottom in each flow field : (a) Z=0mm, (b)Z=1mm, (c)Z=2mm, (d)Z=3mm, (e)Z=4mm (f)Z=5mm, (g)Z=6mm, (h)Z=7mm

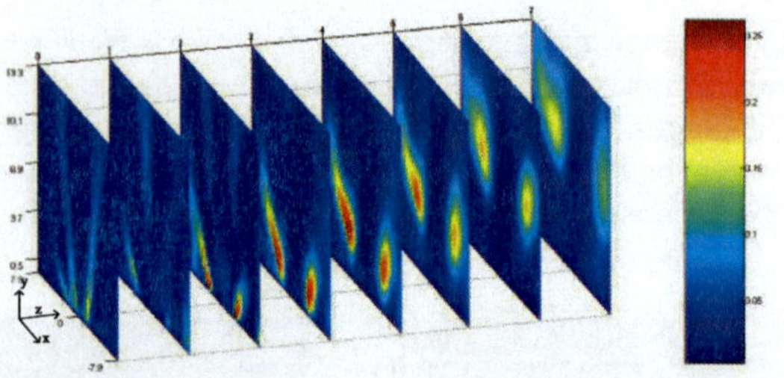

Fig. 6 3D temporal representation of flow field according to 2D slice measurements of valve

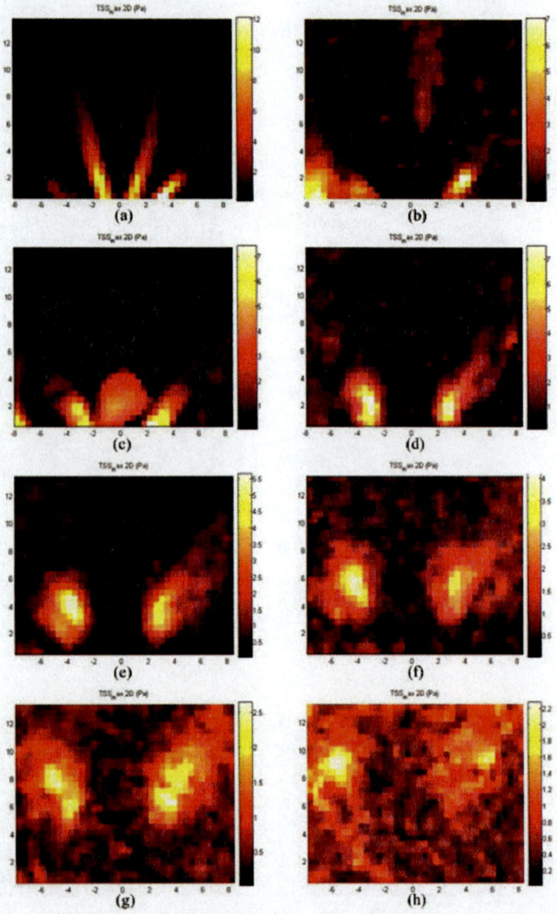

Fig. 7 TSS_2d of flow field: (a)Z=0mm, (b)Z=1mm, (c)Z=2mm, (d)Z=3mm, (e)Z=4mm, (f)Z=5mm, (g)Z=6mm, (h)Z=7mm

As the slices moving to 1mm and 2mm, the velocity of hinge jets dramatically decreased to 0.05m/s shown in Fig.5 (b),(c). On the other hand, the velocity of peripheral leakage jets increased to 0.28m/s, but its spread was limited in a small region at -3 and 3mm away from central plane. Finally, the hinge jets gradually vanished on the slice at 3mm, illustrated in Fig.5 (d). In the contrast, the peripheral leakage jets kept strange tendency and spread to a larger region with more structured shapes. From slices 4-7mm shown as Fig.5(e)-(h), the maximum velocity of two peripheral jets slightly decreased from 0.27 to 0.16 m/s. Additionally, the peripheral jets started to move along y axis from 0mm to 4mm on these slices. The 3D representation of flow field for valve measurement was shown in Fig.6. TSS_2d fields were shown in Fig.7 (a)-(h) for each slice. TSS of hinge leakage jests were decreased from 8 Pa to 0.1 Pa as slices changing from 0mm to 7mm. While TSS of peripheral leakage jests gradually decreased from 12Pa to 4Pa from slices 0mm to 7mm.

C. *Self-preservation*

The concept of a self-preservation indicates that the flow has reached a dynamic equilibrium or asymptotic state in which the mean and higher-order moments evolve together (Townsend, 1976). Mean moments indicate the mean velocity of flow and higher-order moments are referred as Turbulent Shear Stress and the mean kinetic energy of the turbulent velocity fluctuations. Typically, some scaled dependent variables are applied to represent self-preservation. These variables show a universal aspect over a fully developed region. In previous works, two general characteristic scales were frequently taken used to evaluate the self-preservation, which are the mean centerline maximum velocity and a jet's half width in a round shear jet[3].

However, these methods are reserved in the mean velocity field for the evaluation and lose the information of velocity fluctuation in the individual runs. Concerning in this aspect, we introduced the spatial correlations in the velocity fluctuation for the evaluation of jet self-preservation. In statistics, spatial correlation is the specified correlation in the signal along nearby locations in the space. The normalized autocorrelation function of the y component (parallel to the jet direction) of the velocity fluctuation can be defined as equation 6[11].

$$C_y(r) \equiv \langle u_y(0) \cdot u_y(r) \rangle / \langle u_y(0)^2 \rangle \qquad (6)$$

where $\langle \ldots \rangle$ indicates an ensemble average from 1000 runs taken over long periods of time. We calculated the spatial correlation along the distance r either in the direction parallel to the forward direction of jets, $C_y(y)$, or perpendicular to it, $C_y(x)$.

The inset of Fig.8 (a) displayed the direction of spatial correlation for nozzle measurements. We evaluated the spatial correlation of the jet central line on each scanned slice, shown in Fig.8 (b). In addition, we evaluated perpendicular direction of spatial correlation at the downstream of 0.5mm, 3.5mm, 6.5mm, shown in Fig.8 (c). These correlations exhibited a roughly exponential decay as a distance function. In Fig.8 (b), the correlations were convergent at 0.1 for all scanned slices. On the contrary, the perpendicular correlations were eventually convergent to 0 with less distance than parallel correlations. In combination, these indicated that the nozzle jets continuously kept a high level velocity at parallel directions, while they presented a rapid decay in the perpendicular directions. For three locations of perpendicular correlations, 0.5mm correlation showed the fastest decay and 6.5mm correlation showed the slowest decay along the distance. Owing to these observations, the explanation was that the velocity fluctuations gradually became more correlated along the radial direction, at increasing distance from the nozzle in Fig.8 (c). Reynolds Number for the central jet is 3.5×10^3.

In the set of Fig.9 (a), the spatial correlations for valve leakage jet were demonstrated. We evaluated two correlations along the parallel directions to the spread of hinge leakage jets as Cy(Y)1 and Cy(Y)2, shown in Fig.9(b). Furthermore, three downstream locations at 3mm, 7mm and 11mm were extracted for spatial correlations Cy(X) to assess the spread shape of leakage jet, illustrated in Fig.9(c). In comparison with parallel correlations of nozzle measurements, the correlations were convergent closely to 0 in Fig.9 (b). This observation illustrated that the valve leakage jets did not consistently keep as high level velocity as nozzle jets submitted. This evidence could be seen in Fig.9(a) as well. In Fig.9(a), a manifest velocity decay was shown around 5mm downstream locations. Through analyzed perpendicular correlations in Fig9(c), significant differences at three locations were supplied. At downstream location 3mm as Cy(X)3, the correlation curve exhibited a fast convergence at distance 1mm. At downstream location 7mm as Cy(X)2, the correlation curve was convergent at distance 2mm. In contrast, the correlation curve at location 11mm shown a very slow tendency of convergence. Compared to correlation of nozzle jets, valve leakage jets shown much tighter shape at the orientation and much looser shape at downstream away from the orientation. Reynolds Number for the hinge jet 1 is 0.42×10^3 and Reynolds Number for the hinge jet 2 is 0.35×10^3.

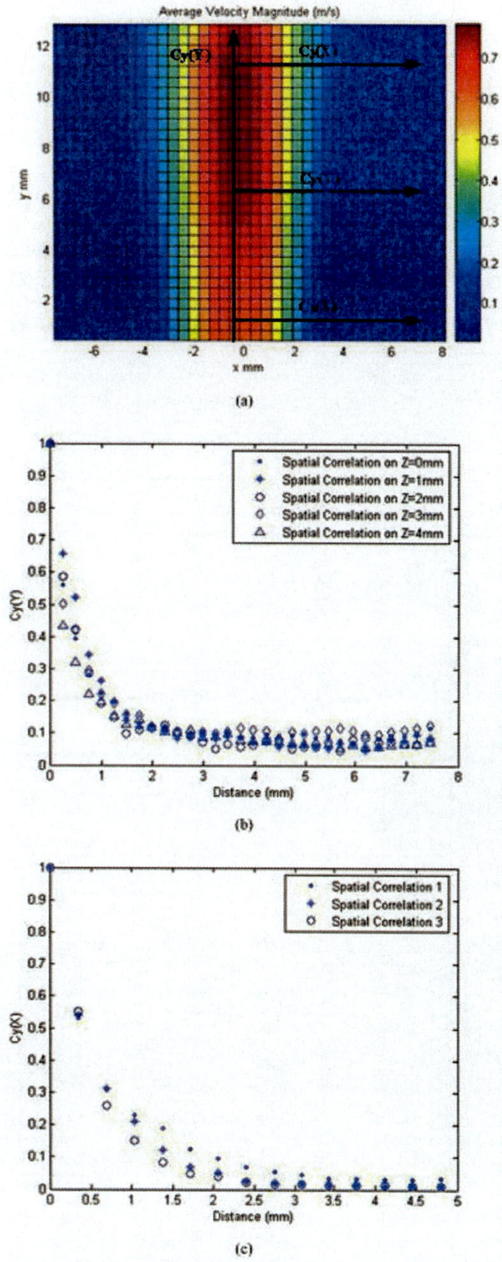

Fig. 8 Spatial correlation functions of the y component of the velocity fluctuation as a function of distance for nozzle measurement:(a)Illustration of correlation directions (b) parallel results (c)perpendicular results

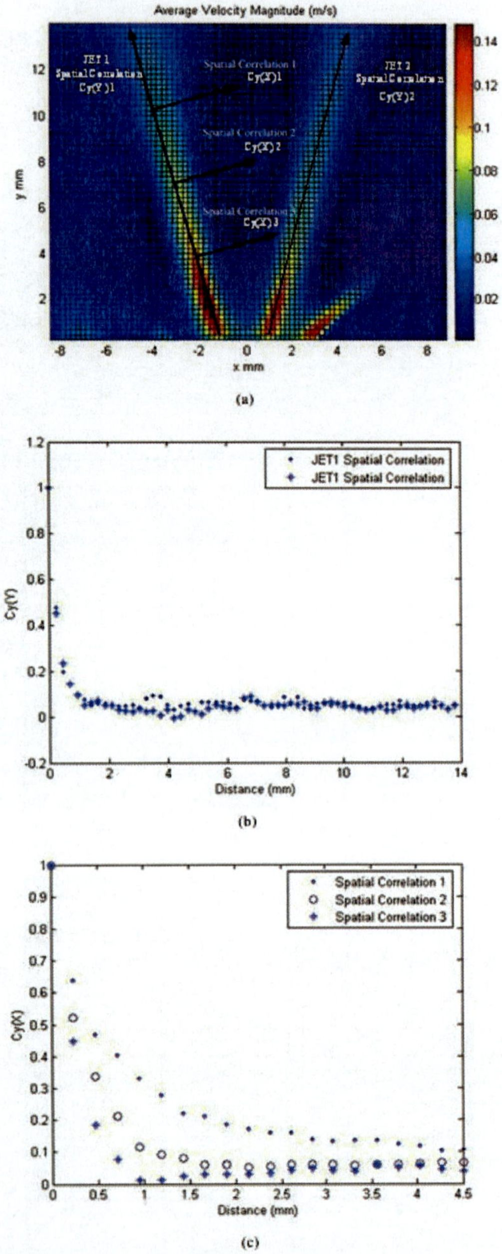

Fig. 9 Spatial correlation functions of the y component of the velocity fluctuation as a function of distance for valve measurement:(a)Illustration of correlation directions (b) parallel results (c)perpendicular results

IV. CONCLUSIONS

Particle image velocimetry was applied to investigate leakage flow fields associated with a standard nozzle and a mechanical bileaflet valve. The images revealed the leakage levers at correspond scanned slices. And these 2D slices were assembled together to present a 3D flow field. From it, a clear changing tendency of leakage jets was illustrated. According to average velocity measurements, the quantitative properties of flow field of leakage jets were supplied, especially the values of TSS_2D for evaluation of possibilities of blood components' damages. TSS_2D around the regurgitant jets were no higher than 12Pa, suggesting that they would not induce hemolysis in the PIV observable part of the spatial distribution of velocities and stresses. According to reference [11], TSSmax up to 180Pa in regurgitant jets of valve may contribute to hemolysis.

In this work, we also introduced the method of spatial correlation to characterize the development of leakage jets. This method was proved as an effective evaluation for the comparison of different kinds of leakage jets.

ACKNOWLEDGMENT

This research is funded by the European Commission, through the MeDDiCA ITN (www.meddica.eu, PITN-GA-2009-238113), Marie Curie actions under FP7, People Pro-gramme. The authors are very grateful to the MeDDiCA project and Marie Curie actions for financial support. Our grant agreement is PITN-GA-2009-238113.

REFERENCES

1. Manning KB, Kini V, Fontaine AA, Deutsch S, Tarbell JM. Regurgitant Flow Field Characteristics of the St. Jude Bileaflet Mechanical Heart Valve under Physiologic Pulsatile Flow Using Particle Image Velocimetry. Artificial Organs 2003;27(9):840-846
2. Sachin KD, Mayur JS, Jyeshtharaj BJ. Investigation of flow and temperature patterns indirect contact condensation using PIV, PLIF and CFD. Chemical Engineering Science 2010;65: 4606-4620
3. Choo YJ, Song CH. PIV measurements of turbulent jet and pool mixing produced by a steam jet discharge in a subcooled water pool. Nuclear Engineering and Design 2010;240:2215-2224
4. Silva G, Leal N, Semiao V. Micro-PIV and CFD characterization of flows in a microchannel: Velocity profiles, surface roughness and Poiseuille numbers. International Journal of Heat and Fluid Flow 2008;29:1211-1220
5. Chung SK, Kim SK. Digital particle image velocimetry studies of nasal airflow. Respiratory Physiology & Neurobiology 2008;163:111-120

6. P.J.Bryanston-Cross and K.S. Chana, PIV Measurements Made in the Stator Trailing Edge Wake Rotor Region in an Annular Transonic Cascade, Proc SPIE 3172.90, Paper presented at the SPIE Conference Optical Technology in Thermal andCombustion Flow, San Diego 27 July - 1st August 1997 SPIE Vol 3783 pp332-8
7. Towers, CE; Towers, DP; Jones, JDC Absolute Fringe Order Calculation Using Optimised Multi-Frequency Selection in Full Field Profilometry. Optics and Lasers in Engineering, vol. 43, pp.788-800.
8. Ellis JT, Travis BR, Yoganathan AP. An In Vitro Study of the Hinge and Near-Field Forward Flow Dynamics of the St. Jude Medical® Regent™ Bileaflet Mechanical Heart Valve. Annals of Biomedical Engineering 2000;28: 524–532
9. Grigioni M, Daniele C, D'Avenio G, Barbaro V. On the monodimensional approach to the estimation of the highest Reynolds shear stress in a turbulent flow. Journal of Biomechanics 2000;33:701-708
10. Segrè PN, Herbolzheimer E, Chaikin PM. Long-range correlation in sedimentation. Physical Review Letters 1997;79:2574-2577
11. Meyer RS, Deustch S, Bachmann CB, Tarbell JM. Laser Doppler velocimetry and flow visualization studies in the regurgitant leakage flow region of three mechanical heart valve. *Artif Organs* 2001;25:292–9.

Author: Guanglei Wang
Institute: Istituto Superiore di Sanità
Street: 299 Viale Regina Elena 00161
City: Rome
Country: Italy
Email: guanglei.wang@iss.it

Peripheral Vascular Measurement Using Electrical Impedance Plethysmography

C. Corciova[1], R. Ciorap[1], R. Matei[2], and A. Salceanu[2]

[1] "Gr. T. Popa" University of Medicine and Pharmacy Iasi, Romania
[2] "Gheorghe Asachi" Technical University of Iasi, Romania

Abstract— **Impedance plethysmography (IPG) is a safe and noninvasive method for measuring peripheral hemodynamics. This paper's aim is to develop a medical device system for a continuous monitoring of this parameter using the impedance technique. This system combines the advanced analog amplifier with the calculation power of digital signal processing to acquire real time monitoring. Our bioimpedance measuring system uses a generator which is under microprocessor control. Experimental investigations are conducted in order to determine the optimum interrelation between current injector output characteristics, power supply and electrode spacing.**

Keywords— **impedance plethysmography, microcontroller, resistivity, algoritm, blood flow.**

I. INTRODUCTION

Numerous attempts and methods have been developed in order to measure pathologic and functional vascular change in the extremities. Little attention has been given to the quantitative changes recordable by various methods of electrical plethysmography, which until recently have been difficult to formulate.

The electrical conductivity method gives a physical measure of the ionic conduction of a given body segment in contrast with electronic conduction characteristic of metallic substances. An attempt will be made to restate and formulate the laws of electrical conduction as they appear applicable to changes within a body segment which is being studied by the passage of a radio frequency current. As has been shown in literature, transient and static values of electrical conductivity are associated, respectively, with dynamic and balanced conditions of arterio - venous blood volume differences within a given segment. [3]

The electrical impedance pulsation represents a changing number of ions brought to the segment by the arterial stream at a rate exceeding the venous outflow during the cycle. The overall change in volume of a segment is the differential effect of expansion and emptying of the vascular components of the entire segment. It may be possible to account segment ally for the volumetric shift of blood by considering its effect as a variable parallel electrical shunt. [5]

Equations for the effect of parallel resistance are well known. The total electrical conductance of the extremity segment is probably equal to the sum of the paralleled conductances of the blood and the corresponding segment. Each additional pulse of blood represents another path through which electrical current will flow. The effective parallel resistive value of the added or displaced blood may be derived by substitution of measured values in the expression:

$$R_{blood} = \frac{R_{total} R_0}{R_0 - R_{total}} \text{ or } R_{blood} \approx \frac{R_0^2}{\Delta R} \tag{1}$$

in which R_0 represents the original resistance of the segment, R_{total}, the new total resistance. R_0- R_{total}, which is equal to ΔR, represents the change in resistance incurred by the change in blood volume of the segment, by pulsation or otherwise.

When small volume and resistive changes occur, then R_0^2 expresses essentially value of the product $R_{total}R_0$. The volume of blood within the segment is a direct and linear function of electrical conductivity. This is true within wide limits of expansion of elastic cylinders such as arteries, veins, intestines, and rubber tubes. The inclusion of ground meat, long bone, or ground bone changes the slope of the relationship but does not destroy the lineal effect. The change in volume of blood uniformly distributed within a segment may then be calculated from the derived expression of the volume of a cylindrical conductor:

$$V_b = \rho \frac{l^2}{R_b} \tag{2}$$

where (ρ) represents the specific resistivity of the segmental blood, (l) the length of the segment being measured, and R_b the calculated effective parallel resistance of the blood related to the change.

The volume of blood pulsed into a peripheral segment is usually equal to the volume of blood leaving during the cycle. If one had an accurate measure of either the input or output, or of both volumes, it is probable that valuable data covering vascular responses could be scientifically expressed. If the measured value is closely proportional to the

S. Vlad and R.V. Ciupa (Eds.): MEDITECH 2011, IFMBE Proceedings 36, pp. 136–139, 2011.
www.springerlink.com

absolute pulse volume, it would not be necessary to know either phase of the volume change. [7]

II. Material and methods

Impedance plethysmography is the measurement of volume changes through the measurement of the electrical impedance of the body tissues of interest. When measuring in limbs, these volume changes are mainly due to the blood flow. Common plethysmography measurements in a limb use four band electrodes placed around it to obtain a homogeneous electric field in the target volume. The signal detected was more affected by electrode placement, which may be important to estimate blood flow, but it is not essential for heart rate estimation. The distance (D) between the current injection electrodes should be as large as feasible, so as to cover as much area as possible of the leg segment being measured. The optimal separation distance between voltages - detection electrodes is not known in advance.

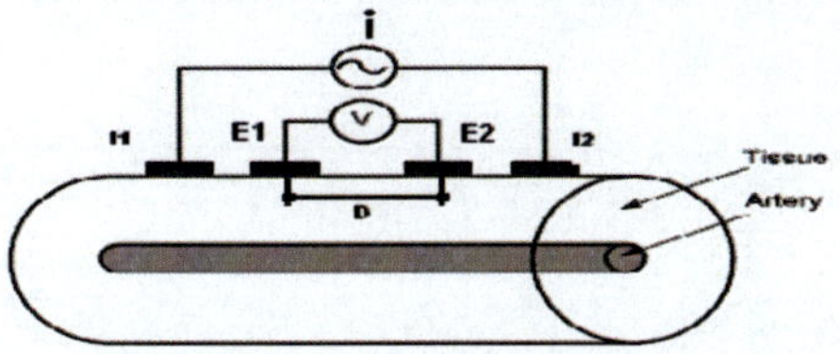

Fig. 1 The tetrapolar arrangement. The inner electrodes are designated E_1, E_2, and the outer electrodes I_1, I_2. The segment D (cm) in length between E_1 and E_2 is the effective resistance (R_0).

Artery expansion during heart systole increases its cross section area from S_0 to $S_0+\Delta S$ and the arterial impedance of this segment decreases correspondingly from Z_0 to $Z_0 - \Delta Z$. The fractional variation is thus $\frac{\Delta Z}{Z_0} = -\frac{\Delta S}{S_0}$.

The functional diagram of the implemented experimental embodiment is illustrated in Figure 2

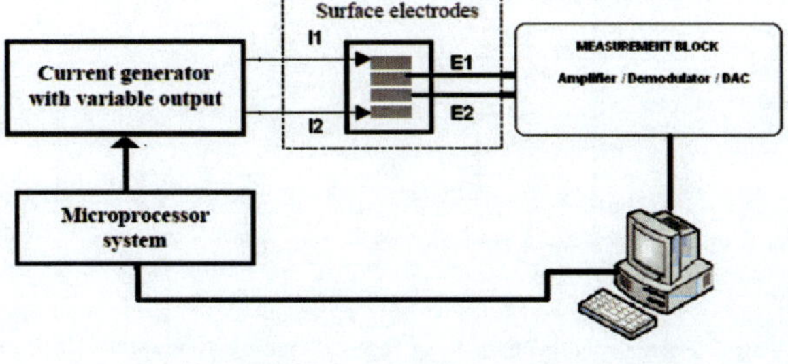

Fig. 2 System for bioimpedance measurement

Essentially, it represents the bioimpedance measuring system device for the purposes of the present paper. Surface electrodes applied to peripheral vasculature form the interconnection to the patient. The measurement method involves providing of a periodic current excitation signal to the body part via a pair of current injecting electrodes. The flow through the body electrical current develops a potential corresponding to the pulsatile impedance variation of the tissues in accordance with the blood circulation. A second electrode pair, positioned between the first electrodes, senses the pulse wave related voltage changes.

The bioimpedance measuring system includes a microprocessor controlled current generator with variable output. The microprocessor system drives the generator's output stage under program control. We used microcontroller MSP430F169 because the total power consumption is reduced. The implemented microprocessor program control provides easy frequency adjustment in the range between 10 kHz to about 200 kHz. The MSP430 has 16 registers, and the ability to perform arithmetic directly on values in the memory. C compilers can be from this and produce more compact, efficient code. Texas Instruments issued a competitive benchmarking document that contains a comparison of the MSP430 with a range of other microcontrollers. Their performance was measured by compiling and executing the source code of a bunch of frequently used applications. [20]

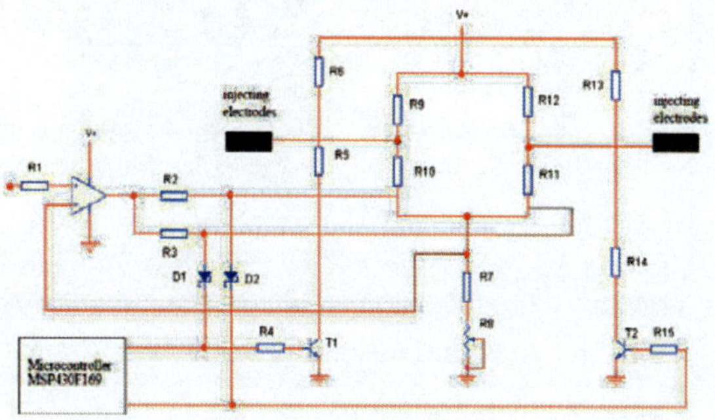

Fig. 3 Circuit diagram of the current generator with microcontroller

An amplitude detector follows and rectifies the modulated signal. After removing the DC component from the signal, an amplification of the resulting pulse wave takes place. The output signal thus obtained, which is indicative of the bioimpedance being measured and hence of the blood flow in the body segment, is visualized by a computer. [12]

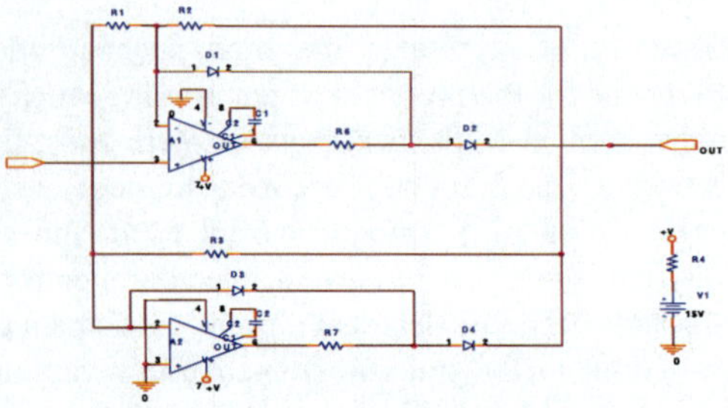

Fig. 4 Schematic diagram for demodulator

The algorithm for calculating the hemodynamic variables was implemented in MATLAB. The majority of the data processing and analysis was done in MATLAB to simplify software development. The system prompts for data inputs and then calculates hemodynamic parameters from the impedance signal. A measurement was made on 20 volunteers, clinically normal. Limb blood flow was calculated as flow / min, flow / beat and as a percentage of the stroke volume from the contemporaneously measured cardiac output. Every calculation was performed as an average over 6 heart beats.

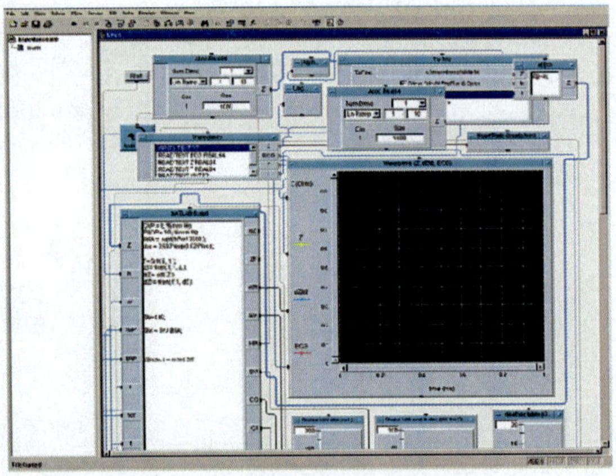

Fig. 5 User interface in MATLAB

III. Results and discussion

The recorded pulse volume is directly proportional, but not necessarily equal to the sum of the true arterial inflow and the venous outflow from a given segment. It follows that the mean height measurement of the pulse wave should be a valid index of this proportional volume. In our study, twice the mean height for the area under the curve is chosen to represent both input and output volume. In effect, this represents a sequestration of the total segmental input without occlusion or run - off of the venous return.

In practice, one obtains the mean height of the pulse volume by planimetric integration over the entire pulse distance. The measurement of several pulses serves to reduce the error. This value is multiplied by two, since the recorded volume increase served both as a measure of input and of output volumes during the entire pulse cycle. The four electrode method basically eliminates the reactive skin and leaves one with a better measure of the internal tissues, including the blood, which appear predominantly resistive to alternating current. The pulsatile volume should probably be calculated on the resistive values obtained by tetrapolar leads, if a closer approximation to the true proportional pulse volume is desired.

The figures below present plethysmography waveforms obtained for different values of distance (D) between the collecting electrodes E_1 and E_2 from the same volunteer.

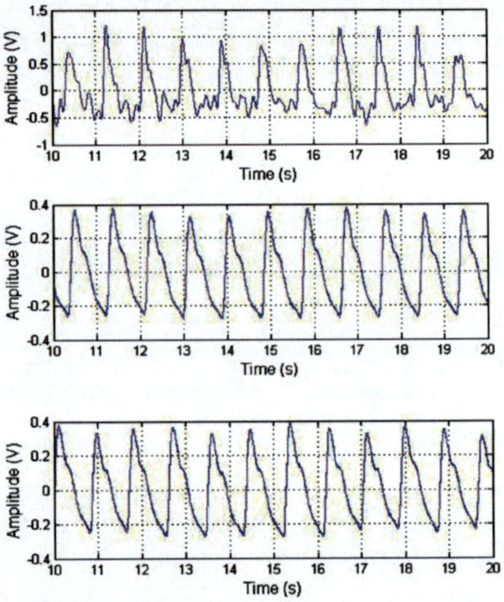

Fig. 6 Plethysmography wave function of the distance (D/2), (D) and (2D) between the E_1 and E_2

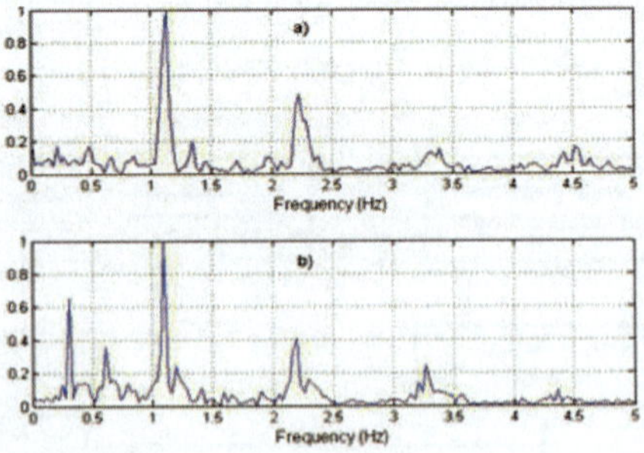

Fig. 7 Power spectral density of the plethysmographic wave: in figure a) for distance (D) and in figure b) for distance (2D)

Table 1 Limb blood flow measurements in normal limbs

	Arm	Leg
% stroke volume	1.10 – 5.00	1.70 – 6.56
Flow / beat (ml)	0.95 – 4.30	1.58 – 7.10
Flow / minute (ml)	92 - 283	103 - 480
Z_0 (ohm)	42.0 - 81	37 - 70

The table 1 shows some calculated values of blood flow.

The system needs to be calibrated in order to be able to calculate meaningful hemodynamic parameters from this data. In order to calibrate this system, a fixed resistor with known impedance would be measured by the system.

IV. Conclusions

In this paper, an impedance plethysmography system for real time noninvasive cardiac output monitoring is developed and evaluated. This system combines the advance analog amplifier with the calculation power of digital signal processing; it is capable of detecting events in the incoming signal under different circumstances. Measurements of limb blood flow may be expressed as flow/min, flow/beat, as proportion of the flow in a control limb. The relative ease of impedance plethysmography facilitates the measurement of limb blood flow relative to the stroke volume or to flow in a control contra lateral limb, when changes in blood flow due to local factors are being monitored.

References

1. Bronzino J.D. (2000), The Biomedical Engineering Handbook – second edition, CRC Press in cooperation with IEEE Press, Boca Raton, Florida
2. Woltjer H., Van der Meer, (1995) Comparison between spot and band electrodes and between two equations for calculation of stroke volume by means of impedance cardiography, Medical Biological Engineering Computing
3. *Payne R. A, D. Isnardi, P. J. D. Andrews, S. R. J. Maxwell and D. J. Webb (2007),* Similarity between the suprasystolic wideband external pulse wave and the first derivative of the intra-arterial pulse wave, British Journal of Anesthesia 99(5).
4. Semmlow, J. L. (2004), Biosignal and Biomedical Image Processing. MATLAB – Based Applications. Marcel Dekker Inc., New York.
5. Brown, B. H., Wilson, A. J., and Bertemes-Filho, P. (2000): Biomedical electronics Bipolar and tetrapolar transfer impedance measurements from volume conductor, Electron. Lett., 36.
6. Ederle J., Blanchard Susan, Bronzino J.D. (2005), Introduction to Biomedical Engineering, ELSEVIER Academic Press.
7. Tanaka S., M. Nogawa and K. Yamakoshi (2006), Fully automatic system for monitoring blood pressure from the volume-oscillometric method, in 27th Annual International Conference of the IEEE Engineering in Medicine and Biology Society.
8. Webster J. G, Medical Instrumentation: Application and Design.
9. Bertemes-Filho P., B. H. Brown, R. H. Smallwood, A. J. Wilson,(2001), Tetrapolar Probe Measure Sensitivity Distribution be Improved, Proc. 11th Int. Conf. Electrical Bio-Impedance (ICEBI).
10. Kubicek W. G., (1989), On the Source of Peak First Time Derivative (dZ/dt) During Impedance Cardiography, Annals of Biomedical Engineering, 17: 459 - 462.
11. Ferreira J., J. J. Sanchez, (2007) Electrical Bioimpedance Measurement System for Limb Edema Monitoring, School of Engineering. Vol. MSc Borås: University College of Borås.
12. Corciova C, Zaharia D, Ciorap R, Matei R (2010), A non invasive system for cardiac output measurements and hemodynamic analysis, Conf Proc. EPE 2010, International workshop on Electromagnetic Compatibility and Engineering in Medicine and Biology, Iasi, 139 - 142;
13. Seoane F, Ferreira J, Sánchez J J, Bragós R (2008), An analog front-end enables electrical impedance spectroscopy system on-chip for biomedical applications, Physiological Measurements 29.
14. Ifeachor, E. C (2002), Digital signal processing: a practical approach, Addison Wesley Publishing Company.
15. Arpinar, V.E. Eyuboglu, B.M., (2001), Microcontroller controlled, multifrequency electrical impedance tomograph,Engineering in Medicine and Biology Society, Proceedings of the 23rd Annual International Conference of the IEEE,25-28.Oct.2001, vol 3, pp. 2289 – 2291.
16. M.Min, T. Parve, V. Kukk, V., A. Kuhlberg, (2002), An implantable analyzer of bio-impedance dynamics: mixed signal approach, IEEE Transactions on Instrumentation and Measurement, Aug.2002, vol. 51, issue 4, pp. 674 – 678.
17. O.Märtens, (1999), DSP-Based Device for Multifrequency Bio- Impedance Measurement, Medical & Biological Engineering & Computing, vol.37,1999, Supplement 1: Proc. Of the 11th Nordic- Baltic Conf.on Biomedical Engineering NBCBME'99, Tallinn, Estonia, 6-10 June 1999, pp.165-166.
18. Gabriel S, Lau RW, Gabriel C. (2004), The dielectric properties of biological tissues, Parts I–III. Phys Med Biol: 2231–93. Online database at: http://niremf.ifac.cnr.it/tissprop/.
19. Areny R.P., Webster J.G., (1993) Bioelectric Impedance Measurements Using Synchronous Sampling, IEEE Trans. On Biomed. Eng., Vol. 40, No.8, pp. 824-829.
20. T. Palko, F. Bialokoz ,J. Weglarz, (1995) Multifrequency Device for Measurement of the Complex Electrical Bio-Impedance Design and Application, Proceedings RC IEEE-EMBS & 14th BMESI.
21. Texas Instruments Datasheets available at: http://www.ti.com/msp430

Autor: Calin Corciova
Institute: "Gr. T. Popa" University of Medicine and Pharmacy, Faculty of Medical Bioengineering
Street: M. Kogalniceanu, No 9-13
City: Iasi

Portable Complex PCG Signal Analyzer

S. Gergely[1], M.N. Roman[2], and R.V. Ciupa[2]

[1] National Institute for Research and Development of Isotopic and Molecular Technologies, Cluj-Napoca, Romania
[2] Technical University, Cluj-Napoca, Romania

Abstract— **There are many *PCG* signal characterization methods and algorithms but their practical implementation are yet limited. This paper presents a device that is more than a classical electronic stethoscope. Therefore the main goal is to make a signal characterization of the *PCG* signal and on that basis to be able to classify the acquired pathology information in accordance with a stored set of reference signals.**

Keywords— **Wavelet transforms, DSP, Shannon entropy, Phonocardiography, Euclidian distance.**

I. INTRODUCTION

The frequency domain spectral properties of the *PCG* signals are a good start for a pathology analysis and classification. The first important parameter is the frequency band of the *PCG* signal which is to be found between 80Hz and 750Hz. As an example shown in figure 1, the spectrum is mainly distributed at the low boundary of the frequency band and in the middle of it, regardless of the pathology signal.

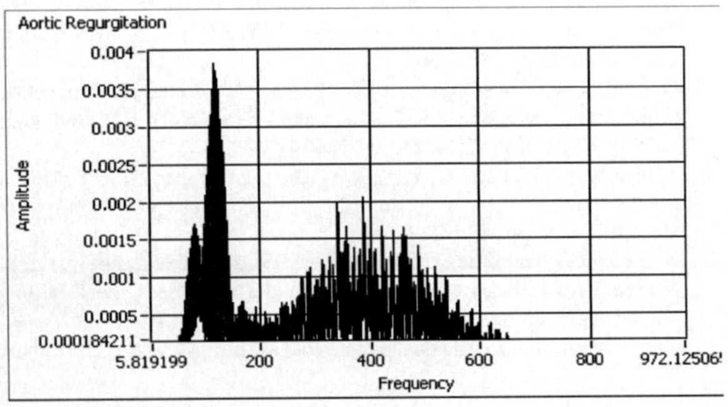

Fig. 1 Aortic regurgitation pathology signal frequency spectrum distribution

Therefore it turns out that a cepstrum analysis is not able to discriminate between let us say a "*(AR) Aortic Regurgitation*" activity and the "*(MR) Mitral Regurgitation*" pathology, even if the signals shown in figure 2, are different in an obvious manner.

What makes the difference is the distribution of the signal energy in the time domain proved by *Parseval's* relation presented in equation 1. This way the same spectral

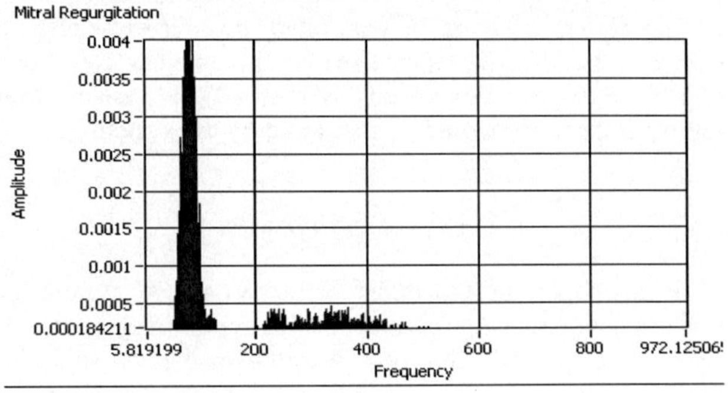

Fig. 2. Mitral regurgitation pathology signal frequency spectrum distribution

distribution of the signal may be generated by a different enve lope, involving only a limited amount of samples, but scattered at a different signal index.

$$\sum_{i=0}^{N-1} x[i]^2 = \frac{2}{N}\sum_{k=0}^{N/2} MagX[k]^2 \tag{1}$$

The best approach is to analyze the non-stationary properties of the signal combined with a time-frequency representation. The need for a combined time-frequency representation derives from the inadequacy of either time domain or frequency domain analysis to fully describe the nature of non-stationary signals. A time frequency distribution of a signal provides information about how the spectral content of the signal evolves with time, thus providing an ideal tool to dissect, analyze and interpret non-stationary signals. This is performed by mapping a one dimensional signal in the time domain, into a two dimensional time-frequency representation of the signal.

The wavelet representation is given in the space-frequency domain, opposite to the *Fourier* analysis that gives only a frequency representation. Compact supports of wavelets provide a space, and their oscillatory nature provides a frequency representation of a transformed function. It is clear that such representation is essential for the non stationary signal. The wavelet representation of a function has the multiresolution property, which means that it is given on several resolution scales. Trough the wavelet transform, the dimension of the data set that store information about the

S. Vlad and R.V. Ciupa (Eds.): MEDITECH 2011, IFMBE Proceedings 36, pp. 140–143, 2011.
www.springerlink.com

function is considerably decreased, while the most important information is not lost. This is very important for a good compression that saves storage in a system memory. Complementary to the wavelet transform is the computing of the *Shannon* Entropy resulted from each decomposition level in the multiresolution representation. The entropy information is important because the information density is different and specific as regards the decomposition level and the pathology signal. The simplest way to build a pattern recognizing algorithm is promising by using the *Euclidian* distance between all decomposition levels and a reference signal which is stored in a SD card memory. This approach has two major drawbacks. First of all the amount of required memory for the signal reference tend to grow with each decomposition level. Secondly it is known that the *Euclidian* distance is sensitive to the time shifting that may occur between the acquired signal and the reference one. The proposed algorithm overcomes the above mentioned issues.

II. Description of the method

Equation 2 shows the matrix of the decomposed signal using the L level wavelet transform applied to N samples. Each row of the matrix represents a decomposition level where the obtained coefficients are half the amount of the original signal or level. The device is acquiring 32768 samples with 16 bit/samples at an 8 KHz sampling rate.

$$W_L(N)=\begin{vmatrix} c_{11} & c_{12} & c_{13} & c_{14} & \cdots & c_{1N} \\ c_{21} & c_{22} & c_{23} & \cdots & c_{2N/2} & 0 \\ c_{31} & c_{32} & \cdots & c_{3N/4} & 0 & 0 \\ c_{41} & c_{42} & \cdots & c_{4N/8} & 0 & 0 \\ \cdots & \cdots & \cdots & \cdots & \cdots & \cdots \\ c_{L,1} & c_{L,2} & \cdots & c_{L,N-2^L} & 0 & 0 \end{vmatrix} \quad (2)$$

where $W_L(N)=W_{ap\_coef}(N)+W_{\det\_coef}(N)$ (2.a)

Equations 2.a shows that the signal matrix includes at the end both approximation and detail coefficients. All these coefficients are taking part in the *Shannon* entropy evaluation.

The first step of proposed algorithm does compute the Shannon entropy of wavelet decomposition shown in figure 3, throughout the *PCG* signal. The results are stored as Shannon coefficients which are specific to a given *PCG* pathology signal. Also the decomposition level of the signal which contains a dyadic number of coefficients is reduced to a single coefficient. All signal specific, computed *Shannon* coefficients are forming together a unique fingerprint which offer an information density characterization. Figure 4 show the Shannon entropy of the approximation coefficients for 10 reference pathology signals. As we can observe in figure 4 the *Shannon* entropy of the approximation coefficients does not characterize in a favorable way our pathology signals, because after the level 4 the usable information amount is very low. Although the approximation coefficients are embedded in the matrix of the signal because of the 300 percent variation of information amount, along all 10 analyzed pathological signals.

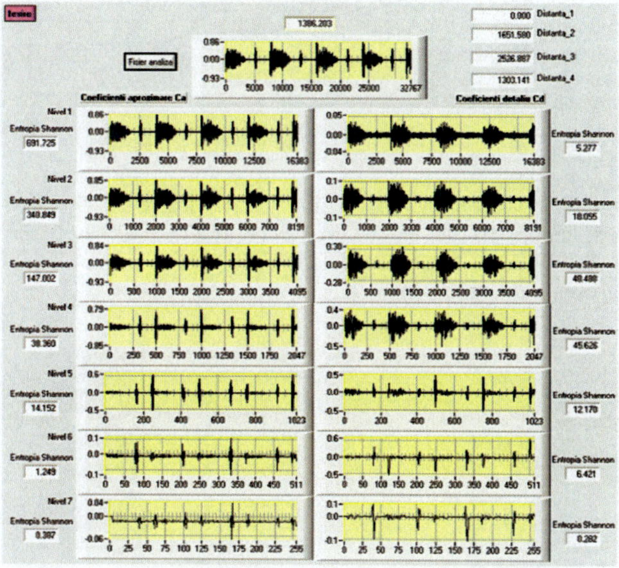

Fig. 3 Wavelet decomposition of a 7 level pathological *PCG* signal

The Shannon entropy of a signal is computed using equation 3.

$$H(X)=\sum_{i=1}^{n} p(x_i)I(x_i)=-\sum_{i=1}^{n} p(x_i)\log p(x_i) \quad (3)$$

After the signal decomposition, the level iterations contain a specific narrow signature reflected in the information content in a specific frequency band. At the end of the decomposition, the signal is characterized by a number of *Shannon*'s entropy coefficients. These coefficients represent the compressed form of the pathology signals. The pathology signals have therefore specific spectral fingerprints which are used to compute the *Euclidian* distance between the stored reference vectors and the currently analyzed signal. The *Euclidian* distance is computed using equation 4.

$$d(p,q)=\sqrt{\sum_{i=1}^{n}(q_i-p_i)^2} \quad (4)$$

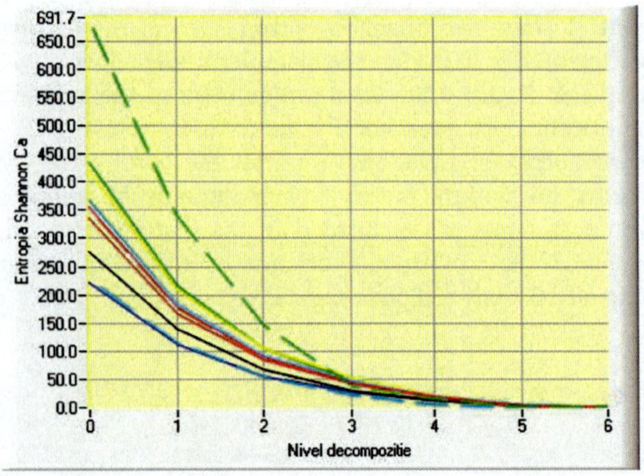

Fig. 4 Shannon entropy of approximation coefficients

Things are very different in figure 5, where the detail coefficients are making the difference. It turned to be obvious that the high frequency details have a high impact on the signals characterization.

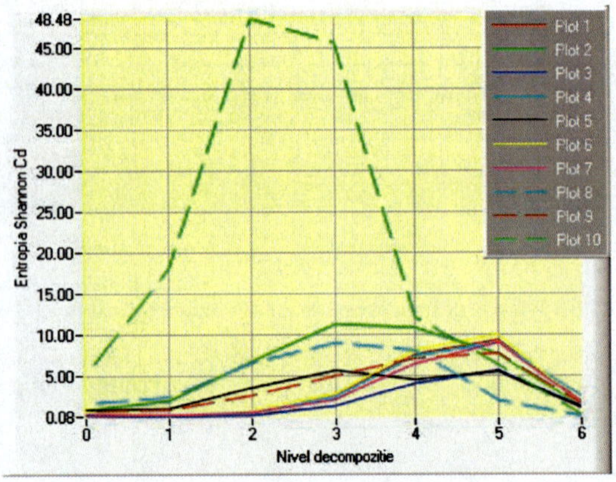

Fig. 5 Shannon entropy of detail coefficients

If there are an enough amount of acquired samples, consequently the *Euclidian* distance is less time shift sensitive. This was tested using a time shifted *PCG* signal which has a close *Shannon* distance value, to another pathology signal. The result is a slight increasing of the Euclidian distance, but after all the algorithm is searching for distance minima between the stored signal coefficients and the currently acquired signal. Figure 5 suggests also that the necessary highest decomposition level does not need to exceed the value 7 (level 6 in figure 5). The reason for that assumption is justified because after level 7 the entropy level drops very low with the meaning of low information content similar to the approximation details. The major advantage of using the *Shannon* entropy instead of the complex matrix shown in equation 2 is the reduction of the original signal to a set of coefficients which equals the number of decomposition level.

The second step of the signal characterization algorithm is a pattern recognition method, based on an envelope cross-correlation. The reference envelope signals are stored in the *SD* card. By using the data compression property of the wavelet's decomposition, the computed reference envelope does contain a reduced number of signal samples in comparison with the original signal. The main difficulty in the characterization of the *PCG* signal is the finding of a useful time/frequency reference. For that reason the reference envelope is a previously cropped signal cycle, from the specific *S1* and *S2* along with all murmur noises. The resulted reference signal cycle is translated over the whole signal at the certain decomposition level to create a cross-correlation. A simple sorting algorithm searches the resulted maxima through each pathology correlation coefficient. At the end of this process all reference envelops are characterized relatively to the acquired signals, therefore we obtain a pathology classification.

The above described algorithms are embedded in a microcomputer that has been build using two microcontrollers. Figure 6 shows the structural design of the device.

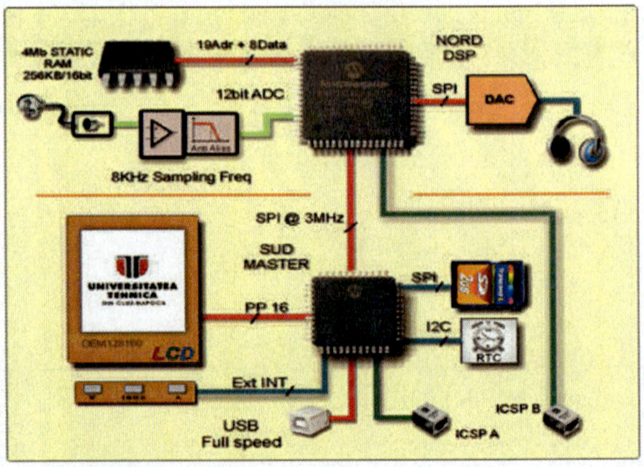

Fig. 6 Bloc diagram of the PCG signal analyzer

The hardware of the device is split in two functional modules. The acquired signal is filtered in real time to the necessary bandwidth by menu selection, using the north side dsPIC30F6012 *DSP* microcontroller. The acquired data is normalized to 16 bit by using the scaling factor given by the FBCL instruction. A very efficient way to suppress the unwanted alias signals is to increase the sampling frequency to at least 10 times the useful bandwidth. This way the order of anti-alias filter can be minimized. Using a clock speed of

120 MHz the 512 *FIR* taps filtering convolution is completed during 512 *MAC* instructions and spending a total time of 18us/sample. After bandwidth filtering, the same *DSP* engine is used for the wavelet transform by using the previously computed *Daubeschies 4* wavelet coefficients. Each filtering iteration is done first for the approximation coefficients and secondly for the detail coefficients. After the iteration is done the sample number is half of the original signal or the previous approximation coefficients. All decomposition levels are stored in the *SD* card. During acquisition the filtered *PCG* signal can be listening, using a pair of headphones. The resulted samples are stored in the 4Mbit static *RAM* for the necessary off line computing of the wavelet transform. The *Shannon* entropy and the Euclidian distance are computed by request after the displaying of the time-frequency scalogram. The signal scalogram is displayed as shown in figure 7 which consists of all wavelet decomposition levels displayed through interpolated values at level borders. Each signal amplitude is corelated to a given color at a certain decomposition level.

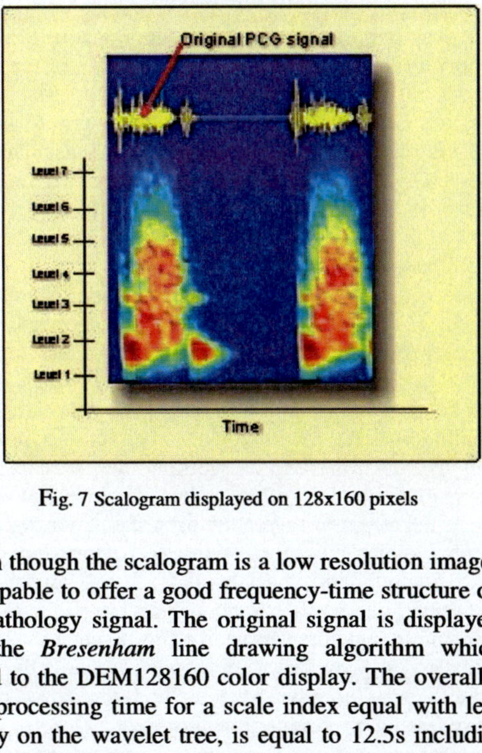

Fig. 7 Scalogram displayed on 128x160 pixels

Even though the scalogram is a low resolution image it is fully capable to offer a good frequency-time structure of the PCG pathology signal. The original signal is displayed by using the *Bresenham* line drawing algorithm which is adapted to the DEM128160 color display. The overall necessary processing time for a scale index equal with level 7 but only on the wavelet tree, is equal to 12.5s including al the read and write time to the *SD* memory card. Hence it is known that the *SD* card is a rather time-consuming storage device; condition that may possibly be improved by using a larger *RAM*, to reduce the *SD* card number of writing-reading cycles.

The second step of pattern recognition method using an envelope cross-correlation algorithm is extremely complex and therefore both microcontrollers are involved in computing the acquired heart rate, the signal envelope and the reference multirate sampling. The requirement of a fast communication channel between the two microcontrollers is solved by using the internal hardware high speed *SPI* module.

III. CONCLUSIONS

Although this new medical imaging device suffers from a low spatial and temporal resolution; it could be proved to be a good choice for low-cost and mobility strategy in cardiac imaging, rather than the expensive ultrasound imaging devices. The classical auscultation technique benefits from a great quality improvement by using a device which is capable to offer a time-frequency representation. The expected research direction is to be guided to improve the image quality on a larger display and increase the number of wavelet decomposing levels to avoid the necessary interpolations used in present. The cross-correlation algorithm is although under construction but the proved interest for this device among the cardiologists, makes us confident for the utility of the newly designed features.

REFERENCES

1. Abbas K. Abbas, Rasha Bassam, Phonocardiography Signal Processing, Morgan & Claypool Publishers, ISBN: 9781598299762
2. Desenka P. Radunovic, Wavelets from Math to Practice, 2009, Springer, ISBN 978-86-7466-345-5
3. Donoho D., 1992, Wavelet shrinkage and WVD: A 10 minute tour. Presented on the International Conference on Wavelets and Applications, Toulouse June 1992
4. Dwight F. Mix, Kraig J. Olejniczak, Elements of Wavelets for Engineers and Scientists, 2003, Willey-Interscience, ISBN 0-471-46617-4
5. Gilbert Strang, Truong Nguyen, Wavelets and Filter Banks, Wellesley-Cambridge Press, 1996, ISBN 0-9614088-7-1
6. Hans-Georg Stark, Wavelets and Signal Processing, 2005, Springer, ISBN 3-540-23433-0
7. Ingrid Daubeschies, Ten Lectures on Wavelets, 1992
8. Lokenath Debnath, Wavelet Transforms and Their Applications, 2002, Birkhauser, ISBN 0-8176-4204-8
9. James S. Walker, A Primer on Wavelets and their Scientific Applications, 1999, CRC Press, ISBN 0849382769
10. Mark Thuillard, Wavelets in Soft Computing, 2001, World Scientific, ISBN 981-02-4609-9

Author: Stefan Gergely
Institute: National Institute for Research and Development of Isotopic and Molecular Technologies
Street: Donath 65-103
City: Cluj-Napoca
Country: Romania
Email: stefan.gergely@itim-cj.ro

Monitoring Neuromuscular Fatigue – A Noninvasive Approach

M. Tarata[1], W. Wolf[2], D. Georgescu[1], D. Alexandru[1], and M. Serbanescu[1]

[1] University of Medicine and Pharmacy of Craiova/Medical Informatics, Craiova, Romania
[2] Universitaet der Bundeswehr Muenchen/ Institute for Communication Engineering, Neubiberg, Germany

Abstract— **Monitoring neuromuscular fatigue in fighter pilots is important and with serious effects. The limit conditions the pilots are submitted to during flight missions make monitoring neuromuscular fatigue (NMF) extremely important, as NMF significantly increases the risk of accident and affects the pilot's efficiency. We developed a noninvasive approach of monitoring the NMF, based on the compression of the Power Spectral Density of the SEMG (surface electromyogram) toward lower frequencies with increasing NMF, quantified by the Instantaneous Median Scale (IMedS) computed from SEMG via the Wavelet Transform. The Raa parameter (Area/Amplitude Ratio) computed on the SEMG was concurrently used. The paper provides a functional, practically oriented overview.**

Keywords— **Neuromuscular fatigue, Area/Amplitude Ratio, Wavelet, Electromyogram, Blood Oxygen Saturation.**

I. Introduction

Due to the limit conditions the pilots are submitted to during flight missions, monitoring NMF in pilots is important and with serious effects, due to the limit conditions the human organism is submitted to. Recent research [1,2,3,4,5], explored NMF in pilots flying on different military aircraft both on land and in mission. Neck, back leg and hand muscles - posturally and functionally important - were studied as mostly exposed to fatigue. The results confirmed the installation and development of fatigue in all studied muscles, mostly in the neck muscles, which may significantly increase the risk of accident and decrease the pilot efficiency in mission.

The power and resistance in the neck muscles are mostly important in fighter and test pilots, therefore extreme attention has to be paid to the recovery following the mission. Cervical injury is directly related to the fatigue in the neck muscles, which involve mostly the Trapezius and Sternocleidomastoidian muscles. These are accessible for surface electromyographic (SEMG) recordings.

Our approach of monitoring the development of NMF is based on the compression of the Power Spectral Density of the SEMG toward lower frequencies, from the beginning of the voluntary contraction, due to the reduction of the conduction velocity in direct relation with the muscular fiber membrane excitability and with neural adaptation. SEMG recorded simultaneously from the same muscles under steady contraction, shows a compression of the spectra toward lower frequencies since the beginning of the contraction [6,7].

In order to study transitional phenomena in muscle contraction and to monitor fatigue in steady and also in dynamic contraction, the use of WT has been investigated, via Instantaneous Median Frequency (IMedS) [8]. The use of WT was shown until now on a rather limited scale only for epochs where Fast Fourier Transform can also be consistently used, i.e. windows of signal where no acceleration or deceleration occurs, situation just occurring in the steady contraction or in some isokinetic exercise.

The Raa parameter (Area/Amplitude Ratio) [9,10,11,12] was concurrently used together with IMedS in the dynamic exercise.

Besides the electromyogram, the oxygen saturation (SaO2) is important in characterizing the muscular activity. Under physical stress the brain is supported only by an increase of blood flow while the circulation is distributed toward vital organs, therefore the price is the installation of the 'central' generalized fatigue.

Under physical stress, the muscle enters a status of oxygen debt. Under conditions of ATP exhaustion and a lack of oxygen, metabolic anaerobic mechanisms act to produce heat and toxic products e.g. the lactic acid, which in its turn has to be metabolized and this leads to muscle fatigue.

While it was reported that a direct effect of hypoxia in reducing the motor drive to the working muscles exists but this effect is moderate [13] other studies conclude that across the range of normoxia to severe hypoxia, the major determinants of central motor output and exercise performance switches from a predominantly peripheral origin of fatigue to a hypoxia-sensitive central component of fatigue, probably involving brain hypoxic effects on effort perception [14].

The paper provides a functional, practically oriented overview, briefly showing the method, the signal recording and processing, preliminary results.

II. Methods

An experimental chair demonstrator was built (Figure 1) to accommodate the subject in different positions. The

S. Vlad and R.V. Ciupa (Eds.): MEDITECH 2011, IFMBE Proceedings 36, pp. 144–147, 2011.
www.springerlink.com

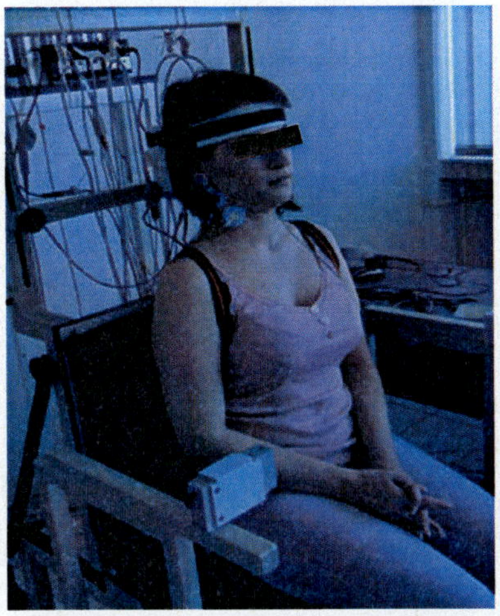

Fig. 1 Experimental chair demonstrator

subject was asked to push (extension) or pull (flexion) with the head against a force transducer.

The signal processing was run through an original application MTT_AEXON_PROC (Figure 2).

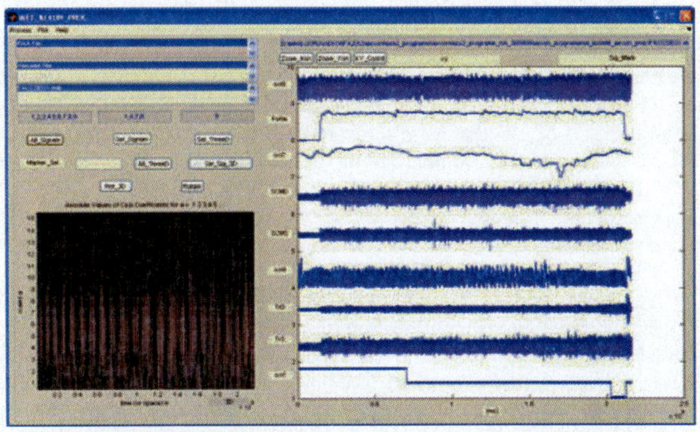

Fig. 2 MTT_AEXON_PROC - Signal processing window

The SEMG was recorded by a BIOPAC system (BIOPAC SYSTEMS INC, USA) from the left and right TR (Trapezius) and SCM (Sternocleidomastoidian) muscles in sustained contraction both in extension and flexion, at 50% MVC Maximum Voluntary Contraction). In the beginning of any experiment the MVC was estimated, the subject having been asked for a maximal contraction sustained for 5 seconds.

As the heart rate generally increases with sustained effort, it was also monitored together with the SaO2. The SaO2 and cardiac rate were taken with a probe fixed on the left ear (OXY100C, BIOPAC SYSTEMS INC, USA). The SEMGs, force signals, SaO2, heart rate were acquired at 1000 samples/second.

A. *Parameters*

The Raa, IMS and IMedS were computed from the original signals (SEMG) on successive epochs, for all the subjects, using a rectangular window.

Raa – Average Area /Amplitude Ratio, with a dimension of time [ms], is computed from the signal in the time domain, calculated from the SEMG between consecutive transversals of the isoelectric line, called 'phases' [9,10]:

$$Raa = \frac{1}{n}\sum_{i=1}^{n}\frac{S_i}{A_i} \qquad (1)$$

with:

n - the number of phases within the considered epoch,
S_i - current phase area, the integral of the ith phase of the signal within the current signal segment,
A_i - the maximal amplitude of the ith phase of the signal within the current signal segment, selected on all m samples within the current phase.

In the simplest case *n* equals 1 and the result is the instantaneous *Raa*.

IMedS (Instantaneous Median Scale) is computed via the Continuous Wavelet Transform (CWT). We use CWT as it works at every scale and preserves all the information in the signal *x(t)*. With CWT, the Ψ(t) MW is smoothly shifted over the full domain of the analyzed signal x(t):

$$CWT_x(s,\tau) = \int x(t)\Psi^{*}_{s,\tau}(t)dt \qquad (2)$$

where

s - scale,
τ - translation (time or space shifting),
$\Psi^{*}_{s,\tau}$ - is obtained by scaling the Ψ(t) MW over τ and s; higher scales correspond to lower frequencies.

The power density function or Scalogram is:

$$SCAL(\tau,s) = |CWT_x(s,\tau)|^2 \qquad (3)$$

where

$SCAL(\tau,s)$ - is the time dependent scalogram.

IMedS - The instantaneous median scale of the scalogram, localized at a specific time moment, computed as the actual scale value separating the scalogram into two surface regions, of equal area:

$$\int_{0}^{IMedS} SCAL(\tau, s)ds = \int_{IMedS}^{scal} SCAL(\tau, s)ds \quad (4)$$

where

scal - is the considered maximal scale considered within the transform.

B. *Data Processing*

The parameter vectors were linearly interpolated (slope+-standard deviation; intercept+-standard deviation). The two-way analysis of variance (ANOVA) was applied to the parameters.

C. *Results*

Raa and IMedS showed positive slopes in all subjects, from the beginning of the contraction, for all the epochs. This proved the central component of the fatigue.

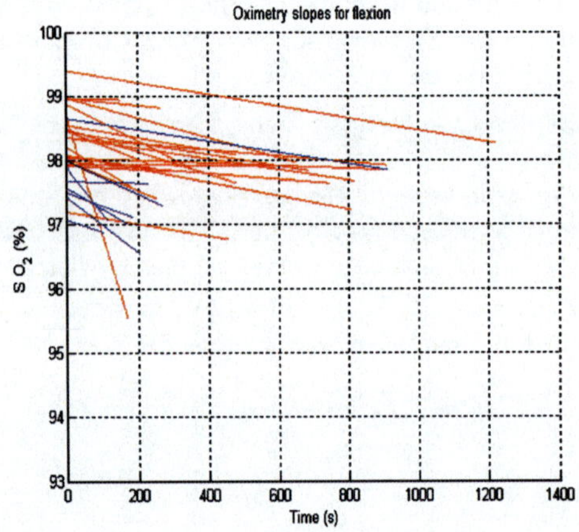

Fig. 3 The evolution of Blood Oxygen Saturation with advancing fatigue

Preliminary experiments run on 12 female and 12 male subjects showed a steady increase (positive slopes) of IMedS (2.88e-003+-3.89e-003 [scale/s]; 3.95 +-0.29 [scales]), Raa (1.23e-003 +-1.491e-003 [ms/s]; 1.88+- 0.22 [ms]) and a decrease (negative slope) of SaO2 (-3.3299e-003+- 9.5170e-003 [%/s]; 97.92+- 0.95 [%]) from the beginning of the contraction up to exhaustion. There are no significant differences between sexes. There are no significant differences between extension and flexion.

III. Conclusions

The preliminary results show that

(i) Exhausting contraction is associated with an increase of IMedS, Raa and a decrease of SaO2

(ii) SaO2 acts as an index of fatiguability, yet SEMG derived parameters are more sensitive than SaO2

Acknowledgment

This work was performed under research grant 82108/2008/CNMP/Romania and aimed a proper choice of techniques to address monitoring of neuromuscular fatigue of pilots. I am thanking the human subjects; without their commitment, these results could not have been possible.

References

1. Netto KJ, Burnett AF (2006), Neck muscle activation and head postures in common high performance aerial combat maneuvers, Aviation Space and Environmental Medicine, Volume 77, Issue 10, October 2006, 1049-1055
2. Ang B, Harms-Ringdahl K (2006), Neck pain and related disability in helicopter pilots: A survey of prevalence and risk factors, Aviation Space and Environmental Medicine Volume 77, Issue 7, July 2006, 713-719
3. Caldwell JA (2005), Fatigue in aviation, Travel Med Infect Dis. 2005 May;3(2):85-96
4. Thuresson M, Ang B, Linder J, Harms-Ringdahl K (2005), Intra-rater reliability of electromyographic recordings and subjective evaluation of neck muscle fatigue among helicopter pilots, J Electromyogr Kinesiol. 2005 Jun;15(3):323-31
5. Green NDC, Brown L (2004), Head positioning and neck muscle activation during air combat, Aviation Space and Environmental Medicine, Volume 75, Issue 8, August 2004, 676-680
6. Tarata MT. Mechanomyography versus Electromyography, in monitoring the muscular fatigue. Biomed Eng Online, 2:3, 2003
7. Madeleine P, Jørgensen LV, Søgaard K, Arendt-Nielsen L, Sjøgaard G. Development of muscle fatigue as assessed by electromyography and mechanomyography during continuous and intermittent low-force contractions: effects of the feedback mode. Eur J Appl Physiol, 87:28–37, 2002
8. Beck TW, Housh TJ, Johnson GO, Weir JP, Cramer JT, Coburn JW, Malek MH. Comparison of Fourier and wavelet transform procedures for examining the mechanomyographic and electromyographic frequency domain responses during fatiguing isokinetic muscle actions of the biceps brachii. J Electromyogr Kinesiol, 15(2):190-199, 2005
9. Tarata M. The Average Area/Amplitude Ratio (Raa), A Consistent Parameter in the Quantitative Analysis of the Electromyogram. In Proceedings of the International Conference on Medical Physics & Biomedical Engineering MPBE'94 May 5-7. Nikos Milonas, Nicosia, 1994, 53-57
10. Tarata M. Monitoring the Evolution of the Muscular Fatigue, Via New Parameters developed from the SEMG Signal. In Proceedings of the ECSAP'97 The First European Conference on Signal Analysis and Prediction 1997 June 24-27. ICT Press, Prague 431-434

11. Tarata M, Monitoring Neuromuscular Fatigue through the Area/Amplitude Ratio Computed on the SEMG and MMG, International conference on Advancements of Medicine and Health Care through Technology, MEDITECH 2009, 23-26 September 2009, Cluj Napoca, in S. Vlad, RV Ciupa and AI Nicu (Eds): MEDITECH 2009, IFMBE Proceedings 26, www.springerlink.com, 215-218
12. Tarata M, Advantages of the Area/Amplitude Ratio over Wavelet-derived Parameters, in Monitoring Neuromuscular Fatigue, Proceedings of The 7th Conference on Measurement, May 20-23 2009, Smolenice, Slovakia, ISBN 978-80-969672-1-6, Komprint sro, Bratislava, 95-98
13. Millet GY, Aubert D, Favier FB, Busso T, Benoît H, Effect of acute hypoxia on central fatigue during repeated isometric leg contractions, Scand J Med Sci Sports. 2009 Oct;19(5):695-702
14. Amann M, Romer LM, Subudhi AW, Pegelow DF, Dempsey JA, Severity of arterial hypoxaemia affects the relative contributions of peripheral muscle fatigue to exercise performance in healthy humans, J Physiol 2007, Volume 581, Number 1, 389-403

Author: Mihai T Tarata
Institute: University of Medicine and Pharmacy of Craiova
Street: Petru Rares 2-4
City: CRAIOVA
Country: ROMANIA
Email: mihaitarata@yahoo.com

Applications of the Hybrid Shielding in Biomagnetometry

M.C. Rau[1], O. Baltag[2], and I. Rau[2]

[1] "Gh. Asachi" Technical University, Faculty of Electrical Engineering, Iasi, Romania
[2] "Gr.T. Popa" University of Medicine and Pharmacy, Faculty of Biomedical Engineering, Iasi, Romania

Abstract— **The work presents the experimental results concerning the electromagnetic characteristics of a shielded room destined to bio-electromagnetism researches and to the utilization of a high resolution SQUID magnetometer. The experimental results regarding the dependence of the shielding factor on the applied field frequency are presented. The phase of the residual magnetic field and its dependence on the frequency of the external field has also been determined. The experimental results confirm the theoretical studies on the utilization of the non-ferromagnetic material shields.**

Keywords— **shielded room, passive shielding, active shielding.**

I. Introduction

The bioelectromagnetism measurements require special conditions, given the very small signals and detected fields.

These conditions concern the necessity to reduce the magnetic fields external to the space where the measurements are carried out, i.e. both the geomagnetic field and the interfering magnetic fields existing outside this space. This kind of space where the magnetic activity is almost zero can be obtained by combining various compensation means. These are represented by the magnetic shields which can be used independently or with other means meant to increase the shielding factors.

The magnetic shields [1], [2], [3] built for bio- electromagnetism measurements are shaped as cubes or parallelepipeds and are made of various materials, using various assembling and construction methods.

The materials they are made of can be ferromagnetic [4], [5] or non-ferromagnetic [6].

For the shielded rooms made of ferromagnetic materials, the shielding is due to the fact that the magnetic flux prefers the path with the highest value of the magnetic permeability. The utilization of multiple layers increases the shielding factor [7] – [9], [10], [11]. To increase the permeability of a ferromagnetic material, an alternating magnetic field produced by coils wound on the shield is applied, i.e. the shaking method is used. In order to increase the shielding factor of the passive shields, other systems can be added. The mostly used method is to add systems of uniaxial or tri-axial coils disposed in Helmholtz configuration, as well as negative feedback assembly, see Table 1.

Table 1 Classification of the hybrid shielding methods

Materials	Combination	Compensation	Applications
Ferromagnetic with high magnetic permeability - mumetal, permalloy, Fe-Si alloy	Cube or parallelepiped shape with 1, 2, 3, 6 or 8 layers with uniaxial or triaxial coils system and negative feed-back assembly	For passive shielding and active shielding	Based on high magnetic permeability of permalloy, for ELF shielding
Non-ferromagnetic (Al, Cu) materials with high electric conductivity	Walls with 1, 2, 3 layers and uniaxial or triaxial coils system and negative feed-back assembly	for passive and dynamic shielding	Based on Lenz law, for EHF shielding
Ferromagnetic and non-ferromagnetic	sandwich type with air, wood, plastic, glass in combination, permalloy with Al or Cu and with uniaxial / triaxial system coils and negative feedback assembly	passive shielding and dynamic shielding	Biomagnetism, satellites, generally equipment using electron guns or ions sources and mass spectrometry, masers and atomic clocks, electron microscopy and transmission electron microscopy, SEM (scanning electron microscopy), SQ UID, MRI, electron beam instruments

The feedback loop closes through a field sensor, a magnetometer, power amplifiers and compensation coils. The sensor is located in the area which must have a minimum magnetic field, inside the compensation coils. These systems are especially used to obtain minimum magnetic fields within different volumes in which the equipment are due to operate. Several terms are used in literature, such as: active compensation [12], [13], [14], active shielding [15], [16], [17], dynamic shielding [18], magnetic field cancellation [19], magnetic field stabilization [20] or hybrid technique [21]. All these terms define compensation systems which operate in negative feedback regime.

S. Vlad and R.V. Ciupa (Eds.): MEDITECH 2011, IFMBE Proceedings 36, pp. 148–151, 2011.
www.springerlink.com

II. THEORETICAL CONSIDERATIONS

In order to measure the biomagnetic fields, besides using these systems for compensating and shielding the disturbing fields, the utilization of high resolution magnetometers is also necessary. The high resolution biomagnetometers have very good sensitivity, but also some limitations, such as:

- reduced immunity to electromagnetic and magnetic fields;
- sensitivity to shocks, vibrations and microseisms;
- necessity of periodic refilling with liquid He or nitrogen.

The disturbing magnetic fields are both continuous, like the geomagnetic field, and natural slowly variable components produced by geomagnetic field micro pulsations or geomagnetic storms. Among the low frequency fields, those produced by the supply networks and by the 50 Hz supply plants and power transformers are the most present in the disturbing signal spectrum, both by their basic frequency, and the second and third order harmonics especially. These frequencies are hard to eliminate from the complex signal of the biomagnetometer, because they can be found in the spectrum of both bioelectric and biomagnetic signals.

Usually, the high frequency fields in the long wave and microwaves range do not penetrate in the channel of biomagnetic signal of the SQUID, but they easily penetrate in the end processing modules. That is why it is recommended to ensure the compatibility of the SQUID installation with the electromagnetic medium within the entire spectral range. The compatibility can be reached by classical shielding methods, introducing the SQUID installation together with the subject in a room shielded against both continuous and alternating magnetic fields. There are several shielding techniques and methods, depending on the SQUID performances, destination and location, and the designer and builder imagination. In the case of shielded rooms made of non-ferromagnetic (Cu, Al) materials, the shielding is based on the Lenz's law. The time-variable electromagnetic field induces in the conducting material eddy currents which, in turn, generate a field opposite to the external field. Simultaneously, the energy absorption through eddy currents occurs. The shielding effect is determined by the skin depth δ, which represents the distance at which the electromagnetic wave is attenuated by a factor 1/e. For the calculation of the skin depth, an electromagnetic wave with intensity H_0 is considered, with normal incidence to the external surface of a plane conducting wall with the thickness z, the intensity Hx of the wave emerging from the conducting wall being diminished. Let's consider a plane wall made of a non-ferromagnetic high conductivity material - Cu or Al, having the separation surface from the medium in the Oyz plane, Figure 1.

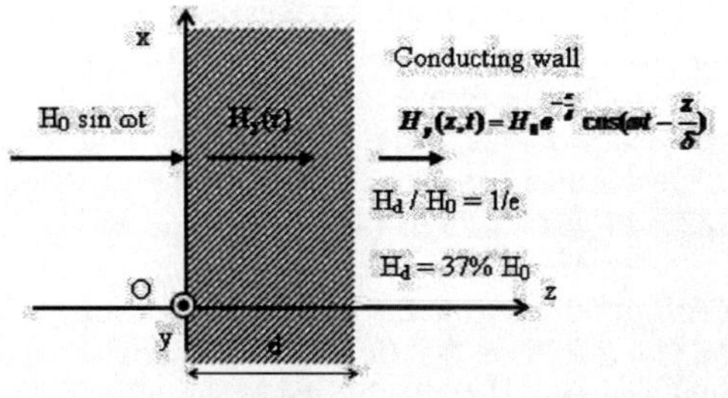

Fig. 1 Wave propagation through conducting wall

The plane wave of intensity H_0 travels in the space on a direction perpendicular to the incidence plane of the conducting wall. Under these conditions, only the field component along the *Ox* axis will be considered Ddeveloping the Helmholtz equation, the component H_Z becomes:

$$H_y(z,t) = H_0 e^{-\frac{z}{\delta}} \cos(\omega t - \frac{z}{\delta}) \tag{1}$$

From the above equation, the field amplitude and phase at the distance *z* result. The phase angle is also useful to express the standard skin depth. At the standard skin depth, the phase angle is about 57^0. Therefore, the electromagnetic field intensity decreases exponentially with the distance run through a conducting medium. In the electromagnetic field theory, it was demonstrated that the skin depth δ depends on the electric and magnetic properties of the medium and the electromagnetic wave frequency according to the relation:

$$\delta = \sqrt{\frac{2}{\omega\sigma\mu}} = \sqrt{\frac{1}{\pi f \sigma\mu}} \tag{2}$$

where: f – field frequency; σ– conductivity of the conducting medium; μ – magnetic permeability of the medium. One can notice that the skin depth decreases as the field frequency increases.

III. EXPERIMENTAL SETUP

The theoretical and experimental researches are dealing with the theoretical calculation and experimental determination of the global shielding factor within different frequency ranges, from the continuous fields to medium frequency fields. At low frequency, the phenomena occur in the magnetic induction zone and, consequently, only the magnetic component of the field is of interest. Even though, from the theoretical results, it follows as obvious that there are two components of the field inside the room, a real and an imaginary one, the studies were not directed to the determination of the field phase variation inside the system. The knowledge of the field phase within the shield is important in the utilization of the systems for magnetic field dynamic control, systems which operates in the negative feedback loop. The realized installation makes use of mixed shielding, consisting of a room made of non-ferromagnetic material, and a group of external coils which control, through the negative feedback, the residual magnetic field inside the room. The shielded room has a parallelepiped shape, sized 2x2x3 m, and has 0.012 m thick aluminum walls. It is located in the centre of a system of coils disposed in a Helmholtz configuration, with the side of 4 m each. From the equation (1), it follows that inside the shielded room there is a residual complex signal with the amplitude smaller than the external field. One can notice that the phase of this signal strongly depends on the field frequency. That is why it is difficult to apply a negative feedback loop, given the fact that it is necessary to correct the phase of the current through the Helmholtz coils in terms of the applied frequency. In order to determine the shielding factors and the residual signal phase, the installation presented in Figure 2 is used.

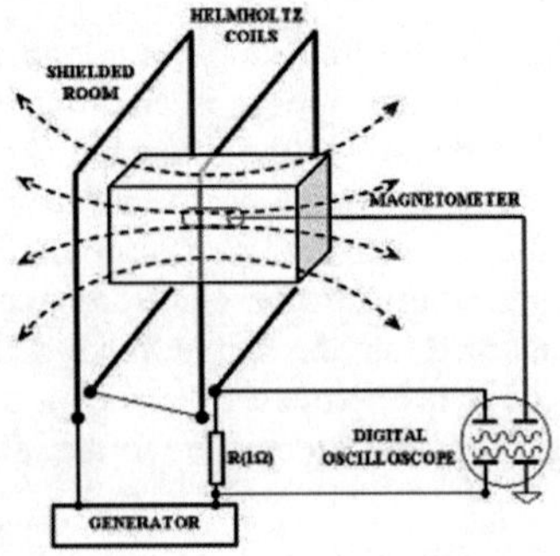

Fig. 2 The block diagram of the installation for electromagnetic characterization of the shielded room

An alternating current is injected through the Helmholtz coils, its intensity being measured by means of a standard resistor and a digital oscilloscope. The residual magnetic field within the shielded room is measured with a saturable fluxgate magnetometer. For the determination of the residual field phase, the current injected through the Helmholtz coils was taken as reference. It has been experimentally found that the shielding factors along the three directions are different Figure 3.

The variation of the shielding factors along the three directions results from the coupling factors between the three pairs of Helmholtz coils and the room walls, which are different along the three directions. Still another reason that the shielding factors are different along the three directions is the presence of the walls which are parallel with the field lines, and within which residual fields are induced by the components of the external magnetic induction produced by the coils.

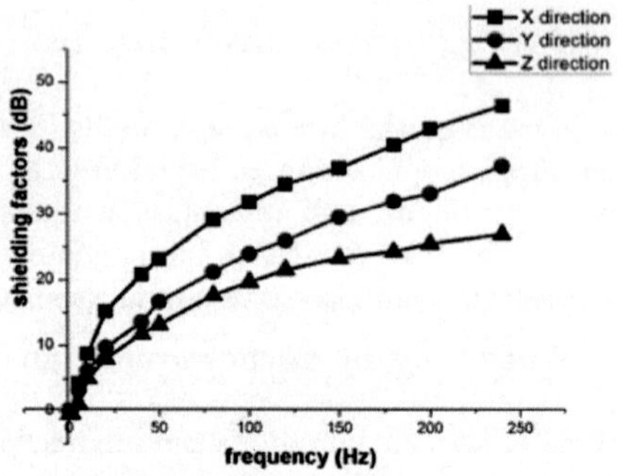

Fig. 3 Shielding factors along the three directions

In fact, one can consider that the wave front of the disturbing induction meet the shielded room walls under different, well defined angles, in each of the three main planes.

The field phase inside the room changes differently along the three directions, within the volume comprised between the geometrical centre and the room walls. Figures 4, 5, 6 present the variation of the residual field phase inside the room along the three orthogonal directions Oxyz.

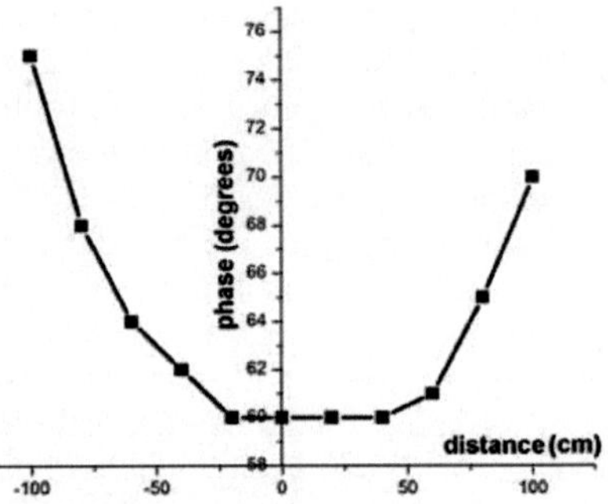

Fig. 4 Variation of the residual field phase inside the room along the Ox direction

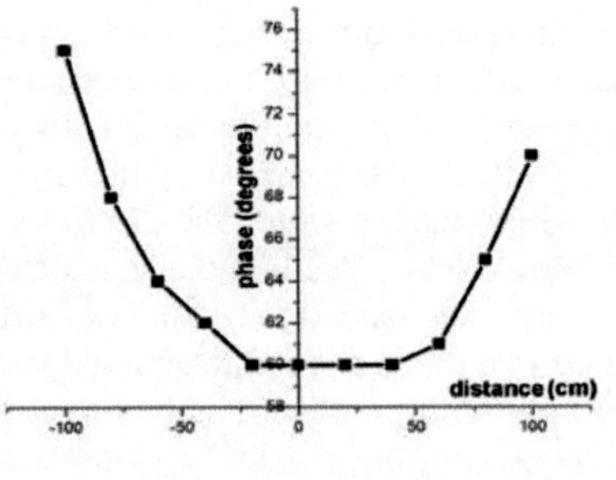

Fig. 5 Variation of the residual field phase inside the room along the Oy direction

The phase variation inside is important because the system that works in the negative feedback loop, having sensors on the inside, must maintain a phase relation corresponding to a stable regime. Consequently, in the negative feedback loop of the power amplifier which injects compensating currents through the external coils, a phase correcting circuit must be introduced depending on frequency. In this way, the negative feedback coefficient can become constant within the entire frequency range, and equal to the negative feedback coefficient for the external fields.

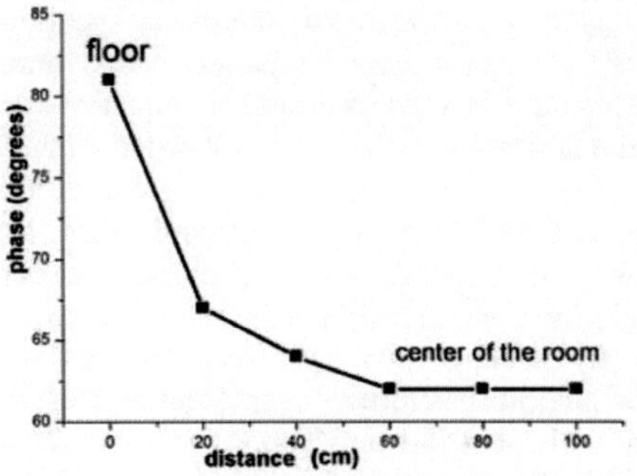

Fig. 6 Variation of the residual field phase inside the room along the Oy direction

IV. Conclusions

The shielded room presents different shielding factors along the three main directions.

The phase of the residual field inside the room depends on the frequency and on the position of the measurement point.

A phase variation was found within the volume of the shielded room along all the three directions.

For the active control/shielding, a phase corrector must be introduced, which maintain the system within the stable zone of the negative feedback loop.

References

1. Mager A.J. (1970) Magnetic shields, IEEE Trans. on Mag. 1:67-75
2. Sterne T.E., (1935) Multi lamellar cylindrical magnetic shields, R.S.I., 6:324-326
3. Schweizer F., (1962) Magnetic shielding factors of a system of concentric spherical shells, J. of Appl. Phys., 33:1001-1003
4. Okada Y.C., Shan B., Huang Jin-Chu, (1994) Ferromagnetic high permeability alloy alone can provide sufficient low-frequency and eddy-current shielding for biomagnetic measurements, IEEE Trans. on Biomedical Engineering, 41:688-697
5. Cohen D., Schlapfer U., Ahlfors S., M. Hamalainen and E. Halgren (2002) New six layer magnetically shielded room for MEG, Proc. of the 13th Int. Conf. on Biomagnetism 2002, Jena, Germany, Berlin: VDE Verlag GmbH,
6. Stroink G., Blackford B., Brown B., Horacek M., (1981) Aluminium shielded room for biomagnetic measurements, Rev. Sci. Instr. 52:463-468,
7. Suzuki A., Ishii K., Yamazaki H., Matsuyama S., Aikawa K., et all (2001) Development of the magnetic aluminum shield for MEG, CYRIC Annual Report, 2001, Tohoku Univ. http://www.cyric.tohoku.ac.jp/english/report/repo2001/Annual2001
8. Nowak,H., Haueisen,J., Gießler,F., Huonker,R., (2002) Magnetic shielding designs, http://www.vitatech.net/magnetic_shielding.php4
9. *** Magnetically shileded rooms, http://www.imedco.ch/englisch/products/frameset_products.html
10. *** Low frequency magnetic shielding, http://www.shieldingsystems.eu/index.php
11. Cohen D., Schlapfer U., Ahlfors S., et all (2002) New six layer magnetically shielded room for MEG, Proc.of the13[th] Int. Conf.on Biomagnetism, Jena, Germany
12. Bork J., Hahlbohm H.D., Klein R., Schnabel A. (2000) The 8 layered magnetically shielded room of PTB: design and construction, Proc. of the 12th Int. Conf. on Biomagnetism, 2000, Espoo, Finland, pp. 970-973
13. K. Kato, K. Yamazaki, T. Sato, A. Haga, et al, (2004) Active magnetic compensation composed of shielding panels, Neurology and Clinical Neurophysiology, **68**:1-4
14. H.J.M. ter Brake, R. Huonker, H. Rogalla, (1993) New results in active noise compensation for magnetically shielded rooms, Meas. Sci. Technol., **4**:370-1375
15. H.J.M. ter Brake, H.J. Wieringa, H. Rogalla, (1991) Improvement of the performance of a mu-metal magnetically shielded room by means of active compensation, Meas. Sci. Tech., **2**:596-601
16. D. Platzek, H. Nowak, F. Giessler, J. Rother, M. Eiselt, (1999) Active shielding to reduce low frequency disturbances in direct current near biomagnetic measurements, Rev. Sci. Instrum. 70: 2465-2470
17. K. Kato, K. Yamazaki, H. Matsuba, C. Sumi, S. Sato, (2001) Active magnetic shield for biomagnetic measurements, Proc. of the 12th Int. Conf. on Biomagnetism 2000, Helsinki University of Technology, Espoo, Finland, Ed. J. Nenonen, R. Ilmoniemi, T. Katila (Vantaa, Finland: Dark) pp. 965-967
18. C. Holmlund, M. Keipi, T. Meinander, A. Penttinen, and H. Seppa, (200) Novel concepts in magnetic shielding, Proc. of the 12th Int. Conf. on Biomagnetism 2000, Helsinki University of Technology, Espoo, Finland, Ed. J. Nenonen, R. Ilmoniemi, T. Katila, (Vantaa, Finland: Dark) pp. 968-969
19. D. Costandache, A. Banarescu, O. Baltag, I. Rau, M. Rau, S. Ojica, (2009) Dynamic shielding in biomagnetism, IFMBE Proceedings, 26:121-124
20. Gu C., Zou S., Han Z., Qu T. M., (2010) Passive magnetic field cancellation device by multiple high-Tc superconducting coils, Review of Scientific Instruments, 81:045101- 045101-5
21. Brys T., Czekaj S., Daum M., Fierlinger P., George D., Henneck, Hochman Z., Kasprzak, Kohlik M., K., K.Kirch, Kuzniak M., Kuehne G., Pichlmaier A., Siodmok A., Szelc A., and Tanner L., (2005) Magnetic Field Stabilization for Magnetically Shielded Volumes by External Field Coils, J. Res. Natl. Inst. Stand. Technol. 110:173-178
22. Kuriki S., Hayashi A., Hirata Y., (2001) Hybrid technique for reduction of environmental magnetic field noise, Proc. of the 12th Int. Conf. on Biomagnetism 2000, Helsinki University of Technology, Espoo, Finland, Ed. J. Nenonen, R. Ilmoniemi, T. Katila (Vantaa, Finland: Dark) pp. 957-960

Author: Miuta Rau
Institute: "Gh. Asachi" Technical University
Street: Blvd. D. Mangeron
City: Iasi
Country: Romania
Email: miuta.carmina@gmail.com

Linear Active/Passive Upper Limb Exerciser

B. Chetran, D. Mandru, S. Noveanu, and O. Tatar

Technical University of Cluj-Napoca, Department of Mechanisms, Precision Mechanics and Mechatronics, Cluj-Napoca, Romania

Abstract— **The objective of this paper is to describe an original linear type exerciser for the rehabilitation of the upper limb. The presentation of the most important characteristics of rehabilitation is followed by an analysis of different variants of exercisers. The core issues on design of exercisers are emphasized and the studied exerciser is described. It has one degree of freedom, is pneumatically actuated, being dedicated to both active and passive exercises.**

Keywords— **Rehabilitation, exerciser, active–passive.**

I. Introduction

The rehabilitation is that component of the medicine (beside the prophylactic and curative medicine) which implies complex assistive processes in order to prevent, compensate, relieve or neutralize impairments, disabilities or handicaps [1]. The main aim of the rehabilitation is the reintegration of the disabled in their families and society. The key concept in rehabilitation is that of residual function; the rehabilitation procedures must maintain, develop and exploit the residual capabilities of the patients.

It is recognized that any rehabilitation procedure is more efficient as is aided by simple devices or complex systems whose role is the replacement or assisting of the totally lost or partially reduced motor or/and sensorial functions [2]. The prosthesis, orthosis, telethesis, mobility aids, wheelchairs, aids for daily living, environmental control systems, augmentative/alternative communication systems, aids for vision and hearing impaired, functional electrical / magnetical stimulation systems, assistive robots, etc. are few examples of currently used rehabilitation equipments [3]. To the above-mentioned products, different types of rehabilitation systems used for motion-based rehabilitation approaches are added. The motion is the basic element of the rehabilitation in walking, running, grasping, gym, and games and training with specialized equipment. The motion-based therapy is recommended in posttraumatic affections of the locomotion apparatus, diseases of the nervous system and rheumatic illnesses [4]. Its task is the amelioration of disabilities by performance's improvement (force, precision, mobility).

In this paper some equipment for performing repetitive exercises (so-called *exercisers*) by the upper limbs are analyzed and an original one degree of freedom, linear type exerciser is presented.

II. Analysis of different exercisers

Manually assisted kinetotherapy is limited by fatigue of the therapists, while with an exerciser, the duration and number of training sessions can be increased, more than that with the aid of the sensors mounted on the exercisers the biomechanical parameters can be recorded or/and analyzed in a real time or later on manner. The great variety of exercises causes a variety of dedicated equipment. Many high-performance exercisers were developed, experimentally tested and now are currently used.

Paper [5] presents a robot for arm therapy which has a semi-exoskeleton structure with six degrees of freedom. It is equipped with position and force sensors. The system is applicable to the training of activities of daily living. The development of a rehabilitation machine for upper limb including a wearable muscle suit without metal frame, actuated by McKibben actuators is presented in [6]. A therapeutic wrist rotator for the passive pronation - supination of the forearm is described in [7]. The device, fixed to a stationary table, includes an electric actuator and a mechanical transmission to provide relatively slow rotational speed and relatively high torque to the output shaft and handgrip. A therapy and training device for the shoulder joint with a trunk base having two shoulders extensions on a rotational joint base was introduced [8]. It has a modular structure and can produce passive motions of the shoulder and elbow. In [9] were studied the possibilities to use the shape memory actuators in the structure of intelligent exercisers. An actuation device based on active element made of CuZnAl alloy with two-way shape memory effect was developed. The device is suited to rehabilitation of the fingers joints. Paper [10] presents a continuous passive motion device including an upper arm support suitably fixed to a drive actuator and an adjustable forearm support. Other similar exercisers are described in [11] – [14]: a two degrees of freedom robotic arm used to play arm wrestling game with human for entertainment or physical exercises, a three degrees of freedom exoskeleton system attached to the upper limb to assist its motion, a therapeutic exerciser equipped with two angular voice – coil actuators coupled by levers to the upper limb, two compact, portable devices for elbow and knee rehabilitation. The resistive forces and torques may be assured by weights, elastic elements and various type of brakes (mechanic, magnetic and based on smart fluids) [15], [16].

S. Vlad and R.V. Ciupa (Eds.): MEDITECH 2011, IFMBE Proceedings 36, pp. 152–155, 2011.
www.springerlink.com

In our previous work, several exercisers were developed. The exerciser shown in [17] was designed to be used in passive and active upper limb's mobilization, aiming the reattainment of the shoulder articulations, elbow's and wrists movement capacity and patient's motor skills. A stationary two degrees of freedom powered therapeutic devices, based on parallel mechanisms, that provide active or passive motions of the hand is presented in [18]. It trains the wrist in flexion-extension and abduction-adduction movements. The device given in [19] provides continuous controlled passive motion and adjustable range of motion for the rehabilitation of the upper extremity forearm. In [20] a cost effective, efficient, easy to use, lightweight and portable exerciser, containing simple mechanisms for regaining the wrist joint's functions, is described.

There is a need for new therapeutic exercisers that integrate various mechanisms, different actuation systems and information technology to develop high performance robotic exercisers.

III. The core issues on exercisers' design

The design of the exercisers is largely influenced by the motion type. The *passive* mobilizations are those movements imposed to a patient's articulation by an exterior intervention, without its neuromuscular system to be involved. They are characterized by the range and speed of motion, both forward and backward. For passive exercises, the specific parameters are: the motion's sequence, the range of motion, the velocity and acceleration. The *active* mobilizations or movements are the ensemble of exercises performed by the patient, voluntarily putting in function his/her neuromuscular system. Active movements may be assisted, free – without exterior resistance or active movements against a resisting force. For the active exercises, the therapist learns the patient about the motions to be executed, and the exerciser gives the necessary resisting force, displaces trajectories and targets to be accomplished.

Both in active and passive exercises, the upper limb's motions can be monitored by using electrogoniometers or other measurement systems; the patient's parameters could be displayed, in order to diagnose.

Due to the fact that the kinethoterapeutic programs are made for each patient, taking into consideration the disease and the patient's characteristics, the exercisers should allow setting multiple functional parameters: range of motion, speed, acceleration, resistive force or torque, a.s.o. The geometrical features of the exercisers are established according to the anthropometric data of the limbs and body and their biomechanics. Since the exercisers are useful aids for therapist in repetitive tasks of the treatment program, their design must take into account the requirements and needs of the therapists as well as those of the patients [21].

IV. The studied exerciser

The studied rehabilitation robotic system is dedicated to both active and passive exercises. User, in the sitting or standing position, has its upper limb "linked" with a mechanical structure (the hand is in direct contact with a handle). Thus the upper limb is mobilized, actively or passively, in front or side work positions (Fig. 1). The exercises further diversify if the system works in a tilted plane from the horizontal, even in a vertical plane [22].

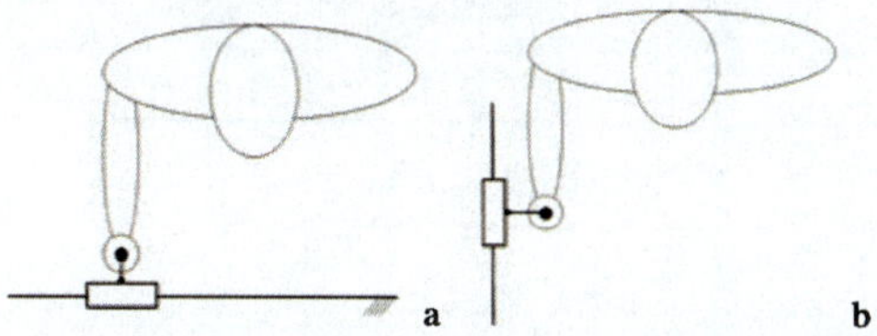

Fig. 1 The work positions of the exerciser

The mechanical structure of the system is made by joining two pneumatic cylinders. The rod ends of the cylinders are fixed and for stabilizing this configuration an additional sliding join was added to increase the mechanism stability. The handle was mounted on the top of the mechanism. A control system was developed. The block diagram of the system is given in Fig. 2.

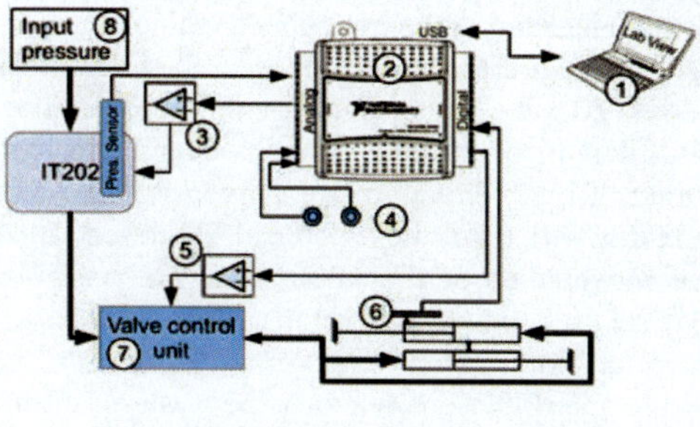

Fig. 2 The block diagram of the exerciser

The control system is made from a PC (1) that communicates with a NI USB 6009 console (2) via USB. The IT202 pressure regulator is controlled by the analog output of (2) through an amplifier (3). The valves are activated digitally with the aid of the amplifier (5). The push and pull forces are measured by two force sensors (4) and for measuring the traveled distance an encoder (6) was added. The pressure is read by the built in pressure sensor of the IT202.

The pneumatic components (Fig. 3) are: two C85E25-160 pneumatic cylinders, one EVZ512 3/2 solenoid valve, one IT202-F003B, electrically adjustable pressure regulator, one EVZ5320 5/3 solenoid valve and one manual pressure regulator. An electronic part has been designed staring from the considerate to use a galvanic separation, with CNY74-4 optotransistors, between the NI – USB 6009 control board and the power connections. The power amplifiers L293 were used to drive the solenoid valves. The IT202 pressure regulator it's controlled in 0-20mA DC current range through one BC639 transistor. The NI – USB 6009 read two Force Sensing Resistors (FSR) sensors mounted in the handle as well as one optical linear encoder.

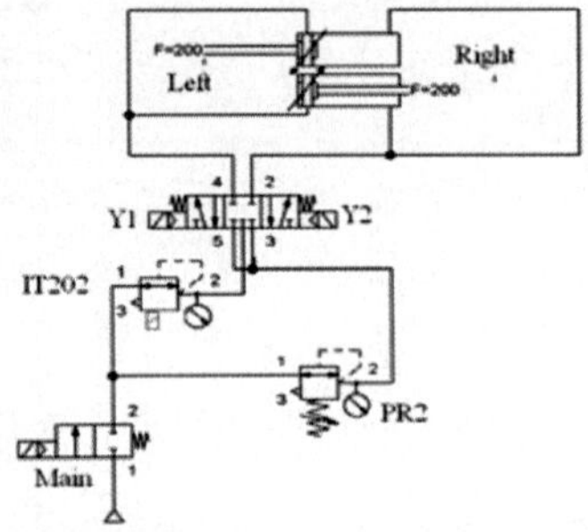

Fig. 3 The pneumatic diagram of the exerciser

When the "Main" solenoid valve is open the compressed air set at 4 bar flows into the IT202 and PR2 pressure regulators. The PR2 is set at 2 bar and is connected at the 3 and 5 terminals of the 5/3 valve and the output of the IT202 is connected at the terminal 1 of the 5/3 valve. The terminals 2 and 4 are connected at the cylinders terminals.

By adjusting, through the "Pressure" slider of the interface, the IT202 at a higher pressure than PR2 and activating the Y1 solenoid of the 5/3 valve, both cylinder roods will travel into "Right" direction. By activating the Y2 solenoid, the cylinder will travel in the "Left" direction. In this way passive movements of the upper limb are possible. When the "MIN Dist." (minimum value of the distance set on the interface) or "MAX Dist." (maximum value of the distance set on the interface) are reached the system automatically changes the direction.

If the patient doesn't want his exercise to be assisted by the program behind the interface, he can manually adjust the force by setting up the IT202 through the "Pressure "slider. For the active assisted movements the pressure of IT202 has to be adjusted in accordance with the force read by the FSR in the handle and the desired force set in the interface.

In the case of exercises with desired low amplitudes and forces the system reads the encoder, the "MAX Dist." and "MIN Dist" and "Force desired" edit boxes of the interface. The system activates the Y1 and Y2 in accordance with the displacement and direction and adjusts the force by setting the IT202 output according with the exercises requirements.

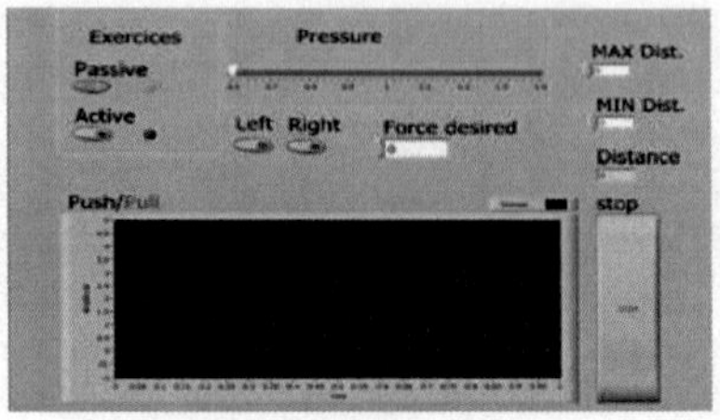

Fig. 4 The developed interface of the exerciser

Lab View 8.6 was used to build the interface of the system (Fig. 4). The program which runs behind this interface allows setting up the type of the exercises, the pressure of the IT202, the force and the distance desired. The interface also offer a visual feed back for the patient by displaying the force and distance read by the sensors. The buttons "Active" / "Passive " allow the user to set the desired exercises with an interlock function that do not allow the execution in the same time. For counting the traveled distance the program reads a digital input port with a linear encoder attached. Also the program has the capacity to limit the traveled distance. The system will change the travel direction automatically if the maximum or minimum distance, from the afferent edit boxes, is reached. The force desired edit box is used to set a maximum force allowed for the exercise. If this force is reached the system will adjust the IT202 in order to lower the handle force.

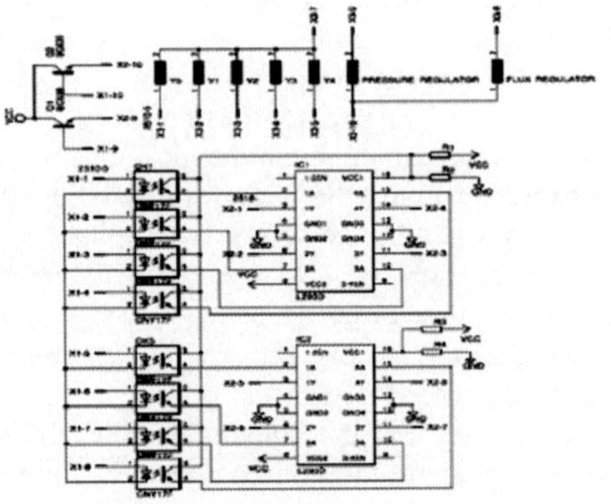

Fig. 5 The amplifiers electrical wiring

Between the NI USB 6009 and the valve control unit, one digital multichannel amplifier is needed (Fig. 5). It consists from eight optocouplers that allows eight digital inputs and a galvanic separation. At the optocouplers output, two

L293 push/pull drivers were used in order to feed the solenoid valves coils with the proper current. The control signal of the IT202 pressure regulator has to be between 0 – 20 mA, according with its data sheet. A BC639 transistor and a different source are used in order to obtain the current signal. The transistor base is connected at the "A0" analog output channel and the collector is feed with 10V DC from the source.

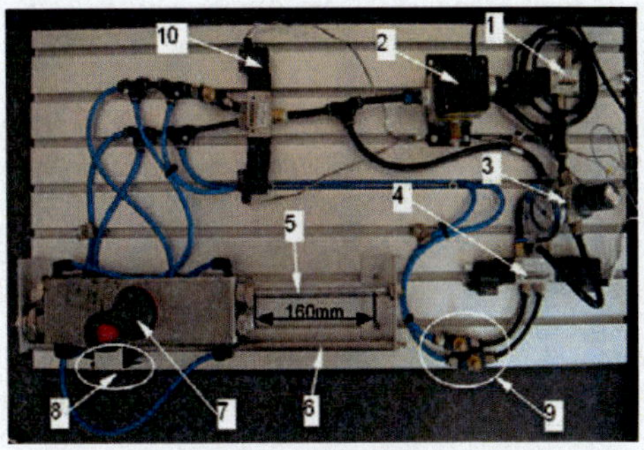

Fig. 6 The prototype of the exerciser

The prototype of the system is given in Fig. 6; the following notations were used: 1 – main valve, 2 – IT202 pressure regulator, 3 – manual pressure regulator, 4 – EVZ5320 5/3 valve, 5 – cylinder rod 160mm stroke, 6 – stability circular beam, 7 – handle mounted on the cylinders case, 8 – encoder and slip bush case, 9 – pneumatic speed controllers, 10 – EVZ5320 5/3 valve. Using the buttons in the interface, the user can switch between active and passive exercises and has the possibility to stop the system in any functioning phase. The forces are varied according to the muscle capabilities of each user. The described exerciser permits a large variety of exercises, which can be accomplished automatically, and modified by programming them.

V. Conclusions

The pneumatically actuated exerciser offers multiple adjustment options for a large variety of exercises. Its benefit is that it reduces the need for a clinical therapist. The patients can use the exerciser independently. The above-described functional prototype is experimentally tested and used as valuable educational tool. Following the preliminary experiments, the exercises that can be done with this exerciser are: in the front work position: active and passive exercises for arm abduction/adduction and flexion/extension of the forearm; in the side and vertical work positions: active and passive exercises for flexion/extension of the forearm combined with anterior/posterior projection of the arm.

Acknowledgment

This paper is supported by the project "Doctoral studies in engineering sciences for developing the knowledge based society-SIDOC" contract no. POSDRU/88/1.5/S/ 60078, project co-funded from European Social Fund through Sectorial Operational Program Human Resources 2007-2013.

References

1. Bronzino, J.D. (ed.) (1995) Biomedical Engineering Handbook, CRC Press & IEEE Press, Hartford
2. Mândru, D. (2001) Ingineria protezării şi reabilitării, Editura Casa Cărţii de Ştiinţă, Cluj-Napoca
3. Kiss, I. (1999) Fizio-kinetoterapie şi recuperarea medicală în afecţiunile aparatului locomotor, Editura Medicală, Bucureşti
4. Plas, H.F. (2001) Kinetoterapie activă, Editura Polirom, Iaşi
5. Nef, T. et al (2006) ARMin – Robot for Rehabilitation of the Upper Extremities, Proc of the 2006 IEEE Int Conf on Rob and Aut, Orlando, 2006, pp. 3152-3157
6. Kobayashi, H., Suzuki, H. (2005) Development of a New Shoulder Mechanism for a Muscle Suit, Proc of the IEEE Int Conf on Mechatronics & Aut, 2005, pp. 1051-1056
7. Baker, E. (1998) Therapeutic wrist rotator, U.S. Patent no. 5.788.607
8. Hassler, A. (2004) Therapeutic and training device for the shoulder joint, U.S. Patent no. 6.824.498
9. Bernardo, D., Martinet, N., Hean, C. (1996) Shape memory intelligent actuator for application to rehabilitation, Proc. of the 9th Int Conf on Mechanics in Medicine and Biology, 1996, pp. 127 – 131
10. Saringer, J., Culhane, J. (1999) Continuous passive motion device for upper extremity forearm therapy, U.S Patent no. 5.951.499
11. Gao, Z., et al. (2006) Design, Fabrication and Application of Arm Wrestling Robot, Int Conf on Mechatronics and Automation, Luoyang, China, 2006.
12. Kiguchi, K., et al. (2003) Design and Control of an Exoskeleton System for Human Upper-Limb Motion Assist, International Conference on Advanced Intelligent Mechatronics, 2003.
13. Gallasch, E., Taferner, H.P., Fend, M. (2000) Stiffness controlled voice coil actuator for muscoskeletal studies, Proc. of the Int. Conference ACTUATOR 2000, pp. 639 – 642
14. Mavroidis, C. et al. (2005) Smart portable rehabilitation devices, ASME 2005 International Design Engineering Technical Conferences, Long Beach, California, 2005
15. Lee, T.C., Tai, H. (2010) Magnetic resistance device for exerciser, U.S. Patent 2010/0069205 A1
16. Kikuchi, T, Ying, J. (2008) Hybrid-PLEMO rehabilitation system for upper limbs with active / passive force feedback mode, Proceedings of the 30th Annual International IEEE Conference, pp. 1973 - 1976
17. Mândru, D., Rusu, C., Noveanu, S. (2004) Research concerning the development of a robotic system for rehabilitation exercises, Mecatronica, 2:. 55 – 60
18. Mandru, D., Stan, S. (2006) Exerciser for Wrist Kinetotherapy, Acta Technica Napocensis, 49, II: 325 – 330
19. Mandru, D. et al (2007) Development of Robotic Systems for Upper and Lower Limbs Kinetotherapy, Acta Electrotehnica, 48/ 4:. 13-18
20. Chetran, B. et al (2010) Wearable Exerciser, Solid State Phenomena 166-167:.115-120, doi:10.4028/www.scientific.net/SSP.166-167.115
21. Lee, M. et al (2005) Design Issues for Therapeutic Robot Systems: Results from a Survey of Physiotherapists, Journal of Intelligent and Robotic Systems 42: 239-252
22. Chetran, B. et al (2011) Rehabilitation exercisers – from simple devices to robotic systems, Acta Technica Napocensis, in press

Model of the Current-Voltage Relation for a Skin Pore

N.M. Birlea[1], S.I. Birlea[2], and E. Culea[1]

[1] Physics Department, Technical University of Cluj-Napoca, Cluj-Napoca, Romania
[2] Electrical & Electronic Engineering, School of Engineering and Informatics, National University of Ireland Galway, Galway, Ireland

Abstract— **Using a Nernst-Planck model, we show that the current density in a membrane's pore as a function of voltage has three types of behavior: a quasi-ohmic behavior at low voltages, with a small slope, a non-ohmic linear dependence at large voltages, with a large slope, and a nonlinear transition region at intermediate voltages. The magnitude of the quasi-ohmic current from low voltages depends mainly on the height of energy barrier inside the pore, *w*, through an exponential term, e^w. The low voltages domain is experimentally accessible and almost unexplored, despite the fact that it can offer direct information about the energy barrier inside a pore. The model has only two assumed parameters, the energy barrier height, *w*, and the relative size of the entrance region of the pore, *r*, with a clear physical meaning, an important advantage for fitting and interpreting experimental data. This simple model for the current-voltage nonlinearity is a good starting point for explaining the electrical behavior of the skin at low voltages.**

Keywords— **skin, current-voltage characteristic, Nernst-Planck model, nonlinearity, pore.**

I. Introduction

An important practical problem is the current-voltage characteristic of the skin. Usually, the skin's electrical response is descriptively presented by Bode plots, Nyquist representations, Cole diagrams/formulae or equivalent electrical circuits. The skin and particularly the outer stratum corneum is a porous membrane with a strong nonlinear response to applied voltages or currents [1-3], but also it has a measurable asymmetry [2-4]. In order to understand the dramatic changes in skin resistance at large voltages, Weaver and his group developed an explicative theoretical model for the skin electrical behavior [5], starting from electroporation studies in bilayer lipid membranes [6]. The electroporation phenomenon consistently describes skin behavior at large voltages, but for low voltages, we have not yet an accepted theory for skin's electrical nonlinearity. The current models for skin at low voltages take into account the appendageal macropores [7], or sweat pores [2]. However a more consistent theoretical base is necessary to understand and to control new drug delivery systems like low voltage iontophoresis [8].

The Nernst-Planck (NP) model provides an explanation for the nonlinearity of current-voltage characteristic of a pore in a membrane [6]. The NP equation is difficult to solve analytically because it is nonlinear. Assuming a linear profile for the ion potential energy inside the pore, it becomes easy to find the solution of the NP equation and an analytic formula for the current-voltage relationship, a very useful feature for experiments.

The current-voltage relation of the membrane's pores is essential for describing the way in which epithelia function, especially the skin's epidermal stratum corneum [9], for predicting electroporation [5], for designing synthetic nanopores or nanodevices [10] and for studying ion channels [11]. All these research interests have strong reverberations into the bio-medical field.

II. Model

Our model assumes the existence of a cylindrical pore, with length *h* and radius *R*, in a plane membrane bathed by an electrolyte with only one ionic species. The generalization for multiple ionic species is straightforward [12, 13] but unnecessary here, for the sake of clarity.

The transport equation states that the stationary flux density of ions through the pore in the membrane is given by the product of the ionic concentration *c*, the mobility *b*, and the gradient of the electrochemical potential μ [12, 14]:

$$F = -cb\frac{d\mu}{dx} \tag{1}$$

The mobility *b* is related to the diffusion coefficient, *D*, of the ion by the Einstein's relation, $b=D/(kT)$, where k = Boltzmann constant, T= absolute temperature. The electrochemical potential μ is:

$$\mu = kT\ln c + qV(x) + W(x) + \mu_o. \tag{2}$$

where the first term is the concentration dependence (the ionic activity are assumed to be unity), the second is the contribution of the applied potential (q= elementary charge), the third is the work *W* required to transfer the ion from a distant point in the aqueous phase to the point *x* in the pore, and the last term is the standard chemical potential in the pore–independent of position.

The potential energy $W(x)$ of an ion inside the pore, the key element for pore conductance, has two distinct parts [15]:

S. Vlad and R.V. Ciupa (Eds.): MEDITECH 2011, IFMBE Proceedings 36, pp. 156–159, 2011.
www.springerlink.com

1. the self-energy due to the induced charges on the dielectric boundary – surrounding liquid and pore wall (Born energy and image or reaction-field energy), proportional to q^2, is *always repulsive*, and
2. the Coulomb interaction with the charges on the pore's wall (if the pore's wall is charged), proportional to q, can be either *attractive* or *repulsive* depending on the signs of interacting charges.

The potential energy $W(x)$ will be treated here as an assumed parameter of the model.

Using relation (2) in equation (1) we obtain the stationary current density J – the flux of ions F, multiplied by ion charge, as:

$$J = -qD\left(\frac{dc}{dx} + c\frac{dE}{dx}\right) = -qDe^{-E}\frac{d(ce^{E})}{dx} \quad (3)$$

where the dimensionless energy variables are:

$$E(x) = \frac{qV(x)}{kT} + \frac{W(x)}{kT} \quad (4)$$

The exponential factor in front of the derivative is transferred to the left side,

$$Je^{E} = -qD\frac{d(ce^{E})}{dx} \quad (5)$$

and the equation is integrated after x,

$$J\int_0^h e^{E}dx = -qD\int_0^h \frac{d(ce^{E})}{dx}dx \quad (6)$$

giving the standard expression for stationary current density [12, 15]:

$$J = -qD\frac{\int_0^h \frac{d(ce^{E})}{dx}dx}{\int_0^h e^{E}dx} = \frac{-qDc(e^{E(h)} - e^{E(0)})}{\int_0^h e^{E}dx} \quad (7)$$

We suppose that the ionic concentration in bulk solution and at the pore's extremities is the same: $c(0)=c(h)=c$. The denominator can be integrate if we know the energy as a function of position. For this purpose the profile of the potential energy inside the pore, $W(x)$, is approximated here by a trapezium as in [6, 13]:

$$\frac{W(x)}{kT} = \begin{cases} w\frac{x}{d} & 0 \le x \le d \\ w & d \le x \le h-d \\ w\frac{h-x}{d} & h-d \le x \le h \end{cases} \quad (8)$$

where w is the adimensional height of the energy barrier (maximum of potential energy) and d is the length of the pore entrance, the length of the pore domain where the potential energy varies.

The externally applied potential $V(x)$ is assumed to vary linearly across the membrane and along the pore as:

$$\frac{qV(x)}{kT} = u\left(1 - \frac{x}{h}\right) \quad (9)$$

where u is the adimensional transmembrane potential:

$$u = q\frac{V(0) - V(h)}{kT} \quad (10)$$

III. RESULTS

We introduce two dimensionless quantities, the relative size of the entrance region of the pore:

$$r = \frac{d}{h} \text{ for } 0<r\le 1/2 \quad (11)$$

and the adimensional current density:

$$j = \frac{Jh}{cqD} \quad (12)$$

With the energy profile (8), the denominator from (7) is:

$$\int_0^h e^{E(x)}dx = \frac{h}{u}\left(\frac{we^{w-ru} - ru}{w - ru}e^{u} - \frac{we^{w+ru} + ru}{w + ru}\right) \quad (13)$$

After further calculations, we obtain the formula of adimensional current density as:

$$j(u) = \frac{u(e^{u} - 1)}{\frac{we^{w-ru} - ru}{w - ru}e^{u} - \frac{we^{w+ru} + ru}{w + ru}} \quad (14)$$

This formula for current density is antisymmetric: $j(u)=-j(-u)$. Thus, it can explain the current nonlinearity but it cannot explain rectification through a pore.

The current density from equation (14) has some properties worth noting. Without an energy barrier, i.e. for w=0, the behavior of the current density is ohmic, the same as in the bulk solution. The bulk current density, j_0, is:

$$j_0(u) = u \tag{15}$$

At large voltages when u>>w, the asymptotic behavior of the current is also linear, but non-ohmic, as seen in Fig. 1. It is not a pure ohmic behavior, because the asymptotic line is shifted relative to the bulk current density j_0, but it has a constant dynamic conductance (resistance) dj/du. This non-ohmic linear current density can be obtained from equation (14) in the limit u>>1, neglecting the terms that contain e^{-u} or e^{-ru}:

$$j(u) \cong u - \frac{w}{r} \tag{16}$$

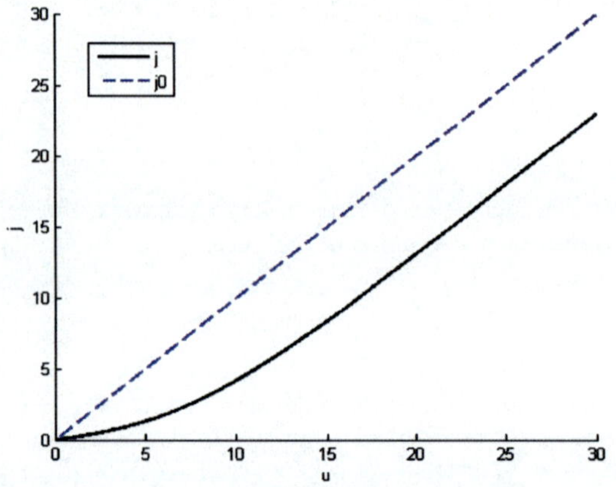

Fig. 1 The linear plot of current density, j, versus applied voltage, u, for a height of the energy barrier of w=2.1, and a relative size of the entrance region of r=0.3. This plot shows mainly the linear behavior of the current density at large voltages. The current density, j, and the bulk current density j_0, has the same slope at large voltages.

Another ohmic behavior is found for r=0, i.e. for a very abrupt energy profile. In this case the current density has a very low value, determined only by the height of the energy barrier: $j(u)=u/e^w$. A similar behavior arises for small applied voltages, when u<<w. Using a Taylor series expansion and neglecting higher order terms, we found the current density formula at small voltages as:

$$j(u) \cong \frac{u}{e^w\left(1-2r+\frac{2r}{w}\right)-\frac{2r}{w}} \tag{17}$$

This fact is confirmed in Fig. 2 – a double logarithmic plot of current versus voltage. At low voltages the slope of current logarithm is similar with that of the bulk current density j_0, this means that j~u, i.e. there is a linear dependence between current and voltage.

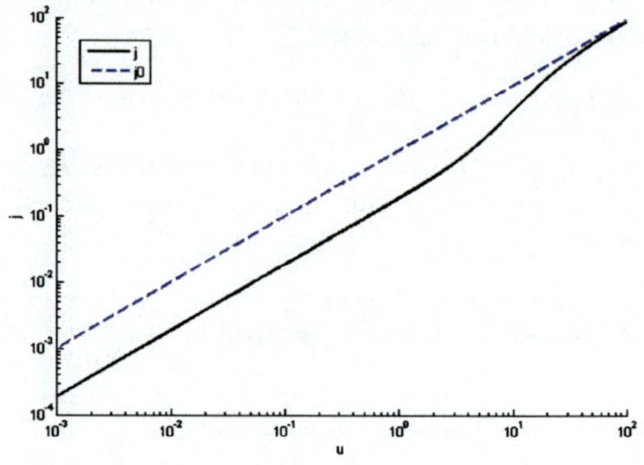

Fig. 2 The double logarithmic plot of current density, j, versus applied voltage, u, for a height of the energy barrier of w=2.1, and a relative size of the entrance region of r=0.3. This plot presents mainly the linear behavior of the current density at small voltages. The current density, j, and the bulk current density j_0, has the same dependence on u at low voltages.

IV. Conclusions

This Nernst-Planck model for the current density in a membrane's pore shows that the current as a function of voltage has three types of behavior, three patterns:

1. a quasi-ohmic linear current at low voltages, with a small slope, whose magnitude depends mainly on the height of the energy barrier, w, through the exponential term e^w;
2. a nonlinear transition region at intermediate voltages, and
3. a non-ohmic linear current at large voltages, with a large slope, whose magnitude depends on the bulk solution conductivity.

The domain of quasi-ohmic linear current is usually below u≈1, which means a voltage below ~25 mV for a normal temperature of 300K. From a practical point of view, this is an experimentally accessible domain. The height of the energy barrier, w, can be obtained from current-voltage measurements made in this domain, using equation (17).

The domain of non-ohmic linear current is above a certain voltage which depends strongly on the energy barrier height, w, and the relative size of the entrance region of the pore, r. Even for small energy barrier heights, w=2.1, and a smooth energy gradient at the pore entrance, r=0.3, the beginning of the high-level linear current is above u=20, which means above ~500mV. These membrane voltages are close to the voltage range in which electroporation starts occurring.

The electroporation phenomenon drastically changes the electrical behavior of the skin. The applied voltage modifies the number of pores into the skin, with a rate of pore creation that depends exponentially on the squared voltage [7]. However for low voltages across the skin (U<5V), where electroporation is supposed to be absent, the current-voltage characteristic of the skin [7] has a strong similarity with our model.

There are other experimental situations where these two linear regimes are clearly visible. In some lipid pores the I–V curve is linear from about −150 to +150 mV, and the I–V relation becomes non-linear over |V|>150 mV [16]. At intermediate pH values of the bathing solution, in polymeric-synthetic nanopores (double-conical pore) the I-V curve deviates from the linear behavior in a similar fashion with our model [10]. In cellular ion channels the I–V relation, usually asymmetric [11, 17], for direct currents has a similar aspect with ours.

The presented model has only two assumed parameters, the energy barrier height, w, and the relative size of the entrance region of the pore, r, with a clear physical meaning. This is an important advantage for fitting and interpreting experimental data. The low voltage domain of linear current, revealed by this model, is experimentally almost unexploited, despite the fact that it can offer direct information about the energy barrier inside a membrane's pore. For all these arguments, we believe that this simple model for the current-voltage nonlinearity is a good starting point for explaining the electrical behavior of the skin at low voltages.

References

1. van Boxtel A (1977) Skin resistance during square-wave electrical pulses of 1 to 10 mA, Medical and Biological Engineering and Computing 15(6), 679-657
2. Grimnes S (1983) Skin impedance and electro-osmosis in the human epidermis. Med Biol Eng Comput. 21(6):739-49.
3. Lochner GP (2003) The voltage-current characteristic of the human skin, MEng (Bio-Engineering) dissertation, University of Pretoria, Pretoria, http://upetd.up.ac.za/thesis/available/etd-09212005-093111/ viewed 2011.
4. Birlea NM, Birlea SI, Tosa V (2009) The skin's electrical asymmetry, Journal of Physics: Conference Series 182, 012020, doi:10.1088/1742-6596/182/1/012020
5. Weaver JC, Vaughan TE, Chizmadzhev Y (1999) Theory of electrical creation of aqueous pathways across skin transport barriers, Adv Drug Deliv Rev. 35(1):21-39.
6. Glaser RW, Leikin SL, Chernomordik LV et al (1988) Reversible electrical breakdown of lipid bilayers: formation and evolution of pores, Biochimica et Biophysica Acta 940:275-287
7. Chizmadzhev YA, Indenbom AV, Kuzmin PI et al (1998) Electrical properties of skin at moderate voltages: contribution of appendageal macropores, Biophysical Journal 74(2) 843–856
8. Xu Q, Kochambilli RP, Song Y et al (2009) Effects of alternating current frequency and permeation enhancers upon human epidermal membrane. Int J Pharm. 372(1-2):24-32.
9. Chizmadzhev YA, Zarnitsin VG, Weaver JC et al (1995) Mechanism of electroinduced ionic species transport through a multilamellar lipid system, Biophysical Journal 68(3): 749-765.
10. Ramirez P, Gomez V, Cervera J et al (2007) Ion transport and selectivity in nanopores with spatially inhomogeneous fixed charge distributions. J Chem Phys. 126(19):194703.
11. Cervera J, Komarov AG, Aguilella VM (2008) Rectification properties and pH-dependent selectivity of meningococcal class 1 porin, Biophysical Journal Volume 94, 1194–1202
12. Neumcke B, Läuger P (1969) Nonlinear electrical effects in lipid bilayer membranes: II. Integration of the generalized Nernst-Planck equations, Biophys. J. 9(9) 1160-1170
13. DeBruin KA, and Krassowska W (1999) Modeling electroporation in a single cell. II. Effects of ionic concentrations. Biophys J. 77(3): 1225–1233.
14. Kilic MS, Bazant MZ, Ajdari A (2007) Steric effects in the dynamics of electrolytes at large applied voltages: I. Double-layer charging, Phys. Rev. E 75, 021502.
15. Kuyucak S, Andersen OS, Chung SH (2001) Models of permeation in ion channels, Rep. Prog. Phys. 64(11): 1427–1472
16. Heimburg T (2010) Lipid ion channels, Biophysical Chemistry 150: 2–22
17. Clapham DE, Runnels LW, Strübing C (2001) The TRP ion channel family, Nat Rev Neurosci. 2(6):387-396.

Author address:

Author: Nicolae-Marius Bîrlea
Institute: Technical University of Cluj-Napoca
Street: Daicoviciu 15
City: Cluj-Napoca
Country: Romania
Email: mbirlea@phys.utcluj.ro

The Skin's Electrical Time Constants

N.M. Birlea[1], S.I. Birlea[2], and E. Culea[1]

[1] Physics Department, Technical University of Cluj-Napoca, Cluj-Napoca, Romania
[2] Electrical & Electronic Engineering, School of Engineering and Informatics, National University of Ireland Galway, Galway, Ireland

Abstract— **We have studied the electrical time constants of human skin potential in response to a constant current impulse (0.4-16 mA/100 μs). Three relaxation times were found with values in the range 10-40 μs, 100-200 μs and 1-4 ms. By correlating the time constants with the electrical model of the skin we found that stratum corneum has a nonlinear resistance that diminishes and a nonlinear capacitance that increases with increasing values of current.**

Keywords— **bioimpedance, impedance spectroscopy, skin, pulsed current, relaxation time.**

I. Introduction

In order to find the relevant electrical parameters describing the skin condition, we studied the relaxation times of the skin potential in response to a constant current impulse. We believe that the time constants could be the characteristic signatures of processes in the skin or markers of the skin's structure.

Bioimpedance spectroscopy studies the frequency dependence of the electrical impedance of biological material. Bioimpedance methods enable on-line monitoring of biological tissues and require low-cost instrumentation. Bioimpedance spectroscopy has many applications as stated in a tutorial book of Grimnes and Martinsen [1]. The skin can be assessed in a noninvasive and fast manner with this method. The electrical impedance of the skin can be used to measure skin moisture [2], to monitor skin irritations or allergic reactions [3], to detect skin cancer [4], to investigate the electroporation of the skin [5] or the transdermal drug delivery [6] and others.

Usually, bioelectrical measurements are performed using sinusoidal currents at several frequencies and the experimental data are fitted with electrical models or empirical formulas. The electrical models of the tissues are combinations of electrical resistances and electrical capacitances. Behind these electrical components are physical mechanisms that generate an interesting behavior of the components versus frequency.

An alternative approach uses square-wave electrical pulses [7] to measure the skin's electrical properties. This time-domain analysis has the advantage of being able to measure instantaneously all the frequency characteristics of the impedance. The method is well suited to obtain the biological impedance which changes with time [8], allows to measure the skin characteristics in a more direct and explicit fashion, and can discriminate more phenomena occurring into the skin. [9].

II. Experimental method

We have measured the electrical properties of human skin using constant current electrical impulses. The measuring system comprises of a signal generator (made in the laboratory) with a 9 V battery as power supply, a bipolar transistor as an open collector current drive, and an oscilloscope (Picoscope 2203) connected to a laptop, which visualizes the potential difference between the electrodes and the current through the electrodes. The signal generator supplies an electrical current impulse (width T= 0.1 ms, the "ON" period) followed by a pause (1 s, the "OFF" period) when the electrodes cannot receive or transmit current (high impedance status).

The oscilloscope (Picoscope 2203, maximum sampling rate 20 MS/s) displays the potential difference between the two electrodes on channel A, the oscilloscope's "ground" being connected to the common point of the electrode with the 100Ω resistance, used for measuring the current on channel B. The signal on channel B triggers the data acquisition and the display.

Two metallic electrodes (Ag), covered by wet cotton (saline solution), deliver the electrical signal to the skin. The reference electrode has a greater surface (25 cm^2) than the active electrode (2 cm^2). To avoid influences on the results from sweat, the skin is wiped using ethanol before the measurement, and after 5 minutes the electrodes are attached on the desired place using elastic ribbons. The reference electrode was placed on the middle of the ventral side of the forearm (hairless skin), and the measuring electrode on the forearm close to the wrist. During the impulse the active electrode was positive in relation to the reference electrode.

A constant current I (0.4, 1, 4, 8, 16 mA) is injected with the described device for a period of T=0.1ms, then the skin's potential is recorded. Figure 1 shows the oscilloscope traces for the potential difference between the electrodes (upper trace, channel A) and the potential difference on the 100Ω resistor (lower trace, channel B), generated by an electrical current of 1 mA.

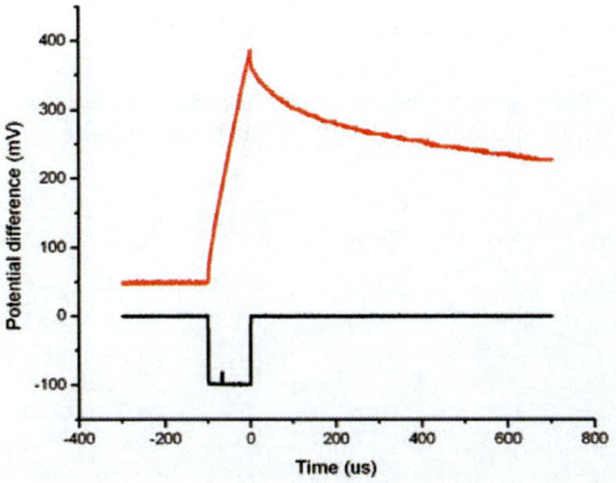

Fig. 1 Oscilloscope traces for the potential difference between the electrodes (upper trace) and the potential difference on the 100Ω resistor (lower trace), generated by an electrical current of 1 mA.

III. Model

The electrical signal applied to the skin is an impulse of constant current with the width T, the "ON" period, followed by a pause when the skin can not receive or transmit current, the "OFF" period. The electrical model for the skin we have chosen consists of 3 parallel R_i C_i configurations (i=1, 2, 3) that are all connected in series and a series resistance R_s ($Z= R_s+\Sigma R_i \| C_i$). In order to find the equation for the potential difference between electrodes a system of 7 equations with 7 unknowns has to be solved. The Kirchhoff equations for the 3 nodes, for the 3 loops formed by R_i and C_i and for the largest loop of the circuit are:

$$R_i \cdot I_i = \int I_i' dt / C_i \text{ (3 equations for the 3 } R_iC_i \text{ loops)}$$
$$I = I_i + I_i' \text{ (3 equations for the currents in 4 nodes)} \quad (1)$$
$$U = U_s + U_1 + U_2 + U_3 \text{ (equation for largest loop)}$$

where:

- I = current through circuit for the time period [0,T];
- I_i = current trough parallel resistance Ri;
- I_i' = current trough capacitor Ci;
- U = potential difference between the two electrodes;
- U_i = potential difference on R_i (I_iR_i) and C_i ($\int I_i' dt/C_i$);
- U_s = potential difference on R_s (IR_s).

The equations for the R_iC_i loops can be written in a differential form as:

$$R_i \cdot dI_i/dt = I_i'/C_i \text{ or } I_i' = \tau_i \cdot dI_i/dt \quad (2)$$

where $\tau_i = R_iC_i$ is the time constant of the i-th loop.

We rewrite the Kirchhoff equations as:

$$I_i' = \tau_i \cdot dI_i/dt$$
$$I = I_i + \tau_i \cdot dI_i/dt \quad (3)$$
$$U = IR_s + I_1R_1 + I_2R_2 + I_3R_3$$

The solutions of the system for $t \in [0,T]$ are:

$$I_i = I \cdot [1 - \exp(-t/\tau_i)]$$
$$U = IR_s + I \cdot \Sigma R_i[1 - \exp(-t/\tau_i)] \quad (4)$$

For $t>T$, the OFF period, the skin does not receive any electrical current so the potential drop on the series resistance is equal to 0 and we are left with a simple exponential decay of the potential for each R_iC_i configuration:

$$U_i = U_i' \exp(-t/\tau_i) \quad (5)$$

where:

$$U_i' = IR_i[1 - \exp(-T/\tau_i)] \quad (6)$$

is the potential difference on the R_iC_i group at the end of the ON period. For sake of simplicity we adopted the time value t=0 at the beginning of the decay. Thus the potential difference between the electrodes will be:

$$U = \Sigma U_i = I \cdot \Sigma R_i[1 - \exp(-T/\tau_i)] \exp(-t/\tau_i) \quad (7)$$

Dividing the potential difference by the excitatory current we obtain the skin response function (dimensionally an impedance)

$$U/I = \Sigma R_i[1 - \exp(-T/\tau_i)] \exp(-t/\tau_i) \quad (8)$$

IV. Data analysis and results

We extracted from the measured data only the points corresponding to the free exponential decay of the electrical potential of the skin ($t>0$). Each time series of the potential difference between the electrodes was divided by the injected current. The resulting response function (dimensionally an impedance), was fitted with a third order exponential decay function using the Microcal OriginPro8 software package. An example of the numerical analysis is presented in figure 2 and the results for all the measurements are given table 1.

Fitting to multiple exponentials is more difficult than fitting to a single exponential. Because the fitting function is:

$$y = y_0 + \Sigma A_i \exp(-t/t_i) \quad (9)$$

the parameter y_0 must be zero, as imposed by the physics of our problem (at very long time the skin potential vanishes). The numerical problem has instability, i.e. its solution may not be unique or may not depend continuously on the data. To avoid to some degree such a situation we begin our nonlinear fitting session with A_1=0, A_2=0 and t_1 and t_2 fixed, in order to estimate t_3. Unfortunately this was possible only by imposing our supposed value for t_3. A better approach was by fixing only A_1=0 and t_1. When the program attains a stable value for fitting parameters we permit A_1 and t_1 to vary. The chi-square values were monitored all the time, in order to see if it is necessary to increase or decrease the number of time constants used.

Table 1 Time constants and amplitudes for the exponential equation used to fit the measured skin response function.

I (mA)	0.4	1	4	8	16
A_1 (Ω)	173	173.1	296	592	743
t_1 (μs)	9.84	20.8	28.0	40.7	25.4
A_2 (Ω)	640	614	800	1185	835
t_2 (μs)	104.6	120	177.9	188	95.1
A_3 (Ω)	3634	2915	2303	1272	703
t_3 (μs)	3772	2883	2828	1829	1182

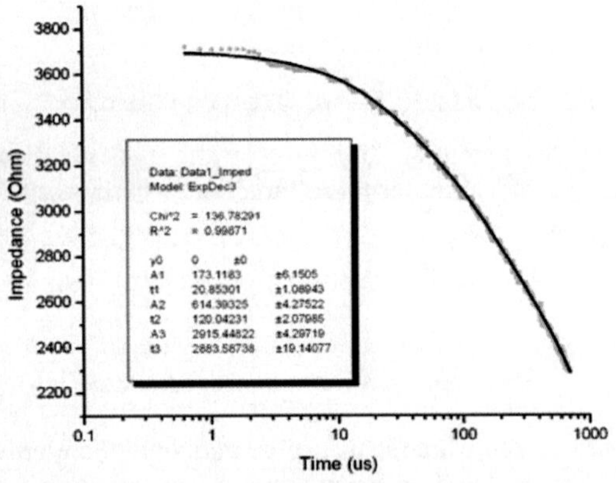

Fig. 2 Non-linear curve fit with a third order exponential decay function of the skin potential divided by the applied current (response function, an impedance) after a pulse of 1mA and 100μs. Logarithmic scale on time axis helps to see the correctness of fitting, especially for short times.

The data from table 1 can be used to find the values of the resistances and the capacitances corresponding to the R_iC_i groups. Comparing equation (8) of the response function to the fitting function (9) we get:

$$A_i = R_i[1-\exp(-T/\tau_i)] \tag{10}$$

Thus we can calculate the resistance R_i and the capacitance C_i as:

$$R_i = A_i/[1-\exp(-T/\tau_i)] \tag{11}$$

$$C_i = \tau_i/R_i = \tau_i\,[1-\exp(-T/\tau_i)]/A_i \tag{12}$$

The values for R_i and C_i are given in table 2.

Table 2 Values of skin resistances and capacitances correspondig to the amplitudes and time constants resulted from the fitting procedure.

I (mA)	0.4	1	4	8	16
t_1 (μs)	9.84	20.8	28.0	40.7	25.4
R_1 (Ω)	173	174.5	304.6	647.5	757.8
C_1 (nF)	56.9	119.2	91.9	62.8	33.5
t_2 (μs)	104.6	120	177.9	188	95.1
R_2 (Ω)	1040	1086	1403	2872	1283
C_2 (nF)	100.6	110.5	126.7	65.4	74.1
T_3 (μs)	3772	2883	2828	1829	1182
R_3 (Ω)	138899	85505	66287	23906	8666
C_3 (nF)	27.1	33.7	42.7	76.5	136.4

V. Discussion and conclusion

Biological materials have unusual dielectric spectra, enormous static polarization and anomalously large relaxation times. All of these complex effects result from physical processes at the solid–liquid interface of a porous system [10]. The existing theories of polarization of heterogeneous media can be useful for the interpretation of frequency spectra of porous materials, or to determin the connectivity of water-bearing channels [11]. Knowledge of the characteristic signatures of these physical mechanisms is important for the correct interpretation of experimental data.

The chosen theoretical model has 3 time constants, electrical equivalent of the three RC groups connected in series. This choice was made because the electrical current flows successively across the different layers of the skin. This specific type of measurement, with constant current excitation, offers the advantage of a simple link between the measured data and the model parameters.

From table 2 we can see that an increase in the excitatory current yields a reduction in magnitude in the largest relaxation time t_3 and the corresponding resistance R_3, which diminishes drastically from 139 kΩ at 0.4mA to 8.7 kΩ at 16mA. These results are consistent with earlier findings of van Boxtel [7] or recent measurements of Kuhn [12, 13]. Capacitance C_3 has an interesting behavior, it increases with the increase in current values. Because the epidermal stratum corneum is credited with the largest value of the resistance among skin layers, it is natural to consider that capacitance C_3 belongs to it. From these measurements we can draw the conclusion that stratum corneum has a nonlinear resistance that diminishes at larger currents and a nonlinear capacitance that increases at larger currents.

The shorter relaxation times t_1 and t_2 have a similar behavior versus current, they increase with increasing current until 8 mA, then decrease at 16mA. R_2 resistance has the same behavior, but the R_1 resistance only increase. If we take into account that the skin potential difference is above 30V only at 16mA, it seems that electroporation phenomenon [14, 15] influences R_2, but not R_1.

Our measurements show the presence of 3 relaxation times in the free decay of skin electrical potential. Resistances R_1 and R_3 have a monotonic behavior, but not resistance R_2. Unexpectedly, capacitance C_3 of stratum corneum is nonlinear and it increases together with the current.

References

1. Grimnes S, Martinsen OG (2000) Bioimpedance & Bioelectricity Basics. Academic Press, London.
2. Martinsen Ø G, Grimnes S and Karlsen J (1995) Electrical methods for skin moisture assessment. Skin Pharmacol. 8 (5), 237-245
3. Nyren M, Kuzmina N and Emtestam L (2003) Electrical impedance as a potential tool to distinguish between allergic and irritant contact dermatitis. J. Am. Acad. Dermatol. 48 (3) 394-400
4. Åberg P (2004) Skin cancer as seen by electrical impedance (PhD thesis), Karolinska Institutet, Stockholm, Sweden
5. Chizmadzhev Y A, Indenbom A V, Kuzmin P I, et al (1998) Electrical properties of skin at moderate voltages: contribution of appendageal macropores. Biophysical Journal 74 843–856
6. Grimnes S 1983 Skin impedance and electro-osmosis in the human epidermis. Med. Biol. Eng. Comput. 21 739-749
7. van Boxtel A (1977) Skin resistance during square-wave electrical pulses of 1 to 10 mA. Med, & Biol. Eng. & Comput., 15, 679-657
8. Yamamoto Y, Isshiki H, and Nakamura T (1996) Instantaneous measurement of electrical parameters in a palm during electrodermal activity. IEEE Transactions on Instrumentation and Measurement, 45(2) 483-487
9. Bârlea N M, Bârlea S I, Culea E (2008) Maxwell-Wagner Effect in the Human Skin, Rom. J. Biophys. 18(1):87-98 http://www.biophysicsnet.ro/rjb/articles/202/nmba.pdf
10. Chelidze T L, Gueguen Y (1999) Electrical spectroscopy of porous rocks: a review - I. Theoretical models, Geophys. J. Int. 137, 1-15
11. Scott J B T (2003) Low-frequency electrical spectroscopy of sandstone, PhD thesis, University of Birmingham
12. Kuhn A (2008) Modeling Transcutaneous Electrical Stimulation, PhD thesis, Swiss Federal Institute of Technology (ETH) in Zurich, Switzerland, http://control.ee.ethz.ch/index.cgi?page=publications&action=details&id=3144
13. Kuhn A, Keller T, Prenaj B, Morari M (2006) The relevance of nonlinear skin properties for a transcutaneous electrical stimulation model. In International Functional Electrical Stimulation Society Conference, vol. 11, 100-102, Zao, Japan. http://control.ee.ethz.ch/index.cgi?page=publications&action=details&id=2484
14. Chizmadzhev Y A, Indenbom A V, Kuzmin P I, Galichenko S V, Weaver JC, Potts RO (1998) Electrical properties of skin at moderate voltages: contribution of appendageal macropores, Biophysical Journal 74, 843–856
15. Pliquett U, Langer R, Weaver J C (1995) Changes in the passive electrical properties of human stratum corneum due to electroporation, Biochimica et Biophysica Acta 1239:111-121

Author address:

Author: Nicolae-Marius Bîrlea
Institute: Technical University of Cluj-Napoca
Street: Daicoviciu 15
City: Cluj-Napoca
Country: Romania
Email: mbirlea@phys.utcluj.ro

Experiments in Electrotherapy for Pain Relief Using a Novel Modality Concept

P. Cevei[1], M. Cevei[2], and I. Jivet[1]

[1] University 'Politehnica' Timisoara, Electronics and Telecommunications Faculty, Timisoara, Romania
[2] University of Oradea, Faculty of Medicine and Pharmacy, Oradea, Romania

Abstract— **The paper presents a novel concept in the modality of electrotherapy use for pain relief. The classical electrotherapy physiological effects for pain relief are critically reviewed with the objective to complement it with a systemic whole human body pain perception management model. A novel chaotic oscillator based modality concept is introduced and objectives defined. For two clinical sample cases chaotic models are associated for clinical trials on the basis of estimated best synchronization. The double pendulum chaotic oscillator model use in electrotherapy for energy waveform delivery is detailed.**

Keywords— **Electrotherapy, pain relief, modality, chaotic oscillators.**

I. INTRODUCTION

The use of electrotherapy in pain relief is a very active ongoing researched topic, a method accepted in medical profession as a palliative mean of pain relief, in the practice of rehabilitation and of physiotherapy medicine [1].

The acute interest on the subject is due to electrotherapy use with success in multiple directions like pain management, tissue repair, treatment of neuromuscular dysfunction, acute and chronic edema and others.

Transcutaneous electrical stimulation (TENS) is the most frequently used method of electrotherapy. The impulses are generated by a device and delivered through electrodes on the skin in direct proximity to the target tissue or nerve bundles to be stimulated.

The electric current induced charges mimic the action potential coming from the central nervous system, offering a mean to intervene in the system function to reduce suffering and also stimulate normal body mechanisms to react to a temporary distressful situation.

II. ELECTROTHERAPY BASICS IN PAIN RELEIF

A. *Pain Physiological Basis*

The human body perceives pain, as a complex system reaction to an out of optimal physiological experience. The brain perception is a sting, burn or ache. Receptors in the body participate in pain building, through generation of electrical impulse originating in the injured area, relayed then by the spinal cord to the brain.

The spinal cord function is as gate and temporary relay where pain can be blocked or its intensity modified before it is transmitted to the brain. The location in the brain where pain signals assembles is the thalamus and then it is relayed to the cognitive cortex.

Different neurotransmitter channels exist in the human body, with some playing a role in human disease causing pain others active in transmitting external receptor signals due to the exterior environment, all acting in combination to determine the painful sensation.

Traditional medicine uses drug doses to alleviate pain sensation or reduce it in the mild case. Intense or severe pain can be or suppress as well by different mechanisms. Morphine and other analgesic drugs are effective in locking on to the corresponding receptors, opening the pain-inhibiting pathways and in this way blocking pain.

A nociceptor is formed by thin nerve fibers in body tissues like skin and muscle, that carry pain signals to the spinal cord and brain. Normally, nociceptors only respond to strong stimuli such as a therapeutic external stimulus [1].

Many kinds of pain can be experienced like the spontaneous pain and abnormal sensitivity following a nerve injury, result from a traumatic injury. Disease or infection, surgery related pain, can persist longer even beyond the time the injury has healed. This forms and may other pain clinical cases are in high need of an efficient pain relief modality as well.

All of the above mentioned pain common clinical cases support the need for a broader insight and a solution for pain management supporting a improved modalities in electrotherapy.

B. *Electrotherapy Novel Modality Premises and Objectives*

According to recent pain relief related literature smart TENS (transcutaneous electrical neural stimulation) is called for to device more efficient energy delivery.

Electro-medicine is based on scientific interdisciplinary studies putting together electrical technology and recent progress in biophysics or biochemistry [3].

S. Vlad and R.V. Ciupa (Eds.): MEDITECH 2011, IFMBE Proceedings 36, pp. 164–167, 2011.
www.springerlink.com

One such recent advancements is the micro-current levels stimulation, a modality that the people involved in the present work adhere to [8].

The present work embraces this idea that minimal energy is sufficient to help remedy the causes of pain in the human body. In general terms classical electrotherapy and drug based therapy use 'substantial overdose' of agents forcing physiologic mechanisms pyramidal structure to come into effect. The success of the treatment relays on the activation of higher level mechanism to react and to correct the pain causing situation.

Our new modality of electric energy injection in the human body can be loosely compared to alternative medicine modality in which minute doses are used and the delivery format are the essential differences from standard medicine.

Homeopathy starts from the assumption that the energetic residual of the chemical has a closer more profound effect than allopathic medicine known to be an excessive invasive procedure.

The physiological premise of the proposed new form of the energy delivery modality is appealing due to the known fact that physiological subsystem in normal healthy subjects indicates a natural variation of the chaotic type. Epilepsy seizure and heart rhythm are the most often cited examples of chaotic evolving physiological cases and best modeled and understood [8].

The novel electrotherapy modality proposed addresses mainly the electric energy delivery format.

We propose the use of chaotic oscillator waveforms as a means of electric energy delivery to the pain region. It is postulated that appropriately chosen chaotic waveform will synchronize the natural rhythm of the physiological subsystem in distress causing pain. The injected energy end effect is thus directed to the objective of the system stabilization and pain removal.

The electric energy effective needed for synchronization we expect to be very small compared with the current density used in electrotherapy nowadays.

III. Chaotic oscillators as models for energy delivery

The use of chaotic oscillators as an energy propagating modality received scarce recent research solutions, as reflected by literature.

It is one of the objectives of the reported research to probe this promising capacity at the energy transfer level among systems using their inherent spread spectrum characteristics of the chaotic oscillators.

Starting from the recent studies on the chaotic oscillators properties, the present paper focused on matching the energy transmitting capacities to the application receptivity, to cover its need for a profound energy transfer.

We have found that the task of matching the transfer is very much depending on a deep understanding of the characteristics of the oscillator behavior relative to the body part undergoing a therapeutic procedure for pain relieve.

Chaotic oscillators are the result of research of the behavior of dynamical systems and are known to be particularly sensitive to initial conditions.

Used in energy transfer application area among autonomous systems, the chaotic oscillators synchronizing capacities occupies a central role. We consider the synchronization level instrumental to the success of the experiments. According to literature there is more then one way the synchronization process can take place among chaotic oscillator:

Direct synchronization – a straightforward form of synchronization among two structural very similar chaotic oscillators when put contact by a medium.

Generalized synchronization – is a type of synchronization happening in between two or more chaotic oscillators that have a different detailed structure.

The synchronization can also be observed related to the generalized phase of the oscillator defined in relation to the equilibrium state of the oscillator.

IV. Method set-up for target clinical cases

There are several recently promoted electrotherapy modalities that are using the same minimal electric energy injection method in the system, like we propose. The *interferential electro-stimulation therapy* is a therapeutic treatment method targeting pain reduction and soft tissue healing. Low energy electrical impulses are injected into the tissues at low-frequencies. This stimulation method prompts the body to secrete endorphins, natural pain killers and inhibitors to relieve pain [6].

The electrotherapy pain management literature to date, indicates initiatives to re-explore the profound mechanisms that traditional therapies a based on, to device methods for their modernizing by using new technological advances for the benefit of efficiency of the clinical practice.

In the modality of electric energy injection, on magnitude and waveform there is little information to develop a solid starting point.

The recent results reported in literature on the options of the devices used to deliver electric energy to the human body, selectively indicate *frequency and amplitude 'windows'*, as more efficient in delivering targeted energy in spatial-temporal forms.

The proposed modality of electrotherapy delivery is different from the use of natural non-conventional electric current generators to deliver electric energy, not as simplified common electric generators sending impulses with amplitude and frequencies, but in a extended form of chaotic oscillator energy waveform.

Two clinical cases have been selected for the first batch of clinical trials of our new electrotherapy modality. Following a search for similarities with respect to internal structure we selected two types of chaotic oscillators for use in electrical energy transfer to relieve pain.

Case 1. Diffused pain caused by inflamed tissue (spondilita anchilozanta for example) develops in a well defined anatomical volume. We considered a matching oscillator for direct synchronization, to match the diverse but localized structure to the body tissue causing the pain. The energy injected in the area by a series of electrodes along a line or curve must have a 'massage type' action, variable amplitude, with no preferential direction or frequency.

The chaotic quasi-oscillatory trajectory model of the double pendulum was chosen as a matching oscillator aiming for direct synchronization in cases of diffused pain.

Case 2. When the anatomy is known but the physiological mechanism is more subtle, like generalized back pain, a spatially structured chaotic oscillator will more likely couple efficiently energy to the body by generalized synchronization.

In neural pain formation the nerve bundle role and interaction can be expressed by nervous pulse interrelation that takes place in time, as a whole relative amplitude complex rather then on individual basis [5].

A two dimensional area splitting and re-mapping chaotic oscillator like the Arnold's chaotic oscillator was chosen as model for electrotherapy delivery in the back pain case.

The *double pendulum* exhibits chaotic behavior if the initial displacement is large as presented in Fig. 1. The position equations are given by the relations below:

$$x1 = L1 \sin \theta 1 \quad (1)$$

$$y1 = -L1 \cos \theta 1 \quad (2)$$

$$x2 = x1 + L2 \sin \theta 2 \quad (3)$$

$$y2 = y1 - L2 \cos \theta 2 \quad (4)$$

where the coordinates of the centers of the arms are x and y respectively , L the arm length and θ vertical inclination.

The solution of equations of motion for the double-pendulum is possible to be obtained analytically if one adds the force balance equations to equations (1)- (4):

$$\sin \theta 1\ (m1\ y1'' + m2\ y2'' + m2\ g + m1\ g) = -\cos \theta 1\ (m1\ x1'' + m2\ x2'') \quad (5)$$

$$\sin \theta 2\ (m2\ y2'' + m2\ g) = -\cos \theta 2\ (m2\ x2'') \quad (6)$$

where supplemental parameters are the mass of the arm m, g the gravitational acceleration, while " represents the second time derivative.

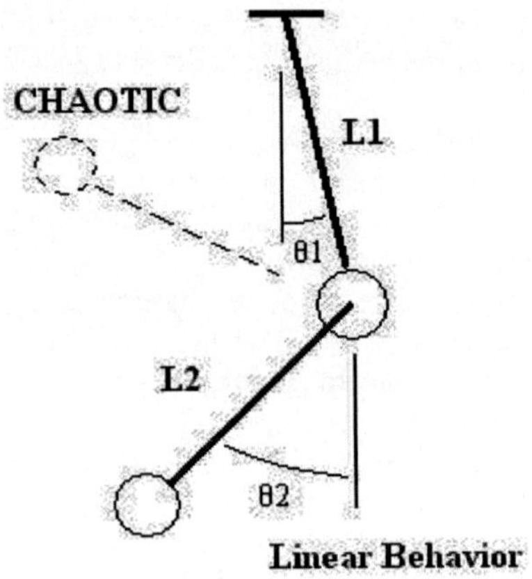

Fig. 1 Motion of a double-pendulum depending on the initial position

Table 1 presents a few approximated values extracted from the calculated trajectory sequence as indicated in Fig. 2., for the initial cycle and the first flip o the second pendulum (the sixth cycle).

Table 1 Sample positions during motion in selected cycles

Position	First	Second
First cycle		
1	90.0	0.0
2	65.5	2.4
3	28.5	5.6
4	-32.2	9.2
5	-60.6	2.3
6	-85.3	-4.5
7	-75.3	-16.3
8	-30.2	-30.2
The sixth cycle [the fist flip]		
1	-3.2	-175.7
2	20.4	-105.1
3	35.3	-24.1
4	21.6	48.8
5	-4.6	100.5

The two one dimensional coupled system model can be easily implemented and applied in two channels as separate variable amplitude over time waveforms from a set of tables with stored values.

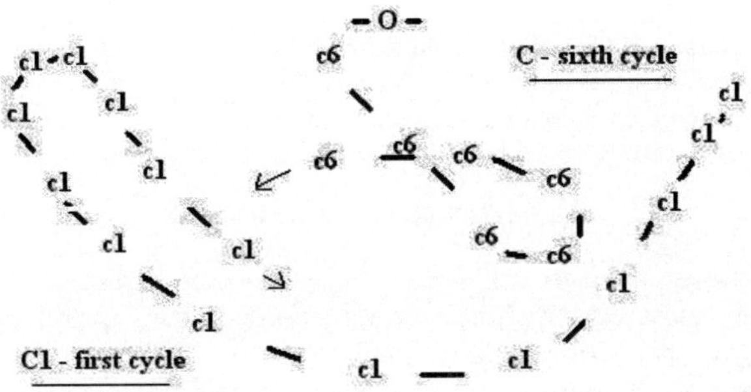

Fig. 2 Sample motion of a double-pendulum exemplifying chaotic behavior.

According to the proposed new electrotherapy modality, the coupled pair of pendulum motion time waveform will be used as a model of chaotic oscillator for energy delivery modulation. The choice of the magnitude of the parameter extracted from the motion model and the electric current to be modulated was not addressed at this stage.

V. Preliminary results and clinical experiments plan

The research results presented in the paper cover the phase of concept validation and planning for clinical trials. A experimental model to be used has been built from previously computed chaotic oscillator data stored in memory.

The current delivering electrodes is a set of five classical electrotherapy electrodes and the power amplification interface.

The power over time histogram has been found to be very uniform over the whole line of electrodes compared to diadynamic current waveforms, clustered over a few values. The amplitude range is quasi-continuos spread over a decade and has just occasional critical points.

The clinical validation trials are under way at University Hospital, University of Oradea.

VI. Conclusions

A novel concept in the electrotherapy delivery modality for pain relief is presented. A novel chaotic oscillator based delivery modality concept is introduced and objectives defined.

It is shown that energy transfer among autonomous systems is mainly dependent of the chaotic oscillators synchronizing capacities. By similarity two chaotic oscillators models have been chosen for the use in a two case clinical trial.

The chaotic energy waveform was found to be more evenly spread over a range of values computing its local histogram.

Following the presently ongoing clinical trials a range of aspects of the newly proposed electrotherapy delivery modality must be addressed. Multiple model interaction, time regime experiments and energy delivery efficiency estimation, to enumerate just a few.

References

1. Watson, T. (2010) Key concepts with electrophysical agents, Physical Therapy Reviews 15(4): 351-359
2. Johnson, M. (2008). TENS In : Electrotherapy: Evidence Based Practice. Ed. Watson. T. Elsevier
3. Elwakil A.S., Kennedy M.P. (1999) Chaotic oscillator Configuration using a Frequency Dependent Negative Rezistor, Journal of Circuits, Systems and Computers, Vol. 9, pp.229-242
4. Brosseau, L., et al. (2002). "Efficacy of the transcutaneous electrical nerve stimulation for the treatment of chronic low back pain." Spine 27(6): 596-603.
5. Hurley,D.A. et al. (2004) A randomized clinical tral of manipulative therapy and interferential therapy for acute back pain, Spine 29 (20), pp 2207-2216)
6. Kulkarni, A.D.et al. The use of microcurrent electrical therapy and cranial electrotherapy stimulation in pain control. Clinical Practice of Alternative Medicine. 2(2):99-102, 2001
7. Jeong-Kyu H.,et al. Chaotic indices and canonical ensemble of heart rate patterns in small-for-gestational age fetuses, J. Perinat. Med. 35 (2007) pp. 210–216
8. http://www.myphysicslab.com/dbl_pendulum.html/(Kinematics of the Double Pendulum)

Author: Ioan Jivet
Institute: University 'Politehnica' Timisoara
Street: V Parvan No 2
City: Timisoara
Country: Romania
Email: ioan.jivet@etc.upt.ro

Double Stimuli Paradigms Should Be Careful Interpreted When Applying Lumbar Magnetic Stimulation

L. Darabant[1], M. Krenn[2], K. Minassian[2,3], M. Cretu[1], W. Mayr[3], and R.V. Ciupa[1]

[1] Electrotechnics Department, Technical University of Cluj-Napoca, Romania
[2] Center for Medical Physics and Biomedical Engineering, Medical University of Vienna, Austria
[3] Institute for Analysis and Scientific Computing, Vienna University of Technology, Austria

Abstract— **We investigated the neural structures activated by transcutaneous magnetic stimulation of the spinal cord, by examining the evoked electromyographic responses in the lower limbs. Only in few cases posterior root-muscle reflexes could be elicited by activation of the posterior root afferent fibers. Generally, M-waves were elicited by direct stimulation of the alpha motoneuron axons within the anterior root or spinal nerve. However, the double stimulus paradigm in the interpretation of the muscle responses by their refractory behavior was inefficient. Stimulation evoked strong contraction of the paraspinal muscles by the first stimulus and increased the distance between magnetic coil and the back during the second pulse, i.e. changed the stimulation conditions. To avoid misinterpretation of the data, additional neurophysiological methods to condition the responses should be applied for their identification, such as tendon vibration.**

Keywords— **Electrical and magnetic stimulation, spinal cord, muscle response, double stimulus paradigm.**

I. Introduction

Magnetic stimulation of the lumbar cord and cauda equina was repetitively investigated. Previous studies have shown that magnetic fields applied over the lowest thoracic and lumbar vertebrae predominantly stimulate alpha-motoneurons. The evoked responses are electromyographically recorded as M-waves in several muscles of the leg. The direct stimulation of the alpha-motoneurons results in short and constant latencies of the compound muscle action potential (CMAP). Some recent developments prompted us to revisit the effect of lumbar magnetic stimulation in the present study. First, Gerasimenko et al. [4] demonstrated a new type of spinal response by magnetic train stimulation in humans with intact nervous system. The availability of stimulators that can apply repetitive magnetic impulses allows us to introduce a double stimulus paradigm. Second, transcutaneous electrical stimulation of the lower thoracic vertebrae level shows sensory fiber activation within the posterior roots [1]. Here we applied a double stimuli paradigm to investigate possible reflex responses to magnetic stimulation of the spinal cord. We further chose tendon vibration as an independent method to identify potential reflex responses.

II. Materials and Method

Seven subjects with intact nervous system participated in the study. EMG signals were recorded from quadriceps (QM), hamstring (HM), tibialis anterior (TA) and triceps surae (TS) muscles, with subjects in the prone position. Electrical and magnetic stimulation was applied to several stimulation sites, at maximum tolerable intensity.

The experiment consisted in stimulating the rostro-caudal area of several subjects, both by electric and magnetic means. We used a computer controlled electrical stimulator and a Magstim Rapid² (The Magstim Company Ltd, Whitland, UK) magnetic stimulator. First electrodes and then a stimulation figure-eight coil were placed above the vertebral column. The nervous impulses generated in this area are propagated through nervous pathways to the lower limbs, producing muscular contractions. The compound muscle action potentials were recorded through electromyography, using an eight channels electromyograph.

Electrical stimulation was used as a reference method, since EMG signals recorded from leg muscles due to spinal stimulation are, by now, well known in the literature. Electrodes were placed above T11T12 vertebras, as one can see in Figure 1.

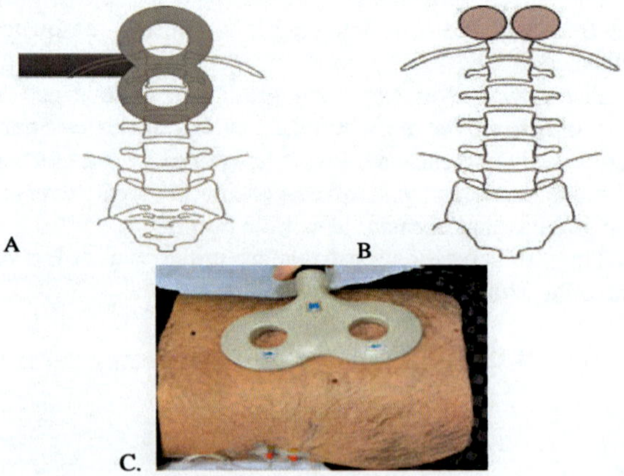

A B C.

Fig. 1 Position of magnetic coil (A) and electrodes (B) for stimulation of the spinal cord. Photo of coil (C)

S. Vlad and R.V. Ciupa (Eds.): MEDITECH 2011, IFMBE Proceedings 36, pp. 168–171, 2011.
www.springerlink.com

We established a certain experimental protocol, starting by defining the coil position with respect to the vertebral column. The abbreviations we used are L- for left, R- for right and M for middle, followed by the angle between the spine and the handle of the stimulation coil, i.e. M90 means that the coil is placed centered above the vertebras and its handle is perpendicular to the spine (Fig. 1 – A and C).

The experimental protocol first tried to recreate the stimulation conditions from electrical stimulation. Therefore, we were looking for short-latency responses, with the coil in the M0 position, above T11-T12 vertebras. We slowly incremented the intensity of the stimulus (by 5% per step), until we reached first the threshold, and then the maximum (tolerable) intensity. This was limited by the subjects' discomfort (while being motivated by the examiner).

We started our stimulation by applying a single stimulus. Afterwards, in order to identify the nature of the muscle response, we applied both the double stimuli paradigm (a good indicator of the refractory period of the response) and the vibration method of modifying the muscle response. Due to the stimulator limits, the shortest interstimulus interval we could apply (and still be able to use magnetic stimulation intensities of 80-100%) was 67 ms (corresponding to a frequency of 15 Hz). Even in these conditions, the stimulator sometimes failed to deliver the second pulse, due to coil heating.

For each position assumed, we repeated our stimulation 10 times (to check for repeatability) and recorded the response of 4 muscles (as mentioned in the introduction, but on both legs). Between each couple of stimuli, we provided our subjects with at least 10 second break, to ensure that their muscles are relaxed and ready for the following pulse.

We also investigated the effect of changes of rostro-caudal stimulation site at maximum tolerable intensity, by assuming several other positions, above a range of vertebras from T9 to L2. We also changed the coil orientation, by assuming the R270 (coil placed above the right side of the spine, and the angle between the spine and the coil's handle has 270 degrees), L90, M0 and M90 positions. For the middle cases, we were extra careful to adjust the position of the coil to elicit bilateral symmetrical responses.

III. Results and discussions

At first, we applied double stimuli on the T11-T12 level, both by electrical and magnetic stimulation. For all stimulation types, the interstimulus interval was set to 67 ms (limited by the magnetic stimulator). The intensity of the stimulus applied by electrical stimulation was 31V, and the CMAP recorded are plotted in Figure 3-A. For magnetic stimulation, the position of the coil was M0, and the intensity of the stimulus was set to 90% of the maximum intensity achievable with our stimulator; simulation results are plotted in Figure 2-B.

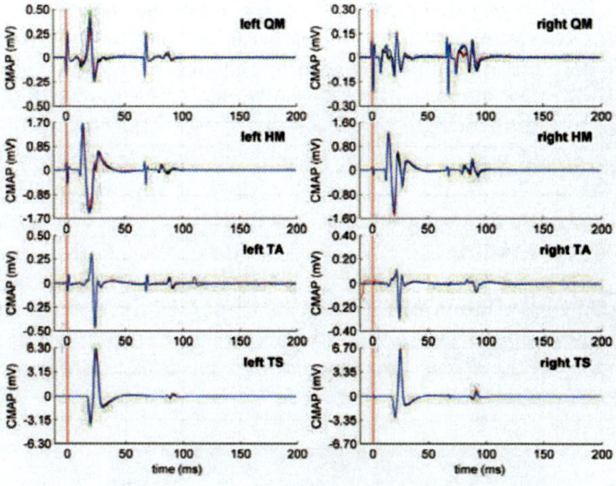

A.

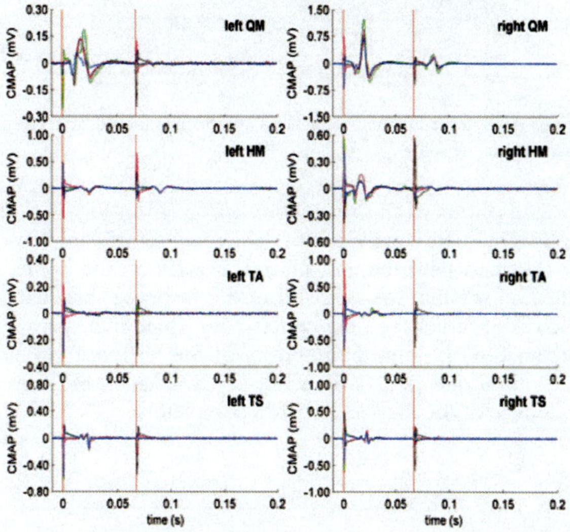

B.

Fig. 2 CMAP recorded for electrical stimulation (A) and magnetic stimulation (B) of the spinal cord on level T11-T12.

For both recordings, one can notice that the CMAP response to the second stimulus applied has much lower amplitudes or is completely suppressed. This is usually an indication of the refractory period similar to the posterior roots muscle (PRM) reflexes.

However, when we took videos during the magnetic stimulation and applied an accelerometer to the spine, we found that during the double-stimulus paradigm, the first stimulus produces a considerable movement/bending of the spine due to a strong contraction of the paraspinal muscles. The largest acceleration takes place 25-30 ms after the stimulus application.

The largest distance of the spine from the magnetic coil can be estimated to be about 1-1.5 cm and was approximately assumed at 90 ms after application of the first pulse. With other words, when the 2nd pulse is applied, stimulation conditions have changed and a less effective field is generated within the spine (due to the distance between the coil and the body). This problem doesn't appear with electrical stimulation, since the electrodes are attached to the subject's skin.

Due to this very important remark, we decided to investigate the variation of the muscles responses for a single stimulus applied and several distances between the magnetic coil and the spine. For this purpose, we used spacers of 1, 1.3, 1.5 and 2 cm – see Figure 3.

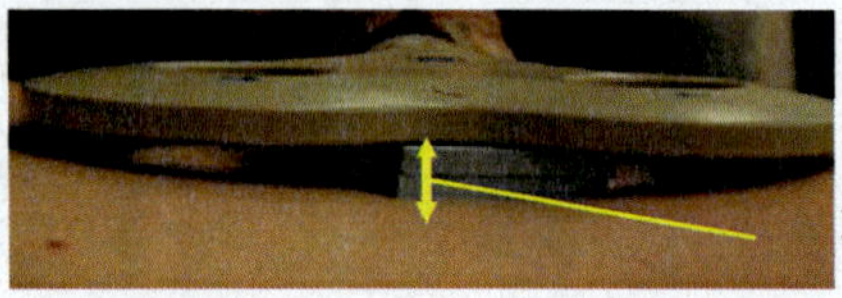

Fig. 3 Applying a stimulus with the magnetic coil placed at a given distance above the spine

The results of this evaluation are given in Fig. 4, for left and right quadriceps muscles. The most significant response is obtained with the coil placed directly on the spine, and one can see that the intensity of the response diminishes as the distance between the coil and the spine increases. For a distance of 1.3 cm, the intensity of the response is 63% of the initial one (left QM), while for a distance of 2 cm, the response of the muscle vanishes completely.

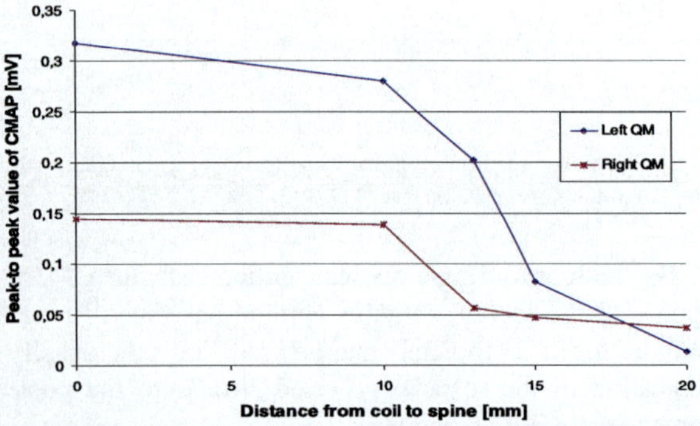

Fig. 4 Peak to peak value of CMAP for a magnetic stimulus of 90% amplitude, applied on level T12L01, position M270, for different distances between the coil and the spine

Therefore, we concluded at the moment that double stimulus paradigms cannot be interpreted unequivocally in case of magnetic stimulation. The inefficiency of the second magnetic pulse to elicit a second response could be due to neuro-physiological mechanisms (refractoriness) or just due to changes of stimulation conditions. In that case, the short-latency responses recorded would be just M waves.

However, we observed that the latencies of CMAP responses to magnetic stimulation are somewhat smaller than the responses to electrical stimulation on the same level (T12) and for the same subject – Table 1.

Table 1 Latencies for electrical and magnetic stimulation

Muscle	Latency for electrical stimulation - ms	Latency for magnetic stimulation - ms
left QM / right QM	14.3 / 6.1 (M-Wave)	8.1 / 9
Left HM / right HM	11.6/ 10.7	11.2 / 9.7
Left TA / right TA	19.6 / 17	17 / 15.6
Left TS / right TS	18 / 17.3	15.1 / 16

This is why we continued our investigation by applying tendon vibration, another known technique of suppressing PRM reflexes [2]. For electrical stimulation, we were able to suppress the evoked response for all subjects - Fig. 5, indicating this way again that the responses are indeed PRM reflexes.

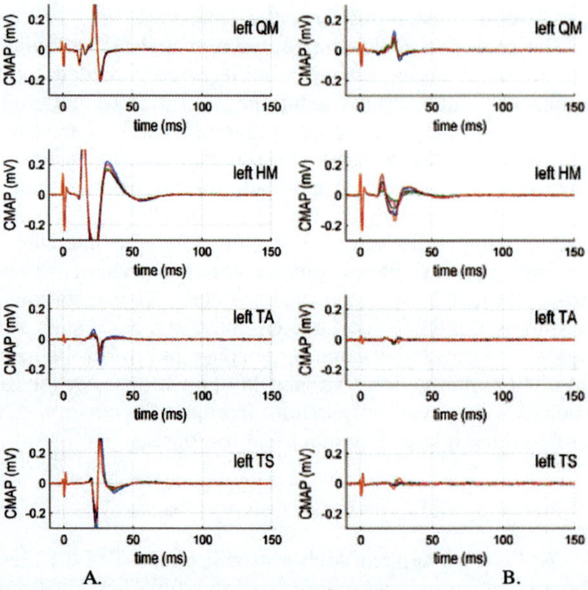

Fig. 5 Muscles responses to a single 28V stimulus applied by electrical stimulation, with electrodes on T11-T12 position, no vibration applied (A) and vibrations applied to left Achilles tendon (B)

One can observe that the responses of left TA and TS is almost gone due to vibrations applied, while the responses of left QM and HM have lower intensities. The responses of the right leg muscles are not influenced by vibrations (same results for single stimulus applied, with or without vibrations of the left Achilles tendon). This is a confirmation of the results obtained by applying the double stimulus paradigm.

If we can easily elicit PRM reflexes by electrical stimulation in all subjects, we could only generate them in one of our subjects by magnetic stimulation. We were able to confirm the nature of these responses by tendon vibrations.

The coil was in position M90, above the L1-L2 vertebras. The stimulus was set to a 90% intensity of the maximum that the magnetic stimulator can deliver. First we applied a single pulse, no vibrations, and then we were able to diminish the intensity of the response in the muscles by vibrating the right Achilles tendon – Fig. 6.

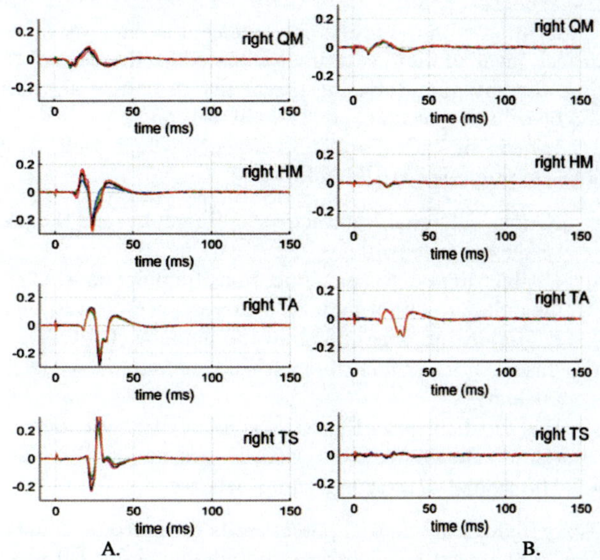

Fig. 6 Muscles responses to a 90% magnetic stimulus, with coil on L1L2 position, no vibration applied (A) and vibrations applied to right Achilles tendon (B)

For all the other subjects of our study, the short-latency responses evoked by magnetic stimulation could not be suppressed by tendon vibration – Fig. 7.

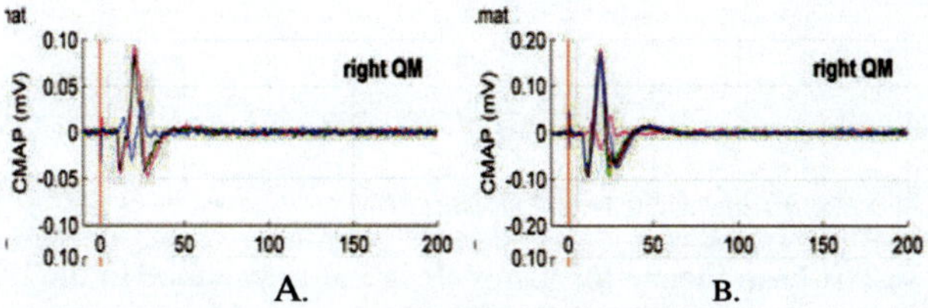

Fig. 7 CMAP for a magnetic stimulus of 100% intensity, with coil on level T12L01, position M0, without vibration (A) and with vibrations applied to right Achilles tendon (B)

IV. Conclusions

Both electrical and magnetic stimulation of the spinal cord can activate efferent fibers, generating M-waves in the lower limb muscles. PMR reflexes can also be elicited by both stimulation techniques, due to activation of afferent fibers. However, we could easily elicit PRM reflexes by electrical stimulation in all subjects, while by magnetic stimulation we were only able to generate them in one out of seven subjects.

Vibration effectively suppresses PRM reflexes. The vibration-induced suppression is due to various mechanisms, probably including presynaptic inhibition and a partial occlusion of input to the spinal cord carried via large afferents, and is a further indication for the reflex origin of the responses evoked by transcutaneous spinal cord stimulation.

The double stimulus paradigms were inefficient in the interpretation of muscles responses due to magnetic stimulation. The first stimulus produces a considerable bending of the spine, due to back muscles contractions, changing this way the stimulation conditions.

Acknowledgment

This research was founded in the frame of an Austria-Romania bilateral project, *"Functional Stimulation of the spinal cord"*, 229 / 2009 and a post-doc grant PD_611/2010.

References

1. Minassian K, Persy I et al. (2007) Posterior root–muscle reflexes elicited by transcutaneous stimulation of the human lumbosacral cord. Muscle&Nerve 35(3):327–336
2. Hofstoetter US, Minassian K, et al. (2008) Modification of reflex responses to lumbar posterior root stimulation by motor tasks in healthy subjects. Artif. Organs, 32(8):644-648
3. Kitano K, Koceja DM (2009) Spinal reflex in human lower leg muscles evoked by transcutaneous spinal cord stimulation. Neuroscience Met. J 180(1):325–329
4. Gerasimenko Y, Gorodnichev R, Machueva E, Pivovarova E, Semyenov D, Savochin, A, Roy RR, Edgerton VR. (2010) Novel and direct access to the human locomotor spinal circuitry. J Neurosci.; 30(10): 3700-3708.

Author: Laura Darabant
Institute: Technical University of Cluj-Napoca
Street: 26-28 Baritiu Str.
City: Cluj-Napoca
Country: Romania
Email: Laura.Darabant@et.utcluj.ro

Virtual Instrument for Early Detecting Dental Decay Using Electrical Impedance Measurement Method

S.G. Lacatusu[1], M. Branzila[1], M. Cretu[1], D. Lacatusu[2], and S. Lacatusu[3]

[1] Technical Univ. "Gh. Asachi" of Iasi, Faculty of Electrical Engineering Energy and Applied Informatics, Iasi, Romania
[2] University of Medicine and Pharmacy "Gr. T. Popa" Iasi, Faculty of Pharmacy, Iasi, Romania
[3] University of Medicine and Pharmacy "Gr. T. Popa" Iasi, Faculty of Dentistry,
Department of Odontology and Periodontology, Iasi, Romania

Abstract— **The paper presents a virtual instrument used to diagnose dental caries using NI data acquisition device and LabVIEW software. We also use two electrodes to measure dental impedance and the results respect the domain literature. Combining hardware and software resources we obtain a flexible and versatile interface very friendly with the user (dentist). Compared to known methods our method is very simple, fast and does not cause pain to patient.**

Keywords— **Virtual Instrument, impedance measurement, dental decay, LabVIEW.**

I. Introduction

The technique of electrical impedance measurements is a promising alternative method for occlusal caries diagnosis that was described as original method in 1951 and it was achieved and used in 1980: Vanguard Electronic caries detector (Massachusetts Manufacturing Corporation, Cambridge) and Caries Meter L (GC International Corportion Interleuvenlaan, Leuven). In 2001 there was made a device for measurement of electrical impedance of the hard dental tissues which was verified in the Cariology laboratory from University of "Medicine & Pharmacy" Iasi, with which occlusal caries in incipient phases could be detected [6].

In dentistry, by clinical examination, occlusal caries are highlighted in a late stage, often even in the complications stage. This is due to specifics of these types of injuries development that lead to the enamel undermining and cause the collapse of cavities in an advanced stage.

In literature, there are concerns for early detection of occlusal caries in order to be able to treat them conservatively and noninvasive, being possible in that way to prevent complications.

Clinical and radiological diagnosis of occlusal caries is difficult. Measurements of resistance in trenches and pits have a better sensitivity than conventional diagnostic methods.

Electrical impedance measurements represent a promising alternative method for diagnosis of occlusal caries.

Correct diagnosis of fissure damage in the early stages is difficult when using only traditional inspection and palpation methods. In the present tactile method should be avoided and use visual method and complementary means of diagnosis [7].

Palpation with the probe can damage the integrity of the surface layer of carious enamel witch cover the enamel lesion, thereby compromising the ability of remineralization and favoring expansion and lesion in dentin [5].

Methods of caries detection based on their underlying physical principles are the following:

- X-rays - Digital subtraction radiography and Digital image enhancement
- Visible light - Fiber optic transillumination (FOTI), Quantitative light-induced fluorescence (QLF) and Digital image fiber optic transillumination (DiFOTI),
- Laser light - Laser fluorescence measurement (DiagnoDent)
- Electrical current - Electrical conductance measurement (ECM) and Electrical impedance measurement
- Ultrasound - Ultrasonic caries detector

Every biological material possesses its own electrical signature. When a current is passed through this biological material will produce an effect on the impedance [9, 10].

Mayuzumi and Suzuki measured impedance teeth and found that at crossing of a sinusoidal current with 400 Hz frequency, the impedance of intact teeth, measured between the trenches, ditches and oral mucosa, is more than 600 kΩ; between dentin and oral mucosa varies between 250 - 600 KΩ, between the flesh and the oral mucosa is less than 15 KΩ. [8]

II. System description

A. *Method and System Architecture*

The base theory for using electrical impedance in the diagnosis of caries consists in that the enamel is a good insulator, but during the caries process, porosity formed in the tissue will be filled with water and saliva ions. These

S. Vlad and R.V. Ciupa (Eds.): MEDITECH 2011, IFMBE Proceedings 36, pp. 172–175, 2011.
www.springerlink.com

moisture-filled pores act as conductive network, causing lower electrical impedance.

The study contents a lot of 23 patients who had extracted premolars and molars teeth for orthodontic or periodontal reasons. All patients gave their consent to be included in the research lot, after they have been informed of the contents of this study. The teeth were cleaned by professionall brushing, isolated and dried. The clinical examination was realized by inspection and palpation with a flexible probe, avoiding excessive pressure exercising that could destroy the integrity of enamel. Sites were examined occlusal grooves and pits. Diagnosis (enamel caries, dentine caries, complicated caries, no decay) was recorded for each site examined.

Virtual Instrument - is based on measurement of electrical impedance of dental tissues, which decreases as the dentin caries progresses, so, the impedance of enamel caries presents 250-600KΩ values, in dentin caries impedance varies between 250 and 15 KΩ, depending on the depth of decay as showed in Fig.1.

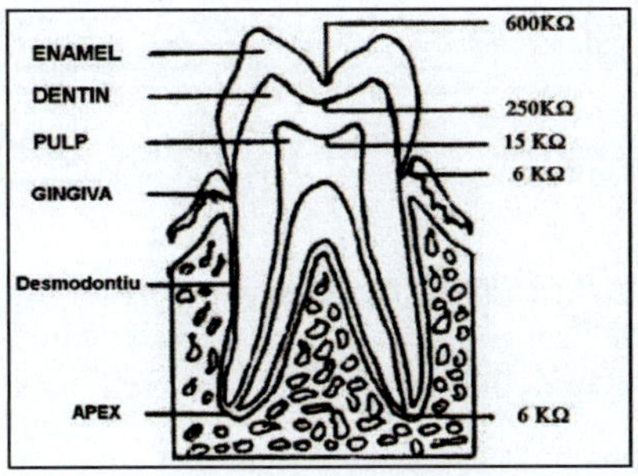

Fig. 1 Impedance values representation depending on the decay type.

Total circuit impedance is the sum of the components in phase (real) and switched phase (imaginary). In our made instrument, the frequency being low, out-of-phase component is negligible and can use the term resistance.

For electronic caries detection at each level, we developed a system (Fig. 2), that consists of the following elements [1, 2]:

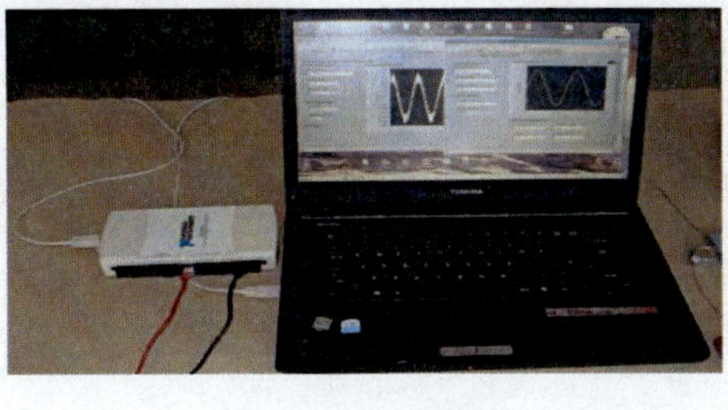

Fig. 2 System image

- signal collection electrodes
- data acquisition board NI-USB,
- laptop with LabVIEW programming environment

For resistance measurement of dental hard tissues in the oral cavity, active electrode is applied to the tooth, and the neutral electrode on the lip (Fig. 3). At the terminals there are attached single use sterile probes. Using the device would not produce any pain due to extremely low generated current.

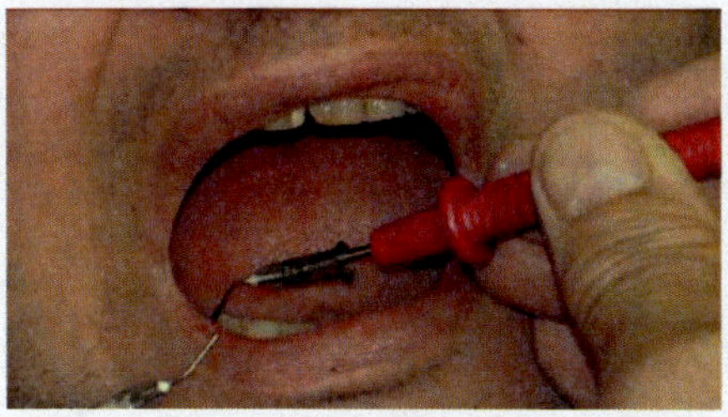

Fig. 3 Impedance measurement at patient level

B. Virtual Instrument

The virtual instrument use a sine wave signal generator with a frequency of 400 Hz and effective value of 1 μA on analogical output channel from the data acquisition board [3, 4]. These values are set in the configuration tab of the virtual instrument, and the analog input channel is also configured as showed in Fig.4. On the output analog channel we use a resistance with 4.5 MΩ to keep a constant current value.

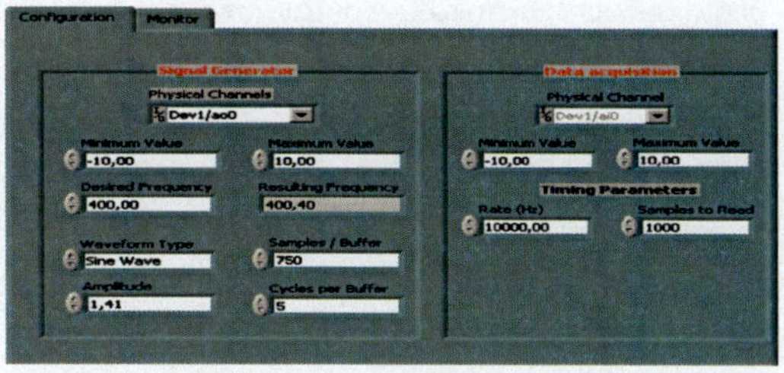

Fig. 4 Front panel-Configuration tab.

Measurement electrode is connected at the analog input channel and the neutral electrode is connected at ground. Collected signal and impedance value are displayed on the monitor tab.

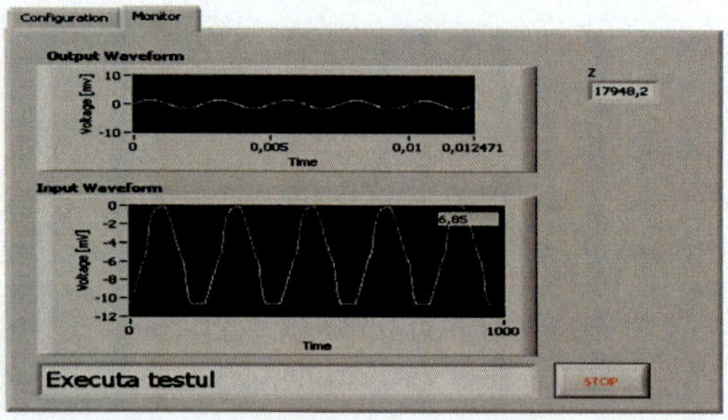

Fig. 5 Front panel-Monitor tab.

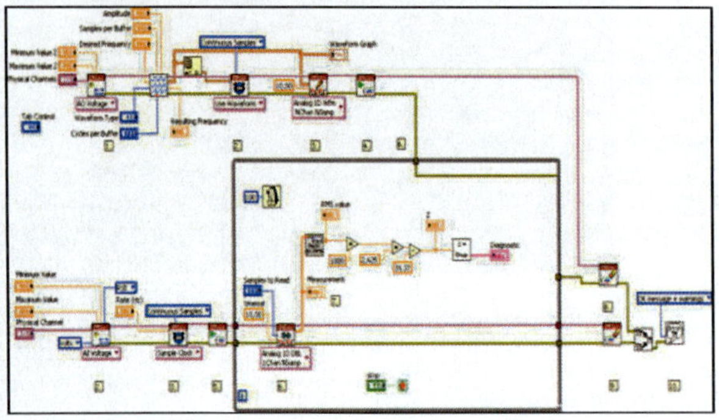

Fig. 6 Diagram block of virtual instrument

To create the virtual signal generator we fallowed the next steps:

- Create an analog output voltage channel.
- Set the waveform generator: amplitude, desired frequency, waveform type, sample per buffer
- Set the sample clock rate based on the input waveform parameters, and set the sample mode to Continuous.
- Write the waveform to the output buffer.
- Start VI.

For voltage signal acquisition and impedance calculation, a few steps must be done:

- Create analog input voltage channel.
- Set the rate for the sample clock-continuous samples.
- Start acquiring samples.
- Read the data in a loop until the user hits the stop button or an error occurs.
- Calculate RMS value
- Calculate impedance value (Z) using predetermined coefficients.
- Use Z to Diagn sub-VI to suggest the diagnostic
- Stop acquiring samples.
- Use the popup dialog box to display an error if any.

The instrument was calibrated against standard resistors. The reproducibility as percent of full scale respects the literature. The linearity of the resistance versus read-out voltage is relatively constant over the full scale. The relationship for calculating the impedance is:

$$Z = c_1 \cdot U + c_2 \quad (1)$$

where c_1 = 2.625 and c_2 = 39.37 obtained from the instrument calibration. Calculated impedance value is used to classify carries types as showed in Fig. 7.

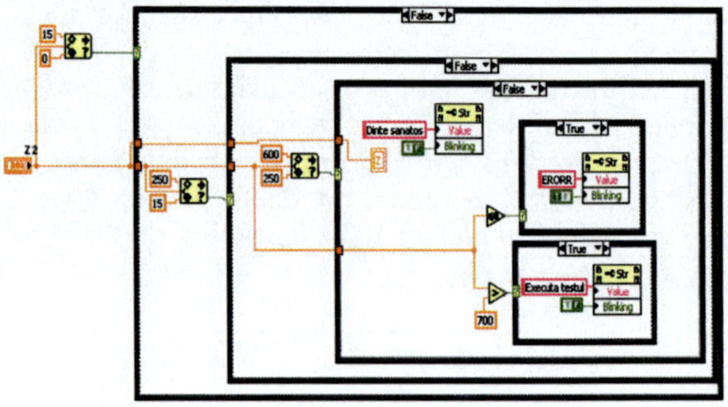

Fig. 7 Block diagram of the sub-VI to extract useful features for diagnosis

The extracted features results are showed in monitor tab as a text that suggests the diagnosis. Depending on the impedance value Z, the displayed text will be:

- Z = 0÷15 kΩ - "Caries at pulp level"
- Z = 15÷250 kΩ - "Caries at dentine level"
- Z = 250÷600 kΩ - "Caries at enamel level"
- Z > 600 kΩ - "Healthy tooth"
- Z < 0 – "Error".

An example is showed in Fig. 8.

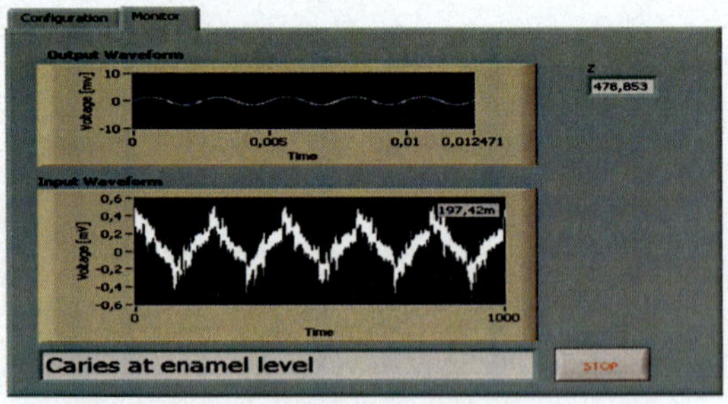

Fig. 8 Diagnostic monitor

III. Conclusions

After experiments at different values of frequency, we find that out-of-phase component is negligible and we can use the term resistance.

Using the technique presented above, we developed a system for resistance measurement and we suggested a diagnosis based on measured values.

The method of measurement using electrical impedance compared with other caries detection methods, such as x-ray or laser light, is noninvasive and does not produce harmful secondary effects. Performance of this method consists in that it can detect caries depth and closeness of the pulp chamber. All other methods show only the presence of dental decay. However the method has some limitations: the decay on the interdental sides, when the decay does not interrupt the marginal ridge. Measuring the electrical resistance of the tooth, errors can occur due to following factors:

- if the tooth is not well insulated (dry air) from the salivary environment;
- contact resistance of electrodes;
- electrode resistance is negligible compared with the measured resistance (in the order miliohm)
- errors from the AC generator

For early diagnosis of fissure caries types, the achieved device has given better results than clinical examination itself.

The dentist tried our instrument in the laboratory. Comparing with other diagnosis methods our methods for electrical resistance of tooth measurement seem to offer the biggest hope for achieving reliable, accurate detection of the earliest stages of enamel demineralization.

The new research directions in caries diagnosis field use virtual instrumentation due to computing power and flexibility. By using virtual instrumentation, our instrument reduces the costs, increases the flexibility and systems modularization.

We believe that this device can be used to record the initial lesions, the ones that have stopped evolving, the remineralized or the progressive.

The virtual instrument that we developed stands to serve as an important tool by helping dentist to speed up diagnoses.

Acknowledgment

This paper was supported by the project PERFORM-ERA "Postdoctoral Performance for Integration in the European Research Area" (ID-57649), financed by the European Social Fund and the Romanian Government.

References

1. Branzila M., Donciu C. (2010) „Distributed System Architecture Using a Prototype Web E-Nose, Advances in Biomedical Sensing, Measurements, Instrumentation and Systems", Ed. Springer, pp. 1-16.
2. Branzila, M , Alexandru, C, Donciu, C, et al. (2007) Virtual environmental measurement center based on remote instrumentation, Environmental Engineering and Management Journal, Volume: 6, Issue: 6, 517-520
3. Cepisca, C., Negoescu, R. (1998) Instrumentatie pentru biosemnale, Editura ICPE.
4. Cretu, M. (2001), Tendinte novatoare in instrumentatie si masurari electrice, Editura Sedcom Libris, Iasi.
5. Lacatusu St. (1996) Caria dentara exploziva, Ed. Cronica, Iasi.
6. Lacatusu St. G., Leanca V. (2002) Aparat electronic pentru detectarea cariilor dentare fisurale, Supliment al Revistei Medicina Stomatologica, Editura Apollonia, vol.1: 389-394
7. Lacatusu St., Gheorghe A., Iovan G., Andrian S. (1997) Posibilitati de diagnostic ale cariilor fisurate, Medicina Stomatologica, nr.3:16-20.
8. Mayuzumi, Y., Suzuki, K. & Sumada, J. (1964) A method of diagnosing incipient caries in pits and fissures by measuring electric resistance. J. Dent. Res. 43, 941.
9. Pretty Iain A., (2006) Caries detection and diagnosis: Novel technologies, Journal of Dentistry, 34: 727–739
10. Ricketts D.N.J., Kidd E.A.M., Beighton D. (1995) Operative and microbiological validation of visual, radiographic and electronic diagnosis of occlusal caries in non-cavitated teeth judged to be in need of operative care, Br. Dent J., 179: 214-220.

Author: Stefan Gruia Lacatusu
Institute: Technical Univ. "Gh.Asachi" of Iasi
Street: Bd. Dimitrie Mangeron, nr.21-23, 700050
City: Iasi
Country: Romania
Email: glacatusu2004@yahoo.com

Analysis of a Dexterous Instrument for Minimally Invasive Procedures, Based on Bellows Actuators

C. Dudescu and D. Mandru

Technical University of Cluj-Napoca / Faculty of Mechanics, Cluj-Napoca, Romania

Abstract— **The paper presents a review of actuators for bio-mechatronic applications, especially for regionally bending of smart instruments for minimally invasive procedures. More detailed analysis was done for bellows actuators. For a system of three parallel bellows an FEM study was performed. For different acting pressures FE results were compared with the analytical solution. The obtained results validate the methods and open future works in this field.**

Keywords— **bellows actuators, FEM analysis, endoscope.**

I. Introduction

The actuators for bio-mechatronic applications require properties like softness, lightness and safety as well as small size, simple design, integrability of sensors and control possibilities. The pneumatic actuators based on deformation of flexible elements fulfill these requirements. In this paper the possible inclusion in the structure of a medical instrument of the bellows actuators is investigated.

The minimally invasive procedures are less invasive than open surgery used for the same purpose. These can be surgical or otherwise, performed with instruments and/or viewing equipment inserted into the body through natural orifices or small incisions. In spite of several absolute advantages (less operative trauma for the patient, shorter hospitalization time, less pain and scarring, and reduced incidence of post-surgical complications), there are disadvantages due to the reduced dexterity, workspace, and sensory input to the surgeon. Suturing and knot tying are difficult to perform, mainly due to the lack of ability to orient the tip of the tools. Endoscopes and colonscopes diagnose various diseases of the gastrointenstinal tract. Due to the stiffness of their body, generate pain and discomfort to patients, especially in angulated regions of intestine where may not easily advance and pass.

The instruments for minimally invasive procedures could be further improved if they could bend regionally along their length as well as their tips.

Conventional, the bending capability is achieved through a series of articulated rigid segments, designed for being passively / actively bent with respect to one another. In [1], a bending mechanism of a manipulator for intrauterine surgery that includes ball joints driven using four wires, is presented. An inner hole through all ball joints makes easy to develop many types of surgical applications. A highly articulated robotic system for cardiac surgery is described in [2]. The prototype is 12 mm in diameter and 300 mm in length, and can achieve a 75 mm radius of curvature. It consists of two concentric tubes, each can alternate between being rigid or limp. Both tubes consist of rigid cylindrical links connected by spherical joints. In [3] a hybrid variant is proposed: a variable flexible endoscope (called Smart Endoscope) in which an outer stem is comprised of rigid hollow tubular segments that can be locked or released with respect to each other to control its stiffness. Four longitudinal tendons are used to accomplish a reliable lock and release mechanism. The *active endoscope* with multiple degrees of freedom, driven by a shape memory alloy (SMA) servo actuator [4] consists of several unit segments, the bending angles of which are independently controlled with a joystick through electric-resistance feedback. A different simple structure is given in [5]: the actuator has two SMA coils springs which bend the tip through two pull wires. In [6] a new polymer-links *active catheter* with integrated interface, within a 2 mm diameter is presented. It is actuated by three SMA helical coils. The catheter introduced in [7] is consisting of a central tube actuated by a single SMA tendon enclosed by an outer sleeve. A bending module for a gastro-intestinal intervention robotic system is proposed in [8]. It is composed of multiple vertebra type elements actuated by SMA wires. Stacking several elements can form an active bending structure. The new design concept of an active catheter is considered in [9] whose operation is based on SMA small diameter tube patterned by laser cutting so active bending in two orthogonal directions is enabled. In our previous works [10], [11] some simple structures with 1 and 2 DOF were proposed. These are based also on SMA wires and helical springs to control the bending angle of the mobile part in respect to the fix one.

The instruments for minimally invasive procedures whose actuation systems are based on different flexible fluidic actuators presents advantageous characteristics for use inside the human body [12]: absence of electrical voltage or high temperature parts, possible operation in presence of radioactivity or magnetic field, delicate manipulation of objects, adaptability to environment during contacts

S. Vlad and R.V. Ciupa (Eds.): MEDITECH 2011, IFMBE Proceedings 36, pp. 176–179, 2011.
www.springerlink.com

thanks to their own compliance, small number of components and miniaturization.

II. The operation principle of bellows actuators

Generally, the sylphon bellows are used as sensorial elements in cases where considerable deflections and rectilinear characteristics are required [13].

In [14], the development of a bellows type gas-liquid phase-change actuator is presented. The actuator is made of a nickel bellows while the operating fluid, enclosed in the bellows, is non-chlorine perfluorcarbon ($C_5F_{11}NO$), with low vaporization energy. When the liquid is vaporized, the volume of the phase-change gas is more than 100 times that of the liquid. The stretching / shrinking motion of the bellows is used for the inching motion of an in-pipe minirobot. Another way to ensure the pressure for bellows actuators is represented by hydrogen absorbing alloys in the structure of metal hydride (MH) actuator [15]. Hydrogen absorbing alloys are capable of storing an amount of hydrogen gas equal to 1000 times their own volume. By heating the alloy, hydrogen is desorbed and the bellows expands, whereas by cooling the alloy, hydrogen is absorbed and the bellows contracts. The mechanical energy of the hydrogen gas pressure is converted into a useful action through a bellows element. A miniature bellows actuator driven by a pair of PWM controlled electrorheological (ER) valves is presented in [16]. The bellows can be expanded and contracted by alternately changing the flow rate into and out of it utilizing the ER valves. A circular micro bellows is the most important element of a thermopneumatic microactuator, [17]. By applying power to a resistive heater in a liquid chamber, the micro bellows can deflect more than a flat membrane of the same dimensions. A bellows manipulator is given in [12]: inside a bellows there are two balloons which are alternatively inflated and generates bending of the structure. Each finger of a dexterous subsea hand, described in [18], consists of three cylindrical metal bellows placed in a parallel arrangement; different pressures in bellows cause finger bending. In the field of minimally invasive procedures, a smart colonoscope system that includes one or more bellows is described in [19]. When three bellows are used, the steering motion of the colonoscope is implemented with different pressure control of each bellows. A similar actuation system composed of three plates interconnected with bellows is analyzed in [20].

The actuators with multiple bellows (Fig. 1) have a symmetrical structure which is asymmetrically powered (asymmetrical energy input into two, three or four independent bellows).

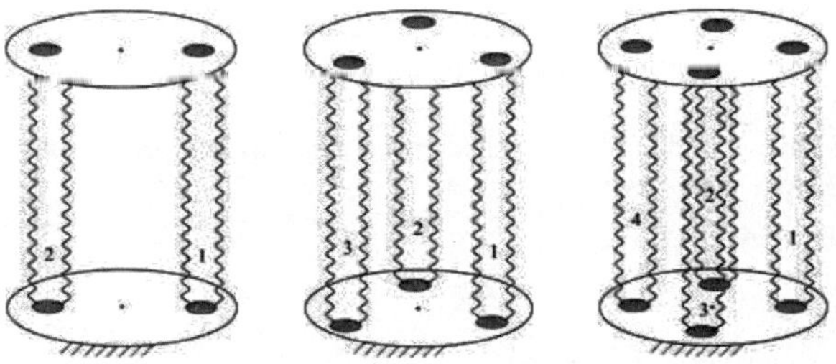

Fig. 1 Actuators with multiple bellows

The proposed actuation system is simple, compact and lightweight. It can generate motion with three degrees of freedom based on the elastic deformation of the bellows, placed at equal intervals about a central axis and constrained by end plates, according to the solution described in [20].

III. The FEM analysis

Simulation of the bellows endoscope by Finite Element Method (FEM) was done on a model of endoscope with three bellows. For these purpose a Sigma-Netics [21] A10x6-28 model of bellows was chosen based on its dimensions and maximum extension. The bellows made of brass ($E=1,05x10^5$ N/mm^2, $v=0.35$) have the following dimensions: outside diameter 10 mm, inside diameter 6 mm, number of active corrugation 28, length corrugated 31 mm, wall thickness 0,090 mm. Maximum deflection of one bellows given by the producer is 8 mm and the effective surface of bellows of 0,5 cm^2.

The three bellows were placed on a radius of 10 mm with an angle of 120° between their longitudinal axes (Fig. 2).

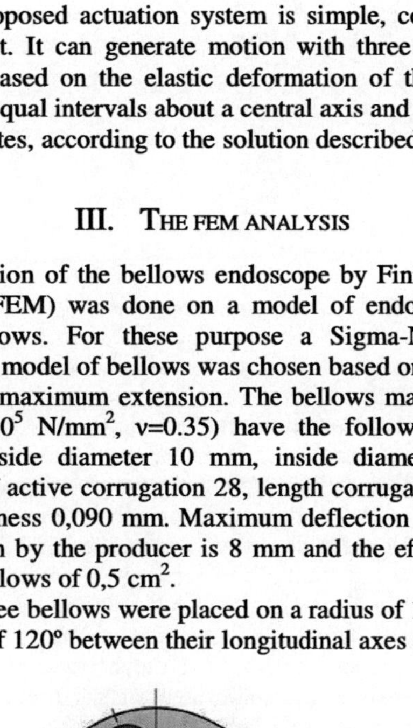

Fig. 2 Position of the three bellows (top view)

The bellows were clamped at their ends by two annular plates (r_i=5mm, r_e=15mm) from the same material but higher thickness (1 mm). One plate was fixed, the other one being oriented by the pressure difference between the individual bellows. Changing the pressure one can move and rotate the free end of the endoscope. The orientation is defined by the angles between the initial and the final position of the free plate. The rotation around *y* axis is given by the angle α_1 measured with respect to *x* axis passing to one bellows and angle α_2 describing rotation about *w* axis as shown in Fig. 3.

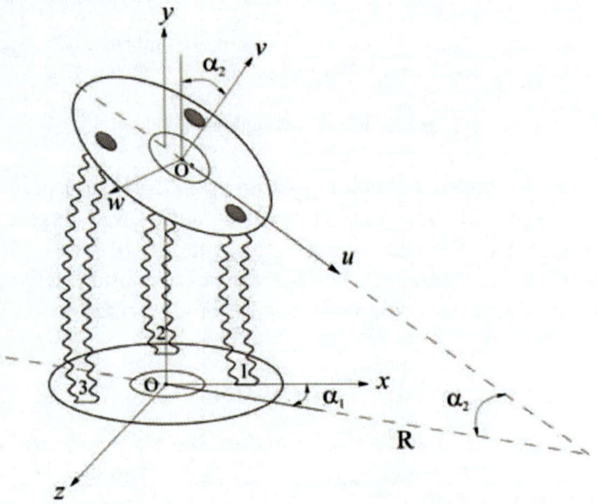

Fig. 3 Orientation of the bellows endoscope free end

The analysis of different positions was limited to a maximum pressure of 0,4 N/mm^2 even if maximum working pressure of one bellows given by the producer is 2 N/mm^2. On one hand the working pressure do not consider the bellows with closed ends and on the on the hand in the above presented set-up the bellows undergo a combined loading (axial and bending) [22]. The higher stress level given by this mounting solution should be avoided by limiting the maximum pressure applied on the bellows. It should be mentioned that in the static analysis performed upon bellows do not act external forces.

Fig. 4 shows vertical extension obtained by FEM (ANSYS software) for a set of three pressures (0,1-0,4-0,2 N/mm^2).

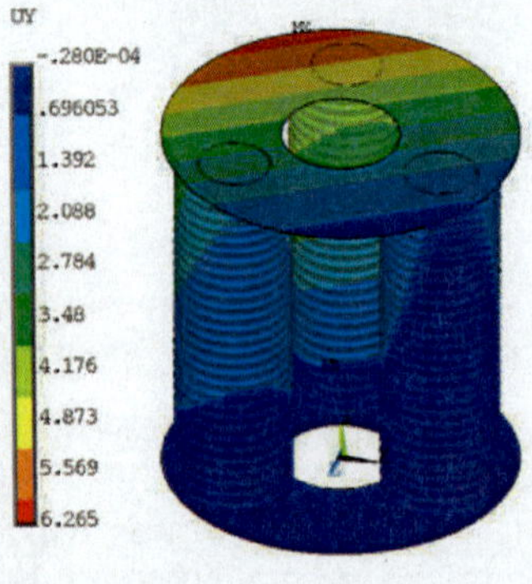

Fig. 4 Vertical extension obtained by FEM of the endoscope with three different pressured bellows

Parallel bellows with fixed end connection determine a reciprocal influence in terms of stiffness of one below upon the other. One observed aspect is the linearity between applied pressure differences and deflection and rotation of the free end, which is an important propriety of this set-up.

The final position having the pressure values calculated by FEM was compared with the analytical solution given by the literature [23] in form of a simplified model. The analytical expression of deflection and orientation of the free plate as a function of the internal pressures can be expressed:

$$\alpha_1 = \arctan\frac{\sqrt{3}(P_2 - P_3)}{2P_1 - P_2 - P_3} \tag{1}$$

$$\alpha_2 = -\frac{S}{3rk}\cdot\frac{2P_1 - P_2 - P_3}{\cos\alpha_1} \tag{2}$$

$$h = \frac{S}{3k}\cdot\sum_{i=1}^{3} P_i + L_0 \tag{3}$$

$$\Delta h_y = \frac{h\cdot\sin\alpha_2}{\alpha_2} - L_0 \tag{4}$$

where:

α_1 - first angle of rotation [rad];
α_2 - second angle of rotation [rad];
P_i - pressure in the bellows *i* [N/mm^2];
S- effective surface of bellows [mm^2];
k - stiffness of the bellows [N/mm];
L_0 - initial length of the unloaded bellows [mm];
r - radius of the circle on which the bellows are placed [mm];
h - length of the arc between the center of plates [mm];
Δh_y – change in length between the centers of plates [mm].

Table 1 Comparison between FEM and analytical values

Pressure [N/mm^2]			Angle α_1 [°]		Angle α_1 [°]		Deflection Δh_y [mm]	
P1	P2	P3	FEM	Eqs. 1	FEM	Eqs. 2	FEM	Eqs. 4
0,05	0,05	0	60	60	2,23	2,34	0,45	0,40
0,1	0,1	0	60	60	4,45	4,68	0,91	0,78
0,2	0,2	0	60	60	8,84	9,36	1,81	1,49
0,3	0,3	0	60	60	13,14	14,05	2,72	2,12
0,4	0,4	0	60	60	17,28	18,73	3,63	2,66
0,4	0,4	0,05	60	60	15,22	16,39	3,85	3,00
0,4	0,4	0,1	60	60	13,13	14,05	4,08	3,33
0,4	0,4	0,2	60	60	8,85	9,36	4,53	3,93
0,4	0,4	0,3	60	60	4,44	4,68	4,99	4,46
0,4	0,4	0,4	0	0	0	0	5,45	4,90
0,05	0,4	0,1	-54,37	-52,41	14,31	15,35	2,50	1,85
0,1	0,4	0,2	-40,68	-40,89	11,63	12,39	3,18	2,60
0,2	0,4	0,3	-30	-30	7,67	8,11	4,08	3,56

Comparison between analytical and FEM results presented in Table 1 revealed that even if a simplified model is used the above expressions give appropriate results in terms of angles and deflection. The bellows stiffness used for calculations (k=4,16 N/mm) is the axial one. In reality, due to bending, this undergoes changes and influences the displacement and angle of the upper plate. The higher the angle α_2 the higher deviations between the analytical solution and FEM calculations occur.

IV. Conclusions

The development of dexterous and compact systems for minimally invasive procedures requires active bending mechanisms with a high degree of miniaturization.

Generally, the actuators based on deformable elements (including the bellows actuators) are recommendable for large strokes and forces within a low volume. Obviously, a problem that occurs is the pressure control.

The proposed system overcomes the lack of dexterity by controlling the bending of a medical instrument for minimally invasive procedures. It has characteristics such as simple structure, perfect flexibility, direct driving and high power/weight ratio.

The FE method proved to be reliable for design of complex bellows actuators and open future works in this field.

References

1. Harada, K., et.al. (2006) Bending Laser Manipulator for Intrauterine Surgery, Proc. of the First IEEE/RAS-EMBS Int. Conf. on Biomed. Robotics and Biomechatronics, BIOROB, Pisa 2006
2. Degani, A., et al. (2006) Percutaneous Intrapericardial Interventions Using A Highly Articulated Robotic Probe, Proc. of the first IEEE/RAS-EMBS Int. Conf. on Biomed. Robotics and Biomechatronics, BIOROB, Pisa 2006
3. Choi, J.H., Sturges, R.H. (2004) Design and Simulation of a Smart Endoscope, Proc. of the Int. Conf., Geneva, 2004, pp. 180-185
4. Ikuta, K., Tsukamato, M., Hirose, S. (1988) Shape memory alloy servo actuator system with electric resistance feedback, Proc. IEEE Int. Conf. on Robotics and Automation, 1988, pp. 427-430
5. Shigeo, M. et al. (1996) Active Endoscope with SMA Coil Springs, Proc.IEEE Int. Conf. on Robotics and Automation., 1996, pp.290-295
6. Park, K.T. et al. (1996) An integrated communication and control system for a multi link active catheter, J.of Micromech. Microeng., vol. 6, pp. 345-351
7. Veeramani, A.S. et al. (2008) Modeling and dynamic behavior of a shape memory alloy actuated catheter, Smart Materials and Structures, 17: 1-14
8. Reynaerts, D.,Peirs, J., Van Brussel, H. (1999) Shape memory microactuation for a gastro-intestinal intervention system, Sensors and Actuators, 77: 157-166
9. Langelaar, M., Van Keulen, F. (2004) Modelling of a shape memory alloy active catheter, Proc. of the 45th AIAA Structures, Structural Dynamics & Material Conf. Palm Springs, 2004, pp. 1-16
10. Mândru D., Tătar, O., Crişan, R. (2002) Aplicaţii ale actuatorilor pe bază de aliaje cu memoria formei în chirurgia minimal invaziva, The Romanian Review of Precision Mechanics, Optics and Mechatronics, Vol. 1-20a, pp. 385-390
11. Mândru, D., Lungu, I., Noveanu, S., Tătar, O. (2008) Applications of Shape Memory Alloy Actuators in Engineering, Annals of the University of Oradea, Fascicle of Management and Technological Engineering, Volume VII (XVII), pp. 922-927
12. De Greef, A., Lambert, P., Delchambre, A. (2009) Towards flexible medical instruments: Review of flexible fluidic actuators, Precision Engineering 33: 311-321
13. Davidson, A. (1990) Handbook of Precision Engineering, Macmillan, London
14. Kato, S., et al. (1998) A bellows type gas-liquid phase-change microactuator for microrobots in thin pipes, Proceedings of MOVIC '98 Conference, vol.1, 1998, pp. 177-182
15. Wakisaka, Y. et al. (1997) Application of Hydrogen Absorbing Alloys to Medical and Rehabilitation Equipment, IEEE Transaction on Rehabilitation Engineering, vol. 5, no.2, pp. 148-157
16. Jolly, M.R. (1998) Properties and applications of commercial controllable fluids, Proc. of the ACTUATOR '98 Int. Conf., Bremen, 1998, pp.414 – 418
17. Yang, X. et al. (1997) Micro Bellow Actuators, Proc. of the Int. Conf. on Solid State Sensors and Actuators, TRANSDUCERS '97, Chicago, 1997 pp. 45 – 48
18. Lane, D.M.et al. (1999) The AMADEUS Dextrous Subsea Hand: Design, Modeling and Sensor Processing, IEEE Journal of Oceanic Engineering, 24: 96-111
19. Kim, B. et al. (2002) Smart Colonoscope System, Proc. of the 2002 IEEE/RSJ Int. Conf. on Intelligent Robots and Systems, Lausanne, 2002, pp. 1367-1372
20. Chen, G. et al. (2003) Identification of the Flexible Actuator of a Colonoscope, Proc. of the 2003 IEEE/RSJ Int. Conference on Intelligent Robots and Systems, Las Vegas, 2003, pp. 3355-3360
21. Sigma-Netics at http://www.sigmanetics.com
22. Dudescu, C. (2005) Strength calculation of bellows (in Romanian), U.T.Press, Cluj-Napoca
23. Thomann, G., Gang Chen, G., Redarce T. (2008) Design and control of an autonomous bendable tip for colonoscopy, J. Micro-Nano Mech. 4:103–114

The Contribution of Technology in Cholangiocarcinoma Treatment

H.C. Neagos[1,2], F. Graur[1,2], O. Neagos[3], R. Elisei[1], A. Szasz[1], A. Muresan[1], L. Furcea[1,2], C. Iancu[1,2], N. Al Hajjar[1,2], O. Bala[1,2], D. Munteanu[1,2], C. Puia[1,2], and L. Vlad[1,2]

[1] Surgical Clinic III, Cluj-Napoca, Romania
[2] University of Medicine and Pharmacy "Iuliu Hatieganu", Cluj-Napoca, Romania
[3] III rd Medical Clinic, Cluj – Napoca, Romania

Abstract— **This article is a review of the principals techniques used in the treatment of cholangiocarcinoma, aims at identifying the most frecquent and valuable methods of therapy. Moreover, its purpose is to emphasize the role of high technology in the cholangiocarcinoma treatment. The multimodal approach in cholangiocarcinoma treatment, involving hepatobiliary surgery, endoscopic biliary or percutaneous drainage as well as other controversial therapies (for example, the photodynamic therapy) represent a progress in biliary tract tumors but at the same time requires an highly developed technology being an essential circumstance for obtaining good results. Thus, it can be said that top technology associated with highly qualified human resources opens a new horizon in cholangiocarcinoma treatment.**

Keywords— **cholangiocarcinoma, technology, curative treatment.**

I. Introduction

Cholangiocarcinoma implies a complex treatment which involves not only tumor resection but also obstructive jaundice remission, the most common clinical sign developed by patients suffering from biliary tract tumors.

Surgery represents the primordial therapeutic option and the only curative treatment which can be used with good results in the palliative treatment of biliary tumors. Hepatobiliary surgery, obtain encouraging results, long-term survival and even cure is considered difficult, highly risky, implying laborious training and presently is supported by optimal technology.

The obstructive jaundice accompanied by pathophysiological context involves the use of therapeutic methods exclusively addressed to cholestasis remission. Moreover, the obstructive jaundice treatment represents in most of the cases the only therapy available to the patient with cholangiocarcinoma, curative surgery being reserved only to 10-20 % of the patients. At the same time, biliary drainage might be the first line emergency therapy consequently followed by radical surgery. The hepatic alteration determined by obstructive jaundice, the comorbidities and the technological progress have assessed the use of minimally invasive biliary drainage procedures asking for high level technology.

The multimodal approach in cholangiocarcinoma treatment, involving hepatobiliary surgery, endoscopic biliary or percutaneous drainage as well as other controversial therapies (for example, the photodynamic therapy) represent a progress in biliary tract tumors but at the same time requires an highly developed technology being an essential circumstance for obtaining good results. Thus, it can be said that top technology associated with highly qualified human resources opens a new horizon in cholangiocarcinoma treatment.

II. Writing paper

A. *Surgery in Cholangiocarcinoma Treatment*

Surgery is the gold standard in cholangiocarcinoma (1) and the sole which allows tumor resection with negative margins. Cholangiocarcinoma surgery implies major visceral resections with extended lymphadenectomy and in selected cases vascular resections but resulting in considerable morbidity and mortality.(2)

Cholangiocarcinoma can be localized intrahepatic, perihilar – Klatskin tumor or distal, extrahepatic. In consequence, the type of surgical procedure is chosen regarding the intraoperatory circumstances in connection to tumor location. Hepatic resection in intrahepatic cholangiocarcinoma is the only type of radical surgical procedure considered by liver surgery centers (3) allowing tumor resection with negative margins.

Klatskin tumors, according to Bismuth classification, require the resection of the main biliary duct with regional lymphadenectomy followed by Roux-en-Y hepaticojejunostomy (type I and II) and in type IIIa and IIIb a combined approach with right or left hepatectomy. Distal cholangiocarcinoma is an indication for cephalic pancreaticoduodenectomy, the only radical procedure possible in this tumor location. Many discussions regarding surgical treatment of cholangiocarcinoma are determined by liver transplantation possibility in selected cases. Even though in the

S. Vlad and R.V. Ciupa (Eds.): MEDITECH 2011, IFMBE Proceedings 36, pp. 180–183, 2011.
www.springerlink.com

past it was a contraindication for liver transplantation, in the last years there have been elaborated treatment protocols by the Mayo Clinic and the University of Nebraska which include liver transplantation and neoadjuvant chemotherapy(4). As a first conclusion, it can be said that in cholangiocarcinoma, hepatobiliopancreatic surgery has a curative intent, is very extended and aggressive, requires medical centers and highly specialized staff and it can't be successfully performed without optimal medical logistics.

Surgical techniques, less complex and depending on technology, are used for the palliative treatment of cholangiocarcinoma and are indicated for biliary decompression and obstructive jaundice remission. A very important role, especially for avoiding useless laparotomy, comes for diagnostic laparoscopy in assessing tumor operability and in detecting hepatic and peritoneal metastasis, but with a variable accuracy in detecting lymph node metastasis and vascular invasion.

B. Technology in Cholangiocarcinoma Surgery

Hepatobiliopancreatic surgery, the only therapy with curative intention in cholangiocarcinoma treatment, is a highly developed department of general surgery and requires substantial technique. Liver, pancreatic or bile duct resection are surgical procedures which are improved by high technology determining a decrease in complications.

Initially laparoscopy and afterwards, robotic surgery became the main technologies used for hepatobiliopancreatic surgery. Liver laparoscopic surgery has developed in the last years so that presently is frequently used in liver surgery centers. However it requires the accomplishment of two conditions: high-advanced technology and specialized medical centers with great experience in such procedures. The results in tumoral liver pathology using laparoscopic surgery are encouraging, studies revealing the great value of laparoscopy even in extended liver resections.(5)

Having an elevated technological endowment is a compulsory condition for successful results in liver laparoscopic surgery. The difficult surgical circumstances, the increased hemorrhagic risk, the necessity of quick and precise maneuvers require high technology. Surgical procedure involves surgical instruments such as: ultrasonic scalpel, LigaSure device, intraoperative ultrasound with laparoscopic transductor, vascular clips.

The ultrasonic scalpel is a device used for liver mobilization by ligaments sectioning, also useful during liver transsection. Even though it has similar effects with those produced by a electrocautery, its benefits consisting in generating a small amount of heath and a limited impact on adjacent structures.(6) Praxis in using this type of scalpel reports minimal blood loss, no biliary leaks and its efficiency in liver resection on cirrhosis (7).

Intraoperatory ultrasound examination presents the advantage of better estimating the extension of the primary tumor being also able of highlighting the resection limits. It is known the importance of intraoperatory ultrasound initially being used in colorectal liver metastasis detection and then proved to be efficient in liver resection for other focal lesions. In 2006, Dr.Bjorn Skjoldbye, during a conference in Sankt Petersburg regarding laparoscopic techniques, claimed the importance of intraoperatory ultrasound in intrahepatic, perihilar and distal cholangiocarcinoma diagnosis and treatment, contributing to subsequent staging and therapeutic protocol.

The easiness to handle, the quality of images and the possibility of 3-dimensional visualization are the main advantages of this method of investigation. The trocar through which the laparoscopic transductor is placed must be 10 mm large while navigation area 30-50 cm and the approach can be accomplished from any point of the operatory field. Moreover, in the last years there were developed 3D systems based on virtualization which allowed not only the tridimensional reconstruction of tumoral liver but also the virtual preoperatory evaluation of the liver remnant after liver resection. In fact, these systems have realized a half- or full-automatic liver-segmentation procedure to visualize liver segments, vessel trees, resected volumes or critical residual organ volumes, either for preoperative planning or intraoperative visualization. At first there were three systems used in the clinical practice: 3-D HepaVision2 (MeVis GmbH, Bremen), LiverLive (Navidez Ltd, Slovenia) and OrgaNicer (German Cancer Research Center, Heidelberg) (10).

A capital acquisition of the technology used in laparoscopic liver surgery is represented by LigaSure, usable in the small and medium vessels ligature during laparoscopic liver transsection. The activity percept of this surgical instrument lays in sealing and cloating the medium gauge vessels, while the ligature of the great-sized vessels (for example the hepatic vein) requires vascular clips adjunction. Many studies assert the usability of Ligasure for the transsection of the liver that has no major structural alteration, but mainly its usability for the cirrhotic liver transsection, as the incidence of intraoperatory bleeding is extremely decreased (11).

C. Robotic Surgery in Cholangiocarcinoma Treatment

Used at first for cholecystectomy, robotic surgery has the glory of developing the minimally invasive surgery concept, as currently there is a large series of surgical interventions that benefits from the robotic technology. The advantages of robotic surgery primarily regard the tridimensional visualization of the operatory field, the decrease in malpraxis and a high convenience for the surgeon during the procedure. The tridimensional visualization of the operatory field, furthermore the haptic system which allows the tactile perception of various structures represents advantages of robotic technology. The tridimensional visualization of the operatory field instantly determines not only the increase in the maneuverability of surgical instruments but also the circumstance that in perceiving structures it does not limit to their depth. Adding to all these the utility of the haptic system, the robotic surgery creates the same conditions for the operatory field as open surgery, unlike laparoscopic surgery, and in addition increases the extent of the operatory circumstance. The benefits regarding the value of the surgical intervention brought by robotic surgery mentioned are the exclusion of the surgeon's trembling, the possibility of vocal guidance and the storage of sites of interest of the operatory area.

Medical history reports multiple cases of robotic liver resection fact that enforces the advantages of robotic surgery over the purely laparoscopic approach. Technology used in such cases was based on the Da Vinci robotic surgical system , post-operatory parametres (blood loss, operation time, hospitalisation period, perioperatory morbidity and mortality) being encouraging (12). Studies emphasize the importance of tridimensional visualisation of the operatory field (13) in robot-assisted hepatic resection and also the learning curve and necessity of developing highly specialized centers of surgery for this type of intervention. Robot-assisted liver resections are possible in cholangiocarcinoma, experienced hands in robotic surgery being capable of accomplighing even a biliary reconstruction.(14)

D. The Contribution of Technology in the Palliative Treatment of Cholangiocarcinoma

The palliative treatment of cholangiocarcinoma mainly consists in biliary drainage, patients with this type of pathology being diagnosed with obstructive jaundice syndrome. Biliary decompression with icteric syndrom remission characterized by decrease of cholestasis represents the primordial objective of this treatment.

The approach of the biliary duct for biliary drainage can be made not only percutaneously – percutaneous drainage – but also through the insertion of stents in biliary tract.

The endoscopic drainage of the biliary duct through the insertion of a stent using the endoscopic approach is most frequently used yet not entirely safe. Medical studies debate on placing either metallic or plastic biliary stents and, also the unilateral or bilateral stents. Placing bilaterally the stents determins an increase in complications, their insertion being more difficult than that of unilateral stents.(15) The quality of stents – metallic or plastic – has been proved to be important in patients's life quality considering the fact that their low level of hope of life after following the palliative treatment. Plastic stents , need to be replaced after aproximately 6 months while metallic stents have a longer existence and determine less complications but present the inconvenient of a higher price.(16) Presently, the therapeutic attitude regarding patients suffering from cholangiocarcinoma associated with obstructive jaundice syndrome consists in inserting the initial plastic stent, followed by a surgical procedure and in the case of not accoplishing the radical treatment, the solution is exploratory laparotomy followed by the insertion of the metallic stent.

Radiofrequency ablation of cholangiocarcinoma represents a modern method in the palliative treatment of this tumor especially of types III and IV according to Bismuth's classification. Under ultrasound or CT guidance, a needle of puncture percutaneously inserted, enters into the tumor mass emiting a high-frequency electromagnetic wave which generates heat and coagulation necrosis and thus, the tissue around the tumor will form a reactive area. The advantages of using the method described above consist in minimal invasiveness, insignificant postoperatory morbidity and mortality and the possibility of repeating the method (17).

Other new method used in palliative treatment of cholangiocarcinoma are photodynamic therapy (PDT), which consists in the administration of a photosensitizer followed by a photoradiation. This process induces the apoptosis and tumor necrosis to a depth of 4 mm to 6 mm (18). All trials are needed to confirm the results of photodynamic therapy associated with stenting and also to identify the best choice for these patients.

III. Conclusions

The hilar, intrahepatic and distal cholangiocarcinoma treatment includes a multidisciplinary approach addressed both to surgery - the only curative option - and to the endoscopic minimally invasive or percutaneous treatment used for the obstructive jaundice remission. Surgical treatment as well as the miniinvasive approach benefits from the development of high technology because of its complexity.

References

1. Dreyer C et al. Cholangiocarcinoma: epidemiology and global management. Rev Med Interne 2008 Aug;29(8):642-51
2. Zografos GN, Farfaras A, Zagouri F, Chrysikos D, Karaliotas K. Cholangiocarcinoma: principles and current trends. Hepatobiliary Pancreat Dis Int 2011 Feb;10(1):10-20.
3. Nuzzo G et al . Intrahepatic cholangiocarcinoma: prognostic factors after liver resection. Updates Surg 2010 Aug;62(1):11-9.

4. Gatto M, Alvaro D. New insights on cholangiocarcinoma. World J Gastrointest Oncol. 2010 Mar 15;2(3):136-45
5. Abu Hilal M et al. Single-Centre Comparative Study of Laparoscopice Versus Open Right Hepatectomy . J Gastrointest Surg. 2011 Mar 5
6. Gagner M. Laparoscopic liver resection: benefits and controversies. Surg Clin N Am 84 (2004) 451–462
7. Abbasoglu O, Sayek I. Parenchymal transection with ultrasonic scalpel in liver resection. . HPB (Oxford). 2003;5(3):167-9
8. Wilheim D, Feussner H, Schneider A, Harms J. Electromagnetically navigated laparoscopic ultrasound. Surg Technol Int. 2003;11:50-54.
9. Bao P, Warmath J, Galloway R Jr, Herline A. Ultrasound-to-computer-tomography registration for image-guided laparoscopic liver surgery. Surg Endosc. 2005;19(3):424-429
10. Grenacher et al. The role of 3-D imaging and computer-based post-processing for surgery of the liver and pancreas. Rofo. 2005 Sep;177(9):1219-26.
11. Liu TQ et al. Application of Ligasure vessel sealing instrument in laparoscopic hepatectomy for liver cancer. Nan Fang Yi 2010 Jul;30(7):1705-6.
12. Sae Byeol Choi et al. Early Experiences of Robotic-assisted Laparoscopic Liver Resection .Yonsei Med J. 2008 Aug;49(4):632-638
13. Lai EC, Tang CN, Yang GP, Li MK. Multimodality laparoscopic liver resection for hepatic malignancy - From conventional total laparoscopic approach to robot-assisted laparoscopic approach. Int J Surg. 2011 Feb 18.
14. Giulianotti PC, Sbrana F, Bianco FM, Addeo P. Robot-assisted laparoscopic extended right hepatectomy with biliary reconstruction. J Laparoendosc Adv Surg Tech A. 2010 Mar;20(2):159-63.
15. De Palma GD, Galloro G, Siciliano S, Iovino P, Catanzano C. Unilateral versus bilateral endoscopic hepatic duct drainage in patients with malignant hilar biliary obstruction: results of a prospective, randomized, and controlled study. Gastrointest Endosc. 2001;53:547–553
16. Katherine Nguyen and James T Sing. Review of endoscopic techniques in the diagnosis and management of cholangiocarcinoma. World J Gastroenterol. 2008 May 21; 14(19): 2995–2999.
17. Wei-Jun Fan et al. Radiofrequency ablation as a treatment for hilar cholangiocarcinoma. World J Gastroenterol. 2008 July 28; 14(28): 4540–4545.
18. Richter JA, Kahaleh M. Photodynamic therapy: Palliation and endoscopic technique in cholangiocarcinoma. World J Gastrointest Endosc. 2010 November 16; 2(11): 357-361

Author: Dr. Horatiu Ciprian Neagos
Institute: Universitatea de Medicina si Farmacie, Cluj-Napoca
Street: Emil Isac
City: Cluj-Napoca
Country: Romania
Email: ciprianneagos@yahoo.com

Non-invasive Steatosis Assessment through the Computerized Processing of Ultrasound Images: Attenuation versus First Order Texture Parameters

M. Lupsor[1], R. Badea[1], C. Vicas[2], S. Nedevschi[2], H. Stefanescu[1], M. Grigorescu[1], C. Radu[1], and D. Crisan[1]

[1] 3rd Medical Clinic, University of Medicine and Pharmacy, Cluj-Napoca, Romania
[2] Technical University, Cluj-Napoca, Romania

Abstract— **Steatosis is a frequent histological finding in patients with chronic hepatitis C virus (VHC) infection. Usual ultrasonography (US) cannot accurately detect the steatosis grade, nor can it always discriminate between steatosis and fibrosis. An improvement of usual US examination is currently under research. A possible approach might be the computerized processing of the data comprised in the US image. In the present paper we set out to compare the performance of two computerized methods for the steatosis assessment on the US images: the attenuation coefficient and the first order textural parameters (FO): Mean, Standard Deviation and Skewness. The attenuation coefficient correlated significantly with steatosis (r=-0.444, p<0.0001), but not with fibrosis (r=-0.046, p=0.395) or necroinflammatory activity (r=-0056, p=0.211). Of the FO parameters, only the FO mean correlated significantly with steatosis (r=0.300, p<0.0001), but also with necroinflammatory activity (r=0.128, p=0004). The present study proves that, in patients having chronic hepatitis C, the attenuation coefficient, but also the FO mean, can discriminate between different steatosis grades; however, the attenuation coefficient has a better performance than the FO mean, being influenced only by steatosis, not by fibrosis or necroinflammatory activity. The area under the ROC curve is significantly better for the attenuation coefficient as compared to the FO mean for the prediction of steatosis regardless of the grade (0.741 vs 0.652, p=0.001), as well as for the prediction of moderate/severe steatosis (0.791 vs 0.719, p=0.043).**

Keywords— **steatosis, chronic hepatitis C, noninvasive, computerized methods, ultrasonography.**

I. Introduction

Hepatic steatosis is a frequent histological finding in patients with chronic hepatitis C virus (VHC) infection, occurring in more than 50% of cases [1]. There is increasing evidence that steatosis is an independent risk factor associated with liver necroinflammatory activity and progression of fibrosis in patients with chronic HCV infection [2]. Therefore, reliable and early diagnosis of hepatic steatosis is crucial to monitor disease progression and therapeutic intervention. The gold standard for assessing diffuse liver disease, including steatosis, is liver histology. However, liver biopsy is an invasive procedure associated with potential complications, as well as sampling error and interobserver variability [3].

Hence, reliable non-invasive methods to assess steatosis in patients with chronic HCV infection are needed. Among these, imaging methods have an important role, and of them, ultrasonography (US) is the best choice from the point of view of the cost, accessibility and lack of side effects. The ultrasonic alterations of fatty liver appear when the fatty load of the hepatocytes exceeds 15–20%. These alterations are represented by hepatomegaly, increased parenchymal echogenicity ("bright liver"), attenuation of the ultrasounds in the subcapsular strata, difficult visualization of the portal vein walls, of the gallbladder wall and of the hepatic capsule, the apparent dilatation of the vessels (especially of the suprahepatic ones) and the false transonic aspect of the parenchyma of the right kidney as opposed to that of the liver. However, the performance characteristics of conventional grey-scale US may vary considerably among studies, ranging from good to poor [4]. One reason might be that concomitant liver pathology (inflammation, fibrosis) may alter the ultrasonographic diagnosis of steatosis [5]. Fibrosis may also appear hyperechoic, but most of the time, fibrosis and fatty infiltration coexist, which is why the term "fatty-fibrotic pattern" is used to define the resulting aspect [6]. It is for this very reason that an improvement of usual ultrasonographic examination is currently under research. A possible approach might be the computerized processing of the data comprised in the ultrasonic image, taking into consideration that all the information concerning the characteristics of the tissue already exist in the echoes returned by the transducer. This is based on the principle according to which the pathological tissular modifications due to a specific disease (such as steatosis or fibrosis) lead to alterations of the physical and micro architectural features (density, thickness, elasticity, homogeneity, etc.). These are very difficult to visualize, but, because they affect the propagation of ultrasounds, they can be perceived through the complex image analysis (the ultrasonic tissular characterization) as a different textural pattern from the healthy one [7]. The ultrasonic tissular characterization can be achieved either by methods based on the study of parenchymal echogenicity and on the attenuation of the ultrasounds (attenuation and backscattering coefficients), or by

S. Vlad and R.V. Ciupa (Eds.): MEDITECH 2011, IFMBE Proceedings 36, pp. 184–189, 2011.
www.springerlink.com

methods based on the quantification of some textural parameters [8-10].

We have proven in our previous research that, between the attenuation and backscattering coefficients, the attenuation coefficient has the best performance in the assessment of steatosis in nonalcoholic steatohepatitis (NASH) patients [11]. Furthermore, the attenuation coefficient is also superior to biochemical methods (such as adiponectin) for the non-invasive assessment of steatosis [12]. At the same time, the attenuation coefficient performs better than the textural parameters derived from the gray level co-occurrence matrix (GLCM) in NASH patients. However, of all tested textural parameters, only the entropy computed on GLCM only correlates to steatosis; even then, it only discriminates between a normal aspect and NASH, but not between grades of steatosis [13].

In the present paper we set out to compare the performance of two computerized methods for the steatosis assessment on the ultrasonographic images in chronic hepatits C patients: the attenuation coefficient and some textural descriptors based on the first order (FO) statistics.

II. Patients and methods

A. *Patients*

526 consecutive patients with chronic HCV infection examined at the 3rd Medical Clinic, University of Medicine and Pharmacy Cluj-Napoca, were prospectively included in this study. All of them were HCV-RNA positive and underwent percutaneous liver biopsy (LB) for the grading and staging of diseases. All patients were referred to an ultrasound exam 1 day prior to LB. Besides the epidemiological data, the following biological parameters were determined for all patients on the same day as LB: aspartate aminotransferase (AST), alanine aminotransferase (ALT), gamma-glutamyl-transpeptidase (GGT), total bilirubin, alkaline phosphatase, fasting blood glucose, fasting serum cholesterol and triglycerides. The study was approved by a local ethical committee of the University of Medicine and Pharmacy Cluj-Napoca. The nature of the study was explained to the patients and they each provided written informed consent before the beginning of the study, in accordance with the principles of the Declaration of Helsinki (revision of Edinburgh, 2000).

B. *Ultrasound Exam*

Each studied patient was submitted to an abdominal ultrasound exam by means of a GE Logiq 7 device, using a 5.5 MHz convex probe, one day before the LB. The examination protocol was built so as to acquire the maximum amount of information from the tissue level, with as little „noise" as possible over added to this process. In order not to change the textural elements, the amount of digital post-processing must be as small as possible, and thus, all the post-processing parameters were set at minimum, and in order to exclude movement artifacts, the tissue image „Freeze" took place as quick as possible (by using a „Frame rate" which must be as high as possible). We worked with harmonic (because it increases the quantity of information coming from tissues). The "Time Gain Compensation" curve was adjusted to a neutral position. The device was set so as to stand on all these principles, and once the setting took place, it was used for all the examined patients. For each patient, US images were acquired from the right lobe through intercostals spaces. Depth was set at 16 cm. The images were saved on the ultrasound machine hard disk in DICOM format and further processed using a special soft designed by the Technical University of Cluj-Napoca.

C. *Computing the Image Coefficients*

On each US image, a straight line was fitted so as to avoid artifacts. This line represents the ultrasound beam path into the liver tissue and it has to be as parallel as possible to the US rays, preferably vertical. The fitted line is the region of interest (fig.1).

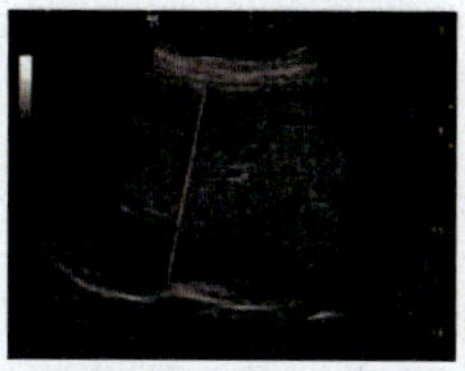

Fig. 1 Ultrasound image from right lobe at 16 cm. The mean gray levels are computed along the white line.

The grey level values for each point along this line are calculated by averaging 7 horizontal pixels (the pixel below the line and three more pixels from each side) [14]. For each point on the line, two values were stored: the average grey level computed as above and the depth (fig.2). As a measure of ultrasonic attenuation, linear regression by least-squares approximation was applied to this dataset. The slope represents the attenuation coefficient.

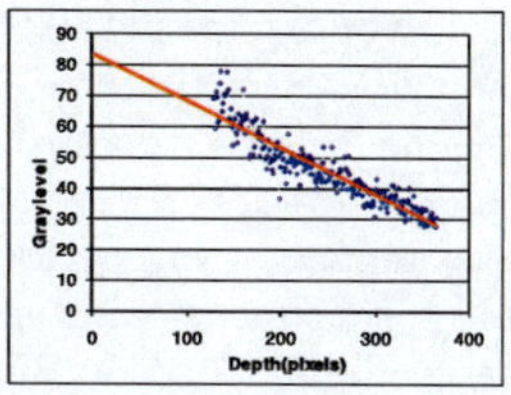

Fig. 2 The graphical representation of the average gray level with respect to depth.

Furthermore, FO parameters were calculated in each patient. The first order parameters employed here belongs to the large family of statistical texture descriptors. The texture is viewed as a collection of pixels, each pixel having a grey level that follows a normal distribution. First order parametrs measure the pixel distributions and report various measures that depend on how much the actual distribution is deviated from the assumed Gauss distribution. In present paper the following coefficients are computed: Mean, Standard Deviation, Skewness. In the following is described how these statistics are computed [15, 16]. First, the histogram of the image is computed. The histogram basically counts how many pixels of each grey level are in the image. Because the grey levels are assumed to follow a Gaussian distribution this histogram should have the classical bell shape. Let *h(i)* be the number of pixels that have a grey value equal to i. Let $p(i) = h(i)/N$ where N are the total number of pixels. We define the following:

Mean:

$$\mu = \sum_{i=0}^{G} i p(i) \tag{1}$$

Standard Deviation:

$$\sigma^2 = \sum_{i=0}^{G} (\mu - i)^2 p(i) \tag{2}$$

Skewness:

$$\mu_3 = \sum_{i=0}^{G} (i - \mu)^3 p(i) \tag{3}$$

It is important to note is that FO Mean represents the mean grey level of the pixels in the image, the FO Standard Deviation measures the width of the histogram (or the contrast). The skweness measures the simmetry of the histogram. If the histogram is assimetrical (i.e. tilted to one side) the skewness will measure this deviation. Skewness is sensitive to the deviations from assumed normal distribution.

These statistics are not computed on the entire image but only on a 64x64 pixels squared region. This region is manually placed on the ultrasound image by a trained radiologist. The area where this region of interest is placed has to be clear of artifacts and as close as possible to the center of the image. Moreover, this squared region should be placed at 1 cm below the upper liver capsule in full liver tissue (fig 3).

D. *Histological Study*

A liver biopsy examination was performed for all the patients. Only biopsy specimens with more than 6 intact portal tracts were eligible for evaluation [17]. Liver fibrosis and necroinflamatory activity were evaluated semiquantitatively according to the METAVIR scoring system [17]. Fibrosis was staged on a 0-4 scale as follows: F0 – no fibrosis; F1 – portal fibrosis without septa; F2 – portal fibrosis and few septa; F3 – numerous septa without cirrhosis; F4 – cirrhosis. Necroinflammatory activity was graded as follows: A0– none; A1 – mild; A2 – moderate; A3 – severe. Steatosis was categorized by visual assessment as: 0- none; 1- steatosis in <33% of hepatocytes; 2- steatosis in 33% to 66% of hepatocytes; and 3- steatosis in > 66% of hepatocytes.

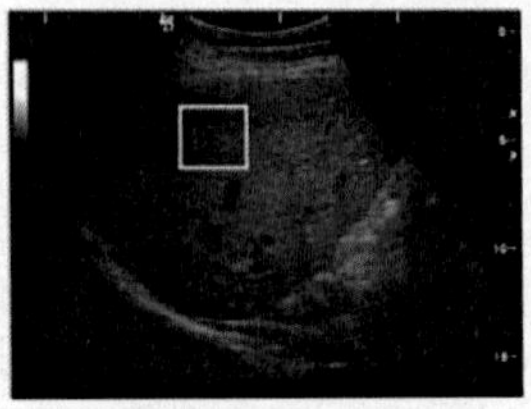

Fig. 3 A squared region of interest established on a right lobe ultrasound image avoiding the major artifacts like blood vessels, shadows, etc.

E. *Statistical Analysis*

The statistical analysis was performed using the SPSS software version 15.0 (SPSS Inc., Chicago, IL, USA). The relationships between the parameters were characterized using the Spearman correlation coefficients. The attenuation coefficient (AC) and FO data were expressed as median values. Differences in mean values were tested by one-way analysis of variance (ANOVA) and Kruskal-Wallis test; The diagnostic performance of AC was assessed using sensitivity (Se), specificity (Sp), positive predictive value (PPV), negative predictive value (NPV), accuracy, likelihood ratios (LR) and receiver operating characteristic (ROC) curves. Optimal cut-off values for AC were chosen to maximize the sum of sensitivity and specificity, and positive and negative predictive values were computed for these cut-off values.

III. Results

526 patients were enrolled in the study with a mean age of 46.79 ± 10.08 years; the majority was women (63.1%). The mean size of the bioptic speciment was 11.10 mm (±2.8), with a mean number of 12.15 (±4.1) portal spaces.

The histopathological analysis found no steatosis in the majority of patients (54.2%), mild seatosis in 34.4% (5-33% fatty content), moderate steatosis in 8% (34-66% fatty content), and severe steatosis in 3.4% (>66% fatty content). Because of the low number of patients in the last two categories, we grouped the patients in 3 groups for further analysis: S0 (patients without steatosis), S1 (patients with mild steatosis) and S2-3 respectively (patients with significant steatosis, either moderate or severe).

In addition, the patients had different fibrosis stages: no fibrosis - F0 (6.7%), mild fibrosis - F1 (36.1%), significant fibrosis - F2 (33.5%), severe fibrosis - F3 (16.5%) and cirrhosis respectively (F4) (7.2%), and various degrees of necroinflammatory activity: A0 (4.9%), A1 (20%), A2 (51.4%), A3 (23.7%).

A. *Correlation between the Attenuation Coefficients and Different Histological Parameters*

The attenuation coefficient correlated significantly with steatosis, but not with fibrosis or necroinflammatory activity. Of the FO parameters, only the FO mean correlated significantly with steatosis but also with necroinflammatory activity (table 1).

We found a significant variability of the attenuation coefficient, but also of the FO mean, in relation to the different steatosis grades (table 2, fig 4).

The area under the ROC curve is significantly better for the attenuation coefficient as compared to the FO mean, for both the prediction of S0 vs S123 and that of S01 vs S23 (table 3, fig 5).

Table 1 Spearman Correlation Coefficient between different coefficients computed on the ultrasound image and steatosis, fibrosis and necroinflammatory activity

	Steatosis		Fibrosis		Necroinflammatory activity	
	r	p	r	p	r	p
Attenuation coefficient	-0.444	<0.0001	-0.046	0.295	-0.056	0.211
FO mean	0.300	<0.0001	0.032	0.468	0.128	0.004
FO standard deviation	0.006	0.897	0.033	0.454	0.065	0.150
FO skewness	-0.075	0.087	0.074	0.090	-0.018	0.686

Table 2 Mean value of the attenuation coefficient and FO mean for different grades of steatosis in HCV patients

	S0 (0-4% fatty content)	S1 (5-33% fatty content)	S2-3 (>33% fatty content)	p
AC	-0.0082±0.065	-0.0586±0.0676	-0.1021±0.0716	<0.0001
FO mean	48.640±10.055	52.917±11.069	58.732±10.404	<0.0001

Table 3 Areas under the ROC curve for the attenuation coefficient and FO mean in the prediction of each steatosis grade

	AUROC for AC	AUROC for FO mean	Difference between AUROCs	Standard error	*p*
S0vsS123	0.741	0.652	0.089	0.026	0.001
S01vsS23	0.791	0.719	0.072	0.036	0.043

Because of the better performance of the attenuation coefficient in the grading of steatosis than that of FO mean, we further tested the performance of the attenuation coefficient in the prediction of steatosis regardless of the grade (S0 vs S123), as well as the prediction of moderate/severe steatosis (S01 vs S23) in HCV patients. The resulting values are displayed in table 4.

Table 4 Attenuation coefficient cutoff values for the diagnosis of steatosis grades ≥ S1 and S2-3

	S 0 vs S 1-2-3	S 0-1 vs S 2-3
AC cutoff value	-0.0471	-0.0564
Se (%)	67.63	86.67
95% CI	61.3-73.5	75.4-94.0
Sp (%)	70.18	64.81
95% CI	64.5-75.4	60.3-69.1
+LR	2.27	2.46
-LR	0.46	0.21
PPV	65.7	24.1
NPV	71.9	97.4
AUROC	0.741	0.791
SE	0.021	0.025
95% CI	0.700-0.777	0.754-0.825

IV. Discussions

An ever-developing field of research is the non-invasive assessment of steatosis occurring in diffuse liver diseases. Of all non-invasive methods, ultrasonography plays an important part thanks to its widespread availability and relatively low cost. However, usual ultrasonography has its limitations: it cannot accurately detect the steatosis grade, nor can it always discriminate between steatosis and other histopathological alterations frequently occurring in diffuse liver diseases (such as inflammation or fibrosis). This justifies the constant attempt to improve the ultrasonographic examination, one method being the computerized processing of images (ultrasonic tissular characterization). The ultrasonic tissular characterization can be achieved by methods based either on the study of parenchymal echogenicity and on the attenuation of the ultrasounds (attenuation and backscattering coefficients), or on the quantification of some textural parameters [8, 9].

Gaitini et al [9] tried to compare textural to attenuation/backscatter indices to suggest the better approach for an objective noninvasive ultrasonic "biopsy".

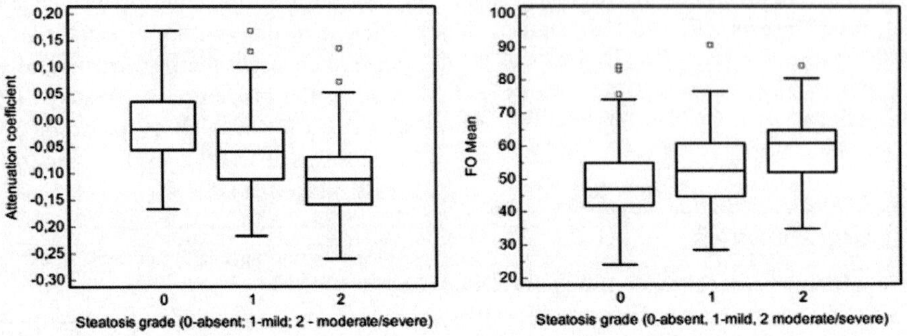

Fig. 4 Values of attenuation coefficient and FO mean according to different grades of steatosis (median and interquartile ranges)

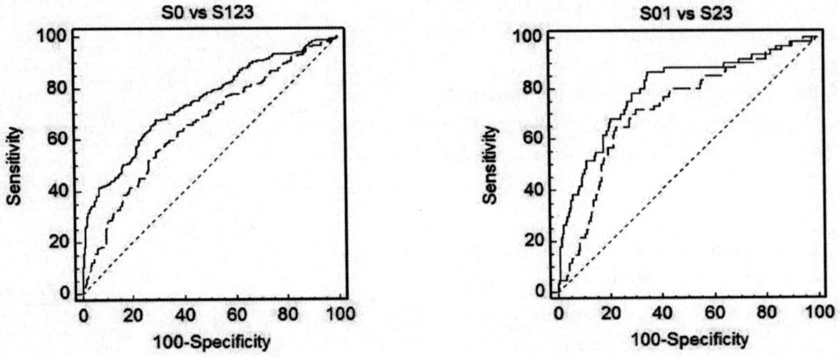

Fig. 5 ROC curves for the attenuation coefficient (continuous line) and FO mean (fragmented line) used for the prediction of each steatosis grade: S0 vs S1-2-3; S0-1 vs S2-3

The attenuation/backscatter indices were superior to textural indices in differentiating between the categories studied. By using the attenuation coefficient, Gaitini [20] obtained an ideal area under the ROC curve (AUROC=1) for the differentiation of patients having „pure" steatosis (without superimposed inflammation of fibrosis) from the healthy liver, but, the patients without severe steatosis had been excluded from that particular study. This approach has, however, a rather limited practical value since the ultrasonographic changes of severe steatosis can be discriminated from the normal liver by the „classic" examination.

Some authors [18] suggest that steatosis can be detected through imaging methods when it is moderate or severe (>33%). In the present study we have found AUROCs of 0.741 and 0.791 respectively for the prediction of the presence of steatosis and respectively for the prediction of moderate/severe steatosis. Such discrimination could not have been possible through the visual inspection alone, especially since these patients have, alongside steatosis, other histologic alterations that may distort the ultrasonographic image (fibrosis, inflammation or ballooning).

In the context of co-existing steatosis, necroinflammatory activity and fibrosis, first of all we wanted to see how much the attenuation coefficient and FO parameters were influenced by the histopathological aspects found in patients with VHC infections. The attenuation coefficient in patients with VHC was found to correlate significantly with steatosis, but there was no significant correlation with activity or fibrosis. This finding might be the first step in the further study of this coefficient used for the differentiation between steatosis and fibrosis on the ultrasonic image. In exchange, the FO mean is influenced by both steatosis and necroinflammatory activity.

Webb et al [19] have also tried to quantify the intrahepatocyte fat content by using the hepatorenal index, defined as the difference between the echogenicity of liver and that of the right kidney. The main limitation of this approach is the reference system (the kidney), whose echogenicity could also be altered by intrinsic conditions. On the other hand, the parameter quantified in these cases is the echogenicity of the liver and kidney respectively; the echogenic pattern of the liver can also result from superimposed fibrosis or inflamation, not only steatosis. In the present study we have proven that the attenuation coefficient correlates only to steatosis, not with inflammation or fibrosis, while the FO mean correlates both to steatosis and inflammation. Indeed, the performance of the attenuation coefficient was better than that of the FO mean, both for the prediction of steatosis (regardless of the grade), as well as for the prediction of moderate steatosis, at least.

The FO Mean parameter measures the average grey level intensities in the designated area. Knowing that, for an ultrasound image, a higher intensity means higher amplitude of the receiving echo one can assume that a higher value for "FO Mean" means that the underlying tissue has an enhanced echogeneity. This echogeneity, in the case of parenchimatous organs can be explained by the presence of numerous but small interfaces. This phenomenon has been observed by many authors in the case of steatosis, so, the correlation of FO Mean with the steatosis is not a surprise. In the case of necroinflamatory activity the liver tissue suffers a local swelling, hence the densities varies and produces numerous interfaces. The other parameters from FO statistics are not sensible to global changes to the image characteristics but to relative changes between pixels. It seems that liver steatosis can be detected only using a global measure of the ultrasound behaviour. A method that focuses only on a small region is less sensitive to steatosis alterations (AUROC for attenuation is better than AUROC for FO Mean).

In conclusion, the present study proves that, in patients having chronic hepatitis C, the attenuation coefficient, but also the FO mean, can discriminate between different steatosis grades; however, the attenuation coefficient has a better performance than the FO mean, being influenced only by steatosis, not by fibrosis or necroinflammatory activity.

Acknowledgment

This work was funded by the Romanian Authority for Scientific Research through the CEEX Program [grants nr 71/2006 and 94/2006] and the 2007-2013 National Program for Research, Development and Innovation [grant PNCDI2-Partnership nr 41-071/2007].

References

1. F Ramalho (2003). Hepatitis C virus infection and liver steatosis. Antiviral Res 60:125-127.
2. G Leandro, A Mangia, J Hui et al (2006. Relationship between steatosis, inflammation, and fibrosis in chronic hepatitis C: a meta-analysis of individual patient data). Gastroenterology 130:1636-1642.
3. A Regev, M Berho, LJ Jeffers (2002). Sampling error and intraobserver variation in liver biopsy in patients with chronic HCV infection Am J Gastroenterol 97: 2614-2618.
4. H Osawa, Y Mori (1996). Sonographic diagnosis of fatty liver using a histogram technique that compares liver and renal cortical echo amplitudes J Clin Ultrasound 24: 25-29.
5. E Caturelli, MM Squillante, A Andriulli et al (1992). Hypoechoic lesions in the "bright liver":a reliable indicator of fatty change. A prospective study. J Gastroenterol Hepatol 7: 469-472.
6. MJ Hepburn, JA Vos, EP Fillman, EJ Lawitz (2005). The accuracy of the report of hepatic steatosis on ultrasonography in patients infected with hepatitis C in a clinical setting:a retrospective observational study. BMC Gastroenterol 5: 14.
7. M. Meziri, WC. Pereira, A. Abdelwahab, C. Degott, P. Laugier (2005). "In vitro chronic hepatic disease characterization with a multiparametric ultrasonic approach", Ultrasonics 43: 305-313,
8. D. Gaitini, M. Lederman, Y. Baruch, et al (2005). Computerised analysis of liver texture with correlation to needle biopsy. Ultraschall Med, 26 (3): 197-202
9. D. Gaitini, Y. Baruch, E. Ghersin, et al (2004). Feasibility study of ultrasonic fatty liver biopsy: texture vs. attenuation and backscatter. Ultrasound Med Biol. 30 (10): 1321-1327
10. Y. Fujii, N. Taniguchi, K. Itoh, et al (2002). A new method for attenuation coefficient measurement in the liver. Comparison with the spectral shift central frequency method J Ultrasound Med 21: 783-788
11. M Lupsor, R Badea, C Vicaş, et al (2008). Ultrasonographic diagnosis of nonalcoholic steatohepatitis based on the quantitative evaluation of the ultrasound beam behavior into the liver. Proceedings of 2008 IEEE-TTTC International Conference on Automation, Quality and Testing, Robotics 3: 112-117
12. M Grigorescu, C Radu, M Lupşor, et al (2008). Comparison between attenuation coefficient computed on the ultrasound image and a biological marker, adiponectin, in the diagnosis of steatosis in non-alcoholic fatty liver disease. Proceedings of 2008 IEEE-TTTC International Conference on Automation, Quality and Testing, Robotics 3: 118-122
13. M. Lupşor, R. Badea, C. Vicaş, et al (2010). Noninvasive steatosis assessment in NASH through the computerized processing of ultrasound images: attenuation versus textural parameters. Proceedings of 2010 IEEE-International Conference on Automation, Quality and Testing, Robotics 2: 333-338
14. A. Szebeni, G. Tolvaj, A. Zalatnai (2006). Correlation of ultrasound attenuation and histopathological parameters of the liver in chronic diffuse liver diseases. Eur J Gastroenterol Hepatol 18 (1): 37-42
15. A. Materka, M. Strzelecki (1998), Texture analysis methods a review," Technical University of Lodz, Institute of Electronics.
16. U. Abeyratne, A. Petropulu (1997) Higher-order statistics for tissue characterization from ultrasound images. IEEE Proc. Signal Processing Workshop Higher-Order Statistics 72-76.
17. Bedossa P, Poynard T (1996). An algorithm for the grading of activity in chronic hepatitis C The METAVIR Cooperative Study Group. Hepatology 24: 289-293
18. S Saadeh, ZM Younossi, EM Remer, et al (2002). The utility of radiological imaging in nonalcoholic fatty liver disease Gastroenterology 123: 745-750
19. Webb M, Yeshua H, Zelber-Sagi S, Santo et al (2009). Diagnostic value of a computerized hepatorenal index for sonographic quantification of liver steatosis AJR Am J Roentgenol 192(4): 909-914

Author: Dr. Monica Lupsor
Institute: Ultrasonography Dept.,3rd Medical Clinic, University of Medicine and Pharmacy Iuliu-Hatieganu
Street: 19 – 21 Croitorilor Str
City: Cluj-Napoca
Country: Romania
Email: monica.lupsor@umfcluj.ro

ECG Signal Baseline Wander Removal Using Wavelet Analysis

Z. German-Sallo

"Petru Maior" University/Electrical Engineering Department, Targu-Mures, Romania

Abstract— **One of the most common problems of ECG recordings is the baseline wandering in the ECG signals during data collection. Baseline wander elimination is considered as a classical problem in ECG signal filtering. This paper presents two wavelet analysis (WA) based ECG signal baseline wander removal methods, The Discrete Wavelet transform based method uses a high level decomposition and eliminates the lowest frequency component. The wavelet packet based searching algorithm uses the energy of the signal in different scales to identify the baseline wander. The algorithm calculates the corresponding energy of wavelet packet coefficients at each scale. After a comparison, the branch of the wavelet binary tree corresponding to higher energy wavelet spaces is chosen. These procedures are tested using specific data records.**

Keywords— **wavelet analysis, ECG signal processing, baseline wander removal.**

I. INTRODUCTION

The ECG signal is obtained by recording the potential difference between two electrodes placed on the body surface. A single normal cycle of the electrocardiogram represents the successive atrial and ventricular depolarization and repolarization. Baseline (isoelectric line) wander is considered as a perturbation which produces artifacts when in ECG parameters measuring, especially the ST segment and the R peak measurements could be strongly affected. The main causes of the baseline wandering are the respiration, electrode impedance change due to perspiration and increased body movements. Therefore, the elimination of the baseline wander can improve very much the accuracy of the clinical information. Figure 1 presents together a clean and a baseline wander affected ECG signal, the differences between them are obvious and the necessity of filtering also.

During the last years, the wavelet analysis has proven to be a very useful tool in many application areas for evaluation of non-stationary signals such as biomedical signals, the ECG signal in particular. The wavelet transform provides a time-frequency representation of the signal, and thus permits the inspection of characteristic waves of the ECG signal at different scales with different resolutions. This paper presents two wavelet analysis (WA) based ECG signal baseline wander removal methods, The Discrete Wavelet Transform (DWT) based method uses a high level decomposition and eliminates the lowest frequency component. The Wavelet Packet (WP) based searching algorithm uses the energy of the signal in different scales to identify the baseline wander.

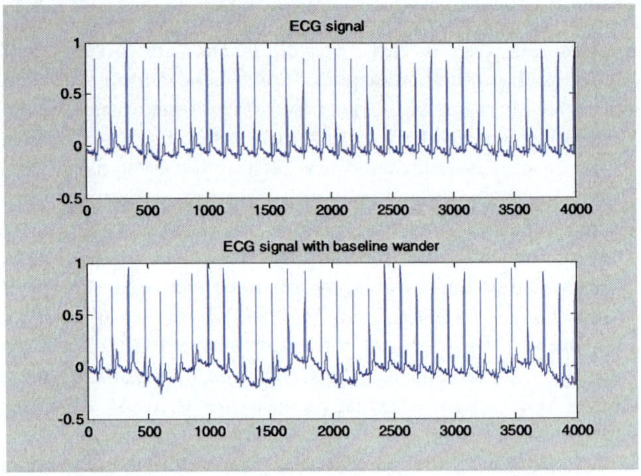

Fig. 1 ECG signal without and with baseline wander

The proposed algorithms are carried out in Matlab environment and are tested using specific ECG recordings from the MIT-BIH Arrhythmia Database, taken from a web-based resource for free access to study of physiological signals.

II. WAWELETS AND WAVELET PACKETS

The wavelet transform (WT) of signal x(t) is defined as a combination of a set of basis functions, obtained by means of dilation a and translation b of a mother wavelet [1].

$$W_a x(b) = \frac{1}{\sqrt{|a|}} \int_{-\infty}^{+\infty} x(t)\,\psi\!\left(\frac{t-b}{a}\right) dt \tag{1}$$

The Discrete Wavelet Transform (DWT) is defined as a discrete dilations and translations of the mother (or analyzing) wavelet. In its most common form, the DWT employs a dyadic grid and orthonormal wavelet basis functions [4].

The Discrete Wavelet Transform (DWT) decomposition of the signal into different frequency bands can be obtained

S. Vlad and R.V. Ciupa (Eds.): MEDITECH 2011, IFMBE Proceedings 36, pp. 190–193, 2011.
www.springerlink.com

by successive high-pass and low-pass filtering of the time domain as shown in figure 2. Translation is accomplished by considering all possible integer translations of ψ(t) and dilation is obtained by multiplying t by a scaling factor which is usually factors of 2.

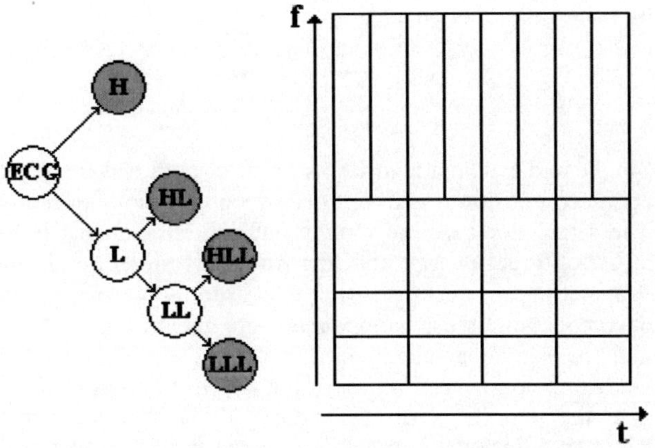

Fig. 2 FIR filter structure for dyadic scale decomposition

The following equation shows how wavelets are generated from the mother wavelet:

$$\psi_{j,k}(t)=2^{j/2}\psi\left(2^{j/2}t-k\right) \quad (2)$$

Wavelet decomposition is a linear expansion and it is expressed as

$$x(t)=\sum_{k=-\infty}^{\infty}c_k\varphi(t-k)+\sum_{k=-\infty}^{\infty}d_k\psi\left(2^{j/2}t-k\right) \quad (3)$$

where φ(t) is called the scaling function and c_k and d_{jk} are the coarse and detail level expansion coefficients. In the field of signal processing, the implementation of wavelet theory is performed using filter banks. In applications one never has to deal directly with the scaling functions or wavelets, only with the coefficients of the associated filters in the filter banks. In a wavelet transform system, the signal is convolved with a pair of maximally decimated quadrature mirror filters (QMF). These filters are related to wavelet and scaling functions as expressed below [4]:

$$\varphi(t)=\sqrt{2}\sum_{k=-\infty}^{+\infty}h_k\varphi(2t+k) \quad (4)$$

$$\psi(t)=\sqrt{2}\sum_{k=-\infty}^{+\infty}g_k\varphi(2t+k) \quad (5)$$

The coefficients are ordered using two dominant patterns, one that works as a smoothing filter (like a moving average), and one pattern that works to bring out the data's detail information. These two orderings of the coefficients are called a quadrature mirror filter pair in signal processing language [6].

The wavelet packet method is a generalization of wavelet decomposition that offers a richer range of possibilities for signal analysis but is more redundant also. In wavelet analysis, a signal is split into an approximation and a detail [5]. The approximation is then itself split into a second-level approximation and detail, and the process is repeated. For n-level decomposition, there are n+1 possible ways to decompose or encode the signal. A single decomposition using wavelet packets generates a large number of bases.

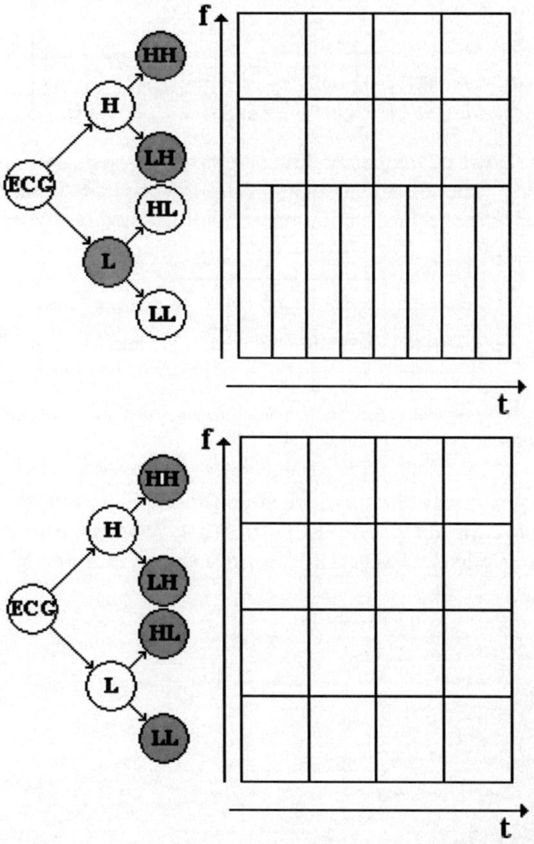

Fig. 3 Time-frequency blocks for second order dyadic wavelet packet decomposition and the corresponding filtering structure

III. Baseline wander removal procedure

The baseline wandering removal is carried out by identifying the low (or the lowest) frequency (large scale) components in the ECG signal. The proposed algorithms are based on the assumption that the baseline wandering and the ECG signal constitute a mixture of two independent signals, obtained as a linear superposition. Usually the typical baseline variation means 15 percent of peak-to-peak ECG amplitude variation of 0.15 to 0.3 Hz [3]. There are two methods to be presented, the first is based on an n-level Discrete Wavelet Transform (DWT) and the other uses the Wavelet Packet (WP) decomposition to find the low frequency component corresponding to the baseline variation. In first case the DWT level is estimated starting from the sampling frequency (which gives the maximum frequency of the ECG signal) and the estimated baseline wander frequency domain $\Delta f_{BLw} = f_{BWH} - f_{BWL}$ as follows:

$$n = round\left[\frac{1}{2}\left(\log_2 \frac{f_{\max}}{\Delta f_{BWH}} + \log_2 \frac{f_{\max}}{\Delta f_{BWL}}\right)\right] \tag{6}$$

The lowest frequency components are identified in the DWT structure and extracted from the original signal (after a selective reconstruction) in a way illustrated on figure 4.

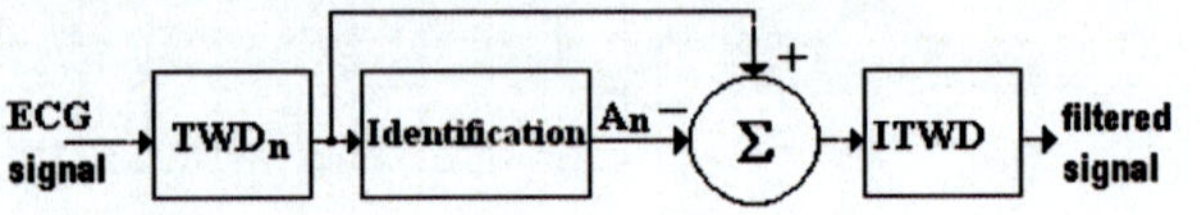

Fig. 4 Wavelet decomposition, lowest frequency component identification and baseline wander removal

A full wavelet packet decomposition binary tree for three scale wavelet packet ECG Baseline Wander Elimination using Wavelet Packets transform is shown in figure 5.

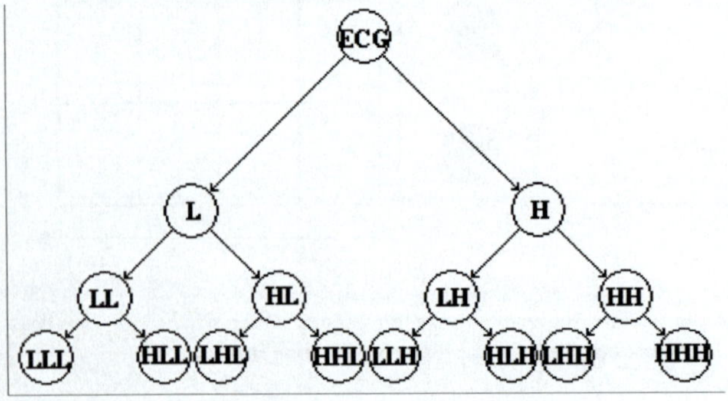

Fig. 5 Third level Wavelet Packet decomposition

In wavelet packet analysis, the details as well as the approximations can be split. This yields more than 2^{2^n-1} different ways to encode the signal [1]. Usually the components which are responsible to baseline wander placed in low frequency but have a relatively great energy. The energy of the signal is given in terms of the WT coefficients by Parseval's relation as:

$$E = \int |x(t)|^2 = \sum_{k=1}^{n} |a_k|^2 + \sum_{k=1}^{n} |d_k|^2 \tag{7}$$

Simple and efficient algorithms exist for both wavelet packets decomposition and optimal decomposition selection. The search for baseline wander causing components is focused on great energy and low frequency structures in the Wavelet Packet decomposition tree. After a Wavelet Packet decomposition these components are identified, removed and the signal is recomposed without them. In fact, that leads to a nonlinear filtering of the signal. In order to eliminate the baseline drift, the estimated baseline wander is subtracted from the original data record and a baseline wander free ECG signal is identified [5].

IV. Results

The test database was extracted from the MITBIH database, results are shown in figures 6 to 8. The measurements were carried out by adding gaussisan white noise to the test signal. The used analyzing function was "db4" (Daubechies) both for DWT or WP decomposition. The estimation is relatively good, the results after extraction show a satisfying baseline wander elimination .

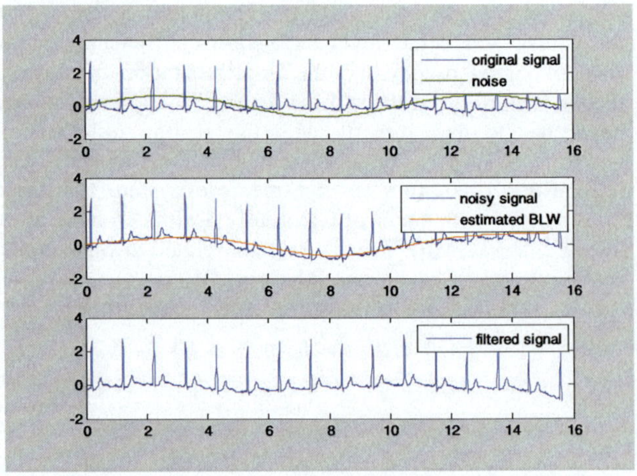

Fig. 6 Results obtained with n[th] level DWT decomposition

In the corresponding wavelet packets situation, each detail coefficient vector is also decomposed into two parts using the same approach as in approximation vector splitting. Results obtained with the use of this method and applied to the same signal, are represented in figures 7.

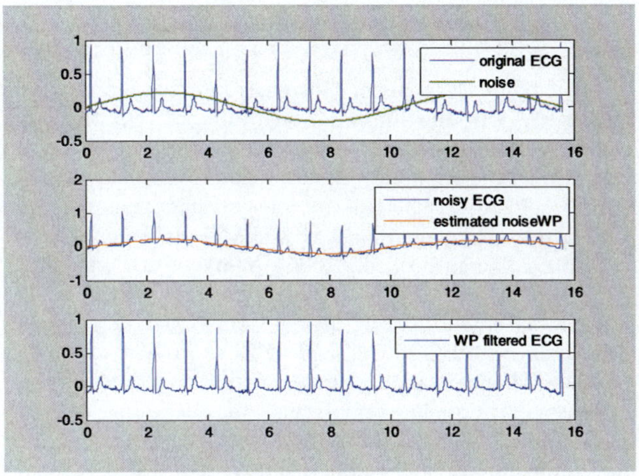

Fig. 7 Results obtained with nth level WP decomposition

To estimate the obtained results the following parameters were measured, :

$$SNR = \log_{10} \frac{P_{filteredECG}}{P_{originalECG} - P_{filteredECG}} \qquad (8)$$

$$Gain = SNR - \log_{10}\left(\frac{P_{originalECG}}{P_{noise}} \right) \qquad (9)$$

The obtained results are presented on figure 8 and 9.

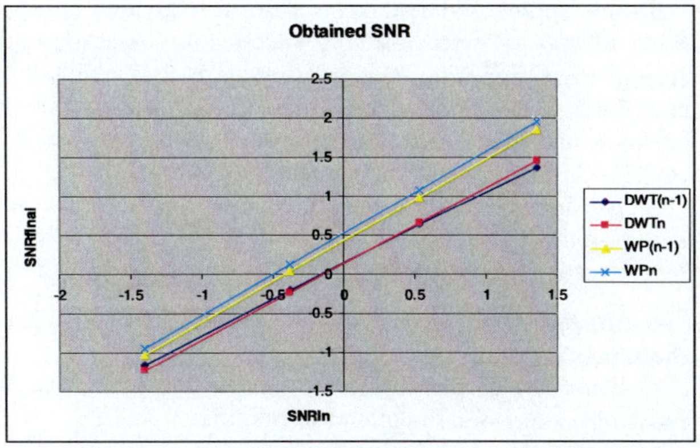

Fig. 8 Results obtained on test signal (Signal to Noise Ratio)

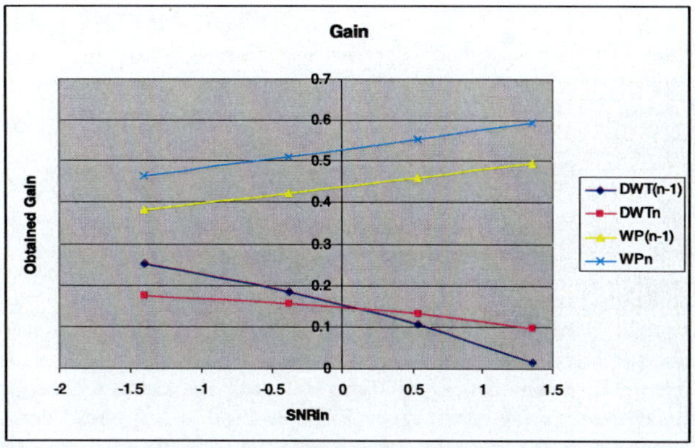

Fig. 9 The obtained gain for different DWT and WP decompositions

V. Conclusions

In this paper were presented two algorithms based on wavelet analysis for canceling baseline wandering in ECG signals. The WP decomposition offers a particular way of decomposing signals, the filtering results are slightly better than in case of using classical DWT. The presented algorithms can eliminate ECG baseline wandering without introducing major deformation in the signal structure. As further work is possible to study the opportunity to extract a weighted sum of low frequency components from the DWT or WP decompositions in order to have a better accuracy of baseline wander identification.

References

1. Aldroubi, A., Unser, M.: Wavelets in Medicine and Biology. CRC Press New York 1996
2. Walnut, D.,F.: An Introduction to wavelet analysis. Birkhäser Boston 2002
3. J. A. Van Alste and T. S. Schilder, "Removal of base-line wander and power-line interference from the ECG by an efficient FIR filter with a reduced number of taps," IEEE Transactions on Biomedical Engineering, vol. 32, no. 12, pp. 1052–1060, 1985.
4. Mallat, S.: A wavelet tour of signal processing. Academic Press London 2001
5. Burrus, S., Gopinath, R.A., Guo.H.: Introduction to wavelets and wavelet transforms. 1998 Prentice Hall, Inc., New Jersey
6. C. R. Meyer and H. N. Keiser, "Electrocardiogram baseline noise estimation and removal using cubic splines and statespace computation techniques," Computers and Biomedical Research, vol. 10, no. 5, pp. 459–470, 1977.

Emotion Investigation Based on Biosignals

E. Lupu, S. Emerich, and R. Arsinte

Technical University of Cluj-Napoca, Communications dept., Cluj-Napoca, Romania

Abstract— **In this paper a physiological signal-based emotion recognition approach is presented. The input biosignals are electromyogram, electrocardiogram, skin conductivity and respiration change. The feature vector is extracted from each signal type by using the same technique based on wavelets and TESPAR DZ method. A Support Vector Machine (SVM) classifier was employed to distinguish among four emotional states: joy, anger, sadness and pleasure. The database employed in our experiments is the AuBT corpus.**

Keywords— **emotional states, TESPAR DZ, biosignals, feature selection, SVM.**

I. Introduction

"Everyone knows what an emotion is, until asked to give a definition" [Beverly Fehr and James Russell]. Emotions play an important role in: motivation, perception, cognition, coping, creativity, attention, planning, reasoning, learning, memory and decision making. Research in emotions is pursued in several scientific disciplines, such as neuroscience, cognitive sciences and psychology. The progress in the aforementioned sciences determines much of the success in the development of affective multimodal systems. In advanced human-machine interaction, emotion recognition is one of the key steps towards emotional intelligence. One way to differentiate emotions is by their being short-term (seconds /minutes), whereas moods are long-term (some days) emotional states, typically global and very variable over the time, dominating the intensity of each short-term emotional state. Moreover, temperaments and personalities are very long-term (months/years/a lifetime) and are much more complex by their including personality factors and moods (Jenkis & et al., 1998).

Various experiments on human judgment on still photographs of posed facial behavior were conducted by Ekman and his colleagues who concluded that there are six basic emotions which can be recognized universally, respectively: happiness, sadness, surprise, fear, anger and disgust (Ekman, 1982). This theory of universality is the most widely used theory in affect sensing by machines [1]. The human communication modes are affected by emotional states through facial expression, body gestures, tone of voice, respiration rate, skin temperature, skin conductance etc.

Biosignals

Physiological signals or biosignals refer to: brain signals measured via functional Near Infrared Spectroscopy (fNIRS), scalp signals measured via electroencephalogram (EEG), and peripheral signals: cardiovascular activity (ECG); electrodermal activity or galvanic skin response (GSR), electromyogram activity (EMG) (Changchun& et al., 2005; Savran& et al., 2006). While visual modalities such as facial expressions and body gestures provide a visible/external understanding of one's emotional state, biosignals such as EEG, ECG and fNIRS provide an invisible/internal understanding of the emotion phenomenon [1].

Psychophysiology establishes the relation between physiological signals and arousal/valence and argues that the activation of the autonomic nervous system changes while emotions are elicited (Levenson, 1988). Consequently, different emotional expressions produce different changes in autonomic activity (for instance, happiness: decreased heart rate, no change in skin temperature; anger: increased heart rate and skin temperature; fear: increased heart rate, decreased skin temperature [6].

However, in general, an optimal set of bio-potential cues that can assist in reliably discriminating among various affective states has not yet been identified.

II. Methods and tools

A. *TESPAR DZ Approach*

In this approach the signal waveform is divided into periods or epochs determined by successive passes through zero of the signal. The time information is thus maintained combined with a simple approximation of the waveform between two successive passes through zero. An overview of TESPAR coding method can be found in [3], [4].

This paper employs a version of the described method based on the TESPAR DZ matrices, by using three descriptors so as to describe each epoch:

- ***duration (D)*** between two successive zero crossings of the signal, expressed in samples;
- ***shape (S)*** of the signal between two successive zero crossings, expressed in number of minima;
- ***amplitude (A)*** which represents the maximum value found amongst the samples of an epoch [5].

S. Vlad and R.V. Ciupa (Eds.): MEDITECH 2011, IFMBE Proceedings 36, pp. 194–197, 2011.
www.springerlink.com

A comparison is made between pairs of epochs by using the TESPAR DZ coding procedure. The descriptors from each epoch pair are compared and symbols are produced. The symbol indicates the differences between the individual D, S and A, features of the two epochs under testing. Different lags can be considered when performing the epochs' descriptors comparison. For example, for a lag=1, comparisons will be made between epoch E and epoch E-1, and for a lag=2, comparisons will be performed between epoch E and epoch E-2. The flowchart of the entire coding procedure or symbol assignation can be found in Fig 1.

For each epoch descriptor, a three-stage vector comparison is generated in the case of each individual epoch pair comparison. Hence, for a lag = 1, when comparing D, S and A, for epochs E1 and E2, the following value results are provided. For D2 versus D1: if D2=D1, the resulted value is 0, if D2<D1, the resulted value is -1 or if D2>D1, the resulted value is +1. In the case of the other descriptors (S, A), the procedure is the same.

From any paired descriptors comparison of D, S and A, one of the 27 possible difference options may be derived by using this algorithm, this indicating the nature of the difference between the pair of epochs being tested The symbols were arbitrarily assigned to the numbers 1 to 27 of the 27 symbol DZ TESPAR alphabet. The TESPAR coder issue represents a string of symbols provided by the comparison of the three descriptors corresponding to each epoch. By converting this simple series of symbols, fixed length structures can be obtained.

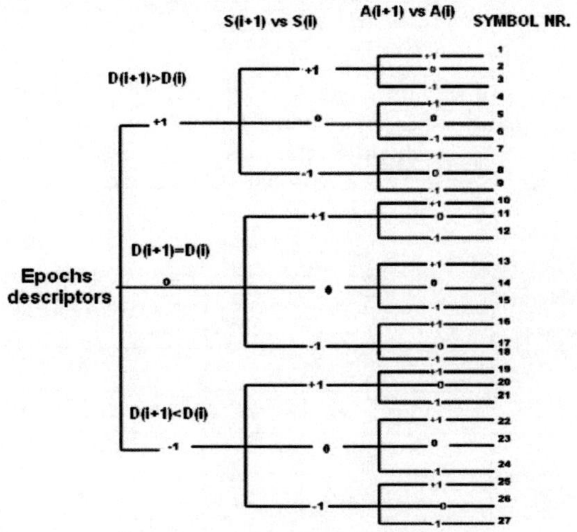

Fig. 1 TESPAR DZ symbols assignation

A one-dimensional vector which counts the number of occurrences of alphabet symbols in an instance can be used, this leading to a histogram [5]. Symbols may be found missing in the coding process, depending on the type of processed signal. Consequently, the alphabet may be redefined with a lower number of symbols.

B. Classification Method

The technique of SVM, developed by Vapnik [10], is a powerful, widely used technique for solving supervised classification problems due to its generalization ability. In this study, the LIBSVM software is used. LIBSVM is an integrated software package for support vector classification, regression and distribution estimation [9]. From the available kernels, the radial basis function was selected for our experiments.

C. AuBT Corpus (Augsburger Database of Biosignal)

The AuBT corpus contains physiological data taken from a single user in four different emotional states: joy, anger, sadness and pleasure. It was recorded while the subject was listening to four music songs, which were previously picked by the user himself according to the four targeted emotion classes. The subject had been advised to select songs which could trigger special memories to him and he was also asked to make an effort to enter the particular needed affective state. Criteria for song selection:

- song1: enjoyable, harmonic, dynamic, moving;
- song2: noisy, loud, irritating, discord ;
- song3: melancholic, reminding of sad memory;
- song4: blissful, slow beat, pleasurable, slumberous.

In order to record electromyogram, electrocardiogram, skin conductivity (SC) and Respiration change (RSP), four-channel biosensors were used. 25 recordings (in 25 days) for each emotion were overall collected.

The length of the recordings is dependent on the length of the songs, but it was later cropped to a fixed length of 2 minutes per session and per emotion. ECG was sampled at 256 Hz, while the other signals were sampled at 32 Hz [7].

III. EXPERIMENTS AND RESULTS

A. Feature Extraction and Selection

The emotional status of a person is inherently reflected in the activity of the nervous system. In psychophysiology, traditional tools for the investigation of human emotional status are based on the recording and on the statistical analysis of biosignals.

This paper proposes a new feature extraction method, based on wavelets and on the TESPAR DZ techniques.

Further, different waveforms extracted from ECG and RSP signals are presented, in the case of the emotional states of joy and, respectively, sadness. The waveforms can be divided into epochs whose dimension, amplitude and shape is variable in time. Consequently, TESPAR DZ algorithm can be applied, this providing the first 27 coefficients of the feature vector.

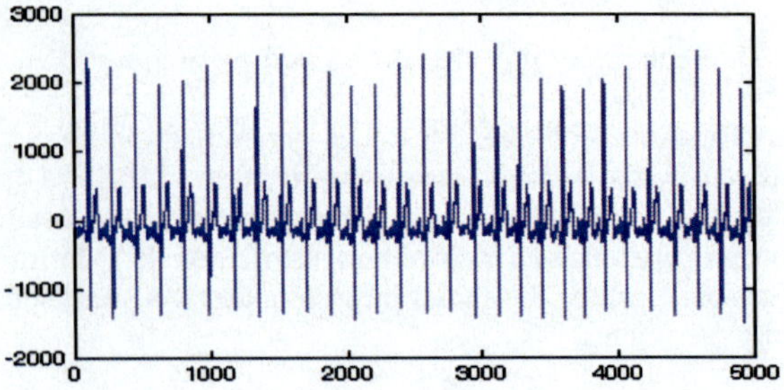

Fig. 2 The first 5000 samples from an ECG signal, emotional state of joy

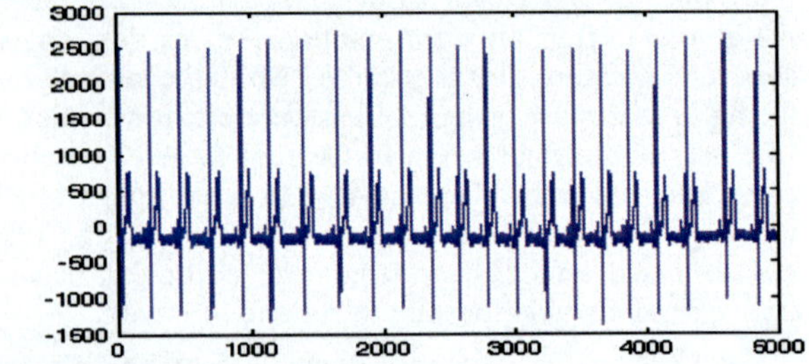

Fig. 3 The first 5000 samples from an ECG signal, emotional state of sadness

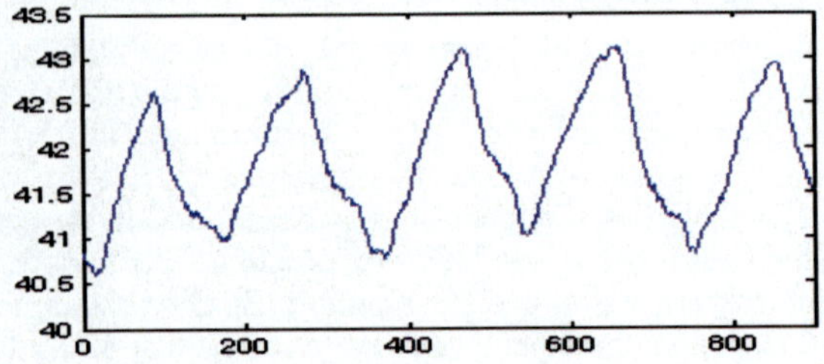

Fig. 4 The first 900 samples from a RSP signal, emotional state of joy

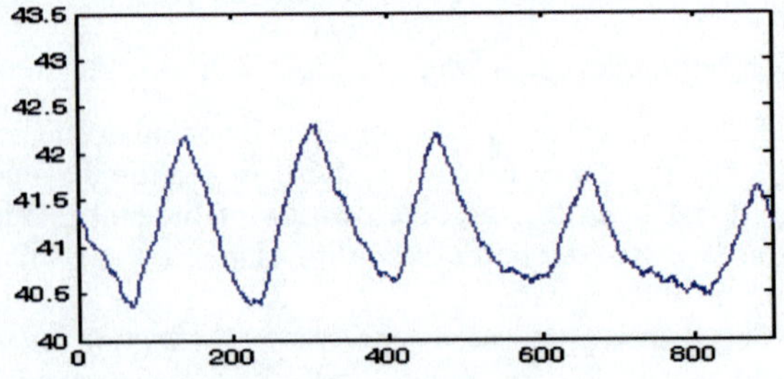

Fig. 5 The first 900 samples from a RSP signal, emotional state of sadness

The signal under test is also decomposed in up to 4 levels, by using the Discrete Wavelet Transform (DWT) and Db4 (Daubechies) as mother function. By passing a signal x through a series of filters, its DWT can be calculated. First, the samples are passed through a low pass filter with impulse response g, while the signal is being decomposed simultaneously using a high-pass filter h. The outputs are giving the detail coefficients (from the high-pass filter) and the approximation coefficients (from the low-pass). The two filters must be related to each other, hence they are known as a quadrature mirror filter. The decomposition is repeated to further increase the frequency resolution and the approximation coefficients are decomposed with high and low pass filters and then down-sampled. The process is represented as a binary tree with nodes representing a sub-space with different time-frequency localization. The tree is known as a filter bank, Fig. 6. The energy content of the approximation and details is then computed, resulting into the next 5 coefficients of the feature vector.

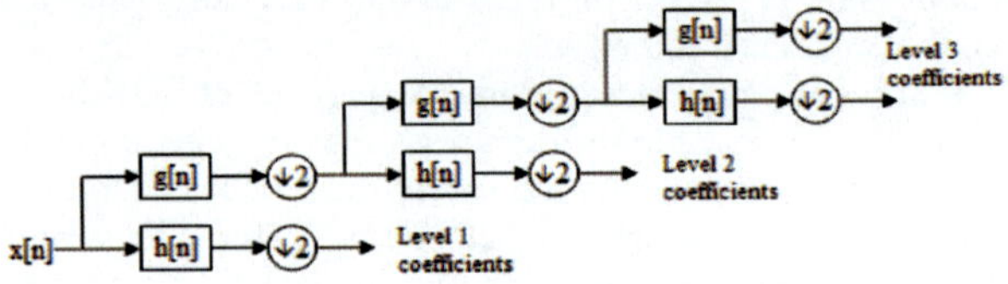

Fig. 6 Wavelet decomposition tree

Finally, the obtained feature vector for each employed biosignal has a fixed dimension equal with 32 coefficients.

$$\underbrace{v1, v2, ..., v27}_{\substack{\text{TESPAR DZ} \\ \text{coefficients}}}, \underbrace{v28, ..., v32}_{\substack{\text{Wavelet Energy} \\ \text{coefficients}}} \qquad (1)$$

The goal of feature selection is that of reducing the dimensionality of input patterns to make the computation feasibl and to meanwhile retain the most relevant features which would reflect the emotional state changes.

Table 1 Feature selection

Feature vector	TESPAR DZ	Wavelet Energy
ECG6	v1,v6, v12	v30,v31
EMG12	v1,v3,v4,v6,v22,v24,v25	v28,v29,v30,v31,v32
SC4	v1,v27	v28,v29
RSP7	v1,v3,v7,v21,v25	v32

The ChiSquaredAttribute Evaluation method was used for this purpose. It evaluates the worth of an attribute by computing the value of the chi-squared statistic with respect to the class. In the next table, the reduced feature vector is described for each biosignal category.

B. AuBT Experiments

The AuBT GUI supports four different signals by default, thier being SC, EMG, RSP and ECG. For each signal, a number of preprocessing steps are applied, such as low pass filtering and normalization. Several statistical features such as mean and standard deviation are then calculated from the preprocessed signals, along with different transformations of the signals, e.g. respiration rate and heart rate variability. The approximation of the first and second derivation is used so as to obtain the same statistical features. 81 features were extracted from the ECG signal, 65 features were extracted from the RSP signal, 19 features were extracted from the SC signal and 21 features were extracted from each of the EMG signals. The complete list of the features extracted can be consulted in [6], [7].

C. Results

Figure 7 shows the results generated by our system. Based on the four selected biosignals, we classified four emotional states: joy, anger, sadness and pleasure. The vectors coresponding to the selected signals are fused at the feature level and provide a 128 coefficients vector. After the feature selection step the vector became shorter, with a length of 29 coefficients, and the resulted acurracy was 89.33%, for the SVM classifier, RBF kernel (C=100 and γ =0.01). By comparing our results with the ones obtained by using AuBT toolbox [7], one can notice that for the AuBT final vector (resulted after feature level fusion and selection – Aubt_40), the classification rate was equal to merely 76.66%, in the case of the same classifier.

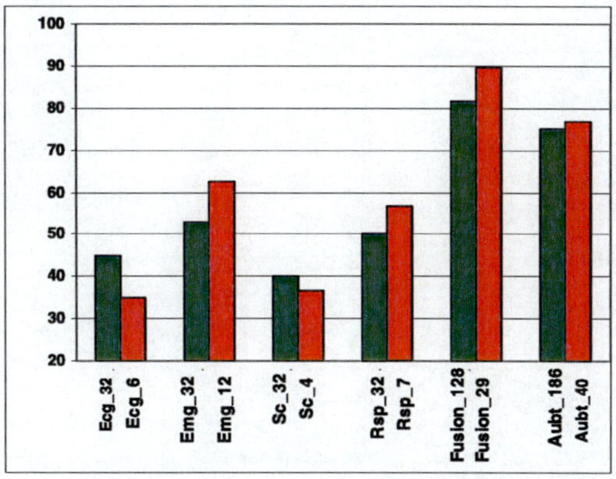

Fig. 7 Classification rates provided by experiments

The results are promising and support the general conclusions in the automatic recognition of emotions that distinguishing emotional states is not an easy task. At this point, experiments have distinguished four emotional classes with good accuracy, compared with the current results found in references.

IV. Conclusions and outlook

The approach we propose offers a better performance than that provided by the AuBT attempt, experiments being made on the same database. Also, SVMs based classification methods represents a major support in pattern recognition research.

Future work will be concentrated on including the validation of the performance on a much larger population of test subjects, on comparing/combining with other biosignals and also on extending the range of emotions under study.

Furthermore, we are aware that recognizing emotions very accurately using one signal source only is a difficult task. Therefore, we also plan on pursuing research in this area by employing the fusion of visual modalities such as facial expressions, body gestures and speech, which provide an external understanding of the emotions phenomenon.

References

1. Hatice Gunes, Massimo Piccardi and Maja Pantic Affective Computing, Focus on Emotion Expression, Synthesis and Recognition cap.10 pp. 185-218
2. Christos D. Katsis, Nikolaos Katertsidis, George Ganiatsas, and Dimitrios I. Fotiadis Toward Emotion Recognition in Car-Racing Drivers: A Biosignal Processing Approach IEEE Transactions on systems, man, and cybernetics—PART A: Systems and humans, Vol. 38, No. 3, May 2008, pp 502-512
3. R. A. King, T. C. Phipps, (1998), Shannon, TESPAR and Approximation Strategies, ICSPAT., Toronto, Vol. 2, pp.1204-1212,.
4. E. Lupu, V.V.Moca, P.G. Pop, TESPAR coding study for speaker recognition, The 30th session of scientific presentations Modern technologies in the XXI Century, Bucharest 2003, pp. 214-221
5. E. Lupu, S. Emerich, F. Beaufort, On-line signature recognition using a global features fusion approach, Acta Tehnica Napocensis Electronics and Telecomunications Vol.50, nr .3/2009, pp 13-20
6. Johannes Wagner, Jonghwa Kim, Elisabeth André, (2005) From Physiological Signals to Emotions: Implementing and Comparing Selected Methods for Feature Extraction and Classification, In IEEE International Conference on Multimedia & Expo (ICME 2005)
7. Johannes Wagner, Augsburg Biosignal Toolbox (AuBT), University of Augsburg
8. HUMAINE: http://emotion-research.net/
9. Chih-Chung Chang and Chih-Jen Lin, LIBSVM - A Library for Support Vector Machines, http://www.csie.ntu.edu.tw/~cjlin/libsvm/
10. Vapnik, V., (1995), The Nature of Statistical Learning Theory, New York, Springer-Verlag.

Author: EUGEN LUPU
Institute: TECHNICAL UNIVERSITY OF CLUJ-NAPOCA
Street: 26-28 BARITIU
City: CLUJ-NAPOCA
Country: ROMANIA
Email: eugen.lupu@com.utcluj.ro

Multirate Sampling in PCG Signal Correlation

S. Gergely[1], M.N. Roman[2], and C. Fort[2]

[1] National Institute for Research and Development of Isotopic and Molecular Technologies, Cluj-Napoca, Romania
[2] Technical University, Cluj-Napoca, Romania

Abstract— **The complex structure of *PCG* signals make them difficult to characterize in an automated approach. This paper presents a method used in *DSP* signal cross-correlation by using a pattern recognizing algorithm based on multirate signal processing of the acquired *PCG* signals.**

Keywords— **Wavelet transforms, DSP, Phonocardiography, Multirate Sampling, Signal Processing.**

I. Introduction

By using the wavelet transform in the analyzing of *PCG* signals it is possible to compress and to preserve all time – frequency characteristics of the signals. On the other hand either time domain or frequency domain analysis does not fully describe the nature of non-stationary signal. A pathological *PCG* signal is dominated by the high frequency components along with the low frequency *S* type pulses. The statistical characterization method is usable only for a primary signal classification. A precise *PCG* signal characterization done with the intention of pathology recognizing, is possible only if a reference signal is in fact compared with the fully or partially acquired signal. The frequency content in the multilevel wavelet transform may well be evaluated by the information content of each level defined by the Shannon entropy. The estimation of the signals envelope as a final characterization is a difficult task that involves intensive computing resources. Therefore the algorithm was designed to be implemented on a device using *DSP* engine for the signal processing.

II. Description of the Method

The first and very important step in the correlation algorithm is to limit the frequency band of the signal to avoid the aliases that could be generated by the interpolation during the multirate signal processing. At the next step it is computed the multilevel wavelet transform of the previously filtered signal. At this stage the main reason for that is to get a signal compression and therefore to lessen the computing tasks. Additional using of the wavelet transform is shown immediately. The total acquisition time is 4.096 seconds at a sampling rate of 8 KHz; as a result the samples amount is this way is decreasing from 32768 to 4096 without any noticeable distortions. The wavelet transform is done by using *Daubeschies 4* coefficients along with signal decimation by two. Another benefit of using the wavelet transform is due the filtering effect which reduces the bandwidth of the signal. Therefore the correlation algorithm is using only the approximation coefficients of the wavelet transform. So that the question is which decomposition level is to be used for the signal correlation algorithm? The envelope of the signal must contain as much as possible amount of samples to get a specific pathology characterization. The property of the multilevel wavelet transform is that, each computed approximation coefficients do contain both the approximation and detail coefficients of the next level. [1][2] The information content of 10 specific *PCG* pathologies is represented by the Shannon entropy showed in figure 1. Consequently the used decomposition level is given by the first decreasing of *Shannon* entropy which is applied to the detail coefficients. Because this level must remain constant for all signals, we are using level 3 (level 2 in figure 1). [3][4]

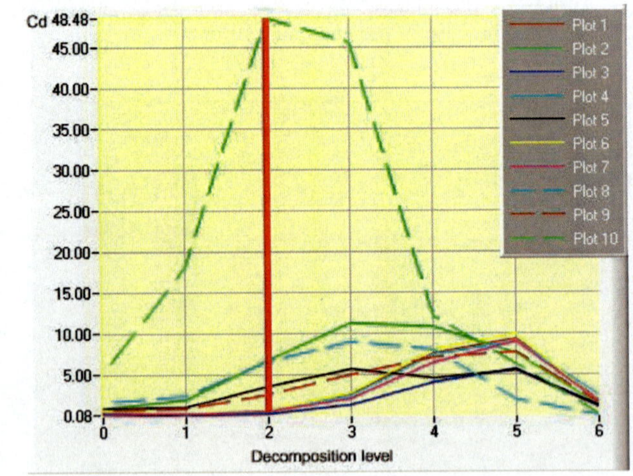

Fig. 1 Selection criterion for the used decomposition level

After all the issue is to test the correlation between the acquired signal and all reference signals that are stored in the *SD* card. Our studies showed that the heart beat rate

S. Vlad and R.V. Ciupa (Eds.): MEDITECH 2011, IFMBE Proceedings 36, pp. 198–201, 2011.
www.springerlink.com

does not affect the overall spectral density of the *PCG* signal. The Shannon sampling theorem has been extended to allow for sampling times which are not uniformly spaced. Several slightly different versions of the non-uniform sampling theorem have arisen. The differences lie in the spaces of functions being considered and the different classes of sampling times which are permitted. The theorem essentially says that a bandlimited signal $x(t)$ is uniquely determined by knowledge of its samples $\{an = x(tn)\}$ as long as the sampling times $\{tn\}$ occur at a rate which is on average higher than the *Nyquist* rate. By compressing or expanding the signal the envelope remains shift able and also the resulted frequency variations are not important. The frequency content is previously analyzed by the using of the *Shannon* entropy which classifies the *PCG* signal at each wavelet decomposition level. [5]

The main difficulty in doing the correlation task is that, the acquired signal is never at the same *HBR* (Heart Beat Rate) like the reference signals; as a result the time shift will corrupt the peak value of the cross-correlation. All reference signal envelopes are recorded at a known *HBR*, information which is stored in the file of each reference signal. The correct cross-correlation is complete after the re-sampling of the acquired signal in accordance to the reference signal *HBR*. After re-sampling, the signal is interpreted as having the same original sampling rate. The complete cross-correlation algorithm is shown in figure 2. The *HBR* ratio is a direct indication of the re-sampling ratio that we need, assuming that equally the analyzed signal and all the references were digitized using the same sampling rate. Therefore:

$$\frac{Sr\uparrow x(tn)}{Rr\downarrow x(tn)}=\frac{L}{M}=RRatio \qquad (1)$$

where *Sr* is the *HBR* of the signal and *Rr* is the *HBR* of the reference.

Obviously the value of *RRatio* possibly will be grater or smaller than 1, otherwise the re-sampling is undefined. The algorithm does support an RRatio from 0.6 up to 1.4 which means a *HBR* that spans from 50BPM to 120BPM, using a 70BPM reference signal. The value for *M* constantly equals the value 100 and *L* may take any value between 60 and 140.That value means that there are inserted a number of *L-1* zeros along the samples. All references are cropped to one cycle pathology signals. These signals are stable without any further processing. The envelope detection algorithm is shown in figure 3. It consists of a squaring and low pass filtering. The reason for choosing this algorithm is that is less computational intensive in comparison to the *Hilbert* transform. This envelope detection method involves squaring the input signal and sending this signal through a lowpass filter. Squaring the signal demodulates the input by using the input as its own carrier wave. This means that half the energy of the signal is pushed up to higher frequencies and half is shifted down toward DC. A *FIR* decimation is used which applies a low pass filter before downsampling the signal. Finally the signal is passed through a minimum-phase lowpass filter, to eliminate the high frequency energy.

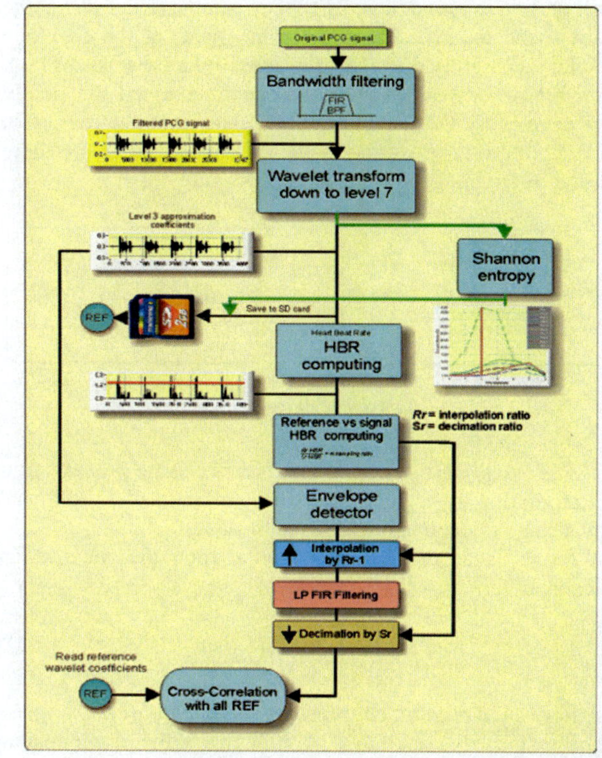

Fig. 2 Multirate sampling in PCG signal corelation

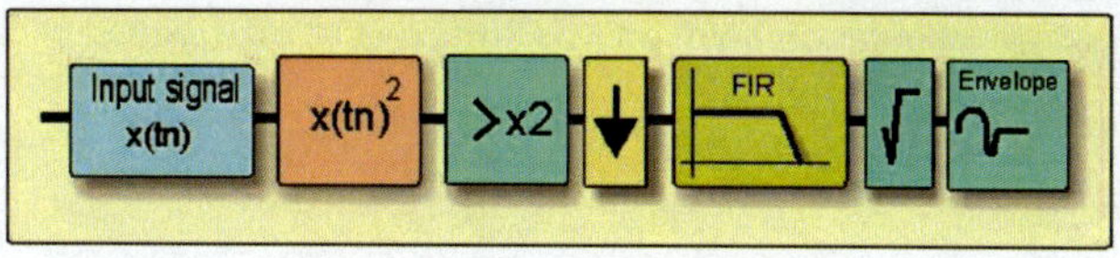

Fig. 3 Envelope detection algorithm

To maintain the correct scale, there are performed two additional operations. First, the signal is amplified by a factor of two. Since the algorithm keeps only the lower half of the signal energy, this gain matches the final energy to its original energy. Second, it is taken the square root of the signal to reverse the scaling distortion that resulted from squaring the signal.

The sensitive part of the overall algorithm is the interpolator used in the multirate sampling module. It turned out that the final correlation operation is obviously sensitive to the unmatched envelope. The issue is the lag over the real envelope of the computed values in case of a under filtering or a diminishing and distorted wave form in case of an over filtering. Therefore it was necessary to calculate the interpolation filter. The first step of using a linear interpolator has the ability of good low frequency attenuation but on the other hand the high frequency components of the envelope signal are slightly distorted. Because of the presence of high frequency components over the envelope signal the finally we have chose the *sinc* interpolator. The general structure of a sampling rate converter by the rational factor *L/M* (where *L* and *M* are integers) is presented in figure 4. [6]

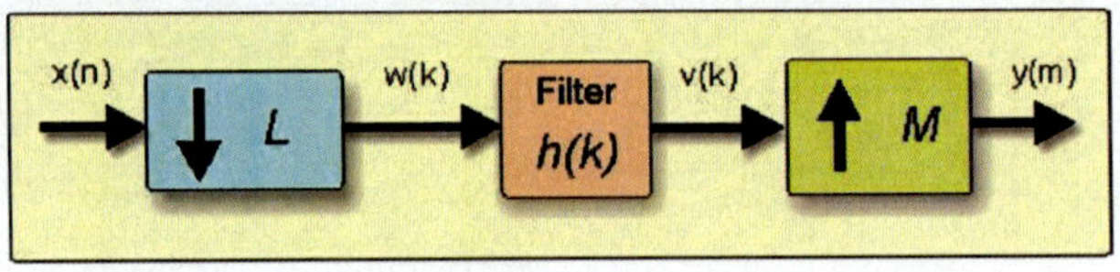

Fig. 4 Sampling rate converter

The sampling rate conversion is a two stage process; first interpolating the signal sequence by a factor of *L*, followed by a stage of decimation by a factor of *M*.

For a linear shift-invariant digital filter the convolution ecuation 2,

$$v(k)=\sum_{-\infty}^{\infty} h(r)w(k-r) \tag{2}$$

specifies the relation between *w(k)*, *v(k)* and *h(k)*. Using the phase linearity advantage of *FIR* filters the implemented filter structure of the *h(k)* filtering stage is next defined.

The interpolation filter is a unified low pass filter which has the cut-off frequency as a minimum between the necessary interpolation point and decimation cut off frequency as shown in equation 3:

$$\theta_c=\min\left(\frac{\pi}{L},\frac{\pi}{M}\right) \tag{3}$$

The first tested interpolation method was the linear interpolation which is the most familiar of all. The linear interpolation filter encompasses two values of the input signal so that we get:

$$N=2L-1 \tag{4}$$

Equation 4 represents the overall length of the filter, where *L* is the interpolation factor. In case of a polyphase filter where h(m) is defined for $-L < m < L$ we obtain for the extension of h(m) to *2L-1* samples:

$$y(nL+\rho)=y_p(nL+\rho)=p_p(0)x(n)+p_p(-1)x(n+1) \tag{5}$$

As the extension of the polyphase filter for *2L-1* samples each polyphase filter has two taps, $p_p(-1)$ and $p_p(0)$. [6]

Based on the definition of the linear interpolator, we get:

$$y(nL+\rho)=x(n)+\Delta x(n)\left(\frac{\rho}{L}\right) \tag{6}$$

$$=x(n)+[x(n+1)-x(n)]\left(\frac{\rho}{L}\right) \tag{7}$$

From equation 5 and 7 we get:

$$p_p(-1)=\frac{\rho}{L},\quad \rho=0,1,...L-1 \tag{8}$$

$$p_p=1-\frac{\rho}{L},\quad \rho=0,1,...L-1 \tag{9}$$

If we represent the sampling rate expanded set as:

$$\hat{p}_p=\begin{cases} p_p\left(\frac{k}{L}\right) & k=0,\pm L,\pm 2L \\ 0 & otherwise \end{cases} \tag{10}$$

Then h(k) can be reconstructed from equation 10:

$$h(m)=\begin{cases} 1-\frac{|m|}{L}, & -L<m<L \\ 0, & otherwise \end{cases} \tag{11}$$

It turned out that equation 11 is in actuality the impulse response of the linear interpolation filter.

To check the frequency response of the interpolator, by applying the *Fourier* transform over equation 11 we get: [6][7][9]

$$H\left(e^{j\omega}\right)=\frac{1}{L}\left[\frac{\sin(\omega L/2)}{\sin(\omega/2)}\right]^2 \tag{12}$$

Thus the *PCG* signal envelope was a mixed frequency combination the linear interpolator was not capable to attenuate sufficiently the created aliases, due to the only 20db attenuation which resulted from the graphical representation of equation 12.

The second method is the direct application of the sampling theorem, where a window bounds the *sinc* function.

All periodic sampled signals can be *sinc*-interpolated exactly using the equation 13.

$$x(t)=\sum_{n=-L}^{M-1} x_n(-1)^n\left[(-1)^{N+1}\tan\left(\pi\frac{t-n}{2N}\right)+\cot\left(\pi\frac{t-n}{2N}\right)\right] \tag{13}$$

where $N=L+M$, the sampling rate is normalized to be T=1 and the period $N=L+M$ samples. By rearranging in a much more visible form we get the equation 14: [10][12]

$$x(t)=\sum_{n=-\infty}^{\infty} x_n \frac{\sin[\pi(t-n)]}{\pi(t-n)}=\frac{\sin(\pi t)}{\pi}\sum_{n=-\infty}^{\infty} x_n \frac{(-1)^n}{t-n} \qquad (14)$$

This form is used to construct a table based sinc interpolation algorithm in which the function *1/t* is sampled, windowed and stored in a table over a small range *t*.

The cross-correlation is shown in figure 5. For each reference envelope, the resulted correlation values are stored in the *SD* memory card. A sorting algorithm searches between the correlation coefficients and obviously the memory sectors where the highest values are find, represents a correct pathology match. The green envelope in figure 4 is the corrected by the *HBR* ratio of the initial (red) signal envelope.

In signal processing, cross-correlation is a measure of similarity of two waveforms as a function of a time-lag applied to one of them. This is also known as a sliding dot product or inner-product. It is commonly used to search a long duration signal for a shorter, known feature. By definition the cross-correlation is presented in equation 15:

$$(f*g)(t)=\int_{-\infty}^{\infty} f^*(\tau)g(t-\tau)d\tau \qquad (15)$$

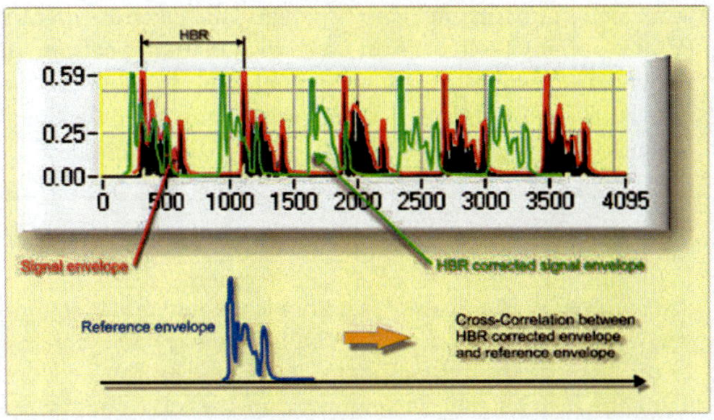

Fig. 5 HBR corrected PCG signal cross-correlation

Similarly, for discrete functions [7], the cross-correlation is defined:

$$(f*g)(t)=\sum_{-\infty}^{\infty} f^*[m]g[n+m] \qquad (16)$$

The amplitude of each sample in the cross-correlation signal is a measure of how much the acquired signal resembles the *PCG* reference signal, at that given location. This means that a peak will occur in the cross-correlation signal for every reference signal that is present in the acquired signal. Except for this noise, the peak generated in the cross-correlation signal is symmetrical between its left and right.

III. Conclusions

A digital resampling method has been described which is convenient for bandlimited characterization of pathological *PCG* signals with smoothly varying sampling rates, and which is attractive for hardware implementation especially for a *DSP* microcontroller. This algorithm is under construction with the intention of using it in a portable complex *PCG* signal analyzer device. The preliminary testing under *LabWindows CVI* simulation are encouraging, to implement the resulted *C* algorithms. All algorithms are written and adapted for a direct implementation in to the designated embedded dsPIC30F6012 microcontroller which is equipped with a 33ns/ *MAC* instruction *DSP* engine.

References

1. Gilbert Strang, Truong Nguyen, Wavelets and Filter Banks, Welles-leyCambridge Press, 1996, ISBN 0-9614088-7-1
2. Hans-Georg Stark, Wavelets and Signal Processing, 2005, Springer, ISBN 3-540-23433-0
3. Mark Thuillard, Wavelets in Soft Computing, 2001, World Scientific, ISBN 981-02-4609-9
4. R. Todd Ogden, Essential Wavelets for Statistical Applications and Data Analysis, 1997, Birkhauser, ISBN 0-8176-3864-4
5. Samuel D. Stearns, Digital Signal Processing using MATLAB, 2002, CRC Press, ISBN 0-8493- 1091-1
6. *Ronald E. Crochiere, Lawrence R. Rabiner, Multirate Digital Signal Prentice-Hall, ISBN 0-13-605162-6*
7. J. P. Lewis, "Fast Template Matching", Vision Interface,p. 120-123, 1995.
8. S. K. Mitra and J. F. Kaiser, Handbook for Digital Signal Processing, New York: Wiley, 1993.
9. R.W. Hamming, Numerical Methods for Scientists and Engineers, Dover Publications, ISBN 0-486-65241-6
10. N. J. Fliege, Multirate Digital signal Processings, John Wiley and Sons, ISBN 0-471-93976-5
11. Robert Oshana, DSP Software Development Techniques for Embedded and Real Time Systems, Elsevier, ISBN 0-7506-7759-7
12. Richard G. Lyons, Understanding Digital Signal Processing, Prentice-Hall, ISBN 0-201-63467-8

Author: Stefan Gergely
Institute: National Institute for Research and Development of Isotopic and Molecular Technologies
Street: Donath 65-103
City: Cluj-Napoca
Country: Romania
Email: stefan.gergely@itim-cj.ro

EMG Signals Case Study: A Time and Frequency Domain Analysis

M. Munteanu[1], C. Rusu[2], D. Moga[3], R. Moga[3], and G. Tont[4]

[1] Technical University of Cluj-Napoca, Faculty of Electrical Engineering, Cluj-Napoca, Romania
[2] Technical University of Cluj-Napoca, Faculty of Electronics, Telecommunications and Information Technology, Cluj-Napoca, Romania
[3] Technical University of Cluj-Napoca, Faculty of Automation and Computer Science, Cluj-Napoca, Romania
[4] University of Oradea, Faculty of Electrical Engineering and Information Technology, Oradea, Romania

Abstract— **The paper aims to analyze the time and frequency domains of EMG signals coming from healthy patients and from patients with muscular disorders (muscular myopathy and neuropathy). The study of these signals can reveal some features in time domain or in frequency domain, that can serve as a basis for diagnosis.**

Keywords— **EMG signals, time/frequency domain analysis, autocorrelation, STFT.**

I. Introduction

The EMG signal records the electrical activity of muscles, being used to detect their abnormal electrical activity, which is a characteristic for certain diseases such as muscular dystrophy, inflammation of muscles, herniated disc or nerve problems of the limbs [1]. When a muscle is active, it will produce an electric current proportional with the level of activity (stimulus) that the muscle may have been subjected [1].

The EMG signal is often used, through its amplitude and frequency information, to determine the effort developed by a muscle or its degree of fatigue, during isotonic/isometric exercises [2].

The simplest way to record the signal that comes from the muscles of certain areas is the SEMG (Surface EMG), the electrodes being applied on the skin covering the muscle under investigation [1]; such a signal can be seen in Figure 1.

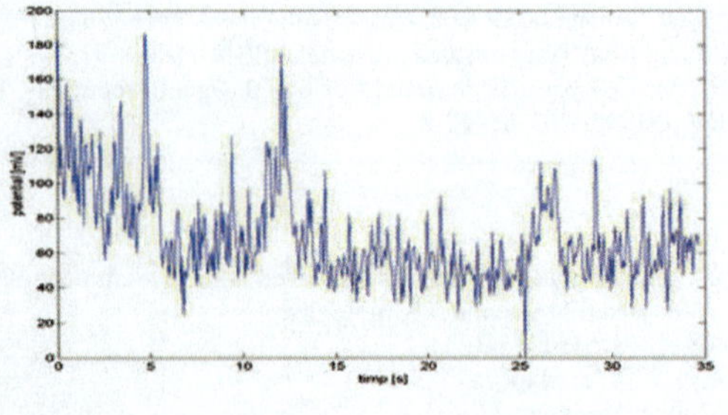

Fig. 1 SEMG signal recorded from a biceps subjected to a load of 2 kilograms [3]

However, the "classical" recording mode of this signal, which offers a high degree of accuracy, is the intra-muscular EMG, that involves placing a needle type electrode into the investigated muscle. Even though this measurement has a high degree of accuracy, it also has the disadvantage of the invasive nature (it causes discomfort and pain) [1].

A fundamental problem in biomedical signal analysis consists in finding information of interest: there are biological signals (ECG) for which this is codified in the form (morphology / amplitude of PQRST complex) and in its spectrum, but there are signals where the information of interest is "hidden" in the signal's amplitude and spectrum [3]. In addition, for some medical signals, a certain periodicity or rhythmicity can be detected.

So, when analyzing the effort of the muscle, one is interested in its time response - namely, the electromyogram (EMG) amplitude for effort, but also in the spectrum of frequencies resulting from the stimulation. In addition, we will study the existence of a rhythmic nature within a signal obtained by invasive measuring from the leg muscles.

II. Medical considerations

Neuropathy is a disease that affects the nervous system (a single nerve or several nerves simultaneously) and can lead to diseases of the motor fibers. Its diagnosis can be done using the EMG signal, in conjunction with nerve conduction velocity study [4].

Myopathy represents a disease that occurs in muscle fibers and which leads to a decrease of muscle tone in hands and legs (and in the abdomen or thorax). Also, in its case, the EMG investigation can help in diagnosing the disease and to estimate its evolution in time [4].

Given what was mentioned previously, this paper aims to investigate some EMG signals (obtained through invasive recordings, for a healthy patient and two subjects suffering of certain muscle diseases) with time and frequency analysis methods.

S. Vlad and R.V. Ciupa (Eds.): MEDITECH 2011, IFMBE Proceedings 36, pp. 202–205, 2011.
www.springerlink.com

III. METHOD AND RESULTS

The analyzed EMG signals were taken from the database [5].

As mentioned in [5], the signals were acquired while the investigated subject executed flex motion of the leg, using concentric electrodes of 25 mm, which were placed into the "tibialis anterior" muscle of the patient (the electrode was repositioned until a satisfactory EMG record was obtained).

The acquired signals were collected from three adults (one without muscle problems and the other two with different diseases) [5] and their appearance in time domain is presented in Figure 2:

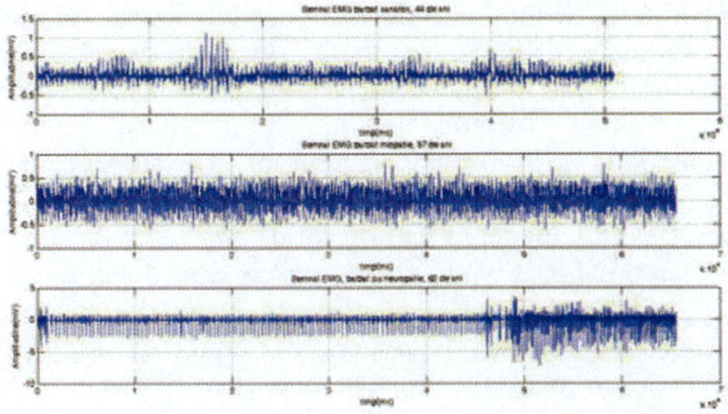

Fig. 2 The EMG signal taken from the three patients

A. *EMG Signal Analysis in Time Domain*

The EMG recording offers the sampled version of a continuous signal, given by the potentials collected from the muscle under investigation. If that muscle will be subjected to regular exercise (in this case flexing the leg), normally every flex of the leg (so every maximum effort) will have an associated peak within the EMG signal. The regular sequence of these maximum potentials (given by the periodical flexing of the leg) could be considered as a periodicity (rhythmicity) characteristic inside the signal.

Three typical EMG signals were subjected to a time domain analysis, to detect the existence of a periodicity (rhythmicity). In order to perform this analysis, a constant section from the beginning of each signal was chosen, for which peak values were detected (depending on the chosen threshold) and their position on the time axis.

The results, obtained with techniques of Virtual Instrumentation - LabVIEW programming [6], [7], are presented in the following figures.

If the local maxima points seen between two absolute maxima are ignored, and considering that the peak pairs corresponding to the moments on the time axis (99.66 and 101.86, 798.98 and 802.08;) come, each one from a leg flexing, then a certain periodicity (rhythmicity) is noticed in the development of high value potentials, at the same time with the leg flexing: they occur on the time axis around values with indexes 100, 450, 800 and 1200. In fact, these highlight a period of about 400*(sampling interval), which is actually the period that the patient is flexing his leg (the maxima related around the samples with index 1000 on the time axis were not considered, as the form of the signal was indicating an artifact of low frequency, probably due to the movement of the patient body).

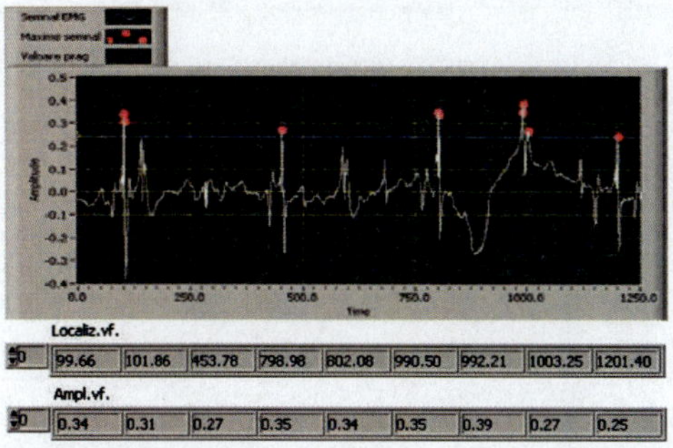

Fig. 3 The positioning and the amplitude of the signal's peak in the EMG taken from a healthy patient

This periodicity (rhythmicity) can also be highlighted by the vectorogram obtained using the Hilbert transform [8]

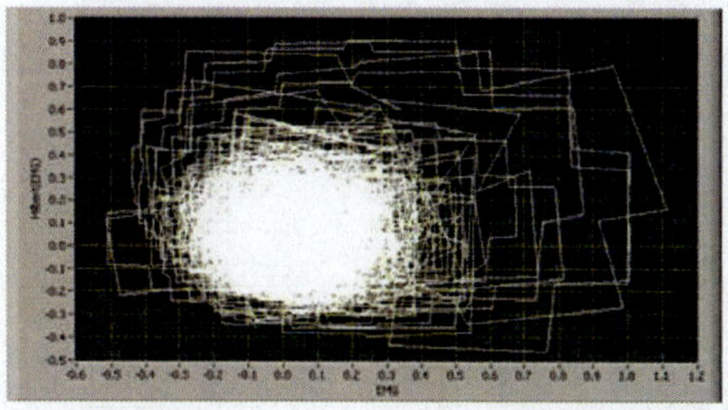

Fig. 4 The Vectorogram of the EMG signal taken from a healthy patient

It can be noticed that the trajectory is concentrating on a limit circle, except for some eccentricities, which can be explained taking into account the evolution of the EMG signal due to fatigue over time. This is also the reason why we selected only the first 1250 samples of signal, the muscular fatigue and the stress influencing both the amplitude of the maximum potentials and also the periodicity (rhythmicity) of their occurrence.

Analysis of signals from patients with muscle problems are presented in figures 5 and 6.

It is noticed that for these signals one can not identify any more the periodicity (rhythmicity) of the potential maxima' occurrence, that was highlighted for the signal taken from a healthy patient.

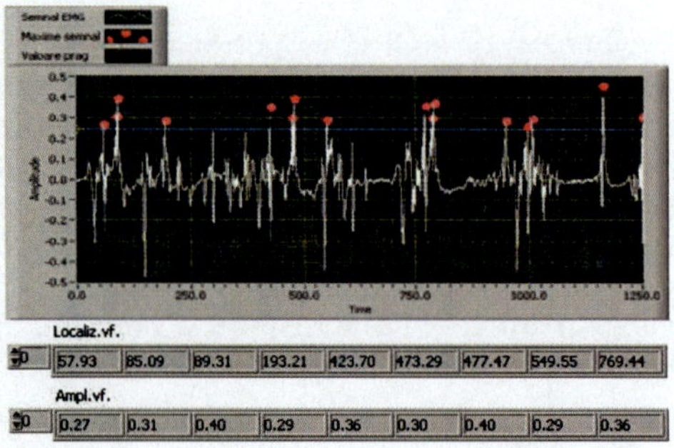

Fig. 5 The positioning and the amplitude of signal's peaks, of the EMG taken from a patient with myopathy

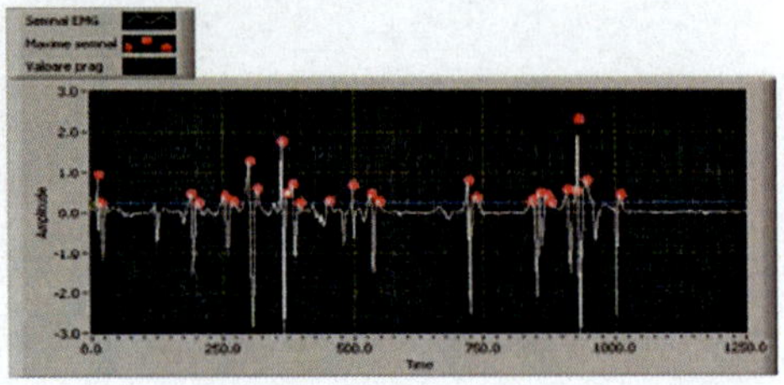

Fig. 6 The EMG signal taken from a patient with neuropathy

The periodicity (rhythmicity) of the occurrence of potential maxima was further investigated by performing the autocorrelation (in Matlab) for each of the studied sequences.

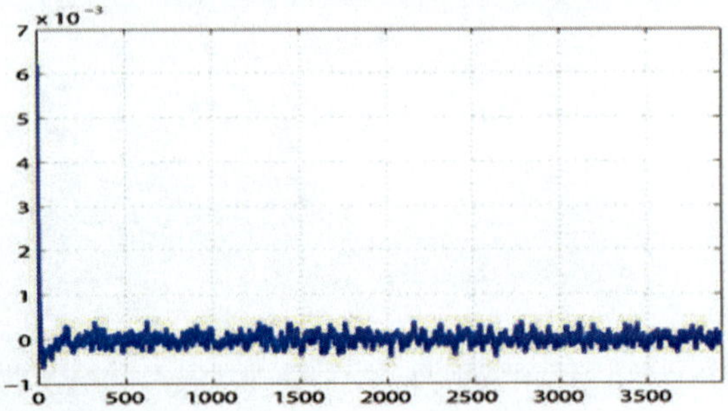

Fig. 7 Autocorrelation of the EMG sequence (miopathy case)

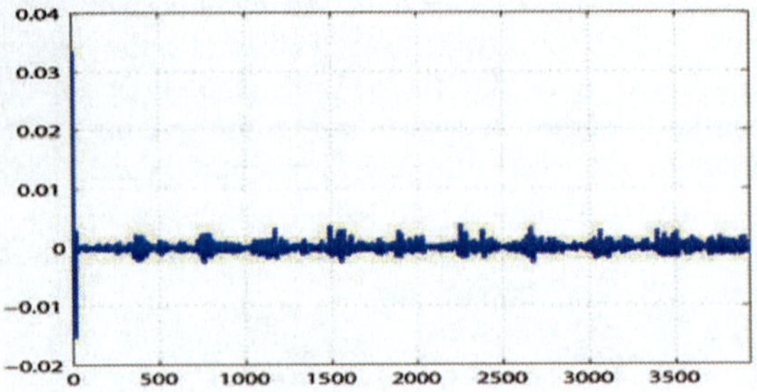

Fig. 8 Autocorrelation of the EMG sequence (neuropathy case)

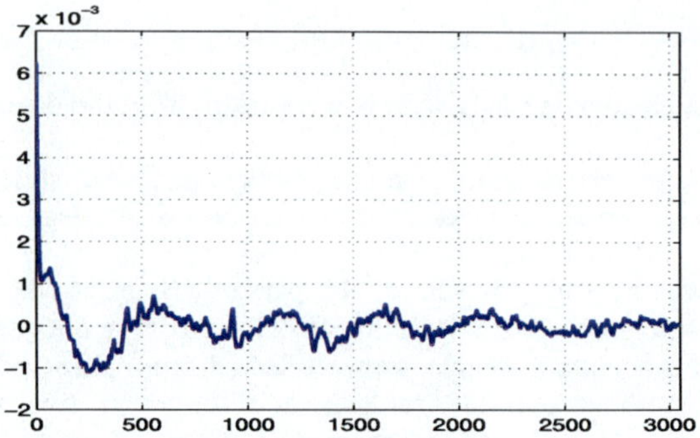

Fig. 9 Autocorrelation of the EMG sequence for a healthy patient

Therefore, healthy muscles subjected to a periodic effort, will provide an EMG signal that, at least for its early phase (where muscular fatigue and stress are not yet installed), presents some features of periodicity (rhythmicity) - demonstrated also by the results of the autocorrelation performed on the first third of the sequence of samples (see figure 9). This tendency is not valid any more for signals taken from patients with various muscle disorders (muscular myopathy, neuropathy – see figures 7 and 8), the peaks corresponding to the effort period (flexing the leg) being "drowned" in many different potentials of different values determined by the abnormal neural activity of muscles.

B. The Analysis of the EMG Signals in the Frequency Domain

Besides the analysis in time-domain, these signals were further investigated through a more complex analysis, such as time-frequency type. This was performed in Matlab [9], using a Short Time Fourier Transform (STFT) in 1024 points and a Hamming Window in 256 points [8].

The results of this analysis are presented as follows:

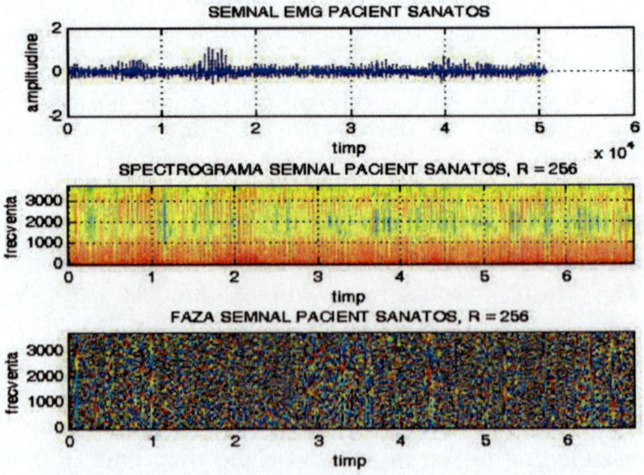

Fig. 10 The EMG signal/ Spectrogram/ Phase Diagram, healthy patient

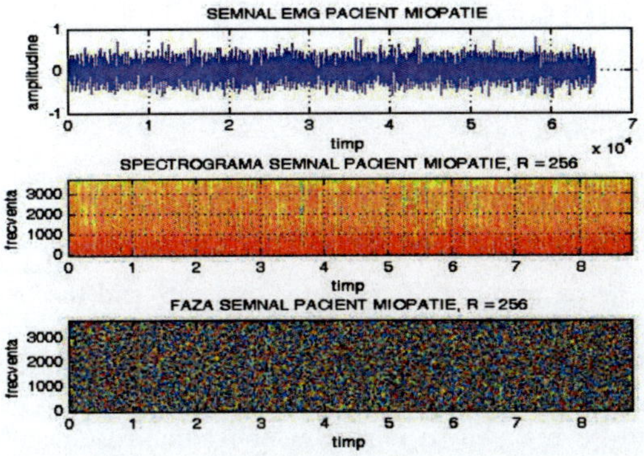

Fig. 11 The EMG signal/ Spectrogram/Phase Diagram, patient with myopathy

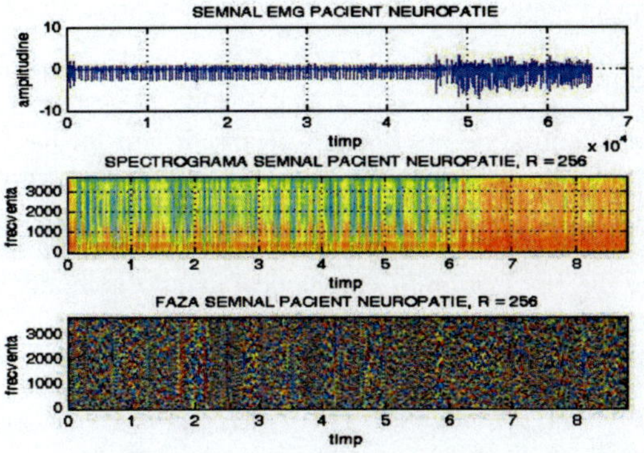

Fig. 12 The EMG signal/ Spectrogram/Phase Diagram, patient with neuropathy

IV. Comments and further developments

The investigation in time domain of the three EMG signals revealed that for healthy muscles subjected to regular exercise, it can be highlighted the existence of a rhythmicity in the occurrence of maximum potentials, at least for the early part of the EMG signal, where muscular fatigue and stress hasn't been installed yet. For signals collected from patients with diseases (in this case muscular myopathy and neuropathy), this trend can no longer be observed, the maximum potentials (corresponding to the period of exercise - flexing the leg) being "drowned" in the multitude of potentials determined by the abnormal neural activity of muscles.

In frequency domain, the results are consistent with those of time domain, meaning that the periodicity is noticeable at the healthy patient.

For the signal acquired from the healthy patient, there are two frequency bands in which the components are significant; these frequency bands are not varying dramatically (very much) in time. We are talking about a low frequency band (up to 1 kHz) and then a higher frequency band (around 3 kHz).

In signals collected from patients with diseases, the high frequency band doesn't have anymore permanent significant components, sudden variations between the amplitude of the components being observed. The phase characteristics are not eloquent, due to the fact that the character of the EMG is closer to a noise type signal than to a harmonic signal.

A further validation of the results would involve the use of signals acquired from a greater number of patients (healthy, but also with muscular disorders) and the study of larger blocks of signals from time domain in order to detect the rhythmicity characteristics for healthy patients.

The validation of these results on larger groups of patients can promote and facilitate the diagnosis of some disorders for the investigated muscles, based on the absence of a certain rhythmicity in the early part of an EMG signal.

Acknowledgment

The work presented in this paper was supported through the research project PN II IDEI – "Arhitectură complexă de monitorizare şi transfer a datelor medicale".

References

1. Rangayyan, M J (2000) A case-study approach to solve problems in Biomedical Signal Analysis, (draft of the book to be published by the IEEE Press, Piscataway, NJ, march 13, 2000)
2. Munteanu M, Rafiroiu D, Curaj A, Velea L, Dobra P, Mitroi D (2008) Study on the relationship EMG-temperature, for the forearm muscles, under isotonic effort, în ISI Proceedings of the 9th WSEAS Int. Conf. on MATHEMATICS & COMPUTERS IN BIOLOGY & CHEMISTRY (MCBC '08), WSEAS Press, ISSN 1790-5125, Bucharest, Romania, June 24-26, 2008, pp 148-153
3. Munteanu M (2007) Simularea, procesarea si transferul datelor medicale prin tehnica Instrumentatiei Virtuale, Mediamira, Cluj-Napoca
4. Manualul Merk (1999), Editia a XVII-a, Bucuresti
5. www.physionet.org
6. Johnson G W (1994) LabVIEW Graphical Programming - Practical Applications in Instrumentation and Control, McGraw-Hill Inc
7. Hedesiu H, Munteanu R Jr (2003) Introducere în programare grafica instrumentala, Mediamira, Cluj-Napoca
8. Rusu C (2000) Prelucrari digitale de semnale,Mediamira, Cluj-Napoca
9. Rivoire M, Ferrier J-L (2000) Matlab, Simulink, Stateflow avec des exercises d'automatique résolus, Ed. Technip

Author: Mihai MUNTEANU
Institute: Technical University of Cluj-Napoca, Faculty of Electrical Engineering, Cluj-Napoca, Romania
Street: Memorandumului 28
City: Cluj-Napoca
Country: ROMANIA
Email: Mihai.Munteanu@et.utcluj.ro

Medical Image Diagnosis Based on Rough Sets Theory

A.L. Ion and S. Udristoiu

University of Craiova/Software Engineering Department, Craiova, Romania

Abstract— **This paper proposes the utilization of rough set theory for modeling the medical images to help physicians in diagnosing. The rough set theory is a powerful approach that permits the searching for patterns in medical images using the minimal length principles. Searching for models with small size is performed by means of many different kinds of reducts that generate the decision rules capable for identifying the medical diagnosis.**

Keywords— **medical image diagnosis, rough sets, image colour, image texture, image shape.**

I. INTRODUCTION

The rough set theory was discovered by Zdzislaw Pawlak [1], [2] and is a powerful mathematical tool for modeling the imperfect and incomplete knowledge, which is an issue debated for a long period of time, by logicians, mathematicians, philosophers and computer scientists [3].

Among methods proposed for modeling the imperfect knowledge [4], the rough set theory is an interesting attempt to solve this problem. This theory is based on an assumption that objects are recognized by partial information about them and some objects can be indiscernible. From this fact it follows that some sets cannot be exactly described by the available information about objects [1], [3].

The methods based on rough set theory have an important utilization in many real life applications. Among the rough set based software systems are ROSETTA [5], RSES [6], and LERS [7], which have been applied to knowledge discover problems.

In this paper we use the rough set approach to discover patterns from medical images for establishing their diagnosis. In the medical domain, a lot of researches were developed to investigate automated techniques for extracting the low–level features that could generate semantic descriptions of the medical image content. Among these techniques are the methods based on machine learning that manually annotate the test image datasets. In the medical domain, algorithms that recognize specific organs with different structures of the medical images are studied in [9]. FIRE [8] application and IRMA [10] use with good results the sub-symbolic processing of images. Though, the actual methodologies of medical image analysis are not generically sufficient for interpreting different diseases. Their major problems are:

1. The description of semantic concepts and the problem understanding-the relationships between the low-level features and semantic concepts are unclear in the actual developed methods. So detailed tests and analysis have to be realized to ensure which combinations of low-level features capture the best the semantic concepts.
2. The generality of the application-in some of the previous researches, only certain semantic concepts could be learned, or the rule were generated of a fixed set of visual features.

The medical applications with automatic diagnosis capacity imply unique challenges, but at the same time new opportunities. Unless we are not physicians, it is a lot harder to understand a medical image than an image taken from nature. On the other hand, there are a lot of formal representations of the medical knowledge that could be exploited to realize the automation of the medical diagnosis in any medical domain.

II. IMAGE REPRESENTATION AND DISCRETIZATIONS

The diagnosis of medical images is directly related to the visual features (colour, texture, shape, position, dimension, etc.), because these attributes capture the information about the semantic meaning. A set of dominant colour regions is obtained from each image by segmentation after the colour characteristic [12]. The HSV colour space quantized to 166 colours is used to represent the colour information [12], [15]. The extraction of colour regions is realized by the colour set back projection algorithm [11]. The specialist selects the representative colour set *C* for the sick regions of medical images from the digestive domain. The algorithm detects the regions having the colour in the colour set *C*. The results of the segmentation algorithm applied to an image diagnosed with gastric ulcer can be visualized in Figure 1.

The visual features of a sick region are represented by 14 parameters [12]:

- The colour, which is represented in the HSV colour space quantized at 166 colours.
- The spatial coherency, which measures the spatial compactness of the pixels of the same colour.

S. Vlad and R.V. Ciupa (Eds.): MEDITECH 2011, IFMBE Proceedings 36, pp. 206–211, 2011.
www.springerlink.com

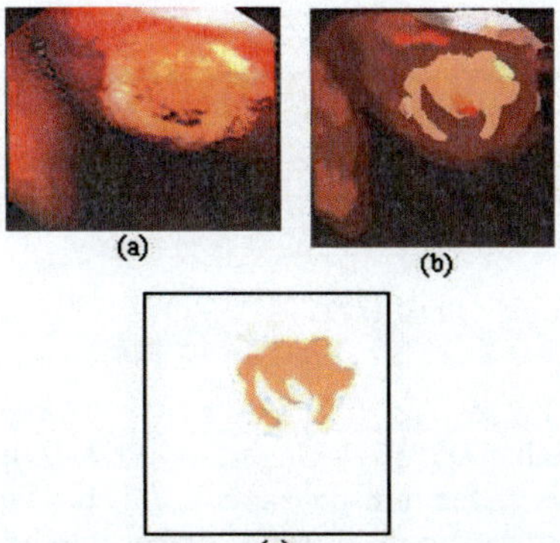

Fig. 1 Segmentation results from an image diagnosed with gastric ulcer: (a) The original image; (b) The quantized image; (c) The sick region.

- A seven-dimension vector (maximum probability, energy, entropy, contrast, cluster shade, cluster prominence, correlation), which represents the texture characteristics.
- The region dimension descriptor, which represents the number of pixels from region.
- The spatial information which is represented by the centroid coordinates of the region and by minimum bounded rectangle.
- A two-dimensional vector (eccentricity and compactness), which represents the shape feature.

The visual features of sick regions were discretized over intervals, using the concept of semantic indicators, which are visual elements: the colour (colour-light-red, etc.), spatial coherency (spatial coherency-weak, spatial coherency-medium, spatial coherency-strong), texture (energy-small, energy-medium, energy-big), dimension (dimension-small, dimension-medium, dimension-big), position (vertical-upper, vertical-center, vertical-bottom, horizontal-upper, etc.), shape (eccentricity- small, compactness-small, etc.).

The values of each semantic descriptor are mapped to a value domain, which corresponds to the mathematical descriptor [12]. At the end of the mapping process, a medical image is represented by means of the terms *figure(ListofRegions)*, where *ListofRegions* is a list of images' sick regions.

III. Modeling image diagnosis using rough sets

A. Rough Sets Foundations

Rough sets theory is an intelligent mathematical tool and it is based on the concept of approximation space [1], [2], [13].

In rough sets theory, the notion of information system determines the knowledge representation system. In this section, we recall some basic definitions from literature [1], [2], [3], [13].

Let U denote a finite non-empty set of objects (sick image regions) called the universe. Further, let A denote a finite non-empty set of attributes. Every attribute $a \in A$, there is a function $a: U \to V_a$, where V_a is the set of all possible values of a, to be called the domain of a. *A pair* $IS = (U, A)$ *is an information system.* Usually, the specification of an information system can be presented in tabular form. Each subset of attributes $B \subseteq A$ determines a binary *B–indiscernibility* relation *IND(B)* consisting of pairs of objects indiscernible with respect to attributes from B like in (1):

$$IND(B) = \{(x, y) \in U \times U : \forall a \in B,\ a(x) = a(y)\} \quad (1)$$

IND(B) is an equivalence relation and determines a partition of U, which is denoted by *U/IND(B)*. The set of objects indiscernible with an object $x \in U$ with respect to the attribute set, B, is denoted by $I_B(x)$ and is called *B-indiscernibility* class.

Thus,

$$I_B(x) = \{y \in U : (x, y) \in IND(B)\} \quad (2)$$

$$U / IND(B) = \{I_B(x) : x \in U\} \quad (3)$$

Table 1 Medical Information System

U	Colour	Texture-entropy	Diagnosis
R_1	light-red	Small	gastric-ulcer
R_2	light-red	Small	gastric-ulcer
R_3	light-red	Small	gastric-ulcer
R_4	light-red	Big	gastric-ulcer
R_5	light-yellow	Big	gastric-ulcer
R_6	light-yellow	Medium	duodenal-ulcer
R_7	light-yellow	Medium	duodenal-ulcer
R_8	medium-yellow	Small	duodenal-ulcer
R_9	medium-yellow	Small	duodenal-ulcer
R_{10}	dark-yellow	Small	duodenal-ulcer
R_{11}	dark-yellow	Small	duodenal-ulcer

Table 2 Partitions Defined by Indiscernibility Relations

IND(B)	Partitions U/IND(B)
IND({Colour})	$\{R_1, R_2, R_3, R_4\}, \{R_5, R_6, R_7\}, \{R_8, R_9\}, \{R_{10}, R_{11}\}$
IND({Colour,Texture-entropy})	$\{R_1, R_2, R_3\}, \{R_4\}, \{R_5\}, \{R_6, R_7\}, \{R_8, R_9\}, \{R_{10}, R_{11}\}$

It is said that a pair $AS_B = (U, IND(B))$ is an approximation space for the information system $IS=(U, A)$, where $B \subseteq A$.

The information system from Table 1 represents the sick regions of images from different diagnoses represented in terms of semantic indicators values, as described in Section II. For simplicity we consider only two semantic indicators as attributes, namely the colour and texture-entropy.

So, our information system is IS = (U, B), where $U = \{R_1, R_2, R_3, R_4, R_5, R_6, R_7, R_8, R_9, R_{10}, R_{11}\}$ *and B = {colour, texture-entropy}.* Some examples of partitions defined by indiscernibility relations for the information system in Table 1 are given in Table 2.

In rough sets theory, the approximations of sets are introduced to deal with inconsistency. A rough set approximates traditional sets using a pair of sets named the lower and upper approximations of the set.

Let $W = \{w_1,\dots,w_n\}$ be the elements of the approximation space $AS_B=(U, IND(B))$. We want to represent *X*, a subset of U, using attribute subset *B*. In general, *X* cannot be expressed exactly, because the set may include and exclude objects which are indistinguishable on the basis of attributes *B*, so we could define *X* using the lower and upper approximation.

The *B-lower approximation X*, $\underline{B}X$, is the union of all equivalence classes in *IND(B)* which are contained by the target set *X*. The lower approximation of *X* is called the positive region of *X* and is noted *POS(X)*.

$$\underline{B}X = \cup\{w_i \mid w_i \subseteq X\} \tag{4}$$

The *B-upper approximation* $\overline{B}X$ is the union of all equivalence classes in *IND(B)* which have non-empty intersection with the target set *X*.

$$\overline{B}X = \cup\{w_i \mid w_i \cap X \neq \varnothing\} \tag{5}$$

Example: Let $X = \{R_1, R_2, R_3, R_4, R_5, R_6, R_7, R_8\}$ be the subset of *U* that we wish to be represented by the attributes set *B={colour, texture-entropy}*. We can approximate *X*, by computing its *B-lower approximation*, $\underline{B}X$ and *B-upper approximation*, $\overline{B}X$.

So, $\underline{B}X = \{\{R_1, R_2, R_3\},\{R_4\},\{R_5\},\{R_6, R_7\}\}$ and $\overline{B}X = \{\{R_1, R_2, R_3\},\{R_4\}, \{R_5\},\{R_6, R_7\}, \{R_8, R_9\}\}$.

The tuple ($\underline{B}X$, $\overline{B}X$) composed of the lower and upper approximation is called a rough set; thus, a rough set is composed of two crisp sets, one representing a *lower boundary* of the target set X, and the other representing an *upper boundary* of the target set X.

The accuracy of a rough set is defined as: cardinality($\underline{B}X$)/cardinality($\overline{B}X$). If the accuracy is equal to 1, then the approximation is perfect.

B. Dispensable Features, Reducts and Core

An important notion used in rough set theory is the decision table. Pawlak [1], [2] gives also a formal definition of a decision table: an information system with distinguished conditional attributes and decision attribute is called a decision table. So, a *tuple DT = (U, C, D), is a decision table. The attributes C = {colour, texture-entropy} are called conditional attributes, instead D = {diagnosis} is called decision attribute.*

The classes *U/IND(C)* and *U/IND(D)* are called condition and decision classes, respectively.

The *C-Positive* region of *D* is given by:

$$POS_C(D) = \bigcup_{X \in IND(D)} \underline{C}X \tag{6}$$

Let $c \in C$ a feature. It is said that *c* is dispensable in the decision table *DT*, if $POS_{C-\{c\}}(D)= POS_C(D)$; otherwise the feature *c* is called indispensable in *DT*. If *c* is an indispensable feature, deleting it from *DT* makes it to be inconsistent.

A set of features *R* in *C* is called a reduct, if *DT'= (U, R, D)* is independent and $POS_R(D)=POS_C(D)$. In other words, a reduct is the minimal feature subset preserving the above condition.

The set of all features indispensable in *C* is denoted by *CORE(C)*. In other words, *CORE(C)* is the set of all reducts of *C*.

C. Producing Rules by Discernibility Matrix

We transform the decision table into discernibility matrix to compute the reducts. Let *DT = (U, C, D)* be the decision table, with $U = \{R_1, R_2, R_3, R_4, R_5, R_6, R_7, R_8, R_9, R_{10}, R_{11}\}$. By a discernibility matrix of *DT*, denoted *DM(T)*, we will mean *nxn* matrix defined as:

$$m_{ij}^{a(R_i)} = \{(a \in C: a(R_i) \neq a(R_j)) \text{ and } (d(R_i) \neq d(R_j))\} \tag{7}$$

where *i, j=1,2,...,11*.

We construct the discernibility matrix, *DM(DT)* as in Table 3, where the colour and texture-entropy are denoted by *C*, respectively *T*. The items within each cell are aggregated disjunctively, and the individual cells are then aggregated conjunctively.

To compute the reducts of the discernibility matrix we use the following theorems that demonstrate equivalence between reducts and prime implicants of suitable Boolean functions [3], [13].

For every object $R_i \in U$, the following Boolean function is defined:

$$g_{R_i}(Colour, Texture) = \bigwedge_{R_j \in U} \left(\bigvee_{a \in m_{ij}} a \right) \tag{8}$$

The following conditions are equivalent [3]:

1. $\{a_{i1}, \dots, a_{in}\}$ *is a reduct for the object* R_i, $i=1,\dots,n$.
2. $a_{i1} \wedge a_{i2} \wedge \dots \wedge a_{in}$ *is a prime implicant of the Boolean function* g_{R_i}.

Table 3 Discernibility Matrix

	R_1	R_2	R_3	R_4	R_5	R_6	R_7	R_8	R_9	R_{10}	R_{11}
R_1	-	-	-	-	-	$C^{\text{light-red}}$ T^{small}	$C^{\text{light-red}}$ T^{small}	$C^{\text{light-red}}$	$C^{\text{light-red}}$	$C^{\text{light-red}}$	$C^{\text{light-red}}$
R_2	-	-	-	-	-	$C^{\text{light-red}}$ T^{small}	$C^{\text{light-red}}$ T^{small}	$C^{\text{light-red}}$	$C^{\text{light-red}}$	$C^{\text{light-red}}$	$C^{\text{light-red}}$
R_3	-	-	-	-	-	$C^{\text{light-red}}$ T^{small}	$C^{\text{light-red}}$ T^{small}	$C^{\text{light-red}}$	$C^{\text{light-red}}$	$C^{\text{light-red}}$	$C^{\text{light-red}}$
R_4	-	-	-	-	-	$C^{\text{light-red}}$ T^{big}	$C^{\text{light-red}}$ T^{big}	$C^{\text{light-red}}$ T^{big}	$C^{\text{light-red}}$ T^{big}	$C^{\text{light-red}}$ T^{big}	$C^{\text{light-red}}$ T^{big}
R_5	-	-	-	-	-	T^{big}	T^{big}	$C^{\text{light-yellow}}$ T^{big}	$C^{\text{light-yellow}}$ T^{big}	$C^{\text{light-yellow}}$ T^{big}	$C^{\text{light-yellow}}$ T^{big}
R_6	$C^{\text{light-yellow}}$ T^{medium}	$C^{\text{light-yellow}}$ T^{medium}	$C^{\text{light-yellow}}$ T^{medium}	$C^{\text{light-yellow}}$ T^{medium}	T^{medium}	-	-	-	-	-	-
R_7	$C^{\text{light-yellow}}$ T^{medium}	$C^{\text{light-yellow}}$ T^{medium}	$C^{\text{light-yellow}}$ T^{medium}	$C^{\text{light-yellow}}$ T^{medium}	T^{medium}	-	-	-	-	-	-
R_8	$C^{\text{medium-yellow}}$	$C^{\text{medium-yellow}}$	$C^{\text{medium-yellow}}$	$C^{\text{medium-yellow}}$ T^{small}	$C^{\text{medium-yellow}}$ T^{small}	-	-	-	-	-	-
R_9	$C^{\text{medium-yellow}}$	$C^{\text{medium-yellow}}$	$C^{\text{medium-yellow}}$	$C^{\text{medium-yellow}}$ T^{small}	$C^{\text{medium-yellow}}$ T^{small}	-	-	-	-	-	-
R_{10}	$C^{\text{dark-yellow}}$	$C^{\text{dark-yellow}}$	$C^{\text{dark-yellow}}$	$C^{\text{dark-yellow}}$ T^{small}	$C^{\text{dark-yellow}}$ T^{small}	-	-	-	-	-	-
R_{11}	$C^{\text{dark-yellow}}$	$C^{\text{dark-yellow}}$	$C^{\text{dark-yellow}}$	$C^{\text{dark-yellow}}$ T^{small}	$C^{\text{dark-yellow}}$ T^{small}	-	-	-	-	-	-

Next, from each decision matrix we form a set of Boolean expressions, one expression for each row of the matrix.

For the gastric ulcer we obtain the following rules based on the table reducts:

1. $(C^{\text{light-red}} \vee T^{\text{small}}) \wedge (C^{\text{light-red}})$
2. $(C^{\text{light-red}} \vee T^{\text{small}}) \wedge (C^{\text{light-red}})$
3. $(C^{\text{light-red}} \vee T^{\text{small}}) \wedge (C^{\text{light-red}})$
4. $(C^{\text{light-red}} \vee T^{\text{big}}) \wedge ((C^{\text{light-red}}))$
5. $(T^{\text{big}}) \wedge (C^{\text{light-yellow}} \vee T^{\text{big}})$

For the duodenal ulcer we obtain the following rules based on the table reducts:

1. $(C^{\text{light-yellow}} \vee T^{\text{medium}}) \wedge T^{\text{medium}}$
2. $(C^{\text{medium-yellow}}) \wedge (C^{\text{medium-yellow}} \vee T^{\text{small}})$
3. $(C^{\text{dark-yellow}}) \wedge (C^{\text{dark-yellow}} \vee T^{\text{small}})$

On Boolean expression the absorption Boolean algebra rule is applied. The absorption law is an identity linking a pair of binary operations.

For example: $a \vee (a \wedge b) = a \wedge (a \vee b) = a$

By applying the absorption rule on the prime implicants, the following rules are generated:

1. Rule 1: (Colour = light-red) → gastric ulcer;
2. Rule 2: (Texture-entropy = big) → gastric ulcer;
3. Rule 3: (Texture-entropy = medium) → duodenal ulcer;
4. Rule 4: (Colour = dark-yellow) → duodenal ulcer.

D. Evaluation of Decision Rules

Decision rules can be evaluated along at least two dimensions: performance (prediction) and explanatory features (description). The performance estimates how well the rules classify new images. The explanatory feature estimates how interpretable the rules are [3].

Let be our decision table *DT = (U, C, D)*. We use the set-theoretical interpretation of rules. It relates a rule to data sets from which the rule is discovered [3]. Using the cardinalities of sets, we obtain the 2x2 contingency table representing the quantitative information about the rule *if features then diagnosis*.

In table 4 the number of images that have a certain feature set and a certain diagnosis is computing. Using the elements of the contingency table, we may define the support (s) and accuracy (a) of a decision rule by:

$$s(rule) = cardinality(featureSet \cap diagnosisSet) \tag{9}$$

$$a(rule) = \frac{cardinality(featureSet \cap diagnosisSet)}{cardinality(featureSet)} \tag{10}$$

where the set $featureSet \cap diagnosisSet$ is composed from image regions which have a certain *featureSet* and a certain *diagnosis*. In term of set theory, the accuracy is the degree in which the set of features rule is included in the set of diagnosis rule.

The coverage(c) of a rule is defined by:

$$c(rule) = \frac{cardinality(featureSet \cap diagnosisSet)}{cardinality(diagnosisSet)} \tag{11}$$

The coverage of a rule is the degree in which the set of diagnosis rule is included in the features set of rule. For the generated Rule 1, the contingency table Table 5 is obtained. For the Rule 1, the support is 4, accuracy is 4/4 and coverage is 4/5. Ryszard et al [14] suggests that high accuracy and coverage are requirements of decision rules.

IV. DECISION RULE EXTRACTION USING ROUGH SETS MODELS AND EXPERIMENTS

In this paper we present the application of rough set to discover the medical diagnosis of images from digestive apparatus. To establish the medical diagnosis the following tasks are carried out:

- selection of the most relevant condition attributes in our case 14 image visual semantic indicators,
- application of rough set based on reduced data,
- discovery of decision rules characterizing the dependency between values of condition attributes and decision attribute.

A rule has the form:

if (colour is red and texture-entropy is small) then the diagnosis is ulcer.

Decision rules are generated from reducts. So in order to compute decision rules, reducts have to be computed first.

This method finds all reducts by computing prime implicants of a Boolean function, as described in Section III.

The rule generation algorithm can be resumed as:

- construct the decision table and discernibility matrix,
- obtain the discernibility function and the prime implicants,
- apply the Boolean algebra rules,
- compute the reducts,
- produce the rules using the reducts.

The image classification algorithm can be resumed as:

- collect all the decision rules in a classifier,
- compute for each rule the support, accuracy and coverage,
- eliminate the rules with the support less than the minimum defined support,
- order the rules by accuracy, than by coverage,
- if an image matches more rules select the first one: an image matches a rule, if all the semantic indicators, which appear in the body of the rule, are included in the characteristics of the image regions.

The image collections used in our experiments were taken from free repositories on the Internet [16], [17]. Two image databases are used for learning and diagnosing process. The database used to learn the correlations between images and digestive diagnoses, contains 200 images. The learning database is categorized into the following diagnoses: duodenal ulcer, gastric ulcer, gastric cancer, esophagitis, and rectocolitis. The system learns each concept by submitting about 20 images per diagnosis. For example, we analyze the performance of the proposed method for colon cancer diagnosis. The rule generation algorithm produces 12 semantic rules that recognize this diagnosis. The test database contains 450 images, from which 67 are relevant for duodenal ulcer diagnosis. A part of images used for learning the diagnoses can be analyzed in Figure 2.

Table 4 General Contingency Table Representing the Quantitative Information about the Rule

	Diagnosis	not(Diagnosis)	
Features	cardinality(features and diagnosis)	cardinality(features and not(diagnosis))	card(features)
not(Features)	cardinality((not)features and diagnosis)	cardinality((not)features and (not)diagnosis)	card(not(features))
	cardinality(diagnosis)	card(not(diagnosis))	card(U)

Table 5 Contingency Table Representing the Quantitative Information about the Rule 1

	diagnosis = gastric ulcer	not(diagnosis = gastric ulcer)	
colour = light-red	card(*colour = light-red and diagnosis = gastric ulcer*) = 4	card(*colour=light-red and not(diagnosis=gastric ulcer)*) = 0	card(*colour=light-red*)= 4
not(colour= light-red)	card(*not(colour=light-red) and diagnosis= gastric ulcer*)=1	card(*not(colour=light-red) and not(diagnosis= gastric ulcer)*)=6	card(*not(colour=light-red)*)=7
	card(diagnosis = gastric ulcer) =5	*card(not(diagnosis = gastric ulcer))*= 6	card(U)=11

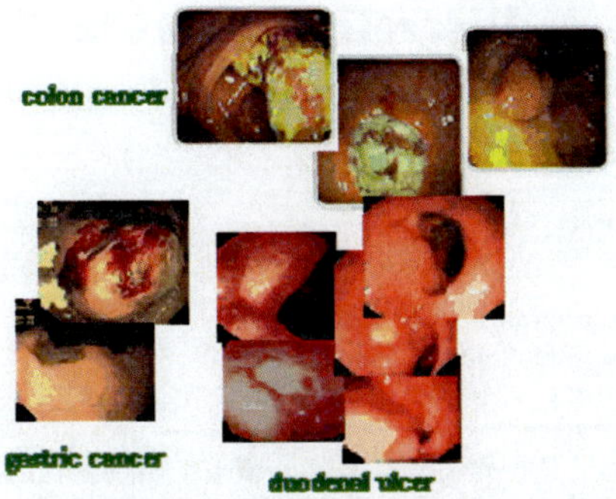

Fig. 2 Medical diagnosed images

After classification, we counted: the number of true positives (images correctly diagnosed with the colon cancer diagnosis) and we found 56 images; the number of false positives (images incorrectly diagnosed with the colon cancer diagnosis) and we found 9 images; the number of true negatives (images correctly diagnosed with a different diagnosis) and we found 377 images; the number of false negatives (images incorrectly diagnosed with a different diagnosis) and we found 8 images. The accuracy, which measures the proportion of true results, is 96.2%. The specificity, which measures the capability of colon cancer rules not to miss the colon cancer images, and not to diagnose images with a different diagnosis, is 97.6%. In our case, this set of rules is very specific.

For the other diagnoses, the counted results are presented in Table 6.

Table 6 Results recorded for different diagnoses

Diagnosis	Accuracy(%)	Specificity(%)
Duodenal Ulcer	96.3	95
Gastric Ulcer	96.7	95.1
Gastric Cancer	95.9	93
Rectocolitis	96.3	95.2

V. Conclusion

Methods proposed and developed in this study could assist physicians by doing automatic diagnose based on visual content of medical images. An important improvement of this paper is in the generation of rules with very high specificity using the rough set theory. The language used for rules representation is Prolog. The advantages of using Prolog are its flexibility and simplicity in representation of rules. The results of the presented method are very promising, being influenced by the complexity and number of endoscopic images.

Acknowledgment

This work was supported by the strategic grant POSDRU/89/1.5/S/61968, Project ID 61968 (2009), co-financed by the European Social Fund within the Sectorial Operational Program Human Resources Development 2007 - 2013.

References

1. Pawlak Z (1986) Rough Relations. Bulletin of the Polish Academy of Sciences, Technical Sciences 34(9-10): 587–590
2. Pawlak Z, Skowron A (1994) Rough Membership Functions. Advances in the Dempster-Shafer Theory of Evidence pp.251-271. John Wiley and Sons, New York
3. Stepaniuk J (2008) Rough Granular Computing in Knowledge Discovery and Data Mining. Springer-Verlag, Germany
4. Zadeh LA, Kacprzyk J (eds.) (1999) Computing with Words in Information/Intelligent Systems 2. Physica-Verlag, Heidelberg
5. Ohrn A, Komorowski J, Skowron A, et al. (1998) The Design and Implementation of a Knowledge Discovery Toolkit Based on Rough Sets - The Rosetta System. Polkowski, L., Skowron, A. (eds.) Rough Sets in Knowledge Discovery 1, Methodology and Applications, pp. 376–399. Physica-Verlag, Heidelberg
6. Bazan J, Szczuk M (2005) The Rough Set Exploration System. Transactions on Rough Sets III, LNCS 3400: 37–56
7. Grzymala-Busse JW (2005) LERS-A Data Mining System. Springer, US
8. Deselaers T, Keysers D, Ney H (2004) FIRE – flexible image retrieval engine: ImageCLEF 2004 evaluation. Multilingual Information Access for Text, Speech and Images 3491:688–698
9. Hong W, Georgescu B, Zhou XS, et al. (2006) Database-guided simultaneous multi-slice 3D segmentation for volumetric data, Proceedings of 9th European Conference on Computer Vision, Graz, Austria; 2006, pp. 397-409
10. Lehmann T, Güld M, Thies C, et al. (2004) Content-based image retrieval in medical applications. Methods Inf Med.43(4): 354-361
11. Smith JR, Chang SF. VisualSEEk: a fully auto-mated content-based image query system. The Fourth ACM International Multimedia Conference and Exhibition, Boston, MA, USA, 1996, pp. 87-98
12. Ion AL, Udristoiu S (2010) Image Mining for discovering medical diagnosis. Information Technology and Control 39(1): 123-129
13. Hassanien AE, Abraham A, Peters JF et al. (2008). Rough Sets in Medical Imaging: Foundations and Trends. Computational Intelligence in Medical Imaging: Techniques and Applications, pp. 47-87. CRC Press, USA
14. Ryszard A, Michalski S (1993) A Theory and Methodology of Inductive Learning. Readings in knowledge acquisition and learning: 323-348
15. Stanescu L, Burdescu D, Ion AL, et al. (2008) Imagistic Database for Medical E-learning, Proceedings of 21st IEEE International Symposium on Computer-Based Medical Systems, Jyvaskyla, Finland; 2008, pp.427-429.
16. Jackson Siegelbaum Gastroenterology at http://gicare.com/Endoscopy-Center/Endoscopy-images.aspx.
17. The Gatrolab Image Library at http://www.gastrolab.net/

On the Stability and Convergence Rate of Some Discretized Schemes for Parametric Deformable Models Used in Medical Image Analysis

A.I. Mitrea[1], O.M. Gurzau[1], and P. Mitrea[2]

[1] Technical University of Cluj-Napoca, Dep. of Mathematics, Romania
[2] Technical University of Cluj-Napoca, Dep. of Computer Science, Romania

Abstract— **In order to find the energy-minimizing surface and to reduce the computational requirements, we consider an associated simplified model, [1], and we derive an algorithm for solving numerically the corresponding Euler-Gauss-Ostrogradsky equation of Calculus of Variations. The stability and the convergence of the algorithm are discussed, together with some aspects regarding the statistical modeling, applied in medical imaging.**

Keywords— **parametric deformable model, EGO-Equation, EGO-Algorithm, stability, convergence rate.**

I. INTRODUCTION

The theory of deformable models is an interdisciplinary scientific domain, which has appeared and developed in the last two decades, in strong connection with practical problems of medicine, image processing and physics; the deformable models represent a promising and vigorously researched model–based approach to computer–assisted medical image analysis.

Deformable models are viewed as curves or surfaces that can move under the influence of the internal forces, which are defined within the curve or surface itself, and external forces, which are computed from the image data [2],[3],[4].

Two basic type of deformable models were pointed out: the parametric or variational models and geometric models. The parametric models start from the original snake introduced by M. Kaas, A. Witkin and D. Terzopoulos [2] and represent curves and surfaces explicitly in their parametric form during the deformation process; the original snake was proposed as an interactive method, which requires expert guidance on the snake initialization and the selection of correct deformation parameters. In the last two decades, a set of deformable variational models have been proposed in order to improve the original snake, such as: *the balloon-snake model* of I. Cohen and L.D. Cohen [1], which add to the internal and external energies the so called balloon-energy and enables the initial contour to be located far from the desired boundary (the curve or surface is viewed as a balloon which is inflated); the *topology-snake* of T. Mc. Inerney and D. Terzopoulos [5], that designed a set of topology changing rules to be used during the balloon deformation; the *distance snake* of I. Cohen and L.D. Cohen [1], which is a deformation strategy implemented by means of the finite element method.

By means of some variational principles an energy-minimizing model (curve or surface) is achieved in this framework, by solving the Euler-Gauss-Ostrogradsky (EGO) Equation using discretized methods [1],[2],[6].

Regarding the geometric deformable models, we notice that these models were proposed independently by Cassels et al. [7] and Malladi et al. [8] and they are based on the curve evolution theory; in this framework the evolving models can be represented implicitly, as a level set of a higher-dimensional function [3,4].

This paper is concerned with the parametric deformable models; our goals are to derive an algorithm for obtaining the energy-minimizing surface, to establish its approximation error and to discuss the corresponding conditions of convergence and stability.

The paper outline is as follows. The next section defines the notion of deformable 3D-model and describes a method for obtaining a simplified 3D model as a sequence of plane curves. In the third section we point out an EGO-Algorithm of implicit type for the 3D simplified model. The convergence and the stability for an explicit-type discretized scheme derived from the algorithm of the third section are discussed in the fourth section. It must be mentioned that similar approaches regarding EGO-Algorithms and their approximation-error (but not their stability) can be found in [9]; however, the EGO-Algorithm presented in this paper is obtained in a different way, by using a general implicit discretization scheme and its convergence rate is better, because the inequalities involving the approximation-error are refined. The fifth section presents some aspects regarding the behavior of prosthetic surgical methods and prosthetic medical materials, based on Software tools, which implement the above mentioned mathematical methods.

II. ENERRGY MINIMIZING-SURFACES

From mathematical point of view, a 3D variational deformable model is emphasized by a family $\mathcal{A}$ of parameterized smooth surfaces with given boundary condition, named

S. Vlad and R.V. Ciupa (Eds.): MEDITECH 2011, IFMBE Proceedings 36, pp. 212–219, 2011.
www.springerlink.com

admissible surfaces, and an associated *energy-functional*. More exactly, let $D = [0,1] \times [0,1]$ be the unit square of $\mathbb{R}^2$ (or a compact plane domain) and

$$(S): v: D \to \mathbb{R}^3,\ v = v(s,r), \quad v = (x,y,z)^T; 0 \le s,r \le 1 \quad (1)$$

an arbitrary surface of class $C^2(D,\mathbb{R}^3)$ In this paper we use the notations $|v|^2 = x^2 + y^2 + z^2$, $v_s = \frac{\partial v}{\partial s}$, $v_r = \frac{\partial v}{\partial r}$, $v_{ss} = \frac{\partial^2 v}{\partial s^2}$, $v_{sr} = \frac{\partial^v}{\partial s \partial r}$, $v_{rr} = \frac{\partial^2 v}{\partial r^2}$. The family $\mathcal{A}$ of *admissible deformations* consists of all parameterized surfaces (1), subject to the boundary conditions $v(s,r) = g(s,r)$ and $\frac{\partial v}{\partial n}(s,r) = h(s,r)$ on the boundary ∂D of D, where $g \in C^2(\partial D, \mathbb{R}^3)$ and $h \in C^1(\partial D, \mathbb{R}^3)$ are given functions and n is the unit normal vector with respect to surface (1). Further, let us consider the following functions: the image intensity functions $I \in C^2(\mathbb{R}^3)$; *the potential function associated* to the external forces $P(v) = -\lambda |\nabla I(v)|^2$, $\lambda > 0$; *the control functions* corresponding to the internal forces acting on the shape of the surface, namely the elasticity functions $w_{10}(s;r)$and $w_{01}(s;r)$; the rigidity functions $w_{20}(s;r)$ and $w_{02}(s;r)$, and the *twist resistance function* $w_{11}(s;r)$.

The energy functional $E: \mathcal{A} \to \mathbb{R}$ incorporates the internal, external and balloon-energy, as follows:

$$E(v) = E_{int}(v) + E_{ext}(v) + E_{bal}(v)$$

$$\begin{aligned} E_{int}(v) &= E_{els}(v) + E_{rig}(v) + E_{twr}(v) \quad (2) \\ E_{els}(v) &= \iint_D \left(w_{10}|v_s|^2 + w_{01}|v_r|^2 \right) \mathrm{ds\,dr} \\ E_{rig}(v) &= \iint_D \left(w_{20}|v_{ss}|^2 + w_{02}|v_{rr}|^2 \right) \mathrm{ds\,dr} \\ E_{twr}(v) &= 2\iint_D w_{11}(s,r)|v_{sr}|^2 \mathrm{ds\,dr} \end{aligned}$$

$$E_{ext}(v) = \iint_D P(v(s,r))\,\mathrm{ds\ dr} \quad (3)$$

$$E_{bal}(v) = \iint_D \det(c_0 v, v_s, v_r)\,\mathrm{ds\,dr};\ c_0 > 0. \quad (4)$$

We notice that $E_{int}(v)$ represents *the internal energy*, $E_{ext}(v)$ is *the energy associated to the external forces* and $E_{bal}(v)$ is named *the balloon-energy*, which can be added, optionally, by the users (the term including $\det(c_0 v, v_s, v_r)$).

The triple $(\mathcal{A}, I, E)$ is said to be a *3D deformable model*, sometimes a *deformable surface*. The basic goal of a deformable model is to minimize its energy functional, which leads to the *energy-minimizing* surface, e.g. the optimal deformable model, provided by the Euler-Gauss-Ostrogradski (EGO) Equation of the Calculus of Variations:

$$\begin{aligned} &\frac{\partial F}{\partial v} - \frac{\partial}{\partial s}\left(\frac{\partial F}{\partial v_s}\right) - \frac{\partial}{\partial r}\left(\frac{\partial F}{\partial v_r}\right) + \frac{\partial^2}{\partial s^2}\left(\frac{\partial F}{\partial v_{ss}}\right) + \\ &+ \frac{\partial^2}{\partial v \partial r}\left(\frac{\partial F}{\partial v_{sr}}\right) + \frac{\partial^2}{\partial r^2}\left(\frac{\partial F}{\partial v_{rr}}\right) = 0 \end{aligned} \quad (5)$$

together the Legendre-Sylvester minimum criterion.

Roughly speaking, the partial derivative (5) may have many solutions, so there may exist many local minimum energy-surfaces. But, in medical imaging the goal of the user is to find a good 3D-contour in a given area. Consequently, a rough prior estimate of the surface is provided (it is at hand of the user); further, this initial surface undergoes a deformation until reaching a local minimum of the energy-functional, according to (EGO) Equation (5). We achieve this deformation process by using the method of the Evolution Equation,[1].

Now, let us describe, mathematically, the method of Evolution Equation. Let

$$(S^0): v = v_0(s,r),\ (s,r) \in D \quad (6)$$

be an initial surface and denote by

$$(S^t): v = v(t,s,r);\ t \ge 0,\ (s,r) \in D \quad (7)$$

a family of surfaces, where the parameter $t \ge 0$ stands for the evolution in time of the model.

Denoting by $G(v, v_s, v_r, v_{ss}, v_{sr}, v_{rr})$ the left-hand member of (5), the *Evolution Equation* associated to the static model $(\mathcal{A}, I, E)$ is:

$$\frac{\partial v}{\partial t} + G(v, v_s, v_r, v_{ss}, v_{sr}, v_{rr}) = 0 \quad (8)$$

together with the initial estimate (condition)

$$v(0,s,r) = v_0(s,r),\ (s,r) \in D \quad (9)$$

and some boundary dynamic conditions.

A solution of the static problem is obtained when the solution $v(t,s,r)$ of (8) becomes stable, which means that $\frac{\partial v}{\partial t}$ approach to 0 (for $t \to \infty$) and the dynamic equation (8) reduces to the static equation (5)at infinity, see also [1].

However, the problem of finding the solutions of (EGO) Equation (8) is not practically possible, because these solutions contain long and complicated expressions or their explicit forms are inaccessible. On the other hand, by using discretized schemes for solving (8), we get a system of algebraic equations with a high computational level. These drawbacks are eliminated by passing to a 2D modeling problem, [1]. More exactly, the third component z of (S) is constrained to depend only on r, by setting $z(s,r) = r$. So, the surface that we seek is given as a sequence of plane curves, named *slices*, and the parameter r of (7) becomes the index of the corresponding slice. In this approach, the

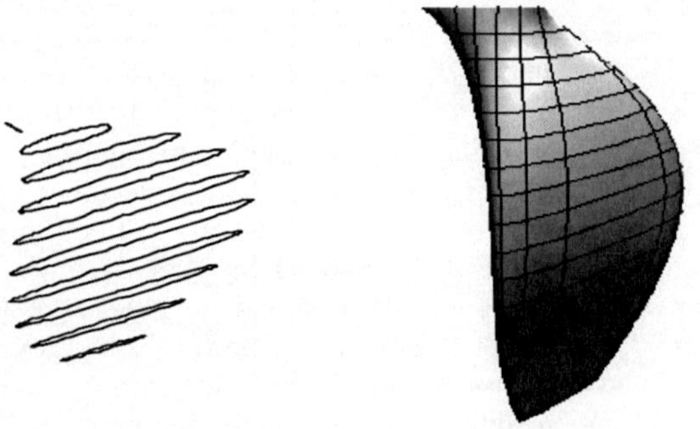

(a) (b)

Fig. 1 (a) Slices, (b) Surface

surface that we seek is viewed as a sequence of a planar curves (slices), indexed by the parameter r, so that each fixed value of r provides a closed curve, lying in a slice of the 3D-image. Consequently, let

$$(\gamma_r) : v(s) = (x(s), y(s)),\ s \in [0,1] \quad (10)$$

be the 2D curve obtained by applying this reconstruction method, for a given r.

In what follows, we suppose that the control functions w_{ij} are positive constants and we set by the seek of simplicity, $w_1 = w_{10}$, $w_2 = w_{20}$. The (EGO) Equation (5), which corresponds to (γ_r), is:

$$2w_2 \tfrac{d^4 v}{ds^4} - 2w_1 \tfrac{d^2 v}{ds^2} - c_0 J_2 \tfrac{dv}{ds} + \nabla P = 0, \quad (11)$$

where $J_2 = \begin{bmatrix} 0 & 1 \\ -1 & 0 \end{bmatrix}$.

Example 1. If we consider in (11) $c_0 = -0.2$, $w_1 = 2.5$, $w_2 = 0.4$ $P = r(x^2 + y^2)$ and $r = 0.1, 0.2, ..., 1$ with boundary conditions $x(0) = x(1) = 1 + \frac{r^2(1-r)^2}{25}$, $x'(0) = x'(1) = \frac{r(1-r)}{10}$, $y(0) = y(1) = 0 + \frac{r^2(1-r)}{25}$, $y'(0) = y'(1) = \frac{2r(1-r)}{5}$ we obtain the graphs of the slices and a 3D reconstruction of the surface, as it can be seen in the figure 1a and 1b.

III. An EGO-Algorithm of Implicit Type for the 3D Simplified Model

Firstly, we use the method of finite differences in order to obtain the discretized version of the EGO-Equation , in dynamic form, see (8) and (11). Let δ and h be the time and the space discretization steps, respectively and denote the plane net of discretization by $\mathcal{R} = \{(t_k, s_i), k \geq 0,\ 0 \leq i \leq N\}$, with $N \in \mathbb{N}^*$, $Nh = 1$, $t_k = k\delta$ and $s_i = ih$. The following notations will be used, too: $v_i^k = v(t_k, s_i)$, $v^k = (v_i^k)_{0 \leq k \leq N}$, $v^k = (x^k, y^k)$, $k \geq 0$, $g^k = (g_1^k, g_2^k)$, with $g_1^k = -\frac{1}{2}\left(\frac{\partial P}{\partial x}\right)(v^k)$, $g_2^k = -\frac{1}{2}\left(\frac{\partial P}{\partial y}\right)(v^k)$; since (γ_r) is a closed curve, it results $v_i^k = v_{i+N}^k$, $i \in \mathbb{Z}$. We approximate the partial derivatives at the nodes of $\mathcal{R}$ as follows:

$$\begin{cases} \frac{\partial v}{\partial t}(t_k, s_i) \simeq \frac{1}{\delta}\left(v_i^k - v_i^{k-1}\right); \frac{\partial v}{\partial s}(t_k, s_i) \simeq \frac{1}{h}\left(v_i^k - v_{i-1}^k\right) \\ \frac{\partial^2 v}{\partial s^2}(t_k, s_i) \simeq \frac{1}{h^2}\left(v_{i+1}^k - 2v_i^k + v_{i-1}^k\right) \\ \frac{\partial^4 v}{\partial s^4}(t_k, s_i) \simeq \frac{1}{h^4}\left(v_{i+2}^k - 4v_{i+1}^k + 6v_i^k - 4v_{i-1}^k + v_{i-2}^k\right) \end{cases} \quad (12)$$

with $k \geq 1; 0 \leq i \leq N$. With the notations $\gamma = \frac{c_0}{h}$, $a_1 = 2\frac{w_1}{h^2} + 6\frac{w_2}{h^4}$, $a_2 = -\frac{w_1}{h^2} - 4\frac{w_2}{h^4}$, $a_3 = \frac{w_2}{h^4}$, we consider the N-th order square matrices K (named *stiffness-matrix*) and L, defined as circular matrices with the first row $(a_1, a_2, a_3, 0, 0, \ldots, a_3, a_2)$ and $(1, -1, 0, 0, \ldots, 0)$, respectively. Thus, the differential equation (11) turns into the algebraic system

$$\tfrac{v^k - v^{k-1}}{\delta} + Kv^k + \gamma L\left(J_2 v^k\right) = g^k, \quad k \geq 1. \quad (13)$$

Denote by $V^k = \left(X^k, Y^k\right)^T$, $k \geq 1$, the solution of the system (13), which approximate the values of $v(t,s)$ at the nodes of $\mathcal{R}$; these solutions satisfy the *discrete EGO-Equation* of implicit type

$$(I_N + \delta K) V^k + \delta\gamma L\left(J_2 V^k\right) = V^{k-1} + \delta g^k,\ k \geq 1, (14)$$

namely

$$\begin{cases} (I_N + \delta K) X^k + \delta\gamma L Y^k = X^{k-1} + \delta g_1^k \\ (I_N + \delta K) Y^k - \delta\gamma L X^k = Y^{k-1} + \delta g_2^k \end{cases} : k \geq 1 \ (15)$$

where I_N is the unit matrix of order N.

The equalities (14) and (15) define a totally implicit discretized scheme, since the unknown $V^k = \left(X^k, Y^k\right)^T$ appears in many terms of these relations (V^k is incorporated in g^k, too). Taking into account that g^k has a complicated expression (generally of non-linear type), we approximate the term g^k (which corresponds to the external forces) by its value of the previous iteration g^{k-1}, obtaining the following discretized scheme:

$$\frac{V^k - V^{k-1}}{\delta} + KV^k + \gamma L\left(J_2 V^k\right) = g^{k-1} \quad (16)$$

The relation (16) defines a *semi-implicit discretized scheme*, because it is explicit with respect to the terms containing V^{k-1} and g^{k-1}, but it is implicit with respect to V^k.

Further, let us approximate both g^k and the matrix terms KV^k and $L\left(J_2V^k\right)$ of (14) by their values of the previous iteration, which leads to the *explicit discretized scheme* given by:

$$V^k = \left(I_N - \delta K - (\gamma\delta L)J_2\right)V^{k-1} + \delta g^{k-1}\ ; k \geq 1. \quad (17)$$

Remark that the EGO-Algorithm of explicit type (17) is a linear approximation of the corresponding implicit algorithms defined by (14) or (16).

Indeed, we derive from (14), with g^{k-1} instead of g^k, or from (16):

$$V^k = \left(I_N + \delta K\right)^{-1}\left(V^{k-1} + \delta g^{k-1} - \delta\gamma L\left(J_2V^{k-1}\right)\right) \quad k \geq 1. \quad (18)$$

On the other hand, from the matrix identity

$$\begin{array}{c}\left(I_N + \delta K\right)\left(I_N + \sum_{j=1}^{m}(-1)^j\delta^jK^j\right) = \\ = I_N + (-1)^m\delta^{m+1}K^{m+1},\end{array} \quad (19)$$

it is easily seen that for $\delta \geq 0$ sufficiently small, by omitting the terms containing $\delta^m, m \geq 2$,the linear approximating formula

$$\left(I_N + \delta K\right)^{-1} \simeq I_N - \delta K \quad (20)$$

holds. Now, the relations (18) and (20) lead to (17).

IV. THE CONVERGENCE AND THE STBILITY OF THE EXPLICIT EGO-ALGORITHM

In this section we refer to the explicit EGO-Algorithm (17). Setting $\alpha = \frac{w_1}{h^2}$ and $\beta = \frac{w_2}{h^4} = a_3$, we derive from (17):

$$\begin{array}{c}V_i^k = -\beta\delta V_{i+2}^{k-1} + \delta\left(\left(\alpha + 4\beta\right)I_2 + \gamma J_2\right)V_{i+1}^{k-1} + \\ + \left(\left(1 - 2\alpha\delta - 6\beta\delta\right)I_2 - \gamma\delta J_2\right)V_i^{k-1} + \\ + \delta\left(\alpha + 4\beta\right)V_{i-1}^{k-1} - \beta\delta V_{i-2}^{k-1} + \delta g_i^{k-1}, \\ k \geq 1, 0 \leq i \leq N-1\end{array} \quad (21)$$

with $V_i^k = \left(X_i^k, Y_i^k\right)^T$, $V_j^k = V_{j+N}^k$, $j \in \mathbb{Z}$, $k \geq 0$ and $g_i^k = (g_{1i}^k, g_{2i}^k)$, $g_{1i}^k = -\frac{1}{2}\frac{\partial P}{\partial x}\left(v_i^k\right)$, $g_{2i}^k = -\frac{1}{2}\frac{\partial P}{\partial y}\left(v_i^k\right)$.

In order to establish the convergence of the EGO-Algorithm (21), denote by

$$\varepsilon_i^k = v_i^k - V_i^k, \ k \geq 0, \ 0 \leq i \leq N-1 \quad (22)$$

By using the Taylor expansions of $v(t,s)$ at the point $N_i^{k-1}(t_{k-1}, s_i) \in \mathcal{R}$, namely:

$$\left\{\begin{array}{c}v_i^k = v_i^{k-1} + \delta\frac{\partial v}{\partial t}\left(N_i^{k-1}\right) + \frac{\delta^2}{2!}\frac{\partial^2 v}{\partial t^2}\left(N_i^{k-1}\right) + \cdots \\ v_{i\pm1}^{k-1} = v_i^{k-1} \pm h\frac{\partial v}{\partial s}\left(N_i^{k-1}\right) + \\ + \frac{h^2}{2!}\frac{\partial^2 v}{\partial s^2}\left(N_i^{k-1}\right) \pm \frac{h^3}{3!}\frac{\partial^3 v}{\partial s^3}\left(N_i^{k-1}\right) + \cdots \\ v_{i\pm2}^{k-1} = v_i^{k-1} \pm 2h\frac{\partial v}{\partial s}\left(N_i^{k-1}\right) + \\ + \frac{(2h)^2}{2!}\frac{\partial^2 v}{\partial s^2}\left(N_i^{k-1}\right) \pm \frac{(2h)^3}{3!}\frac{\partial^3 v}{\partial s^3}\left(N_i^{k-1}\right) + \cdots\end{array}\right. \quad (23)$$

we obtain from (21), (22) , (23):

$$\begin{array}{c}\varepsilon_i^k = \left(\left(1 - 2\alpha\delta - 6\beta\delta\right)I_2 - \gamma\delta J_2\right)\varepsilon_i^{k-1} + \\ + \delta\left(\left(\alpha + 4\beta\right)I_2 + \gamma J_2\right)\varepsilon_{i+1}^{k-1} + \\ + \delta\left(\alpha + 4\beta\right)\varepsilon_{i-1}^{k-1} - \beta\delta\left(\varepsilon_{i+2}^{k-1} + \varepsilon_{i-2}^{k-1}\right) + \delta Rv_i^{k-1},\end{array} \quad (24)$$

where Rv_i^{k-1} is the *residue* of the algorithm, see [10], namely:

$$\begin{array}{c}Rv_i^{k-1} = \delta\left(\frac{1}{2}\frac{\partial^2 v}{\partial t^2} + \frac{1}{6}\delta\frac{\partial^3 v}{\partial t^3} + \cdots\right)\left(N_i^{k-1}\right) + \\ + h^2 w_2\left(\frac{1}{6}\frac{\partial^6 v}{\partial s^6} + \frac{127}{5040}h^2\frac{\partial^8 v}{\partial s^8} + \cdots\right)\left(N_i^{k-1}\right) - \\ - w_1h^2\left(\frac{1}{12}\frac{\partial^4}{\partial s^4} + \frac{h^2}{60}\frac{\partial^6 v}{\partial s^6} + \cdots\right)\left(N_i^{k-1}\right) - \\ - c_0J_2h\left(\frac{1}{2}\frac{\partial^2 v}{\partial s^2} + \frac{h^2}{24}\frac{\partial^4 v}{\partial s^4} + \cdots\right)\left(N_i^{k-1}\right); \\ 0 \leq i \leq N-1; k \geq 1\end{array} \quad (25)$$

Let

$$E^k = \max\left\{|\varepsilon_{i\pm2}^k|, |\varepsilon_{i\pm1}^k|, |\varepsilon_i^k|, 0 \leq i \leq N-1\right\}, k \geq 0 \quad (26)$$

the *approximation-error* at the k-th iteration of the EGO-Algorithm. Suppose that the partial derivatives in (25) are uniformly bounded. Thus, the relations (24), (25) and (26), combined with the classical inequality $\sqrt{x^2 + y^2} \leq |x| + |y|, x, y \in \mathbb{R}$, provides the upper estimate:

$$\begin{array}{c}E^k \leq \left(10\beta\delta + 2\alpha\delta + 2\gamma\delta + \left(1 - 2\alpha\delta - 6\beta\delta\right)\right)E^{k-1} + \\ + M_1\delta^2 + M_2|2w_2 - w_1|\delta h^2 + M_3c_0\delta h\end{array} \quad (27)$$

Under the practical assumption (in medical imaging, for example)

$$1 - 6\beta\delta - 2\alpha\delta > 0 \quad (28)$$

we derive from (27):

$$E^k \leq qE^{k-1} + M_1\delta^2 + M_2|2w_2 - w_1|\delta h^2 + M_3c_0\delta h \quad (29)$$

for $k \geq 1$, with:

$$q = 1 + 4\beta\delta + 2\gamma\delta \quad (30)$$

Writing the estimate (29), successively, for $k, k-1, \ldots, 1$ and taking into account that $E^0 = 0$, we get:

$$E^k \leq \frac{q^k - 1}{q - 1} \cdot M(h, \delta), \tag{31}$$

with

$$M(h, \delta) = M_1\delta^2 + M_2|2w_2 - w_1|\delta h^2 + M_3 c_0 \delta h \tag{32}$$

Now, from the previous notations, we get:

$$w_1 = \alpha h^2, \; w_2 = \beta h^2 \text{ and } c_0 = \gamma h \tag{33}$$

In the medical imaging the parameters α, β and γ are viewed as absolutely positive constants, [1], what is in accordance with the hypothesis (28) written in the form:

$$\delta < \frac{1}{6\beta + 2\alpha}, \tag{34}$$

The inequalities $1 + 4\beta\delta \leq q \leq 1 + 6\beta\delta$, which follows from (30) and (33), combined with the relations (31), (32) and $(1+x)^{1/x} \leq e$ for $x > 0$, lead to the estimate:

$$\begin{aligned} E^k \leq \tfrac{\exp(6\beta\delta k) - 1}{4\beta} \cdot \\ \cdot \left(M_1\delta + M_2|2\beta h^2 - \alpha|h^4 + M_3\gamma h^2\right) \end{aligned} \tag{35}$$

With the practical hypothesis:

$$k\delta = O(1) \tag{36}$$

we derive from (35), for $h > 0$ sufficiently small and $\delta = O\left(\frac{1}{k}\right)$

$$E^k = \begin{cases} O(\delta) + O(h^2), \text{ if } c_0 > 0 \\ O(\delta) + O(h^4), \text{ if } c_0 = \gamma = 0 \end{cases} \tag{37}$$

It is easily seen that (36) implies (34) for k sufficiently large. Therefore, $E^k \to 0$ if $\delta \to 0, h \to 0$, so that the following convergence statement holds.

If the condition (36) is fulfilled and the weight-coefficients w_1, w_2 and c_0 are given by (33), then the EGO-Algorithm (21) is convergent and its approximation-error at the k-th iteration satisfies the estimate (37).

Further, let us examine *the stability of the explicit EGO-Algorithm* (21), with $c_0 = 0$, i.e. $\gamma = 0$. The intuitive idea regarding the stability is that small errors in the initial conditions should cause small errors in the solution. On the other hand, the study of the stability is necessary in order to use the Lax Theorem of convergence [11]. By omitting the small term $\delta R v_i^{k-1}$ in (24), the errors $\tilde{\varepsilon}_i^k$ of the EGO-Algorithm (21) with $c_0 = \gamma = 0$ satisfy the relation:

$$\begin{aligned} \tilde{\varepsilon}_i^k = (1 - 6\beta\delta - 2\alpha\delta)\,\tilde{\varepsilon}_i^{k-1} + \\ (\alpha\delta + 4\beta\delta)\left(\tilde{\varepsilon}_{i+1}^{k-1} + \tilde{\varepsilon}_{i-1}^{k-1}\right) - \\ -\beta\delta\left(\tilde{\varepsilon}_{i+2}^{k-1} + \tilde{\varepsilon}_{i-2}^{k-1}\right), k \geq 1 \end{aligned} \tag{38}$$

In order to use the method of von Neumann [10,12],let

$$\begin{aligned} \tilde{\varepsilon}_i^k = \exp(\mu k \delta) \cdot \exp(j\omega i h) \cdot (1,1)^T = \\ = \mu^k \cdot \exp(j\omega i h)(1,1)^T, \mu = \exp(\nu\delta) \end{aligned} \tag{39}$$

where j and $\nu = \nu(\omega)$ are complex number, $j^2 = -1$ and ω denotes the frequency. Taking into account that $|exp(j\omega i h)| = 1$, it is obvious that the error $\tilde{\varepsilon}_i^k$ doesn't increase in time if

$$|\mu| < 1. \tag{40}$$

The inequality (40) is referred to as the *stability criterion of von Neumann.*

In our case, we derive from (38) and (39) with $\eta = \omega h$

$$\begin{aligned} \mu = (1 - \alpha\delta - 6\beta\delta) + \\ +\delta\,(\alpha + 4\beta)\,(\exp(j\eta) + \exp(-j\eta)) - \\ -\beta\delta\,(\exp(2j\eta) + \exp(-2j\eta)) \end{aligned} \tag{41}$$

By using the relations $\exp(j\alpha) + \exp(-j\alpha) = 2\cos\alpha$ and $1 - \cos\alpha = 2\sin^2(\alpha/2), \alpha \in \mathbb{R}$ the equality (41) becomes:

$$\mu = 1 - 4\alpha\delta\sin^2\eta - 16\beta\delta\sin^4\eta \tag{42}$$

Now, combining the relations (40) and (42) we get:

$$2\alpha\delta + 6\beta\delta \leq 1. \tag{43}$$

namely

$$2\frac{\delta}{h^4}\left(w_1 h^2 + 4w_2\right) \leq 1, \tag{44}$$

which represents *the stability condition* of the considered EGO-Algorithm. Remark that the inequality (44) leads to the *necessary stability conditions*:

$$w_2\frac{\delta}{h^4} \leq \frac{1}{8}; \;\; w_1\frac{\delta}{h^2} \leq \frac{1}{2} \tag{45}$$

V. Monitoring the Behavior of Prosthetic Surgical Methods and Prosthetic Medical Materials, Based on Software Implementation

In order to apply the results of the theoretical researches detailed above in the medical imaging domain, a 3D visual software environment –named MoDef- was implemented, aiming to visualize and follow-up the deformation behavior of the surgical (abdominal, maxilla-facial and orthodontic) prosthetic materials. That is performed on three distinct, but convergent levels, as follows:

a) 3d reconstruction visual software component, aimed to tracks the evolution of the prosthetic materials, based on processing the US images of the anatomic context of a lot of surgical patients;
b) deformable prosthetic material's behavior forecasting software component, based on software tools which implements the above described mathematical methods;
c) comparative parallel tracking software component, aimed to simultaneous supervise in time both (a) and (b) levels, in comparison with the results provided by the stochastic analysis component of the 3D visual software environment MoDef.

Concerning the 3D visualizing of the prosthetic meshes by means of the MoDef software environment components, two levels of reconstruction are performed, namely:

1) On the first level, a polynomial interpolation method is applied on each slice of the US image of the prosthetic mesh, acquired based on succeeding positions of the transducer, obtained by rotating them with a constant angle in a same pre-established direction; more exactly, the curves representing the sections of the surgical mesh acquired by the transducer, are extracted from the context of the US image, based on specific image processing methods - namely contour detection methods, that are implemented at the level of the image processing operators of the MoDef environment's image processing library. Starting with this set of basic mesh surface definition curves, extracted from the US images acquired at pre-established moments in time, a complete and consistent collection of 3D generator curve sets is obtained, by means of 3D polynomial interpolation methods, based on Lagrange, Hermite or Birkhoff operators.
2) On the second level, the complete collection of the 3D generator curves obtained at the first level is processed based on Blended Interpolating Methods (BIM), as well as with 3D continuous representation techniques, in order to obtain "solid-view", respectively "wired-view" representations of the prosthetic mesh.

In what follows some preliminary experiments made in 3DS Max7, followed by some relevant results obtained with the 3D Reconstruction component of MoDef 3D Visual environment are presented.

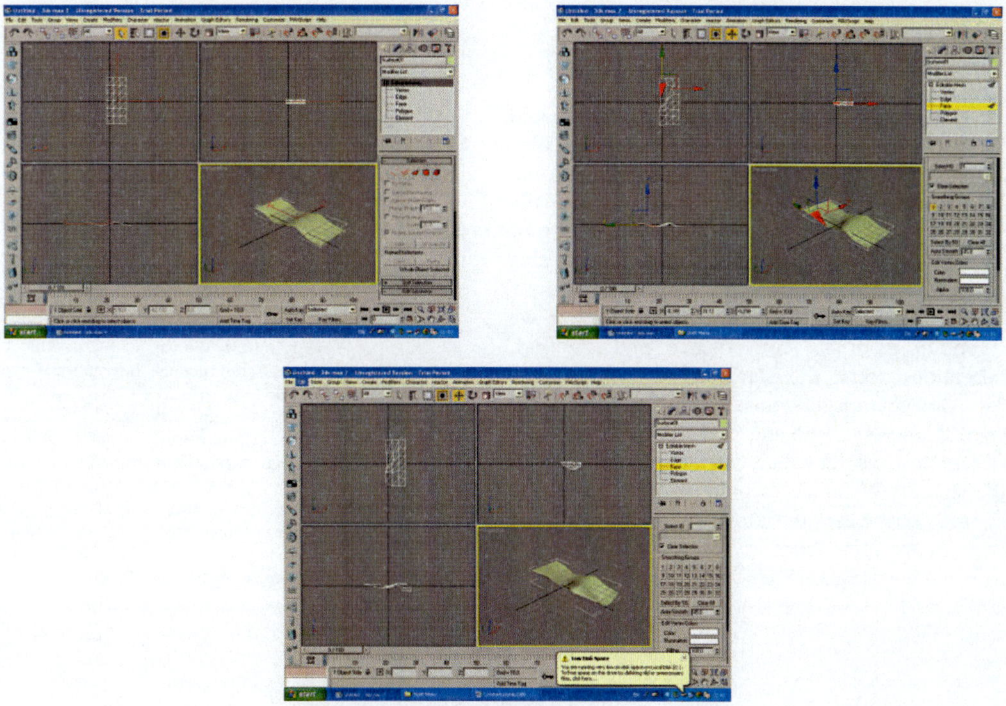

Fig. 2 Preliminary experiments made in 3DS Max7

(a) Initial 3D representation of the deformable surface of the surgical mesh

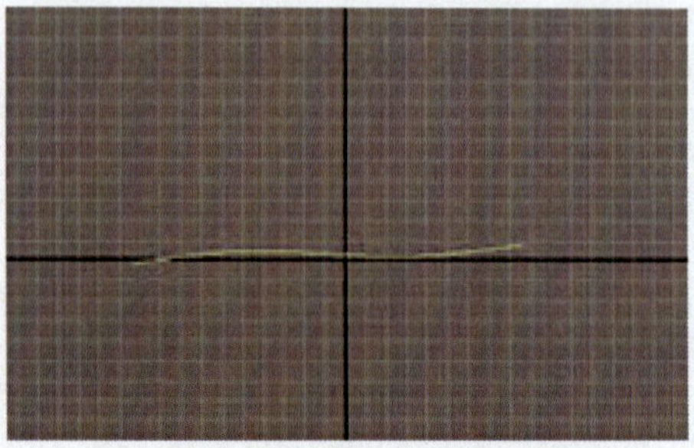

(b) Curve representing a section of the surgical mesh acquired by the transducer, extracted from the context of the US image based on specific image processing methods

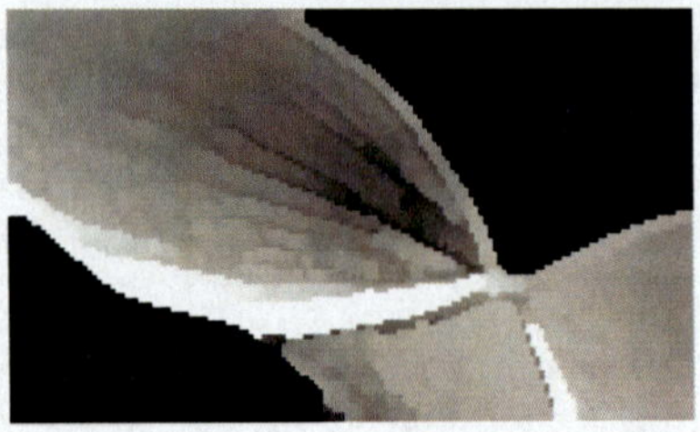

(c) The surface of the prosthetic mesh after the deformations produced in time due to the anatomic assimilation process

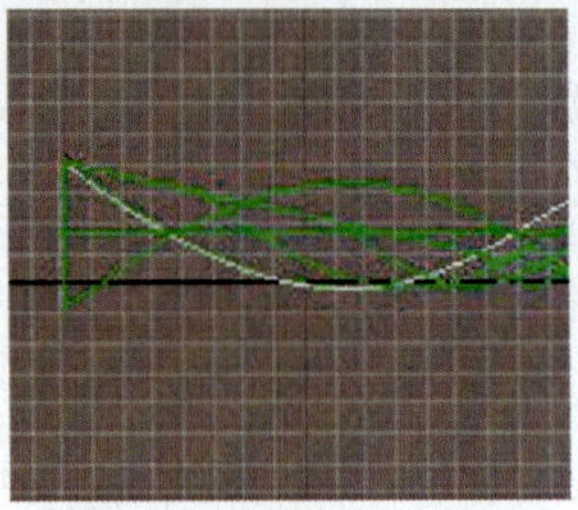

(d) The basic set of generating curves, used to obtain the "solid-view" representations of the prosthetic mesh

Fig. 3 Results obtained with the 3D Reconstruction component of MoDef 3D Visual environment

VI. Conclusions

In this paper we considered a 3D parametric deformable model and we defined, according to [1], its associated simplified model, with the aim to reduce the computational requirements. In order to find the energy-minimizing surface (the optimal model), we derived an EGO-Algorithm of implicit type, starting from the EGO-Equation of Calculus of Variations and using a finite difference method. We estimate the approximation error of this algorithm and we established conditions of its convergence and stability. Some considerations about the statistical modeling were presented, too. Our next target consists in using the discretized scheme of Krank-Nicolson and finite-element methods in order to obtain new results concerning the rate of convergence and the stability of the corresponding algorithms; the probabilistic and geometric deformable models will be considered, too, in our approaches.

Acknowledgment

These researches are supported by the Project PN2-Partnership, No. 11018 (MoDef).

References

1. L.D. Cohen and I. Cohen. (1993). Finite Element Methods for Active Contour Models and Balloons for 2D and 3D Images. IEEE PAMI 15(11), 1131—1150.
2. Michael Kaas, Andrew Witkin, and Demetri Terzopoulos. (1988) Snakes: Active contour models. International Journal of Computer Visison 1, 321—331.
3. L.He, Zh.Pens, B.Guarding, X.Wang, C.Y.Han, K.L.Weiss and W. G. Wer (2008) A comparative study of deformable contour methods on medical image segmentation Image and Comp.Vision, 26, 141-163.
4. R.Hegadi, A. Kop, M. Hangarge (2010) A Survey on Deformable Model and its Applications to Medical Imaging, IJCA Special Issue on "Recent trends in Image Processing and Pattern Recognition.
5. T. McInerney, D. terzopoulos (2000), T-snakes:topologically adaptive snakes,Med. Image Analysis, 4(2),73-91

6. S. Nedevschi and D. Mitrea. (2003) Contour detection based on active contour models. Bull.Appl.Math.Comp.Sci. (Techn. Univ. of Budapest) XCVII(2275), 107—118.
7. V. Caselles, F. Catte, T. Coll, and F.Dibos. (1993).A geometric model for active contours. Numerische Mathematik 66, 1—31.
8. R. Malladi, J. Sethian, and B. Vemuri. (1995) Shape modeling with front propagation: A level set approach. PAMI 17(2), 158—175.
9. A.I.Mitrea, O.M.Gurzau, P.Mitrea, D.Cimpean, Numerical methods for finding the optimal model in the theory of deformable models, Proceedings of 12-th Symposium of Mathematics and its Applications, Ed. Politehnica, Timisoara(2009), pp.446-453, ISSN 1224-6069
10. D. Trif. (1997). Numerical methods for differential equations (in romanian) Ed. Transilvania Press .
11. G. Aubert, P. Kornprobst (2006) Mathematical Problems in Image Processing. Partial Differential Equations and the Calculus of Variations, Springer
12. E.Isaacson , B. Keller, (1966) Analysis of Numerical Methods, New York, Wiley and Sons.

Author: Alexandru Ioan Mitrea
Institute: Technical University of Cluj-Napoca
Street: G. Baritiu 25
City: Cluj-Napoca
Country: Romania
Email: Alexandru.Ioan.Mitrea@math.utcluj.ro

Texture-Based Methods and Dimensionality Reduction Techniques Involved in the Detection of the Inflammatory Bowel Diseases from Ultrasound Images

D. Mitrea[1], P. Mitrea[1], R. Badea[2], M. Socaciu[2], L. Ciobanu[2], A. Golea[2], C. Hagiu[2], and A. Seiceanu[2]

[1] Technical University of Cluj-Napoca, Computer Science Department, Cluj-Napoca, Romania
[2] Clinica Medicala III, Department of Ultrasonography, Cluj-Napoca, Romania

Abstract— **The inflammatory bowel diseases (IBD) are severe, chronic and recurring disorders, requiring continuous patient monitoring. The most reliable methods for the diagnosis of the inflammatory bowel diseases are invasive (endoscopy, colonoscopy, histopathology) or irradiating (CT). We aim to develop computerized methods for the noninvasive assessment of the bowel inflammation level based on information obtained from ultrasound images. In this work, we study the role of the textural parameters in characterizing different types of inflammatory bowel diseases and the colorectal tumors. The dimensionality reduction techniques are taken into consideration in order to obtain the relevant textural features and to improve the result of the automatic diagnosis. The Principal Component Analysis (PCA) method and the Correlation based Feature Selection (CFS) method, as well as their combinations, are assessed for this purpose. The methods of Support Vector Machines (SVM) and Multilayer Perceptron (MLP), which gave very good results in our former experiments, are implemented for the automatic diagnosis. B-mode ultrasound images belonging to biopsied patients, are used. The patients were suffering from the following types of diseases: Crohn's disease, ulcerative recto-colitis, colo-rectal cancer.**

Keywords— **inflammatory bowel diseases (IBD), nonivasive diagnosis, texture, dimensionality reduction techniques, classification performance.**

I. Introduction

The inflammatory bowel diseases (IBD) are a group of disorders that frequently mark the population of the developed countries. Their evolution is frequently chronic, with activation peaks and remission periods, elements that are conditioned by the rapidity of the diagnosis, by the follow-up efficiency and by the therapeutic means. The clinical scores which are mostly used in order to assess the phase of IBD are the Crohn's Disease Activity Index (CDAI) and Truelove Witts. Together with laboratory parameters, they can assess to some degree the activity but are not enough accurate. [1] The standard methods of diagnosis and of inflammatory phase assessment are the endoscopic, radiologic and histopathology exams, but these are too invasive for severe forms and they cannot be permanently repeated in order to monitor the clinical evolution. Also, computer tomography and magnetic resonance imaging are elective imagistic methods, but are less accessible and quite expensive. Ultrasonography has similar potential in diagnosis, but with advantages like: noninvasivity, reduced cost and the possibility of repeatability. Numerous literature studies have proven their role for the examination of the digestive tube pathology [1], [2], [3]. Our aim is to develop new methods of assessment in the case of the inflammatory bowel diseases, based on utrasonography examination, combined with some modern techniques like computer-aided analysis of images. The texture is considered as an important visual feature, able to provide subtle information concerning the structure of the internal organ tissues. In our work, we develop texture-based methods in order to emphasize the features that characterize each inflammatory bowel disease and the digestive tumors. Image enhancement techniques, texture analysis methods, feature selection methods, feature extraction methods and classification methods, are involved in our research. For the detection of the relevant textural features, we implemented the Correlation-based Feature Selection (CFS) method, which retains only the features that are mostly correlated with the class parameter. Concerning the feature extraction methods, the Principal Component Analysis (PCA) technique was considered. The PCA is the best known linear features extractor that searches for the best directions of data projection in order to emphasize the main variation modes. We combined the PCA method with the CFS method in order to add class sensitivity to the feature extraction method. For the purpose of automatic diagnosis, we implemented the methods of Support Vector Machines (SVM) and Multilayer Preceptron (MLP), well known for their performance. The paper presents the state of the art in the approached domain, then the medical aspects concerning the inflammatory bowel diseases and the colonic tumors. The proposed solution is illustrated in details, then the experiments are presented. At the end, we formulate the conclusion and we propose the further development directions.

II. The state of the art

In [4] the authors analyzed the fluorescent images of colonic tissue based on textural parameters derived from the Grey Level Cooccurrence Matrix (GLCM), in order to distinguish the colonic healthy mucosa versus

S. Vlad and R.V. Ciupa (Eds.): MEDITECH 2011, IFMBE Proceedings 36, pp. 220–225, 2011.
www.springerlink.com

adenocarcinoma. A modified version of Multiple Discriminant Analysis was used for dimensionality reduction, such that only four final features resulted. A minimum Mahalanobis Distance, a Linear Discriminant Classifier and a simple evaluation 'score' method were used for performing two-category classification and provided 95% accuracy. In [5] the authors used the Grey Level Cooccurrence Matrix (GLCM), together with morphological features (shape, orientation), in order to characterize the malignant and benign tissues from biopsy slides of patients with colon cancer. The Support Vector Machines (SVM) method with a 3rd degree polynomial kernel provided very satisfying classification results (above 90%). Textural features such as the skewness and kurtosis of the image intensity histogram, together with RGB color features, were used in order to perform the segmentation of the bowel lumen from endoscopic images [6]. Concerning the implementation of the dimensionality reduction methods in the medical domain, in [7] the authors combined the method of Principal Component Analysis with genetic algorithms for the selection of the gene expressions. The new method leaded to the improvement of the classification accuracy, in comparison with the case when only the PCA method was applied. However, in the case of the inflammatory bowel diseases, the computerized techniques are poorly implemented and there is no systematic study of the relevant features that characterize each type of disease. We aim to perform this in our work.

III. THE DESCRIPTION OF THE INFLAMMATORY BOWEL DISEASES AND OF THEIR ASPECT IN ULTRASOUND IMAGES

Bowel inflammation is a common disorder found in several intestinal diseases, ranging from malignant to purely inflammatory ones. However, there is a group of intestinal diseases, known as Inflammatory Bowel Diseases (IBD), that have unknown causes, but are important because of their severity. The Crohn's disease (CD) and Ulcerative colitis (UC) are the most frequent forms. The inflammation extends to all the layers in Crohn's disease [8], and only to the mucosa in the Ulcerative colitis [9]. Concerning the case of the Ulcerative colitis, hypoechoic lesions that correspond to the ulcerations appear. Patients with inflammatory bowel diseases present, alternatively, periods of remission and of activity. The evolution is marked by the onset of complications, among which some of major severity, so that the assessment of the inflammation activity and of the treatment response is crucial in monitoring these patients. Inflammatory or neoplastic bowel pathology is associated with thickening of the bowel wall. Concerning the features of the Crohn's disease, the characteristic visual appearance is that of a "target" image, corresponding to the transversal bowel section taken on a specific part of the bowel, where the walls are thickened over 4 mm, having a circumferential, stenotic aspect (Fig.1). In the acute form, a removal of the separation between the composing layers, because of edema and inflammation, is being noticed. In the chronic form, the composing layers appear as being distinct. The dominant process at histological layer is that of fibrosis, which generates an echo-poor halo surrounding a central echogenic zone [8]. Ulcerative colitis is a mucosal disease, the halo being a less prominent feature (Fig. 2). [9] The colorectal tumors (Fig.3), although distinct from inflammatory bowel diseases, share a lot of characteristics with the latter, like wall thickening and increased vascularity. However, like every tumor, they are characterized by the heterogeneity of the tissue structure and by the complexity and irregularity of the vessel structure. [10] Eloquent examples of these diseases are illustrated in the figures given bellow:

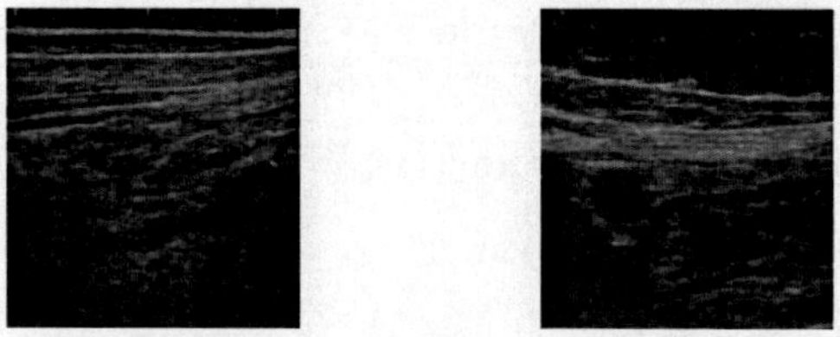

a. b.

Fig. 1 Crohn's Disease; (a) Before treatment, (b) After treatment (chronic form)

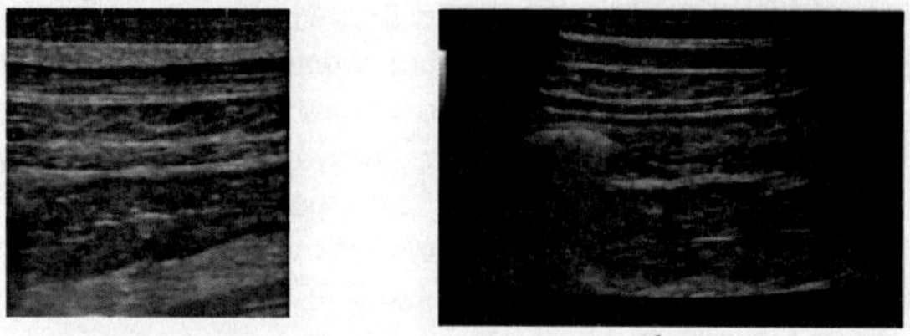

a. b.

Fig. 2 Ulcerative recto-colitis (a) before treatment, (b) after treatment (chronic form); thickened walls natively visible

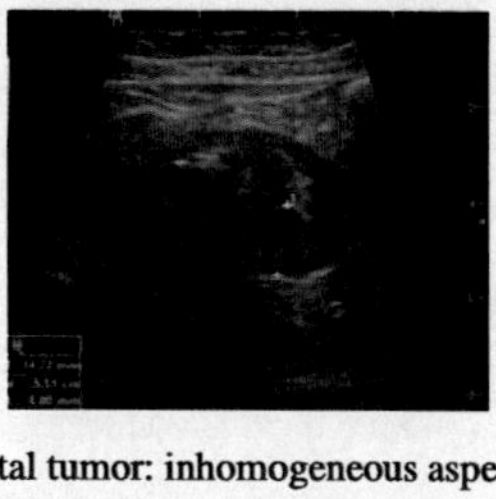

Fig. 3 Colo-rectal tumor: inhomogeneous aspect of the bowel wall

IV. DESCRIPTION OF THE PROPOSED SOLUTION

In our work, we compute a number of potentially useful textural parameters, we apply dimensionality reduction methods in order to find the most appropriate representation for the textural features and to select the relevant ones; then we assess the possibility of the automatic diagnosis, by providing the features at the inputs of some powerful classifiers – Support Vector Machines (SVM) and Multilayer Perceptron (MLP) [11].

A. *Methods for the Computation of the Textural Parameters*

First order statistics, as well as second order statistics, edge-based statistics, fractal based methods and multi-resolution methods were considered. *The first order statistics* of the grey levels, appropriate for liver and tumor tissue characterization, were the mean grey level, the maximum and minimum of the grey levels and the autocorrelation index, a texture granularity measure [12]. *The second order statistics* of the grey levels were provided by the *Grey Levels Co-occurrence Matrix (GLCM),* which computed, for each possible pair of grey levels ($g1$, $g2$), the number of pairs of pixels, of intensities $g1$ and $g2$, being in a spatial relationship given by a specified displacement vector $(\vec{dx}, \vec{dy})$, as results from (1):

$$\begin{aligned} C_D(g_1,g_2) = \#\{&((x,y),(x',y')): \\ &f(x,y)=g_1, f(x',y')=g_2, |x-x'|=dx, \\ &|y-y'|=dy, \operatorname{sgn}(dx\cdot dy)=\operatorname{sgn}((x-x')\cdot(y-y'))\} \end{aligned} \tag{1}$$

We computed the GLCM probability matrix [13] in our implementation. The parameters that we computed from GLCM, considered as being able to best characterize the tumor tissue, were: contrast, variance, local homogeneity, correlation, energy and entropy [13]. The maximum distance considered between the pixels from the same pair was two, all the possible orientations of the displacement vector being adopted. The final values of the Haralick parameters resulted by averaging the values of all the resulted cooccurrence matrices. The Third Order GLCM (TOGLCM) [14] matrix was also computed, in a similar way, the intention being that of improving the accuracy of the IBD characterization and detection, by considering the co-occurrence of three pixels with three distinct values of the grey levels, situated in a certain spatial relation, given by two displacement vectors, as described by (2):

$$\begin{aligned} C_D(g_1,g_2,g_3) = \#\{&((x,y),(x',y'),(x'',y'')): \\ &f(x,y)=g_1, f(x',y')=g_2, f(x'',y'')=g_3, \\ &|x-x'|=dx_1, |x'-x''|=dx_2, |y-y'|=dy_1, \\ &|y'-y''|=dy_2, \operatorname{sgn}(dx_1\cdot dy_1)=\operatorname{sgn}((x-x')\cdot(y-y')), \\ &\operatorname{sgn}(dx_2\cdot dy_2)=\operatorname{sgn}((x''-x')\cdot(y''-y'))\} \end{aligned} \tag{2}$$

The edge-based statistics (edge-frequency and edge-contrast) also provided useful information, as they computed the relative number of separations between regions with different intensity values (edge frequency) and also the relative amount of the difference between these regions (edge contrast). *The variability in edge orientations* was also taken into consideration [12]. *The fractals* provided a measure of the grey level structure complexity in the region of interest. We computed the Hurst coefficient, as described in [15]. For emphasizing *the textural microstructures* of the liver and tumor tissue, we applied the Laws convolution filters in order to detect: *levels, edges, spots, waves, ripples* [12]. We used *the Wavelet transform* for the decomposition of the signal spectrum in components having various frequencies, and for the analysis of the signal at various resolutions. We considered the Haar function as basis for the Wavelet transforms [15]. The decomposition was performed at two levels: at the first level, the signal components were computed on the original image; at the second level, the decomposition was performed for each component obtained on the first level: low-low, low-high, high-low and high-high. Then, we computed the Shannon entropy value of the grey levels for each component [15], using (3):

$$Entropy = -\sum_{x=1}^{N}\sum_{y=1}^{M} |I(x,y)| \log_2 |I(x,y)| \tag{3}$$

All the features were computed on regions of interest having 30x30 pixels in size. They were independent of orientation and scaled with the size of the region of interest.

B. *Dimensionality Reduction Methods*

a.) Principal Component Analysis (PCA)

The method of Principal Component Analysis (PCA) is designed to reduce the dimensionality of the data by projecting the feature vectors on a space of lower dimensions, putting in evidence the main data variation modes. In mathematical terms, the purpose of PCA is to find a linear mapping M, that maximizes the quantity $M^T cov(X)M$, where cov(X) is the covariance matrix [16] of the dataset X. It can be demonstrated that M, the linear mapping matrix, is formed by the first d eigenvectors that correspond to the first d largest eigenvalues, denoted by λ, of the covariance matrix built on the mean subtracted data. The equation that provides the solution is:

$$cov(X)M=\lambda M \tag{4}$$

The transformation is a linear one, consisting in the computation of the covariance matrix and of the eigenvectors and eigenvalues of this matrix. The transformation of the original data is described by the following mathematical expression:

$$Y = (X - \bar{X})M \tag{5}$$

In the above formula, X represents the original data set, M is the linear mapping matrix, $\bar{X}$ is the mean subtracted data and Y is the transformed data set. The following steps are due in order to perform a reliable PCA [16]: 1. Subtract the mean of the original data, for each considered feature; 2. Compute cov(X), the covariance matrix of the mean subtracted data. 3. Determine the eigenvalues and the eigenvectors of cov(X); 4. Sort the eigenvalues ascending and choose the first eigenvectors, corresponding to the highest eigenvalues, that cover a specified amount of variance in the data; 5. Transform the data according to the mathematical relation given by (5). PCA maximizes the variance of the extracted features, these features being also uncorrelated.

b.) Correlation-based Feature Selection (CFS)

From the class of feature selection methods, we applied the Correlation Based Feature Selection (CFS), combined with genetic search, this being an efficient search method that generates a complete set of feature subsets [17]. The CFS method eliminates the redundant, dependent features, choosing those that are the most correlated with the class parameter. For each analyzed subset s, a merit is computed [17].The subset that has the highest merit is finally selected.

c.) Combinations of the above-mentioned methods

The dimensionality reduction methods described above were used individually and in combination. Both variants of successive application (first CFS, then PCA; respectively first PCA, then CFS) were considered for assessment.

C. *Performing Automatic Diagnosis of the Inflammatory Bowel Diseases*

During the validation phase, the most appropriate pattern classification methods were used in order to make an objective evaluation concerning the set of relevant features. The Multilayer Perceptron method [11] was implemented, as our experiments demonstrated that it was the most accurate. The method of *Artificial Neural Networks* is inspired from the human perception, so it aims to build a structure that respects the model of a human neural network. In the majority of cases, it has three or more layers, being called Multilayer Perceptron. Usually, for training the network, the classical *backpropagation* algorithm is used [11]. The method of Support Vector Machines (SVM) uses a mapping function that performs the projection of the initial data into a new space, where the data is linearly separable and the identification of the support vectors is possible [D]. The classification performance was evaluated using specific accuracy parameters like the recognition rate (percent of correctly classified instances), the sensitivity (TP rate), the specificity (TN rate) and the area under ROC (AUC) [11].

V. EXPERIMENTS AND DISCUSSIONS

A. *Training Set Structure*

The training set contained the following classes: Crohn's disease (CD), active form, ulcero-hemoragic rectocolitis (UHRC), active form, colo-rectal tumors and healthy patients. A number of 50 patients per class were considered. For each patient, 3 B-mode ultrasound images were retained, corresponding to various orientations of the transducer. The images were acquired using the same settings of the ultrasound machine: frequency of 9.0 MHz, gain of 56, depth of 4 cm. The textural features were computed on regions of interest having 30x30 pixels in size, situated on the superior bowel wall in the cases of the CD and UHRC, and inside the tumor in the case of colo-rectal cancer. The dimensionality reduction methods were applied and the textural features which were relevant for the differentiation of the diseases were detected and analyzed. Also, the role of the TOGLCM parameters was studied in comparison with that of the second order GLCM matrix parameters. During the validation phase, the classification accuracy of the original dataset was assessed, and then the contribution of the applied dimensionality reduction methods on the classification performance was evaluated. Our specific Visual C++ application, based on the Weka 3.5 library [18], was used in order to compute the textural features, to separate the relevant textural features from the non-relevant ones, and to assess the classification performance.

B. *The Relevant Textural Features for the Differentiation between the Considered Diseases*

The method of Correlation-based Feature Selection (CFS) combined with the genetic search method and also that of Principal Components, in conjunction with the ranker method, both from the Weka 3.5 library, were applied for dimensionality reduction. The textural features selected by the CFS method, as well as those that had the highest weights inside the principal components, are illustrated in the tables bellow. Those textural features that were selected by both methods are formatted with italic.

Features like edge frequency, edge contrast, directional gradient magnitude, and directional gradient variance emphasize the visual separation of the bowel wall layers, which is more obvious in the case of the CD, and less obvious in the case of UHRC. The GLCM contrast, TOGLCM contrast, GLCM variance, and the entropy computed at multiple resolutions after applying the Wavelet transform, illustrate the difference in homogeneity that occurs in the aspect of the bowel wall for the considered diseases. The frequency of the edges and textural microstructures is more increased in the case of the CD from the same reason. When differentiating the active form of these diseases from the healthy bowel wall, the autocorrelation index appears as being important as well, indicating a variation in the tissue granularity, because of the inflammation.

Table 1 The relevant textural features for the differentiation between Chron's disease and ulcero-hemoragic recto-colitis

Feature selection method	Relevant textural features
Correlation-based Feature Selection (CFS), $Merit_s = 0.494$	*GLCM contrast*, GLCM variance, *Edge contrast*, *Hurst fractal index*, Wavelet_entropy3, Wavelet_entropy8_lh, Wavelet_entropy5_hl, Laws spot mean, TOGLCM contrast
Principal Component Analysis (PCA)	Directional gradient magnitude, Directional gradient variance, Edge frequency, *Edge contrast*, *Hurst fractal index*, Laws spot frequency, Laws wave frequency, *GLCM_contrast*, *TOGLCM contrast*, *Wavelet_entropy2*, Wavelet_entropy8_ll

Table 2 The relevant textural features for the differentiation between the colo-rectal tumors and other tissues

Feature selection method	Relevant textural features
Correlation-based Feature Selection (CFS), $Merit_s = 0.777$	GLCM homogeneity, Minimum of the grey levels, Wavelet Entropy3, Wavelet Entropy7_hh, directional gradient variance, Laws spot mean, Laws spot frequency, Laws wave mean, *Laws wave frequency*, *TOGLCM homogeneity*
Principal Component Analysis (PCA)	Hurst fractal index, *Laws wave frequency*, Laws edge frequency, Laws wavelet frequency, edge contrast, *TOGLCM homogeneity*, GLCM correlation, TOGLCM variance

The textural features determined as being relevant in the case of differentiation between the colo-rectal tumors and kinds of tissues refer to the decreased homogeneity and increased entropy of the tumor tissue (GLCM homogeneity, TOGLCM homogeneity, the entropy computed after applying the Wavelet Transform), to the chaotic structure (directional gradient variance) of the tumors, and to the complexity of the tumor tissue (the mean and frequency of the textural microstructures obtained after applying the Laws convolution filters, the Hurst fractal index). We can notice, in all the considered cases, of differentiation between various types of IBD and between the colo-rectal tumors and the other affections, that both the GLCM and TOGLCM parameters are important.

C. *The Automatic Diagnosis of the IBD and Colo-Rectal Tumors*

a.) The role of the GLCM and TOGLCM features in the classification process

First, we study the importance of the TOGLCM textural features, in comparison with that of the GLCM textural features. We assess the classification accuracy by considering each group of features separately, as well as by taking into account both groups of features in the same time. The Multilayer Perceptron (MLP) classifier and the Support Vector Machine (SVM) classifier of Weka 3.5 were implemented for this purpose. The values of the parameters were the following: the recognition rate was 0.2 the momentum was 0.8 and the number of nodes within the single hidden layer was a = (number of features + number of classes)/2 for the MLP method; the polynomial kernel of third degree was considered for the SVM method in the case of tumor recognition, as it provided the best results in this case, while the Radial Basis Function (RBF) kernel, provided the best results in the case of differentiation between CD and UHRC.

Table 3 The accuracy of differentiation between CD and UHRC for various groups of textural features

Classification method	Recognition rate	TP rate	TN rate	AUC	Time
The GLCM parameters and the other textural features					
Multilayer Perceptron	68%	64%	72%	64.6%	3s
Support Vector Machines	72%	90%	54%	72%	0.05s
The TOGLCM parameters and the other textural features					
Multilayer Perceptron	60%	52%	68%	62.6%	2.83s
Support Vector Machines	76%	88%	64%	76%	0.06s
The GLCM parameters, the TOGLCM parameters and the other textural features					
Multilayer Perceptron	70%	68%	72%	65.1%	3.58s
Support Vector Machines	78%	98%	58%	78%	0.05s

Table 4 The classification accuracy of the colorectal tumors for various groups of textural features

Classification method	Recognition rate	TP rate	TN rate	AUC	Time
The GLCM parameters and the other textural features					
Multilayer Perceptron	75%	64.3%	85.7%	88.3%	1.7s
Support Vector Machines	75%	57.1%	92.9%	75%	0.16s
The TOGLCM parameters and the other textural features					
Multilayer Perceptron	71.42%	57.1%	85.7%	86.2%	1.56s
Support Vector Machines	78.57%	64.3%	92.9%	78.6%	0.05s
The GLCM parameters, the TOGLCM parameters and the other textural features					
Multilayer Perceptron	82.14%	78.6%	85.7%	89.8%	2.11s
Support Vector Machines	85.71%	78.6%	92.9%	85.7%	0.03s

As Table 3 and Table 4 show, the classification performance is best when using both the GLCM and TOGLCM parameters. Although a cubic correlation was found between each pair of corresponding GLCM and TOGLCM parameters, this correlation is considered weak, unable to induce redundancy.

b.) The effect of the dimensionality reduction methods on the classification accuracy

Table 5 The effect of the dimensionality reduction methods on the classification accuracy, assessed by the SVM method

Classification method	Recognition rate	TP rate	TN rate	AUC	Time
Differentiation between CD and the UHRC					
CFS	74%	94%	54%	74%	0.05s
PCA	81.42%	85%	78%	81.5%	0.25s
CFS + PCA	82.91%	89.2%	76.7%	82.9%	0.17s
PCA + CFS	73.26%	80.8%	66%	73.4%	0.03s
Differentiation between the colo-rectal tumors and other kinds of tissues					
CFS	85.71%	85.7%	85.7%	85.7%	0.03s
PCA	84.61%	84.6%	84.6%	84.6%	0.38s
CFS + PCA	88.46%	92.3%	84.6%	88.5%	0.19s
PCA + CFS	87.46%	84.6%	92.3%	88.5%	0.28s

In the case of the SVM classifier, the CFS method generated a decrease in the classification accuracy, when it was applied on the original dataset, as well as after the PCA method. The PCA method improved the classification performance, in comparison with the original case, when it was applied individually, and also after the CFS method. The application of the CFS method after PCA altered the natural order of the principal components, some of those corresponding to the highest eigenvalues being eliminated.

VI. Conclusions

The considered textural features were efficient concerning the characterization and the automatic diagnosis of the inflammatory bowel diseases and the colo-rectal cancer. The implementation of the TOGLCM matrix was benefic, as in combination with the second order GLCM matrix, it leaded to an increase in the classification performance. The textural features which were detected as relevant for the differentiation between CD and UHRC referred to the contrast and heterogeneity in the gray levels structure of the bowel wall, as well as to the frequency of edges and textural microstructures, indicating the increased visibility of the bowel wall layers in the case of the Crohn's disease. In the case of the colo-rectal tumors, the relevant textural features emphasize the heterogeneity, the chaotic structure and the complexity of the tumor tissue. The applied dimensionality reduction methods had a benefic role on the classification performance, the resulted classification accuracy being situated around 80% in the case of the differentiation between the Crohn's disease and the ulcerative recto-colitis, respectively around 85% in the case of the diagnosis of the colo-rectal tumors. The best method proved to be the combination CFS+PCA. Our solution will be improved by considering a larger number of items/class and by applying classifier combinations.

Acknowledgment

The study was founded by Romanian Ministry of Education and Research, CNCSIS Project PNII-ID-PCE-833/2008 *(SonoDig).*

References

1. L. Ruess, A.R. Blas, "Inflammatory bowel disease in children and young adults: correlation of sonographic and clinical parameters during treatment", *AJR Am J Roentgenol.* 175 (1): 79-84, 2000
2. G. Maconi, L. Carsana, "Small bowel stenosis in Crohn's disease: clinical, biochemical and ultrasonographic evaluation of histological features". *Aliment Pharm.* 18 (7): 749-756, 2003
3. N. Lassau, S. Koscielny, "Evaluation of contrast-enhanced color Doppler ultrasound for the quantification of angiogenesis in vivo". *Invest Radiol.* 36(1): 50-55, 2001
4. V. Atlamazoglou, D. Yova, "Texture analysis of fluorescence microscopic images of colonic tissue sections",*Medical and Biological Engineering and Computing, 39(2)*: 145-151, 2001
5. K. Masood, "Co-occurrence and morphological analysis for colon tissue biopsy classification" , Procs of the 4th International Workshop on Frontiers of Information Technology 2006, Pakistan, pp. 211-216.
6. S. Bejakovic, R. Kumar. "Analysis of Crohn's Disease Lesions in Capsule Endoscopy Images", International Conference on Robotics and Automation, ICRA 2009, pp.2793-2798
7. M. Pechenizkiy, A. Tsymbal, S. Puuronen, "PCA based feature transformation for classification: Issues in medical diagnosis", *Proceedings of the 17th IEEE Symposium on Computer-Based Medical Systems*, June 2004, pp. 535 – 540
8. M. Migaleddu, E. Quaia "Inflammatory activity in Crohn disease". *Abdom Imaging*, 33: 589–597, 2008
9. J. A. Pradel, X.R. David "Sonographic assessment of the normal and abnormal bowel wall in nondiverticular ileitis and colitis." *Abdom Imaging*, 2:167 – 172, 1997
10. G.R. Schmutz, "Echographie". In: Schmutz GR, & co. Ecographie et endosonographie du tube digestif et de la cavite abdominale". Ed. Vigot, Paris, pp. 15 – 21, 1994
11. R. Duda, *Pattern Classification (2nd ed)*, Wiley Interscience, 2003
12. A.K. Jain, *Fundamentals of Digital Image Processing*, Engl. Cliffs, NJ: Prentice Hall, 1989
13. D.A.Clausi "An analysis of co-occurrence texture statistics as a function of grey level quantization". *Canadian Journal of Remote Sensing.*; 28(1):45-62 , 2002
14. M. Huber, J. Carballido-Gamio, Developing and testing of texture discriminators for the analysis of trabecular bone in proximal femur radiographs, Med. Phys. 36 (11): 5089 – 5098, November 2009.
15. J.R. Parker, Algorithms for Image Processing and Computer Vision, Wiley Computer Publishing, 1996
16. L.J.P. Van der Maaten, E.O.Postma, H.J. van der Herik, „Dimensionality reduction: A comparative review", January 2008, pp. 1-22.
17. M. Hall "Benchmarking attribute selection techniques for data mining". *IEEE Trans Knowledge & Data Eng.,* 15(3): 1-16, 2003
18. Weka 3, Data Mining Software in Java, http://www.cs.waikato.ac.nz /ml/weka/, 2004

Computer Assisted Optical Podoscope for Orthostatic Measurements

S. Crisan[1],V.D. Zaharia[1], C. Curta[2], and E.D. Irimia[3]

[1] Technical University of Cluj-Napoca, Department of Electrical Measurements, Faculty of Electrical Engineering, Cluj-Napoca, Romania
[2] Technical University of Cluj-Napoca, Department of Electrotechnics, Faculty of Electrical Engineering, Cluj-Napoca, Romania
[3] Infectious Diseases Clinical Hospital Cluj-Napoca, Integrated Ambulatory, Cluj-Napoca, Romania

Abstract— **Extensive static and dynamic foot problems in modern society are a direct consequence of the lifestyle of the individual. Continuous use of transportation, prolonged periods of physical inactivity and obesity modify the normal structure of the foot in time leading to traumatic disorders. Correctly assessing the plantar pressure and identifying potential problems due to the unbalance of the foot structure are key factors in preventing and treating of conditions such as flat feet, high arches, bunions, hammertoes, plantar fasciitis etc. The research leading to this paper addresses these issues and a low cost computer aided optical podoscope is presented. With the use of modern computational systems, patient information can be analyzed, stored and compared and visual pressure points can be linked to accurate measurements when body mass and touch areas are known. By combining different illumination techniques the effects of the environment lighting are diminished and the images of the foot interaction with the sensitive surface increase in resolution. Moreover, the pressure information can be separated or connected to the rest of the foot allowing for better visual identification of the static and dynamic foot problems.**

Keywords— **plantar pressure, podoscope, digital measurements, optical sensing, tactile systems.**

I. Introduction

The human foot is a complex biomechanical structure with 26 bones, 33 joints, 107 ligaments and 19 muscles. When standing, the healthy human foot supports the body on the ground at three points:

- the joint of the 1st metatarsal
- the 5th metatarsal
- the heel bone

Arches span between these three main points and when the feet are placed side by side, the arches form a dome-like structure. This structure is not rigid, and the tendons and joints between the bones dampen the shocks when an individual is walking, making the gait cycle smooth [1]. When correctly in balance, this system can withstand large forces and the arch-type pattern is one of the strongest models found in nature.

Modern age has altered the requirements originally intended for the human foot. Increased motorized mobility leads to sedentarism, the life style of the average individual has become more inactive and the muscles lose their tonus. In addition, incorrect pressure on the plantar area can be applied in the case of overweight people or even by fitting the wrong shoes.

When the foot's biomechanical structure goes out of balance this will eventually lead to several types of foot disorders, such as different degrees of flat-feet or high arches. Such disorders may then lead to related problems in the ankles, knees, hips and lower back. These problems can also be responsible for sore muscles and aching nerves in many areas of the body. If the foot problems are not corrected in due time, there is a risk of damage to the upper joints and spinal vertebrae. [1]

While many problems should be corrected in childhood years, due to the lack of information or lack of medical devices capable of aiding the physician in correctly diagnosing a plantar pressure disorder, many children reach maturity with severe feet problems. Moreover, barefoot running on hard, artificial surfaces or selecting improper footwear for children may lead to early complications affecting gait and general mobility.

Both static and dynamic assessments of the foot structure can be made by analyzing the interaction between the supportive surface –the sole- and hard surfaces. Slight differences between pressure points and the general posture of the foot can provide valuable information for correctly diagnosing foot related disorders [2].

In a normal foot, the pressure of the heel bone and the capitulum of the first metatarsal are stronger than that of the fifth metatarsal. This effect is caused by the position of the centre of gravity and the weight distribution. The *weight line* represents the median line of the maximum pressure area. The weight line of a healthy foot runs through the inner edge of the heel bone, thus increasing the load of the inner part (medial) of the foot. The force on the inner arch from the longitudinal arch is extensive, and the ligament system is only able to sustain the arch to the limit of its flexibility. [1]

S. Vlad and R.V. Ciupa (Eds.): MEDITECH 2011, IFMBE Proceedings 36, pp. 226–229, 2011.
www.springerlink.com

Figure 1 reveals the weight line –calculated from the intensity of pressure points derived from the interaction between the foot and a rigid surface- in a normal foot and the modifications of the weight line direction that span in a flat foot.

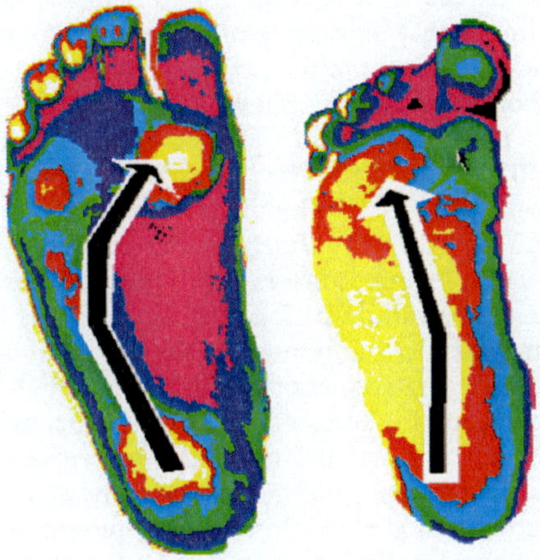

Fig. 1 Weight line calculated from pressure points in a normal foot (left) and a flat foot (right) [1]

Static problems are the main cause of foot disorders. Dropped arches or the changes arising in the process of the arches dropping can cause the majority of common foot problems [1].

II. Plantar pressure assessment

The use of plantar pressure assessment devices for diagnosis support is well known in medical literature. The advent of new scanning technologies and the increased computational power of microprocessors allows for complex detection and storing of patient's data for future reference and comparison.

The foot pressure measurement devices or *podoscopes*, generally provide an accurate way of identifying pressure problems and can serve in the diagnosis of multiple conditions. These devices are capable to offer visual information on disorders such as flat feet (all degrees), high arches, excessive pronation, early detection of bunions (Hallux Valgus), hammertoes etc.

In order to design and implement an accurate podoscope, the sensing area of the device has to allow a good discrimination between different pressures and to withstand the applied forces in both static and dynamic working regimes. Plantar pressures can be measured using a variety of instruments and methods, including force-sensing resistors (FSRs),[3] hydrocells,[4] microcapsules,[5] projection devices,[6] optical podoscopes,[7] and capacitive transducers,[8] as well as by critical light deflection.[9].Newer pressure detecting techniques include optical linear scanning[10] and CCD/CMOS camera based systems[1].

Figure 2 shows a compilation of different podoscope devices and the measurement techniques the systems are based upon.

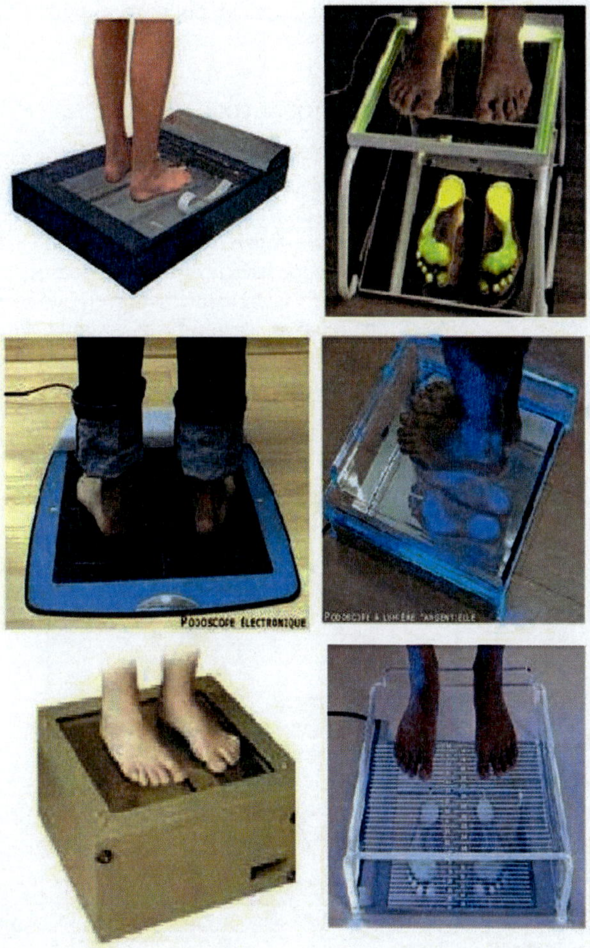

Fig. 2 Podoscope types: analogue with mirror and side illumination, structured visible lighting, optical scanner, capacitive sensing, camera based acquisition

In comparison with capacitive, resistive or microcapsule devices, optical podoscopes carry the advantage of increased visibility of the patient's foot since the pressure point information is fused with the rest of the plantar surface. However, mirror type podoscopes force the physician to observe the image from a certain angle and they are highly dependant on the environmental lighting. Optical line scanning produces high contrast images at the expense of speed and the patient has to remain immobile for the duration of the acquisition –an issue when analyzing the foot pressure of small children. The camera based podoscopes offer the best contrast to acquisition time ratio but, in most implementations, noise from external lighting and the difficulty in separating pressure data from the rest of the body diminish the quality of the final image.

This paper is focused on the design and implementation of a low cost dual lighting optical podoscope with CCD camera scanning.

III. Low cost optical podoscope design

In order to reduce the environmental effects –as discussed in the previous chapter- a dual illumination system with triggered acquisition has been developed. By applying radiation emission concepts from multi-touch optical systems [11], the implemented optical podoscope uses near infrared sources at a wavelength of 850nm where CCD cameras have a good response. In figure 3, the relative sensibility of common video cameras based on Charged Couple Device technology along with the spectral response of the human eye is presented.

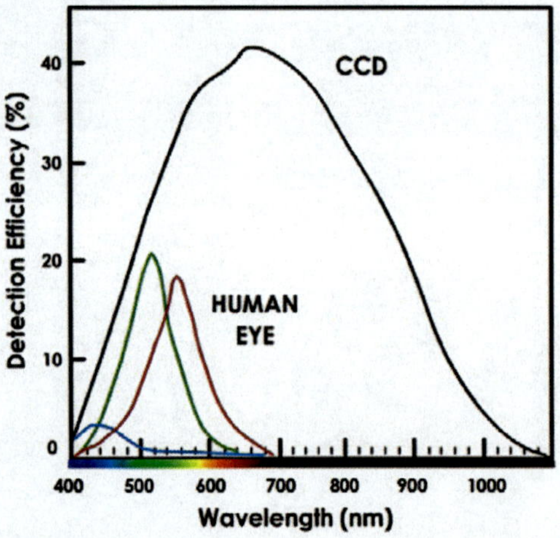

Fig. 3 Comparison response of the human eye cone cells and a CCD camera[11]

The IR blocking filter of the camera is removed in order to allow the full spectrum of visible and near infrared radiation to be perceived by the camera. To reduce the effects of external lighting, a band-pass filter with a central wavelength of 850nm and an acceptance band of 10 nm is mounted on the lens system.

The optical podoscope design is based on authors' previous work related to optical touch systems [13]. The primary illumination source consists of 2 IR LED arrays placed inside a baffle that holds a 12mm thick acrylic plate. According to the FTIR principle described in [12], the radiation enters the acrylic and is reflected internally with minimum loss. When a foot is placed on the plate, radiation is scattered towards the camera because of the higher refractive index of the new medium. The camera acquires the resulting image that represents only the pressure points with a magnitude correlated to the pixel intensity.

The secondary lighting system is constructed with 2 IR illuminators placed inside the device facing downwards at an angle of 30...40 degrees. The design of the radiation source is based on the diffused illumination method used in many multi-touch devices [11] .Each illuminator holds 48 LEDs placed at variable angles inside the illuminator case. In order to prevent uneven illumination, a diffuser foil is applied on the illuminators and the bottom of the device holds a white plate to insure the reflection of the radiation towards the acrylic foot support. Infrared light gets reflected from the bottom of the foot and the relative pixel values in the acquired image will represent the distance towards the camera. This lighting method also allows the capture of snapshots of the lower and upper limb area as seen by the camera.

A transparent transfer foil is applied on the acrylic plate in order to improve the sensitivity of the device. The foil has micro bumps that replace the air when a foot is placed on the podoscope. A protection foil is also applied on top of the device in order to maintain the integrity of the acrylic and prevent scratches on the sensitive transfer foil surface.

The whole system is enclosed in a box, 50 cm wide and 45 cm long constructed from 1.8 cm thick chipboard. The case also holds the LED power supply, the USB and power connectors

The design concept of the low cost computer aided optical podoscope is presented in figure 4.

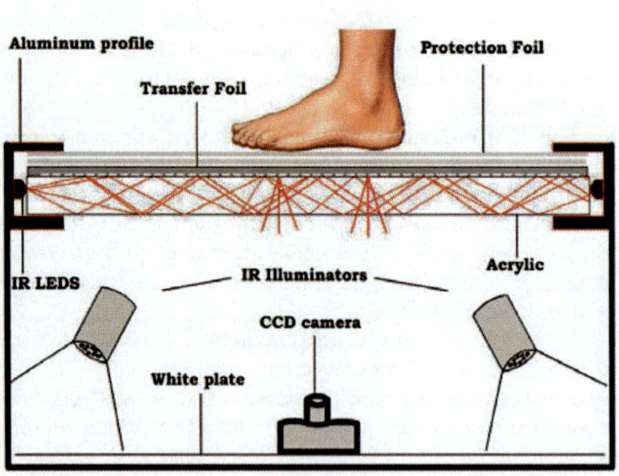

Fig. 4 Schematic of the proposed computer assisted optical podoscope

The camera is calibrated on startup and the coordinates of the tactile area are mapped to the actual pixel dimensions of the final image. Since the scanning procedure is not time critical –an important requirement in multi-touch systems- a camera with higher resolution and low frame rate is preferred. The optimal values for the resolution are at least 640 x 480 pixels and a minimum of 15 frames per second in terms of acquisition speed.

Both sources of radiation are turned on in sequence and 4 images are recorded, one for each state(no radiation, first source only, second source only, both sources on) The whole scanning procedure lasts less than 500ms and the pressure data can be interpreted by fusing multiple images. By subtracting pixel values from opposing state images, the effects of radiation noise from the environment are cancelled and only the pressure information can be displayed if necessary. Patient data along with pressure information is stored in a database for future comparison, the information can be brought on screen or printed and the application is modular and can be extended to include more options.

62 experiments were performed and images of 42 individual feet (both left and right) were extracted and processed with various load conditions.

When inspecting the visual data, false color imagery is available and the edge detection algorithms implemented in the software application are capable of calculating various lengths and angles of the foot. Based on the weight line determination, the application has been trained with several qualitative diagnostics meant to aid the orthopedist.

IV. Conclusions

Plantar pressure analysis is an efficient way to diagnose several foot disorders such as flat feet, high arches, bunion and hammertoes formation etc. While there are numerous implementations of foot pressure scanning devices or podoscopes, this paper aims to present a low cost computer assisted optical podoscope with synchronized dual illumination that reduces environmental effects and improves scanning resolution as well as providing an increased number of visualization options.

The results of the preliminary tests are encouraging and while extensive clinical studies are required to assess the viability of such a system in medical practice, the scanning capabilities of the proposed podoscope exceed other existing systems.

Future research will focus on analyzing both static and dynamic behavior for high-strain athletes as well as gait studies related to individuals with low physical activity.

Acknowledgment

This paper was supported by the project "Development and support of multidisciplinary postdoctoral programmes in major technical areas of national strategy of Research - Development - Innovation" 4D-POSTDOC, contract no. POSDRU/89/1.5/S/52603, project co-funded by the European Social Fund through Sectoral Operational Programme Human Resources Development 2007-2013.

References

1. Pedikom technical considerations (2011)- http://www.pedikom.co.uk/
2. Orlin MN, McPoil TG. Plantar pressure assessment. Phys Ther. (2000);80:399–409
3. Cavanagh PR, Hewitt FG, Perry JE. In-shoe plantar pressure measurement:a review. The Foot. (1992);2:185–194.
4. Schaff PS. An overview of foot pressure measurement systems. Clin Podiatr Med Surg. (1993);10:403– 415.
5. Brand PW, Ebner JD. Pressure sensitive devices for denervated hands and feet: a preliminary communication. *J Bone Joint Surg Am.* 1969;51:109 –116Sydney, Australia, (2003), pp 789–792
6. Harris RI, Beath T. *Army Foot Survey: An Investigation of Foot Ailments in Canadian Soldiers.* Ottawa, Ontario, Canada: National Research Council of Canada; (1947). NRC 1574.
7. Fromhertz WA. Examination. In: Hunt GC, ed. *Physical Therapy of the Foot and Ankle.* New York, NY: Churchill Livingstone Inc; (1988): 59–90.
8. Nicol K, Henning EM. Time-dependent method for measuring force distribution using a flexible mat as a capacitor. In: Komi PV, ed. *Biomechanics V-B.* Baltimore, Md: University Park Press; (1976):433–440.
9. Alexander IJ, Chao EYS, Johnson KA. The assessment of dynamic foot-to-ground contact forces and plantar pressure distribution: a review of the evolution of current techniques and clinical applications. *Foot Ankle.* (1990);11:152–167.
10. Gray, Henry. *Anatomy of the Human Body.* Philadelphia: Lea & Febiger, 1918; Online edition Bartleby.com, (2000)
11. Çetin, G., Bedi, R., Sandler, S. (Eds.) Multi-touch Technologies. 1st edition at http://www.nuigroup.com 2009
12. Jeff Han. "Low-Cost Multi-Touch Sensing through Frustrated Total Internal Reflection" *Proceedings of the 18th Annual ACM Symposium on User Interface Software and Technology* 2005
13. Septimiu Crisan, Ioan G. Tarnovan, Combined illumination method for finger and object tracking in multi-touch systems, Technological Development in a Sustainable Economy symposium, Technical University „Gheorghe Asachi" Iaşi, 2011

Author: Septimiu Crisan
Institute: Technical University of Cluj-Napoca
Street: Memorandumului 28
City: Cluj-Napoca
Country: Romania
Email: septimiu.crisan@mas.utcluj.o

Digital Microscopy Used in Synthetic Structures Analysis of Dental Prosthesis

M. Baritz[1] and D. Cotoros[2]

[1] University Transilvania Brasov, Precision Mechanics and Mechatronics Department, Brasov, Romania
[2] University Transilvania Brasov, Mechanics Department, Brasov, Romania

Abstract— **In this paper we present some aspects related to the analysis of surface quality for dental materials obtained by synthesis (photo-polymerization) process. In the first part, the biocompatibility properties of dental materials are analyzed, using Fuzzy logic. In the second part, the images obtained from the digital microscope with 500x of magnification, of some samples made of photo-polymerizable materials are analyzed using software dedicated to image processing. The results of these procedures are presented in a modular way in order to highlight both the 3D surface profile and the shape of roughness variation along the selected direction on the studied surface. In the final part of the paper we presented some conclusions both concerning the mechanical tests and images analysis.**

Keywords— **dental prosthesis, photo-polymerization process, digital microscopy.**

I. Introduction

There was a substantial revolution in dental prosthetics along with the occurrence of oral implant procedures. Oral implant allowed dental prosthetics to use additional pillars that might be inserted where necessary in the prosthetic area. Thus, some edentations that not so long ago could be approached only by mobilizable or entire mobile systems can be obtained today by fixed prosthetic restorations. [1]

The prosthetic parts can be exclusively attached on implants (pure implanter) or mixed (dental-implanter). Prosthetic works on implant may replace from a single tooth to an entire arcade.

They can be made of various materials: metal-acrylate or metal-ceramics. Many times though, the metal-ceramics works (with porcelain antagonists) are preferred especially for the integral rigidity of the structure. On the other hand the acrylic works present the benefit of shocks damping but are not resistant enough both to the forces developed during the food fragmenting and to the effect of fluid substances upon surfaces. [2]

Also on the implantation support they can perform the so called *implant prosthesis*, which is a mobilizable prosthetic device. Based on a small number of implants they may manufacture a prosthesis including special aggregation systems that provide the prosthesis an increased stability and support in comparison to the normal one. Thus, by help of implants dentists may create abutments to apply different super-structures, connected to the fixed telescopic prosthesis. At this time the fixed implantology prosthetics is dominated by bolting but prosthetic works can also be cemented. The prosthetic abutment applied on the implant represents the trans-mucous component while the implant body is covered in order to display a natural tooth aspect. But one of the most difficult issues related to the mobilizable or mobile prosthesis construction and technology is represented by the manufacturing of artificial dental arcades, with the most possible individualized occlusal shape and within its frame, selecting and assembling artificial teeth.

Artificial teeth are usually included in one of the following situations: *single maxillary mobile prosthesis*, with noble alloys as antagonists, acrylic or diacrylic resins, in order to prevent their accelerated wear; *alveolar ridge*; *rubbed off or periodontal antagonists*; or the case when there is a *single dental prosthesis or a metal bridge on the antagonist arcade*. Wherever would be the position of teeth in prosthetic area, artificial teeth are defined by characteristics related to color, shape, dimension and occlusal shape.

Besides for the frontal teeth the order of importance is color, shape, height and width. From the material point of view, the used artificial teeth are consisting of PMMA co-polymerized by reticular agents and usually these are provided with an increased resistance to cracking by using a greater amount of reticular agents. In the contact zone with the prosthetic basis, a lower concentration of reticular agent is recorded, than in the incisal, respectively occlusal areas, in order to facilitate the chemical connection to the polymers in the prosthetic basis. To provide the most physiognomic aspect, artificial teeth use a large range of pigments and to increase resistance of teeth, they are treated with inorganic micro-particles. For long-term successful performance of all dental implant types the following general factors should be considered: biomaterials, biomechanics, dental evaluation, medical evaluation, surgical requirement, healing processes, prosthodontics, laboratory fabrication, post insertion maintenance. All practitioners involved in patient care should be knowledgeable regarding these factors and their interrelationships. Standards of dental practice would suggest the following general contra-indications for the above three categories of dental implants: debilitating or uncontrolled disease, pregnancy, lack of

S. Vlad and R.V. Ciupa (Eds.): MEDITECH 2011, IFMBE Proceedings 36, pp. 230–233, 2011.
www.springerlink.com

adequate training of practitioner, conditions, diseases or treatment that severely compromise healing, e.g., including radiation therapy, poor patient motivation, psychiatric disorders that interfere with patient understanding and compliance with necessary procedures, unrealistic patient expectations, unattainable prosthodontic reconstruction, inability of patient to manage oral hygiene, patient hypersensitivity to specific components of the implant.[12]

II. Determination of Biocompatibility Based upon Fuzzy Logic

In order to determine the biocompatibility of the materials used in dental prosthetics and implantology a questionnaire was conceived, which was filled in by a human subjects' sample with prosthetic works made of the same type of acrylic material. [6]

Based on the questionnaire's answers and suing a module of the software developed on Fuzzy logic, we accomplished the analysis of the materials used in manufacturing dental prosthetics works. The result of the analysis is materialized in a graphic presenting the analyzed dental material biocompatibility by means of percents, for the selected subjects' sample.

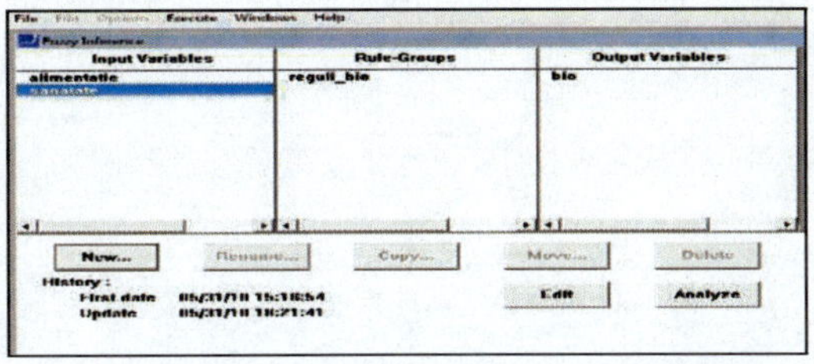

Fig. 1 Main window of biocompatibility analyze software

The first stage in biocompatibility analysis by Fuzzy logic consisted in introducing initial data in a main window (fig.1) considering the most two important causes leading to materials incompatibility: *nourishment and health* of the studied human subjects. Graphics were made at a percent scale of 1 to 100. To each of the two variables may correspond from 2 to n concepts. Thus, for the "nourishment" variable we considered as valid the following concepts: "soft food", "hard food", "acid food" and "sweets".

For the second variable "health" we established the concepts: "bad", "average" and "good". Fuzzy analysis continued with the second stage, introducing final data, where we established a single variable as being biocompatibility ("bio") with the following concepts: "null", "partial" and "total".

Fuzzy type analysis assumes the compilation of initial data and of the final ones based upon the definition of certain rules that are presented in a separate window.

Prior to starting the fuzzy analysis process we checked on software basis all the introduced data and rules to avoid the errors during analysis.

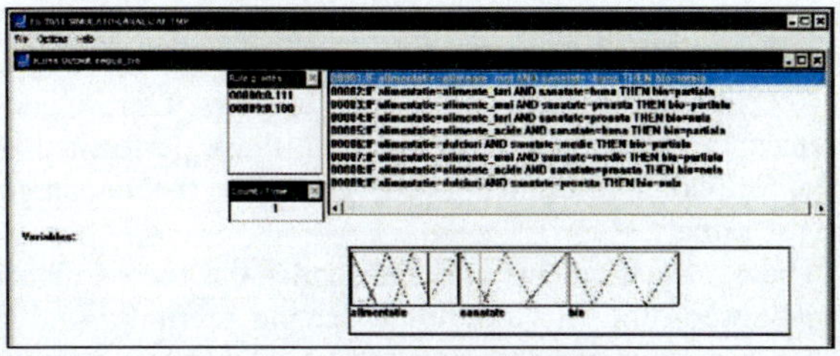

Fig. 2 Biocompatibility results

The last stage, concerning the analysis results was performed using a soft simulator that calculated based upon the numerical values and established rules, the percentage of studied material biocompatibility.

The obtained results reveal the fact that the biocompatibility level stage is remaining at low values due to the health state and nourishment style of the investigated sample of subjects.

III. Polymerization Process of Restoration Materials Samples – Microscopic Analyzes

Most of the restoration materials should withstand forces during manufacturing or mastication, so the mechanical properties are important in understanding and predicting the material behavior under load.

Because a single mechanical property cannot represent a quality measure, the application of the involved principles in a range of mechanical properties is essential, especially considering the human factor implication.

One of the most important applications in dentistry is the study of the forces applied to teeth and dental restorations. The maximum forces recorded by strain gauges and telemetry devices reach 250 to 3500N. [3]

The forces developed in the dental occlusion for an adult subject decrease from the molar area towards the incisors, reaching forces values from 400 to 800N, upon the first and second molar. Of the same importance for the study of forces developed in the natural teeth occlusion, is the determination of stress and strains in the restoration type works, such as insertions, fixed connections, partial and total prostheses.

One of the first investigations of the occlusion forces shows that average biting force in patients with replacement of first molar is determined at 250N for the restored part and 300N for the opposite side, in comparison to the average biting forces for permanent teeth, reaching 665N for molars and 220N for incisors.

In a different study, forces measured for patients with partial prostheses are form 67 to 235N. Generally, the force in women bite is 90N smaller than the one applied by a man. These studies indicate that the mastication force on the first molar with a fixed connection is approx. 40% of the force exerted by the patients with natural teeth. [10, 11]

Recent measurements performed by help of strain gauges devices are much more accurate than those performed with other previous equipments, but generally the conclusions are the same.

These measurements concluded that the forces distribution between the first premolar, second premolar and first molar in a complete dentition can be established as approx. 15%, 30%, and 55% of the normal force.

IV. EXPERIMENTAL SETUP

The experimental setup used for the microscopic analysis of the polymerizable dental materials samples consists of a digital microscope Keyence VHX-600 type, with objective magnification between 500x and 5000x, an object field of 0, 25 mm^2 and software suitable for the assessment studies and surface quality measurements, roughness, 3D representations. [4,5] The used samples were manufactured in the same conditions and assessed according to the same procedures.

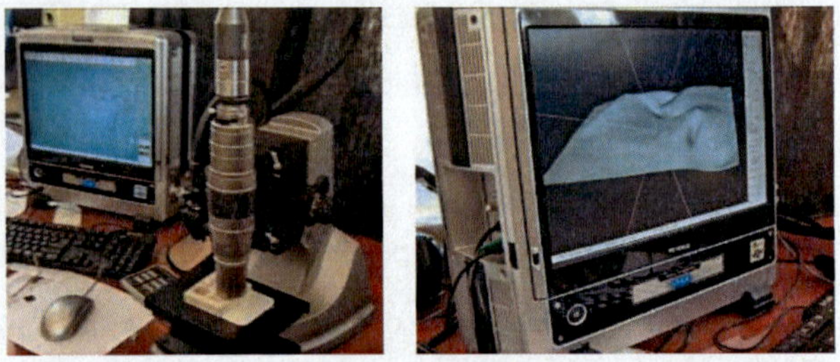

Fig. 3 Keyence VHX-600 digital microscope

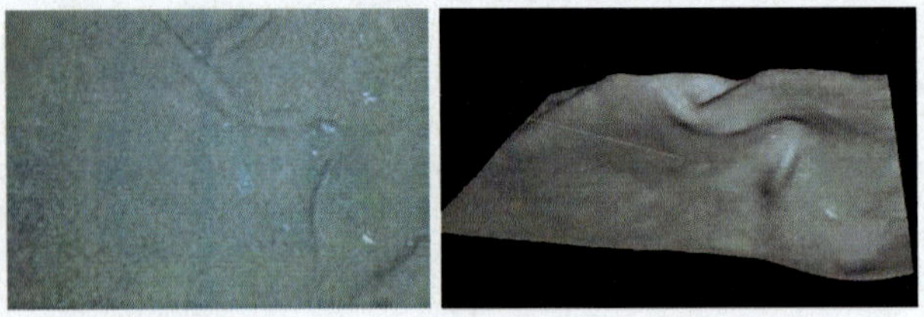

Fig. 4 Samples images from Keyence VHX-600 digital microscope

From the point of view of the polymerization process, an important aspect is represented by the polymerization time, which is a parameter affecting the mechanical characteristics of the prosthesis teeth, dental restorations or implants.

Polymerization time for the composite diacrylic polymerizable resins cannot be measured based on viscosity changes. Approximately 75% of the process takes place in the first 10 minutes, the reaction continues slowly for 24h. The sub-polymerized layer at the surface has an internal conversion ratio of approx. 25%. By comparing some materials used for artificial teeth construction we may notice that in the case of *dental acrylate* (having the following characteristics – compressive strength of 84 MPa, elastic modulus of 1700 MPa and elasticity limit of 55 MPa) this is used in dental technique offices in 80% situations unlike the ceramics materials which are present only in 20% of situations.

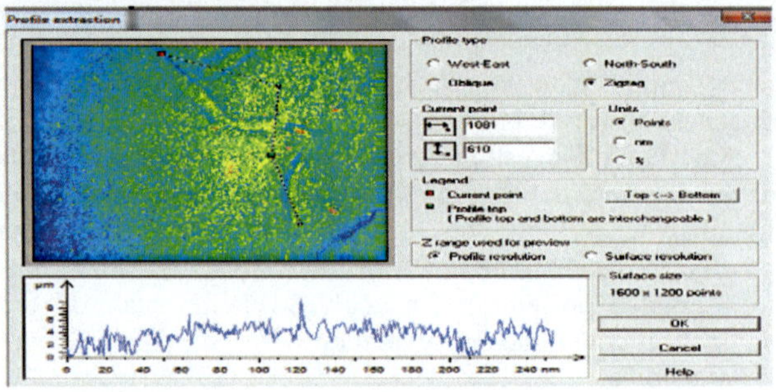

Fig. 5 TE-ECONOM structure, photo-polymerization time of 5 min (images from measurement software)

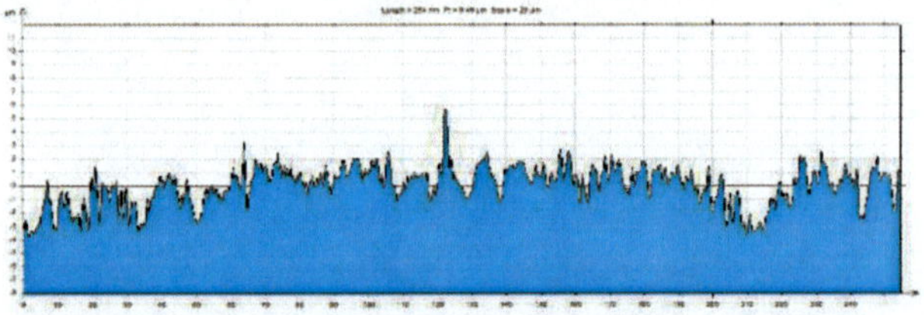

Fig. 6 Roughness profile variation

Duropont composite material (having the characteristics – compressive strength of 90 MPa, elastic modulus of 1600 MPa and elasticity limit of 45 MPa) presents a highly superior hardness to the presently used acrylates. Unlike these, the *duropont* composite polymerizes in 6 atm external pressure conditions and even if it does not show the *cromasit* hardness, the favorable price makes it the most used material for dental prosthetic works. During the performed tests we manufactured some working samples with the same size and volume. First working samples were made of **TE-ECONOM** material and were polymerized for various time intervals (5 min, 6 min, 7 min and respectively 9 min) and monitoring the photo-polymerization process in order to avoid other environmental influences.

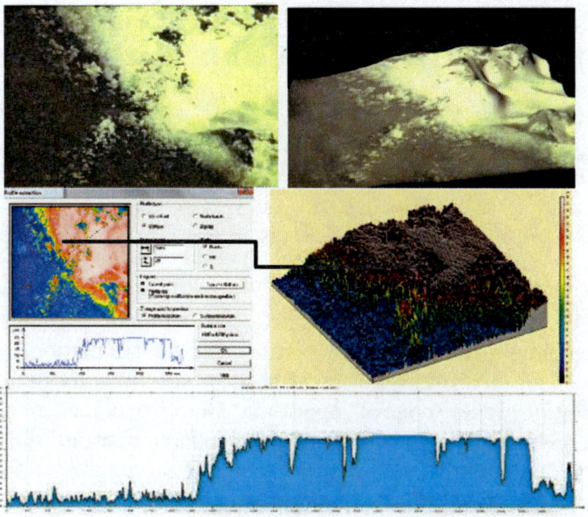

Fig. 7 **VALUX-PLUS** structure– photo-polymerization time of 5 min (500x digital microscope) and roughness profile variation in the area marked for sample **VALUX-PLUS** [9]

The second material was **VALUX-PLUS**, in photo-polymerization process, time 5 minutes and (fig.7.) From the performed measurements presented above we may observe the following: according to the materials polymerization degree we notice changes in their aspect depending on the photo-polymerization time interval; for VALUX PLUS material we observed an incomplete polymerization due to the white spots upon the material surface, while for all the TE-ECONOM samples, the photo-polymerization was uniform, there were no white spots on the material surface; the tested materials withstand very well the applied forces considering that: these analyzed materials resisted up to a 2300 N force, the equivalent of a 117 MPa strain for VALUX PLUS, respectively 2500 N, the equivalent of a 127 MPa strain for TE-ECONOM.

V. Conclusions

All these results are determined considering that the bite force of a human being may reach the maximum value of 270 N. We also noticed based on the surfaces profile analysis that the photo-polymerization process determining the best surface quality must take place along a 6 min time interval for TE-ECONOM material and respectively along 9 minutes for VALUX Plusmaterial. Analyzing the benefits of composite materials based upon resins, used as dental materials, we may find the following: they do not include Hg; due to a suitable edge adjusting and a volume constant in time they do not allow deposits in the contact area between the two materials (root and tooth); there is a biocompatibility with the human organism; they obtain very hard materials with high mechanical resistance and consequently at least 20 years life cycle; the hardness of these materials being below the one of the dental enamel it does not scratch the antagonist teeth during mastication; the hardening reaction of these materials used in the dental office for root canals takes place in a few minutes, which proves to be very comfortable to the patient.

In the future, these researches will be continued with evaluations from clinical practice and taking into account different substances which these materials get in contact.

Acknowledgment

These researches are part of the Grant PNII-IDEI 744 with CNCSIS Romania and we've developed the investigations with equipments in Mechatronic Researches Department from University Transilvania of Brasov, Romania.

References

1. Grosu L., s.a. (1983) Biosistemul orofacial,Cluj- Napoca, Ed.Dacia,.
2. Bratu D., s.a. (1994) Materiale dentare-Materiale utilizate în cabinetul de stomatologie; Editura Helicon;
3. Regenio M, et al. (2009), Stress distribution of an internal connection implant prostheses set, Stomatologija, Baltic Dental and Maxillofacial Journal, 11;
4. Cotoros, DL. et al. (2009) Aspects concerning impact tests on composites for rigid implants, WORLD CONGRESS ON ENGINEERING, London England, Pages: 1658-1661;
5. Cotoros D. (2010) Analyses by image processing of surface quality of mobile skeletal dental prosthesis, International Conference on CNC Technologies, Bucharest, Romania, May 05-07;
6. Stanciu A., Cotoros D., Baritz M., Florescu M. (2008), Simulation of Mechanical Properties for Fibre Reinforced Composite Materials, Theoretical and experimental aspects of continuum mechanics, WSEAS Cambridge, UK, ISBN 978-960-6766-38-1;
7. Ieremia L., Dociu I., (1987) Functia si disfunctia ocluzala, Editura Medicala, Bucuresti, Romania;
8. Lussi A., (2006) Dental Erosion From Diagnosis to Therapy, Copyright 2006 by S. Karger AG www.karger.com, ISSN 0077–0892;
9. http://www.digitalsurf.fr/en/index.html accessed oct.2010;
10. Albrektsson, T. Et al., (2004), Oral implant surfaces: part 1—review focusing on topographic and chemical properties of different surfaces and in vivo responses to them. Int. J. Prosthodont. 17, 536–543;
11. Anders Palmquist et al. (2010), Titanium oral implants: surface characteristics, interface biology and clinical outcome, J. R. Soc. Interface 2010, **7**, S515-S527 doi: 10.1098/rsif.2010.0118.focus;
12. Navaro M. Et al. (2008), Biomaterials in orthopaedics, *J. R. Soc. Interface* 2008 **5**, 1137-1158;

Author: Mihaela Baritz
Institute: University Transilvania from Brasov
Street: B-ul Eroilor nr.29
City: Brasov
Country: Romania
Email: mbaritz@unitbv.ro

Robustness Tests of a Model Based Predictive Control Strategy for Depth of Anesthesia Regulation in a Propofol to Bispectral Index Framework

C.M. Ionescu[1], I. Nascu[1,2], and R. De Keyser[1]

[1] Ghent University, Dept. Electrical energy, Systems and Automation, Technologiepark 913, B9052, Gent, Belgium
[2] Technical University of Cluj-Napoca, Dept. Automation, Cluj-Napoca, Romania

Abstract— **This paper verifies the robustness of a model based predictive control scheme for depth of anesthesia (DOA) regulation. The manipulated variable is Propofol, which is used in a Model based Predictive Control (MPC) algorithm for automatic induction and control of DOA. In turn, DOA is evaluated by means of the Bispectral index (BIS). The simulation tests are performed on a set of 17 virtually generated realistic patients with significantly varying sensitivity to Propofol infusion. The results show a high-efficiency, optimal dosage and robustness of the MPC algorithm to induce and maintain the desired BIS reference while rejecting typical disturbances from surgery.**

Keywords— **Depth of anesthesia (DOA), predictive control, propofol, bispectral index, robustness.**

I. Introduction

General anesthesia plays an important role in surgery and intensive care unit and requires critical assessment of induced quantities of drugs into the patient. There are three major interactive parts in anesthesia: sedation, analgesia and neuromuscular blockade.

Usually, anesthesiologists control the drug dosing during anesthesia by monitoring hemodynamic signals. This open-loop technique reaches the target level of sedation fast, but it may result in minimal values (undershoot) which are not safe for the patient. On the other hand, if the drug delivery regulation is done automatically, anesthesiologists will have more time to concentrate on critical issues that may threaten the safety of the patient. Control of anesthesia poses a manifold of challenges: multivariable characteristics, variable time delays, inter- and intra-patient variability, dynamics dependent on anesthetic substances and stability issues [1], [2]. Numerous PID controllers have been designed during decades, but since these controllers cannot anticipate the response of the patient and do not have any prior knowledge of the drug metabolism, the performances were sub-optimal. Therefore, model based strategies using fuzzy [3], adaptive [1] and predictive [4] control algorithms have been developed and applied in clinical trials.

In this paper, we present a single input (Propofol) – single output (bispectral index) model based predictive control (MPC) algorithm for controlling the depth of anesthesia. The patient models for prediction and for simulation purposes are given in the second section, followed by the description of the control algorithm. The closed loop results are given in the fourth section for induction of anesthesia and for maintenance within clinically acceptable values while rejecting typical surgery disturbances.

II. Patient Model

A. *The Pharmacokinetic-Pharmacodynamic Model*

Propofol is a hypnotic agent, for which the pharmacologic properties have been well described and studied in different kind of patients. Given its beneficial pharmacological profile, Propofol is used as one of the drugs of choice for both induction and maintenance of the hypnotic component of anesthesia and intensive care sedation . This drug is the input of the patient model and the output is the Bispectral Index (BIS), a signal derived from the electroencephalogram (EEG). Using EEG, several derived, computerized parameters like the BIS have been tested and validated as a promising measure of the hypnotic component of anesthesia [5]. BIS values lie in the range of 0-100; whereas 90-100 range represents fully awake patients; 60-70 range and 40-60 range indicate light and moderate hypnotic state, respectively. For the induction phase of DOA, a BIS value of 50 is considered suitable.

In figure 1 the pharmacokinetic (PK) – pharmacodynamic (PD) blocks denote the 4th order compartmental model for Propofol [6], [7]. Compartmental models are used to represent the distribution of drugs in the body, i.e. mass balance. In each compartment the drug concentration is assumed to be uniform, as in a perfect and instantaneous mixing:

$$\begin{aligned}
\dot{x}_1(t) &= -[k_{10}+k_{12}+k_{13}]\cdot x_1(t)+k_{21}\cdot x_2(t)\\
&+k_{31}\cdot x_3(t)+u(t)\\
\dot{x}_2(t) &= k_{12}\cdot x_1(t)-k_{21}\cdot x_2(t)\\
\dot{x}_3(t) &= k_{13}\cdot x_1(t)-k_{31}\cdot x_3(t)\\
\dot{x}_e(t) &= -k_{e0}\cdot x_e(t)-k_{1e}\cdot x_1(t)
\end{aligned} \tag{1}$$

S. Vlad and R.V. Ciupa (Eds.): MEDITECH 2011, IFMBE Proceedings 36, pp. 234–239, 2011.
www.springerlink.com

where x_1 [mg] denotes the amount of drug in the central compartment. The blood concentration is expressed by x_1/V_1. The peripheral compartments 2 and 3 model the drug exchange of the blood with well and poorly diffused body tissues. The masses of drug in fast and slow equilibrating peripheral compartments are denoted by x_2 and x_3 respectively. The parameters k_{ij} for $i \neq j$, denote the drug transfer frequency from the ith to the jth compartment and $u(t)$ [mg/s] is the infusion rate of the anesthetic drug into the central compartment. The parameters k_{ij} of the PK models depend on age, weight, height and gender and can be calculated for propofol:

$$V_1 = 4.27\ [l], V_2 = 18.9 - 0.391 \cdot (age - 53)\ [l], V_3 = 2.38\ [l]$$

$$C_{l1} = 1.89 + 0.0456(weight - 77) - 0.0681(lbm - 59) +$$

$$+0.0264(height - 177)\ [l/\min]$$

$$C_{l2} = 1.29 - 0.024(age - 53)\ [l/\min], C_{l3} = 0.836\ [l/\min]$$

$$k_{10} = \frac{C_{l1}}{V_1}\left[\min^{-1}\right], k_{12} = \frac{C_{l2}}{V_1}\left[\min^{-1}\right], k_{13} = \frac{C_{l3}}{V_1}\left[\min^{-1}\right],$$

$$k_{21} = \frac{C_{l2}}{V_2}\left[\min^{-1}\right], k_{31} = \frac{C_{l3}}{V_3}\left[\min^{-1}\right]$$

where C_{l1} is the rate at which the drug is cleared from the body, and C_{l2} and C_{l3} are the rates at which the drug is removed from the central compartment to the other two compartments by distribution. The lean body mass (lbm) for men and women have the following expressions:

$$1.1 \cdot weight - 128 \cdot \frac{weight^2}{height^2}$$

and

$$1.07 \cdot weight - 148 \cdot \frac{weight^2}{height^2},$$

respectively.

An additional hypothetical effect compartment was proposed to represent the lag between drug plasma concentration and drug response. The concentration of drug in this compartment is represented by x_e. The effect compartment receives drug from the central compartment by a first-order process and it is regarded as a virtual additional compartment. Therefore, the drug transfer frequency from the central compartment to the effect site-compartment is equal to the frequency of drug removal from the effect-site compartment: $k_{e0}=k_{1e}=0.456\ [min^{-1}]$. Knowing k_{e0}, the apparent concentration in the effect compartment can be calculated since k_{e0} will precisely characterize the temporal effects of equilibration between the plasma concentration and the corresponding drug effect. Consequently, the equation is often used as:

$$\dot{C}_e(t) = k_{e0} \cdot (C_e(t) - C_p(t)) \tag{2}$$

with C_e called the *effect-site compartment concentration*,. The BIS variable can be related to the drug effect concentration C_e by the empirical static but time varying nonlinear relationship [4], called also the *Hill curve:*

$$BIS(t) = E_0 - E_{\max} \cdot \frac{C_e(t)^\gamma}{C_e(t)^\gamma + EC_{50}^\gamma} \tag{3}$$

where E_0 denotes the baseline value (awake state - without drug), which by convention is typically assigned a value of 100, E_{max} denotes the maximum effect achieved by the drug infusion, EC_{50} is the drug concentration at half maximal effect and represents the patient sensitivity to the drug, and γ determines the steepness of the curve. The constant k_{10}

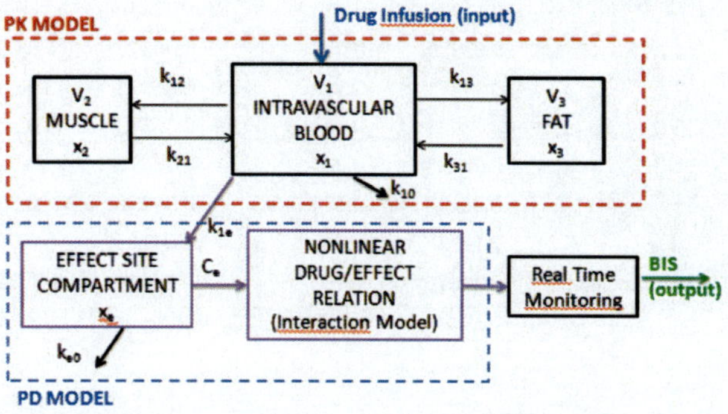

Fig. 1 Compartmental model of the patient, where PK denotes the pharmacokinetic model and PD denotes the pharmacodynamic model.

B. *Nominal Patient Model Parameters*

A set of model parameter values for the nominal patient are necessary in order to be used for prediction purposes. For comparison purposes, the same nominal patient as that given in [8] will be used in this study. The corresponding model parameters are given in Table 1.

Table 1 Values of the parameters for the nominal patient used for prediction of BIS values based on known Propofol infusion rates.

parameter	value
k_{10} (min^{-1})	0.119
k_{12} (min^{-1})	0.112
k_{21} (min^{-1})	0.055
k_{13} (min^{-1})	0.0410
k_{31} (min^{-1})	0.0033
V_1 (l)	5.05
V_2 (l)	30.6
V_3 (l)	191.1

Table 2 Values of the parameters for the 17 patient set arranged in decreasing order of the BIS sensitivity to Propofol infusion [8].

Patient no.	Parameter							
	k_{10}	k_{12}	k_{21}	k_{13}	k_{31}	k_{e0}	EC_{50}	γ
1 (sensitive)	0.08925	0.084	0.06875	0.031425	0.004125	0.459	1.6	2
2	0.14875	0.14	0.04125	0.052375	0.004125	0.239	1.6	2
3	0.14875	0.112	0.04125	0.0419	0.004125	0.239	1.6	3.122
4	0.14875	0.14	0.04125	0.052375	0.004125	0.239	1.6	3.122
5	0.08925	0.084	0.04125	0.052375	0.002475	0.459	2.65	2.561
6	0.08925	0.084	0.06875	0.031425	0.002475	0.349	2.65	2.561
7	0.14875	0.112	0.06875	0.031425	0.002475	0.459	2.65	2.561
8 (nominal)	0.119	0.112	0.055	0.0419	0.0033	0.349	2.65	2.561
9	0.119	0.112	0.055	0.0419	0.0033	0.239	2.65	2
10	0.119	0.112	0.055	0.0419	0.0033	0.239	2.65	2.561
11	0.08925	0.084	0.06875	0.031425	0.002475	0.459	3.7	2
12	0.14875	0.112	0.06875	0.031425	0.002475	0.349	3.7	2.561
13	0.08925	0.084	0.06875	0.031425	0.002475	0.239	3.7	2.561
14	0.08925	0.084	0.06875	0.031425	0.002475	0.239	3.7	3.122
15	0.08925	0.084	0.04125	0.052375	0.002475	0.239	3.7	3.122
16	0.14875	0.14	0.04125	0.052375	0.004125	0.349	3.7	2.561
17 (insensitive)	0.14875	0.14	0.04125	0.052375	0.002475	0.239	3.7	3.122

C. Population Database

The population database consisted of 17 virtually generated realistic patients, whose model parameter values are given in Table 2. These values are taken from [8] and based on a statistical analysis on inter-patient variability. These 17 patient relevant sets are arranged in the decreasing order of their BIS sensitivity to the amount of propofol infusion. For the **insensitive patient**, the rates in the central compartment k_{10}, k_{12} and k_{13} are high (0.149, 0.14 and 0.052, respectively) and the k_{21} and k_{31} rate constants are low (0.041 and 0.002, respectively). In the PD parameters, higher EC_{50} (3.7) indicates the need for more drug to get to the same hypnosis level, higher γ (3.12) indicates higher nonlinearity (slope in the Sigmoid Hill curve) and lower k_{e0} (0.239) indicates sluggishness in response. For the **sensitive patient** k_{10}, k_{12} and k_{13} are low (0.089,0.084 and 0.031, respectively) and the k_{21} and k_{31} rate constants are high (0.069 and 0.004, respectively). In the PD parameters, lower EC_{50} (1.6) indicates the need for less drug to get to the same hypnosis level, lower γ (2) indicates lower nonlinearity (slope in the Sigmoid Hill curve) and higher k_{e0} (0.459) indicates faster response. With these patients at hand, robustness of the control algorithm can be tested.

III. THE EPSAC-MPC APPROACH

A. Controller Design

In the general MPC scheme represented in figure 2, the patient model is used to predict the current value of the output variable (BIS). The difference between the measured BIS from the patient and the model output (residual), serves as feedback signal in the prediction block. With this residual and the input *u*, the prediction block predicts the future values of the output BIS. On the basis of these predicted BIS values, the controller calculates the future optimal infusion rates over a number of samples in the future, called the prediction horizon. However, only the first calculated sample is applied to the process (i.e. principle of receding horizon).

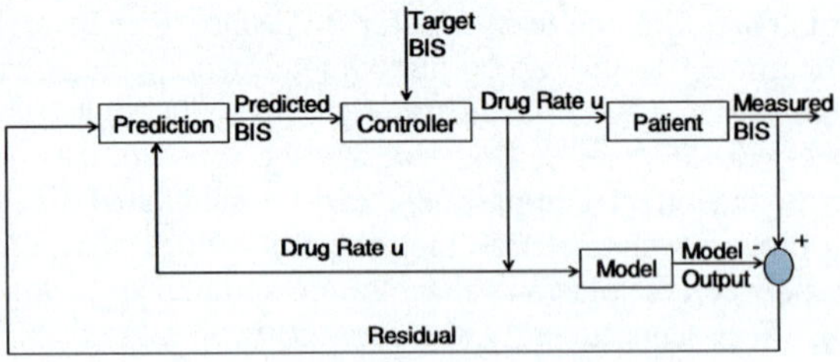

Fig. 2 MPC scheme for closed loop BIS regulation

In this paper, we apply the EPSAC (Extended Prediction Self-Adaptive Control) strategy described in detail in [9]. The EPSAC-MPC is based on a generic process model:

$$y(t) = x(t) + n(t) \tag{4}$$

The disturbance $n(t)$ includes the effects in the measured output $y(t)$ which do not come from the model input $u(t)$ via the available model. These non-measurable disturbances have a stochastic character with non-zero average value, which can be modeled by a colored noise process:

$$n(t) = \left[C(q^{-1})/D(q^{-1})\right]\cdot e(t) \tag{5}$$

with: $e(t)$ - uncorrelated (white) noise with zero mean value; $C(q^{-1})$ and $D(q^{-1})$ - monic polynomials in the backward shift operator q^{-1} of orders n_c and n_d. In this application, the disturbance filter $C(q^{-1})/D(q^{-1})$ is defined as a pure integrator. The relationship between $u(t)$ and $x(t)$ is given by the generic dynamic system model:

$$x(t) = f\left[x(t-1), x(t-2), \cdots, u(t-1), u(t-2), \cdots\right] \tag{6}$$

In our case the input applied to the patient, $u(t)$, represents the Propofol delivery rate. The model output is then represented by:

$$x(t) = m_1 \cdot C_{eProp}(t - T_d) + m_2 \cdot C_{eRem}(t - T_d) \tag{7}$$

The process output is predicted at time instant t over the prediction horizon N_2, based on the measurements available at that moment and the future outputs of the control signal. The predicted values of the output are:

$$y(t+k/t) = x(t+k/t) + n(t+k/t) \tag{8}$$

Prediction of $x(t+k|t)$ and of $n(t+k|t)$ can be done respectively by recursion of the process model and by using filtering techniques on the noise model (5) [9]. In EPSAC for linear models, the future response is considered as being the cumulative result of two effects:

$$y(t+k/t) = y_{base}(t+k/t) + y_{opt}(t+k/t) \tag{9}$$

where $y_{base}(t+k/t)$ represents:

- effect of past control $\{u(t\text{-}1), u(t\text{-}2), \ldots\}$ (initial conditions at time t);
- effect of a *base* future control scenario, called $u_{base}(t+k\,|\,t),\ k \geq 0$, which is defined *a priori*; for linear systems the choice is irrelevant, a simple choice being $\{u_{base}(t+k\,|\,t) \equiv 0,\ k \geq 0\}$;
- effect of future (predicted) disturbances $n(t+k|t)$.

while, $y_{opt}(t+k/t)$ represents:

- effect of the *optimizing* future control actions $\{\delta u(t\,|\,t),\ \delta u(t+1\,|\,t),\ \ldots\ \delta u(t+N_u-1\,|\,t)\}$ with $\delta u(t+k\,|\,t) = u(t+k\,|\,t) - u_{base}(t+k\,|\,t)$. The *design* parameter N_u, called the *control horizon* (a well-known concept in MPC-literature), is considered in this paper equal to 1.

The controller output is obtained by minimizing a cost function. The basic cost function is:

$$J(\mathbf{U}) = \sum_{k=N_1}^{N_2}\left[r(t+k/t) - y(t+k/t)\right]^2 \tag{10}$$

where $r(t+k/t)$ is the desired *reference trajectory*. The cost function (10) is a quadratic form in **U**, which leads after minimization w.r.t. **U** to the optimal solution:

$$\mathbf{U}^* = \left[\mathbf{G}^T\cdot\mathbf{G}\right]^{-1}\mathbf{G}^T\cdot(\mathbf{R} - \bar{\mathbf{Y}}) \tag{11}$$

with **R** the reference trajectory,

$$\bar{\mathbf{Y}} = \left[Y_{base}(t+N_1/t)\ldots\ldots Y_{base}(t+N_2/t)\right]^T$$

$$\mathbf{U} = \left[\delta u(t/t)\ldots\ldots\delta u(t+N_u-1/t)\right]^T \text{ and}$$

$$\mathbf{G} = \begin{bmatrix} gProp_{N_1} \\ \ldots \\ gProp_{N_2} \end{bmatrix}$$

where $gProp_{N_1}$ .... $gProp_{N_2}$ are the coefficients of the unit step response of the PK-PD propofol model. The EPSAC strategy was implemented to control the value of BIS, using the PK-PD model presented from (1), (2) and (3). The prediction model was used with the nominal values from table 1, while each patient was simulated with the values from table 2. In this way, significant modeling errors are introduced in the control scheme, accounting for inter-patient variability. In this study, the MPC control parameters are set to: N_1=1, N_u=1 and a sampling time for control action every 5 seconds.

B. *Performance Evaluation*

In order to evaluate the performance in the closed loop, we introduce the index defined by

$$IAE = \int_0^T \left|R(\tau) - BIS(\tau)\right| d\tau \tag{12}$$

with $R(\tau)$ the desired reference BIS value, in this case set to BIS=50, BIS(τ) the real output of the patient and τ the time instant. Notice that the index defined in (12) is known as the integral absolute error (IAE).

Next, it is important to introduce realistic disturbances, in this case typical surgery stimuli. A standard stimulus profile has been defined as in figure 3, whereas each interval denotes a specific event in the operation theatre. Hence, stimulus A mimics the response to intubation; B represents surgical incision followed by a period of no surgical stimulation (i.e. waiting for pathology result); C represents an abrupt stimulus after a period of low level stimulation; D shows onset of a continuous normal surgical stimulation; E,F, and G simulate short-lasting, larger stimulation within the surgical period; and H simulates the withdrawal of stimulation during the closing period.

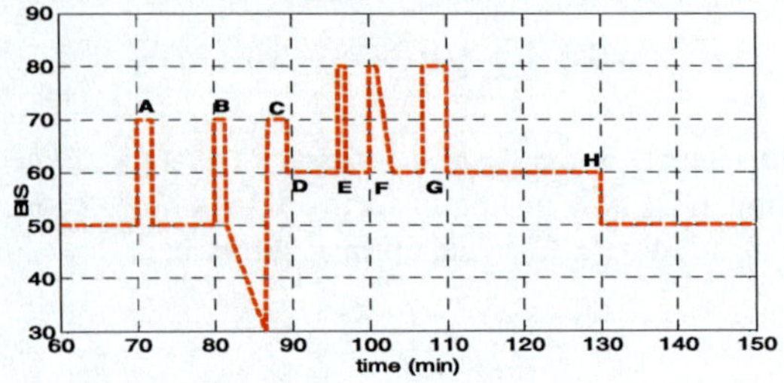

Fig. 3 The artificially generated disturbance signal

IV. Results

A. *Induction Phase*

Ideally, the induction of the patient in an operational DOA is required to be done as soon as possible, such that few time is lost before the surgeon can start. It is therefore desirable that the patient reaches the BIS=50 target and remains within the target value without much undershoot, i.e. values below BIS=30 should be avoided. A predictive controller can be either i) tuned such that the control effort is very significant, thus the response will be fast, at the cost of undershoot values; either ii) tuned in a conservative manner, such that the response is smooth but slow. This is achieved by tuning the value of the N_2 parameter (i.e. the prediction horizon). Figures 4 and 6 depict the results for prediction horizon of N_2=35 and N_2=12, respectively. Since our pool of patients varies significantly in the degree of sensitivity to Propofol according to the nonlinear relation from (3), it is clear that to obtain a good result for the entire set of patients, one must have a conservative controller, which brings smoothly the C_e values to their optimal levels, as concluded from figures 5 and 7, respectively.

The corresponding IAE values for each patient for the induction phase with the aggressive and the conservative MPC strategies have the mean and standard deviation 198.14±84.68 respectively 163.41 ±49.19.

B. *Maintenance Phase*

During the maintenance phase, it is important that the controller rejects the disturbances as soon as possible,

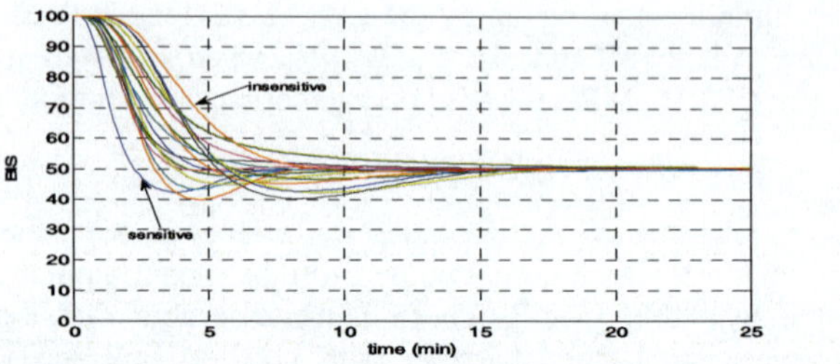

Fig. 4 Closed loop response of BIS during the induction phase with the conservative controller, for N_2=35

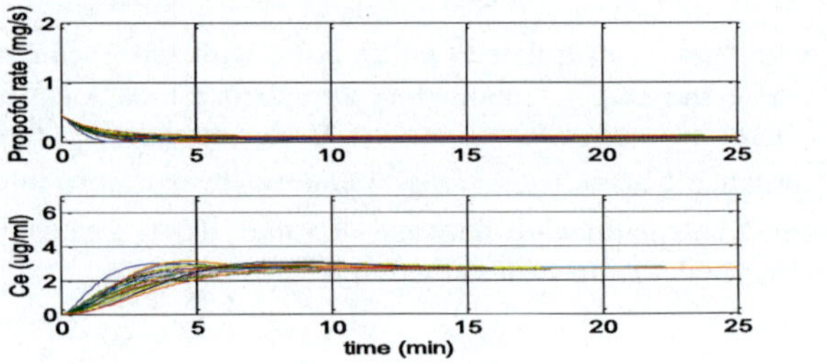

Fig. 5 Closed loop response of propofol rate and $C_e$during the induction phase with the conservative controller, for N_2=35

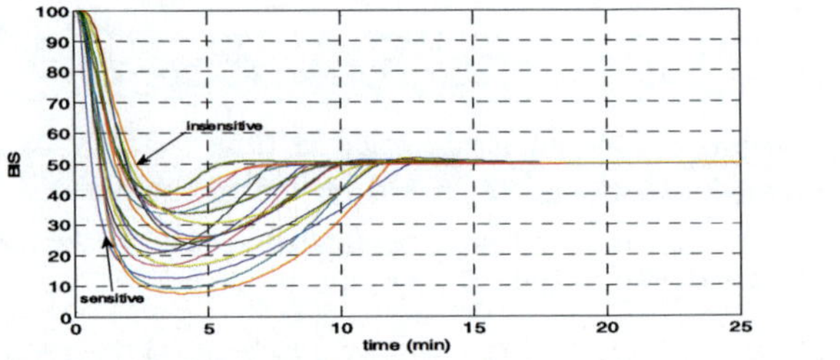

Fig. 6 Closed loop response of BIS during the induction phase with the fast/aggressive controller, for N_2=12

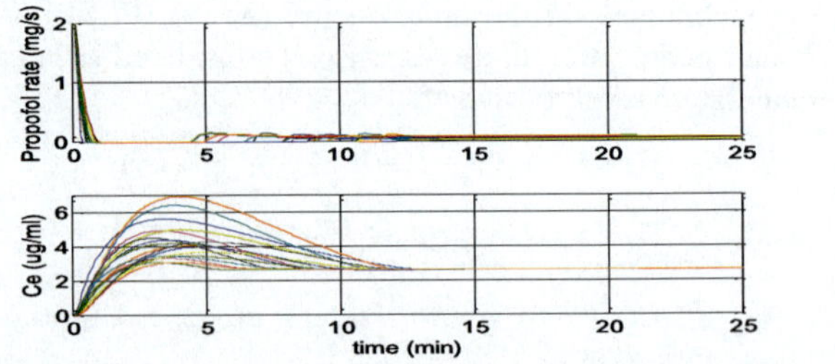

Fig. 7 Closed loop response of propofol rate and C_e during the induction phase with the aggressive controller, for N_2=12

keeping the patient within the BIS interval of 40 to 60 BIS values (thus around the BIS=50 value). The controller results for rejecting the disturbances defined in figure 3, are given in figures 8 and 10, for the two control designs, respectively. It can be observed that the aggressive controller presented in figure 10 has a faster response to the disturbances but the undershoot is considerably greater than the case of the conservative controller. During this phase we can say that we obtain better performances if we use the conservative controller

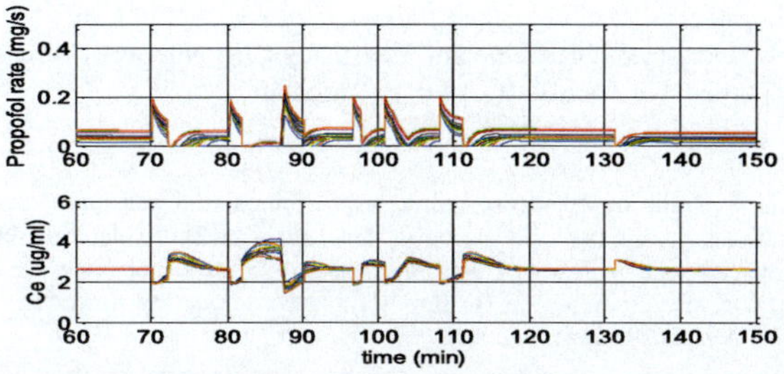

Fig. 8 Closed loop response of BIS during the maintenance phase with the conservative controller, for N_2=35

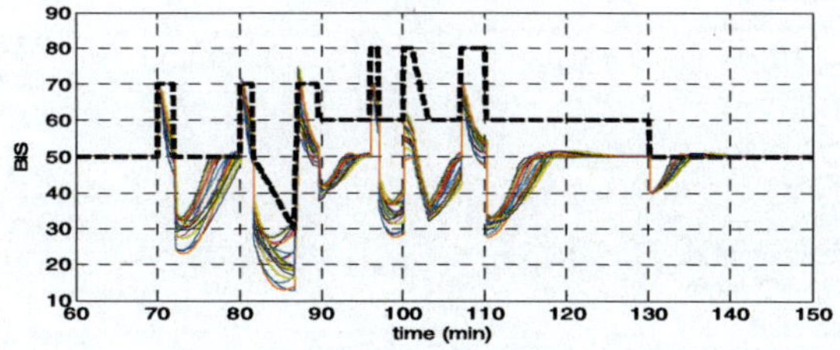

Fig. 9 Closed loop response of propofol rate and C_e during the maintenance phase with the conservative controller, for N_2=35

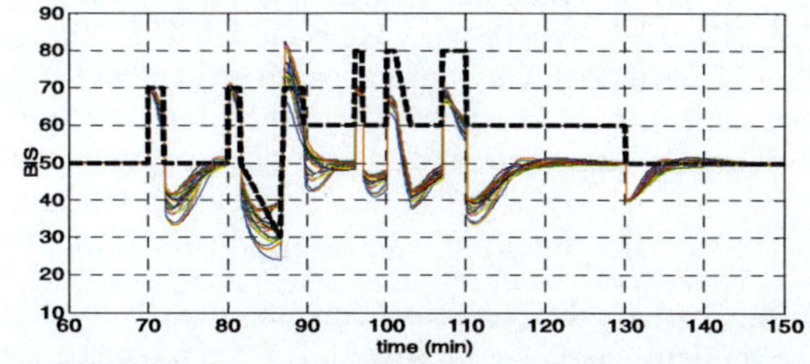

Fig. 10 Closed loop response of BIS during the maintenance phase with the aggressive controller, for N_2=12

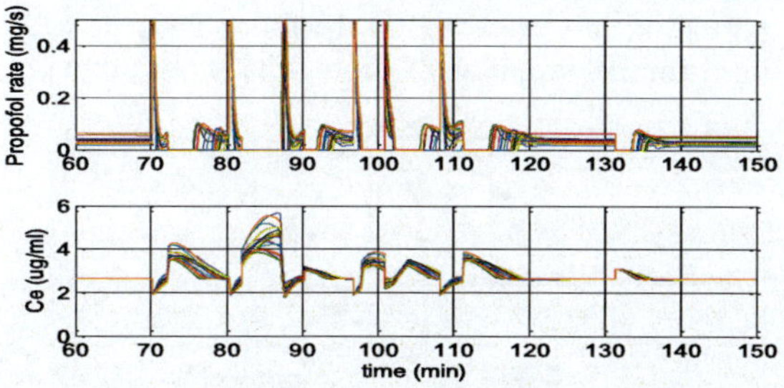

Fig. 11 Closed loop response of propofol rate and C_e during the maintenance phase with the aggressive controller, for N_2=12

V. Conclusions

In this paper, we discuss the robustness of a predictive anesthesia control strategy applied to a set of 17 virtually generated realistic patients. The patients pose significant variation in the sensitivity to the applied drug rates and the controller shows good performance despite strong inter-patient variability. The disturbance rejection tests are done on a realistic disturbance signal, capturing typical events from the operation theatre. Overall, the tuning of the predictive control algorithm in terms of the prediction horizon plays a crucial role in defining the controller speed in the closed loop paradigm.

References

1. W. Haddad, T. Hayakawa, J. Bailey, "Nonlinear adaptive control for intensive care unit sedation and operating room hypnosis", *American Control Conference*, 1808-1813, 2003
2. M. Struys, H. Vereecke, A. Moerman et al;, "Ability of the bispectral index, autoregressive modeling with exogenous input-derived auditory evoked potentials and predicted propofol concentrations to measure patient responsiveness during anesthesia", *Anesthesiology*, 99:802-812, 2003
3. M. Curatolo, M. Derighetti, S. Petersen-Felix, P. Feigenwinter, M. Fisher, A Zbinden, "Fuzzy logic control of inspired isoflurane and oxygen concentrations using minimal flow anesthesia", Bristish J of Anesthesia, 76: 245-250, 1996
4. C.M. Ionescu, R. De Keyser, B. Torrico, T. De Smet, M. Struys, J. E. Normey-Rico, "Robust predictive control strategy applied for propofol dosing using BIS as a controlled variable", *IEEE Trans Biomed Eng*, 55: 2161-2170, 2008
5. M. Struys T. De Smet, S. Greenwald, A. Absalom, S. Binge, E. Mortier, "Performance evaluation of two published closed-loop control systems using bispectral index monitoring: a simulation study", Anesthesiology, 100: 640-647, 2004
6. T. W. Schnider, C. F. Minto, P. L. Gambus, C. Andresen, DB Goodale, EJ Youngs, "The influence of method of administration and covariates on the pharmacokinetics of Propofol in adult volunteers", *Anesthesiology*, 88: 1170-1182,
7. T. W. Schnider, C. F. Minto, P. L. Gambus, C. Andresen, DB Goodale, EJ Youngs, "The influence of age on Propofol pharmacodynamics", *Anesthesiology*, 90:1502-16, 1999
8. S. Yelneedi, L. Samavedham, G.P. Rangaiah, "Advanced control strategies for the regulation of hypnosis with propofol", *Ind Eng Chem Res*, 48: 3880-3897, 2009
9. R. De Keyser, "Model based predictive control for linear systems", UNESCO Encyclopaedia of Life Support Systems, 6.43.16.1, Eolss Publishers Co Ltd, Oxford, 2003

Author: Clara M. Ionescu
Institute: Ghent University
Street: Technologiepark 913, B9052
City: Gent
Country: Belgium
Email: ClaraMihaela.Ionescu@UGent.be

An Overview on Mathematical Models of Human Crystalline Lens

S. Talu[1], S. Giovanzana[2], S.D. Talu[3], and M. Talu[4]

[1] Faculty of Mechanics / Department of Descriptive Geometry and Engineering Graphics, Technical University of Cluj-Napoca, Romania
[2] DAUR – Laboratory of Design Tools and Methods in Industrial Engineering, University of Padova, Padova 35100, Italy
[3] Faculty of Medicine / Department of Ophthalmology, "Iuliu Hatieganu" University of Medicine and Pharmacy Cluj-Napoca, Romania
[4] Faculty of Mechanics / Department of Applied Mechanics, University of Craiova, Romania

Abstract— **To describe the human crystalline lens, mathematical models are required. Advanced mathematical models are applied for human vision studies and biomechanical behavior of the crystalline lens. The accurate modeling of the crystalline lens is important in the development of intraocular lenses. This paper presents an overview of researches for human crystalline lens modeling using mathematical models.**

Keywords— **human visual system, lens geometry, computational lens models.**

I. Introduction

The human lens is an elastic capsule containing cellular tissue of non-uniform gradient index. This index is difficult to measure and makes it difficult to deal with human lens parameters, especially for posterior surface of the lens that is difficult to image since it is seen through all the previous layers. The lens grows continually throughout life, with new epithelial cells forming at the equator, all the properties of the lens, are then age-dependent.

The first one to use spherical curve to describe the lens shape was Gullstrand [1] in his eye model but from the first experimental data [2] it was seen that an aspherical coefficient must be taken into account also for the optical zones of the lens that are about 5 mm diameter for anterior surface and 4 mm diameter for posterior surface.

In Table 1 some experimental results for anterior and posterior human crystalline lens radii and asphericity are given, the earlier experiments present average data, instead more recent experiments are age-dependent.

Moreover it must be underlined that modern imaging techniques, e.g. magnetic resonance [3, 4], are able to produce 3D image of the lens against previous imaging techniques, e.g. Scheimpflug lamp [5], which are able to produce one single meridian image.

The accurate modeling of the crystalline lens is important in research of intraocular lenses, so mathematical models of the lens based on geometrical constraints derived by the images must be used.

In this paper an overview of the most known mathematical models used for modeling the human crystalline lens is presented and analyzed.

Table 1 Radii of curvature, conic asphericities and relevant standard deviations of anterior and posterior lens experimentally determined by various authors. Number of eye and age with related ranges or standard deviation are also reported (D stands for accommodation power in dioptres, A stands for age in years)

Author	Anterior radius (mm)	Anterior asphericity	Posterior radius (mm)	Posterior asphericity
Lowe and Clark (1973), in vivo	11.26	—	—	—
Howcroft and Parker (1977), in vitro	7.3 ± 0.3	-1.08 ± 9.412	-5.4 ± 0.1	-0.12 ± 1.74
Glasser and Campbell (1999), in vitro	$4.32 + 0.068 \cdot A$	-1	$-3.14 - 0.05 \cdot A$	-1
Dubbelman et al. (2001), slit lamp	$12.9 – 0.057 \cdot A$	-4 ± 4.7	$-6.5 - 0.017 \cdot A$	-3 ± 3.6
Rosen et al. (2006), in vitro	$7.5 + 0.046 \cdot A$	-0.8 ± 1.7	-5.5	-1.1 ± 1.5

In chapter II six different models are presented: conic model, figuring conicoid model, Hermans conic patch model, Kasprzak hyperbolic cosine model, Urs 10th–order Fourier series model, Giovanzana parametric model.

In chapter III are presented the results for curvature and longitudinal spherical aberration analysis for each model.

II. Materials and Methods

In this part six different mathematical models used to describe the lens shape are presented. All the models are rotationally symmetric along the optical axis z, and their form is presented in the yz-plane where y is the distance from the optical axis and z is the sagitta.

All the models presented in the following part will refer to Fig. 1 for the definition of geometrical constraints.

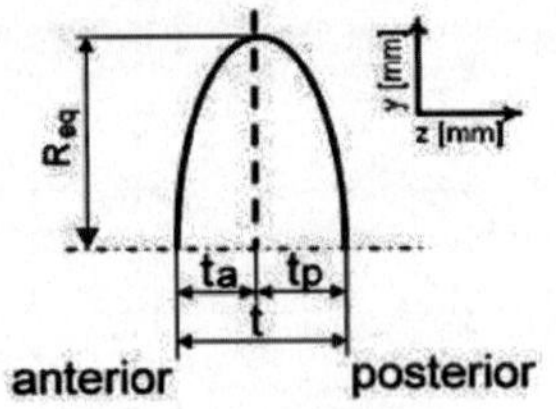

Fig. 1 Reference frame definition for a crystalline human lens

S. Vlad and R.V. Ciupa (Eds.): MEDITECH 2011, IFMBE Proceedings 36, pp. 240–245, 2011.
www.springerlink.com

A. Conic Model

The most known and used model for describing an optics revolution surface, e.g. cornea, crystalline lens and also ophthalmic lens, is the conic model. Its general equation is:

$$y^2 = 2Rz - (1 + q)\cdot z^2 \tag{1}$$

where R is radius of curvature and q is the aspheric coefficient. This formula is really important in optometry since the radius of curvature could be linked to the power of the surface described by equation (1). To derive these two coefficients a least square method is usually adopted to minimize the distance between the model and the raw points extracted by the imaging of the lens.

To describe the lens surface equation (1) must be modified adding two constraints, one for anterior and one for posterior part of the lens. These constraints are the anterior thickness t_a, and the posterior thickness t_p.

These geometrical constraints are derived from the image of the lens if the equator is clearly visible in the image, or are taken as a ratio of the total thickness, t, of the lens.

The final form of the conic model for anterior and posterior part of the lens in term of sagittal z is then:

$$y^2 = 2R_a\cdot(z + t_a) - (1 + q_a)\cdot(z + t_a)^2 \tag{2a}$$

$$y^2 = 2R_p\cdot(z - t_p) - (1 + q_p)\cdot(z - t_p)^2 \tag{2b}$$

where R_a is taken positive and R_p is taken negative; the form used in equation (2a) and (2b) let the lens equator stands on the y-axis.

B. Figuring Conicoid Model

The conic model itself does not provide a smooth junction of anterior and posterior part at the equator. This happens only for particular combinations of R and q.

To overcome this problem the figuring conicoid model is introduced [6]. The addition of figuring terms to equation (1) was already used in lens design [7] to modify the lens out of the optic zone.

In Smith's original formulation the number of figuring terms is not specified in a following work [8] the same work group define a number of three figuring conicoid, two of them linked to the geometry of the lens and one taken as free parameters assessed with least square method.

The final form of the figuring conicoid model for anterior and posterior part of the lens in term of sagittal z is then:

$$y^2 = 2R_a\cdot(z + t_a) - (1 + q_a)\cdot(z + t_a)^2 + v_{1a}\cdot(z + t_a)^3 + v_{2a}\cdot(z + t_a)^4 + v_{3a}\cdot(z + t_a)^5 \tag{3a}$$

$$y^2 = 2R_p\cdot(z - t_p) - (1 + q_p)\cdot(z - t_p)^2 + v_{1p}\cdot(z - t_p)^3 + v_{2p}\cdot(z - t_p)^4 + v_{3p}\cdot(z - t_p)^5 \tag{3b}$$

where v_{1a}, v_{2a} and v_{3a} are the figuring terms for anterior part and v_{1p}, v_{2p} and v_{3p} are the figuring terms for posterior part. Three conditions must be satisfied at the equator of coordinate (0, R_{eq}):

1. y at the equator must be the same for equation (3a) and (3b)
2. to have a smooth joint at the equator dy/dz = 0 when z = 0
3. to have a smooth joint at the equator also $(d^2y/dz^2)_a = (d^2y/dz^2)_p$

Solving these conditions for the equation (3a) and (3b) lead to the following equations for five parameters:

$$v_{1a} = (4R_{eq}^2 + 2q_at_a^2 + v_{3a}t_a^5 + 2t_a^2 - 6R_at_a)/t_a^3 \tag{4}$$

$$v_{2a} = -(3R_{eq}^2 + q_at_a^2 + 2v_{3a}t_a^5 + t_a^2 - 4R_at_a)/t_a^4 \tag{5}$$

$$v_{1p} = -(4R_{eq}^2 + 2q_pt_p^2 - v_{3p}t_p^5 + 2t_p^2 + 6R_pt_p)/t_p^3 \tag{6}$$

$$v_{2p} = -(3R_{eq}^2 + q_pt_p^2 - 2v_{3p}t_p^5 + t_p^2 + 4R_pt_p)/t_p^4 \tag{7}$$

$$v_{3p} = (q_p - q_a + 3v_{1p}(2t_p - t_a) + 6v_{2p}(2t_p - t_a)^2 - 3v_{1a}t_a - 6v_{2a}t_a^2 - 10v_{3a}t_a^3)/(10(2t_p - t_a)^3) \tag{8}$$

as it can be seen the free parameter v_{3a} can be found using a least square method.

C. Hermans Conic Patch Model

Another approach used by Hermans et al. [4] and based on the works of Dubbleman and Van der Heijde [5] and Hermans et al. [9], is to use different conic patches for the different zones of the lens. Their model is a parametric curve with a set of parameters that have a physical meaning; being formed by an anterior and posterior conic function as:

$$y^2 = 2R_a\cdot(z + t_a) \tag{9a}$$

$$y^2 = 2R_a\cdot(z - t_p) \tag{9b}$$

where the anterior part (9a) stops when y = 2.5 mm (the radius of anterior optical zone) and the posterior one (9b) when y = 2 mm (the radius of posterior optical zone). These functions are then linked together by other two conics for anterior and posterior part that join at the equator plane, and are of the form:

$$y = R_{eq} - \frac{cz^2}{1 + \sqrt{1 - kc^2z^2}} \tag{10}$$

where R_{eq} is the equatorial radius of the lens, k = 1 + q is the asphericity coefficient and c is the curvature of the conicoid junction.

To assure a continuous junction and a continuous derivative between the different conic patches c and k for anterior and posterior surface must satisfied the following equations:

$$c = \frac{R_{eq} - y_p}{z_p\left[\frac{(R_{eq} - y_p)y_p}{r} + z_p\right]} \quad (11)$$

$$k = \frac{z_p\left[\frac{2(R_{eq} - y_p)y_p}{r} + z_p\right]}{(R_{eq} - y_p)^2} \quad (12)$$

where the terms (z_p, y_p) stands for the interception of the junction zones with the anterior or posterior optic zone.

D. *Kasprzak Hyperbolic Cosine Model*

Only few authors try to use different equations to describe both the anterior and posterior lens surface by a unique function, such as Kasprzak [10] hyperbolic cosine functions, Urs et al. [11] 10th–order Fourier series.

Kasprzak model is particularly interesting since it has a continuous curvature profile. This model uses a function to describe shape of the lens, given in polar coordinates:

$$\rho(\theta) = \rho_a(\theta) + \rho_p(\theta) + t/2 \quad (13)$$

where $\rho(\theta)$ is the distance from the lens centre along the angle θ that varies from 0 to π. The function has anterior and posterior contribution given by:

$$\rho_a(\theta)=(a_a/2)\cdot[\cosh((\pi - \theta)b_a) - 1]\cdot[1 - \tanh(m(s_a - \theta))] \quad (14a)$$

$$\rho_p(\theta)=(a_p/2)\cdot[\cosh(\theta b_p) - 1]\cdot[1 + \tanh(m(s_p - \theta))] \quad (14b)$$

where a and b can be derived by the lens radius and thickness t and the hyperbolic tangent terms give an appropriate cut–off function that reduce the mutual influence of the anterior and posterior contribution. The coefficients m and s describe the slope and the shift of the hyperbolic tangent. But Kasprzak do not give a method to evaluate the coefficients expect least square method.

E. *Urs 10th–Order Fourier Series Model*

As for Kasprzak model the Urs one [11] uses trigonometric functions instead of conics.

The Fourier series in a general form is written as:

$$\rho(\theta) = \sum_{n=0}^{10} (b_n)\cos(n\theta) \quad (15)$$

where the anterior thickness is found when $\theta = \pi$, the posterior thickness for $\theta = 0$ and the equator for $\theta = \pi/2$.

A remarkable difference between the Urs model and the Kasprzak one is that she was able to assess the b_n coefficients and create then an age-dependent model so that equation (15) becomes:

$$\rho(\theta) = \sum_{n=0}^{10} (A_{n1} + A_{n2} \times age)\cos(n\theta) \quad (16)$$

In Table 2 are given the results of Urs et al. [11] experiment with the values of the A_n parameters, valid in the age range from 20 to 69. Other studies from Schachar [12, 13] demonstrate that out of this range the trend of radius of curvature and thickness may be different from Urs et al.

Table 2 Fourier coefficients of the age-dependent Fourier model

Parameter	A_{n1}	A_{n2}
A_0	2.6466	$812.11\cdot10^{-5}$
A_1	0.2246	$170.62\cdot10^{-5}$
A_2	– 0.97938	$-297.37\cdot10^{-5}$
A_3	0.010573	$-34.901\cdot10^{-5}$
A_4	0.37993	$-26.276\cdot10^{-5}$
A_5	– 0.032321	$1.6647\cdot10^{-5}$
A_6	– 0.16846	$69.192\cdot10^{-5}$
A_7	0.027934	$-9.5571\cdot10^{-5}$
A_8	0.066522	$-42.251\cdot10^{-5}$
A_9	– 0.014232	$1.7295\cdot10^{-5}$
A_{10}	– 0.021375	$18.638\cdot10^{-5}$

F. *Giovanzana Parametric Model*

A parametric model that is able to assess the coefficients according to geometrical and optical constraints is the model proposed by Giovanzana et al. [14] and then revised in [15]. This model is based on Chien et al. model [16] but with the addiction of such constraints the coefficients can be assessed without any numerical method, e.g. least square.

The anterior part of the lens can be described by parametric function for $\pi/2 < u < \pi$ as:

$$z_a(u) = (b_{0a} + b_{1a}(\pi - u)^2 + b_{2a}(\pi - u)^4)\cdot\cos(u) \quad (17a)$$

$$y_a(u) = a_a\cdot\sin(u) \quad (17b)$$

while the posterior part of the lens can be written for $0 < u < \pi/2$ as:

$$z_p(u) = (b_{0p} + b_{1p}u^2 + b_{2p}u^4)\cdot\cos(u) \quad (18a)$$

$$y_p(u) = a_p\cdot\sin(u) \quad (18b)$$

To derive the coefficients a_i and b_i, geometrical constraints was imposed adopting the following considerations:

1. y at the equator must be the same for equation (17b) and (18b)
2. in the optical axis the value of the equation (17a) must be equal to thickness of the anterior side t_a

3. in the optical axis the value of the equation (18a) must be equal to thickness of the posterior side t_p

Also the following optical constraints are considered:

4. in the optical axis the curvature radius of the anterior side R_a must be respected
5. in the optical axis the curvature radius of the posterior side R_p must be respected
6. 6th–order Taylor expansion series of equation (17a) must be equal to the 6th–order Taylor expansion series of equation (2a) of the conic model
7. to ensure continuity in the equatorial plane the same radius of curvature for both sides of the lens was imposed

In particular the constraint number 6 allow to entangle the aspheric coefficient, typical of conic model, into the parametric model. Solving these conditions for the equation (17) and (18) lead to the following equations for eight parameters:

$$a_a = a_p = R_{eq} \quad (19)$$

$$b_{0a} = t_a \text{ and } b_{0p} = t_p \quad (20)$$

$$b_{1a} = 1/2\cdot(t_a - R_{eq}^2/R_a) \text{ and } b_{1p} = 1/2\cdot(t_p + R_{eq}^2/R_p) \quad (21)$$

$$b_{2a} = 5/24\cdot t_a - 1/12\cdot R_{eq}^2/R_a - (1 + q_a)/8\cdot R_{eq}^4/R_a^3 \quad (22)$$

$$b_{2p} = b_{2a} + 2\cdot(\pi^2 + 8)/\pi^4\cdot(t_a - t_p) - 2R_{eq}^2/\pi^2\cdot(1/R_a + 1/R_p) \quad (23)$$

These set of constraints are linked to the contour extremis of the lens profile, by thickness t and equatorial radius R_{eq}; to optical properties of the central zone, by the radius of curvature R; to the whole optical zone, by the aspherical coefficient q; and let the model to have a continuous curvature along the whole profile.

G. Methods

To analyse these models data in Smith et al. [8] are used. In particular are used as reference data of Kasprzak model since as already said the author don't give any method to find the model coefficients. Geometrical and optical constraints of the lens are summarised in Table 3. In particular in [8] the values of the asphericity coefficient q are not provided and are then founded with a least square method.

Table 3 Lens constraints

R_a	q_a	R_p	q_p	t	t_a / t_p	R_{eq}
5.431	– 0.078	– 4.454	– 0.61	5.014	0.874	3.992

Curvature analysis is performed on the different models. The radius of curvature is given by next formula:

$$R = \frac{\left((z_u)^2 + (y_u)^2\right)^{3/2}}{z_u y_{uu} - z_{uu} y_u} \quad (24)$$

where R is the radius of curvature and y and z are parameterised respect to u.

III. Results

Using the data of Table 3 the conic model can be found. Applying formula (24) the curvature data is found and plotted against the distance from the optical axis (Fig. 2).

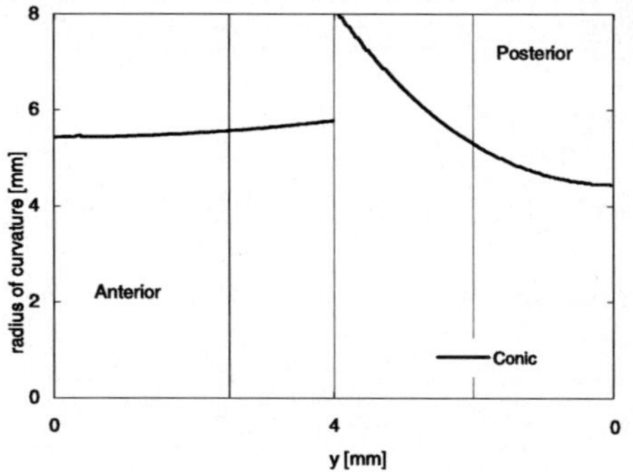

Fig. 2 Radius of curvature of the conic model versus y direction

Anterior part of the lens is on the left side from 0 to 4, and posterior one is on the right part from 4 to 0. The vertical lines in the plot represents from left to right the anterior optic zone, the equator, the posterior optic zone.

Both side of the lens present an increasing value or the radius of curvature according to the asphericity value.

It can be also seen the jump in curvature at the equator.

The figuring conicoid model parameters data are summarised in Table 4. In particular the value of parameter v_{3a} is found with a least square method, the rest of parameters are found according to formula in section II.B.

Table 4 Figuring conicoid model parameters

Parameter	Anterior surface	Posterior surface
q	– 0.078	– 0.61
v_1 (mm^{-1})	– 0.432650	0.179712
v_2 (mm^{-1})	0.142355	0.000749
v_3 (mm^{-1})	– 0.045000	0.000934

In Fig. 3 the radius of curvature for the figuring conicoid model is plotted against the distance from the optical axis.

It can be immediately seen that curvature of the model is continuous and in both optic zones the curvature is close to the conic model.

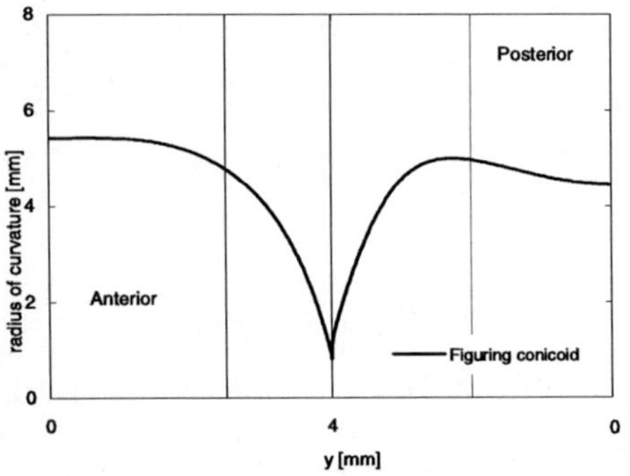

Fig. 3 Radius of curvature of the figuring conicoid model versus y direction

Hermans conic patch model parameters are summarised in Table 5. The parameters are concerning only the junction zones since both the optical zones have q = –1.

Table 5 Hermans model parameters

Parameter	Anterior surface	Posterior surface
c (mm^{-1})	0.786	0.672
k	0.309	0.245

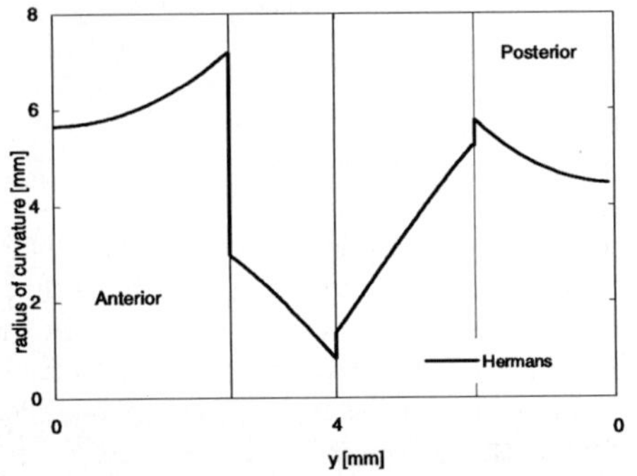

Fig. 4 Radius of curvature of the Hermans model versus y direction

In Fig. 4 the radius of curvature of the Hermans conic patch model is plotted. As it can be seen the curvature in the optical zones does not follow the conic model due to different asphericity coefficient. Moreover the model presents a curvature jump for every zone passage.

In Table 6 the data for Kasprzak model taken from [8] are summarised. All the coefficients are derived by least square method.

Table 6 Kasprzak model parameters

Parameter	Anterior surface	Posterior surface
a	0.965	0.977
b	1.186	1.061
s	1.808	1.669
m	3.706	

In Fig. 5 the radius of curvature for the Kasprzak model is plotted. It can be seen that this model presents a continuous curvature all along the lens shape.

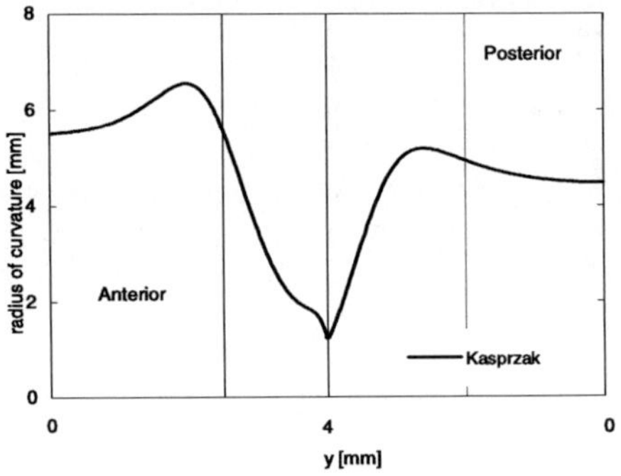

Fig. 5 Radius of curvature of the Kasprzak model versus y direction

Urs model depends only by age from [8] is known that the lens analysed is a seven year old ex vivo lens it is decided to take the age input data for Urs parameters as 7.

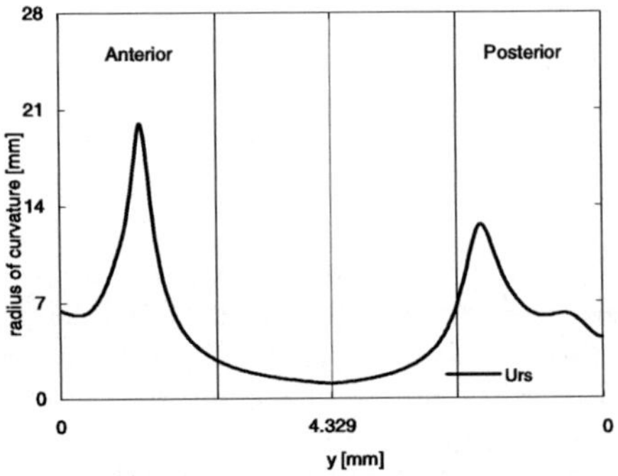

Fig. 6 Radius of curvature of the Urs model versus y direction

Radius of curvature for the Urs model is plotted in Fig. 6. As it can be seen the model presents much more variability against the other models.

In Table 7 the parameters for Giovanzana model are found using as input the data of Table 3.

Table 7 Giovanzana model parameters

Parameter	Anterior surface	Posterior surface
b_0	2.339	2.675
b_1	−0.298	−0.451
b_2	0.060	0.067

Fig. 7 presents the radius of curvature of Giovanzana model.

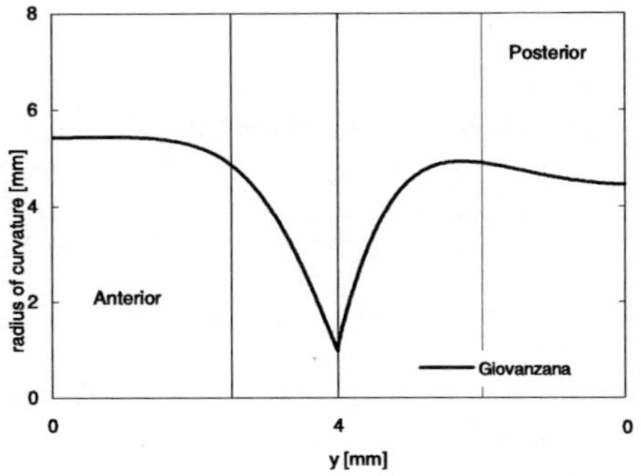

Fig. 7 Radius of curvature of the Giovanzana model versus y direction

As it can be seen the model is continuous all along the surface and is close to the conic model, in the optic zones, and to the figuring conicoid model all over the surface.

IV. Discussion and conclusion

Several simple models such as circular, parabolic or conic sections are usually adopted to describe the surfaces of the human crystalline lens.

More complex mathematical models such as figuring conicoid [8], conic patch [4], hyperbolic cosine [10], Fourier series [11], parametric curves [13] use more coefficients typically derived by fitting techniques [17], or derived from geometrical and optical constraints.

Figuring conicoid, Kasprzak, Urs and Giovanzana models present a continuous curvature that is important for models that can be used in FEM analysis.

Kasprzak and Urs models present more difficulties to assess the coefficients, the first one since is not given by the author any method to find them; the second since they are presented only dependent by age.

Figuring conicoid model and Giovanzana model are very similar in shape and in the way they assess the coefficients. However Giovanzana model is able to fully derive the coefficients by geometrical and optical constraints.

Both these models presents a radius of curvature similar to the conic model in the optical zone and this is for sure important for the optical properties of these two models that can be easily used then to modelised a human lens more similar to the real shape of the lens.

References

1. Gullstrand A (1909) Helmholz's handbuch der physiologichen optic, vol. 1, 3rd edition, english translation edited by Southall JP, 1924 Optical of America
2. Lowe R F, Clark B A (1973) Posterior corneal curvature. J Ophthalmlol 57: 464–470
3. Jones C E, Atchison D A, Meder R, Pope J M (2005) Refractive index distribution and optical properties of the isolated human lens measured using magnetic resonance imaging (MRI). Vision Res 45 (18): 2352–2366
4. Hermans E, Pouwels P J, Dubbelman M, Kuijer J P A, Van der Heijde R G L, Heethaar R M (2009) Constant volume of the human lens and decrease in surface area of the capsular bag during accommodation: an MRI and Scheimpflug study. Invest Ophth Vis Sci 50 (1): 281–289
5. Dubbelman M, Van der Heijde R G L (2001) The shape of the aging human lens: curvature, equivalent refractive index and the lens paradox. Vision Res 41: 1867–1877
6. Smith G (2003) The optical properties of the crystalline lens and their significance. Clin Exp Optom 86 (1): 3–18
7. Atchison D A (1992) Spectacle lens design: a review. Appl Opt 31 (19): 3579–3585
8. Smith G, Atchison D A, Iskander D R, Jones C E, Pope J M (2009) Mathematical models for describing the shape of the in vitro unstretched human crystalline lens. Vision Res 49: 2442–2452
9. Hermans E, Dubbleman M, Van der Heijde R G L, Heethaar R M (2007) The shape of the human lens nucleus with accommodation. J Vision 7 (10): 1–10
10. Kasprzak H T, Iskander D R (2006) Approximating ocular surfaces by generalised conic curves. Ophthal Physl Opt 26: 602–609
11. Urs R, Ho A, Manns F, Parel J (2010) Age–dependent Fourier model of the shape of the isolated ex vivo human crystalline lens. Vision Res 50: 1041–1047.
12. Schachar R A (2004) Central Surface Curvatures of Postmortem-Extracted Intact Human Crystalline Lenses. Ophthalmology 111 (9): 1699–1704
13. Schachar R A (2005) Growth patterns of fresh human crystalline lenses measured by in vitro photographic biometry. J Anat 206: 575–580
14. Giovanzana S, Savio G, Meneghello R, Concheri G (2011) Shape analysis of a parametric human lens model based on geometrical constraints. J Mod Opt: 1–11 DOI 10.1080/09500340.2011.554895
15. Giovanzana S (2011) A Virtual Environment for modeling and analysis of human eye. University of Padova
16. Chien M C, Tseng H, Schachar R A (2003) A mathematical expression for the human crystalline lens. Compr Ther 29: 245–258
17. Xirui Z, Yuhua P, Xiaobing W, Aizhen L (2008) Modeling of human crystalline lens. The 2nd Int. Conference on Bioinformatics and Biomedical Engineering (ICBBE 2008), Shanghai, China, pp. 1787–1791

Author: Stefan Talu
Institute: Faculty of Mechanics / Department of Descriptive Geometry and Engineering Graphics, Technical University of Cluj-Napoca
Street: B-dul Muncii Street no. 103-105
City: Cluj-Napoca
Country: Romania
Email: stefan_ta@yahoo.com

Mathematical Analysis of the Human Crystalline Lens in Giovanzana Parametric Model

S. Talu[1], S. Giovanzana[2], M. Talu[3], and S.D. Talu[4]

[1] Faculty of Mechanics / Department of Descriptive Geometry and Engineering Graphics, Technical University of Cluj-Napoca, Romania
[2] DAUR – Laboratory of Design Tools and Methods in Industrial Engineering, University of Padova, Padova 35100, Italy
[3] Faculty of Mechanics / Department of Applied Mechanics, University of Craiova, Romania
[4] Faculty of Medicine / Department of Ophthalmology, "Iuliu Hatieganu" University of Medicine and Pharmacy Cluj-Napoca, Romania

Abstract— **The objective of this paper is to present a mathematical analysis of the human crystalline lens in Giovanzana parametric model. This model can serve to improve computational modeling, such as finite element modeling of the human crystalline lens.**

Keywords— **human crystalline lens, lens geometry, geometrical constraint.**

I. Introduction

The human crystalline lens, one of the most complex optical elements of the eye appearing as biconvex lens, is located behind iris, pupil and in front of vitreous body [1].

In the last decades an important research effort has been directed to understand different aspects of the human crystalline lens, ranging from the physical to the biological process involved [2, 3, 4, 5].

There is much interest in mathematical analysis and computational modeling of human crystalline lens [6, 7, 8].

Numerical modeling requires access to high quality data on geometric and material properties of the lens.

Chien et al. [8] proposed and analysed twelve different mathematical lens models. The best function that fits Fincham's lens data [9] from photomicrography is:

$$z(u) = (b_0 + b_1 u^2 + b_2 u^4)\cdot \cos u$$
$$y(u) = a \cdot \sin u \tag{1}$$

where u is the parameter ranging between 0 to $\pi/2$, z is the direction along the optical axis, y is the distance from the optical axis in radial direction, a and b_i are coefficients which are found with a fitting process.

The parametric model proposed is based on Chien et al. model [8] in which the coefficients a and b_i are derived from geometrical constraint and not by fitting methods.

This method is close to the one in [10] but it differs for the change of a constraint that implies different optical properties of the final model.

Three real lens images are then analysed comparing the proposed model with the lens shape. Merit function and curvature data are shown.

II. Materials and Methods

Following Chien et al. model [8] the anterior part of the lens can be described considering parametric function for $\pi/2 < u < \pi$:

$$z_a(u) = (b_{0a} + b_{1a}(\pi - u)^2 + b_{2a}(\pi - u)^4)\cdot \cos u$$
$$y_a(u) = a_a \cdot \sin u \tag{2}$$

while the posterior part of the lens can be written for $0 < u < \pi/2$ as:

$$z_p(u) = (b_{0p} + b_{1p}u^2 + b_{2p}u^4)\cdot \cos u$$
$$y_p(u) = a_p \cdot \sin u \tag{3}$$

where u substitution with $(\pi - u)$ allows to mirror the equation (2) and simultaneously gives a continuous passage through the junction zone due to parameterization.

This definition is compatible both with a canonical mathematical definition of the angles where the angles grow counter clockwise and with the lens disposition in an optical system so that the anterior part of the lens stands on the left and the posterior part stands on the right of the equatorial plane that pass through the origin of the axis.

Any geometrical constraints, such as the vertical tangent in the optical axis or the horizontal tangent in the junction zone at the equator, were already proposed by Chien et al. [8] in order to identify possible model equations.

To derive the eight coefficients a_i and b_i, eight geometrical constraints was imposed adopting simple considerations.

We now show the geometrical constraints used to define the first set of parameters.

To respect the lens shape, both the equation at the equator (where $z_a = z_p = 0$) must be equal to the lens equatorial radius R_{eq}, than $y_a(\pi/2) = y_p(\pi/2) = R_{eq}$.

S. Vlad and R.V. Ciupa (Eds.): MEDITECH 2011, IFMBE Proceedings 36, pp. 246–251, 2011.
www.springerlink.com

Taking into account the lens shape, in the optical axis ($y_a = y_p = 0$) the z_a value of the functions (equation 2) must be equal to thickness of the anterior side t_a, while the z_p value of the functions (equation 3) must be equal to thickness of the posterior side t_p, than $z_a(\pi) = t_a$ and $z_p(0) = t_p$.

These set of constraints are linked only to the contour extrema (thickness and equatorial radius) of the lens profile, so that we have call them geometrical constraints.

By the first condition and the second condition the coefficients a_a, a_p, b_{0a} and b_{0p} were easily found:

$$a_a = a_p = R_{eq} \tag{4}$$

$$b_{0a} = t_a \text{ and } b_{0p} = t_p \tag{5}$$

To ensure the optical properties, in the optical axis ($y_a = y_p = 0$) the curvature radius of anterior side R_a and posterior side R_p must be respected, so that $r(\pi) = R_a$ and $r(0) = R_p$.

Calculating $r(\theta)$ by the formula in differential geometry for $u = 0$ and $u = \pi$ the following relationships were obtained:

$$r(\pi) = a_a^2/(b_{oa} - 2b_{1a}),\ r(0) = a_p^2/(b_{op} - 2b_{1p}) \tag{6}$$

Replacing in equation (6) a_a, a_p, b_{0a} and b_{0p}, derived from equations (4) and (5), the coefficients b_{1a} and b_{1p} were found:

$$b_{1a} = \frac{1}{2}\left(t_a - \frac{R_{eq}^2}{R_a}\right) \tag{7}$$

$$b_{1p} = \frac{1}{2}\left(t_p + \frac{R_{eq}^2}{R_p}\right) \tag{8}$$

The previous constraints ensure a good trend of the optical zone only at a first order of approximation.

To let the model follow more the shape of the lens is necessary to expand the equation in Taylor series.

The explicit equation of the model for anterior part can be expressed as:

$$z_a = -\left(b_{0a} + b_{1a} a\sin\left(\frac{y}{a}\right)^2 + b_{2a} a\sin\left(\frac{y}{a}\right)^4\right) \cdot \frac{\sqrt{a^2 - y^2}}{a} \tag{9}$$

where the function (a·sin(y/a)) it has been explicitated.

On the other hand the surfaces of the lens are usually approximated in the literature as conicoid [11]:

$$z_a = \frac{y^2}{R_a + \sqrt{R_a - (1+q_a)y^2}} - t_a \tag{10}$$

where the term t_a gives the displacement of the conicoid so that the equator stands in $z = 0$.

The conicoid has an important feature when that function is used to fit an image that can be the imaging of the crystalline lens or of the cornea or other optical surfaces. The term R is the curvature radius of the central zone when $y = 0$, so that it is linked to a punctual characteristic of the surface. The term q is the aspherical coefficient as already pointed out and it is linked to the size of the fitting zone [6]. This characteristic of the q coefficient is for sure not ideal if we want to fit a single conicoid on the lens surface but for our purpose is important since it is linked to a whole region of the lens and not only to a single point.

If we in fact expand in Taylor's series the aspheric function till the 6th term we obtain:

$$z_a = -t_a + \frac{1}{2R_a} y^2 + \frac{1+q_a}{8R_a^3} y^4 \tag{11}$$

and if we do it for the Chien equation [8] we obtain:

$$z_a = -b_{0a} + \left(\frac{b_{0a}}{2a^2} - \frac{b_{1a}}{a^2}\right) y^2 + \left(\frac{b_{0a}}{8a^4} + \frac{b_{1a}}{6a^4} - \frac{b_{2a}}{a^4}\right) y^4 \tag{12}$$

If we want that in a region close to the optical axis, corresponding to the fitting zone, the model follows the aspheric equation.

We can easily found the coefficients b_{0a} and b_{1a} which meaning is already explained before in (11) and (12).

Instead equating:

$$\left(\frac{b_{0a}}{8a^4} + \frac{b_{1a}}{6a^4} - \frac{b_{2a}}{a^4}\right) = \frac{1+q_a}{8R_a^3} \tag{13}$$

we can find the last coefficient b_{2a}:

$$b_{2a} = \frac{5}{24} t_a - \frac{1}{12} \frac{R_{eq}^2}{R_a} - \frac{1+q_a}{8} \frac{R_{eq}^4}{R_a^3} \tag{14}$$

as it can be seen this constraints is closely linked to the shape of the whole anterior surface, by the use of the q coefficient.

On the other hand if we would like to construct the aspheric function from the model we can find the asphericity coefficient q:

$$q = \frac{5}{3}\frac{R_a{}^3 t_a}{R_{eq}{}^4} - 8\frac{R_a{}^3 b_{2a}}{R_{eq}{}^4} - \frac{2}{3}\frac{R_a{}^2}{R_{eq}{}^2} - 1 \quad (15)$$

where the terms referred to the anterior part of the lens may be substituted with the ones referred to the posterior one to find the posterior asphericity.

We have now to assess the last coefficient, to ensure continuity in the equatorial plane ($z_a = z_p = 0$) the same radius of curvature for both sides of the lens was imposed:

$$r_a(\pi/2) = r_p(\pi/2) \quad (16)$$

in the equatorial plane, where u = π/2, the radius of curvature is given by:

$$r_a\left(\frac{\pi}{2}\right) = \frac{(z_a'(\pi/2))^2}{y_a''(\pi/2)} \quad (17)$$

$$r_p\left(\frac{\pi}{2}\right) = \frac{(z_p'(\pi/2))^2}{y_p''(\pi/2)} \quad (18)$$

One can demonstrate that the derivate of $y_a''(\pi/2) = y_p''(\pi/2) = -R_{eq}$ so that the equation (20) becomes after some substitutions:

$$-b_{0p} - \frac{\pi^2 \cdot b_{1p}}{4} - \frac{\pi^4 \cdot b_{2p}}{16} = -b_{0a} - \frac{\pi^2 \cdot b_{1a}}{4} - \frac{\pi^4 \cdot b_{2a}}{16} \quad (19)$$

and as a result:

$$b_{2p} = b_{2a} + 2\frac{\pi^2 + 8}{\pi^4}(t_a - t_p) - 2\frac{R_{eq}{}^2}{\pi^2}\left(\frac{1}{R_a} + \frac{1}{R_p}\right) \quad (20)$$

These set of constraints are linked to optical properties of the central zone, by the radius of curvature R; to the whole optical zone, by the aspherical coefficient q; and let the model to have a continuous curvature along the whole profile, so that we have call the optical constraints.

Moreover, all the geometrical and optical constraints involved in the model are often available in the literature and frequently several equation relate this parameter with accommodation and age [5, 11].

The proposed model is compared against shadow-photogrammetry imaging provided by R. Urs and F. Manns of the Ophthalmic Biophysics Center, Bascom Palmer Eye Institute, University of Miami Miller School of Medicine, Miami, USA, testing shape, volume and curvature [7].

The images referred to three lenses relevant to a 20, 42 and 63 year old in vitro lens (Fig. 1).

Fig. 1 Human crystalline lens of 42 year old men acquired by shadow-photogrammetry imaging

For the optical constraints a fitting process was used.

Over a defined diameter a conics was fitted on the lens contour using a least square method.

In particular was used the MATLAB built-in function [12] that use the Levenberg-Marquardt algorithm [13, 14, 15] which solves nonlinear least squares curve fitting problems of the form:

$$\min \left\| f(r,q) \right\|_2^2 = 0 \quad (21)$$

as MATLAB optimization toolbox explains, with this method is possible to evaluate the anterior and posterior radius, R_a and R_p respectively, and asphericity coefficient, q_a and q_p respectively. Data for both geometrical and optical constraints are summarised in Table 1.

Table 1 Constraints adopted to define the shape of the lens adopting the proposed model, to compare it with the other models mentioned in the first column (values in mm)

Lens contour	R_a	R_p	q_a	t_a	t_p	t	R_{eq}
20	7.207	4.884	−2.862	1.580	2.465	4.045	4.447
42	8.877	5.968	−1.477	1.908	2.377	4.285	4.662
63	14.015	6.327	2.8	1.869	2.646	4.515	5.108

The shape of the proposed model is compared with the 20, 42 and 63 year old lens contour, plotting above the lens profile the magnified normal deviation. To summarize this results it was adopted the merit function proposed in Smith et al. [6], which is the sum of the square error evaluated along the normal to the lens profile of the proposed model.

The merit function was performed in the optical zones, in the junction zones and in all the lens profile. Moreover, maximum and average distances between models are calculated.

A solid of revolution was created and then the volume is assessed. The radius of curvature was calculated using the method presented in [11] and it is plotted against the radial y–axis.

III. Results

Figure 2 shows the results of the normal deviation, with its sign, of the proposed model fitted on different age lens contour, magnification is 20 times. It can be seen that, 20 year old and 42 year old contours are fitted better then 63 year old contour, if we do not consider the red zone in the 20 year old model that should be arise from an error in the Canny filter the maximum deviation is lower 0.06 mm over the two contours.

The 63 year old model presents much more deviations then the other two with a maximum deviation of 0.175 mm. This may arise due to contour that is less clear then the other models, and also the fact that there are imbalances between the superior and inferior parts of the lens. The superior one presents less deviation then the inferior one. Further investigations have shown that the difficult in the contour evaluation with Canny filter may introduce errors during the rotation of the lens, leading to imbalances and fitting errors.

Looking at the anterior (|y| < 2.5 mm) and posterior (|y| < 2 mm) optic zone (the y amplitude is selected following the work of Hermans et al. [5]) we can see that the proposed model is very close to both the 20 and 42 year old with deviations of the order of 10^{-5}, for the 63 year old model the deviation is one order more.

From an optical point of view this means that the proposed model is close to fitting conics in both the anterior and posterior optic zone leading to a situation where there optic difference between the proposed model and the standard conics model is not as high as it will be seen in the chapter IV.

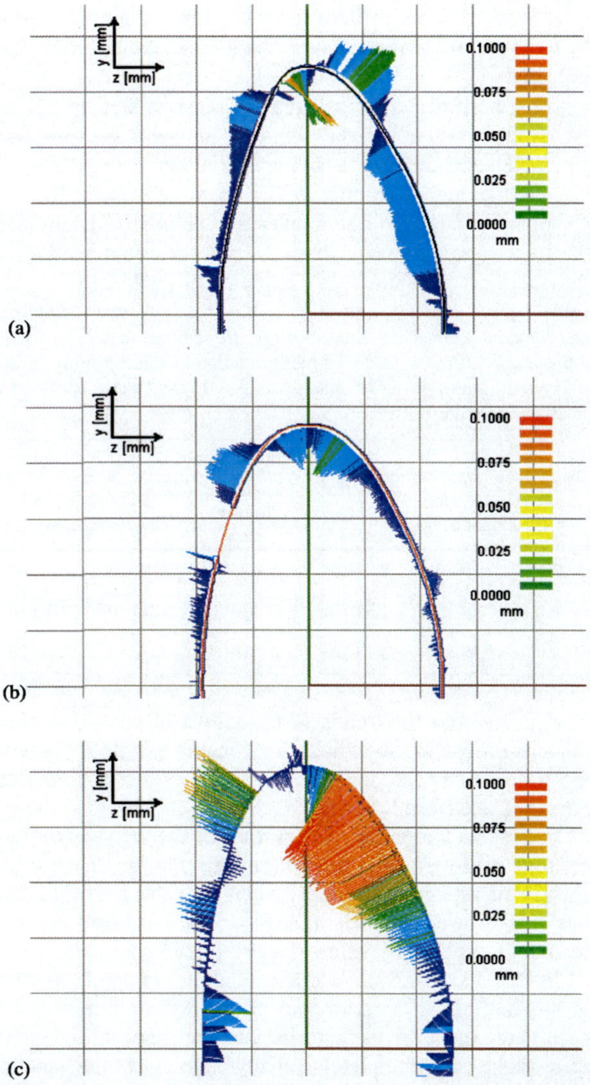

Fig. 2 Comparison between the shape of the proposed model and different lens contours: (a) 20 year old, (b) 42 year old, (c) 63 year old. The normal distances between the models, magnified 20 times, are depicted above the profile of the proposed model.

In Table 2 the deviations between the model and the lens contours are quantitatively summarized by the merit function for the whole lens and for its portions, together with the maximum and the average distances. Analysing these results similar conclusion can be drawn:

- the merit function between the proposed model and the 42 year old lens contour is the least in all the profile, and in the different sections except the anterior optic zone

- the merit function between the proposed model and the 20 year old lens contour is the least in the anterior optic zone
- the merit function assumes highest values in all the profile and in all the sections between the proposed model and the 63 year old lens contour
- average and maximum deviations confirm that the proposed model fits better over the 42 year old lens contour, so mid-age lens, then the other extrema age lenses

Table 2 Merit function, maximum and average deviation of the proposed model against the lens contours. In addition to results for the entire lens profile, the values of merit function in anterior optic zone ($z < 0$, $y < 2.5$ mm), posterior optic zone ($z > 0$, $y < 2$ mm), and the two junction zones are shown

Lens contour	Merit function [mm^2]				
	Anterior optic zone	Anterior junction zone	Posterior junction zone	Posterior optic zone	All profile
20	$1.70 \cdot 10^{-5}$	$2.89 \cdot 10^{-4}$	$3.25 \cdot 10^{-4}$	$6.69 \cdot 10^{-4}$	$4.41 \cdot 10^{-4}$
42	$7.44 \cdot 10^{-5}$	$1.33 \cdot 10^{-4}$	$9.20 \cdot 10^{-5}$	$2.00 \cdot 10^{-5}$	$1.16 \cdot 10^{-4}$
63	$5.88 \cdot 10^{-4}$	$3.17 \cdot 10^{-3}$	$5.26 \cdot 10^{-4}$	$8.08 \cdot 10^{-5}$	$1.50 \cdot 10^{-3}$

Fig. 3 shows the trends of the radius of curvature along the lens profile of the models. The curves are plotted against the y direction, anterior part of the lens from 0 to R_{eq} and posterior one from R_{eq} to 0.

In Fig. 3a the radius of curvature of the 20 year old lens contour is plotted. It is possible to see that the anterior part shown an asphericity value higher then the posterior part, but if it is used the calculation for the posterior part will lead to an asphericity value of $q_p = -1.256$.

In Fig. 3b the 42 year old lens contour radius of curvature is higher both in anterior and posterior part but it can be seen that except for the anterior part the shape of the curvature does not completely change. In fact the anterior asphericity, which drive most the curvature in the optic zones, changes from -2.862 to -1.477 but the posterior asphericity does not present a substantial difference, $q_p = -1.402$, the change is about 11 %.

According to the literature [11] the anterior and posterior radius of curvature increases also in Fig. 3c that show the radius of curvature of the 63 year old lens contour. The interesting thing is again the shape of the curvature which from the anterior junction till the posterior optic zone remains as previous cases.

The posterior asphericity in the last case is $q_p = -1.246$ which present a difference of less the 1 % with the 20 year old lens. So according to the shape of the three curvatures we can say that the curvature at the equator does not show sensible differences, the same does the asphericity coefficient. Much difference instead arises on the anterior surface. As last data the optical power of the lens of the three ages is provided in Table 3, this was calculated using 1.42 as lens refraction index and 1.336 as surrounding media index and using common formulas [11].

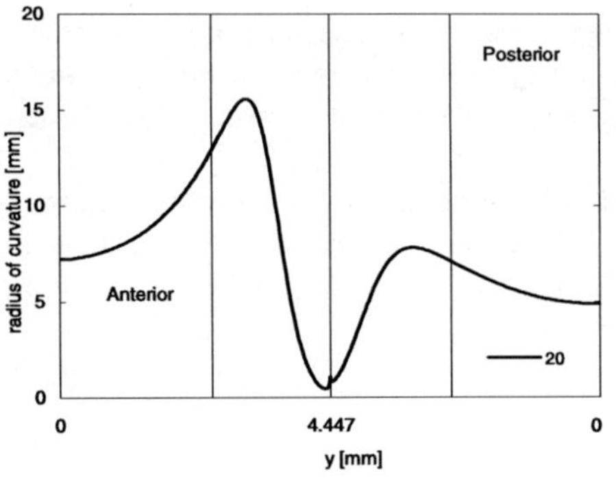

(a)

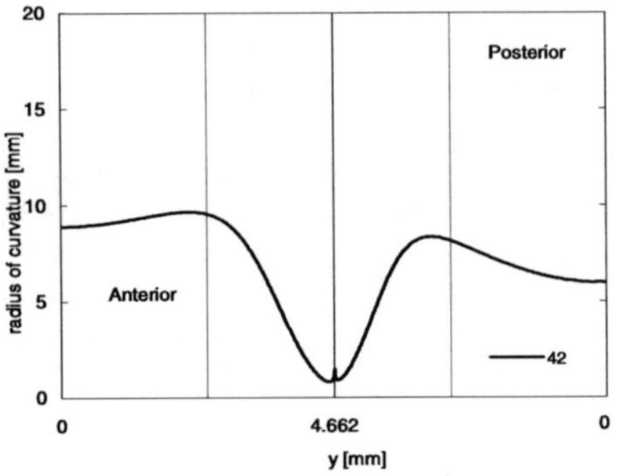

(b)

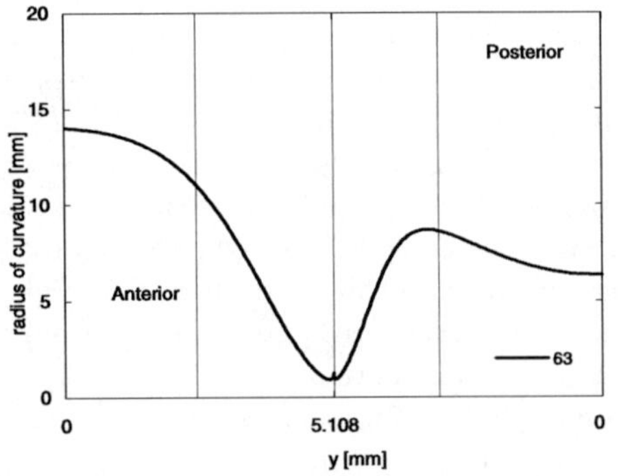

(c)

Fig. 3 Radius of curvature of the proposed model fitted over three lens contours, plotted versus y direction, anterior part of the lens is on the left side from 0 to 4, and posterior one is on the right part from 4 to 0. (a) Proposed 20 year old lens contour; (b) 42 year old lens contour; (c) 63 year old lens contour.

Table 3 Vertex power for the three lens contour

Lens contour	Volume [mm^3]
20	29.42
42	23.94
63	19.49

This data are in agreement with radius founded by Urs et al. [7] and so to vertex power.

From this data the lens paradox that Brown [16, 17] and Dubbelman and Van der Heijde [18] found does not arise. With lens paradox we mean that lens radius decreases with age so that the power of the lens increases, against the total power of the eye that decrease, and this is the well known hyperopisation through age. Of course we would like to remind that the set of data presented here is not as extensive as other studies in the literature.

IV. Discussion and conclusion

Geometrical behaviours of the model are in good agreement with the other ones described in the literature as can be seen in the results.

In particular the merit function MF in the optic zone against real lens contours is higher then MF against models but the value is comparable with other studies in the literature [6]. The best results for the anterior and posterior optic zone were in the mid-age 42 year old lens contour.

Radius of curvature shows a similar trend; these circumstances signify that the model is able to evaluate the optical properties. In particular it has been seen from the lens contour that the posterior radius of curvature, that is the one that drive more the optical properties of the lens remain quite stable over different ages while is the anterior radius of curvature that change much more.

This model can also be applied to lens nucleus shape modeling.

In future improvement of the proposed model, the coefficients will be derived from more imaging data of different age crystalline lenses. Also the internal structure of the lens shall be taken into account for optical analysis, such as optical power and spherical aberration.

References

1. Gullstrand A (1909) Helmholz's handbuch der physiologichen optic, vol. 1, 3rd edition, english translation edited by Southall JP, 1924 Optical of America
2. Smith G (2003) The optical properties of the crystalline lens and their significance. Clin Exp Optom 86 (1): 3–18
3. Hermans E, Dubbleman M, Van der Heijde R G L, Heethaar R M (2007) The shape of the human lens nucleus with accommodation. J Vision 7 (10): 1–10
4. Atchison D A (1992) Spectacle lens design: a review. Appl Opt 31 (19): 3579–3585
5. Hermans E, Pouwels P J, Dubbelman M, Kuijer J P A, Van der Heijde R G L, Heethaar R M (2009) Constant volume of the human lens and decrease in surface area of the capsular bag during accommodation: an MRI and Scheimpflug study. Invest Ophth Vis Sci 50 (1): 281–289
6. Smith G, Atchison D A, Iskander D R, Jones C E, Pope J M (2009) Mathematical models for describing the shape of the in vitro unstretched human crystalline lens. Vision Res 49: 2442–2452
7. Urs R, Ho A, Manns F, Parel J (2010) Age–dependent Fourier model of the shape of the isolated ex vivo human crystalline lens. Vision Res 50: 1041–1047.
8. Chien M C, Tseng H, Schachar R A (2003) A mathematical expression for the human crystalline lens. Compr Ther 29 (4): 245–258
9. Fincham E F (1937) The mechanism of accommodation. Brit J Ophthalmol 8: 5–80.
10. Giovanzana S, Savio G, Meneghello R, Concheri G (2011) Shape analysis of a parametric human lens model based on geometrical constraints. J Mod Opt: 1–11 DOI 10.1080/09500340.2011.554895
11. Giovanzana S (2011) A Virtual Environment for modeling and analysis of human eye. University of Padova.
12. MATLAB, User Guide, The MathWorks, Inc., 1994-2011, at http://www.mathworks.com
13. Levenberg K (1944) A method for the solution of certain problems in least squares. Quarterly Applied Mathematics 2: 164–168
14. Marquardt D (1963) An algorithm for least squares estimation of nonlinear parameters. SIAM Journal of Applied Mathematics 11: 431–441
15. Moré J J (1977) The Levenberg-Marquardt Algorithm: Implementation and Theory, Numerical Analysis. Ed. G. A. Watson, Lecture Notes in Mathematics 630, Springer Verlag
16. Brown N (1974) The change in lens curvature with age. Exp Eye Res 19: 175-83.
17. Moffat B A, Atchison D A, Pope J M (2002) Explanation of the lens paradox. Optom Vis Sci. 79 (3): 148-50
18. Dubbelman M, Van der Heijde R G L (2001) The shape of the aging human lens: curvature, equivalent refractive index and the lens paradox. Vision Res 41: 1867–1877

Author: Stefan Talu
Institute: Faculty of Mechanics / Department of Descriptive Geometry and Engineering Graphics, Technical University of Cluj-Napoca
Street: B-dul Muncii Street no. 103-105
City: Cluj-Napoca
Country: Romania
Email: stefan_ta@yahoo.com

On Approximation of Human Corneal Surface with Superellipsoids

M. Talu[1], S. Talu[2], S.D. Talu[3], and R. Shah[4]

[1] Faculty of Mechanics / Department of Applied Mechanics, University of Craiova, Romania
[2] Faculty of Mechanics / Department of Descriptive Geometry and Engineering Graphics, Technical University of Cluj-Napoca, Romania
[3] Faculty of Medicine / Department of Ophthalmology, "Iuliu Hatieganu" University of Medicine and Pharmacy Cluj-Napoca, Romania
[4] Massachusetts Eye Research and Surgery Institution, Boston, Massachusetts, U.S.A.

Abstract— **The objective of this paper is to present results of the theoretical and experimental researches for determination of an approximation of human corneal surface with superellipsoids using computational geometry. The mathematical formula permits a complex representation and the tool allowing exploring the physical and optical characteristics of the cornea. The spatial shape of the cornea can be described using different mathematical models with particular parameters for different subjects (women and men). These researches are applied in geometric constructions and computer aided design used in corneal refractive surgery, human vision studies, solid modelling and biomechanical behavior of the cornea.**

Keywords— **human corneal surface, superellipsoids, computational geometry.**

I. Introduction

The ophthalmology has benefited from computational methods in computer graphics, used for evaluation of patient data [1].

The visual analyzer is a complex system that has the purpose to receive, analyze and synthetize the informations regarding the shape, the dimension, the colour, the movement and the spatial depth of the objects in the surrounding area (Fig. 1).

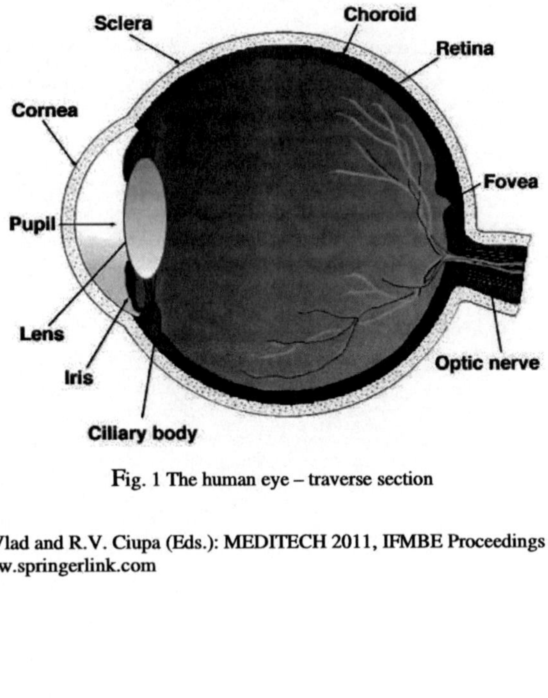

Fig. 1 The human eye – traverse section

The accurately measurement of corneal shape, refractive power and thickness has become important with the continuously growing popularity of refractive surgery procedures. Also, exact measurement of the corneal curvature is essential to properly fit contact lenses [1, 2].

Corneal topography is a modern invaluable tool to assist in the diagnosis and management of keratoconus as well as for the prevention of inappropriate refractive surgery in the patient groups [3 -10].

The cornea is the transparent front part of the eye that covers the iris, pupil, and anterior chamber (placed between air and aqueous humour), providing most of an eye's optical power, with a refraction index of 1.376. The refractive power of the cornea is approximately 43 dioptres, roughly two-thirds of the eye's total refractive power [11 – 21].

The central dioptric power of the cornea (43 dioptres) results from the addition of the dioptric powers of the three optical interfaces (air - tear film = + 43.6 D, tear film - cornea = + 5.3 D, cornea - aqueous humor = - 5.8 D).

The cornea presents the typical aspect of a divergent meniscus, in contact with transparent media having different refractive indices.

Cornea is a transparent and refractive tissue having a fixed curvature. Because transparency is of prime importance the cornea does not have blood vessels; it receives nutrients via diffusion from the tear fluid at the outside and the aqueous humour at the inside and also from neurotrophins supplied by nerve fibres that innervate it.

The adult cornea is normally clear, has a uniform surface, and is comprised of five layers: epithelium, Bowman's membrane, stroma, Descemet's membrane and the endothelium.

The corneal shape is maintained by its elastic properties in conjunction with intraocular pressure, generated by the continuous production and outflow of aqueous humor in the eye.

The adult cornea has a diameter of about 11.5 mm and a thickness of (0.5 - 0.6) mm in the center and (0.6 - 0.8) mm at the periphery.

The normal cornea has an aspheric profile that is more steeply curved in the center relative to the periphery.

The cornea is transparent for radiations comprised between (400 - 760) nm.

S. Vlad and R.V. Ciupa (Eds.): MEDITECH 2011, IFMBE Proceedings 36, pp. 252–255, 2011.
www.springerlink.com

The transparence is influenced by anatomical and biochemical factors, such as: the regularity of the cells, the parallelism of the fibers, the absence of the blood vessels.

From the biochemical point of view, the transparence depends on the hydration, which is arround 80 %. Transparency, avascularity, and immunologic privilege makes the cornea a special tissue.

To model the eye as an optical system it is necessary to describe of the anterior and the posterior corneal surfaces.

The aim of human corneal modelling is to determine the geometrical shape of the corneal surface using algorithms, to convert the input information to a mathematical description of the corneal surface.

The anterior corneal surface has been frequently described in the literature, as it can be measured by widely available techniques [11 – 20]. However, the literature provides only limited data on the posterior corneal surface [14, 15]; asphericity information is particularly scarce.

The asphericity of the posterior cornea has been measured by different authours using a combined approach involving a keratoscope to measure the anterior corneal surface and pachymetry to measure the thickness profile of the cornea [2, 13, 15, 21].

II. Mathematical model

A. *Theoretical Considerations*

The mathematical and computer models provide important possibilities that are not available in the experimental studies, which make it to be a useful supplement to experimental studies of the human cornea.

In recent years, superellipsoids have received significant attention for object modeling with their simple and flexible shape description and efficient computer graphical representation, having an important potential for the 3D objects modeling [22, 23, 24]. Using a superellipsoid makes it easy to create simple 3d primitives.

This type of implicit surface can represent various shapes to a high level of accuracy and allows the user to benefit from a generic graphics library, with the implementation of user objects or modules for specific applications.

A superellipsoid, as an ellipsoid's extension, is the result of the spherical product of two 2D models (two superellipses) [22].

A superellipse, analogous to a circle, can be expressed using next relation:

$$\left(\frac{x}{a}\right)^{2/\varepsilon}+\left(\frac{y}{b}\right)^{2/\varepsilon}=1,\ a>0, b>0. \quad (1)$$

and can be written in next form:

$$s(\theta)=\begin{bmatrix} a\cos^{\varepsilon}\theta \\ b\sin^{\varepsilon}\theta \end{bmatrix},\ -\pi\le\theta\le\pi. \quad (2)$$

where exponentiation with ε is a signed power function such that:

$$\cos^{\varepsilon}\theta = sign(\cos\theta)\left|\cos\theta\right|^{\varepsilon}. \quad (3)$$

Superellipsoids can be expressed by a spherical product of a pair of such superellipses:

$$r(\eta,\omega)=s_1(\eta)\otimes s_2(\omega)=\begin{bmatrix} \cos^{\varepsilon 1}\eta \\ a_3\sin^{\varepsilon 1}\eta \end{bmatrix}\otimes\begin{bmatrix} a_1\cos^{\varepsilon 2}\omega \\ a_2\sin^{\varepsilon 2}\omega \end{bmatrix}= \\ =\begin{bmatrix} a_1\cos^{\varepsilon 1}\eta\cos^{\varepsilon 2}\omega \\ a_2\cos^{\varepsilon 1}\eta\sin^{\varepsilon 2}\omega \\ a_3\sin^{\varepsilon 1}\eta \end{bmatrix},\ -\frac{\pi}{2}\le\eta\le\frac{\pi}{2};\ -\pi\le\omega\le\pi. \quad (4)$$

The a_1, a_2, a_3 parameters are scaling factors along the three coordinate axes. ε_1 and ε_2 are derived from the exponents of the two original superellipses.

This flexibility achieved by raising each trigonometric term to an exponent is of particular interest to us. In simple terms, these exponents, control the relative roundness and squareness in both the horizontal and vertical directions.

The shape of the superellipsoid cross section parallel to the [*xoy*] plane is determined by ε_1, while the shape of the superellipsoid cross section in a plane perpendicular to the [*xoy*] plane and containing *z* axis is determined by ε_2.

A superellipsoid is defined as the solution of the general form of the implicit equation [23]:

$$\left(\left(\frac{x}{a_1}\right)^{2/\varepsilon_2}+\left(\frac{y}{a_2}\right)^{2/\varepsilon_2}\right)^{\varepsilon_2/\varepsilon_1}+\left(\frac{z}{a_3}\right)^{2/\varepsilon_1}=1. \quad (5)$$

All points with coordinates (*x*, *y*, *z*) that correspond to the above equation lie on the surface of the superellipsoid. This is a compact model defined by only five parameters that permits to handle a large variety of shapes.

The exponent functions are continuous to ensure that the superellipsoid model deforms continuously and thus has a smooth surface.

This form provides an information on the position of a 3D point related to the superellipsoid surface, that is important for interior/exterior determination [24].

We have an inside-outside function *F* (*x*, *y*, *z*):

F (*x*, *y*, *z*) = 1 when the point lies on the surface;
F (*x*, *y*, *z*) < 1 when the point is inside the superellipsoid;
F (*x*, *y*, *z*) > 1 when the point is outside.

III. Materials and Methods

In this section we present a description of our dataset and a summary of the experiments performed.

Based on the experimental research and the statistical analysis, we consider the examination data obtained in our previous work [25].

Gender distribution and age distribution by gender don't differ significantly and therefore possible age dependencies are not influenced by gender [25].

These measurements were performed in order to be able to investigate a possible statistical deviation.

The average radius of the anterior corneal surface was 7.85 ± 0.25 mm. The average radius of the posterior corneal surface was 6.45 ± 0.20 mm. The ratio between the posterior and the anterior radius of curvature was 0.82 ± 0.02. Mean corneal thickness was 0.572 ± 0.032 mm [25].

We proposed a method that permits an anatomical correspondence between the original data and the created model. This allows us to visualize both textual information and imaging data of our classification results, aiding in clinical decision support.

The computational geometric analyses in this paper are based on the use of the Wolfram CDF Player 8.0 application for computation that generated the 3D superellipsoids [26, 27]. The choice of the geometrical formulation used in the analysis was chosen to ensure that the performance of the model is satisfactory and also to avoid errors associated with modeling materials.

Multiple view range images were used to capture data for the entire surface of the superellipsoid.

To test the geometrical computation the parameters for superellipsoids were chosen for the anterior corneal surface.

The determined parameters for the approximation of human corneal surface with superellipsoids are:

$a_1 = a_2 = a_3 = 1$; $\varepsilon_1 = 1$, $\varepsilon_2 = 0.8 ... 1.8$.

The fitting of contour points to the superellipsoid was evaluated by a defined distance measure using the inside-outside function [28, 29]:

$$E = \frac{1}{n}\sum_{i=1}^{n}(\sqrt{a_1 a_2 a_3}(1 - F^{\varepsilon_1}(x_i, y_i, z_i)))^2 \qquad (6)$$

where (x_i, y_i, z_i) are the detected contours points.

Because the error function is nonlinear the problem of fitting given contour points to superellipsoid model is a nonlinear estimation problem.

We used the Levenberg-Marquardt algorithm [30, 31, 32] which solves nonlinear least square minimization of the error function.

In this case it is necessary to input a set of initial values. We considered that the origin of object-centered coordinate system is aligned to the center of gravity of all the n contour points. The distance between the outermost contour points along each coordinate axis of the object-centered coordinate system is the site of initial fitting curve.

If it is increased the surface model, it is possible to capture the surface irregularities that distinguish between the different data. In the same time, using the correct superellipsoid parameters ensures that we capture most of the irregularities present over the measured of the corneal surface. We determined that higher superellipsoid parameters that were able to provide a better model fit, but it was also susceptible to noise.

For smooth continuity in the representation of the measured data, a mathematical procedure was developed for interpolating the locations where no measurement had been taken. At locations where no measurement has been taken, the corresponding locations are obtained by interpolation at neighbouring points. In addition, the interpolated values are constrained to be no greater than the measured values.

As an evaluation of the measurement technique, the algorithm was tested.

IV. Conclusions

In the context of corneal surface, models and simulations are used to examine the function, structure and nonlinear dynamics as an accurate and rigorous tool for generating quantitative and qualitative predictions.

Mathematical analysis of corneal models is often restricted to special cases with particular features. Numerical simulations can expand the range and allow understanding how analysis of the special cases relates to more realistic situations.

In this paper are proposed contributions concerning in determination of a new mathematical and graphical model using the superellipsoids for the corneal surface, with particular parameters for different subjects (women and men).

The determined parameters for the approximation of human corneal surface with superellipsoids are:

$a_1 = a_2 = a_3 = 1$; $\varepsilon_1 = 1$, $\varepsilon_2 = 0.8 ... 1.8$.

This analysis predict that the radius and asphericity of the vertical and horizontal meridians of individual subjects differ. As predicted by the analysis the anterior surface of the cornea becomes more curved with age, but more so in the horizontal meridian than in the vertical meridian.

The radii of both anterior and posterior corneal surfaces and asphericity of anterior corneal surface are not significantly age-dependent, but the asphericity of the posterior corneal surface is age-dependent.

There is a stronger relationship between posterior asphericity and anterior asphericity of the cornea.

These results provide the possibility of optimizing the refractive surgery by considering a new mathematical and CAD graphical model for the corneal surface.

In future improvement of the proposed model, deformable surface and advanced techniques for surface extraction will be chosen.

References

1. Saragoussi J J, Arne J L, Colin J, Montard M (2001) Chirurgie refractive. Masson, Paris
2. Anera R G, Jimenez J R, Jimenez L B, Dıaz J A (2002) Corneal asphericity on visual function after refractive surgery. Optik 113 (2): 83–88
3. Guillon M, Lydon D P, Wilson C (1986) Corneal topography: a clinical model. Ophthal Physiol Opt 6: 47-56
4. Carroll J P (1994) A method to describe corneal topography. Optom Vis Sci 71: 259–264
5. Carney L G, Mainstone J C, Henderson B A (1997) Corneal topography and myopia. A cross-sectionnal study. Invest Ophthalmol Vis Sci 38: 311-320
6. Langenbucher A, Viestenz A, Seitz B (2002) Conoidal fitting of corneal topography height data after excimer laser penetrating keratoplasty. J Refract Surg 18: 63–70
7. Kasprzak H T, Jankowska-Kuchta E (1996) A new analytical approximation of corneal topography. J Modern Opt 43: 1135–1148
8. Buehren T, Lee BJ, Collins MJ, Iskander DR (2002) Ocular microfluctuations and videokeratoscopy. Cornea 21: 346-351
9. Buehren T, Collins M J, Iskander D R, Davis B, Lingelbach B (2001) The stability of corneal topography in the post-blink interval. Cornea 20: 826-33
10. Zhu M, Collins M J, Iskander D R (2007) Dynamics of ocular surface topography. Eye 21: 624–632
11. Kiely P M, Smith G, Carney L G (1982) The mean shape of the human cornea. Optica Acta 29: 1027–1040
12. Mainstone J C, Carney L G, Anderson C R, Clem F M, Stephensen A L, Wilson M D (1998) Corneal shape in hyperopia. Clin Exp Optom 3: 131-137
13. Sheridan M, Douthwaite WA (1989) Corneal asphericity and refractive error. Ophthal Physiol Opt 9: 235- 238
14. Patel S, Marshall J, Fitzke E W (1993) The shape and radius of the posterior corneal surface. Refract Corneal Surg 9: 173-181
15. Dubbelman M, Weeber H A, Van der Heijde R G, Volker-Dieben H J (2002) Radius and asphericity of the posterior corneal surface determined by corrected Scheimpflug photography. Acta Ophthalmol Scand 80: 379–383
16. Mandell RB, St Helen R (1971) Mathematical model of the corneal contour. Br J Physiol Opt 26: 185-197
17. Burek H, Douthwaite W (1993) Mathematical models of the general corneal surface. Ophthal Physiol Opt 13: 68–72
18. Kasprzak H T, Iskander D R (2006) Approximating ocular surfaces by generalized conic curves. Ophthal Physiol Opt 26: 602–609
19. Iskander D R, Collins M J, Davis B (2001) Optimal modeling of corneal surfaces with Zernike polynomials. IEEE Trans Biomed Eng 48: 87–95
20. Iskander D R, Morelande M R, Collins M J, Davis B (2002) Modeling corneal surfaces with radial polynomials. IEEE Trans Biomed Eng 49: 320–328
21. Lotmar W (1971) Theoretical eye model with aspherics. J Opt Soc Am 61: 1522–1529
22. Jaklic A, Leonardis A, Solina F (2000) Segmentation and Recovery of Superquadric. Computational imaging and vision. Kluwer Academic Publishers, Dordrecth, The Netherlands
23. Velho L, Gomes J, Figueiredo L H (2002) Implicit Objects in Computer Graphics. Springer-Verlag, New York, Inc.
24. Chevalier L, Jaillet F, Baskurt A (2003) Segmentation and superquadric modeling of 3D objects. Journal of Winter School of Computer Graphics, WSCG'03, 11 (2): 232-239
25. Talu S D, Talu S, Talu M, Shah R (2007) CAD of human corneal surface using new mathematical models. Acta Electrotehnica of the Technical University of Cluj-Napoca, Romania, 48 (4): 299-302
26. Wolfram CDF Player 8.0.1.0 by Wolfram Research, Inc., 100 Trade Center Drive Champaign, IL 61820-7237 USA, 1988-2011, at http://www.wolfram.com/cdf-player.
27. Kragler R, Superquadrics from the Wolfram Demonstrations Project, at http://demonstrations.wolfram.com/Superquadrics/
28. You S, Neumann U (1998) Automatic Object Modeling for 3D Virtual Environment, International Workshop on Non-linear Model Based Image Analysis, Scotland
29. Bhabhrawala T (2004) Shape recovery from medical image data using. Extended superquadrics. Doctoral thesis, Buffalo, New York, USA.
30. Levenberg K (1944) A method for the solution of certain problems in least squares. Quarterly Applied Mathematics 2: 164–168
31. Marquardt D (1963) An algorithm for least squares estimation of nonlinear parameters. SIAM Journal of Applied Mathematics 11: 431–441
32. Moré J J (1977) The Levenberg-Marquardt Algorithm: Implementation and Theory, Numerical Analysis. Ed. G. A. Watson, Lecture Notes in Mathematics 630, Springer Verlag

Author: Mihai Talu
Institute: Faculty of Mechanics / Department of Applied Mechanics, University of Craiova, Romania
Street: Calea Bucuresti Street, no. 165
City: Craiova
Country: Romania
Email: mihai_talu@personal.ro

Numerical Simulation of Thrombus Aspiration Catheter: Preliminary Results

S. Soleimani[1,2], G. Pennati[1], and G. Dubini[1]

[1] Laboratory of Biological Structure Mechanic Department of Structural Engineering, Politecnico di Milano, Milan, Italy
[2] Bioengineering Department, Politecnico di Milano, Milan, Italy

Abstract— **Thrombus aspiration is one of the available therapies for the treatment of thrombosis diseases. It is an effective option especially in the case of early stage clot formation. Nowadays there is a major research effort to improve catheter design in term of ease of use, less drawback and higher efficiency. The main aim of this study is to consider the performance of a standard catheter and predict the behavior of clot during the aspiration phase using computational fluid dynamics (CFD) techniques. Three cases are modeled and compared one another to assess catheter performance and efficiency according to different rheological models for the clot. In the first case the clot is considered to be Newtonian with early stage coagulation and no surface tension, in the second case the clot exhibits higher viscosity due to the longer time passed after coagulation, but it has not surface tension, in the third case the clot is assumed to be Newtonian which is formed in early stages of coagulation but has surface tension. In all cases clot is assumed to behave like a viscous fluid.**

Keywords— **Catheter, Clot, Viscosity, surface tension, CFD.**

I. INTRODUCTION

As a biological pump, heart needs nourishment and oxygen that are supplied through the coronary system. The coronary system can face different diseases like blockage by thrombosis. Vascular injury can lead to platelet aggregation and coagulation which per se lead to thrombin. In this process aggregation is the result of physical process while coagulation is the result of biochemical enzyme reactions [1]. Thrombosis is one of the main causes of mortality. There are different locations for thrombus formation: venous, deep vein, portal vein, renal vein. The thrombus formation process is complex and relates to a number of mechanical, physiological and pathological parameters. Among theses many factors, rheological parameters [2] are the most important and, in turn, mainly relate to the ratio of thrombin to fibrinogen [3] and the time elapsed after clot initiation.

Balloon angioplasty and stenting, fibrinolysis and mechanical thrombectomy are three approaches which can be used for the treatment. In any approach the efficacy of clot removal, procedure-related embolization rate and mortality are main parameters [4].

Angioplasty and stenting include:

Exciter laser coronary angioplasty, Cutting balloon angioplasty, Bare-metal stents, Drug-eluting stents.

Thrombectomy consists of: Simple Aspiration Devices, Hydrodynamic devices, Fragmentation Devices (Directional atherectomy, Rotational atherectomy, Extraction Athrectomy), Ultrasound devices. [4-5]

However cutting balloon and laser angioplasty did not show any improvement in restenosis. The most common used device is Angiojet Rheolytic Thrombectomy System (Possis Medical, Minneapolis,MN, USA) and the Export XT Aspiration Catheter (Medtronic Vascular, Santa Rosa, CA) [5].

However, generally speaking the comparison of the approaches is sometimes hard to judge because of lack of consistent data. There are many different aspiration catheters especially regarding to the tip shape like TVAC (duckbill tip shape) (Nipro,Osaka, Japan), Thrombuster (Kaneka Medix Corporation, Osaka, Japan), Percusurge Export catheter (Medtronics, Minneapolis, MN), Rescue (oblique straight tip shape) (Boston Scientific, Natic, MA) and Diver CE Catheter (Invatec, Roncadelle, BS, Italy). For example TVAC showed high performance for intramural aspiration due to its special tip shape [6-7]. The aim of this work is to describe and better understand the clot behavior during absorption through catheter under different assumptions on clot rheology.

II. MATERIAL AND METHODS

For absorption of clot, the catheter is deployed and guided along the coronary vessel to proximally reach the clot. From case to case the absorption can be even done from the distance of 2 cm. In the normal procedure it is sometimes needed to apply aspiration more than one time to absorb the entire clot.

Usually the catheter for coronary purpose has 2-2.6 mm (6F-7F) diameter, almost half of the size of the coronary. (Figure 1) [7]. The catheter is formed of two lumens which are inside each other: the smaller tube is for guidance while the larger one is for clot absorption. However in this study the catheter assumed to have 0.85 mm diameter.

In reality the applied suction pressure follows a diagram in such a way that it is almost flat in its starting phase, then decreasing to zero with steep slope [8], however in this work a constant negative pressure is applied through the catheter.

S. Vlad and R.V. Ciupa (Eds.): MEDITECH 2011, IFMBE Proceedings 36, pp. 256–259, 2011.
www.springerlink.com

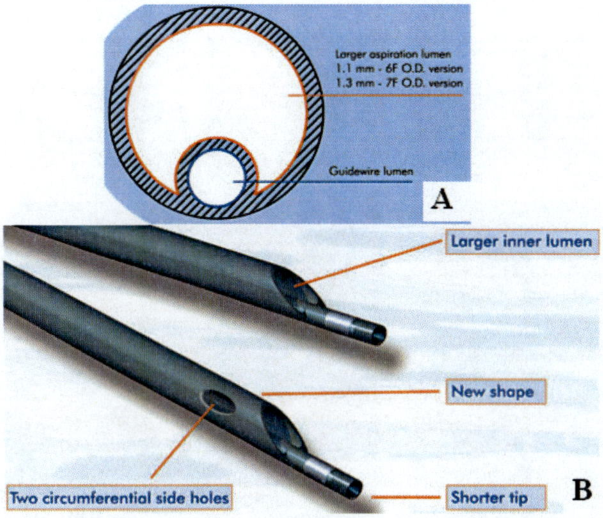

Fig. 1 Schematic view of Diver CE Catheter , A) tip shape and B) body shape for the catheter with and without lateral holes, [7]

A. *Viscosity Models*

The mechanical properties of blood and clot are very important in predicting and modeling their behavior during suction. In this work the blood behavior is considered to be Newtonian. This is mainly true when the velocity is high enough, say the shear rate is higher than 100 s^{-1} [9]. The clot mechanical properties varies radically due to different parameters like blood hematocrit [10], time passed after coagulation, clot type and forming procedure [2]. The properties of blood are given in Table 1 according to Gay et al. [11].

Balossino et al. [8] considered the clot in its early time after coagulation when its viscosity is 0.035 Pa·s, ten times larger than normal blood viscosity and the surface tension is 0.05 N/m.

In this work it is assumed that clot absorption in blood field can be described as a two-phase flow and the Volume-of-Fluid (VOF) technique is used to trace the interface of two phases [8].

Table 1 Mechanical properties of blood and clot

Parameters	Value
Blood density	1060 kg/m^3
Blood viscosity	0.0035 Pa·s
Early clot viscosity	0.035 Pa·s
Late clot viscosity	0.6 Pa·s

B. *Modeling*

In this work the clot behavior is studied in the period after suction start, monitoring how much clot is absorbed and how its shape changes. To the purpose three different cases are applied and considered. In case 1, the clot has just formed and is almost soft enough to be considered with a ten-time viscosity of blood (0.035 Pa·s) [8]. In this case surface tension is not considered between clot and blood, the vessel nor catheter. Furthermore blood density is set to 1060 kg/m^3. In case 2, the clot has a much higher viscosity (0.6 Pa·s) which corresponds to a longer time elapsed since coagulation initiation. In Case 3, the surface tension with amount of 0.1 N/m is considered to play a role while the clot viscosity is Newtonian (viscosity equal to 0.035 Pa·s). The geometry is meshed using ANSYS ICEM CFD 12.1 (ANSYS Inc., Canonsburg, PA, USA) with tetrahedral elements.

C. *Mathematics Background*

The adopted numerical procedure is based on the finite volume commercial package ANSYS/Fluent 12.1. First the domain is discretized to apply the finite volume form of equation of mass, momentum and volume of fluid interface.

The formula of mass and momentum conservation for an incompressible fluid are as follow:

$$\nabla u = 0 \tag{1}$$

$$\frac{\partial \rho \vec{u}}{\partial t} + \nabla\left(\rho \vec{u}\vec{u}\right) = -\nabla p + \nabla \tau + \rho g \tag{2}$$

where u is the fluid velocity, ρ is the fluid density, p is the pressure and τ is the stress tensor. The interface locations between the two fluids (blood and clot) are traced through Volume of Fluid (VOF) approach. In VOF methodology [12], the volume of each fluid is defined in the cell through the formula $F_{vol}=\gamma \cdot V_{cell}$, where V_{cell} is the computational cell volume and γ is the liquid fraction in this cell. When the cell is totally filled by one fluid, γ is equal to 1; if it is filled by the other, γ is equal to 0; if the cell is partially filled by either volume, γ should satisfy the following equation:

$$\frac{\partial \gamma}{\partial t} + \nabla(\gamma u) + \nabla\left[\gamma(1-\gamma)u_r\right] = 0 \tag{3}$$

where u_r is the velocity field at interface [13]. A single momentum equation is solved for the entire domain and the velocity filed is shared between the fluid phases [12].

The generic properties in each cell (θ) are computed according to following equation [13]:

$$\theta = \gamma \cdot \theta_{fluid1} + (1-\gamma)\theta_{fluid2} \tag{4}$$

In this study the adopted solution methods include: PRESTO for pressure, second order upwind for momentum, PISO for pressure-velocity coupling and Geo-reconstruct for volume fraction.

The vessel diameter is 4.2 mm while the inside diameter of catheter is 0.85 mm in this study. A schematic view of the catheter tip in the artery is shown in Figure 2. The applied pressure through catheter is 100.000 Pa.

III. Results

Figures 3 shows snap shots of fluid (blood and clot movement) as time progresses from start of applying negative pressure. Clot is visible as black part in the figure. At early stage of clot absorption (0.005 s after suction start) the amount of absorbed early clot is higher than late clot. This trend will be same until all clot is absorbed, which means that it needs a little bit more time to absorbing late clot comparing to early clot. However this extra time is not significant. Anyhow the shape of clot during absorption is almost the same for theses two cases. Regarding to Figure 3-c the shape of clot during absorption is completely different when the surface tension is considered by the amount of 0.1 N/m. Furthermore it seems that the absorbed clot in this case is almost the least at the same time comparing the two other cases (with no surface tension). However it is clear that the entire clot is absorbed via catheter in all the three cases.

The velocity contours and vectors at the time 0.005 s after suction start for the first case are presented in Figure 4. The velocity differs radically from inside to outside the catheter. Furthermore there is a relatively large area of vortex flow right at the tip, in upper region of the catheter. Inside the catheter the velocity shows a higher amount in the superior aspect of the catheter rather than in its inferior part.

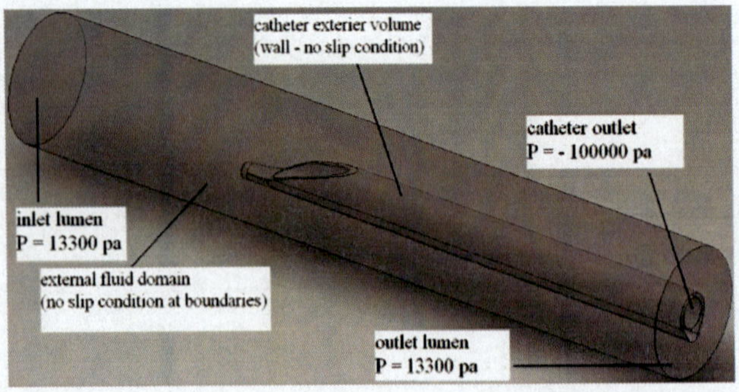

Fig. 2 Schematic view of the vessel, the catheter inside and the imposed boundary conditions

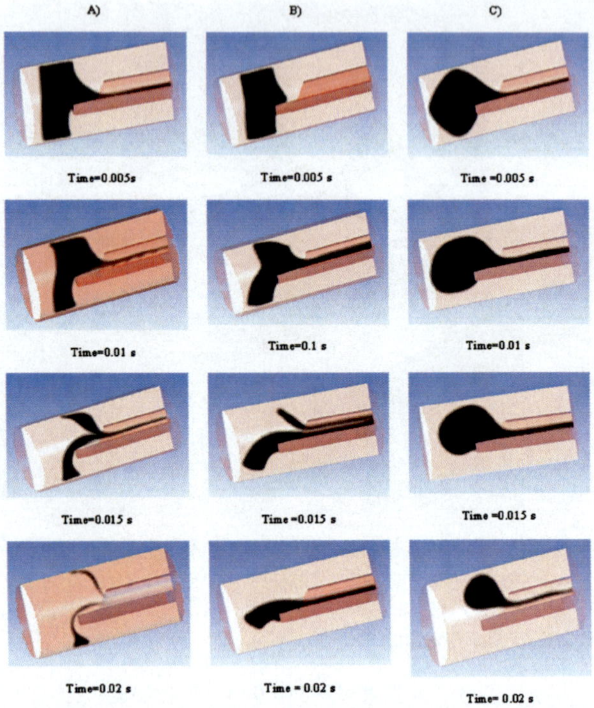

Fig. 3 Snap shots of clot absorption when time progresses, for the three cases of A) early clot, Newtonian, viscosity 0.035 Pa•s, no surface tension; B) late clot, Newtonian, viscosity 0.6 Pa•s, no surface tension; C) Newtonian, viscosity 0.035 Pa•s, surface tension 0.1 N/m

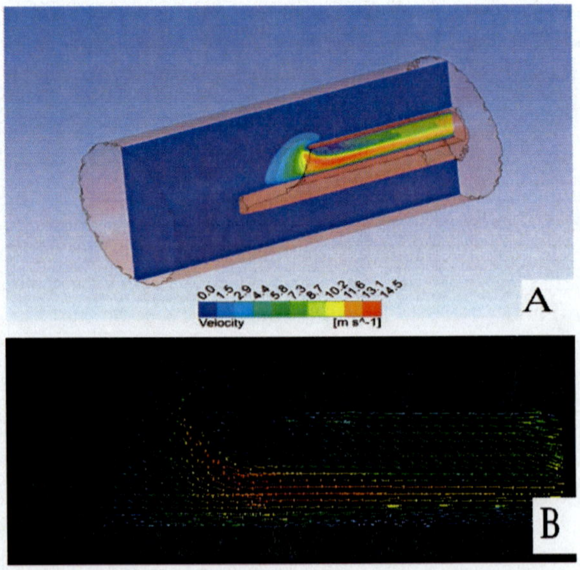

Fig. 4 The A) contour and B) vector maps of fluid velocity in the area around the catheter tip at time 0.005 s after suction start for the case of clot with viscosity 0.035 Pa•s and no surface tension

IV. Conclusions

In this work the behavior of a blood clot is investigated. The preliminary results from modeling show that during the clot absorption procedure, the shape of clot and the absorbed amount vary according to the rheological properties of clot. So far clot properties have been changed in terms of viscosity, whether they are in an early and late phases of maturation. The case with surface tension has also been studied to assess whether it may cause a difference in the absorption rate. In the case of late clot type the amount of clot absorption slightly varies, but the clot shape is more importantly affected when surface tension is modeled.

Numerical simulations clearly indicated that mechanical properties can play role in clot movement, which is different in the three cases. According to the model results in all cases, clot will be completely absorbed in a short time after suction start. It means that the applied pressure is large enough for aspiring all clot inside the vessel in all cases. When the surface tension is taken into account, simulation results indicate that it has a main effect on the clot shape during absorption.

One may consider that manipulating the entrance geometry at the tip of catheter can lead to higher performance in term of less blood, but higher fraction of clot absorption. In future development, the effect of different designs of the catheter tip on clot aspiration will also be investigated. Since there is a limitation for the aspiration pressure mainly due the collapse of the arterial wall, the pressure parameter will also be studied to understand the whole system behavior in terms of the ratio of clot to blood absorption. Furthermore the applied negative pressure curve exhibits an extended plateau at the beginning and then a decrease while the adopted boundary conditions at the vessel wall are steady in the present study. Future work could include the systolic-diastolic change, too.

Acknowledgment

This work has been funded by the FP7, People Programme, Marie Curie Actions with the grant agreement PITN-GA-2009-238113.

References

1. Leiderman K, Ferguson AL (2010) Grow with the flow: a spatial–temporal model of platelet deposition and blood coagulation under flow. Math Med Biol 28:47-84
2. Anand M, Rajagopal K, Rajagopal K R (2006) A viscoelastic fluid model for describing the mechanics of a coarse ligated plasma clot. Theor Comput Fluid Dyn 20: 239–250
3. Collet JP, Woodhead JL, Soria J et al. (1996) Fibrinogen Dusart: Electron Microscopy of Molecules, Fibers and Clots, and Viscoelastic Properties of Clots. Biophys J 70: 500-510
4. Muller-Hulsbeck S and Jahnke T (2003) Peripheral Arterial Applications of Percutaneous Mechanical Thrombectomy. Techniques in Vascular and Interventional Radiology 6: 22-34
5. Hudson PA, Kim MS, Carroll JD (2010) Coronary ischemia and percutaneous intervention. Cardiovascular Pathology 19:12–21
6. Sakurada M, Ikari Y and Isshiki T (2004) Improved Performance of a New Thrombus Aspiration Catheter: Outcomes From In Vitro Experiments and a Case Presentation. Catheterization and Cardiovascular Interventions 63:299–306
7. INVATEC at http://www.invatec.com/tool/home
8. Balossino R, Dubini G, Migliavacca F et al. (2010) Numerical simulation of thrombus aspiration in two realistic models of catheter tips. Artificial Organs 34:301–310
9. Johnston BM, Johnston PR, Corney S, Kilpatrick D (2006) Non-Newtonian blood flow in human right coronary arteries. Transient simulations Journal of Biomechanics 39: 1116–1128
10. Schmitt C, Henni AH and Cloutier G (2007) Characterization of Time-Varying Mechanical Viscoelastic Parameters of Mimicking Deep Vein Thrombi with 2D Dynamic Elastography. IEEE International Ultrasonics Symposium Proc, New York, USA, 2007, pp 1009-1012
11. Gay M and Zhang LT (2009) Numerical studies of blood flow in healthy, stenosed, and stented carotid. Int. J. Numer. Meth. Fluids 61:453–472
12. Hirt CW and Nicols BD (1981) Volume of Fluid (VOF) method for the dynamics of free boundaries. J Comp Phys 39: 201-225
13. Favero JL, Cardozo NS, Secchi AR, Jasakc H (2010) Simulation of Free Surface Viscoelastic Fluid Flow Using the viscoelastic Inter-Foam Solver. 20th European Symposium on Computer Aided Process Engineering – ESCAPE20 Proc, Ischia, Naples, Italy, 2010, pp 31-36

Developing a Lumped Model for the Vestibular Receptors

A. Codrean, A. Korodi, I. Jivet, and T.-L. Dragomir

"Politehnica" University of Timisoara/Faculty of Automation and Computers, Timisoara, Romania

Abstract— **Researching on the vestibular-sympathetic reflex mechanism requires as an intermediary step to establish a lumped model for the Vestibular Receptors. Knowing that Otolith Organs and Semicircular Canals are sensing changes in the head velocity on 6 degrees of freedom, the purpose of this paper is to set up models for both types of receptors, branching them into Regular and Irregular subcategories, and to consider a certain convergence of their outputs. The developed model for the Vestibular Receptors is tested in the sit to stand orthostatic stress scenario.**

Keywords— **vestibular-sympathetic reflex, vestibular receptors, otolith organs, semicircular canals, fractional order systems.**

I. Introduction

The vestibular research domain has shown significant progress in the last decades, with an increasing interest in studying how the Vestibular System mediates several nervous reflex mechanisms. Such a reflex mechanism is the vestibular-sympathetic reflex, which has been highlighted by many scientists for its important role in cardiovascular regulation ([1]).

At a simplistic level, the vestibular-sympathetic reflex can be considered consisting of vestibular receptors (afferents), vestibular nucleus (central processing) and the sympathetic system (efferent). In this context, the aim of the current paper is to develop a lumped model for the Vestibular Receptors (VR), which could be further coupled with existing models for the Vestibular Nucleus ([2]), respectively with models of the sympathetic system (several models exist in the literature, initially built for the baroreflex mechanism – e.g. [3]). The long term goal would be to quantify through a mathematical model the action of the vestibular-sympathetic system on the cardiovascular system during an orthostatic stress scenario.

During an orthostatic stress scenario, like sit to stand, the motion of the head would be detected by both types of VR: Otolith Organs (linear acceleration) and Semicircular Canals (angular acceleration). Thus, it would be important that the model of the VR includes both types of receptors. Moreover, knowing that the frequency range of head movement is between 0.1 and 10 Hz, the frequency domain of interest for the developed models will be set accordingly.

The task of modeling the VR (Otolith Organs and Semicircular Canals) is complex, mainly due to the distributed nature of these receptors. Because of this, one of the most common modeling approaches is to develop models of individual receptors units based on experimental frequency characteristics ([4], [5], [6] and [7]). These studies are usually presenting a classification of the receptor units according to their observed dynamic behavior (regular receptors and irregular receptors). Finally, despite all the progress in this direction, models that can characterize the whole ensemble of VR are still missing.

In this context, the aim of the current paper is to obtain a lumped model that could characterize, at least qualitatively, the dynamic behavior of the VR.

The structure of the paper is as follows. Section II presents the structure of the general model for the VR. It also describes the models for Otolith Organs and Semicircular Canals, respectively the implementation manner. Section III presents the development of the modules designed to generate the excitatory input signals for the models. Section IV discusses how the outputs of the models and the sub-models are converging. Section V shows the simulation results and Section VI draws conclusions.

Because the modeling procedure presented further refers actually to MIMO type systems, with multiple channels, for simplifying the presentation the notation $u \rightarrow y$ will be used when referring to an informational channel having the input signal u and the output signal y. Equivalently, we will refer to the orientation $u \rightarrow y$ of a system or subsystem.

II. Modeling the vestibular receptors

The structure of the proposed model is presented in Fig. 1. Four types of receptor groups are considered: Regular Otolith Organs, Irregular Otolith Organs, Regular Semicircular Canals and Irregular Semicircular Canals. The model for each group is built based on the averaged dynamics (frequency characteristics) of the receptors from that group. The inputs for each model are generated through special modules that transform the measured linear accelerations (a_x, a_y, a_z) and angular velocity components ($\omega_x, \omega_y, \omega_z$) according to the excitatory direction or area of each receptor group. The outputs of the models (n_{OO}, n_{SCC}) are considered to converge linearly through summation or weighted summation. In the end, the n_{VR} output signal is obtained.

S. Vlad and R.V. Ciupa (Eds.): MEDITECH 2011, IFMBE Proceedings 36, pp. 260–265, 2011.
www.springerlink.com

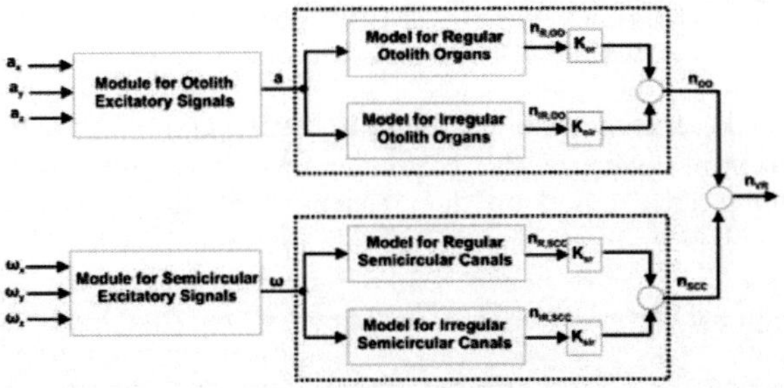

Fig. 1 Model of the Vestibular Receptors

A. *Models for Regular and Irregular Otolith Organs*

Due to the small number of parameters and to the good approximation capabilities, for the Otolith Organs we adopted the model from [4]. The model has the following form:

$$H_{OO}(s) = K \cdot \frac{s^n (1+T_d s)^m}{(1+T_p s)} \quad (1)$$

In [4], the parameters were obtained through identification based on the averaged experimental frequency characteristics ([0.1-10] Hz domain) for Regular Otolith Organs, respectively Irregular Otolith Organs. Thus, two models were obtained: one for the regular otoliths with the parameters {K=0.15, T_p=0.07 seconds, T_d=0.027 seconds, n=0.15, m=1.43}, and one for the irregular otoliths with the parameters {K=0.68, T_p=0.07 seconds, T_d=0.045 seconds, n=0.3, m=1.21}. In respect to Fig. 1, the Regular Otolith Organs model has the orientation a → $n_{R,OO}$, while the Irregular Otolith Organs model has the orientation a → $n_{IR,OO}$.

The dynamic behavior of the Regular Otolith Organs and that of the Irregular Otolith Organs differs in that the Irregular Receptors show a larger phase lead and larger amplitude at high frequency (also a larger slope for the amplitude-frequency characteristics). This can be observed also from the models parameters – the irregular otoliths have larger gain K and derivative time constant T_d.

B. *Models for Regular and Irregular Semicircular Canals*

The model for the Semicircular Canals was adopted from [5] and has the following form:

$$H_{SCC}(s) = K \cdot \frac{s^n (1+T_d s)}{(1+T_{p1} s)(1+T_{p2} s)} \quad (2)$$

The parameters were obtained in [5] through identification based on the averaged experimental frequency characteristics ([0.1-4] Hz domain) for Regular Semicircular Canals, respectively Irregular Semicircular Canals. The model obtained for the Regular Semicircular Canals has the parameters {K=6.3, T_{p1}=7 seconds, T_{p2}=0.003 seconds, T_d=0.004 seconds, n=0.08}, and the model obtained for the Irregular Semicircular Canals has the parameters {K=2.1, T_{p1}=3 seconds, T_{p2}=0.003 seconds, T_d=0.02 seconds, n=0.3, m=1.21}. In respect to Fig. 1, the Regular Semicircular Canals model has the orientation a → $n_{R,SCC}$, while the Irregular Semicircular Canals model has the orientation a → $n_{IR,SCC}$.

Due to the values of the time constants T_{p1}, the differences between the dynamic behavior regarding the regular and irregular receptors are of the same nature as the ones presented for the Otolith Organs.

C. *Model Implementation Issues*

Once the models are identified, the next step is to find the proper procedure for implementing these models in a simulation environment (e.g. Matlab/Simulink). The issue that emerges is that the models described through (1) and (2) fall into the category of fractional order systems. Fractional order systems can not be implemented directly and usually different approximation methods are used to find the equivalent integer order system. The papers in which such models are developed from experiments (e.g. [4], [5], [7]) avoid using such methods by presenting only frequency characteristics or a time domain response for specific input signals (e.g. trapezoid). Our intention here is to implement the models through the use of approximation methods so that it would permit time domain simulations with any type of input signals.

In case of equations (1) and (2), two types of fractional order elements are presented: a fractional derivative - s^n and a fractional power zero $(1+Ts)^m$.

For the fractional derivative quite a few approximation methods can be found in the literature. From a careful comparative analysis of the methods implemented in the Non-Integer Matlab Toolbox ([8]) – Crone method, Carlson's method, Matsuda's method - we reached the conclusion the Crone method produces the best results. Basically, the main idea of the Crone method, initially developed in [9], is to approximate the fractional order derivative over a predefined frequency range through N zero-pole pairs:

$$s^n \approx k \prod_{i=1}^{N} \frac{1+s/\omega_{zi}}{1+s/\omega_{pi}} \quad (3)$$

The order N depends on the frequency range and on the minimal accepted approximation error. The manner in which the zero-pole pairs are computed is further discussed in [8].

For the fractional power zero the most appropriate approximation method found in the literature was that of

Charef ([10]). Charef's method was actually developed for approximating a power pole, but by inversing the sign of the fractional power the power pole becomes a power zero. The specific formula for this method resembles with that of the Crone's method:

$$\frac{1}{(1+Ts)^m} \approx k\prod_{i=1}^{N}\frac{1+s/\omega_{zi}}{1+s/\omega_{pi}} \quad (4)$$

In our case the power exponent m from (4) would be negative, which would lead to an inversion of the nominator and denominator of the approximating integer order transfer function. In the end, for implementing Charef's method a Matlab script was developed based on the algorithm described in [10].

Next, we used Crone method and Charef's method to determine de integer order models corresponding to equations (1) and (2). It should be also mentioned here that for fractional order systems, in the absence of time domain measurements, frequency characteristics represent the only way of assessing the approximation errors of models like (3) and (4).

For the two Otolith Organs models (regular and irregular), both Crone method and Charef's method were used. Based on the comparison between the frequency characteristics calculated analytically for the fractional order model and the computed frequency characteristics of the approximation integer order models, the maximum error in amplitude and phase were below 0.3 dB and below 5 degrees on the frequency domain of interest ([0.1 -10]Hz). The order N for the approximations was 6 for the Crone method and 5 for Charef's method.

For the two Semicircular Canals models (regular and irregular), only the Crone method was needed. The maximum approximation errors in amplitude and phase were below 0.05 dB and below 1.5 degrees on the frequency domain of interest ([0.1 -10] Hz). The order N for the approximations with the Crone method was 6.

Two additional issues surfaced at the implementation of the Semicircular Canals models. The first issue was that in order to reproduce the results presented in [5] the fractional power n was modified from 0.008 to 1.008 for the regular receptors and from 0.3 to 1.3 for the irregular receptors. The second issue relates to the fact that the frequency characteristics presented in [5] are on the frequency domain of [0.1-4] Hz instead of [0.1-10] Hz. However, by comparing the frequency characteristics of the model with the experimental results from [6] on the frequency domain [1-10] Hz it was concluded that the model's extrapolation capabilities on the frequency domain of interest are acceptable.

As a last remark regarding the implementation of the four models, it should be mentioned that the output of each model was summed with the mean firing rate (MRF). The averaged MFR for each group of receptors corresponding to each of the four models were extracted from [11] and [5].

III. Generating the excitatory signals for the otolith and semicircular receptors

The large number of afferents that innervate the motion receptor organs of the vestibular system are divided into two groups, based on their function and location: otolith afferents and semicircular canals afferents. Each afferent has a different response in amplitude and phase to the same stimulus, although differences between afferents that belong to the same group are merely significant. The similarities between these neurons allow the usage of a single mean response for the whole group. This response can be obtained by composing either the inputs of the model, or the outputs.

Composing the excitatory signals (the inputs) is more favorable, since the modeling complexity is reduced by using a single model for all afferents of the same group with mean values for its parameters. These values are calculated for a set of receptors considered representative for each group.

The inputs for each module that generates the excitatory signals for each one of the two models (otolith organs and semicircular canals – shown in Fig.1) are considered to be coordinates of the acceleration and velocity in a fixed spatial frame with origin in the left vestibular labyrinth (stereotaxic coordinates).

Otolith organ afferents are located in the utricular and sacular maculae and they respond to linear head acceleration. Each otolith afferent can be characterized by a polarization vector, which summarizes its directional properties when the head is tilted in various directions with respect to the sagittal, coronal and transverse planes of the human body ([12]). The polarization vector indicates the afferent's maximum sensitivity direction. In [7] Fernandez and Goldberg considered a stimulus excitatory if the force that generated the stimulus and the afferent's polarization vector had the same direction. If these directions were opposite, the stimulus was considered inhibitory. Excitatory and inhibitory stimuli have different effects on the polarity of the afferents which cause pattern modifications in nervous discharges. So it is necessary to determine if a signal stimulating the afferent is excitatory or inhibitory.

Angular head acceleration is detected by afferents situated in the semicircular canals. Each inner ear has a set of three semicircular canals: anterior (AC), posterior (PC) and horizontal canals (HC). The normals to the canal planes can be considered axes of a fixed coordinate frame. When the head is in an upright position, the lateral canal axis is tilted back about 30°, with respect to a vertical axis pointing out of the top of the head. The vertical canal axes (AC and PC) are tilted back about 40° with respect to the negative naso-occipital axis ([12]).

As same as the otolith organ afferents, semicircular canal afferents can be divided into groups, depending on their position in the three canals. Each group can be characterized by a maximum sensitivity vector, described in [5].

In computing the input acceleration and velocity signals for the model two properties are taken into consideration: magnitude and sign. The magnitude is responsible for the amplitude of the response, while the sign indicates if the input signal has an excitatory or inhibitory effect. These two values are calculated using predefined excitatory and inhibitory areas for otolith afferents, respectively the velocity's projections on the maximum sensitivity vectors for semicircular canals. The necessity of using zones (and not vectors) for the otolith afferents is justified by the vast distribution of maximum sensitivity directions for otolith afferents that cannot be replaced by a mean sensitivity direction.

The magnitude of the input signal for the otolith afferents model is calculated with the Euclidean norm:

$$\|a\| = \sqrt{a_x^2 + a_y^2 + a_z^2} \qquad (5)$$

where a_x, a_y, and a_z are the acceleration's coordinates in the fixed spatial frame. The hypothetical predefined areas used for sign computing are shown in Fig. 2. Forward movement ($a_x > 0$) has an excitatory effect, while backward movement inhibits the afferent. The sign and magnitude for the acceleration on the horizontal plane (h_{xy}) are used to calculate the final value for the sign.

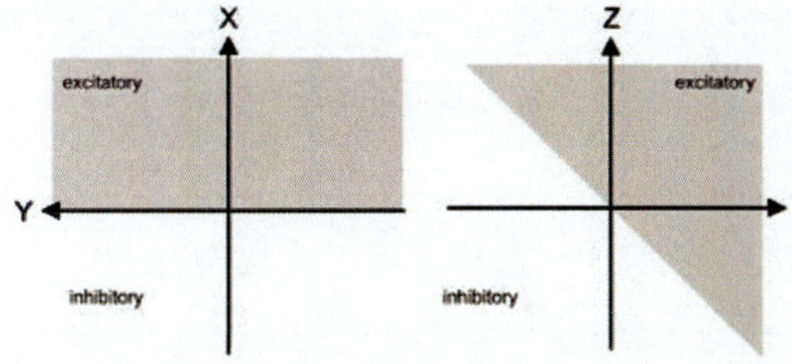

Fig. 2 Predefined excitatory and inhibitory areas for the otolith afferents

Both the magnitude and the sign of the input signal for the semicircular canals model are computed using the projections of the velocity on each canal's maximum sensitivity vector (MSV) – e.g. Fig. 3b. In [5], MSV is described by the direction cosines in stereotaxic space. The projections on MSV are calculated by multiplying the cosines with the stereotaxic coordinates of the velocity:

$$\omega_i = \omega_x \cdot \cos\theta_x^i + \omega_y \cdot \cos\theta_y^i + \omega_z \cdot \cos\theta_z^i\,, \quad i \in \{LHC, RHC, LAC, RAC, LPC, RPC\} \qquad (6)$$

where ω_x, ω_y, ω_z are the velocity's stereotaxic coordinates, LHC stands for left horizontal canals, RHC stands for right horizontal canals, etc. The magnitude and the sign of the input signal are obtained by summing the 6 equations, one for each canal from the right and the left inner ear:

$$\omega = \omega_{LHC} + \omega_{RHC} + \omega_{LAC} + \omega_{RAC} + \omega_{LPC} + \omega_{RAC} \qquad (7)$$

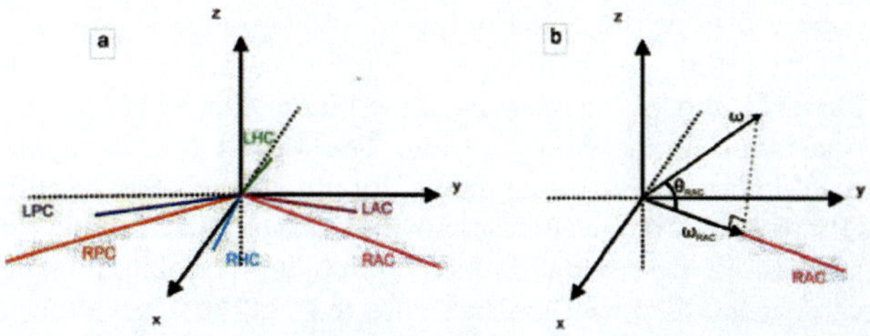

Fig. 3 a) MSV for semicircular canals; b) input velocity's projection on RAC MSV

The resulting magnitude is different from 0, because of the position of complementary semicircular canals in almost coplanar planes: LHRH (171° - angle between MSV), LARP (170)° and RALP (173°) ([5]), as shown in Fig.3.a. The sign of the sum determines whether the signal is excitatory (sign is greater than 0) or inhibitory (sign is less than 0).

IV. Convergence of different types of receptors

The arising issue in manipulating the output of the models (Regular and Irregular Otolith Organs, respectively Regular and Irregular Semicircular Canals) is to establish the type of convergence of the afferents.

Regarding the convergence of Regular-Irregular afferents, the weighted summation is considered, as in other studies from the literature [13]. Considering the perspective of the vestibular-sympathetic reflex mechanism, the inputs will be perceived by groups of regular and irregular afferents. Therefore, the weights corresponding to each type of afferents (k_{or}, k_{oir} - Regular and Irregular Otolith Organs, respectively k_{sr}, k_{sir} - Regular and Irregular Semicircular Canals) are set according to the size of representative group – the receptor groups considered in [11] and [14]. Following this pattern, according to data from [11], the calculated weights are k_{or}=0.6985 (234 regular receptors/335 receptors), and k_{oir}=0.3015 (101 irregular receptors/335 receptors). Considering information from [14], the calculated weights are k_{sr}=0.6709 (212/316), and k_{sir}=0.3291 (104/316).

The last issue is to set the Otolith Organs – Semicircular Canals convergence. According to studies from the literature [15], [16], [13], the result of the convergence will be provided after simple linear summation.

V. Simulation results

The developed models of the VR were tested in an orthostatic stress scenario, particularly body posture change from sit to stand position. The model used in these simulations is the one presented in Fig. 1.

The inputs for models are determined using a Matlab Simulink model that simulates the kinematics of the transition from sit to stand position. The values for the linear accelerations [m/s^2] (as inputs for models of the Otolith Organs) and the angular velocities [deg/s] (as inputs for models of the Semicircular Canals) are validated using various studies from the literature (e.g. [17]). The relevant inputs for the mentioned scenario are presented in Fig. 4. The output of the first excitatory signal generator blocks, providing a resultant linear acceleration a is also illustrated in Fig. 4a (the resultant angular velocity will be the same as the only input for the sit to stand posture change).

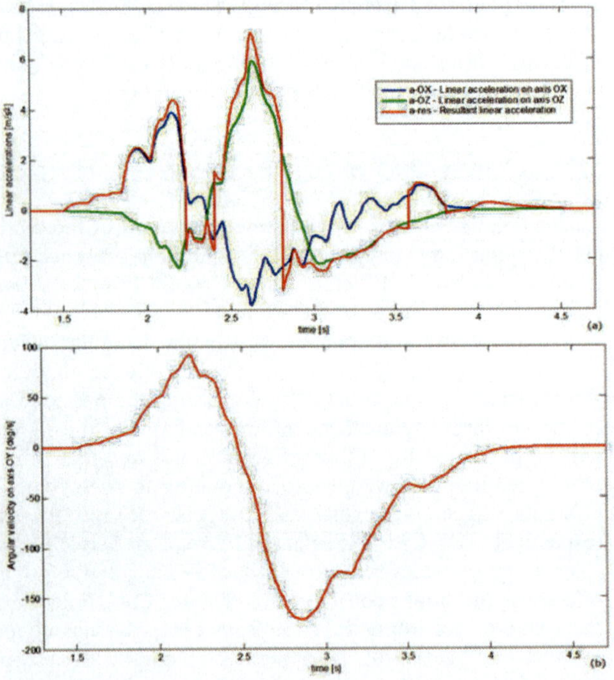

Fig. 4 Linear accelerations (a) and angular velocity (b) for sit to stand scenario

The outputs of the Regular and Irregular Otolith Organs, respectively Semicircular Canals are shown in Fig. 5 and 6. The composite output of the VR model is presented in Fig. 7.

As it can be observed in Fig. 5, 6 and 7, the firing rates of all VR models are corresponding qualitatively to the specifics of the considered scenario. The excitatory and inhibitory behaviors of the afferents are matching with the inputs from Fig. 4.

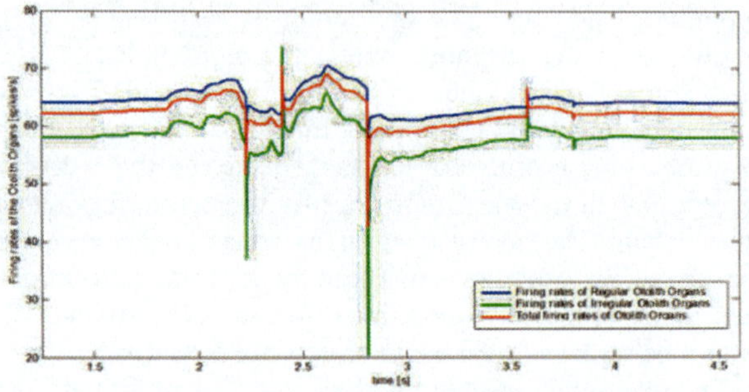

Fig. 5 Firing rates from the models of the Otolith Organs

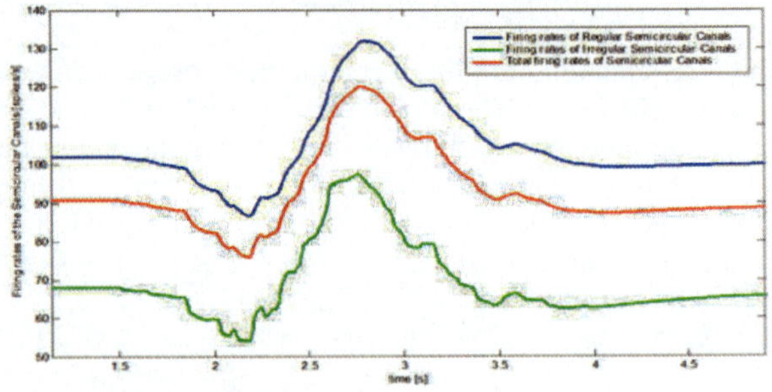

Fig. 6 Firing rates from the models of the Semicircular Canals

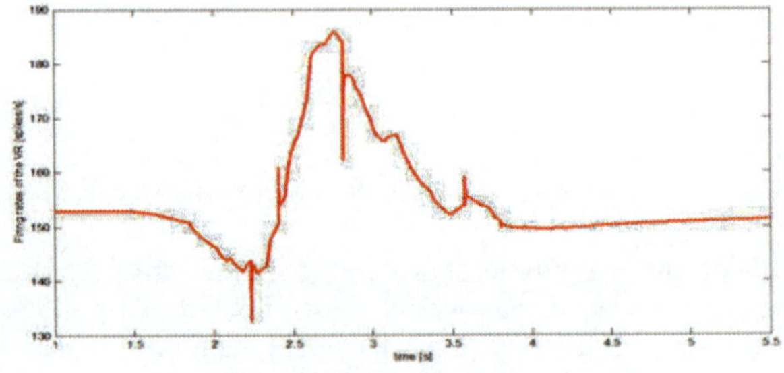

Fig. 7 The composite output of the VR model

VI. Conclusions

The vestibular receptors are the first information processing elements in the vestibular-sympathetic reflex, sensing linear and angular movement. In this direction, the current paper develops a lumped model for the Vestibular Receptors based on the current information available in the literature. The obtained simulation results for orthostatic stress encourage the pursue of a model for the entire vestibular-sympathetic reflex by coupling the VR model with models for the Vestibular Nucleus and for the Sympathetic System.

REFERENCES

1. Carter J.R., Chester A.R. (2008) Sympathetic responses to vestibular activation in humans. Am. J. Physiol. Integr. Comp. Physiol., Vol. 294, pp. 681–688, 2008.
2. Codrean A., Korodi A., Dragomir T.-L., Ceregan V. (2011) Modeling the Vestibular Nucleus. Intelligent Control and Computer Engineering, Lecture Notes in Electrical Engineering, Springer, Vol. 70, pp. 293-306, 2011.
3. Olufsen M., Ottesen J., Tran H., Lipsitz H., Novak V. (2006) Modeling baroreflex regulation of heart rate during orthostatic stress. Am. J. Physiol. Regulatory Integrative Comp. Physiol., Vol. 291, pp.1355-1368, 2006
4. Angelaki D.E. (2000) Spatiotemporal Processing of Linear Acceleration: Primary Afferent and Central Vestibular Neuron Responses. Journal of Neurophysiology, Vol. 84, No. 4, pp. 2113-2132, 2000.
5. Haque A., Angelaki D.E., Dickman J.D. (2004) Spatial tuning and dynamics of vestibular semicircular canal afferents in rhesus monkeys. Experimental Brain Research, Vol. 155, pp.81-90, 2004.
6. Hullar T.E., Santina C.C.D., Hirvonen T., Lasker D.M., Carey J.P., Minor L.B. (2004) Responses of Irregularly Discharging Chinchilla Semicircular Canal Vestibular-Nerve Afferents During High-Frequency Head Rotations. Journal of Neurophysiology, Vol. 93, No. 5, pp. 2777-2786, 2004.
7. Fernandez C., Goldberg J.M. (1976) Physiology of Peripheral Neurons Innervating Otolith Organs of the Squirrel Monkey. III. Response Dynamics. Journal of Neurophysiology, Vol. 39, No. 5, pp. 996-1008, 1976.
8. Valerio D., Sa Da Costa J. (2004) Ninteger: A Non-Integer Control Toolbox for MATLAB. Proceedings of the First IFAC Workshop on Fractional Differentiation and Applications, pp. 208-213, Bordeaux, France, 2004.
9. Oustaloup A. (1991) La commande CRONE: commande robuste d'ordre non entier. Hermès, Paris, 1991.
10. Charef A.; Sun H.H.; Tsao Y.Y.; Onaral B. (1992) Fractal system as represented by singularity function. IEEE Transactions on Automatic Control, Vol.37, No.9, pp.1465-1470, 1992.
11. Fernandez C., Goldberg J.M. (1976) Physiology of Peripheral Neurons Innervating Otolith Organs of the Squirrel Monkey. I. Response to Static Tilts and to Long-Duration Centrifugal Force. Journal of Neurophysiology, Vol. 39, No. 5, pp. 970-984, 1976.
12. Highstein S. M., Fay R. R., Popper A.N. (2004) The Vestibular System, Springer-Verlag, New York, 2004.
13. Wilden A. (2002) Analyse und Modellierung vestibularer Information in den tiefen Kleinhirnkernen. PhD Thesis, Ludwig-Maximilians-Universitat zu Munchen, Munchen, 2002.
14. Goldberg J.M., Highstein S.M., Moschovakis A.K., Fernandez C. (1987). Inputs From Regularly and Irregularly Discharging Vestibular Nerve Afferents to Secondary Neurons in the Vestibular Nuclei of the Squirrel Monkey. I. An Electrophysiological Analysis. Journal of Neurophysiology, Vol. 58, No. 4, pp. 700-718, 1987.
15. Furman J. M., Schor R. H. (2011) Semicircular Canal-Otolith Organ Interaction During Off-Vertical Axis Rotation in Humans. Journal of the Association for Research in Otolaryngology, Vol. 2, No. 1, pp. 22-30, 2001.
16. Crane B. T., Viirre E. S., Demer J. L. (1997) The human horizontal vestibulo-ocular reflex during combined linear and angular acceleration. Experimental Brain Research, Vol 114, No. 2, pp. 304-320, 1997.
17. Bussone W.R. (2005) Linear and Angular Head Accelerations in Daily Life. Master thesis at Virginia Polytechnic Institute and State University, USA, 2005.

Author: Alexandru Codrean
Institute: "Politehnica" University of Timisoara
Faculty of Automation and Computers
Department of Automation and Applied Informatics
Street: Bd. Vasile Parvan no. 2, postal code 300223
City: Timisoara
Country: Romania
Email: alexandru.codrean@aut.upt.ro

Integration of the Calcium Dynamics in the Excitation Contraction Coupling Process within a Multiscale Model of the Left Ventricle

B. Bhattacharya-Ghosh[1], S. Schievano[2], and V. Díaz-Zuccarini[1]

[1] University College London/Mechanical Engineering, WC1E 7JE, London, United Kingdom
[2] University College London/Institute of Child Health, WC1N 3JH, London, United Kingdom

Abstract— **The description of multiscale processes in a unified and coherent fashion is of paramount importance in cardiovascular computational physiology. In this paper, a model that links the process of cardiac excitation and contraction from a cellular to a tissue level will be presented. A model of the action potential will be linked to the kinetics (attachment and detachment) of the proteins responsible for cardiac contraction (via the so called "crossbridges mechanism"). As a final step, the coupling of the calcium-dependent kinetics model to a higher-scale model of the left ventricle will complete the multi-scale model of cardiac dynamics from the sub-cellular level to the organ.**

Keywords— **Multi-scale modeling and simulation, cardiac modeling, action potential, dynamic calcium concentration, crossbridge kinetics.**

I. Introduction

Cardiac muscle can be represented in mathematical terms as a multi-scale model and can be described by different subunits as shown in Fig. 1 [1]. During the excitation-contraction process (ECP), the cardiac muscle cell, also called cardiomyocyte is responsible for the contractile and relaxing periodic behavior of the heart.

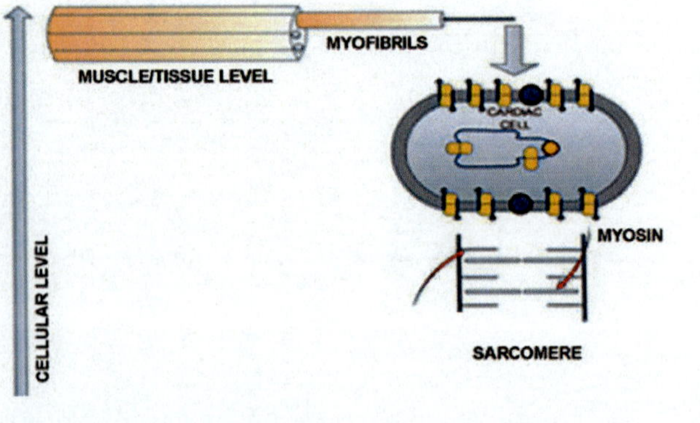

Fig. 1 Multi-scale model of a muscle (top left to bottom right): bundle of muscle fibres (muscle), myofibril, single muscle fibre (cell) and sarcomere (consists of thick (myosin) and thin (actin) filaments)

Within the cardiomyocyte, the sarcomere is the fundamental unit, while calcium (Ca^{2+}) is the key signalling ion of this process. Myocytes contract when the intracellular calcium concentration rises. To describe the ECP at cellular level, the Luo Rudy (LRII) model [2] [3], a well known cardiac cell model, is used and as a result, the action potential (AP) is calculated. A change in the concentration in the intracellular calcium produces a change in the action potential and is directly related to the kinetics at the protein level that are at the basis of cardiac contraction. Two protein chains are responsible for the basic contraction mechanisms in cardiac muscle (as shown in Fig. 1) at the sarcomere level. The actin-myosin sliding mechanism is well documented in cardiac literature [4] and is based on the chemical reactions of attachment and detachment between the actin and myosin protein chains. In this paper, the sliding mechanisms and related kinetics, are determined by a four state model (4SM) [5].

To extend the coupling to a higher scale both models are implemented into a multi-scale model of the left ventricle (LV). The LRII and 4SM will provide a mathematical description of the sub-cellular and cellular scales and this will be an input within the LV model that will connect with the tissue and organ scales (e.g. it calculates, ventricular pressure, energy and force (Fig. 2.)).

This paper will present a multi-physics model of cardiac contraction that includes the ECP with dependence on the intracellular concentration ($[Ca^{2+}]$) at different scales. To simulate the complex interactions between the different scales, from the cellular level to the muscle (tissue) level and organ level, an integrative approach, linking and modifying the existing models that represent these scales will be presented in the sections that follow.

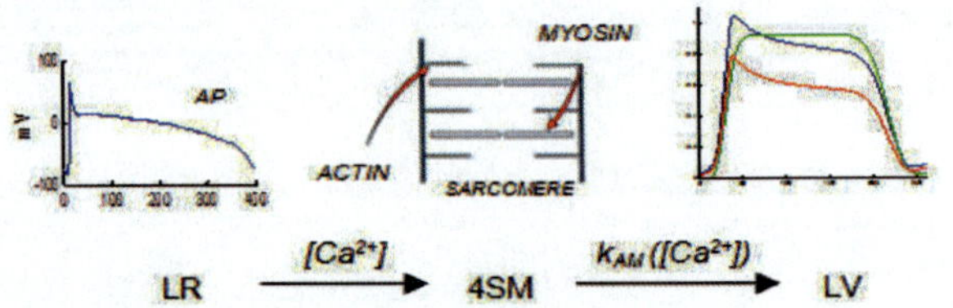

Fig. 2 Coupling process from LR to LV via 4SM; based on $[Ca^{2+}]$ and $[Ca^{2+}]$ dependent actin-myosin-kinetics ($k_{AM}([Ca^{2+}])$)

S. Vlad and R.V. Ciupa (Eds.): MEDITECH 2011, IFMBE Proceedings 36, pp. 266–269, 2011.
www.springerlink.com

II. METHODS

A. Action Potential

All cardiac cells have the ability to generate an electrical impulse, i.e. an AP. Therefore an initiated impulse will trigger the contraction from one cell to the next. Each cardiac cell includes various time and voltage dependent channels [6] which are responsible for the behaviour and characteristics of the AP. The AP is the combined result of all channel-currents and is measured at the membrane surface of a cell. It is divided into different phases as shown in Fig. 3.

The fundamental ions affecting each phase of the AP are Na^+, K^+ and Ca^{2+}. Dependent on their influx, efflux and concentration, all ion-specific channels that are related by the change a particular ion will also have an impact on the AP itself.

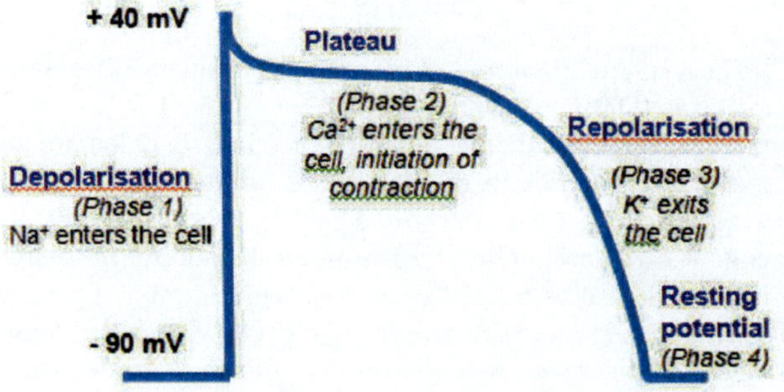

Fig. 3 Phases of AP: Phase 1 – Depolarisation., Phase 2 - Plateau, Phase 3 - Repolarisation, Phase 4 – Resting potential

B. Luo Rudy Model

The LRII model is widely used in the electrophysiology literature to simulate the AP. The numerical reconstruction of the ventricular AP is based on a Hodgkin-Huxley-type approach [7].

An advanced model of LRII by Livshitz and Rudy [8] accounts for the dynamic changes in ionic concentrations and ionic fluxes during the action potential and will be used further on this paper. The LRII model structure according to the processes that control myoplasmic concentrations of Ca^{2+}, Na^+, and K^+ is shown in Fig. 4 [9].

This model will use a set of differential equations based on [7] to calculate the action potential and the currents. The rate of change of membrane potential V is given by

$$\frac{dV}{dt} = \frac{1}{C} \cdot (I_t + I_{st}) \tag{1}$$

where C is the membrane capacitance, I_{st} is a stimulus current and I_t is the sum of all ionic fluxes that carry a specific ion x.

$$I_t = \Sigma I_x \tag{2}$$

The general equation to calculate the current I_x of any channel is provided by

$$I_x = g_x \cdot GAI_x \cdot (V_m - E_x) \tag{3}$$

where x represents the specific ion, e.g. Ca^{2+}, Na^+, and K^+, g_x the total conductance, GAI_x the fraction of channels that are open, V_m the membrane potential and E_x the Nernst potential.

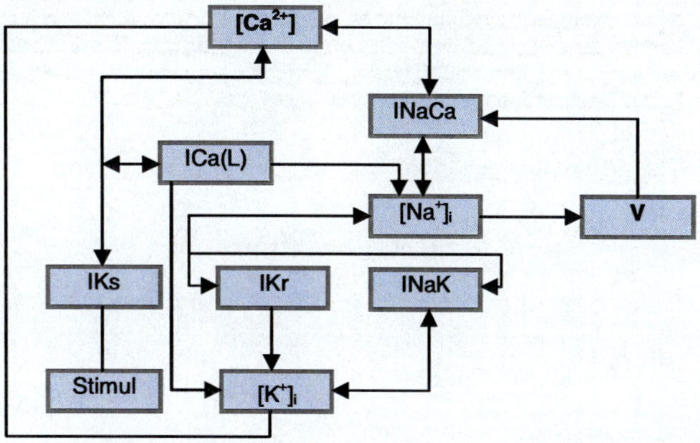

Fig. 4 Calcium dependence in Livshitz-Rudy model

C. Four State Model

A change in the calcium concentration activates the process of crossbridges (acting as extensions of myosin to attach and detach to actin). Changing myoplasmic Ca^{2+} concentration and myofilament sensitivity to Ca^{2+} are mainly attributed to the mechanism of Ca^{2+} contraction coupling in cardiac muscle [10]. A link between the left ventricular pressure and the Ca^{2+} concentration may reveal the impact of regulatory proteins and kinetics of actin and myosin cross-bridge interaction on the calcium contraction coupling [11]. The model represents interactions between the actin and myosin for cross-bridge formation and includes the binding of Ca^{2+} to troponin C (TnC) on the actin myofilament (TnCA).

State 1 and 2 illustrate the phases of myosinheads (crossbridges) (M) not attached to actin (A). No force is generated by myosinheads due to their detached state. State 3 and 4 show the phases of attached crossbridges, when force is generated. During state 2 and 3 calcium is strongly bound to troponin C (TnC) within the troponin complex located on the actin filament (A).

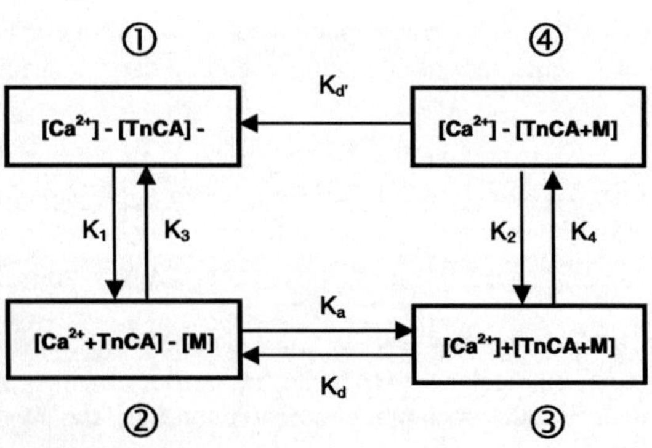

Fig. 5 Four State Model. TnCA, troponin C (TnC) protein molecule on the actin (A) myofilament; M, myosin head. "+" indicate strong bonds while "-" indicate weak bonds. K_1, K_2, K_3 and K_4 are measures of Ca^{2+}-binding affinity. K_a, K_d and $K_{d'}$ are measures of crossbridge association and dissociation rates.

The differential equations that describe this process are:

$$\frac{d[TnCA]}{dt} = -K_1[Ca][TnCA] + K_3[CaTnCA] + K_{d'}[TnCAM] \quad (4)$$

$$\frac{d[M]}{dt} = K_{d'}[TnCAM] - K_a[CaTnCA][M] + K_{d'}[CaTnCAM] \quad (5)$$

$$\frac{d[CaTnCA]}{dt} = K_1[Ca][TnCA] - K_3[CaTnCA] - K_a[CaTnCA][M] + K_d[CaTnCAM] \quad (6)$$

$$\frac{d[CaTnCAM]}{dt} = K_a[CaTnCA][M] + K_2[Ca][TnCAM] - (K_d + K_4)[CaTnCAM] \quad (7)$$

$$\frac{d[TnCAM]}{dt} = -K_4[CaTnCAM] - K_{d'}[TnCAM] - K_2[Ca][TnCAM] \quad (8)$$

The total number of crossbridges will be calculated as follows:

$$[XA] = [CaTnCAM] + [TnCAM] \quad (9)$$

D. *Left Ventricle Model*

To simulate the cardiac dynamics of the left ventricle we use the left ventricular model described in [1]. It represents different biological levels from the mechanisms of contraction up to the hemodynamics of the left ventricle. The model characterizes cardiac function via a set of equations that are described and calculated in [1]. The muscular energy (E_{XA}) of the LV can be calculated based on the attached and detached crossbridges (XA). This will provide a link to calculate the force (F_m) and left ventricular pressure (LVP) as follows:

$$E_{XA} = A_A \cdot XA + B_A \cdot XA \cdot (\log(XA) - 1) \quad (10)$$

$$F_m = F_A \cdot E_{XA} + F_P \cdot (1 - E_m) \quad (11)$$

$$LVP = \frac{F_m \cdot n}{3 \cdot A_T \cdot L_m^2} \quad (12)$$

Full details of this formulation and its constants are presented in [1].

III. Results

The mathematical model includes 25 ordinary differential equations (ODE), of which 18 ODE [8] in the LR II provide the intracellular calcium function needed to calculate the crossbridge kinetics in the 4SM. The kinetics are governed by additional 5 ODE described in [5], that are highly dependent on $[Ca^{2+}]$. The ODE in the 4SM supply an insight on the actin, myosin and troponin concentration at a particular time during an entire crossbridge cycle. Given the shown kinetics dependence on calcium that allows us to determine the proportion of attached and detached crossbridges (XA).

The percentage of attached crossbridges is calculated by the following equation:

$$XA = \frac{[M_{Max}] - [M]}{[M_{Max}]} = \frac{[CaTnCAM] + [TnCAM]}{[M_{Max}]} \quad (13)$$

Distinguishing the LV muscular energy E_{XA} is required to calculate other variables such as, F_m and LVP (Fig. 6). These are calculated using XA from the 4SM and further equations of the LV model [1] in a differential form to implement and present the couple model in its entirety.

Results were calculated, using the MATLAB® ODE suite. The coupling was achieved by a combination of ODE and algebraic equations, following [1] as a blueprint. Data paced at a cycle length of 400 ms is chosen according to related time-dependent initial conditions for the LR II from [8] and time-independent initial conditions for the 4SM and LV model from [5] and [1], respectively.

IV. Discussion

This model has been used successfully to study the mechanism of the cardiac ECP. Simulation results suggest that the coupled model is able to integrate well the calcium dynamics into the whole left ventricular model and is able to show different aspects of cardiac dynamics at different scales. Fig. 6 shows E_{XA}, F_m and LVP, which are exemplary results of the macroscopic level based on the microscopic level as $[Ca^{2+}]$ and $k_{AM}([Ca^{2+}])$ (Fig. 2). The results show that the model exhibit high sensitivity to the crossbridge kinetics. Also, the specific behavior of CaTnCA within the kinetic model is a dominant and sensitive factor to determine the characteristics of other kinetic rates (Fig. 5).

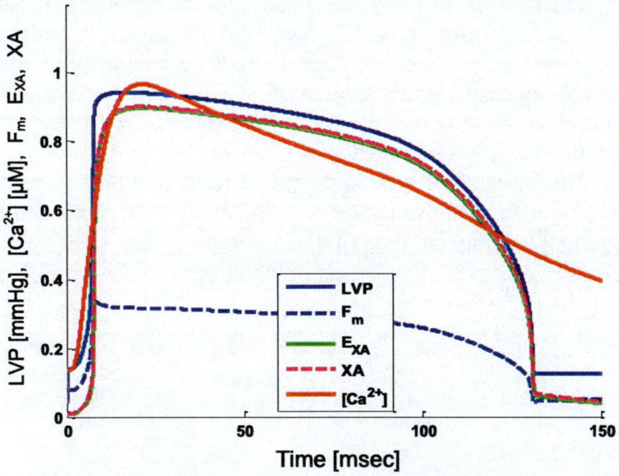

Fig. 6 Normalized results for: LVP (blue line; divided by a factor of 120), F_m (dashed blue line) and E_{XA} (green line) resulting from LV model with dependence on XA (dashed magenta line) from 4SM and $[Ca^{2+}]$ (red line; divided by a factor of 1.6) from LRII model

V. Conclusions

In this paper we presented a multi-physics & multi-scale model of the LV. This model allows us to understand the consequences at the macroscopic (organ) level of the mechanisms of contraction in the heart. The overall coupling of the calcium dynamics that activates and affects the crossbridge kinetics from a protein to a cellular level, at the same time leads us to achieve and distinguish the mechanisms of ECP in the LV at an organ level. For future studies this multi-scale model can be used to understand how pharmacological intervention can affect cardiac contraction at different scales.

Acknowledgments

The authors would like to acknowledge the financial support of the European Commission, FP7, People Programme, Marie Curie actions via the Marie Curie Initial Training Network MeDDiCA (PITN-GA-2009-238113).

References

1. Vanessa Díaz-Zuccarini, Jacques LeFèvre An energetically coherent lumped parameter model of the left ventricle specially developed for educational purposes Computers in Biology and Medicine 37 (2007) 774 – 784
2. CH Luo and Yoram Rudy A model of the ventricular cardiac action potential. Depolarization, repolarization, and their interaction Circulation Research is published by the American Heart Association. 7272 Greenville Avenue, Dallas, Circ. Res. 1991;68;1501-1526
3. CH Luo and Yoram Rudy A dynamic model of the cardiac ventricular action potential. I. Simulations of ionic currents and concentration changes Circulation Research is published by the American Heart Association. 7272 Greenville Avenue, Dallas,Circ. Res. 1994;74;1071-1096
4. B. F. Gray and Gonda The Sliding Filament Model of Muscle Contraction II. The Energetic and Dynamical Predictions of a Quantum Mechanical Transducer Model J. theor. Biol. (1977) 69, 187-230
5. Samhita S. Rhodes, Kristina M. Ropella, Said H. Audi, Amadou K. S. Camara, Leo G. Kevin, Paul S. Pagel and David F. Stowe Crossbridge kinetics modeled from myoplasmic $[Ca^{2+}]$ and LV pressure at 17°C and after 37°C and 17°C ischemia Am J Physiol Heart Circ Physiol 284:H1217-H1229, 2003. First published 12 December 2002
6. Rob MacLeod and Quan Ni Simulation of Cardiac Action Potentials, Background Information January 29, 2006
7. Bernard Victorri, Alain Vinet, Fernan A. Roberge and Jean-Pierre Redrouhard Numerical Integration in the Reconstruction of Cardiac Action Potentials Using Hodgkin-Huxley-Type Models Computers and Biomedical Research 18, 10-23 (1985)
8. Leonid Livshitz and Yoram Rudy Uniqueness and Stability of Action Potential Models during Rest, Pacing, and Conduction Using Problem-Solving Environment Cardiac Bioelectricity and Arrhythmia Center, Washington University in St. Louis, St. Louis, Missouri, Biophysical Journal Volume 97 September 2009 1265–1276
9. Interactive Tool for Cell Model Simulation Leonid Livshitz and Yoram Rudy Poster - Cardiac Bioelectricity and Arrhythmia Center (CBAC)
10. George S.B. Williams, Gregory D. Smith, Eric A. Sobie, M. Saleet Jafri Models of cardiac excitation–contraction coupling in ventricular myocytes Mathematical Biosciences 226 (2010) 1–15
11. Landesberg, A., and S. Sideman Mechanical regulation of cardiac muscle by coupling calcium kinetics with crossbridge cycling: a dynamic model. Am. J. PhysioZ. 267 (Heart Circ. PhysioZ. 36): H779-H795, 1994

Author: Dipl.-Ing. Benjamin Bhattacharya-Ghosh
Institute: University College London
Street: 74Gower Street, LG
City: WC1E 6EG,London
Country: United Kingdom
Email: b.bhattacharya-ghosh@ucl.ac.uk

Biomechanics of Noncarious Cervical Lesions

G. Beresescu[1] and L.C. Brezeanu[2]

[1] University of Medicine and Pharmacy, Tg.Mures, Romania
[2] Petru Maior University of Tg.Mures, Faculty of Engineering, Tg.Mures, Romania

Abstract— **Noncarious cervical lesions (NCCLs) restoration represents a unique clinical situation due to their multifactorial etiology. Though the mechanical theory of cervical lesions formation is widely accepted, its mechanism is not fully understood. The incidence of NCCL refers to the facial and oral aspects of the teeth. Finite Elements Method (FEM) were drawn up, applied with various occlusal forces and analyzed in order to observe the stress distribution. The standard biomechanical unit involves restorative material, tooth structure and interface between the restoration and tooth. The purpose of this study was to examine the NCCL formation caused by occlusal forces and the behavior of restored and unrestored lesions.**

Keywords— **Finite Element Analysis, abfraction, cervical lesion, stress, displacement, restorative material.**

I. Introduction

Noncarious cervical lesions (NCCL) are considerable restorative challenges for the dentist. NCCL are defined as the loss of tooth structure at the cement-enamel junction. However, literature also describes other destructive processes that originate on the external surface of the tooth and affect it causing irreversible damage to the tooth structure, such as erosion, abrasion, attrition and abfraction. Dental erosion represents the physical results of loss of hard tissue caused by acid attack. Abrasion represents the pathological loss of hard tissue through abnormal mechanical processes. The term of dental attrition is used to describe the physiological wear of hard tissue. The term of abfraction describes a special type of wedge shaped defect in the cervical region of the tooth.

The tooth has not a rigid structure, hence it can suffer strains when various forces/loads are applied. Intraoral loads vary from 10N to 430N, the normal clinical values being considered of 70N [9]. One current hypothesis is that the tensile or compressive strains gradually produce micro fractures.

Loads applied under various angles result in different flexures of the tooth: lateral flexure at occlusal loads of 40° or axial flexure at occlusal loads directed axially to the tooth (Fig.1).

Lately, the numerical analysis methods have become indispensable in solving engineering and biomechanical problems mainly due to their increasing reliability and accuracy. In the field of biomechanics, the finite element methods are able to address a wider range of problems than the conventional methods, because of their structural and material complexity. However, both conventional and FEM present a major shortcoming concerning their inability to predict failure by fracture.

The bulk of dentistry papers address the influence of dental materials used in NCCL restorations focusing mainly on their retention/loss rates. Thus, there are few studies treating the biomechanical mechanism of NCCL and the selection of that restoration material which best exhibits the most appropriate elastic characteristics.

The purpose of this study was a radical approach to the behavior of an intact and restored/unrestored tooth undergoing a mechanical load of various values.

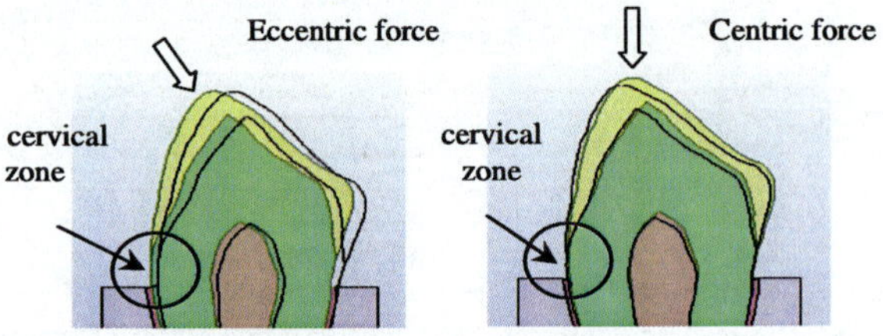

Fig. 1 Diagram of tooth flexure creating cervical stresses

II. Materials and methods

A 2D mathematical finite elements analysis model was generated, using an intact normal human mandibular canine. The quality of the analysis results depends on the accuracy of the model.

Properties of dental tissue are shown in Table 1.

Table 1 Properties of dental tissue

Materials	Young's modulus E [MPa]	Poisson's ratio μ
Enamel	$6{,}9.10^4$	0,30
Dentine	$1{,}67.10^4$	0,31
PDL	12	0,45
Bone	$1{,}47.10^4$	0,3
Pulp	2	0,45

S. Vlad and R.V. Ciupa (Eds.): MEDITECH 2011, IFMBE Proceedings 36, pp. 270–275, 2011.
www.springerlink.com

All materials were considered elastic (right proportion between stresses and specific strain and Hooke law valability) and isotropic (with identical elastic characteristics on all directions). Longitudinal elastic modulus (Young's modulus E) and Poisson's ratio μ values for the materials used in the model were derived from standard texts [3].

Numerical analysis was carried out using ALGOR-Fempro solver. A plan model reproducing a vestibular and lingual section of the lower canine was created. A denser mesh with a large number of EF was built in the area of interest in order to obtain the best replica of the tooth and the most faithful analyses of the situation. To simulate material continuity, all the parts of the dental structure are considered connected and forming whole body (Fig.2).

Two situations of tooth loading were considered:

a. Oblique nodal force at 40 degrees to vertical applied onto the vestibular aspect at h=8.993mm from cervical area of increasing magnitudes: 40, 80, 120, 160, 200N (Fig.1.a.);
b. Vertical nodal force of increasing magnitudes: 40, 80, 120, 160, 200N applied onto the tip of the tooth (Fig.1.b.).

The values of the loadings (40-200N) are considered "study loads" that cover the whole range of the clinical situations. The forces applied were of the same values, both for the vertical and tensile stress, in order to obtain the most accurate results by means of comparison of the two situations.

The study sets off from the working hypothesis that there are various differences in stress profile between healthy teeth and teeth with cervical enamel damage.

III. Results

The results of the present study are shown in the following significant values:

- Equivalent stress Von Mises σ_{ech};
- Stress following tooth direction Z-Z;
- Minimum main stress (compression effect) σ_2;
- Resultant displacement.

Model I. Healthy tooth - lesion mechanism

The present study used simulations of different values and positions of the loads, both vertical and oblique, on a healthy tooth. The result show that the most stress-prone area with the highest risk of mechanic damage is the cervical area of the tooth (Fig.1).

A maximum stress at 0,1mm above the cervical line for an oblique load was noticed (Fig.3).

The variation graphs for equivalent stress values of the eccentric and centric forces applied on the cervical interface were created (Fig.4, Fig.5).

Model II. Tooth with cervical lesion

Loads of different positions and magnitudes applied on a tooth with cervical lesion will lead to an increase of stress in the area.

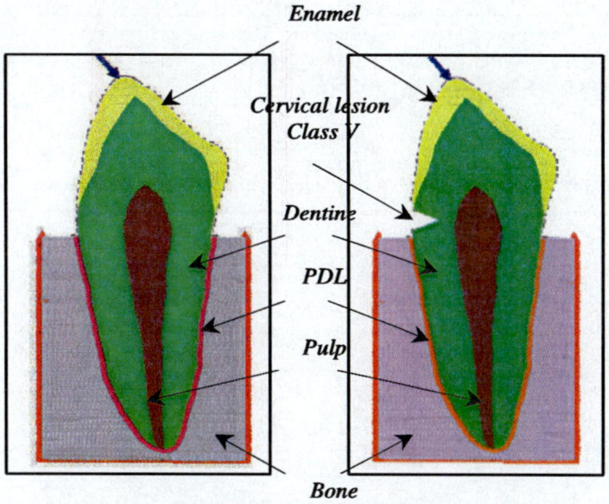

Fig. 2 2D model of the lower canine.
Healthy tooth and tooth with lesion

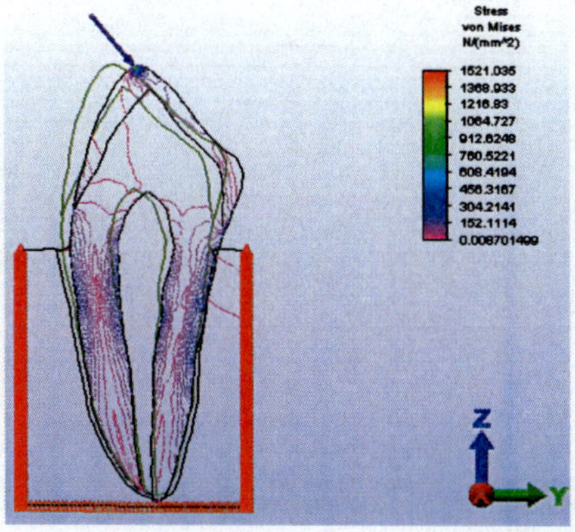

Fig. 3 Equivalent stress Von Mises distribution
Curves of equal values corresponding to an eccentric force of F=160N

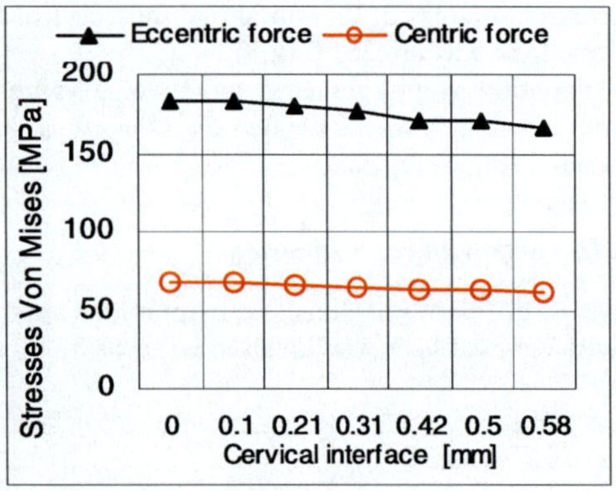

Fig. 4 Von Mises equivalent stress variation in the cervical interface for different position of a force: F=160N

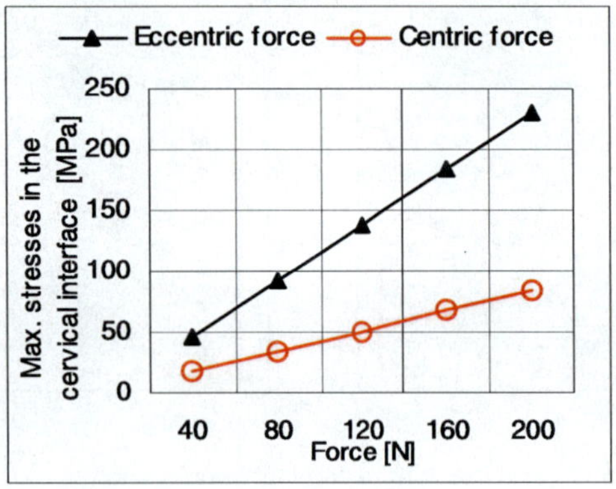

Fig. 5 Comparison: eccentric force – centric force The node with maximum stress Von Mises values (0,1 mm from colet)

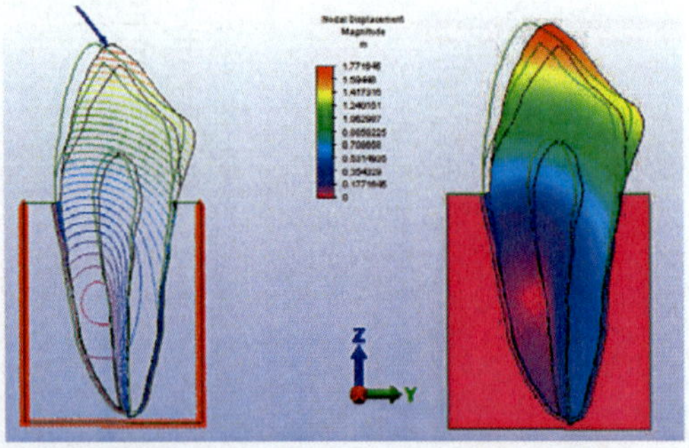

Fig. 6 Displacements distribution for eccentric force of F=160N (curves and areas of equal stress)

The lesion will become a stress concentrator with cracks propagating onto and into the tooth, ultimately leading to tooth fracture (Fig.7).

The results obtained after simulations on a tooth lesionned on the cervical area were compared to those obtained following simulations on a healthy tooth (Fig.8).

The same values were considered of significance both for the healthy and lesionned tooth.

Model III. Tooth with restored lesion

Simulations on a restored lesion showed that after restoration the values of the stress in all the elements of the dental structure exhibited slight differences similar to those noticed in the healthy tooth (Fig.3, Fig.15).

After restoration, both the stress concentrator in the bottom of the cavity (Fig.16) and the displacement of stress towards the apex of the restoration will disappear (Fig.18).

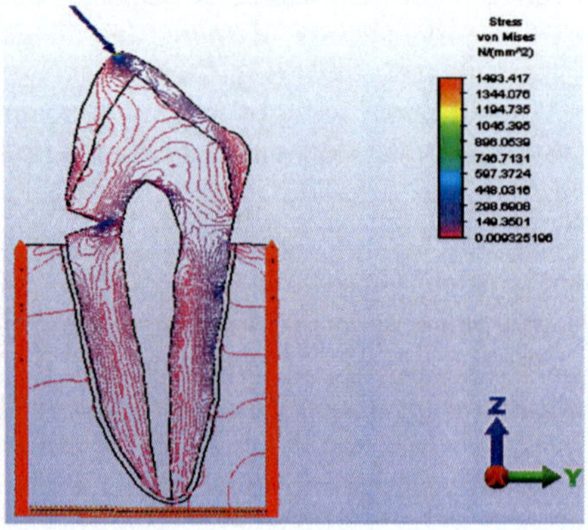

Fig. 7 Equivalent stress Von Mises distribution. Curves of equal values corresponding to a eccentric force of F=160N

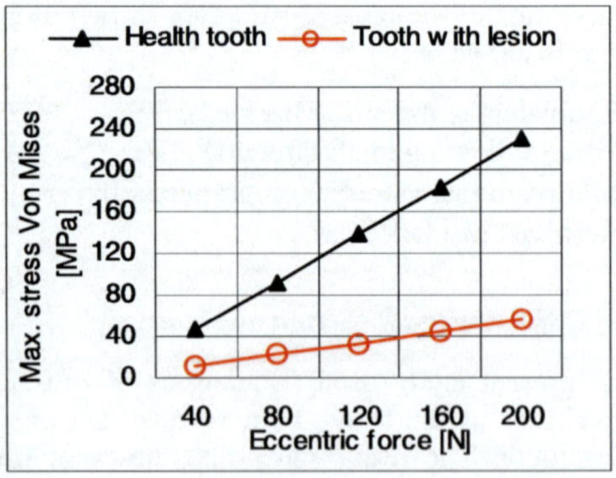

Fig. 8 Comparison: intact tooth – tooth with cervical lesion for oblique force The node with maximum stress Von Mises values (0,1 mm from colet)

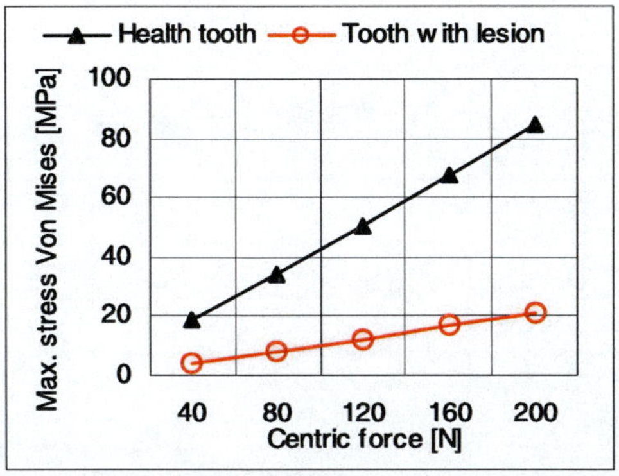

Fig. 9 Comparison: intact tooth – tooth with cervical lesion for centric force. The node with maximum stress Von Mises values (0,1 mm from colet)

Since the quality of any material is defined by its elastic characteristics (Young's modulus E and Poisson's ratio μ) it is extremely important to use those restoration materials which present the closest elastic characteristics to the material to be restored.

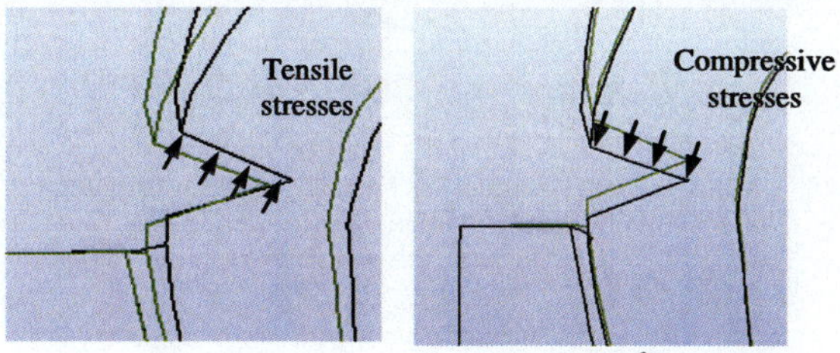

a. b.

Fig. 10 Deformed position of the lesion:
a. eccentric force; b. centric force

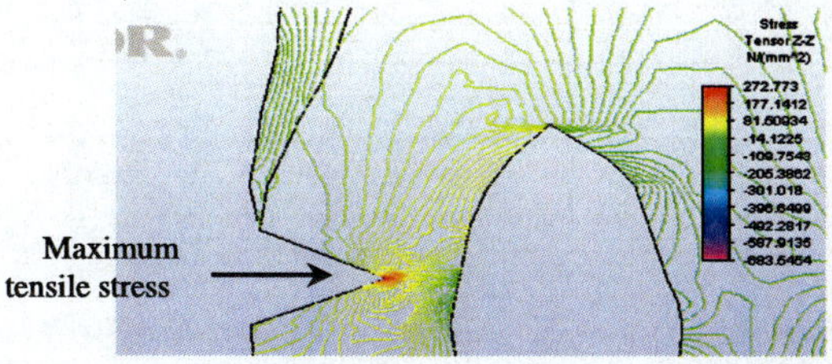

Fig. 11 Stress distribution following vertical axes of the tooth Z-Z in the bottom of the lesion for the eccentric force (positive values in the lesion)

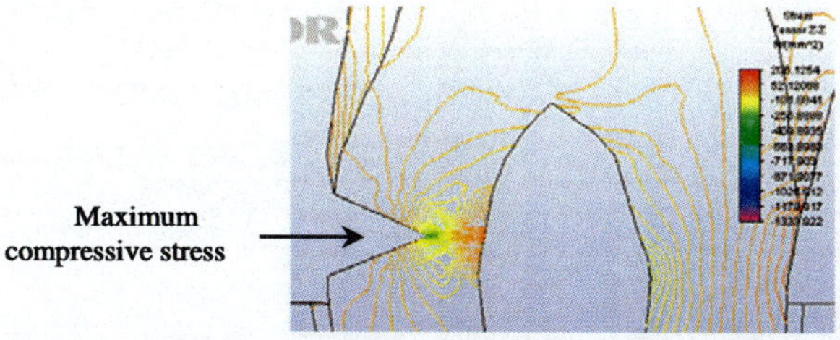

Fig. 12 Stress distribution following vertical axes of the tooth Z-Z in the bottom of the lesion for the centric force (negative values in the lesion)

IV. Discussion

Our study is based on the golden rule in engineering according to which any stress will always follow the direction of the most rigid material, that is to say, of the material with the highest elastic modulus.

The results obtained showed that:

a. *in the healthy tooth*

- maximum stress occurs in the cervical area irrespective of load direction (Fig.4);
- maximum stress values occur at 0,1mm above the cervical line (Fig.4.);
- stress values in the cervical area increase with occlusal loads values (Fig.5);
- Von Mises equivalent stress values for the same value of the load are higher for oblique loads (Fig.5).

The results show that from a mechanic stand point, the maximum strain appears in the cervical area at 0,1 mm above the cervical line, irrespective of load direction. This is the area were the tooth is most exposed to flexure leading to a concentration of stress which increases with the occlusal forces, ultimately leading to cracks.

b. *in the tooth with cervical lesion*

- Von Mises equivalent stress values are higher in the lesionned tooth than in the healthy tooth (Fig.8, Fig.9);
- oblique loads lead to lateral flexure of the tooth and vertical loads lead to axial compression;
- vertical direction of occlusal loads result in higher values of the stress in the lesionned area (Fig.13);
- as a result of load direction on the tooth, a stretching of the tooth appears at oblique loads and a compression of the tooth appears at vertical loads (Fig.10); the phenomenon determines positive maximum strain values in the bottom of the lesion (stretching strain) at oblique loads, respectively negative maximum strain values in the bottom of the lesion (compressive strain) at vertical loads (Fig.13);
- maximum stress values appear in the bottom of the lesion which becomes a stress concentrator (Fig. 11, Fig.12);

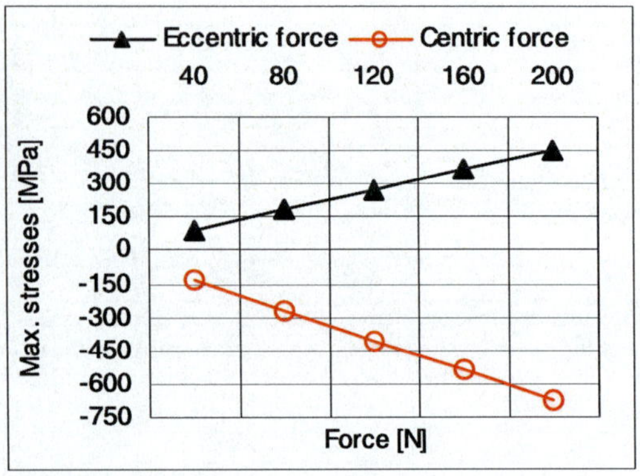

Fig. 13 Variation of maximum stresses Von Mises values stress in the concentrator at various values of both eccentric and centric force

- maximum stress values in the bottom of the lesion increase with external loads (Fig.13);
- the stress concentrator in the bottom of the lesion will lead, in time, to cracks and their propagation onto and into the tooth; this constitutes a risk factor for fracture

c. *in the restored tooth*

- after reconstruction, stress values in all structure elements register differences similar to those in the healthy tooth (Fig.3, Fig.14, Fig.16, Fig.17);
- stress values in the dentine exhibit very close figures, indicating that the reconstruction material undergoes the same distribution of stress as the healthy tooth (Fig.17);
- maximum strain in the reconstruction appears in the bottom of it irrespective of the material used, direction and magnitude of loads;
- considering that maximum strength for dentine ~ $\sigma_{a\ dentine}$ = 105,5 MPa [5], it became clear that loads higher then F=80N will damage the tooth structure.

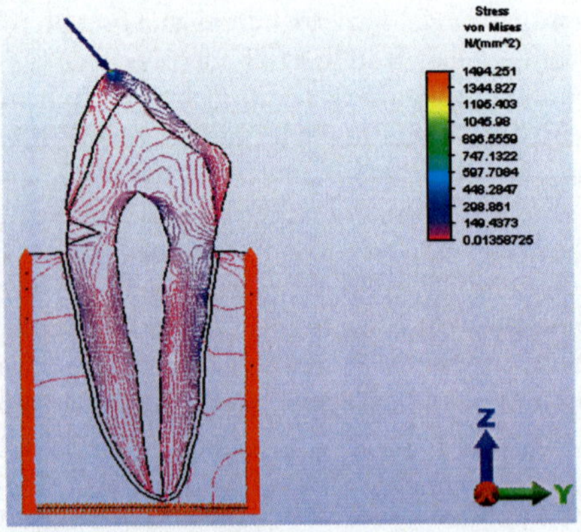

Fig. 14 Equivalent stress Von Mises distribution
Curves of equal values corresponding to a eccentric force of F=160N

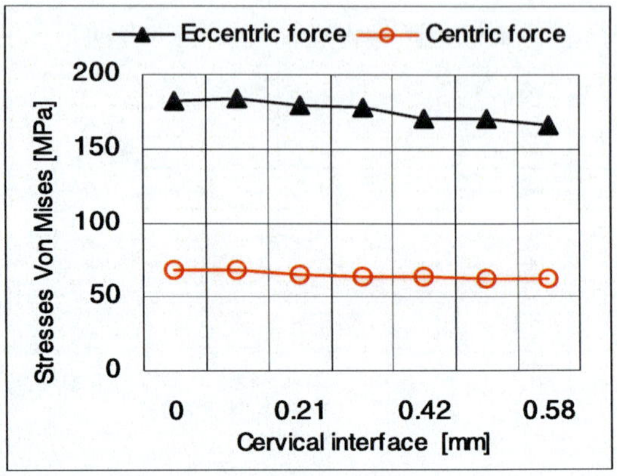

Fig. 15 Von Mises stress values in the cervical interface for different positions of a oblique force F=160N

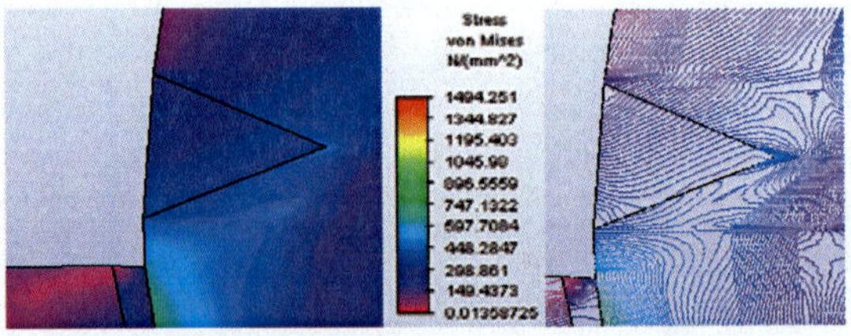

Fig. 16 Detail - Equivalent stress Von Mises distribution
Areas and curves of equal values corresponding to a eccentric force of F=160N

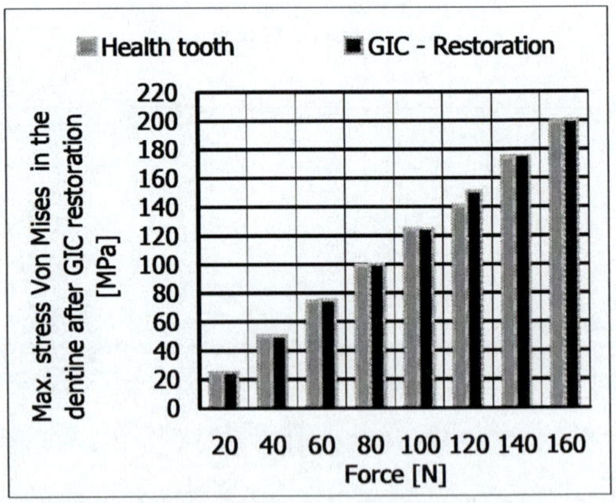

Fig. 17 Maximum Von Mises Stresses in the dentine.
Comparison: healthy tooth - tooth with restoration

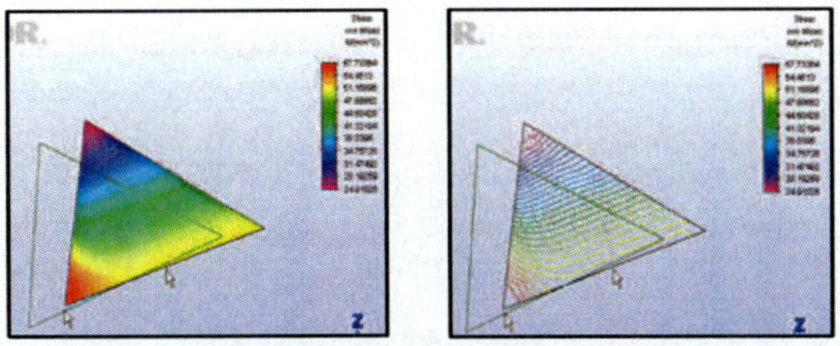

Fig. 18 Von Mises Stress along the cervical interface

V. Conclusions

We appreciate that Finite Elements Method is a valuable complementary method which offers accurate images of the behavior of the structure under study, while being a noninvasive method of analysis.

With all the limitations of this numerical study, the following conclusions can be drawn:

1. Any load applied on a tooth can result in enamel damage on the cervical area leading to NCCL and modest possibilities of preventing the restart of the process.
2. The cervical lesions induce a stress concentrator, which, in turn, leads to cracks on the surface of the enamel and even fracture of the tooth in the cervical area.
3. As materials of choice in NCCL restorations, we suggest those with the closest elastic characteristics to the dental tissues to be replaced.

FEM was also applied to analyze the behavior of restored lesions. The material fracture formulation is based on rotating crack model and propagation. The modeling technique presented offers an insightful understanding of the nonlinear relationship between loading capacity, damage and softening in multiple materials possible.

The finite element code adopted allows an automatic insertion of cracks, enables remeshing and accommodates self contact between the cracked interfaces.

Numeric simulation of a NCCL process and its restorations can became both an alternative to the clinical and experimental studies on the tooth behavior and a major part in selecting and developing biomaterials.

References

1. Beata Dejak, Andrzej Mlotkowski, Maciej Romanowicz, (2005), Finite element analysis of mechanism of cervical lesion formation in simulated molars during mastication and parafunction, The Journal of Prosthetic Dentistry, Volume 94, Number 6, deccember, 2005, pp.520-529;
2. Boitor, C, Frăţilă, A, Ionaş, M, Pascu, A, Oleksik, V, Pârvu, B, (2009), Analiza prin metoda elementului finit a leziunilor de abfracţie dentară, Revista Română de Stomatologie, Vol LV, Nr. 4, 2009, pp.263-267;
3. Gurbuz, T, Sengul, F, Altun, C, (2008), Finite Element Stress Analysis of Short-post Core and Over Restorations Prepared with Different Restorative Materials, Dental Materials Journal, Vol. 27(4), 2008, pp.499-507;
4. Ichim, I, Li, Q, Loughran, J, Swain, MV, Kieser, J, (2007), Restoration of non-carious cervical lesions, Part I. Modelling of restorative fracture", Dental Materials Journal, Vol. 23, 2007, pp.1553-1561;
5. Ichim, I, Schmidlin, PR, Li, Q, Swain, M.V, Kieser, J, Swain, MV, (2007), Restoration of non-carious cervical lesions. Part II. Restorative material selection to minimize fracture, Dental Materials Journal, Vol. 23, 2007, pp.1562-1569;
6. Narayanaswamy, S, Meena, N, Shetty, A, Kumari, A, Naveen, DN, (2008), Finite element analysis of stress concentration in Class V restorations of four groups of restorative materials in mandibular premolar, Journal of Conservative Dentistry, Vol.11 3, 2008, pp121-128;
7. J.E.A. Palamara, D.Palamara, H.H. Messer, M.J. Tyas, (2006), Tooth morphology and characteristics of non-carious cervical lesion, Journal of Dentistry, 2006,.Vol. 34, pp.185-194;
8. S. Rees, M. Hammadeh, (2006), Undermining of enamel as a mechanism of abfraction lesion formation: a finite element study, European Journal of Oral Sciences, No. 112, 2004, pp.347-352;
9. Roberson TM, Heymann HO, Swift EJ, (2006), Art and Science of Operative Dentistry, Mosby, 2006.

Authors:

Gabriela Beresescu
Institute: University of Medicine and Pharmacy
Street: 38, Gheorghe Marinescu
City: Targu Mures
Country: Romania
Email: gabriella_fellicia@yahoo.com

Ligia Cristina Brezeanu
Institute: "Petru Maior" University of Targu Mures
Street: 1, N. Iorga
City: Targu Mures
Country: Romania
Email: ligia.brezeanu@gmail.com

Modelling the Development of In-Stent Restenosis: Preliminary Results of a Structural Model

C.M. Amatruda, D.R. Hose, P.V. Lawford, and A.J. Narracott

Medical Physics Group, Department of Cardiovascular Science, University of Sheffield, Sheffield, UK

Abstract— **Coronary artery diseases are at present commonly treated by minimally invasive treatment such as intravascular stents. However, such treatment is compromised by in-stent restenosis, a re-narrowing of the artery related to the injury of the vessel wall as a result of the local stress caused by the stent struts. The aim of this study is to develop a computational model to represent this tissue growth in a finite element model of a stented coronary artery.**

A 1/6 symmetry section of a coronary artery was generated within ANSYS Mechanical APDL version 12.0 (ANSYS Inc.). A 'ghost' finite element mesh was used to represent the region of the vessel where neointimal growth is expected to occur. The model was used to assess the influence of the initial strain in the neointima on the stress state within the vessel under subsequent pulsatile pressure loading.

These results suggest that, if the proliferative response within the neointima is related to local stress, some initial strain must be present during the deposition of the tissue. A validated model has the potential to reduce the occurrence of restenosis through improved understanding of stent/artery interactions.

Keywords— **Restenosis, Stent, Finite Element Method.**

I. Introduction

Ischemic heart disease represents the most common cause of death in the world, and one of its manifestations is obstructive coronary artery disease [1].

Coronary artery disease is nowadays mostly treated with percutaneous coronary interventional procedures, in particular with stenting operations. One of the main disadvantages of this technique is in-stent restenosis, a re-narrowing of the artery related to the injury of the vessel wall as a result of the local stress caused by the stent struts, which often occurs a few months after this intervention [2].

Arteries have been studied for a long period, from various points of view: morphology, physiology, anatomy, mechanical properties [3], in both animal models and humans.

Vessel models in the literature tend to represent only the passive mechanical properties [4-5] when considering the interaction between the vessel wall and a stent; recent studies have included active vessel behavior, to improve understanding of restenosis [6] [7].

Boyle [7] studied the development of neointima using a cell-centred lattice-based approach. To represent the coronary artery, three types of cells were modeled: smooth muscle cells in their contractile or synthetic phenotype and endothelial cells. Extracellular components taken in consideration were extracellular matrix, matrix degrading factors and growth stimuli. Finite element analysis was used to determine an initial stress-based injury criteria: stent deployment into a hyperelastic cylinder was simulated using ABAQUS (SIMULIA). This model does not take into consideration the evolution of the stress state within the vessel during neointimal growth.

The COAST (Complex Automata Simulation Technique) project [8] developed a multiscale framework: bulk flow, drug diffusion, and smooth muscle cell models were coupled to predict restenosis following stent deployment. After stent deployment, smooth muscle cell proliferation was shown to depend on the blood flow and drug concentration. A limitation of the model, with respect to the representation of the vessel and neointima, is that it considers the behaviour of smooth muscle cells and rupture of the internal elastic lamina without considering the role of the extracellular matrix.

The aim of this study is to propose a finite element model of stent expansion within the coronary artery to evaluate the stress distribution during stent implantation and the evolution of the state of stress during neointimal growth as a result of mechano-biological interactions. The novelty of this study is that the mechanical state of the vessel is updated during the tissue growth response.

It is proposed that a 'ghost' finite element mesh can be used to represent the region of the vessel where neointimal growth is expected to occur.

An example of a first stage of neointima growth is simulated by means of the modification of 'ghost' mesh properties: this represents change in neointimal constituents during the remodelling process. The purpose of this initial work is to assess the feasibility of this approach and the influence of assumptions relating to the initial stress state of the neointimal tissue.

S. Vlad and R.V. Ciupa (Eds.): MEDITECH 2011, IFMBE Proceedings 36, pp. 276–279, 2011.
www.springerlink.com

II. Materials and methods

A. *The Stented Vessel Model with 'Ghost' Mesh*

A section of a coronary artery, with typical dimensions, was generated within ANSYS Mechanical APDL version 12.0 (ANSYS Inc.): the aim of such a model is to capture both the stent expansion inside the vessel, with the subsequent stresses and strains, and the tissue growth inside the lumen, driven by a response to these mechanical stimuli.

The model is symmetric and represents 1/6 of the section of a coronary vessel, as shown in figure 1.

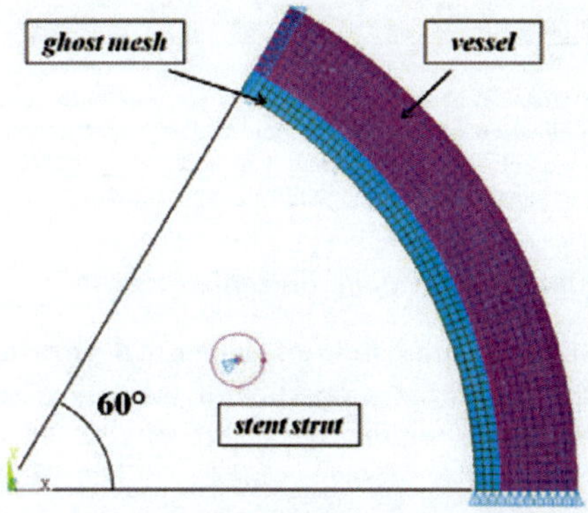

Fig. 1 Symmetric model of 1/6 of a vessel section

It is proposed that a 'ghost' finite element mesh can be used to represent the region of the vessel where neointimal growth is expected to occur. The effect on the stress state within the vessel is first evaluated under three distinct conditions to assess the influence of the 'ghost' mesh behaviour on the solution.

I. Vessel only (no 'ghost' mesh defined)

II. Vessel with a ghost mesh in the inner side of the vessel (to represent potential neointimal tissue) with:

   a. "Killed elements" in the ghost mesh: killing elements allows the software to update the mesh geometry under loading without the generation of a stress state within the material, and in particular does not store any strain information.

   b. Reduced material properties (Elastic modulus = 100 kPa): in this case the ghost mesh will contribute to a degree, dependant on the relative change in these properties, in determining the stress distribution within the vessel and neointima. The state of strain will be updated as the ghost mesh deforms.

Linear elastic material properties (E = 2 MPa) are assumed for the vessel. The geometry of the model is specified as follows:

Vessel radius = 2 mm, Vessel thickness = 0.29 mm,

Neointima thickness = 0.86 mm

A single stent strut, represented by a rigid circular body of radius 0.1 mm, is displaced in the radial direction to a final position of 2.2 mm, as figure 2 shows, with contact defined between the strut and the inner vessel wall. A uniform pressure of 16 kPa is applied to the inner vessel wall to represent the mean coronary artery pressure.

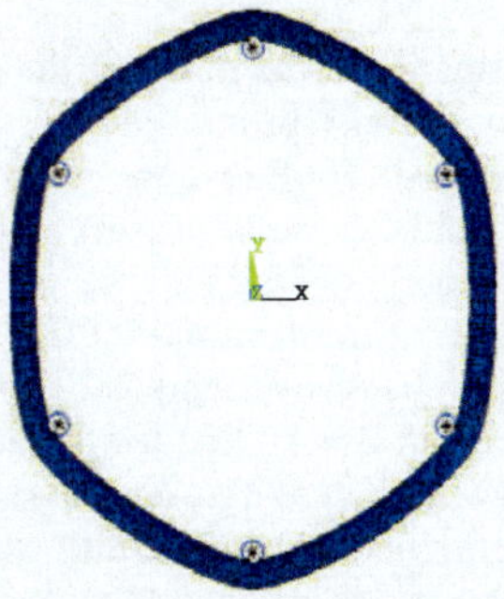

Fig. 2 Expanded symmetry representation of vessel and stent struts

The contact definition ensures that the stent does not "see" the presence of the ghost mesh and interacts only with the vessel inner surface.

B. *Neointimal Growth and Pulsatile Pressure*

Following evaluation of the initial stress distribution resulting from stent deployment, the early stage of neointimal growth has been simulated for both 'ghost' mesh models described above (II.a and II.b). A thinner neointima (0.095mm) is defined to represent the initial stage of restenosis. The test was then conducted in three steps:

i. Stent strut displacement and mean coronary pressure loading applied,

ii. Update of 'ghost' mesh material properties to represent neointimal formation, followed by application of the pressure to the new internal surface of the vessel,

iii. Evaluation of stress distribution at minimum, mean and maximum coronary pressure values.

The neointima is modelled with the same elastic modulus as the vessel and, as an initial simplifying hypothesis, neointimal growth is assumed to be homogeneous. The range of coronary pressures applied is illustrated in figure 3 (mean pressure: 16 kPa, max pressure: 18 kPa, min pressure: 14 kPa).

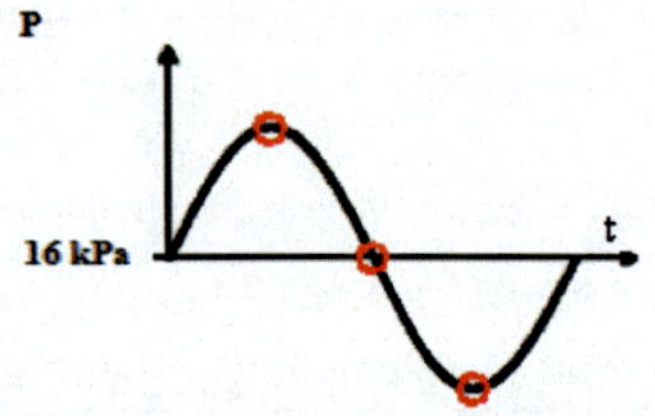

Fig. 3 Variation in coronary pressures during cardiac cycle

III. Results

A. *Influence of 'Ghost' Mesh Properties on Vascular Stress Distribution*

An initial comparison was made of the stress distribution under application of only an internal pressure, with no strut displacement defined. In this case identical results were obtained with model I (simple vessel) and II.a (ghost mesh with killed elements), whilst the use of a reduced elastic modulus for the ghost mesh in model II.b gave very similar results, with peak stresses around 2% lower than the other two simulations (with E = 10 kPa for the ghost mesh).

The stress distribution within the vessel wall after stent strut displacement and with application of a uniform internal pressure is shown for all models in figure 4a.

A small difference is observed in the peak stresses between model I and II.a (<0.8% of the first principal stress), with small differences also between model I and model II.b (<0.3% for the 10 kPa ghost mesh). The distribution of stress within the vessel wall is similar for all three models.

Figure 4b shows results of three simulations in terms of first principal strains: the three models show the same strain through the vessel thickness. The main difference is observed between models II.a and II.b, where the update of strain in the ghost mesh is as expected, with no strain developed within the killed elements in II.a and a non-uniform distribution of strain in the model II.b.

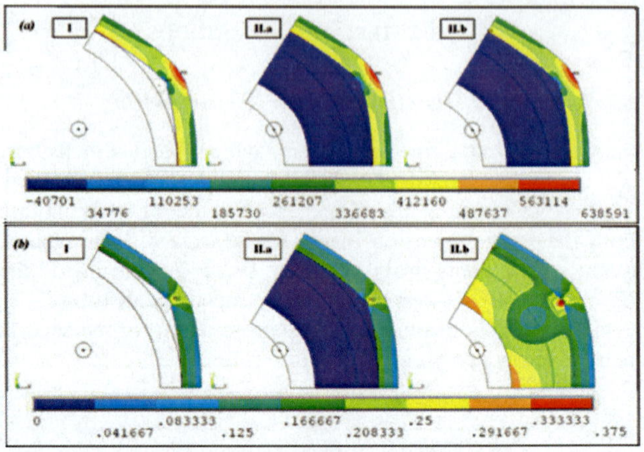

Fig. 4 Comparison of a) stress and b) strain distribution after stent deployment and application of internal pressure to the vessel lumen. I) Vessel only II.a) Vessel with killed element in neointimal 'ghost' mesh II.b) Vessel with reduced stiffness in neointimal 'ghost' mesh

B. *Neointimal Growth and Pulsatile Pressure*

Following the simulation of neointimal growth, through the update of material properties of the 'ghost' mesh, the stress distribution within the vessel was evaluated at the three representative pressure values. Figure 5 shows the distribution of circumferential stress for all three models at the mean coronary pressure value. The most significant variation in stress occurs within the neointima between models II.a and II.b.

The variation of stress on the symmetry plane at the inner and outer surfaces of both the vessel wall and neointimal region at minimum and maximum pressures is reported in table 1.

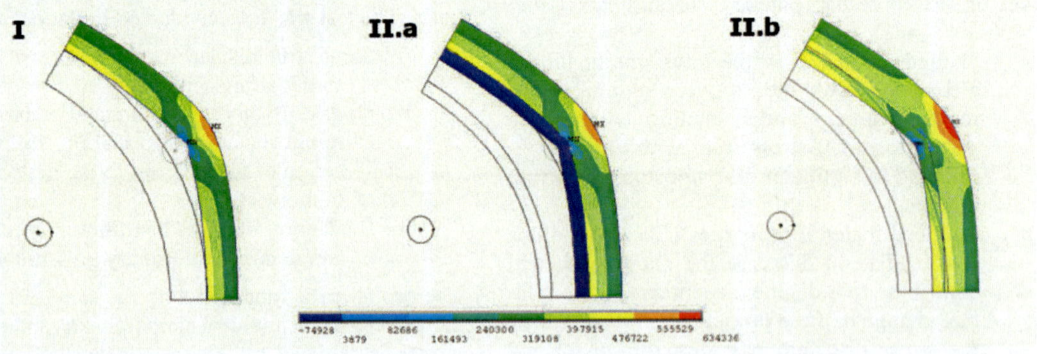

Fig. 5 Variation in circumferential stress for the three model types at the mean coronary pressure loading. I) Vessel only II.a) Vessel with killed elements II.b) Vessel with strain update

These results also demonstrate the large changes in neointimal stress observed between models II.a and II.b, although the stress distribution within the vessel is still comparable.

Table 1 Variation in circumferential stress with model type and applied pressure at the inner and outer surfaces of both the vessel and the neointima regions

		Circumferential stress (kPa)			
		Inner vessel	Outer vessel	Inner neointima	Outer neointima
I	P_{min}	486	265	N/A	N/A
	P_{max}	483	284	N/A	N/A
II.a	P_{min}	485	264	3	-1
	P_{max}	485	282	-1	1
II.b	P_{min}	455	231	448	384
	P_{max}	455	246	445	386

IV. Discussion

The models presented in this study examine variations in the initial strain of neointimal tissue generated in response to vascular injury after stent deployment. It is unknown whether new tissue is generated with a state of strain similar to the tissue in the arterial wall or in a stress free state. In order to understand the relationship between restenosis and the changes in structural response of the vessel during the remodeling process, it is necessary to develop hypotheses which describe the relationship between structural stress and biological processes occurring within the tissue. Development of these hypotheses and implementation within this modeling framework will form the basis of future work. The simple hypothesis of homogeneous neointimal growth along the circumference of the vessel requires sophistication. Spatial localisation of the neointima could be included within the model through a local stress rule. However, regardless of the localisation of growth, the results presented here suggest that, if the proliferative response within the neointima is related to local stress, some initial strain must be present during the deposition of the tissue.

It should also be noted that a purely structural approach is unlikely to fully describe the mechanisms associated with the cessation of the process of restenosis. The ultimate aim of this research is the combination of the modeling framework presented here with existing models of the proliferative response to variations in fluid shear shear stress [6].

V. Conclusions

This study has examined the influence of the initial strain of neointimal tissue on the stress distribution within a stented artery under pulsatile pressure loading. Further work will develop hypotheses of the mechano-biology of neointimal formation, implement these within the modeling framework and validate model outcomes. A validated model has the potential to reduce the occurrence of restenosis through improved understanding of stent/artery interactions.

Acknowledgment

This research is funded by the European Commission, through the MeDDiCA ITN (www.meddica.eu, Marie Curie Actions, grant agreement PITN-GA-2009-238113).

References

1. King, S.B., 3rd, J.J. Marshall, and P.E. Tummala, Revascularization for coronary artery disease: stents versus bypass surgery. Annu Rev Med. **61**: p. 199-213.
2. Arjomand, H., et al., Percutaneous coronary intervention: historical perspectives, current status, and future directions. Am Heart J, 2003. **146**(5): p. 787-96.
3. Ozolanta, I., et al., Changes in the mechanical properties, biochemical contents and wall structure of the human coronary arteries with age and sex. Med Eng Phys, 1998. **20**(7): p. 523-33.
4. Gijsen, F.J., et al., Simulation of stent deployment in a realistic human coronary artery. Biomed Eng Online, 2008. **7**: p. 23.
5. Lally, C., F. Dolan, and P.J. Prendergast, Cardiovascular stent design and vessel stresses: a finite element analysis. J Biomech, 2005. **38**(8): p. 1574-81.
6. Evans, D.J., et al., The application of multiscale modelling to the process of development and prevention of stenosis in a stented coronary artery. Philos Transact A Math Phys Eng Sci, 2008. **366**(1879): p. 3343-60.
7. Boyle, C.J., et al., Computational simulation methodologies for mechanobiological modelling: a cell-centred approach to neointima development in stents. Philos Transact A Math Phys Eng Sci, 2010. **368**(1921): p. 2919-35.
8. Caiazzo, A., et al., Towards a Complex Automata Multiscale Model of In-Stent Restenosis. Computational Science - ICCS 2009, 2009. **5544/2009**: p. 705-714.

Author: Claudia Maria Amatruda
Institute: Medical Physics Group, Department of Cardiovascular Science, University of Sheffield
Street: Beech Hill Road
City: Sheffield, S10 2RX
Country: UK
Email: C.M.Amatruda@sheffield.ac.uk

In Silico Evaluation of Some Dihydroxamic Derivatives of Diphenylether, as Hybrid Hydroxamic-Allosteric Inhibitors for MMP13

T. Petreus, C.E. Cotrutz, B. Stoica, M. Neamtu, P.D. Sirbu, and A. Neamtu

"Gr.T.Popa" University of Medicine and Pharmacy, Iasi, Romania

Abstract— **Synthetic inhibitors for MMPs may become effective if they include a functional group as hydroxamic acid, carboxylic acid, sulfhydryl, etc. capable of binding the catalytic Zn, while at least one functional group provides a hydrogen bond interaction with the enzyme backbone, and one or more side chains will undergo effective van der Waals interactions with the enzyme subsites.**

Due to the fact that previous clinical trials that have used small inhibitor molecules and especially hydroxamic derivatives (batiamastat, marimastat) have shown poor clinical results and even malignancy rebound, the next generation of MMP inhibitors is directed toward exodomain-substrate interactions.

In the present study we have docked both known and experimentally tested inhibitors and also new proposed inhibitor models. Then we have compared binding affinities of the known compounds with those of three newly proposed ligands.

The purpose of this study was evaluate the affinity degree of these hypothetic hybrid inhibitors for the catalytic domain of MMP13.

The docking study performed with the open source software Autodock Vina, have generated promising results regarding the possibility to propose hypothetical but potent hybrid hydroxamic-allosteric inhibitors for the catalytic domain of MMP. Interactions between the newly proposed ligands and the catalytic site of MMP13 show new interesting alternative options for tunnel-like catalytic site enzymes. It appears that the direct Zn ion coordination is not solely responsible for enzyme inhibition but also allosteric inhibition may play an important role.

Our results show that the proposed inhibitors, nominated as ligands 3, 4 and 5, mainly dihydroxamic derivatives of diphenylether, has both hydroxamic potency but als the ability to perform allosteric inhibition at least for MMP13 catalytic site.

Further studies will consider evaluation of these theoretical inhibitors by docking on other MMPs with different S1' pockets.Regarding the proposed extended docking studies, we suppose that the synthesis of ligand-5 and the experimental data should confirm our molecular docking results.

Keywords— **molecular docking, metalloproteinase, synthetic inhibitor, hydroxamate.**

I. INTRODUCTION

Matrix metalloproteinases (MMP) are representative enzymes involved in decomposition of the natural compounds in the extracellular matrix [1,2]. Zn and Ca are essential ions for these enzymes activity. Even if MMP plays important roles in physiological processes, their overexpression plays also crucial roles in pathological processes as multiple sclerosis, arthritis, Alzheimer disease and especially in cancer and metastasis [3].

Interactions between ligands and substrate macromolecules – as enzyme-substrate or enzyme-inhibitor interactions – can be explored by molecular docking. Most of the methods are using molecular modeling algorithms that may involve either the exhaustive exploration of all possible ligand conformations and their relative position according to the enzyme. Some new software as Autodock Vina [4] is faster than the classical ones and are recommended to be used in ligand screening methods. Any MMP catalytic site is characterized by a Zn atom and a conserved zinc binding motif, HExxHxxGxxH. Synthetic inhibitors for MMPs may become effective if they include a functional group as hydroxamic acid, carboxylic acid, sulfhydryl, etc. capable of binding the catalytic Zn, while at least one functional group provides a hydrogen bond interaction with the enzyme backbone, and one or more side chains will undergo effective van der Waals interactions with the enzyme subsites.

II. PURPOSE

Due to the fact that previous clinical trials that have used small inhibitor molecules and especially hydroxamic derivatives (batiamastat, marimastat) have shown poor clinical results and even malignancy rebound, the next generation of MMP inhibitors is directed toward exodomain-substrate interactions [5]. The third generation of MMP inhibitors show no Zn-biding activity and exploits the presence and depth of the S1' pocket in most metalloproteases. As MMP13 shows a peculiar S1' loop on an additional S1' pocket side, and regarding the previous experimental data obtained by Johnson et al [6], we have imagined three new hybrid compounds that conserve either the hydroxamic group and also develop an exosite-compliant domain.

In the present study we have used the technique of molecular docking to explore the possibility of designing new selective inhibitors for MMPs based on an allosteric

S. Vlad and R.V. Ciupa (Eds.): MEDITECH 2011, IFMBE Proceedings 36, pp. 280–283, 2011.
www.springerlink.com

regulation philosophy. Docking experiments were performed either on experimentally tested inhibitors and also on three new proposed inhibitor models. We have compared binding affinities of the known compounds with those of proposed ligands.

The purpose of this study was evaluate the affinity degree of these hypothetic hybrid inhibitors for the catalytic domain of MMP13.

III. Material and methods

We considered useful to apply molecular docking method that attempts to predict non-covalent binding between a macromolecule (here a metallo-enzyme, MMP13) and a small molecule (MMP inhibitors) starting from their unbounded structures. Docking methods are reproducing chemical potentials, and thus the bound conformation preference and the free binding energy [7].

In our docking study we have used a new program for molecular docking and virtual screening, Autodock Vina [4]. Vina uses Iterated Global Local Optimizer [8] which, briefly, uses a succession of steps consisting of a mutation, followed by a local optimization. In each step, the method of Broyden-Fletcher-Goldfarb-Shanno (BFGS) [9] is used for the local optimization that uses not only the scoring value but also its gradient.

While the algorithm implemented in Vina is based on a heuristic scoring function, its computes for at least 4 times faster than other docking software. By using multithreading, Vina can further speed up the execution by taking advantage of multiple CPUs or CPU cores. [4, 7].

Crystal structures for MMP13 (PDBID:830c) and the two known hydroxamic derivative inhibitors were chosen from ProteinDataBank [10]. In our study, for convenience, the known ligands were nominated as ligand-1(4-[4-(4-chloro-phenoxy)-benzene-sulfonylmethyl]-tetrahydro-pyran-4-carboxylic acid hydroxyamide) [11] and ligand-2 (2-{4-[4-(4-chloro-phenoxy)-benzenesulfonyl]-tetrahydro-pyran-4-yl}-n-hydroxy-acetamide) [12]. New proposed ligands were depicted as ligand-3, 4 and 5 respectively, and they represent ideal constructs, nominated as dihydroxamic derivatives of diphenylether (table 1).

Models visualization and graphic representation was performed by VMD, a molecular visualization program for displaying, animating, and analyzing large biomolecular systems using 3-D graphics and built-in scripting [13].

IV. Results

The docking study performed with the open source software Autodock Vina, have generated promising results regarding the possibility to propose hypothetical but potent hybrid hydroxamic-allosteric inhibitors for the catalytic domain of MMPs. In table 1 we are showing the docking results generated for three newly imagined MMP inhibitors, compared to two already described inhibitors with known crystallographic structure in complex with MMP-13 (collagenase 3). Affinity scores obtained for the described inhibitors are correlated with experimental data described in literature [6].

Table 1 Gibbs free energy of binding for 5 different MMP13 inhibitors

MMP Inhibitor	Gibbs free energy of binding (kcal/mol)	Chemical formula
LIGAND 1 4-[4-(4-chloro-phenoxy)-benzenesulfonyl-methyl]- tetrahydro-pyran-4-carboxylic acid hydroxyamide	-9.6	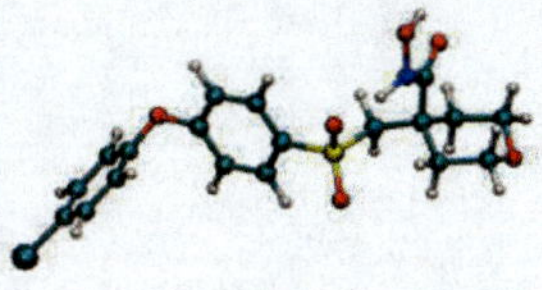
LIGAND 2 2-{4-[4-(4-chloro-phenoxy)-benzenesulfonyl]-tetrahydro-pyran-4-yl}-n-hydroxy-acetamide	-7,2	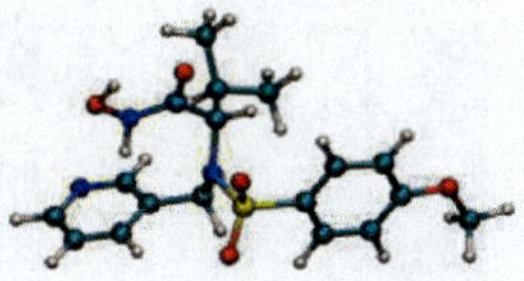
LIGAND 3 PROPOSED MODEL	-9,8	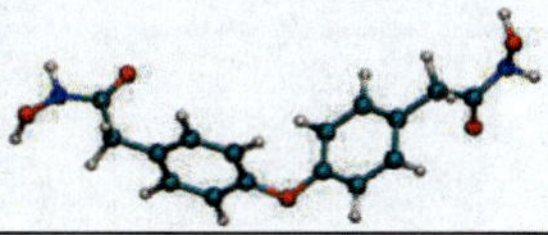
LIGAND 4 PROPOSED MODEL	-10,3	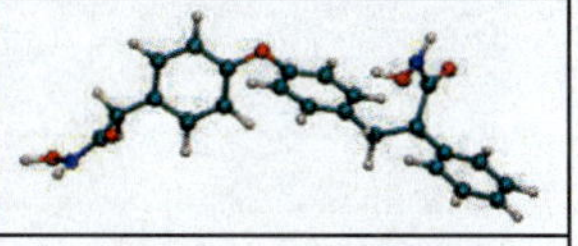
LIGAND 5 PROPOSED MODEL	-11,1	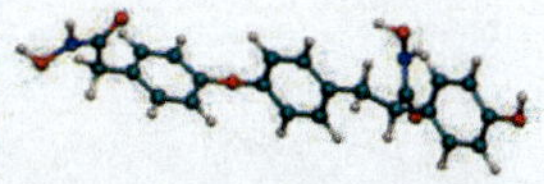

We observed that ligand-5 experimental has the highest affinity for the MMP-13 catalytic site (free Gibbs energy of -11,1 kcal/mol). Moreover, all three newly imagined inhibitors show higher affinities than previous known inhibitors (ligand 3 = -9,8 kcal/mol and ligand 4 = -10,3 kcal/mol, compared to ligand 1 = -9,6kcal/mol and ligand 2 = -7,2 kcal/mol). The proposed structures for docked ligand-3, -4, and -5 are depicted in figure 1; ligand 5 seems to realize an optimal fit into the catalytic site (Fig. 1).

Due to the fact that we have performed a comparative analysis between experimental data that shows only the Ki and not the free energy values, and while conversion between these constants ranges high output errors, we have performed also a docking analysis for the compounds used in the mentioned experiments [11]. Thus, on the MMP13 crystallographic structure we have docked ligand-1 and ligand-2 (see table 1) and we have obtained a good agreement with the experimental structure.

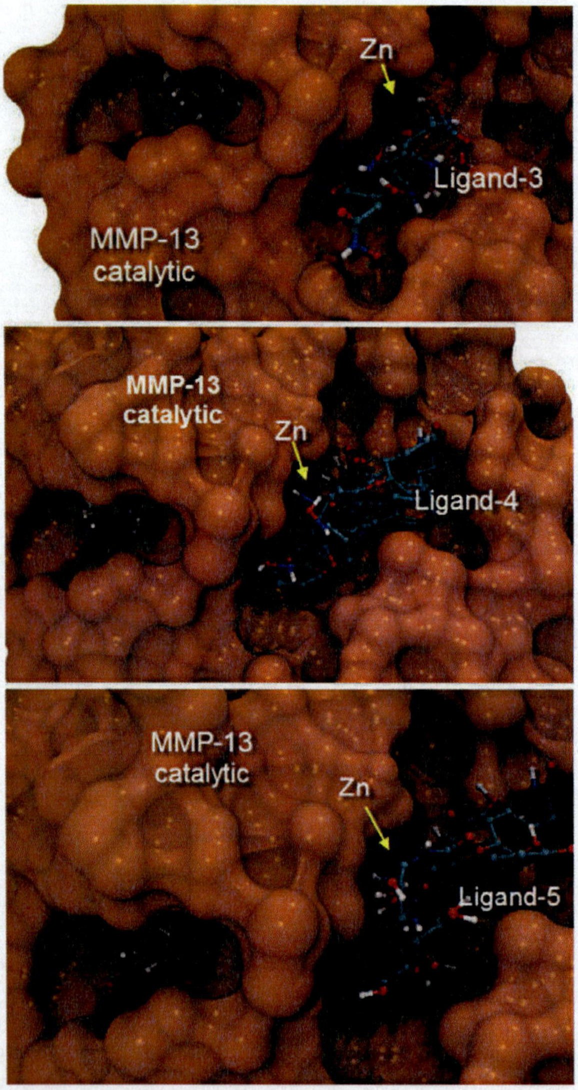

Fig. 1 Docking results for the new proposed dihydroxamic derivatives of diphenylether, nominated as ligand-3, -4 and -5. Posed conformations with the most favorable binding energy are located deep in the catalytic site tunnel

While during docking process, the mobility of the side chains was limited, the position of the hydroxamic group is a little bit displaced from the crystallographic results (Fig. 2). Observing that the inhibitor position in the catalytic site is very close to the crystallographic location, we may consider that the values resulted from docking analysis are reproducible and may be subject for further synthesis process.

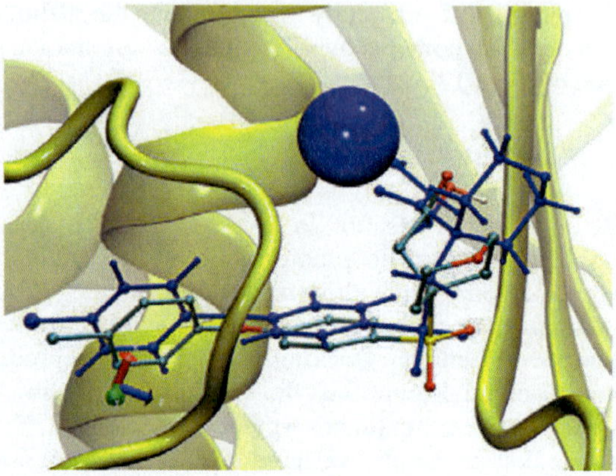

Fig. 2 Ribbon model of MMP13 catalytic site (yellow) with bound ligand 1 (in blue stick-ball model, crystallographic structure; in CPK model, the docked structure by Autodock Vina). Zn ion - large blue sphere.

Then, experimental data [11] shows that ligand-1 (Ki=0.52nM) is a more potent inhibitor than ligand-2 (Ki=1.9nM) [12], fact that is also demonstrated by docking procedure using Autodock Vina.

V. Discussions

Interactions between the newly proposed ligands and the catalytic site of MMP13 show new interesting alternative options for tunnel-like catalytic site enzymes. It appears that the direct Zn ion coordination is not solely responsible for enzyme inhibition but also allosteric inhibition may play an important role. Allosteric control of MMP2 action on gelatin was demonstrated by Ingvarsen et al [14].

As the tested compounds, whose crystallographic structure is described in complex with MMP13 (PDBID:830c), both contain a phenoxy-benzensulfone group, we have imagined a symmetrical molecule (ligand 3) with conserved the aromatic bicyclic structure on which we have grafted two hydroxamic terminal groups, obtaining three dihydroxamic derivatives of diphenylether. Hydrophobic ligand areas are interacting with complementary areas in the MMP13 catalytic tunnel (Fig. 3). Ligand-4 has a supplementary aromatic group, interacting with Pro242 and Ile 243 in the S1' loop. Ligand 5 has a supplementary hydroxyl group that is defining probably hydrogen bridge with Gly 183. The most important aminoacid in this relation between the inhibitor ligand 4 or 5 and the MMP13 catalytic site is

Leu185. The aliphatic non-linear side chain of Leu185 goes parallel with the neighbor aromatic cycle, insuring a large contact surface between the enzyme hydrophobic pocket and the designed inhibitor.

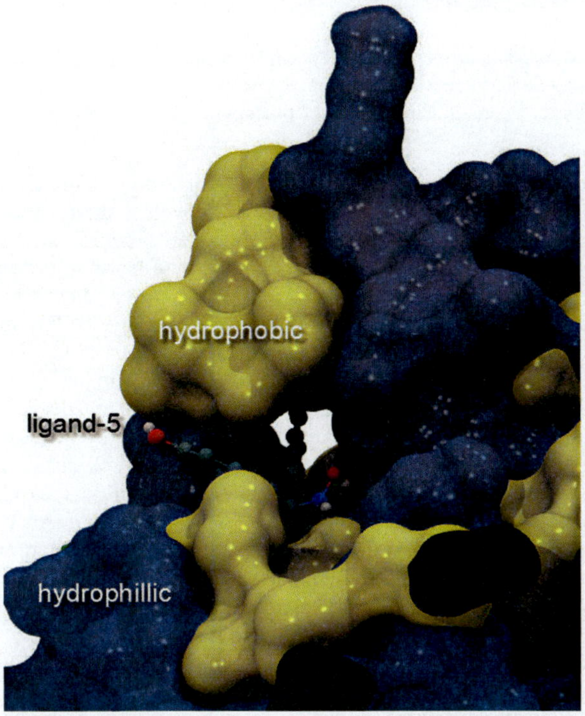

Fig. 3 Surface model of MMP13 catalytic site hydrophobic (yellow) and hydrophillic areas with bound ligand 5 (CPK stick-ball mode).

Ligand-5 is able to form H-bonds with Gly183 and at the same time an extended apolar contact with Leu185, areas that are not classic described as Zn coordination sites. Thus, ligand-5 appears to be the best option prom the proposed inhibitors that will be able to perform allosteric modulation of MMP catalytic site activity.

VI. Conclusions

Our results show that the proposed inhibitors, nominated as ligands 3, 4 and 5, mainly dihydroxamic derivatives of diphenylether, has both hydroxamic potency but also the ability to perform allosteric inhibition at least for MMP13 catalytic site.

Further studies will consider evaluation of these theoretical inhibitors by docking on other MMPs with different S1' pockets.

Regarding the proposed extended docking studies, we suppose that the synthesis of ligand-5 and the experimental data should confirm our molecular docking results.

Acknowledgments

This paper was supported by the grant IDEI 2560/2008.

References

1. Shapiro SD. (2000) A concise yet informative stroll through matrix metalloproteinases and TIMPs. *J.Cell.Sci.*, 113 (19), .3355-3356
2. Woessner JF Jr. (1999) Matrix metalloproteinase inhibition. From the Jurassic to the third millennium. *Ann N Y Acad Sci.*.30(878), 388-403
3. Page-McCaw A,Ewald AJ, Werb Z (2007) Matrix metalloproteinases and the regulation of tissue remodelling, *Nat. Rev. Mol. Cell. Biol.* 8 221–233
4. Trott O, A. J. Olson (2010), AutoDock Vina: improving the speed and accuracy of docking with a new scoring function, efficient optimization and multithreading, *Journal of Computational Chemistry* 31 455-461
5. Overall CM, Lopez-Otin C (2002) Strategies for MMP inhibition in cancer: innovations for the post-trial era, *Nat. Rev., Cancer* 2, 657–672
6. Johnson AR, Pavlovsky AG, Ortwine DF, Prior F, Man CF, Bornemeier DA, Banotai CA, Mueller WT, McConnell P, Yan C, Baragi V, Lesch C, Roark WH, Wilson M, Datta K, Guzman R, Han HK, Dyer RD. Discovery and characterization of a novel inhibitor of matrix metalloprotease-13 that reduces cartilage damage in vivo without joint fibroplasia side effects. J Biol Chem. 2007 21;282(38):27781-91
7. Gilson MK, Given JA, Bush B, McCammon JA (1997) The statistical-thermodynamic basis for computation of binding affinities. A critical review. *Biophys. J.* 72:1047-1069
8. Baxter J (1981) Local Optima Avoidance in Depot Location *J. Oper. Res.* Soc., 32, 815-819
9. Nocedal J, Wright SJ (1999) *Numerical Optimization*; in Springer Series in Operations Research Springer Verlag: Berlin
10. Berman HM, Westbrook J, Feng Z, Gilliland G, Bhat TN, Weissig H, Shindyalov IN, Bourne PE (2000) The Protein Data Bank *Nucleic Acids Research*, 28: 235-24
11. Lovejoy B, Welch AR, Carr S, Luong C, Broka C, Hendricks RT, Campbell JA, Walker KA, Martin R, Van Wart H, Browner MF. (1999) Crystal structures of MMP-1 and -13 reveal the structural basis for selectivity of collagenase inhibitors. *Nat Struct Biol.* 6(3):217-21
12. Zhang X., Gonnella NC, Koehn J, Pathak N, Ganu V, Melton R, Parker D, Hu SI, Nam KY. (2000) Solution structure of the catalytic domain of human collagenase-3 (MMP-13) complexed to a potent non-peptidic sulfonamide inhibitor: binding comparison with stromelysin-1 and collagenase-1. *J Mol Biol.* 11;301(2):513-24
13. Humphrey, W., Dalke, A. and Schulten, K. (1996), VMD - Visual Molecular Dynamics, *J. Molec. Graphics*, 14, 33-38
14. Ingvarsen S, Madsen DH, Hillig T, Lund LR, Holmbeck K, Behrendt N, Engelholm LH (2008) Dimerization of endogenous MT1-MMP is a regulatory step in the activation of the 72-kDa gelatinase MMP-2 on fibroblasts and fibrosarcoma cells, *Biol. Chem.* 389, 943–953

Author: Tudor Petreus
Institute: "Gr.T.Popa"University of Medicine and Pharmacy Iasi
Street: 16, Universitatii
City: Iasi
Country: Romania
Email: biotudor@gmail.com

The Effectiveness of Interactive Clinical Case Study Simulation in Palliative Medicine

M. Florea[1], M. Crisan[2], M. Gherman[1], and R. Manasia[3]

[1] University of Medicine and Pharmacy /Family Medicine Department, Cluj-Napoca, Romania
[2] University of Medicine and Pharmacy / Histology Department, Cluj-Napoca, Romania
[3] University of Medicine and Pharmacy / I Pediatric Hospital, Cluj-Napoca, Romania

Abstract— **The medical students' interest in case studies used as an instructional strategy in modern medical education and in teaching tools like virtual patients and simulations is growing. Professional skills and knowledge about palliative care are widely accepted to be deficient and the medical care in palliative care settings remains inadequate. Reduced student-patient contact times in hospitals, mainly in palliative care settings with progressive chronic disease patients and increasing cost of technology means generated the need to create innovative methods that help students develop their clinical skills. Palliative medicine interactive clinical case study simulation (ICCSS) offers a cost-effective method to measure the impact of this virtual instructional tool on the students' clinical skills. It helped us to measure and compare their choices made in diagnosis, treatment, and follow-up process prior and after an educational module in palliative medicine. By simulating the most common patient profiles for palliative medicine, the interactive clinical case study simulation (ICCSS) allowed us to record the students' diagnostic and treatment decisions, to show the likely outcome of each decision, and to provide individualized feedback to identify and correct inappropriate decisions.**

The interactive clinical case study simulation (ICCSS) can be successfully integrated with face-to-face teaching in palliative medicine like a supportive and innovative educational method.

Keywords— **interactive clinical case study simulations, educational tool, palliative medicine, effectiveness, medical students.**

I. Introduction

Modern computer technology has made possible the creation of simulated clinical experience and the precise measurement of patterns of clinical decision making. [1,2]. Reduced student-patient contact times in hospitals, mainly in palliative care settings with progressive chronic disease patients and increasing cost of technology means generated the need to create innovative methods that help students to develop their clinical skills. In addition the current methods for assessing changes in students' clinical practice by means of direct observation are too costly and time-consuming and this also rise the need to develop alternative to traditional medical education [3,5,7]. E-Learning tools like virtual patients allow to help students to develop their clinical reasoning skills, without putting the patient at risk [4,6,].These educational methods maximize learning process offering exposure to new ideas and active discussions concerning: bad news communication strategy, differential diagnosis of the disease complications, alternative plans, prognostic estimations in quantitative terms. The learner has the opportunity to revise diagnosticand treatment decisions and is guided toward the optimal diagnosis and appropriate treatment. Clinical case simulations can assess clinical competence and provide an individualized educational experience. This kind of interactive technology environment can deliver complex data rapidly and at a lower cost than any other outcome measurement strategy. The interactive case study also appreciates the learner's behavior, it generates data of greate validity and it can be distributed and administered entirely by computer, with a minimal commitment of time and resources. [9,10.11,12]. There is also a potential for collaborative learning to break the isolation of learners realized in computer-based learning technologies. Advances in synchronous distance education and collaborative technologies like Weblogs, message boards, chats, e-mail, and teleconferencing are making such collaborative learning more readily available.

II. Methods

Three interactive clinical case study simulations (ICCSS) were created by loading specific case information into a pre-programmed software template that has been designed to mimic the clinical diagnostic process, treatment selection and follow-up criterias in palliative medicine.

The aim was the improvement of palliative care education of medical students. The goals were to:

(1) improve physician communications skills in regard to breaking bad news and end-of-life care;
(2) integrate evidence-based medicine into patient recommendations; and
(3) enhance clinical skills in the practice of palliative medicine, particularly the symptoms', evaluation, diagnostic decision, new complications recognising and management of total suffering.

S. Vlad and R.V. Ciupa (Eds.): MEDITECH 2011, IFMBE Proceedings 36, pp. 284–287, 2011.
www.springerlink.com

These simulations were constructed as text-only module, as multimedia module with audio, video, and animation for maximum educational impact, and as an intermediate blend of text and multimedia. We recorded a group of 40 students' baseline skills and knowledge in palliative medicine prior to the interactive clinical case study simulations (ICCSS) sessions. We administered the educational module in palliative medicine with interactive clinical case study simulations (ICCSS) to the same group of students and we appreciated the students' improvements in making diagnostic and treatment decisions, the ability to recognise and correct inappropriate decisions and the impact of the educational programme on the students' clinical thinking. Three interactive clinical case study simulations included multimedia material, images, movies and audio recordings about an advanced breast cancer patient, liver chirrosis with hepatocarcinoma and a stage IV chronic obstructive pulmonary disease patient . The interactivity was based on quantitative and qualitative feedback. The interactive clinical case study simulations recorded every decision the students made in diagnosis, treatment, and follow-up.

III. Results

A number of 14 medical students (35%) out of 40 performed correct diagnostic and treatment decisions prior to the educational program in palliative medicine with interactive clinical case study simulations (ICCSS), in comparison with 35 students (87 %) which selected correct the diagnostic, treatment and the follow –up criterias after the ICCSS palliative medicine module.($p<0,004$).

In addition, 21 students (52%) recognised their previous errors and corrected them.

Palliative medicine interactive clinical case study simulations (ICCSS) offers a cost-effective method for measuring the impact of this virtual instructional tool on the students' clinical skills. It helped us to appreciate their choices made in diagnosis, treatment, and follow-up process. By simulating the most common patient profiles for palliative medicine , the interactive clinical case study simulations (ICCSS) allowed us to:

-record the students' diagnostic and treatment decisions,
-show the likely outcome of each decision, and
-provide individualized feedback to identify and correct inappropriate decisions.

Most of the students considered that working with ICCS was fun and the format appealing. They found the cases interesting , the knowledge base of the cases appropriate and the key feature of the cases relevant for their clinical practice.

IV. Discussions

Medical educators are facing different challenges than their predecessors in teaching tomorrow's physicians. The multiple changes in health care delivery and the advances in medicine have increased demands on academic faculty, resulting in less time for teaching than has previously been the case.

For "new" fields such as palliative care, and complementary medicine or genomics and geriatrics, is difficult to find time for teaching when medical school curricula are already challenged to cover conventional materials. Traditional educator-centered teaching is yielding to a learner-centered model and the recent shift toward competency-based curricula emphasizes the learning outcome, not the process, of education.[8]

The Federal Interagency Working Group on Information Technology Research and Development has recommended the establishment of centers to explore "new delivery modes for educating medical practitioners and providing continuing medical education".[11, 16, 20].

Simulation based learning provides:

-interactive learning in virtual clinical setting
-with no risk to patient and
-no liability for error and
-ability to provide real life clinical experience for superior learning.

The more advanced and sophisticated computers simulation based e-learning products allow students and medical professionals:

-to easily learn complex medical processes and
-perfect their skills.

Simulation learning challenge and motivate students and help them to identify their own strengths and weaknesses.

The important benefits of interactive clinical case simulations are:

-better retention, understand better,
-remember longer and decide faster,
-critical thinking development,
-freedom to make mistakes and learn from them,
-schedule learning anytime anywhere.

The interactive clinical case study simulations (ICCSS) in palliative medicine fits that description.

Many studies of collaborative and interactive learning in medicine have shown higher levels of learner satisfaction, improvements in knowledge, self-awareness, understanding of concepts, achievement of course objectives, and changes in practice.[12,13,14,15].

A growing emphasis on competency-based medical education has forced educators to reevaluate their traditional roles. In this changing time, educators no longer serve as the sole distributors of content, but are becoming facilitators of learning and assessors of competency. The interactive clinical case study simulations (ICCSS) in palliative medicine offer the opportunity for educators to evolve into this new role by providing them with new resources to facilitate the learning process.

We have found the interactive clinical case study simulation (ICCSS) to be a useful teaching and assessment tool integrated in the palliative medicine module. It allows learning to be individualized (adaptive learning), enhancing learners' interactions with others (collaborative learning), and transforming the role of the teacher.

Learning enhancement permits greater learner interactivity and promotes learners' efficiency, motivation, cognitive effectiveness, and flexibility of learning style. By enabling learners to be more active participants, the interactive clinical case study simulation (ICCSS) can motivate them to become more engaged with the content. Interactive learning shifts the focus from a passive, teacher-centered model to one that is active and learner-centered, offering a stronger learning stimulus. Interactivity helps to maintain the learner's interest and provides a means for individual practice and reinforcement. Learners have control over the content, learning sequence, pace of learning, time. All these allow them to tailor their learning style and experience to meet personal learning objectives.

The learning process efficiency is likely to translate into improved motivation and performance, resulting in better achievement of knowledge, skills, and attitudes. The interactive clinical case study simulation (ICCSS) and multimedia learning materials offers learners the flexibility to select options to accommodate their diverse learning styles. The integration of this educational method into medical education can favour the shift toward educators more involved as facilitators of learning and assessors of competency. [13, 14] .

The good correlation between the observer's (examiner) and the recipient's (medical student) perception of this educational tool provides evidence of the validity of the assessment. Future areas for research may include assessing contexts for effective use of interactive technology environment in medical education, the adaptation of this educational tools to a wide variety of medical specialties and clinical settings, an exploration of methods for simplifying the e-learning creation process to gain wider acceptance and the use of a multimedia instructional design process by medical educators. In addition. data from ongoing evauation will also allow curricular refinement.

V. Conclusions

The use of the interactive clinical case study simulations in palliative medicine demontrated effectiveness improving the students' scores on case study exercises and offered the oportunity to identify and correct inappropriate decisions. The interactive clinical case study simulations were used as supplementary learning and formative feedback resource for students in a self-directed learning process.

Case-based e-learning can be successfully integrated with face-to-face teaching in palliative medicine like a supportive and innovative educational method.

Virtual patients and computerized teaching methods in palliative medicine education are powerful tools for healthcare educators, offering the right content when and where they need it.

However, because students often focus on the decision to provide or withhold the palliative care intervention, rather than paying attention to the patients' and families' values and concerns , face to face education (student- real patient) must be associated, mainly to improve students' communications skills.

This two educational methods could not be used interchangeably but integrated.

References

1. Bergin R, Youngblood P, Ayers MK, Boberg J, Bolander K, Courteille O, Dev P, Hindbeck H, Leonard II EE, Stringer R, Thalme A, Fors U: Interactive simulated patient: experiences with collaborative e-learning in medicine J Educ Comput Res 2003, 29(3):387-
2. Botezatu M, Hult H, Tessma M, Fors U: As time goes by: stakeholder opinions on the implementation of a virtual patient simulation system. Med Teach 2010, 32(11):509-16.
3. Gesundheit N, Brutlag P, Youngblood P, Gunning WT, Zary N, Fors U: The use of virtual patients to assess the clinical skills and reasoning of medical students: initial insights on student acceptance.Med Teach 2009, 31(8):739-42
4. Gordon J, et al. "Practicing" medicine without risk: students' and educators' responses to high-fidelity patient simulation. Acad Med, 2001;76(5):469-72.
5. Jack P. Freer, Karen L. Zinnerstrom.The Palliative Medicine Extended Standardized Patient Scenario: A Preliminary ReportJournal of Palliative Medicine. March 2001, 4(1): 49-56.
6. Jennifer Weller, MB BS, MClinEd. FANZCA, FRCASimulation in Undergraduate Medical Education: Bridging the Gap Between Theory and Practice JEPM, Vol.: 5, No. I I, July-Dec, 2003
7. Johnson CE, Hurtubise LC, Castrop J, et al . Learning management systems: technology to measure the medical knowledge competency of the ACGME. Med Educ. 2004;38:599–608.
8. Jorge G. Ruiz, Michael J. Mintzer, Rosanne M. Leipzig. The Impact of E-Learning in Medical Education. Acad Med. 2006; 81:207–212
9. Littlejohn A. Issues in reusing online resources. In: Littlejohn A (ed). Reusing Online Resources: A Sustainable Approach to eLearning. London: Creative Print and Design, 2003:1–6.

10. McCowan C et al. Lessons from a randomized controlled trial designed to evaluate computer decision support software to improve the management of asthma. Medical Informatics and the Internet in Medicine, 2001, 26:191–201.
11. MedEdPortal http://www.aamc.org/meded/ mededportal). Accessed 22 November 2005. Association of American Medical Colleges, Washington, DC, 2005.
12. Meigs JB et al. A controlled trial of web-based diabetes disease management: the MGH diabetes primary care improvement project. Diabetes Care, 2003, 26:750–757.
13. Morgan P, Cleave-Hogg D. A Canadian simulation experience: faculty and student opinions of a performance evaluation study. Br J Anaesth, 2000;85(5):779-81.
14. Nair BR, Finucane PM. Reforming medical education to enhance the management of chronic disease. Med J Aust. 2003;179:257– 59
15. Reddy R, Wladawsky-Berger I (co-chairs). President's Information Technology Advisory Committee. Transforming Health Care through Information Technology. Arlington, VA: National Coordination Office for Information Technology Research & Development, 2001.
16. Smith R. Guidelines for Authors of Learning Objects http://www.nmc.org/guidelines/ NMC/20LO/20Guidelines.pdf). Accessed 22 November 2005. The New Media Consortium, Austin, TX, 2004.
17. Ward JP, Gordon J, Field MJ, Lehmann HP. Communication and information technology in medical education. Lancet. 2001;357: 792–96.
18. Walker R, Dieter M, Panko W, Valenta A. What it will take to create new Internet initiatives in health care. J Med Syst. 2003;27: 95–103.
19. Wiecha J, Barrie N. Collaborative online learning: a new approach to distance CME. Acad Med. 2002;77:928–29.
20. Wiecha JM, Gramling R, Joachim P, Vanderschmidt H. Collaborative e-learning using streaming video and asynchronous discussion boards to teach the cognitive foundation of medical interviewing: a case study. J Med Internet Res. 2003;5:e13.

Author: Mira Florea
Institute: University of Medicine and Pharmacy/ Family Medicine Department
Street: Friedrich Hegel 7A
City: Cluj-Napoca
Country: Romania
Email: miraflorea@yahoo.com

Stent Fracture Prediction in Percutaneous Pulmonary Valve Implantation: A Patient-Specific Finite Element Analysis

D. Cosentino[1], C. Capelli[2], G. Pennati[3], V. Díaz-Zuccarini[1], P. Bonhoeffer[2], A.M. Taylor[2], and S. Schievano[2]

[1] Mechanical Engineering Department, UCL, London, UK
[2] Centre for Cardiovascular Imaging, UCL Institute of Cardiovascular Sciences & Great Ormond Street Hospital for Children, London, UK
[3] Laboratory of Biological Structure Mechanics, Structural Engineering Department, Politecnico di Milano, Milano, Italy

Abstract— **The major limitation of the current percutaneous pulmonary valve implantation (PPVI) device is stent fracture. In this study, patient-specific analyses were developed to reproduce the realistic loading conditions experienced by the device *in-situ*, in order to predict fractures.**

Biplane fluoroscopy images of 5 patients who underwent PPVI and experienced fracture were used to reconstruct the 3D *in-situ* device geometry at 3 different times of the procedure and cardiac cycle (end of balloon inflation, early systole and diastole). From the superimposition of these 3 stent configurations, the displacements of the strut junctions of the stent were measured. Asymmetries were calculated in all 3 orthogonal directions for every instant reproduced.

A finite element (FE) model of the stent in the initial crimped configuration was created. The previously measured displacements were applied to 2 nodes of the FE stent model in the corresponding strut junction, and the stent deployment history was reproduced for each patient. A fatigue study was performed using the Goodman method and the Sines criterion. Both these methods were able to predict stent fracture in every analysed case. Furthermore, the zones with the highest risk of fracture predicted by the simulations included the areas where fractures were actually detected from X-rays images.

Keywords— **Percutaneous pulmonary valved stent, Patient-specific, Finite element analysis, Mechanical fatigue, Fluoroscopy images.**

I. INTRODUCTION

Percutaneous pulmonary valve implantation (PPVI) technique is based on the concept that a bovine jugular venous valve sewn inside a balloon expandable stent (Melody™, Medtronic Inc., USA) can be reduced in size by crimping it into a catheter, and then introduced through a peripheral vessel to the desired implantation site in the heart.

Although nowadays PPVI is a successful alternative to surgery for patients with pulmonary valve dysfunction [1], stent fractures remain one of the major limitations of this technique [2]. Both bench experiments and computational analyses developed in the past were not able to predict this problem observed in the clinical experience [3, 4, 5]. This suggests that both experimental set-up and finite element (FE) analysis incorrectly reproduced the stent *in vivo* loading conditions.

The aim of this study was to enhance the modelling of PPVI loading conditions by reproducing patient-specific procedures in order to investigate stent fractures.

II. MATERIAL AND METHODS

A. Patients' Selection

Five patients (named patients I to V) who underwent PPVI and experienced fracture were chosen. Criteria for patients' selection were the following ones: (i) PPVI procedure should have been done in a catheterization laboratory equipped with Axiom Artis Flat Detector system (Siemens, Germany) to eliminate image distortion [3, 6]; (ii) fluoroscopy exams should have been performed with the arm positions in the antero-posterior and perpendicular-lateral projections, in order to have projections of the device in 2 orthogonal planes; (iii) fluoroscopic images should have been visible throughout the balloon expansion procedure and at least one cardiac cycle at implantation completed.

Patients' age at the time of implant was between 10 and 18 years, and fractures were detected from X-rays images between the 1st and the 6th month after device implantation in all patients. The exact location of these fractures was detected from X-rays examination.

Informed consent for research use was given by the patients or by their parents in case of minor age.

B. Reconstruction from Fluoroscopy Images

Orthogonal 2D X-ray projections from PPVI fluoroscopy images allowed 3D *in-situ* stent reconstructions [3]. This was achieved by identifying the intersection points between the struts forming the stent in both fluoroscopy projections; these points were then projected into the 3D space by tracing parallel rays (CAD software Rhinoceros, McNeel, USA). The intersection points between the rays determined the position of the strut junctions in the 3D space. The

S. Vlad and R.V. Ciupa (Eds.): MEDITECH 2011, IFMBE Proceedings 36, pp. 288–293, 2011.
www.springerlink.com

junctions were subsequently joined together by straight segments to reproduce the zigzag wires of the device, and consequently the whole structure. The length of the reconstructed segments (L_{rec}) was measured and compared to the real length (L_{real}) of the stent struts, in order to estimate the reconstruction error as follows:

$$Error[\%] = \frac{|L_{rec} - L_{real}|}{L_{real}} \quad (1)$$

The stent configuration was reconstructed for each patient at the end of balloon inflation (pre-recoil), at early systole and at diastole in every patient. The 3D reconstructions were then aligned according to 2 axes and superimposed in order to eliminate stent rigid displacements due to heart beating and respiration motions. From this, it was possible to calculate the displacements of every junction point from the pre-recoil state through systole to diastole. Circumferential, radial and longitudinal asymmetries [3] were calculated for every patient, at the pre-recoil, systole and diastole steps.

C. Finite Element Modelling

PPVI Melody™ stent is made of 90%Platinum-10%Iridium alloy with golden thick sleeves around the intersection junctions for reinforcement of the wire welding. The stent geometry for FE analysis was reproduced in the initial crimped status of the device into the catheter. A structured hexahedral mesh of 119,360 elements was used to model the platinum-iridium stent structure. To reproduce the golden coverings, an additional set of elements with depth of 0.025 mm was modelled around the junctions and a structured hexahedral mesh was generated using 36,320 elements. Geometrical and material properties [3, 7] are listed in Table 1.

Patient-specific stent deployment configurations were replicated in ABAQUS/Standard (Simulia, USA) using nodal displacement boundary conditions. These displacements were those previously measured from the device reconstructions at different times of the procedure and cardiac cycle (II.B). In particular, nodal displacements (x and y directions) were applied to 2 internal nodes of each strut junction of the stent. Due to this type of displacement condition, some of the elements surrounding these nodes resulted distorted, causing an abnormal stress distribution. Therefore, a set of 128 elements for every internal strut junctions and of 48 elements for every external strut junction, as well as the golden reinforcement elements, was subtracted from the model for the analysis of stress results. This hypothesis can be considered conceivable as fractures

Table 1 Stent geometrical and material properties

Wire diameter	0.33 mm
Initial internal diameter	4.00 mm
Initial overall length	34.32 mm
Central zig-zag segment length	5.78 mm
Terminal zig-zag segment length	5.62 mm
PLATINUM	
Young modulus	224 GPa
Poisson ratio	0.37
Yield stress	285 MPa
Ultimate strength, S_{ult}	875 MPa
Fatigue endurance strength, S_e	263 MPa
GOLD	
Young modulus	80 GPa
Poisson ratio	0.42

at the strut junctions have not been reported for Melody™ stent once the golden reinforcements were introduced.

Three displacement steps were performed spaced by 2 elastic recovering steps for every patient. The first step led the stent from its initial crimped status to the end of balloon inflation; the second and the third ones replicated a cardiac cycle from systole to diastole.

D. Fatigue Analysis

A fatigue analysis was performed using the Goodman method [8] and the Sines criterion [9]. The first plots the mean and the alternating stress (σ_m and σ_a, respectively) during the cardiac cycle for each element as points in a graph that shows also the material strength limits. Using the maximum principal stress component (σ) at systole (sys) and diastole (dia), σ_m and σ_a were calculated as follows:

$$\sigma_m = \frac{\sigma_{sys} + \sigma_{dia}}{2} \quad (2)$$

$$\sigma_a = \frac{|\sigma_{sys} - \sigma_{dia}|}{2} \quad (3)$$

Fatigue safety factor (FSF) was calculated as [10]:

$$\frac{1}{FSF} = \frac{\sigma_m}{S_{ult}} + \frac{\sigma_a}{S_e} \quad (4)$$

where S_{ult} and S_e are the ultimate material strength and the fatigue endurance strength, respectively.

The Sines criterion coincides with the Goodman method if the stress status is uniaxial. If the stress field includes two or three dimensions, the Sines criterion is more appropriate, being a multiaxial fatigue criterion. It uses the Von Mises stress as the controlling parameter, whereas Goodman uses the maximum principal stress component. To evaluate the fatigue resistance with the Sines criterion, the equivalent Sines stress, for the comparison with the material strength S_e, was calculated as in the first member of the following disequation:

$$\sqrt{3}\left(\sqrt{J_2}\right)_a + 3\frac{S_e}{S_{ult}}\sigma_{H,m} \le S_e \qquad (5)$$

where $\left(\sqrt{J_2}\right)_a$ and $\sigma_{H,m}$ are the amplitude of the mean square root of the second deviatoric stress invariant and the hydrostatic pressure respectively.

III. Results

A. *Reconstruction from Fluoroscopy Images*

The device reconstructions at the 3 different instants analysed are shown for one patient in figure 1a, while figure 2 shows the superimposition of the initial crimped status of the stent and its 3 subsequent deployment steps. The maximum percentage error in the stent reconstruction was 8%. The reconstructed stents presented asymmetries in all 3 orthogonal directions. As an example radial, longitudinal and circumferential asymmetries for one patient are shown in figure 3, 4 and 5 respectively. Asymmetries strongly depended on the right ventricular outflow tract anatomy. In all patients, a common trend was observed in the longitudinal direction: minimum values of standard diameter were always measured in the internal rings, between sections 2 and 6.

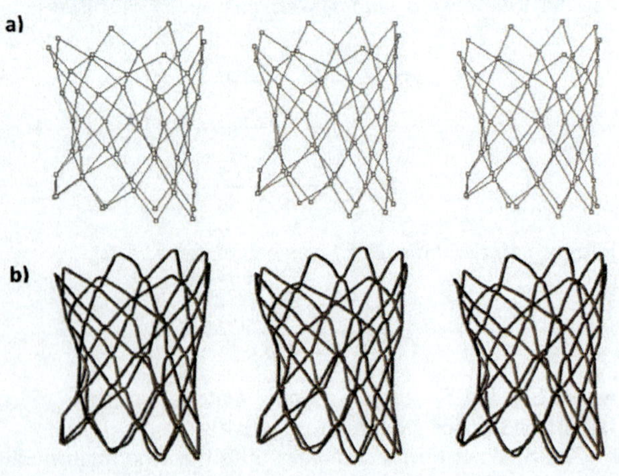

Fig. 1 Stent fluoroscopy reconstructions for patient II at pre-recoil, systole and diastole (a) and corresponding FE deployed configurations

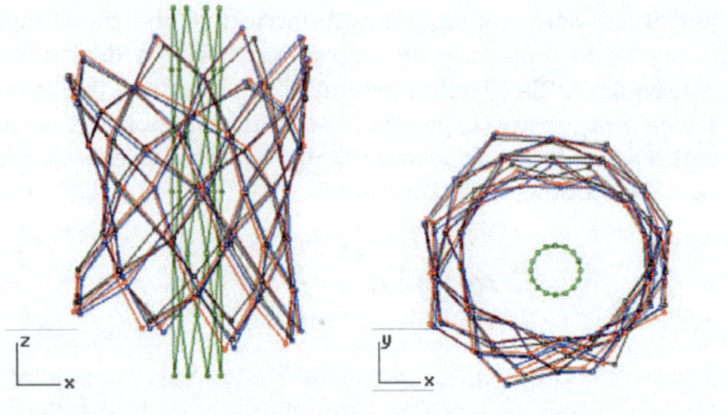

Fig. 2 Superimposition of the stent fluoroscopy reconstructions for patient II at pre-recoil (black), systole (red), diastole (blue) and initial crimped (green) status – lateral and top views

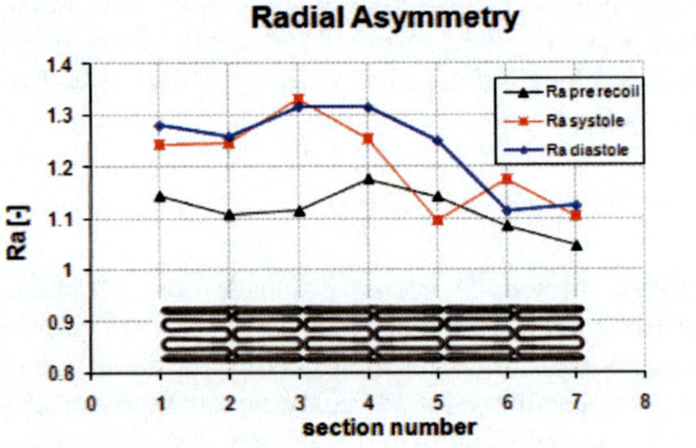

Fig. 3 Radial asymmetry measured in patient III at pre-recoil (black), systole (red) and diastole (blue)

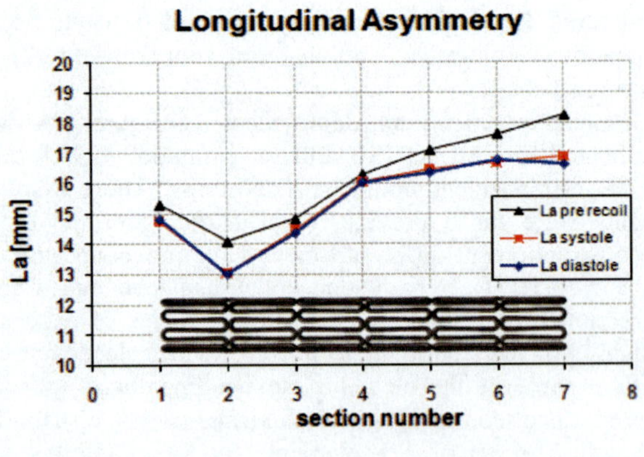

Fig. 4 Longitudinal asymmetry measured in patient III at pre-recoil (black), systole (red) and diastole (blue) compare comparison

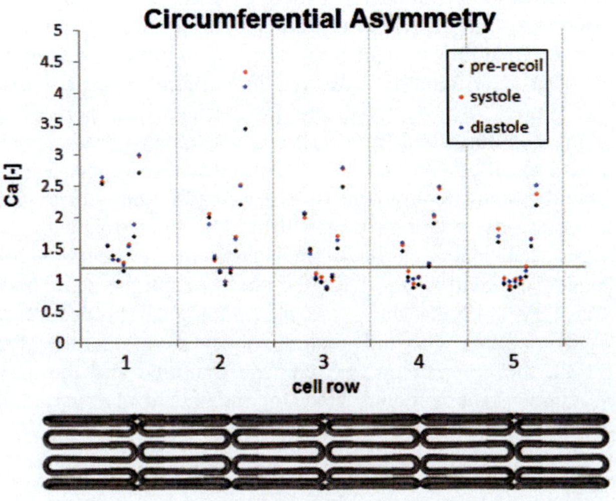

Fig. 5 Circumferential asymmetry measured in patient III at pre-recoil (black), systole (red) and diastole (blue). Comparison with the circumferential symmetric value of 1.18, corresponding to the ratio between the longitudinal and circumferential cell diagonals open in a symmetrically deployed stent of 20 mm internal diameter

B. Finite Element Modelling

Stent deployments was successful in all patients and the stent junction points reached the correct final position in x and y direction (figure 2) as imposed by the nodal displacement boundary conditions (figure 1b). The z coordinate of these points was compared to the corresponding coordinate in the fluoroscopic reconstructions. The maximum difference between the FE analysis and the reconstructed z coordinate was 0.78 mm (2.3% of the entire stent length in the crimped configuration).

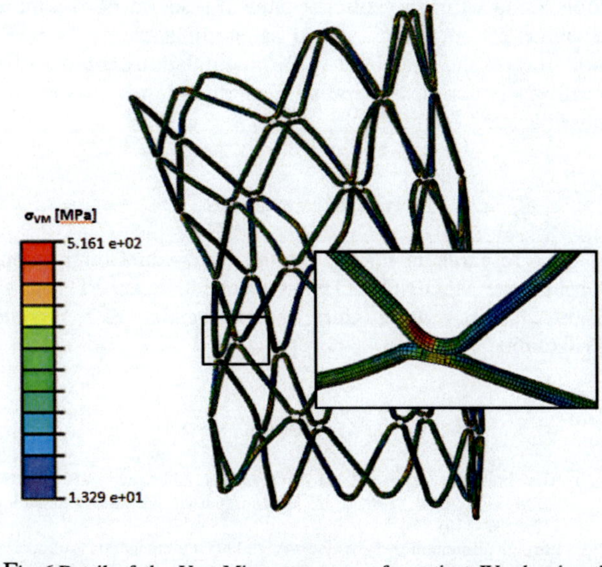

Fig. 6 Detail of the Von Mises stress map for patient IV, showing the location with highest stress

Table 2 reports Von Mises (σ_{VM}) maximum stresses reached in every patient and in every deployment step, as well as the maximum Sines stress (σ_{SIN}) and the inverse of the FSF. The maximum σ_{VM} was not located in the same ring for every patient and in 3 cases out of 5 it changed location at the end of each step. However, in every simulation, the highest stress occurred close to the strut intersections, which are the most highly bent portions of the device (figure 6). The peak σ_{VM} was reached during diastole in every patient, and its values ranged between 516.1 and 612.8 MPa.

Table 2 Maximum Von Mises stresses, at every deployment step, maximum Sines stresses and the inverse of fatigue safety factor in the 5 studied cases

		I	II	III	IV	V
$\sigma_{VM,\,max}$ [MPa]	Precoil	519.8	508.0	528.1	510.0	563.9
	Systole	530.9	521.3	523.0	515.8	593.9
	Diastole	538.1	526.5	535.1	516.1	612.8
$\sigma_{SIN,\,max}$ [MPa]		408.9	412.4	348.6	348.6	394.7
1/FSF		1.38	1.93	1.7	1.51	1.7

C. Fatigue Analysis

Figure 7 shows the Goodman diagram for the II analysed case. The points falling above the Goodman line indicate the elements likely to fracture. Goodman method was a good fracture predictor in all cases.

The Sines criterion predicted fractures in every case, with the most stressed regions in 3 cases out of 5 close to the strut intersections between the first and the second ring from the proximal end; in one case it was at the lowest terminal crown and in the last one it was detected between the second and the third crown. In almost all simulations it was verified that the location of fractures, detected from X-rays images, was included in the zones with the highest risk of fracture predicted by Goodman and Sines criteria (figure 8).

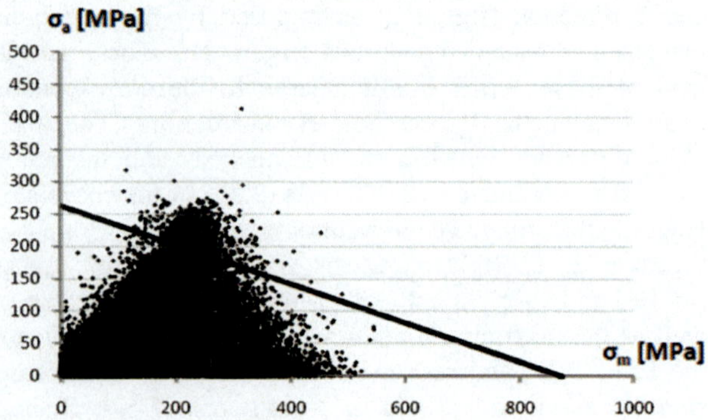

Fig. 7 Goodman diagram for patient II

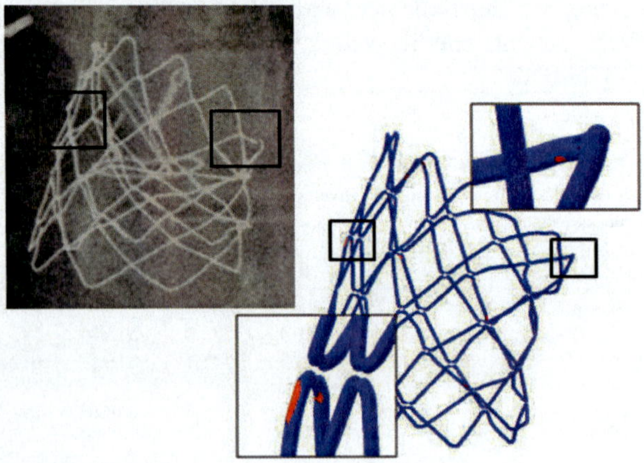

Fig. 8 Stent fractures highlighted in the X-ray image for patient I (left) and Sines stress map in the corresponding simulation with highest risk fracture zones coloured in red (right)

IV. Discussion

PPVI procedure was performed for the first time in 2000 as a less invasive treatment for right ventricular outflow tract dysfunction [11]. An increasing number of patients are benefiting from this technique that proved to be a successful alternative to surgery. Yet, a well recognised complication is stent fracture. Experimental and computational studies in the past have not been able to predict the device fractures, probably because the boundary conditions used were too much simplified. Thus, a patient-specific approach was used in this study, aiming at replicating more realistically the stent deployment history in 5 patients.

Biplane fluoroscopic images were used to reconstruct the device shape *in-situ*, and to measure its deformation during the procedure and the cardiac cycle. This method allowed the calculation of the stent strut displacements, subsequently used in a stent FE model as boundary conditions. Reproducing the stent deployment from its crimped initial status up to its final configuration resulted in a more realistic stress distribution. As a result, the fatigue analysis, performed using Goodman and Sines criteria, was able to predict stent fractures in all 5 cases. This highlights the importance of realistic stent deployment to assess loading conditions and therefore the major role played by the interaction with the right ventricular outflow tract anatomy. Patient-specific analysis should be used for the study of fatigue stent fracture. By applying the same methodological procedure to study patients who did not experience stent fracture after PPVI, the comparison between the fractured and the non-fractured groups could lead to understand the variables which most affect PPVI stent failure.

V. Conclusions

Reproducing patient-specific analysis with more realistic loading conditions can provide more accurate information regarding the stent mechanical performance and its fatigue life.

Furthermore, from a clinical perspective, a better understanding of the phenomena inducing fracture in PPVI stents could help predict success/failure of the procedure for each individual patient before the procedure is performed. Hence, interventional cardiologist would be driven by engineering tools along with conventional clinical assessment to a more accurate patient selection and safer implantation. In addition, the results of patient-specific simulations could aid in settling a better post-operative planning, in order to monitor possible stent fractures.

Acknowledgment

This research is funded by the European Commission, through the MeDDiCA ITN (www.meddica.eu, PITN-GA-2009-238113), Marie Curie actions under FP7, People Programme.

References

1. Khambadkone S et al. (2005) Percutaneous pulmonary valve implantation in humans: results in 59 consecutive patients. Circulation 112:1189–1197
2. Lurz P, Bonhoeffer P, Taylor A (2009) Percutaneous pulmonary valve implantation: an update. Expert Rev Cardiovasc Ther 7(7):1–10
3. Schievano S et al. (2010) Patient specific finite element analysis results in more accurate prediction of stent fractures: application to percutaneous pulmonary valve implantation. J Biomech 43(4):687–693

4. Nordmeyer J et al. (2007) Risk stratification, systematic classification, and anticipatory management strategies for stent fractures after percutaneous pulmonary valve implantation. Circulation 11:1392–1397
5. Vezmar M et al. (2010) Percutaneous pulmonary valve implantation in the young, 2-year follow-up. J Am Coll Cariol Intv 3:439–448 DOI 10.1016/j.jcin.201002003
6. Vano et al. (2005) Dynamic flat panel detector versus image intensifier in cardiac imaging: dose and image quality. Phys Med Biol 50(23):5731–5742
7. Schievano S et al. (2007) Finite element analysis of stent fracture in percutaneous pulmonary valve implantation. J Interv Cardiol 20(6):546–554
8. Beden SN et al. (2009) Fatigue life assessment of different steel-based shell matrials under variable amplitude loading. Eur J Sci Research 29(1), 157–159
9. Sines G, Ohgi G (1981) Fatigue criteria under combined stresses or strains. J Eng Mater Technol 103(2):82–90 DOI:10.1115/1.3224995
10. Marrey RV et al. (2006) Fatigue and life prediction for cobalt-chromium stents: a fracture mechanics analysis. Biomaterials 27(9):1988–2000
11. Raikou et al. (2011) An assessment of the cost of percutaneous pulmonary valve implantation (PPVI) versus surgical valve replacement (SVR) in patients with right ventricular outflow tract dysfunction. J Med Econ 14:47–52

Author: Daria Cosentino
Institute: UCL Mechanical Engineering Department & Great Ormond Street Hospital for Children
Street: Great Ormond Street
City: London, WC1N 3JH
Country: United Kingdom
Email: d.cosentino@ucl.ac.uk

The Study of Massive Trochanterion Fractures

A.I. Botean, I.A. Takacs, and M. Hardau

Technical University of Cluj-Napoca / Strength of Materials Department, Mechanical Faculty, Cluj-Napoca, Romania

Abstract— **This paper studies the massive trochanterion fractures using two methods of fixation, namely DHS (Dynamic Hip Screw) and Gamma rod. The analysis is carried out both experimentally (photoelasticimetry and tests on human femur) and numerically (finite element method).**

Keywords— **femoral bone, massive trochanterion, fracture, photoelasticimetry, finite element.**

I. Introduction

The massive trochanterion fractures are among the most common fractures seen in elderly subjects and can be generated even after minor trauma [1]. This type of fracture requires an emergency surgery, minimally aggressive to speed up mobilization and reduce the subjects immobilization period in the supine position so that the recovery can be initiated early.

Kyle's classification of the trochanteric fractures divides them in four categories (Figure 1) [4]: 1 – stable intertrochanterion fractures without displacement and tearing; 2 – stable intertrochanterion fracture with displacement and minimum tearing; 3 – unstable intertrochanterion fracture with displacement and posterior-medial breakage; 4 – unstable intertrochanterion fracture with posterior displacement, posterior-medial breakage and undertrochnaterion component.

The stability is provided by the bone tissue or by the particular soft parts. This study examines the biomechanics of the osteosynthesis using two methods (Figure 2):

1. DHS (Dynamic Hip Screw) – in which case the force arm unloads the body's weight on a larger length, positioned between the femoral head and the plate fixed to the external cortex;
2. Gamma rod – in which case the rod is inserted in the centre of the shaft, without opening the focus area of the fracture.

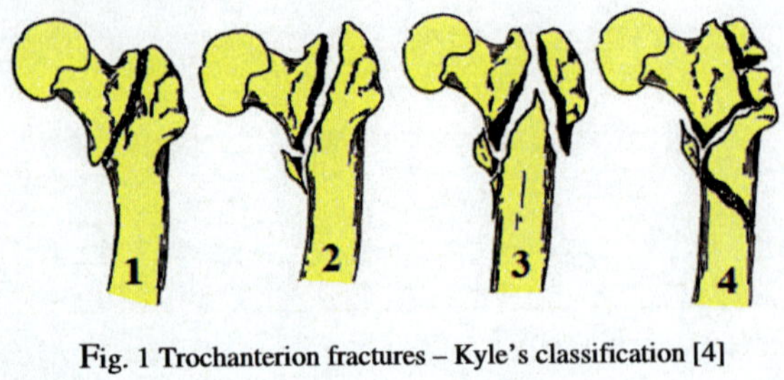

Fig. 1 Trochanterion fractures – Kyle's classification [4]

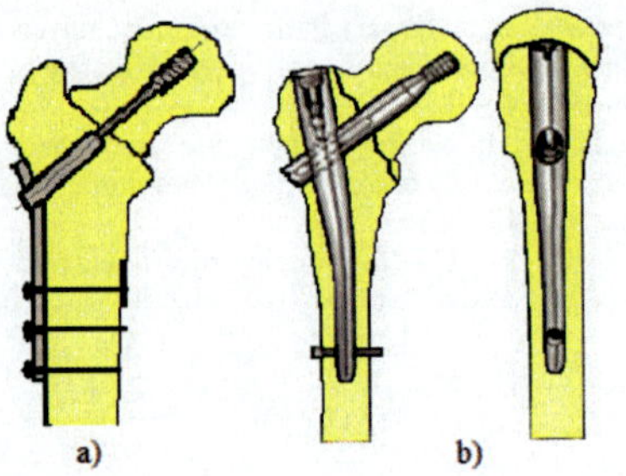

a) b)

Fig. 2 a) The fixation of the massive trochanterion – DHS method; b) The fixation of the massive trochanterion – Gamma rod [4]

II. Experimental analysis

The study is carried out on two femoral bones that represent the massive trochanterion fracture. For osteosynthesis are used a DHS system (Figure 3) and the Gamma rod (Figure 4).

The experimental setup is composed of the following elements: 1 –U2B10kN (HBM) force transducer that establish the value of the load; 2 –WA20mm (HBM) displacement transducer that show the movement of the femoral head on the vertical direction; 3 - Catman Easy (HBM) data acquisition interface; 4 – Spider8 (HBM) data acquisition system.

When using the DHS system the femur is loaded with the maximum mass of 134.6 kg, the recorded displacement of the femoral head being 14.48 mm. The higher values of the loaded weight produce pronounced crushing and the contact areas cannot support the load.

When using the Gamma rod the maximum weight that loads the femoral bone is 161 kg and the displacement recorded in the vertical plane is 7.61 mm. In this case as well, the crushing problem in the contact area becomes extremely important.

To highlight the state of stresses in the contact area between the metal elements and the bone matter we appeal on an optical investigation method, namely the photoelasticimetry. Based on experimental optical principles and mathematical theory of elasticity, photoelasticity was noted from the very beginning to be a simple experimental technique with broad possibilities of application in the state of stresses and strain analysis. Unlike other tensometers methods (mechanical, optical or strain gauges method) which

S. Vlad and R.V. Ciupa (Eds.): MEDITECH 2011, IFMBE Proceedings 36, pp. 294–297, 2011.
www.springerlink.com

provide information in discrete points, photoelasticimetry provides a complete full field of stresses, thereby enabling the determination of stresses (in magnitude and direction) at any point on the tested model [2], [3].

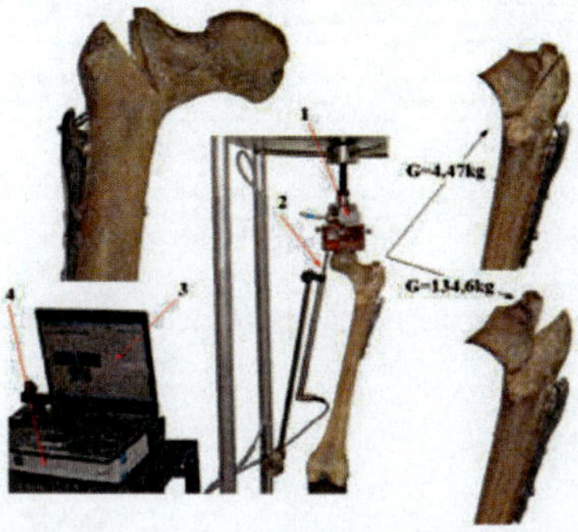

Fig. 3 Experimental setup for testing the model with the DHS system

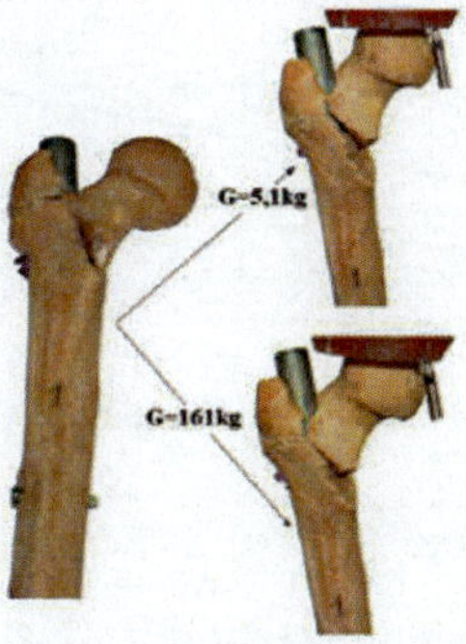

Fig. 4 The loading and straining manner using Gamma rod

Thus, in Figure 5 is shown the three dimensional model obtained by casting epoxy resin, in which a DHS system is introduced.

In terms of quality, analysing the distribution of the isocromates (Figure 5b and 5c), the maximum load areas are in the points O_1 and O_2 as well as O_3 and O_4.

In Figure 6 is presented the three dimensional model made from epoxy resin in which a Gamma rod is inserted.

Qualitatively speaking, analysing the distribution of the isocromates (figure 6b and 6c), we can see that the maximum loaded areas are in the points O_5, O_6 and O_7, O_8.

The results lead to important qualitative indications on the most loaded areas, representing the basis for a numerical analysis (finite element method) or giving information that deserves to be taken into account in choosing the optimum process in regard to the fixation of fractures of the massive trochanterion.

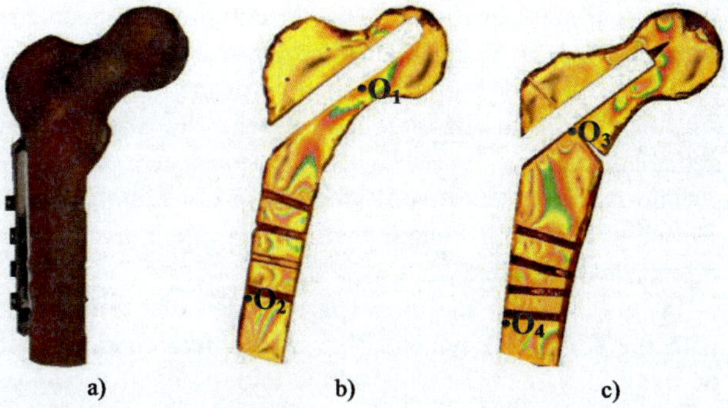

Fig. 5 Model obtained from epoxy resin – DHS system: a) model before loading; b) plane section through model without fracture; c) plane section through the model with fracture

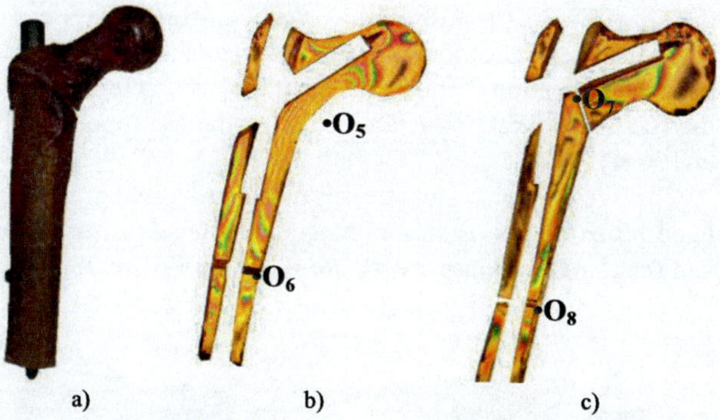

Fig. 6 Model obtained from epoxy resin – Gamma system: a) model before loading; b) plane section through model without fracture; c) plane section through the model with fracture

III. Numerical analysis

For numerical modelling using finite element method the specific features of the osteoarticular system's biomechanics should be considered. Finite element calls as input the actual numerical values of the elastic constants and of the loads. Great difficulties arise from trying to determine them.

The bone is highly anisotropic and inhomogeneous and it is desirable that the idealized structure to be developed as such. If in which regards the development of the structure no unsolved problems remain, the most difficult problems come from determining the variable elastic characteristics of the bone tissue [1].

Difficulties also result in determining the values of the loads and their points of application. In general, the forces are transmitted to a bone through the joint surfaces, which are three dimensional, with a rather complicated geometric configuration. In these circumstances, the consideration of concentrated loads is a rough shaping of the reality, and the consideration of a distribute load requires knowledge of its distribution law.

The method has proved effective in objectively and effectively analysing complex structures, such as those of osteoarticular system.

In this case a plane model is provided that corresponds with the two cases studied. The mechanical characteristics of the epoxy resin suitable to the experimental model are taken into account.

Figure 7a shows the meshing design of the experimental model for the first studied case using triangular finite elements. The material used for the numerical simulation is epoxy resin. The loading is performed with a uniformly distributed load (q = 10 N/mm). Also, in figure 7a are defined the used constrains. Contact elements are not defined in this application. Instead, the material characteristics for the fracture's fixation system of the model are (epoxy resin and steel).

The Tresca equivalent stresses distribution that are calculated based on the maximum tangential stresses distribution and the third resistance theory are shown in Figure 7b.

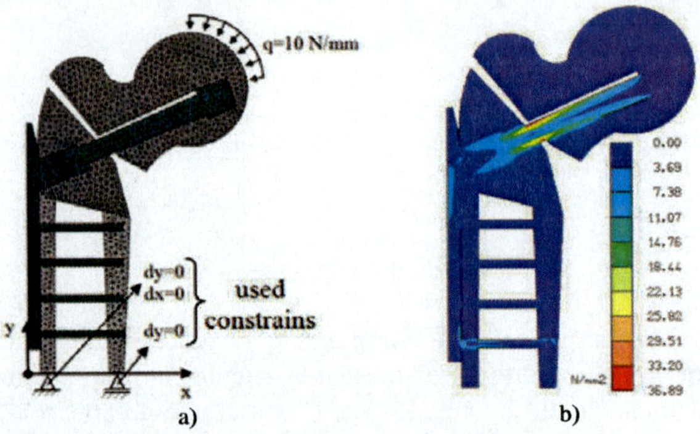

Fig. 7 a) Meshing design for the DHS model
b) The variation of the Tresca equivalent stresses – DHS model

In Figures 8 and 9 we can see the distribution of the Tresca equivalent stresses, both in the fractured section D-D' and in the area where the metal plate is inserted E-E'. The highest stresses are in the section area (D-D'). For the area where the metal plate is inserted, the maximum values of the stresses appear in the fourth screw.

The same analysis regarding the meshing model (Figure 10a) and Tresca equivalent stresses variation (Figure 10b) are done for the Gamma fixation system.

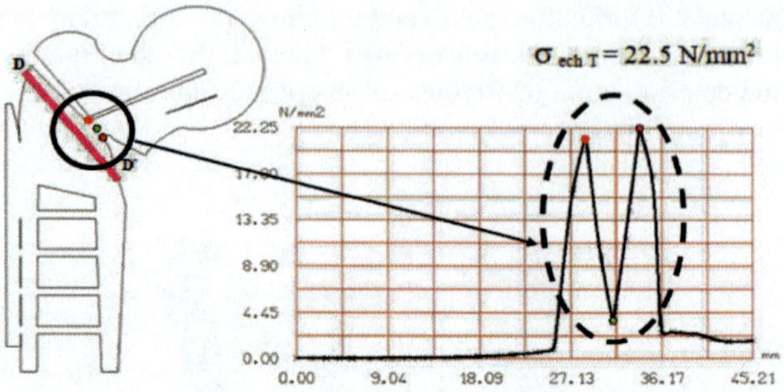

Fig. 8 The variation of the Tresca equivalent stresses in the D-D' section – DHS model

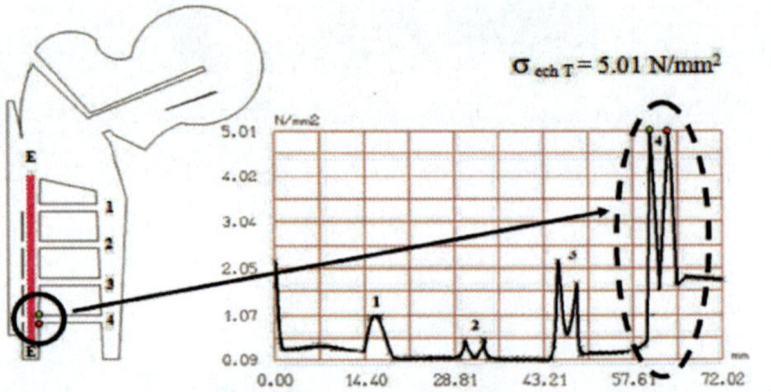

Fig. 9 The variation of the Tresca equivalent stresses in the E-E' section – DHS model

The variation of the Tresca equivalent stresses are presented for the fracture section F-F' (Figure 11) and for the area where the safety screw is inserted H-H' (Figure 12).

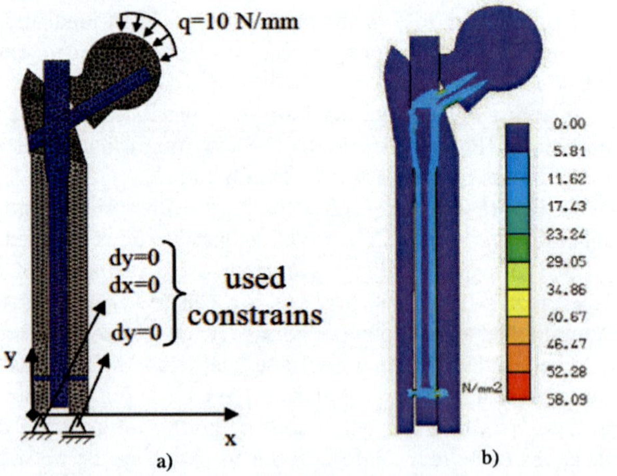

Fig. 10 a) Meshing design for the Gamma model; b) The variation of the Tresca equivalent stresses – Gamma model

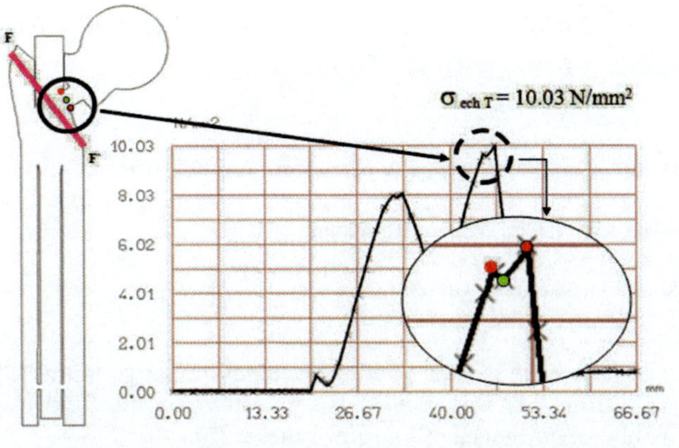

Fig. 11 The variation of the Tresca equivalent stresses in the F-F' section – Gamma model

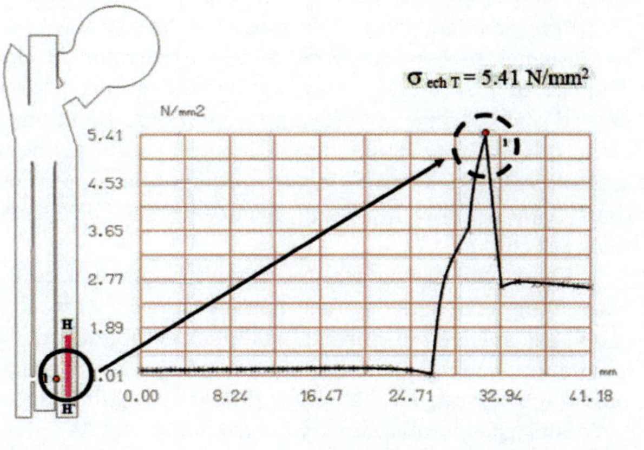

Fig. 12 The variation of the Tresca equivalent stresses in the H-H' section – Gamma model

IV. Conclusions

From the study conducted we can conclude the following:

- the numerical and experimental analysis identifies the area where the load is maximum, which is the surface area between the greater trochanter and the lesser trochanter.
- by using photoelasticimetry it could be underlined the stresses distribution in the fractured area of the femoral bone.
- from Figures 7b and 10b results that the equivalent stresses Tresca have maximum values in the inferior fibres of the rods (DHS and Gamma system), noticing that for the Gamma system they have maximum value ($\sigma_{ech\ T}$ = 58.09 N/mm^2) in comparison with the DHS system ($\sigma_{ech\ T}$ = 36.89 N/mm^2).
- as well, in the contact area between the rod (or screws) and the model (bone tissue or epoxy resin) the loads level could be quantified.
- thus, according with Figures 8 and 11, the DHS system has the maximum stresses ($\sigma_{ech\ T}$ = 22.5 N/mm^2), in comparison with the Gamma system ($\sigma_{ech\ T}$ = 10.03 N/mm^2).
- Figures 9 and 12 show the equivalent stresses Tresca distribution in the areas where the two systems are fixated, the femoral bone shaft. It can be noticed that, in this area, in the case of the DHS fixation system the equivalent stresses Tresca are 5.01 N/mm^2, while for the Gamma system their value is 5.42 N/mm^2.
- the comparative study between the two systems used in the fixation of a trochanteric fracture shows that the Gamma system is the better choice; the disadvantage is that is also the expensive one.

Acknowledgment

This paper was supported by the project "Improvement of the doctoral studies quality in engineering science for development of the knowledge based society-QDOC" contract no. POSDRU/107/1.5/S/78534, project co-funded by the European Social Fund through the Sectorial Operational Program Human Resources 2007-2013.

References

1. Antonescu, D., s.a., 1986, Metode de calcul şi tehnici experimentale de analiza tensiunilor în biomecanica. Bucuresti, Editura Tehnica
2. Cretu, A., 2001, Tensiuni... Stresses... Contraintes... Cluj-Napoca, Editura Mediamira
3. Pastrav, I., 2001, Metode optice de analiza experimentala a tensiunilor. Lucrari de laborator. Cluj-Napoca, Editura U.T.PRES
4. http://www.scritube.com

Author: Takacs Ioana Alexandra
Institute: Technical University of Cluj-Napoca, Strength of Materials Department
City: Cluj-Napoca
Country: Romania
Email: ioana.takacs@rezi.utcluj.ro

Multi-physics and Multi-scale Computational Evaluation of the Thrombogenic and Cavitational Potential of a Bileaflet Mechanical Heart Valve

R.K. Nallamothu[1], Yan Li[1], D.R. Rafiroiu[1], V. Díaz-Zuccarini[2], A.J. Narracott[3], P.V. Lawford[3], and D.R. Hose[3]

[1] Electrical Engineering Dep., Technical University, Cluj-Napoca, Romania
[2] Mechanical Engineering Dep., University College, London, United Kingdom
[3] Medical Physics Group, University of Sheffield, Sheffield, United Kingdom

Abstract— **The paper presents a multi-physics and multi-scale modeling approach on the closure dynamics of a bileaflet prosthetic heart valve. The modeling methodology consists of coupling a multi-scale model of the left ventricle contraction to a 3D CFD-FSI model, in order to investigate the closure of a cavity pivot bileaflet valve. The rebound motion of the valve is also modeled. The potential of the valve to generate thrombogenic and cavitation effects are assessed on the basis of the duration for which the blood flow is exposed to elevated wall shear stress, large negative pressure transients and increased flow vorticity. A special attention is given to the flow field in the hinge region of the valve.**

Keywords— **multi-physics, multi-scale, mechanical heart valve, thrombogenicity, cavitation, hinge flow.**

I. Introduction

Cardiovascular engineering is one of the success stories in Bioengineering. A specific example of successful cardiovascular engineering applications is the design analysis of prosthetic heart valves [1]. Although existing techniques, including mechanical heart valves (MHV) and biological valves (BV), have stood the test of time there are still some important issues to solve regarding thrombogenecity of MHVs and life-time duration for biological valves.

Computational Fluid Dynamics (CFD) has emerged as a promising tool, which, alongside experimentation, can yield insights of unprecedented detail into the hemodynamics of prosthetic heart valves. For CFD to realize its full potential, however, it must rely on numerical techniques that can handle the enormous geometrical complexities of prosthetic devices with spatial and temporal resolution sufficiently high to accurately capture all hemodynamically relevant scales of motion [2]. Valve function is driven by interaction between the blood (fluid) and the motion of solid valve structure. In order to examine such systems computationally it is necessary to consider both the solid and the fluid phases simultaneously, which requires a fluid-structure interaction (FSI) analysis. Recent computational models have used both custom and commercial codes such as ANSYS-CFX, ANSYS-Fluent and LS-DYNA.

These have been applied to the study of native mitral [3] and aortic [4] valves and also to determine the fluid dynamic performance of valve prostheses [5].

The effects of the heart, vasculature and the systemic response to the changing physiological environment are often not included within local 3D models of valve function. Multi-physics and multi-scale modeling brings new insight by allowing such interactions to be investigated *in silico*. The use of FSI analyses has highlighted the need for improved and interactive boundary conditions. Solutions include FSI analyses coupled with lumped parameter boundary condition models which can represent biochemical reactions at the cellular level and electro-mechanical events in the heart [6].

Patients with mechanical heart valve implants need to be under long-term anticoagulant therapy in order to minimize problems related to thromboembolic complications [7]. Cavitation bubble development due to the large negative pressure transients on the inflow side of the leaflet edge and their subsequent collapse may contribute to both the platelet activation and the structural damage of the valve [8].

The bileaflet mechanical heart valve (BMHV) has been used for almost two decades and remains the most widely implanted valve design. Like any other mechanical or bio prosthetic heart valve, bileaflet valves are not free from complications. They can still cause major complications including hemolysis, platelet destruction, and thromboembolic events. Investigations have concluded that the stresses imposed on blood by MHVs during the closing phase, in both mitral and aortic position, can initiate hemolysis and the coagulation cascade. Based on their attempts to investigate the leakage, hinge, and near hinge flow fields of different BMHVs [9], manufacturers continuously try to improve the design of their products.

The work presented here is continuation of our previous efforts in building up multi-physics and multi-scale models of the BMHV's fluid mechanics. Our previous work has been done with the purpose of investigating the hemolytic and cavitation potentials of the valves. Focusing on the most critical points of a valve dynamics, the closure and the rebound motion of its leaflets, we were seeking for evidence of high negative pressure transients, high vorticity and high

S. Vlad and R.V. Ciupa (Eds.): MEDITECH 2011, IFMBE Proceedings 36, pp. 298–305, 2011.
www.springerlink.com

wall shear stress values in the hydrodynamic field [10]. Using the same modeling methodology, we are now extending our study to the whole cardiac cycle, focusing not only on the closure of the valve but also on the acceleration phase during systole and on the reverse flow during diastole. Special attention is given to the hinge flow. The valve is considered in mitral position but other studies currently under development are dealing with both the aortic and the mitral valve simultaneously.

II. COMPUTATIONAL MODEL

A. *CFD Model*

To build up the computer aided design (CAD) model, an explanted 23 mm tissue annulus diameter (TAD) bileaflet mechanical heart valve was laser scanned. The resulted CAD image of the assembled valve is shown in Figure 1.

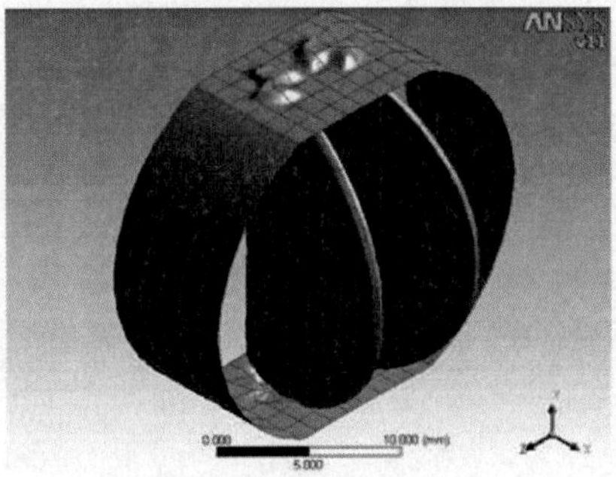

Fig. 1 The CAD image of the scanned valve, consisting of two identical leaflets and the housing

Considering the valve to be mounted in mitral position, the CAD model was completed with two cylindrical chambers representing the atrium and ventricle. The atrial chamber, positioned in the negative OX direction and the ventricular chamber, positioned in the positive OX direction are both 24 mm in length. For saving computational resources, only one quarter of the valve was considered in the 3D model.

Blood was considered as an incompressible Newtonian fluid with density $\rho = 1100\,\text{kg/m}^3$ and dynamic viscosity $\mu = 0.004\,\text{kg/m}\cdot\text{s}$. The unsteady flow field inside the valve is described by the 3D equations of continuity and momentum (1), with the boundary conditions suggested by Figure. 2.

$$\rho \frac{\partial \vec{u}}{\partial t} + (\vec{u}\cdot\nabla)\vec{u}\rho = \nabla p + \mu\Delta\vec{u};\ \nabla\cdot\vec{u} = 0 \qquad (1)$$

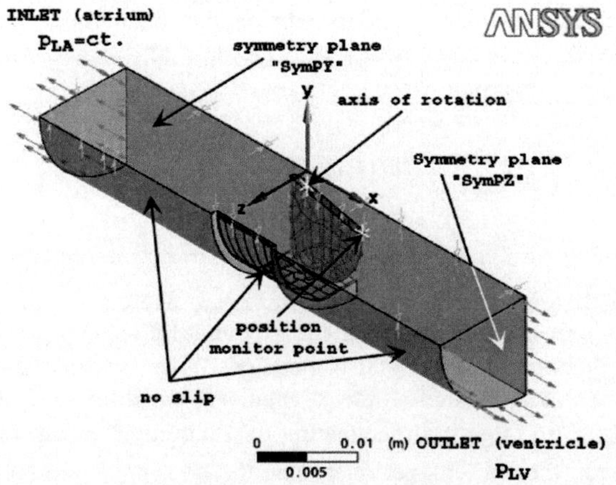

Fig. 2 The overall CAD model and the associated boundary conditions

The fluid-structure interaction model is outlined in Figure 3 and it is based on an explicit incremental method. Each leaflet rotates under the combined effects of hydrodynamic, buoyancy and gravitational forces acting on it. At every time step, the drag $\vec{f}_Q^z$ and lift $\vec{f}_Q^x$ force exerted by the flowing fluid on an arbitrary point Q of the leaflet's surface, are reduced to the centroid, G. The total contribution of drag $\vec{F}_G^z$ and lift $\vec{F}_G^x$ are added to the difference between the gravitational and buoyancy forces $\left(\vec{G} - \vec{A}\right)$.

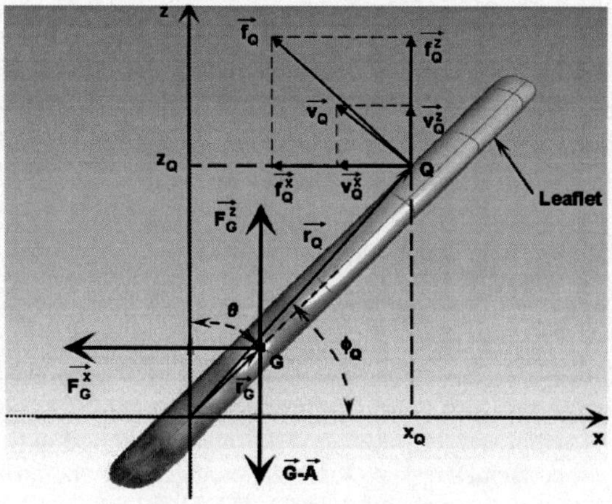

Fig. 3 The fluid-structure interaction model

The condition for the conservation of the kinetic moment is imposed, resulting in the dynamic equation (2) that describes the motion. In equation (2), $d\theta$ represents the leaflet's angular step, dt is the time step, ω_{old} is the angular velocity at the beginning of the current time step and M is the total moment acted on the leaflet from the external forces. The moment of inertia of each leaflet is $I = 7.21\,\mathrm{g}\cdot\mathrm{mm}^2$.

$$d\theta = \omega_{old}\,dt + \frac{M}{2I}dt^2 \tag{2}$$

$$M = r_G\left[F_G^x \sin\theta + \left(F_G^z - G + A\right)\cos\theta\right] \tag{3}$$

As there is speculation that vortexes are likely to occur after the first impact between the leaflet and the valve housing, the rebound motion is also modeled. When, within the current time step, the angular position exceeds $\theta_{reb} = 63.7$ dgrs. corresponding to a minimum distance between the leaflet's periphery and the housing, a virtual torsion spring with constant k is suddenly twisted between the current position and θ_{reb}. The torque exerted by the spring is added to the gravitational force, thus reversing the leaflet's rotation. The dynamic equation changes accordingly:

$$d\theta = \omega_{old}\,dt + \frac{M}{2I}dt^2 - k\frac{\left(\theta_{old} - \theta_{reb}\right)^2}{2I}dt^2 \tag{4}$$

$$\begin{aligned} dx_{new} &= r_Q\left[\cos\left(\varphi_{Q_{old}} + d\theta\right) - \cos\varphi_{Q_{old}}\right] \\ dz_{new} &= r_Q\left[\sin\left(\varphi_{Q_{old}} + d\theta\right) - \sin\varphi_{Q_{old}}\right] \end{aligned} \tag{5}$$

Displacements of every point Q from the surface of the leaflet are calculated as a function of the distance to the center of rotation r_Q, its angular coordinate at the previous time step $\varphi_{Q_{old}}$, and the current angular step $d\theta$, according to equation (5). The current time linear displacements are added to the "old" coordinates and the moving boundary is displaced together with the mesh.

B. *Left Ventricle Model*

For representing the contraction of the left ventricle (LV), a complex boundary condition is used. Figure 4 shows the sub-levels of organization of the ventricle that were taken into consideration. The connection between the LV model and the valve model is also illustrated in figure 4.

A constant pressure source is connected to the atrium via the mitral valve (input model). The blood fills the LV (output model) and ejects a volume of blood into the arterial network via the aortic valve. Contraction in the cardiac muscle is described at a number of levels, starting from the level of the contractile proteins (actin and myosin), following a hierarchical path, from the microscopic level up to the tissue (muscle level) and to the organ level, to finally reach the hemodynamic part of the LV and its arterial load.

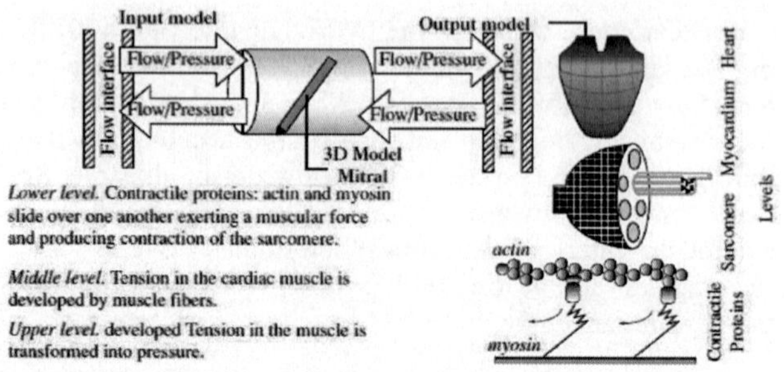

Fig. 4 Representation of the LV model and its different physical scales, coupled to the 3D model of the mitral valve [6].

Complete details of the formulation of the model can be found in [11].

III. Results

The computational model presented in section I has already been used to simulate the valve closure exclusively. The leaflets' movement was separated into two consecutive phases i.e., the *approach* phase and the *rebound* phase. The influences of various parameters of the model, like the intensity of the LV contraction and the resilience coefficient on the cavitation and hemolytic potential of the valve have been investigated [10].

In the current study, we have extended our searching for hemolytic and cavitation evidences to the whole cardiac cycle. Figure 5 shows the pressure and position histories of the valve over a 0.8 s long cardiac cycle. A detail about the bouncing motion of the leaflets is included.

The first to extract from our simulation data was the minimum pressure at the surface of the leaflets. The third panel of figure 5 shows that during the rebound motion, transient negative pressure spikes occur, whilst in the rest of the cardiac cycle the negative pressure values are low. The maximum negative pressure value recorded at the surface of the leaflets during the rebound was of -669.4 mmHg. That is below the vapor pressure of blood of -713 mmHg [12].

This suggests that, under the specifications taken into account in the model and with the inherent influence of the numeric in the results, no cavitation potential exists for this kind of valve. However, it might happen that, at hyperdynamic physiologic states (higher LV pressure rates) [12] and/or grater resilience coefficients [10], the negative pressure spike be greater than the vapor pressure. This demonstrates the potential for cavitation with implanted mechanical valves in vivo.

Yet other mechanisms responsible for cavitation exist. Vortex formations at the atrial side of the valve may contribute to cavitation. In figure 6, the maximum vorticity at the surface of the leaflets was plotted against time for the entire cardiac cycle. The maximum vorticity occurs during the rebound but still relatively high values are present after the valve closure, at mid-systole and later. That is due to the retrograde flow through the closed valve which may increase the hemolytic potential of the valve.

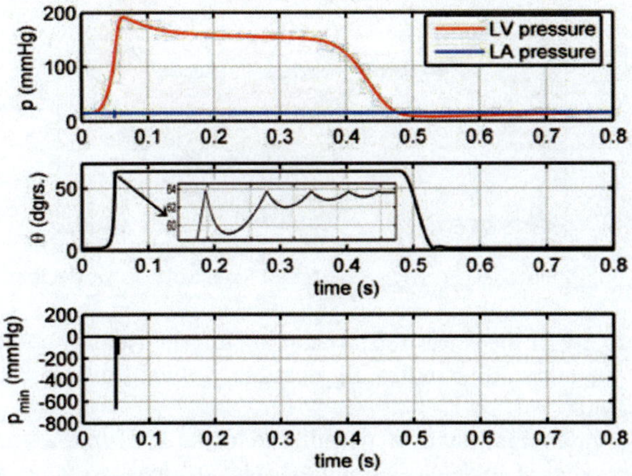

Fig. 5 Total pressure drop on the valve and position history over the entire cardiac cycle, together with the minim pressure at the surface of the leaflets

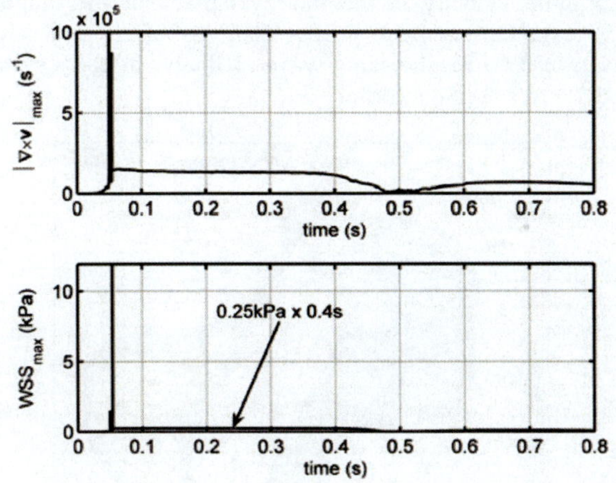

Fig. 6 Maximum vorticity and maximum wall shear stress at the surface of the leaflets vs. time

A relevant indicator of platelet activation is the area under the wall shear stress vs. time curve $\int \tau_{max}(t)dt$. The values of this integral that exceed $3.5\,\mathrm{Pa}\cdot\mathrm{s}$ are indicators for platelet activation [13]. The second panel of figure 6 shows the maximum wall shear stress values at the surface of the leaflet vs. time. Again, the maximum WSS value (14 kPa) occurs during the rebound but still high values are noticed at mid-systole and later. Using a value of 0.25 kPa for the WSS and 0.4 s for the duration of the stress from the plot, WSS-time product of $100\,\mathrm{Pa}\cdot\mathrm{s}$ is obtained as representative of the area under the curve. The value exceeds the magnitudes of $3.5\,\mathrm{Pa}\cdot\mathrm{s}$ suggested for platelet factor 3 release indicative of platelet activation. This extremely high value is due to the leaking valve during late systole and we have to look for its source in the incomplete closure of the valve and in the hinge flow field.

Numerous researches have sought to characterize the flow field inside the hinge region of BMHVs in an effort to better understand the relationship between hinge design and thromboembolic potential. Most of studies have only captured two-dimensional velocity fields at selected locations [15], [17]. Researchers have resorted to numerical studies to obtain further information on the hinge flow fields. However, the relevance of their results seems to be insufficient, due to "lack of spatial resolution or the use of non-physiologic flow conditions". Simon *et al* [16] have implemented a pseudo multi-scale approach to simulate the three-dimensional physiologic flow in the hinge region of BMHV under aortic conditions.

In the present study, we try to have a preliminary insight into the hinge flow field of a BMHV under mitral conditions, using the multi-physics and multi-scale modeling approach described above. The hinge geometry was characterized using the same nomenclature and methodology as in Simon et al. [16]. The leaflet ear was positioned within the hinge recess such that the gap width (distance between the bottom of the hinge recess and the tip of the leaflet ear, defined as (b-a) in Fig. 7) was approximately 200 μm.

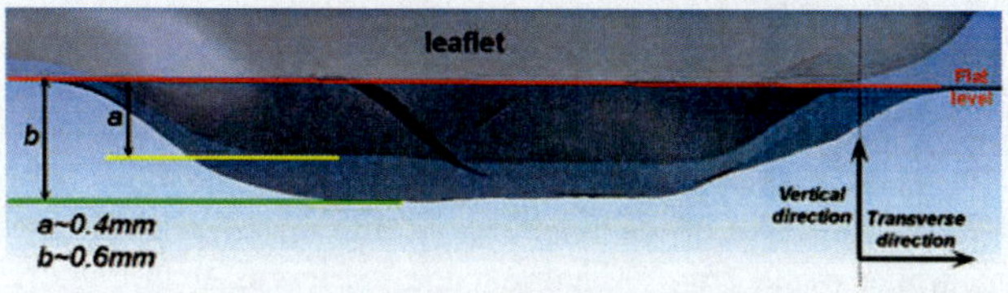

Fig. 7 Side view of the hinge pivot and recess at fully closed position

Figure 8 shows the valvular pressure drop and flow waveforms. The cardiac cycle duration was set to 0.8s which corresponds to a heart rate of 75beat/min. The instances of time corresponding to the beginning of systole, mid-systole, end-systole, mid acceleration and peak-diastole are included. The flow waveform indicates a persistent leakage of the valve after its closure. There are two possible sources for this leakage in our model. One might be a slight deformation of the valve housing either at explantation or during its preparation for scanning, as the housing and the two leaflets had to be separated. Most certainly, the second source for leakage is the small gap (0.1mm) that had to be foreseen between the leaflet and the hosing in order to model the rebound. That gap allows for the virtual spring (which we have introduced in our model to simulate the rebound) to be compressed by the closing leaflet and to render the elastic deformation energy back to the leaflet, throwing it back. At the end of bouncing, the leaflet settles up at 0.1mm away from the housing, thus maintain this gap all along the systole. A double size gap exists within each hinge, between the leaflet ear and the housing recess. The central gap between the two fully closed leaflets is much smaller. It means that most of the leakage is through the hinges and between the leaflets and the housing.

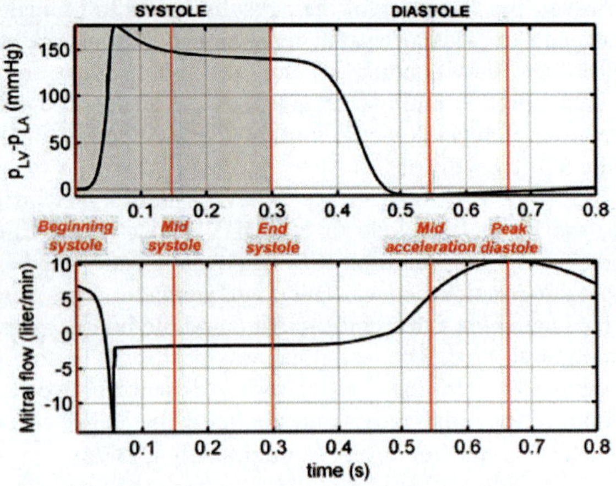

Fig. 8 Valvular pressure drop and flow waveforms

Under these circumstances, which may be thought as inherent somehow, we tried to have a look at the flow fields in the flat level of the hinge, at specific instances during the cardiac cycle (fig. 8 second panel). Figures 9 through 15 illustrate these velocity fields. The direction of the forward flow is from right to left (from left atrium to the left ventricle). Arrows point in the direction of the velocity vector and are colored and sized by velocity magnitude. For a better view of the velocity distribution within the flat level, contour plots were also added to each graph.

Before the onset of the left ventricle contraction (beginning of systole), the valve is still opened and a forward flow passes through the valve. This is the initial state (t=0s) of the valve for the current cardiac cycle. A forward jet at a maximum velocity of 1.26 m/s was found in the central region of the hinge (fig. 9). The same plot shows areas of low

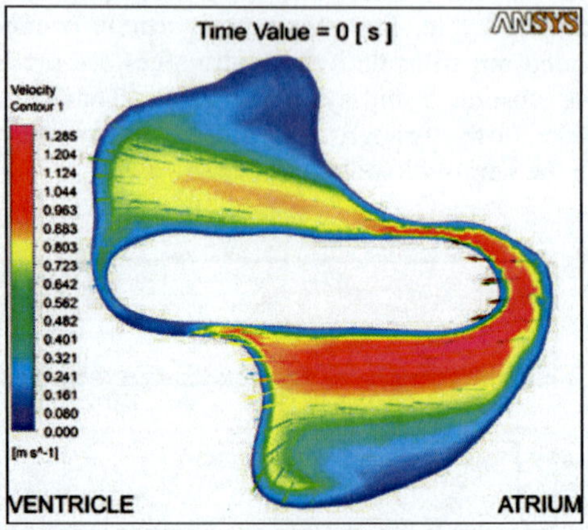

Fig. 9 Flow field at early systole for a 23-mm mitral valve, flat level

velocity in the lower (atrial) and upper (ventricular) corners of the hinge, thus indicating possible vertical flows in those regions.

One hundred and fifty milliseconds later, at mid-systole, the valve is fully closed and a backward leakage jet is observed towards the atrial corner of the hinge (fig. 10). As Leo explains [14], this leakage jet is drawn towards the atrial side of the hinge, at early-systole, by the leaflet. The maximum velocity in the flat-level plane of the hinge is 9.59 m/s. This is much greater than any of the two values given by Leo for the same valve (3.17 m/s at early systole and 2.4 m/s at mid systole).

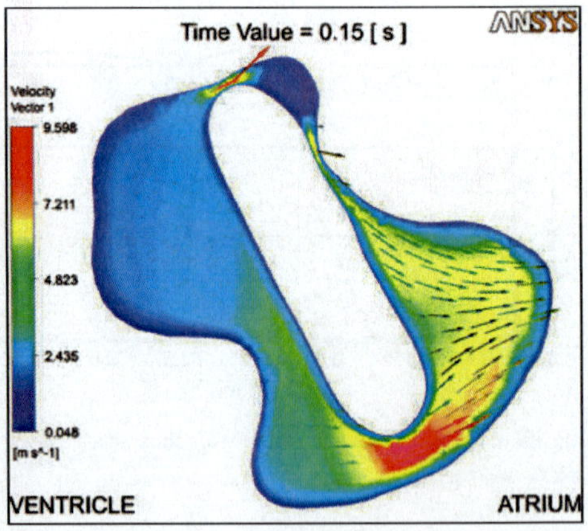

Fig. 10 Flow field at mid-systole for a 23-mm mitral valve, flat level; high backward flow velocities of nearly 10 m/s are present because of the leakage

A similar backward leakage jet was found at the end of systole (fig. 11) but the maxim velocity is a little bit lower (9.37 m/s).

During the diastole, at mid-acceleration, the forward flow recovers, with the same location relative to the leaflet's ear as it was at the end of the previous diastole (fig. 9) but, with a maximum velocity of 0.96 m/s (fig.12) only. The maximum velocity is much lower than the one reported by Leo for the mid diastole. This is the moment when the small flow velocity adjacent to the central forward jet starts generating vortical structures in the corners of the hinge (fig. 13).

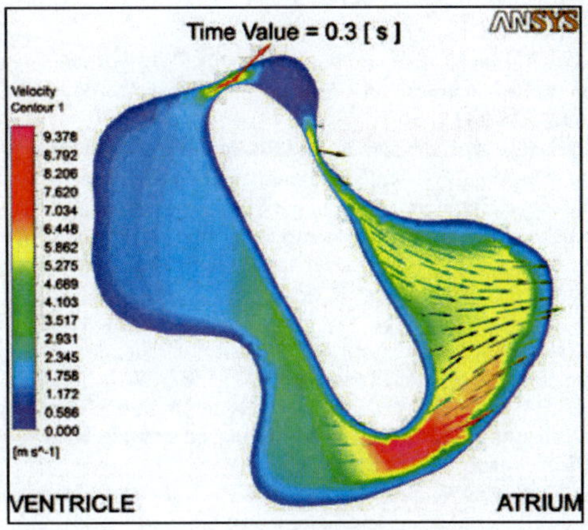

Fig. 11 Flow field at end-systole for a 23-mm mitral valve, flat level; high backward flow velocities of nearly 10 m/s are present because of the leakage.

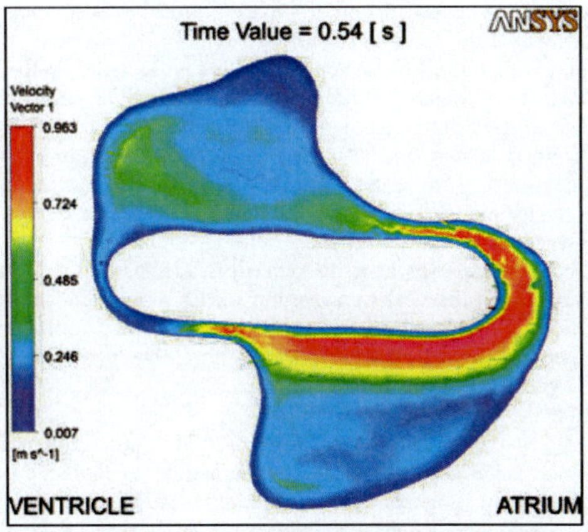

Fig. 12 Flow field at mid-acceleration for a 23-mm mitral valve, flat level

Towards the peak diastole (fig. 14) the forward central jet increases in intensity, the maximum velocity reaching the value of 2.5 m/s. This value is much closer to the one reported by Leo (3.2 m/s). The vortical flow in the corners increases in intensity (fig. 15).

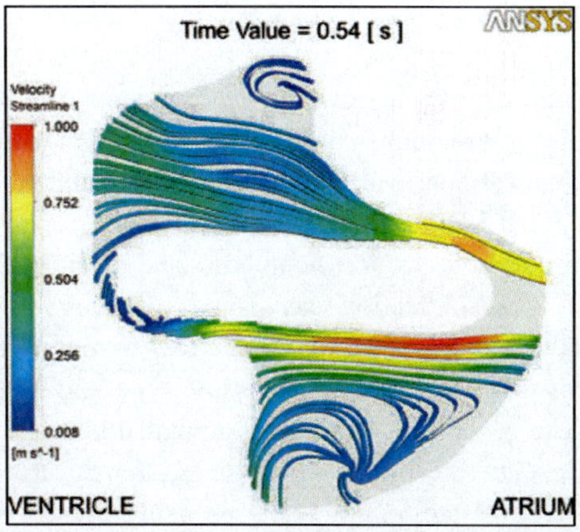

Fig. 13 Streamlines at mid-acceleration for a 23-mm mitral valve, flat level; vortical structures occur in the both the atrial and the ventricular corner of the hinge recess

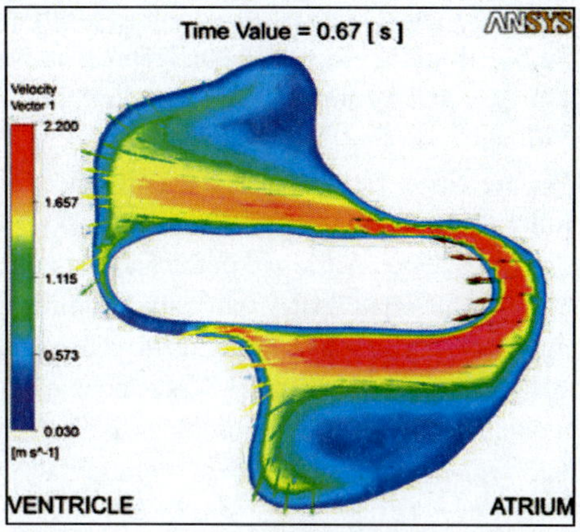

Fig. 14 Flow field at peak-diastole for a 23-mm mitral valve, flat level

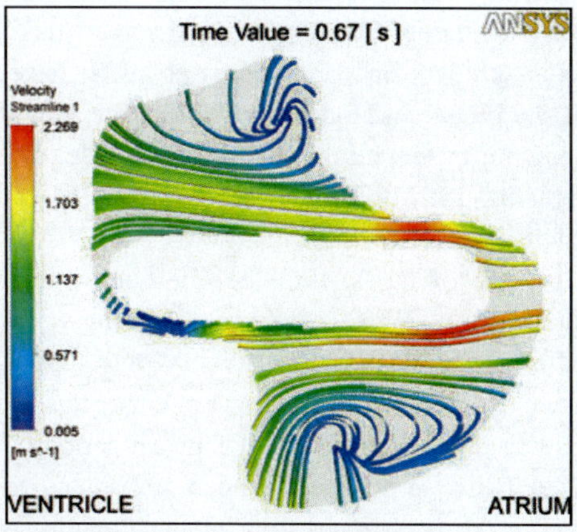

Fig. 15 Streamlines at peak-diastole for a 23-mm mitral valve, flat level; vortical structures develop in the both the atrial and the ventricular corner of the hinge recess

IV. Conclusions

By coupling a multi-scale model of the left ventricle contraction to a three-dimensional CFD-FSI model of a bileaflet mechanical heart valve, we succeeded to draw a picture of the global biomechanics of the valve and to investigate the possible sources for hemolytic and cavitation drawbacks. A special attention was given to the hinge flow field; the results being compared to the experimental ones obtained by other authors for the same type and size of the valve. The general aspects of the simulated hinge flow field were close to the experimental ones, the differences between some of the values being probably caused by some damage that has been brought to the explanted valve and by the rebound model which knowingly leaves a leakage gap between the leaflets and the valve housing.

These results encourage us to continuing our modeling adventure by struggling to get a better representation of the valve geometry and by improving our rebound model such that its influence on the flow field is reduced to the minimum. We are also aware that the two cylindrical chambers representing the atrium and the ventricle are poor approximations of the real geometries of both chambers. Interesting debates about the effects that the actual geometry of the heart might have on the valve functionality are already present in the literature. For example, dependence of the valve-valve interaction on the flow conditions is expected [18]. Therefore, replacing the cylindrical representation of the adjacent chambers of the valves with more realistic geometrical models is also foreseen.

At the same time, the CFD part of the model will be improved by switching from the laminar model that we have used to a turbulent one, thus allowing us to have a better view at the global and local flow fields. We would also be happy to cooperate with our partners from MeDDiCA, particularly with ISS, by comparing their 3D PIV measurements with our numerical simulations.

To the best of our knowledge, the current study is the first numerical modeling insight into the hinge flow field of a bileaflet mechanical heart valve in mitral position. All other studies published so far concern either the aortic position or they are just experimental investigations. We consider MeDDiCA as the best chance we've had to involve ourselves in such an interesting but challenging scientific activity.

Acknowledgment

"This project is funded by the European Commission; Marie Curie Initial Training Networks, FP7-PEOPLE-ITN-MeDDiCA 2008, 238113".

We also take this opportunity to acknowledge the MeDDiCA participants, professors and colleagues for their continuous support.

References

1. Dasi L, Simon H, Sucosky P et al. (2009) Fluid mechanics of artificial heart valves. Clinical and Experimental Pharmacology and Physiology **36**, 225-237
2. Yoganathan A, Chandran K. and Sotiropoulos F. (2005) Flow in Prosthetic Heart Valves: State of-the-Art and Future Directions. Annals of Biomedical Engineering **33**(12) 1689–1694
3. Kunzelman K, Einstein D. and Cochran P (2007) Fluid-structure interaction models of the mitral valve: Function in normal and pathological states. *Phil. Trans. R. Soc. B* **362** 1393–1406
4. Carmody, C., Burriesci, G., Howard, I. and Patterson E (2006) An approach to the simulation of fluid-structure interaction in the aortic valve. *Journal of Biomechanics* **39**, (1) 158-169
5. Watton P, Luo X, et al (2007) Dynamic modeling of prosthetic chorded mitral valves using the immersed boundary method. Journal of Biomechanics **40**(3) 613–626
6. Diaz-Zuccarini V, Hose D, Lawford P, et al (2008) Multiphysics and multiscale simulation: application to a coupled model of the left ventricle and a mechanical heart valve. *International Journal for Multiscale Computational Engineering* **6**(1) 65–76.
7. Johansen, P. (2004) Mechanical heart valve cavitation. Expert Review of Medical Devices **1**(1), 95–104
8. Hwansung L, Yoshiyuki T (2006) Mechanism for cavitation phenomenon in mechanical heart valves. *Journal of Mechanical Science and Technology* **20**(8) 1118–1124
9. Simon H.A. (2005) Influence of the implant location on the hinge and leakage flow fields through bileaflet mechanical heart valves. MS Thesis http://hdl.handle.net/1853/5159
10. Rafiroiu D, Ciupa R (2009) Medical devices design in cardiovascular applications. Current and future trends at Cluj Napoca Technical University. Journal of Nonlinear Optics, Quantum Optics, **39** (2-3) 101-115
11. LeFevre J., LeFevre L. and Couteiro B. (1999) A bond graph model of Chemo-mechanical transduction in the Mammalian left ventricle. *Simulation Practice and Theory* **7**(5–6) 531–552.
12. Dexter EU, Aluri S, Radcliffe RR, et al (1999) In vivo demonstration of cavitation potential of a mechanical heart valve. ASAIO J **45**(5) 436-441
13. Cheng R, Lai Y. and Chandran, K. (2004) Three-dimensional fluid-structure interaction simulation of bileaflet mechanical heart valve flow dynamics. *Ann Biomed. Eng.* **32**(11), 1471–1483.
14. Hwa Liang Leo, (2005) An in vitro investigation of the flow fields through bileaflet and polymeric prosthetic heart valves. PhD Thesis http://hdl.handle.net/1853/11644
15. Ellis JT. and Yoganathan AP. (2000) A comparison of the hinge and near-hinge flow fields of the ST Jude Medical hemodynamic Plus and Regent bileaflet mechanical heart valves, J Thorac Cardiovasc Surg; 119:89-93
16. Simon HA. Ge L. Sotiropoulos F. Yoganathan AP. (2010) Simulation of the three-dimensional hinge flow fields of a bileaflet mechanical heart valve under aortic conditions. Ann Biomed Eng, 38(3): 841-853

17. Simon HA. Leo H-L. Carberry J. Yoganathan AP. (2004) Comparison of the hinge flow fields of two bileaflet mechanical heart valves under aortic and mitral conditions. Ann Biomed Eng, 32(12): 1607-1617
18. Stijnen J.M.A Bogaerds A.C.B de Hart J. Bovendeerd P.H.M. de Mol B.A.J.M. van de Vosse J.M.A. (2009) Computational analysis of ventricular valve-valve interaction: Influence of flow conditions. International Journal of Computational Fluid Dynamics, 23(8): 609-622

Corresponding author:

Rajeev Kumar Nallamothu
Technical University of Cluj-Napoca
26-28 Gerorge Baritiu
Cluj-Napoca
Romania
Rajeev.NALLAMOTHU@et.utcluj.ro

CFD Analysis Study on the Impact of the Coronary Anatomy on the Efficiency of the Coronary Thrombectomy: The Effect of Bend Angles

Yan Li[1], R.K. Nallamothu[1], D. Rafiroiu[1], A. Iancu[2], V. Díaz-Zuccarini[3], A.J. Narracott[4], D.R. Hose[4], and P.V. Lawford[4]

[1] Technical University of Cluj-Napoca/Electrical Engineering, Cluj-Napoca, Romania
[2] University of Medicine and Pharmacy "Iuliu Hatieganu"/Cardiology, Cluj-Napoca, Romania
[3] University College London/Mechanical Engineering, London, England
[4] University of Sheffield/Medical Physics, Sheffield, England

Abstract— **Primary percutaneous coronary intervention is effective in opening the infarct-related artery in patients with myocardial infarction with ST-segment elevation. However, the embolization of athero-thrombotic debris induces micro vascular obstruction and diminishes myocardial reperfusion. The aspiration catheter's design, the coronary anatomy and the physical properties of the thrombi are key factors in ensuring a successful interventional procedure.**

The aim of our study is to use computational fluid dynamics (CFD) analysis for assessing the impact that the thrombus age and the coronary anatomy might have on the aspiration efficiency. A generic model of an Export AP (Medtronic) catheter tip was created and extended with distal shafts having different bend angles. Double bended catheter shafts were also considered. A two-phase CFD model with different values for the viscosity of the thrombus was implemented in ANSYS-CFX, allowing us to explore the impact of the viscosity on the efficiency of the aspiration.

Some authors have recently used *in-vitro* tests for exploring the impact of the bends angles and thrombus age on the efficiency of some commercial thrombus aspiration devices, including the Export catheter.

Based on a fundamental study we tried to look for some explanations for their findings.

Keywords— **percutaneous coronary intervention, catheter, thrombus aspiration, thrombus age, bending.**

I. Introduction

Primary percutaneous coronary intervention (p-PCI) has emerged as the preferred treatment of ST-elevation myocardial infarction. It has been proved to be a very effective procedure in obtaining patency of the infarct-related artery [1]. However, in 5-50% of cases, distal embolism by thrombus or ruptured plaque can occur, thus limiting the effectiveness of myocardial reperfusion and leading to larger myocardial damage and worse prognosis [2]. Despite the epicardial vessels reopening, this distal micro-embolization entrains microvascular obstructions thus leading to the *no reflow phenomenon.*

Numerous thrombus aspiration systems have been devised and intensively studied until today. The "ideal" catheter must be rapidly exchangeable, easily manageable and have a sufficiently large lumen to effectively absorb the thrombi [3]. The manual devices like the Diver® (fig. 1a), the Pronto® (fig. 1b) or the Export® catheter (fig. 1.c) are the most popular. Yet numerical analyses of thrombus aspiration performed by Pennati et al [4] showed that the catheters with side holes have a worse performance if compared with single central lumen catheters.

Probably, one of the most extensively studied thrombus aspiration catheters is the Export catheter manufactured by Medtronic (Minneapolis, Minnesota, USA). Some experimental studies, [3] and [5], have recently tried to identify factors which could explain the success or the failure of thrombus aspiration using Export catheters.

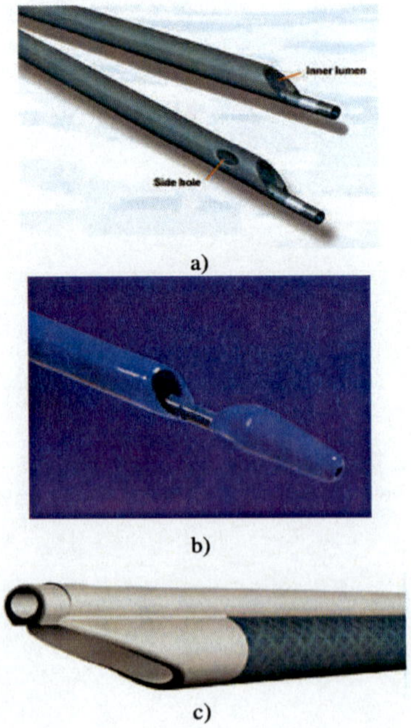

Fig. 1 Thrombus absorption catheter designs: a) Diver, b) Pronto, c) Export

S. Vlad and R.V. Ciupa (Eds.): MEDITECH 2011, IFMBE Proceedings 36, pp. 306–309, 2011.
www.springerlink.com

The efficiency of the Export catheter has been investigated experimentally, in correlation with the diameter and the bending angle of the coronary artery [3]. The same paper also describes how the thrombus age might affect the thrombus aspiration efficiency.

The respective study highlights a strong impact of the coronary artery's diameter on the efficiency of the thrombus aspiration. The efficiency of the thrombectomy has been evaluated according to the percentage of the total clot mass that has been aspired by the catheter. The same authors have noticed that neither the thrombus age nor the bending angle of the vessel has a significant effect on the thrombus aspiration. However, similar works done by Pioud et al. [5] suggest some possible effects of double bended vessels on the thrombus aspiration efficiency.

Even though these kind of experimental studies are extremely useful and reliable, they still raise a lot of questions. A good alternative for straightening out this matter is resorting to numerical simulation. Very few authors have attempted so far to do numerical simulation of the thrombus aspiration. Pennati at al. have recently tried to simulate thrombus aspiration using two realistic models of catheter tips [4]. Starting from the belief that different commercial aspiration catheters could produce different thrombus removals; the authors have tried to clarify the blood clot suction phenomenon in aspiration catheters and identify the main design parameters.

The aim of the present study is to develop a numerical simulation tool meant to investigate the impact of the coronary artery anatomy on the efficiency of the thrombus removal. A CFD analysis of the thrombus aspiration from a 3.6 mm coronary artery using an Export AP catheter tip was simulated. The impact of the thrombus age and of the artery bend angles on the efficiency of aspiration device was investigated.

II. Materials and methods

For the numerical analysis we used the same methodology as Pennati et al. did, except for the 3D reconstruction of the catheter tip. Based on measurements made on the tip of a 0.068" Export AP 6F catheter and some data available in the literature [6], we created the computer aided design (CAD) model illustrated in fig.2.

The tip of the catheter was then introduced in a straight tube, 3.6 mm in diameter, representing the coronary artery from which the thrombus is extracted. The CAD model of the catheter tip was extended with either a straight or a bended distal shaft. The vessel was extended over the same distance. Different bending angles (30°, 60° and 90°) have been introduced, at 3 cm from the tip of the catheter (fig. 3).

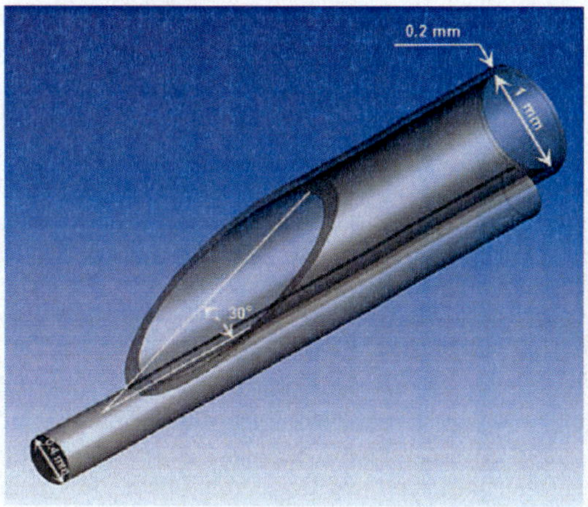

Fig. 2 CAD model of the catheter tip

A second set of CAD models was created for simulating the thrombus absorption from a double bended artery (fig. 4).

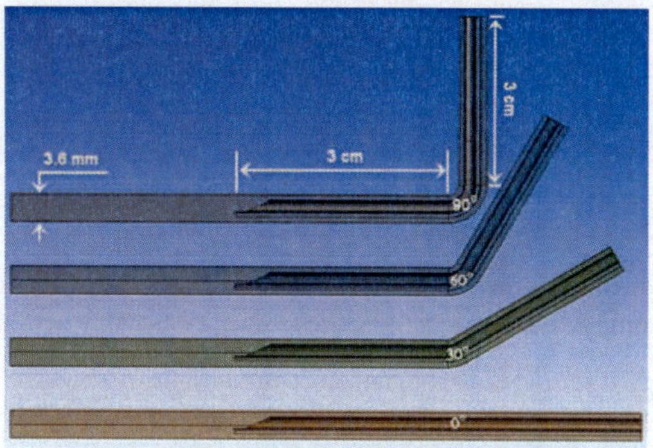

Fig. 3 Single bended catheters

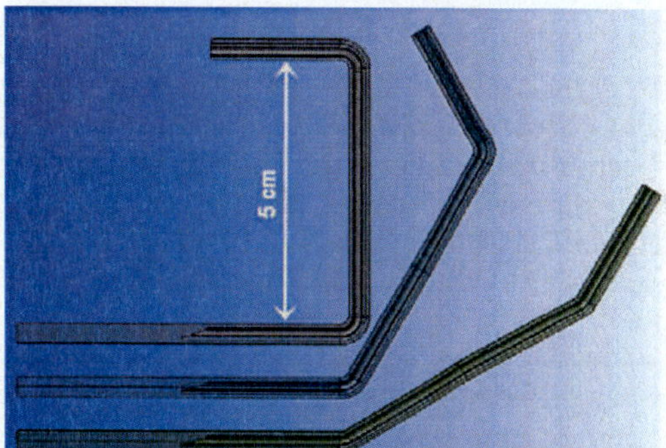

Fig. 4 Double bended catheters

The thrombus absorption was then simulated in ANSYS-CFX, based a two-phase laminar flow model. Blood and thrombus were modeled as two immiscible fluids having a density of 1600 and 1300 kg/m^3 respectively. The dynamic viscosity of blood was set to 0.0035 Pa-s. Two different values of the dynamic viscosity of the thrombus were used, thus mimicking two different ages of the clot: 0.035 Pa-s for a 6-hour thrombus and 0.6 Pa-s for a 12-hour thrombus. The vessel was considered to be fully clotted over a distance of 2 cm. The catheter tip was positioned proximal to the thrombus as indicated in fig. 5.

A constant pressure of -100 mmHg was applied at the outlet of the catheter while at both ends of the vessel the pressure was set to 100 mmHg, thus resulting in a 200 mmHg of difference in pressure between the outlet of the catheter and the interior of the vessel. This looks to be coherent with a vacuum of about 700 mmHg generated with the syringe and pressure drop across the shaft [3]. The whole fluid domain was initialized at 100 mmHg of pressure.

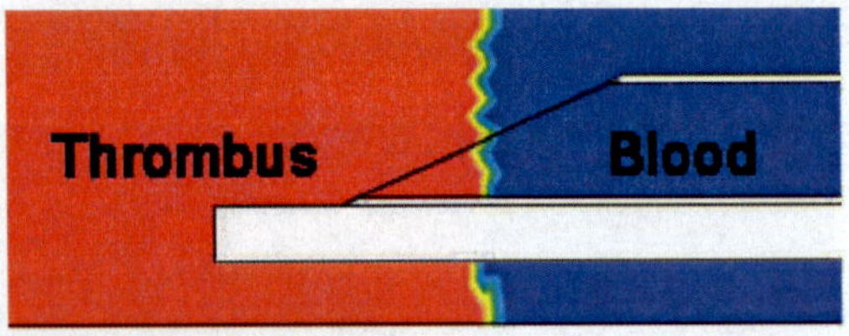

Fig. 5 Position of the catheter tip relative to thrombus before the aspiration

During the transient simulation, the total mass of clot still remaining in the vessel and the catheter was monitored.

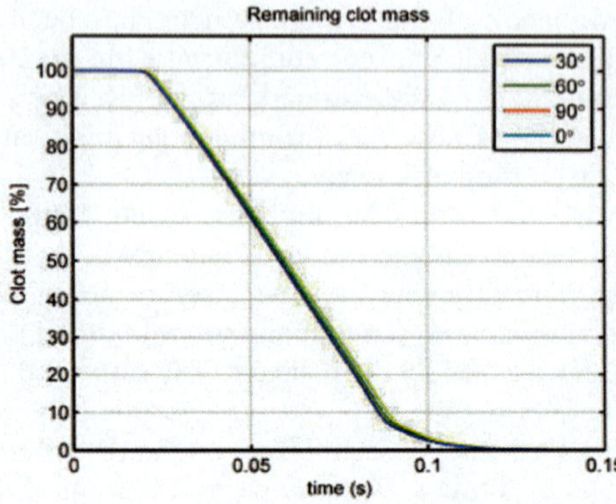

Fig. 6 The 6-hour thrombus mass percentage still remaining in the vessel

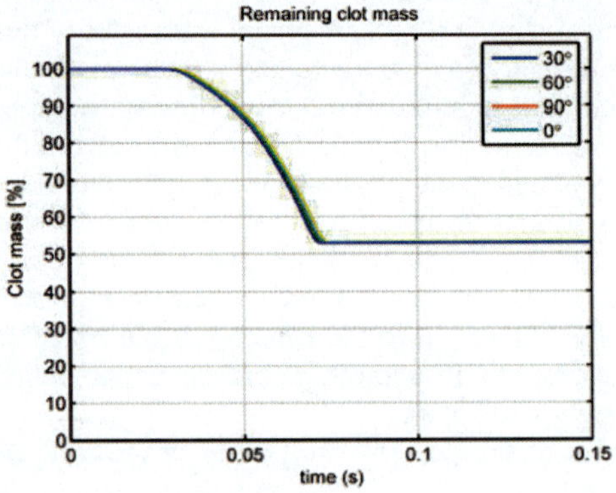

Fig. 7 The 12-hour thrombus mass percentage still remaining in the vessel

III. RESULTS AND DISCUTIONS

Four sets of simulations were carried out. The first set of simulations was carried out for the 6-hour thrombus ($\mu_{thrombus} = 0.035$ Pa-s) and the second one for the 12-hour thrombus ($\mu_{thrombus} = 0.6$ Pa-s).

Figure 6 show the histories of the mass percentages of the 6-hour thrombus still remaining in the vessel during the aspiration through the straight and the bended catheters. It becomes clear that the bends of the catheter and the vessel don't have any effect on the aspiration efficiency. This is consistent with the experimental observations of Leormain [2] and Pioud [4].

The same conclusion concerning the impact of the bend angles on the aspiration efficiency emerges from the next set of simulation results, which have been obtained for the 12-hour thrombus. However, a different behavior of the aspiration process for the more viscous thrombus was observed. Figure 7 suggests that the pressure wave travels slower through the catheter, thus making the effective aspiration begin a little bit later than for the less viscous thrombus.

Another major difference occurred later, when the thrombus was fragmented (fig. 8a). As a consequence, almost 50% of the clot mass was captured between the outer surface of the catheter and the interior of the vessel (fig. 8b), thus remaining there for a long time. Only after approximately 0.5s, the thrombus debris was completely absorbed from the vessel. Thrombus fragmentation may be dangerous due to the risk of distal embolism. However, a better rheological model of the thrombus is necessary for assessing the risk of fragmentation. The surface tensions and the elastic behavior of the thrombus might improve its cohesion.

The third and the fourth set of results were obtained for the double bended catheter. Aspiration of the 6 and 12-hour thrombus at different bending angles was simulated. A comparison between the aspiration characteristics of the single bended and the double bended catheters is illustrated in figure 9, for both the low and the high viscosity thrombus. The delay of the thrombus aspiration with our double bended catheter model is due to a greater length of the distal shaft compared to the single bended catheter; it takes a longer time for the pressure wave to reach the tip of the catheter. Based on the results presented in figure 9 we might say that the second bend of a catheter doesn't have any effect on the aspiration capability of the catheter.

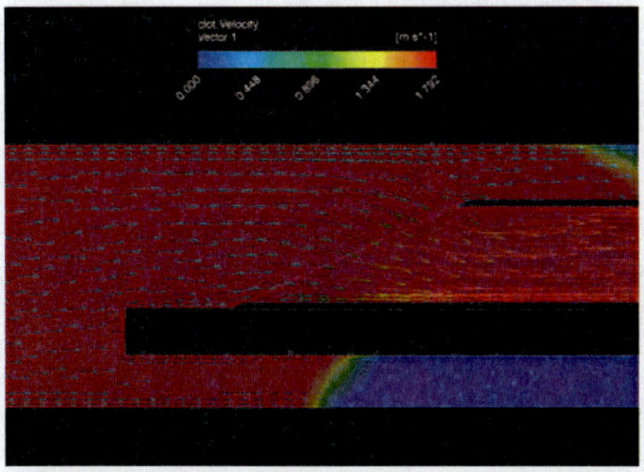

a)

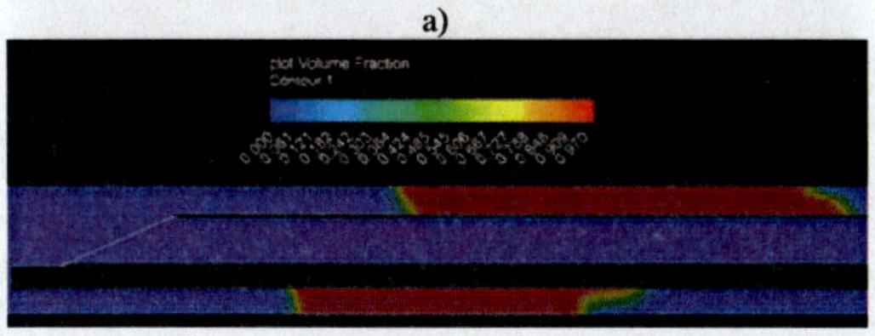

b)

Fig. 8 a) Clot velocity vector plot and volume fraction contour around the inlet of the catheter at the moment when the thrombus was fragmented (t=0.025s), b) Clot volume fraction contour after thrombus fragmentation, when blood was aspired by the catheter (t=0.07s).

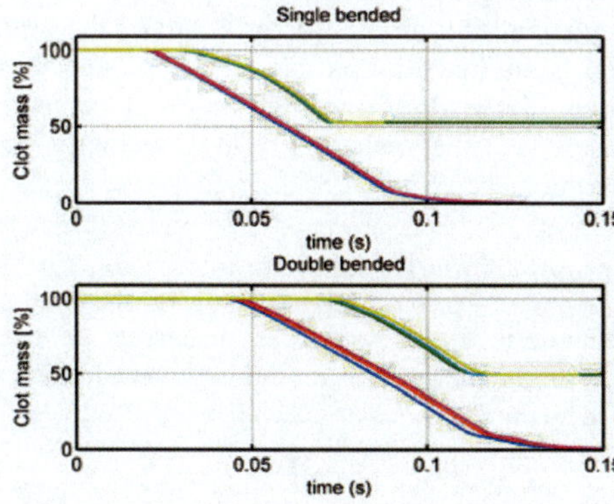

Fig. 9 Comparison between the aspiration characteristics of the single- and the double-bended catheters, at different bending angles and thrombus viscosities

IV. Conclusions

Based on a two-phase CFD model, simulation of thrombus aspiration was carried out. Our aim was to assess the impact that the thrombus age and the bending angles of the distal shaft of the catheter have on the efficiency of the thrombus aspiration. The thrombus aspiration was affected neither by the bending angles nor by the number of bends. However, the length of the catheter seems to be determinative for the delays in aspiration.

On the other hand, the viscosity of the thrombus has a greater impact on the aspiration. The viscous thrombi got fragmented during their aspiration thus creating the premises for distal embolization. According to other simulation tests that we carried out but we didn't include in this paper, not only when partially penetrated by the catheter's tip, the thrombus gets fragmented. Even if the catheter is placed at 2cm relative to the proximal end of the thrombus, fragmentation occurs. However, we are aware that a non-Newtonian model for the thrombus might be more realistic, as would some sort of surface tension effect, thus explaining the differences between our results and those of Pioud et al.

Acknowledgment

This research was funded by the European Commission, through the MeDDiCA ITN (www.meddica.eu) Marie Curie Actions grant agreement PITN-GA-2009-238113.

References

1. Napodano M, Ramondo A, Tarantini G et al. (2009) Predictors and time-related impact of distal embolization during primary angioplasty. European Heart J. 30:305–313
2. Niccoli G, Burzotta F, Galiuto L, Crea F (2009) Myocardial no-reflow in humans. J. of the American College of Cardiology 54(4):281-292
3. Leromain A.S, Fayard M, Lorgis L, (2010) Impact du diamètre et de l'âge du thrombus sur la thromboaspiration. Ètude experimentale. Ann Cardiol Angeiol (Paris) doi: 10.1016/j.ancard. 2010.07.015
4. Pennati G, Balossino R, Dubini G, Migliavacca F, (2010) Numerical Simulation of thrombus aspiration in two realistic models of catheter tips. Artificial Organs 34(4):301-310
5. Pioud V, Lorgis L, Collin B, et al (2010) Coronary thrombectomy, technical comparison of two systems on a laboratory bench: the impact of bends, angles and thrombus age. Euro Interv 6:(in press)
6. Comparison of Dimensions and Aspiration Rate of the Pronto® V3, Pronto® LP, Export® XT, Export® AP, Fetch®, Xtract™, Diver C.E.™ and QuickCat™ Catheters (2009), ML1623 Rev. F 12/09 c2009 Vascular Solutions, Inc.

Author: Yan Li
Institute: Technical University of Cluj-Napoca
Street: Memorandumului, 28
City: Cluj-Napoca
Country: Romania
Email: ally_ly_3@hotmail.com

Magnetically Targeted Drug Transport and Fixation

A.A. Dobre[1], A.M. Morega[1,2], and M. Morega[1]

[1] Faculty of Electrical Engineering, University POLITEHNICA of Bucharest, Bucharest, Romania
[2] "Gheorghe Mihoc – Caius Iacob" Institute of Mathematical Statistic and Applied Mathematics, Romanian Academy

Abstract— **This paper presents a mathematical model and numerical simulation results on transport and targeting of a medical substance carried by magnetic nanoparticles through a high gradient magnetic field. In our study we use simpler yet consistent models for the hemodynamic flow, and more complex, realistic computational domains based on medical images for the iliac arterial branching. The biocompatible drug carrier is injected in the blood. An optimized array of permanent magnets generates the targeting magnetic field, in the process of medication delivery in the region of therapeutic concern.**

Keywords— **magnetic drug targeting, flow – magnetic field interaction, numerical simulation, optimization.**

I. INTRODUCTION

Magnetic drug targeting (MDT) is a noninvasive modern technique to reduce the side effects related to the excessive distribution of powerful medication and improve its efficiency. In this therapy, the medication carried by superparamagnetic nanoparticles and injected in the blood stream interacts with an external magnetic field aimed at targeting the drug and fixing it mostly in the region of interest (ROI) for optimal delivery [1-3].

For example, tumor formations excision may be improved by destroying more of the affected tissue rather than healthy tissue through magnetic drug targeted therapies. Worth noting, magnetic targeting has also industrial applications (*e.g.*, micro stirring devices) [4].

In this paper we focus on the analysis and optimization of a static magnetic field source (*e.g.*, a permanent magnet), able to generate a high gradient magnetic field, to obtain a localized and an as high as possible fluid flow – magnetic field interaction that may prolong the time of residence of the medication in the ROI. To this aim, we investigated, first, simpler two-dimensional idealized models (*e.g.*, channels, ducts) for the interaction of the aggregate fluid (blood *and* magnetic drug) with the magnetic field; next we turned our attention to more complex 3D models.

Finally, using the permanent magnet configuration found through this procedure, we investigate the flow – field interaction in a more realistic, medical image based model for the aorta and iliac arteries. To start with, the magnetic field source optimization process begins with the study of 2D models for the blood flow – magnetic field interaction analysis. Next, a more complex case described by an idealized 3D computational domain was considered. Finally, using the optimized array of magnets we studied the flow – magnetic field for a more realistic model, based on a 3D computational domain built out of medical image. The numerical simulations were performed in the finite element method (FEM) technique.

II. THE MATHEMATICAL MODEL

In this study we neglect the flow-vessel walls structural interactions. Previous studies [6] revealed that, although a problem of concern in many circumstances, it is less so in magnetic drug targeting. The magnetic drug transport and fixation problem is analyzed by coupling the magnetic field model to the fluid flow. The aggregate fluid – blood and medication – has the magnetic properties of the drug carrier (a superparamagnetic material). First, the static magnetic field problem of the permanent magnet is solved for to find the (magnetization) body forces. Next, the fluid-flow interaction is studied. In this two-step approach we neglect the reaction of the flow upon the external magnetic field. The reason is the relatively low velocity field in hemodynamic.

A. The Magnetic Field Model

The magnetic field source is an array of permanent magnets. The magnetic field model is governed then by

Ampère's law

$$\nabla \times \mathbf{H} = 0 \tag{1}$$

Magnetic flux law

$$\nabla \cdot \mathbf{B} = 0 \tag{2}$$

Constitutive law for

$$\mathbf{B} = \mu_0 \mu_{r,mag} \mathbf{H} + \mathbf{B}_{rem} \text{ (permanent magnet)} \tag{3}$$

$$\mathbf{B} = \mu_0 \left[\mathbf{H} + \mathbf{M}_{ff}(\mathbf{H})\right] \text{ (aggregate fluid)} \tag{4}$$

$$\mathbf{B} = \mu_0 \mathbf{H} \text{ (elsewhere)} \tag{5}$$

S. Vlad and R.V. Ciupa (Eds.): MEDITECH 2011, IFMBE Proceedings 36, pp. 310–315, 2011.
www.springerlink.com

Here, μ_0 is the magnetic permeability of air; μ_r is the relative magnetic permeability of the permanent magnet; **H** is the magnetic field strength; **B** is the magnetic flux density; $\mathbf{B}_{rem}$ is the remanent magnetic flux density; and $\mathbf{M}_{ff}$ is the magnetization of the aggregate fluid, a function of **H**. Using the magnetic vector potential **A** (and the divergence free gauge condition)

$$\mathbf{B} = \nabla \times \mathbf{A}, \quad \nabla \cdot \mathbf{A} = 0, \qquad (6)$$

the mathematical model for the magnetic field problem is

$$\nabla \times \left(\mu_0^{-1}\mu_r^{-1}\nabla \times \mathbf{A}\right) = 0. \qquad (7)$$

The computational domain was conveniently extended (*i.e.*, the confining boundary was set "far away") such that magnetic insulation boundary conditions $(\mathbf{n} \times \mathbf{A} = 0)$ may be set. Alternatively, "infinite elements" may be used. Although this latter approach leads to computational domains of smaller sizes, hence fewer elements, the computational effort may be larger than acceptable.

B. The Hemodynamic Model

The magnetic fluid aggregate is assumed Newtonian, with constant properties, and its flow is incompressible and laminar [5], [6]. This approach is consistent with larger arteries, of "resistive" nature [11]. Different rheological models may be needed in different sections of the arterial tree. The flow model is described then by

momentum balance (*Navier-Stokes*)

$$\rho\left[\frac{\partial \mathbf{u}}{\partial t} + (\mathbf{u} \cdot \nabla)\mathbf{u}\right] = -\nabla\left[-p\mathbf{I} + \eta\left(\nabla\mathbf{u} + (\nabla\mathbf{u})^T\right)\right] + \mathbf{f}_{mg}, \qquad (8)$$

mass conservation

$$\nabla \cdot \mathbf{u} = 0. \qquad (9)$$

Here **u** is the velocity field, p is pressure, ρ is the mass density (1,060 kg/m^3]), η is the dynamic viscosity (0.005 Pa·s]), and **I** is the unity matrix. The magnetic body forces in eq. (8), due to the fluid aggregate magnetization in an external magnetic field, are given by $\mathbf{f}_{mg} = \mu_0(\mathbf{M} \cdot \nabla)\mathbf{H}$ [7]. We recall that the ensemble blood and magnetic drug carrier behaves as a magnetizable fluid that bears the magnetic properties of the material of which the magnetic nanoparticles are made of.

The boundary conditions that close the hemodynamic model are either constant or pulsatille inlet velocity, uniform pressure condition at the outlet(s), and no-slip (solid wall) conditions at the vessels walls. Mass transfer through the vessels walls is neglected in this study.

It should be mentioned that the magnetic field used in MDT is too small to impede the flow by interacting with the hematocrit and O_2 concentration in blood. In the absence of the magnetic particles used in MDT much higher magnetic flux densities are needed to influence the blood flow [12].

III. MAGNETIC FIELD SOURCE OPTIMIZATION

Several arrays of permanent magnets were studied in order to find out the highest gradient magnetic field source structure.

A. 2D Models and Analysis

First, a simple 1 cm wide and 0.2 cm tall rectangular permanent magnet is considered. Then, nine configurations of permanent magnet arrays were studied. For each configuration three cases were analyzed. They were obtained by varying the magnets width (d_1) and the spacing (d_2), *i.e.* the aspect ratio (AR) d_1/d_2, while keeping the same overall dimensions of the array: height (0.2 cm) and total length (1 cm).

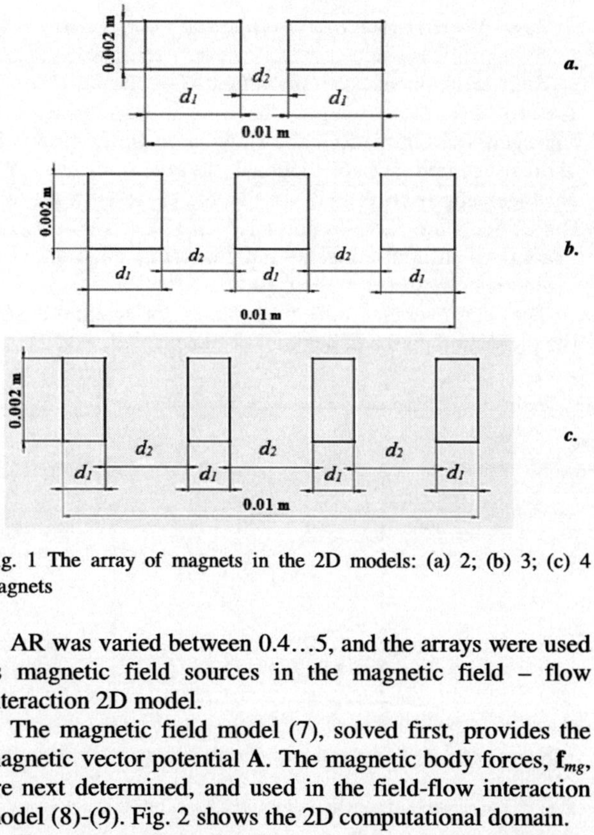

Fig. 1 The array of magnets in the 2D models: (a) 2; (b) 3; (c) 4 magnets

AR was varied between 0.4…5, and the arrays were used as magnetic field sources in the magnetic field – flow interaction 2D model.

The magnetic field model (7), solved first, provides the magnetic vector potential **A**. The magnetic body forces, $\mathbf{f}_{mg}$, are next determined, and used in the field-flow interaction model (8)-(9). Fig. 2 shows the 2D computational domain.

The numerical model (1) – (9) was implemented and numerically solved by Comsol FEM [8]. The simulation results, for pulsatile flow conditions, were analyzed with the aim at finding the optimal array of magnets, judged by the associated magnetic body forces (*i.e.*, the magnetic field gradient). In this 2D model there are two components of $\mathbf{f}_{mg}$: the horizontal (Ox) component, $\mathbf{f}_{mg,x}$, opposite to the flow, and the vertical component (Oy), $\mathbf{f}_{mg,y}$, that attracts the aggregate fluid to the vessel wall. As the magnetic targeting effect is more significant during the minimum mass flow rate interval, we focus on this specific moment.

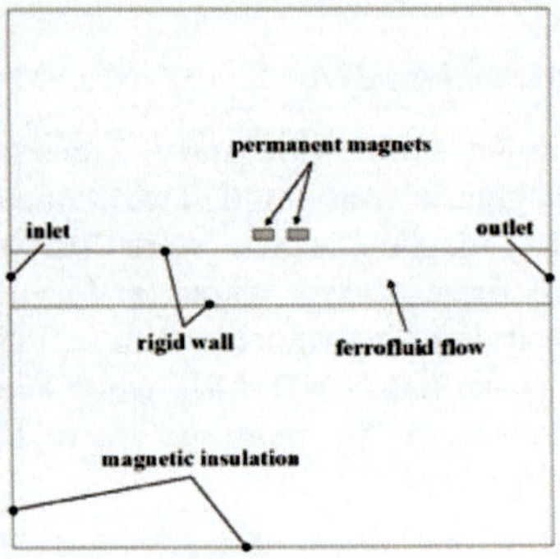

Fig. 2 The 2D computational domain and the boundary conditions

Both components are important: $\mathbf{f}_{mg,x}$ (parallel to the stream) acts into extending the time of residence of the aggregate fluid in the ROI (by slowing down the flow). The vertical component (orthogonal to the stream), $\mathbf{f}_{mg,y}$, enhances approach velocity to the vessel walls in the ROI. Both effects are more important during the low mass flow interval of the pulsating arterial flow. Fig. 3, 4 show the magnetic body forces at the wall.

They are averaged over the span of the magnet shadow (its projection on the upper wall of the channel, Fig. 2).

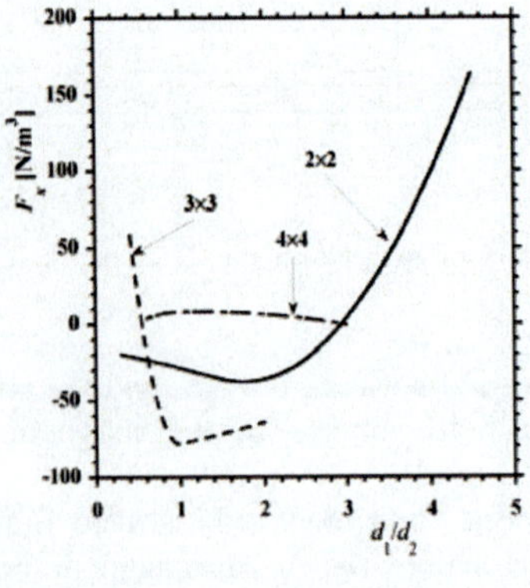

Fig. 3 The average magnetic body force $\mathbf{f}_{mg,x}$ for different arrays and ARs

Sure, at the wall, the flow velocity is zero (in virtue of the no-slip condition) but the magnetic forces reach here their upper (asymptotic) bounds. These values are used for comparison purposes.

The simulation results show that the larger the array the higher is the stream-wise gradient of the magnetic field whereas the vertical component does not vary significantly. We may then conjecture that the smallest array (2 magnets) may be an advisable choice. Moreover, increasing ARs – leading to a compact magnet – provides better design.

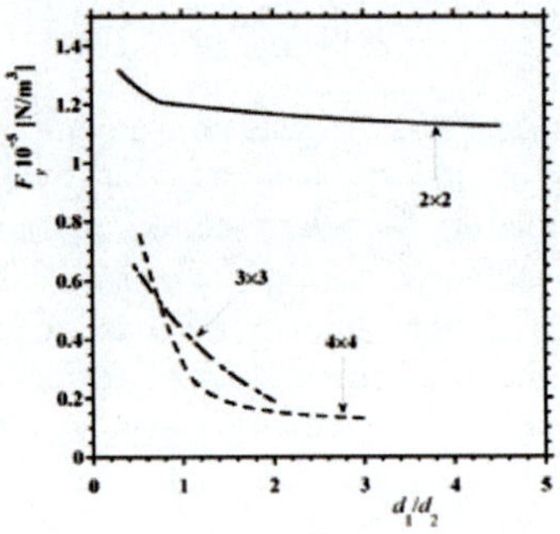

Fig. 4 The average magnetic body force $\mathbf{f}_{mg,y}$ for different arrays and ARs.

It is interesting to notice that the 4 magnets array gives a maximum stream-wise force with respect to the AR for d_1~d_2. In what follows we look at these results in a 3D layout – for this notional flow geometry – to verify whether these 2D results are still valid.

B. 3D Models and Analysis

The three arrays (2×2, 3x3, 4x4 magnets), occupying the same 3D volume, were studied for different ARs. Fig. 5 shows the computational domain for a notional vessel with the 4x4 array of magnets. The non-magnetic mass that embeds the array is not shown here, for better view.

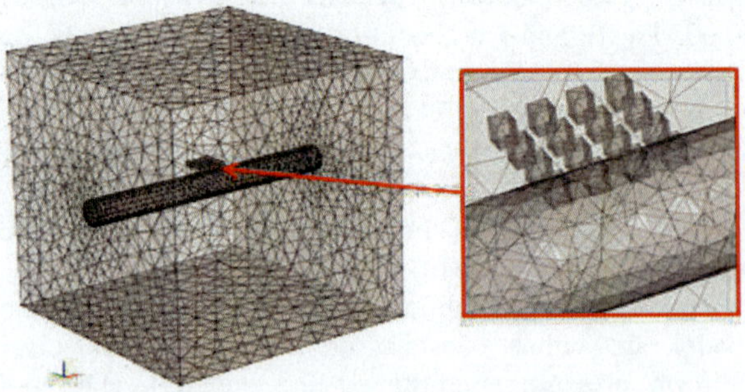

Fig. 5 The FEM mesh for the 4×4 array of magnets –3D notional model.

The mathematical model that describes the static magnetic field and the pulsating hydrodynamic flow coupled problems (1)–(9) were solved numerically [8]. The boundary conditions are consistent to the previous 2D models. The mass flow rate has the same time variation as in the 2D analysis

$$U_{in}(t) = U_0\left[\sin(\omega t) + \sqrt{\sin(\omega t)^2}\right], \qquad (10)$$

where $U_0 = 50$ cm/s, and $\omega = 2\pi f$ rad/s, $f = 60$ beats/minute.

The remanent flux density acts in the Oz direction, and the flow is in Ox direction. Oy direction, typical for 3D here, makes the difference between the 2D and 3D models.

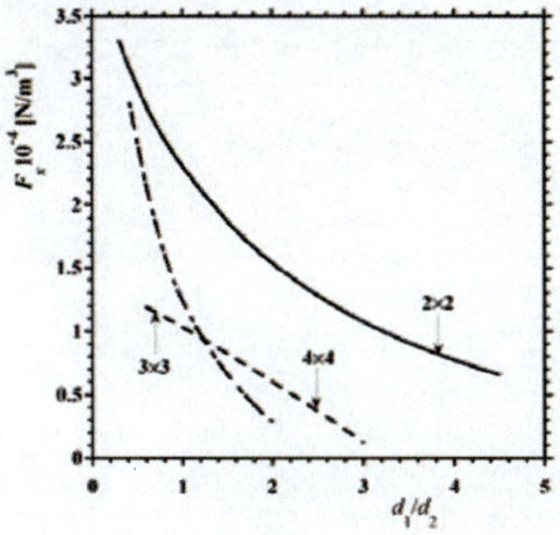

Fig. 6 The body force $\mathbf{f}_{mg,x}$ for different arrays and ARs – 3D models

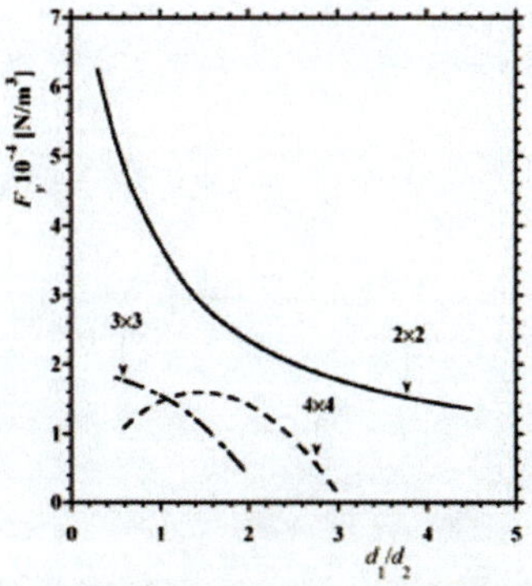

Fig. 7 The body force $\mathbf{f}_{mg,y}$ for different arrays and ARs – 3D models

In our study of the optimal magnet array we pursued the same steps, solving the problem for different ARs. The averaged body forces are shown in Fig. 6, 7, 8. Apparently, the 2×2 array provides for the best solution. However, the body forces decrease with increasing AR. This effect was noticed for the body force transversal to the walls (Fig. 4) in the previous 2D analysis. Here we note the same (decreasing) trend exhibited by the stream-wise component. This behavior, different in the 2D analysis (Fig. 3), may be explained by a 3D effect, not evidenced by the simpler 2D model. As expected, the d_1~d_2 case is best for the 2×2 array.

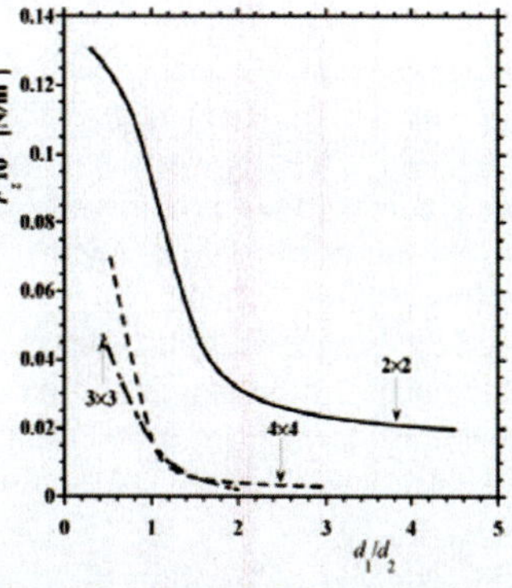

Fig. 8 The body force $\mathbf{f}_{mg,z}$ for different arrays and ARs – 3D models

The magnetic body force component (almost) perpendicular to the wall (responsible for conveying the drug to the vessel wall) has a knee for AR ~ 2. For values of AR less than 2 it exhibits a steep decrease with respect to AR, whereas for larger than 2 values of AR the decrease flattens. Apparently, for AR in the range 3…5 this component is almost constant.

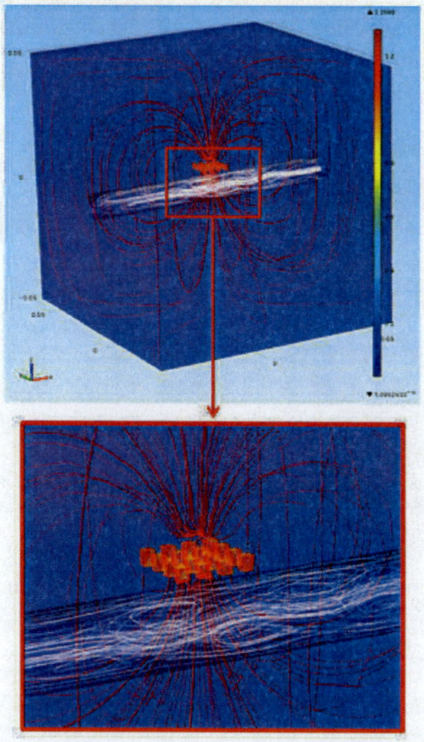

Fig. 9 The magnetic field and flow at the minimum mass flow rate. Recirculation areas build up and magnetic body forces may help conveying the medication to the ROI

It is interesting to notice that, as for the 2D analysis, the 4×4 array (16 permanent magnets, each 0.143 cm wide, equally spaced by 0.143 cm from each other) has a maximum at AR ~ 1 (d_1~d_2), when it provides its highest magnetic field gradient. This arrangement behaves almost as well as the 4x4 array in what concerns the stream-wise body force for lower values of AR.

Fig. 9 shows the magnetic field spectra (field lines and boundary surface map for the magnetic flux density) and the flow field (velocity streamlines) for the 4×4 array of magnets. A magnified view of the region where the magnet is located (presumably, nearby the region of therapeutic interest) provides details of the flow field. Recirculation is generated at every minimum mass flow rate. These particular moments provide for best conditions for the injected medication to diffuse in the therapeutic ROI.

IV. A MORE REALISTIC MODEL

To approach as closer as possible the medical procedure of the magnetically targeted drug transport and fixation a more realistic (*i.e.*, patient related) computational domain is needed. To this aim, we used a DICOM image set acquired by a MRI scanner. Fig. 10 shows several slices in the set that may provide the needed imagistic information. The ROI were segmented out and processed using a set of imagistic reconstruction tools [9].

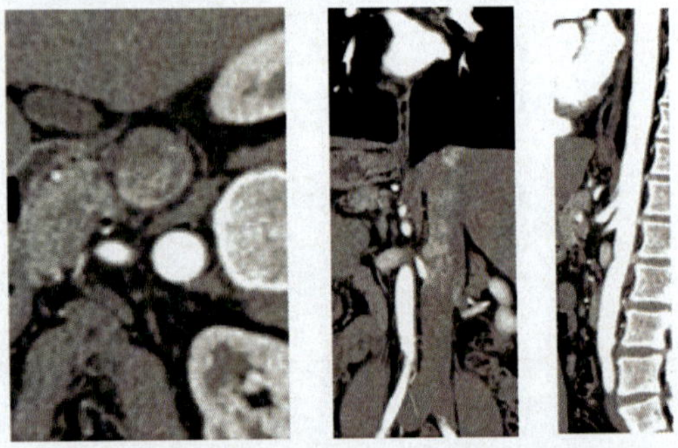

Fig. 10 Slices from a MRI DICOM set containing information about the human arteries in the chest and abdominal region that were used as source for the reconstruction process of the aorta and iliac 3D solid model

The computational "box" that contains the magneto-static problem of the 4×4 permanent magnet array (d_1=d_2) was generated using a CAD program [10] (Fig. 11). Particular care was devoted to model the magnetic array itself. During the meshing process the 16 permanent magnets may suffer unwanted edge fillet processes that would request a much too fine mesh to provide satisfactory numerical accuracy. We overcome this inconvenient by analytically modeling the 4×4 array. The model was FEM discretized (Fig. 12).

The blood vessel is part of the arterial tree called of "resistance" type with diameters in the range 20-25 mm [11], [6]. Therefore the pressures at the inlet and outlet flow ports may vary synchronously. We assume p_I = 11,000 N/m^2, p_2 = 10,000 N/m^2, and p_3 = 10,000 N/m^2, where here $p_i(t) = p_i[1 + K\sin(t + 3/2)]$, i = {I, 2, 3}, and K is a factor of order 10^{-1} (here, K = 0.2).

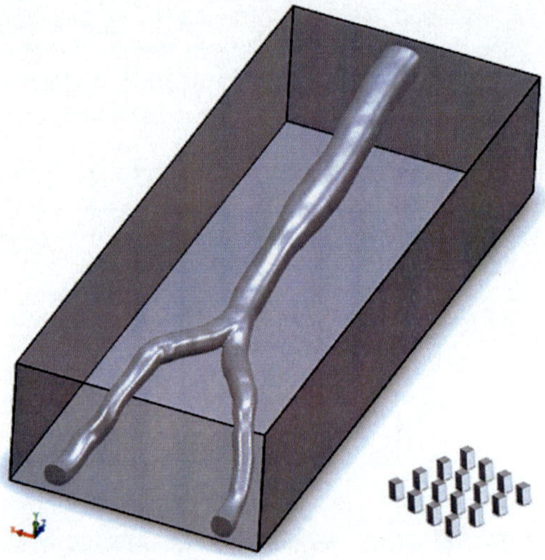

Fig. 11 The reconstructed 3D solid model of the arteries. The 4×4 array of magnets is shown (enlarged) aside. Only the magnets are shown

Fig. 12 The FEM mesh – *approx.* 265,000 Lagrange tetrahedral elements

Fig. 13 The unperturbed, pulsating flow – at minimum mass flow rate

Fig. 13 shows the flow unperturbed by the magnetic field through velocity arrows and pressure surface color map. The non-zero average mass flow rate and the outlet conditions set for the downstream boundaries do not promote recirculation built up.

Fig. 14 shows the magnetic and the flow field at minimum mass flow rate. Here, although less effective than for the flow with zero mass flow rate intervals (section III, Fig. 9), the magnetic body forces do modify, locally, the flow pattern.

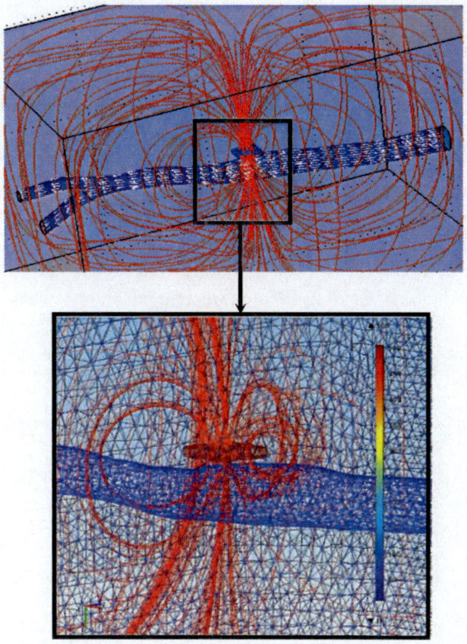

Fig. 14 The flow with magnetic field – at minimum mass flow rate

The investigation of this transport effect combined with the diffusion of medication through the vessel walls makes the object of a future research. In this study the drug that coats magnetic nanoparticles has not a distinct identity and it is rather dispersed in the blood stream.

V. Conclusions

Magnetic drug targeting (MDT) may be an invasive procedure therefore the design of the magnetic field source to best provide for medication delivery is of concern, and this paper reports numerical simulation results on its optimization. Different array-type of magnets were investigated, and compared based on the field gradient they produce. Apparently, the 2D and 3D models lead to different results, and it is suggested that the optimization (hence, the flow-field interaction itself) needs to be conducted on 3D models. Finally, to port these findings to a more realistic situation, a 3D model obtained out of patient specific MRI imagery for a segment of the arterial tree was built. The pulsating arterial flow – magnetic field was then investigated.

Acknowledgments

The work was conducted in the Laboratory for Electrical Engineering in Medicine (IEM) – Multiphysics Models, the BIOINGTEH platform at UPB, and it supported in part by POSDRU/88/1.5/S/61178 and CNCSIS PCCE-55/2008.

References

1. Alexiou C, Jurgons R, Schmid, RJ, Bergemann C, Henke J, Erhardt W, Huenges E, Parak F (2003) Magnetic drug targeting-biodistribution of the magnetic carrier and the chemotherapeutic agent mitoxantrone after locoregional cancer treatment, J. Drug Target. 11, 3:139-149.
2. Voltairas PA, Fotiadis DI, Michalis LK (2002) Hydrodynamics of Magnetic Drug Targeting, J. Biomech., 35:813–821.
3. Plavins J, Lauva M (1993) Study of Colloidal Magnetite Binding Erythrocytes: Prospects for Cell Separation, J. of Magnetism and Magnetic Materials, 122:349-353.
4. Suzuki H, (2003) Development of a Chaotic Micro-Mixer Using Magnetic Beads, PhD Thesis, UCLA, USA.
5. Morega AM, Dobre A, Morega M, Mocanu D (2009) Computational modeling of arterial blood flow, Proc. Second MediTech Conference, 23-26 September 2009, Cluj-Napoca, Romania.
6. Morega A M, Dobre AA, Morega M (2010) Numerical simulation of magnetic drug targeting with flow – structural interaction in an arterial branching region of interest, 17-19 Nov. 2010 Comsol Conference, Versailles, France
7. Rosensweig RE (1997) Ferrohydrodynamics, Dover Publications, NY
8. Comsol Multiphysics, v. 3.5a (2010), COMSOL A.B., Sweden.
9. Simpelware v. 3.2, Simpleware Ltd., UK, 2009.
10. SolidWorks (2009)
11. Feijóo RA (2000) Computational methods in biology, second Summer School LNCC/MCT, Petrópolis, January 2000.
12. Morega, A.M. and Morega, M. (2005) A FEM analysis of Magnetically induced biomagnetic fluid mixing, Rev. Roumaine Sci. Techn. Electrotech. et Energ., 50, 2:239-248.

Author: Alexandru M. MOREGA
Institute: University POLITEHNICA of Bucharest
Street: Splaiul Independenței, nr. 313, sector 6, 060042
City: Bucharest
Author: ROMANIA
E-mail: amm@iem.pub.ro

3D Simulation Analysis of Transcranial Magnetic Stimulation

C. Curta[1], S. Crisan[2], and R.V. Ciupa[1]

[1] Technical University of Cluj-Napoca / Electrotechnics Department, Faculty of Electrical Engineering, Cluj-Napoca, Romania
[2] Technical University of Cluj-Napoca / Electrical Measurements Department, Faculty of Electrical Engineering, Cluj-Napoca, Romania

Abstract— **This paper presents the results from a simulation of Transcranial Magnetic Stimulation using realistic brain model. The simulation found some interesting results regarding the distribution of the electric field induced in the cortex.**

Keywords— **Transcranial magnetic stimulation, coil position, bioelectromagnetism.**

I. Introduction

Transcranial Magnetic Stimulation (TMS) allows direct initiation of cortical activity, adding a new dimension to studies of the human brain. In TMS, the cortical cells are stimulated non-invasively by strong magnetic field pulses that induce a flow of current in the tissue leading to membrane depolarization and thereby to neural excitation.

TMS is used in neurology to determine different conditions by evaluating the cortical-motor threshold or to assess the continuity of nervous pathways.

In recent years TMS has proven its capabilities in treating psychiatric conditions like depression or schizophrenia. There is also a lot of research under development for the treatment of other psychiatric conditions [1][2].

Because psychiatric treatments imply stimulation (or inhibition) of certain cortex gyrus, we designed a simulation to see the effects of a circular stimulation coil on a realistic model of the cortex.

II. TMS – Basic Principles

The neurons are stimulated by applying a rapidly changing magnetic field. In TMS the excitation is obtained through a pulse current that drives a coil situated in the vicinity of the head (figure 1). The source of the activation of neurons is the electric field E induced in the tissue by the varying magnetic field (Faraday's Law) [3]:

$$\nabla \times E = -\frac{\partial B}{\partial t} \tag{1}$$

At cellular level the electric field E affects the transmembrane potential which may lead to local membrane depolarization and firing of the neuron (figure 2) [4].

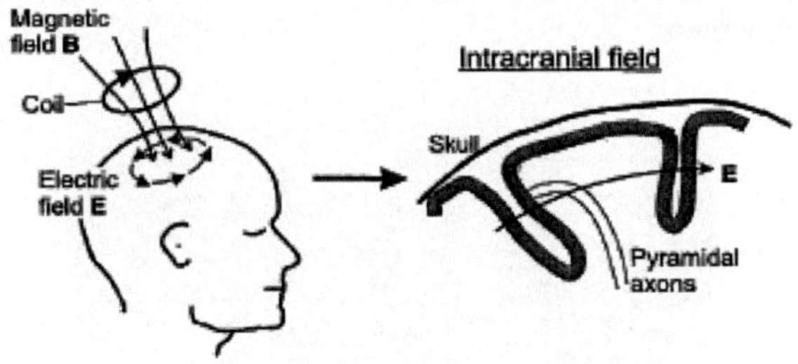

Fig. 1 TMS mechanism – macroscopic view [3]

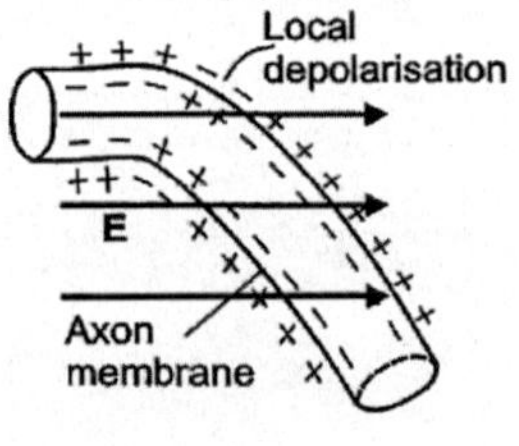

Fig. 2 TMS mechanism – cellular level [3]

III. Simulation Setup

Because the surface of the cortex is highly irregular, we wanted to view the effects of stimulation on a realistic brain model. To reach our goal we designed a double 70 mm copper circular coil placed in the vicinity of the brain. The brain is represented by an anatomically correct right side hemisphere. The simulation was performed using 3D Finite Element Modeling software.

The generally accepted time periods for TMS are 200-500 μs [5]. The coil is excited by a 1 kA current with a frequency of 2.5 kHz, which corresponds to duration of 400 μs for each pulse. This time period is similar to the one used by Magstim Rapid^2 magnetic stimulators.

The brain hemisphere was modeled as a 3D solid object composed of 261 faces (figure 3). It includes white and gray matter, modeled as a homogeneous volume. Conductivity of the brain was set according to recent research regarding brain conductivity. For body temperature and frequencies between 10 Hz-10 kHz the conductivity was set to 1.79 S/m [6].

S. Vlad and R.V. Ciupa (Eds.): MEDITECH 2011, IFMBE Proceedings 36, pp. 316–319, 2011.
www.springerlink.com

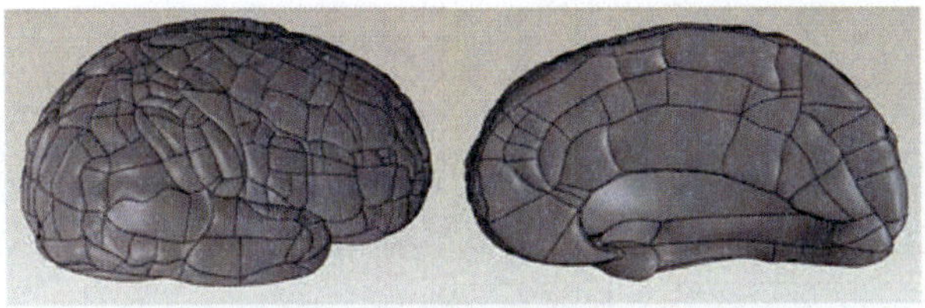

Fig. 3 Cortex views

The exterior diameter of the coil loops is 70 mm, and the interior 40 mm. The coil and cortex objects were included in a sphere containing air. This enclosure has the radius 10 times bigger than the radius of the coil loops. This allows us to set the Neumann condition on the exterior surface of the enclosure without important distortions to the magnetic field produced by the coils.

To mimic realistic stimulation conditions, the coil was positioned parallel to the scalp, rather than to the surface of the cortex. The center of the stimulator is positioned at coordinates (0, 0, 0), and the coils are parallel to the YZ plane (figure 4).

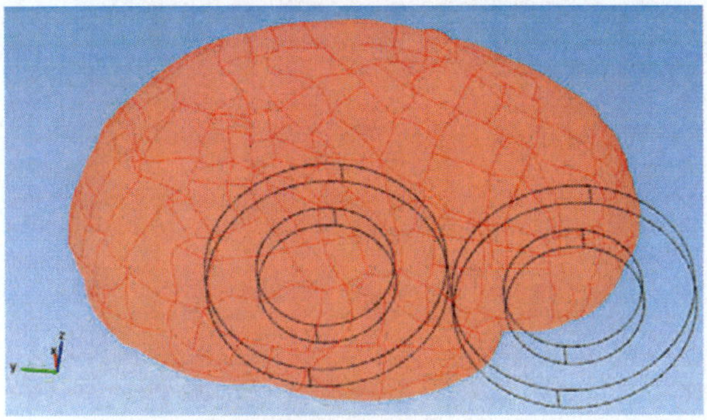

Fig. 4 Overview for the position of the coil relative to the brain

IV. Results and discussions

The solution was solved for 753554 degrees of freedom. The overview of the results is in concordance with our expectations – the biggest values for induces electric field are located under the center of the stimulator. But the distribution is not uniform (figure 5).

Although one could expect bigger values for the electric field in the gyri, since they are closer to the coil, we find higher intensities along the channels (called sulci) between them.

For the vertical slices, their positions were chosen to include the center of the stimulator and the points under the inner margin of the left coil (figure 6).

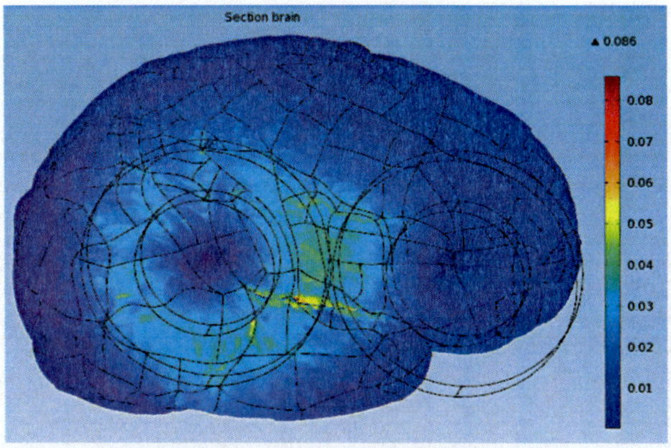

Fig. 5 Overview of induced electric field (V/m)

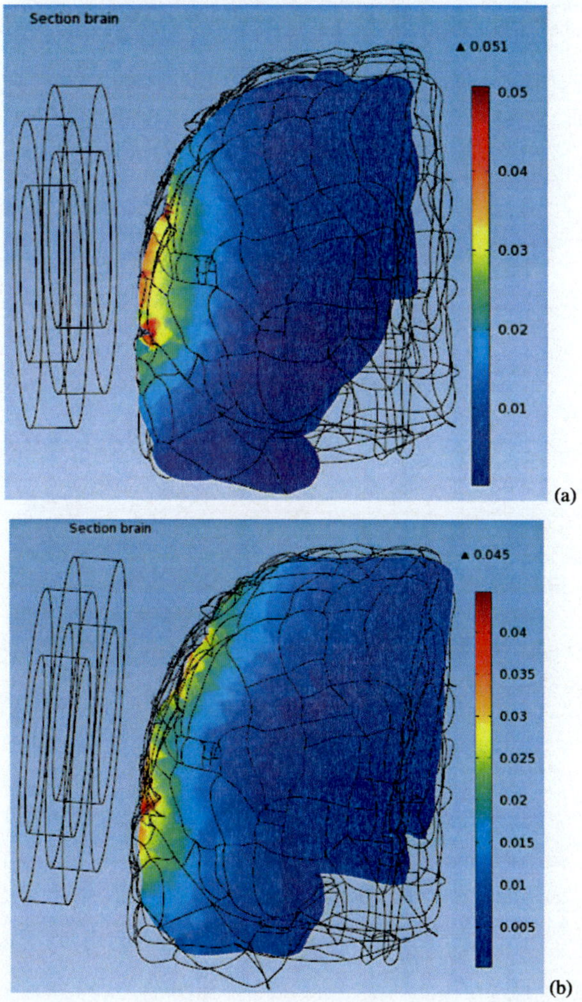

Fig. 6 XZ plane slices for induced electric field (V/m). Slice position: (a) 0 mm; (b) +55 mm

Figure 7 contains vertical slices through the center of each coil. The difference between the maximum values in the two slices is due to the distance between each coil and the brain.

For the horizontal slices, their positions were chosen so that they include the points under the center of the stimulator (figure 8 a), or the points under the inner margin of the coils (figure 8 b and c).

If we look at different slices through the cortex, we realize that the biggest values for the electric field are not necessarily in the closest vicinity of the coil, but rather in the tight sulci (depressions or fissures in the surface of the brain). This means that stimulation can occur at rather smaller power levels in sulci (figure 6, 7 and 8).

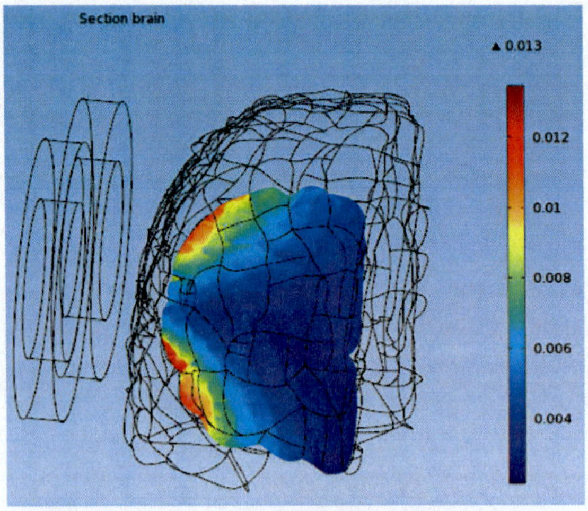

(a)

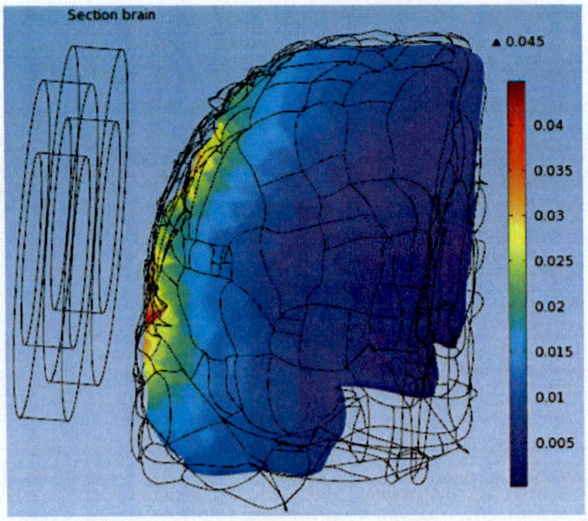

(b)

Fig. 7 XZ plane slices for induced electric field (V/m). Slice position: (a) -36 mm; (b) +36 mm

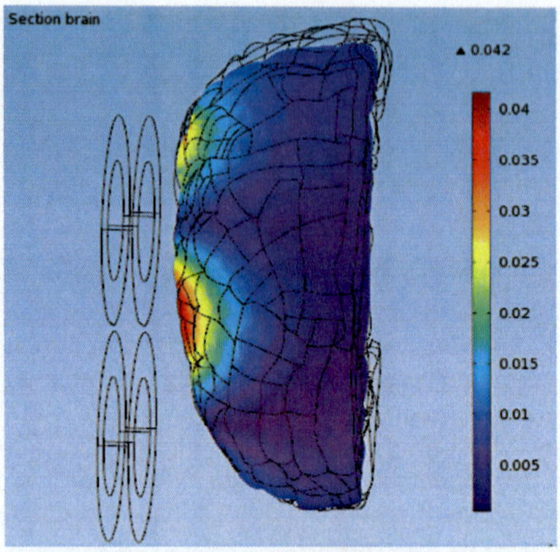

(a)

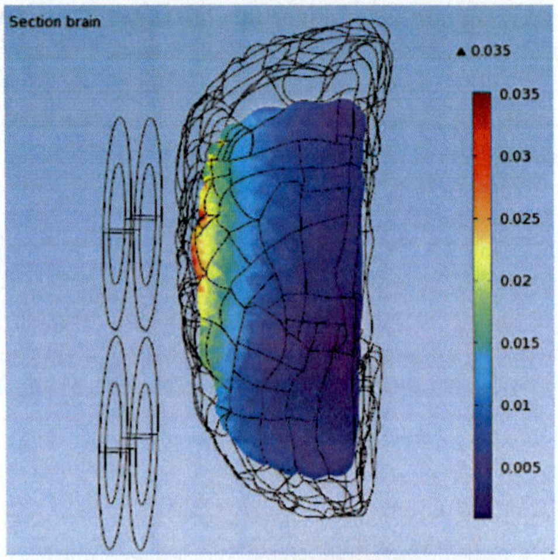

(b)

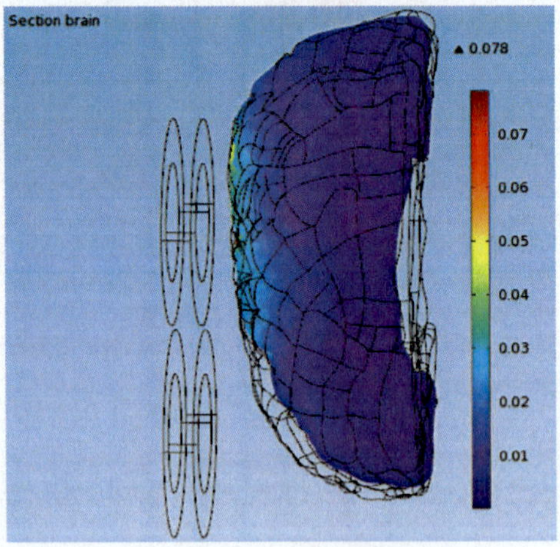

(c)

Fig. 8 XY plane slices for induced electric field (V/m). Slice position: (a) 0 mm; (b) +23 mm; (c) -23 mm

The highest value for the induced electric field is 86 mV/m and is located in the lateral sulcus (also called Sylvian fissure). Electric field values in the gyri are smaller than 50 mV/m.

Analyzing the results obtained we see that the electric field intensity is at least 20-30% higher in the sulci then in the gyri (figure 6, 7 and 8). Also, the highest values of the electric field are found in the lateral sulcus, despite the fact that the it is not perfectly under the center of the stimulator. The maximum value in that sulcus is almost double the value in the gyri located under the center of the stimulator (figure 5 and figure 6a).

V. Conclusions

Our results show that the shape of the targeted area of the cortex greatly influences the distribution of the induced electric field during TMS. This effect has a big impact on the locus of stimulation, and should be taken into consideration by physicians when applying TMS.

More studies are needed to confirm our findings, including more complete models that contain the scalp, skull, cerebrospinal fluid and the cortex.

Acknowledgment

This paper was supported by the project "Development and support of multidisciplinary postdoctoral programs in major technical areas of national strategy of Research - Development – Innovation

4D-POSTDOC, contract no. POSDRU/89/1.5/S/52603, project co-funded by the European Social Fund through Sectoral Operational Programme Human Resources Development 2007-2013.

References

1. Pascual-Leone A, Rubio B, Pallardo F, Catala MD. Rapid-rate transcranial magnetic stimulation of the left dorsolateral prefrontal cortex in drug-resistant depression. Lancet. 1996;348:233-237.
2. Fitzgerald PB, Daskalakis ZJ. A. Review of Repetitive Transcranial Magnetic Stimulation Use in the Treatment of Schizophrenia. Arch Gen Psychiatry. 2003;60:1002-1008.
3. Risto J. Ilmoniemi, Jarmo Ruohonen, Jari Karhu, Transcranial magnetic stimulation – A new tool for functional imaging of the brain, Critical reviews in Biomedical Engineering, 27(3-5) 1999, 241-284
4. Jarmo Ruohonen, Transcranial Magnetic Stimulation: Modeling and New Techniques, PhD Thesis, Espoo, Finland,1998.
5. Eric Wassermann, Charles Epstein and Ulf Ziemann, Oxford Handbook of Transcranial Stimulation, Oxford University Press, 2008
6. Baumann S.B., Wozny D.R., Kelly S.K., Meno, F.M., The Electrical Conductivity of Human Cerebrospinal Fluid at Body Temperature, IEEE Trans. on Biomedical Engineering, Vol. 44, No. 3, March 1997, pp. 220-223

Author: Curta Catalin
Institute: Technical University of Cluj-Napoca
Street: Baritiu 26
City: Cluj-Napoca
Country: Romania
Email: Catalin.Curta@et.utcluj.ro

The Romanian Public's Perception of Electromagnetic Fields Risk

D. Curseu, M. Popa, and D. Sirbu

University of Medicine and Pharmacy /Environmental Health Department, Cluj-Napoca, Romania

Abstract— **Potential health risks from exposure to electromagnetic fields (EMF) are a hot topic in recent environmental health research. The aim of this survey is to help understanding of the perception that the Romanian public has upon the links between electromagnetic fields and health. Nervousness/restlessness, headache and sleep disturbances were the most often reported symptoms and power lines, mobile phones, microwave ovens, computers and TV screens were perceived as the main EMF sources suspected to be associated to symptoms. The general results show that the level of knowledge about the event increases the concern about risks incurring from it. Population's information level was strongly related to level of perceived risk, growing concern about EMF.**

Keywords— **Electromagnetic fields, risk perception, environment, health.**

I. INTRODUCTION

In today's world, there is increasing concern about various suspected environmental illnesses. The electric and magnetic fields are unavoidably produced wherever electrical energy is generated, transmitted or used, and are thus inherent in modern societies. These fields surround all electrical wires and all electrical equipment, such as computers, televisions, electrical stoves, telephones, transformers, electric heaters, cell phones, fluorescent lights, all types of wireless devices and more. The health consequences of technological developments can be difficult to predict and manage, and the public reaction is largely hesitant and in many cases tends to develop a negative attitude toward this matter.

Despite evidence that there are no adverse health effects within guideline limits, a number of individuals have reported a variety of health problems that they have attributed to exposure to electromagnetic field (EMF). These individuals have often been described as being "electromagnetic hypersensitive" (EHS), and EHS is considered a new illness of industrialized society. Although there is little scientific evidence to support the idea of EHS, the World Health Organization (WHO) says the "symptoms are certainly real" and "can be a disabling problem for the affected individual". Reacting to this rising tide of claims of a new illness, the WHO issued a fact sheet on the allergies, which it dubbed "electromagnetic hypersensitivity" and likened it to multiple chemical sensitivities [1]. The understanding of risk perception is of fundamental importance not only to improve the communication between scientists and the general public, but also to evaluate the plausibility and relevance of claimed effects such as "electromagnetic hypersensitivity".

The aim of this work was to assess the Rumanian Public's perception upon the links between electromagnetic fields and health and to looks at what proportion the population has received information on EMF and whether receiving this information affects levels of concern about possible health risks.

II. MATERIAL AND METHOD

In order to assess and understand in detail the public's perception of electromagnetic fields risk and to investigate population's information upon this issue, a questionnaire was handed out to 500 of adult residents from Cluj-Napoca. The reply rate was 65 % (325 responders). The survey's effective respondents were almost gender balanced (with 55% male and 45% female). More than half of the interviewed were above 45 years old (48% aged between 20-45 years old and 52% between 46-70 years old), and 60% of them have high school diploma and above. The current jobs of the respondents are in the industries of service (24%), students (20%), retired (17.2%), manufacturing (12%), health or education (10.8%) freelance (8%) and house persons (8%).

III. RESULTS

The particular questions were designed to assist in determining the levels of environmental health concern, risk perception and general information about EMF.

The first part of the study looks at the place of EMF between the other environmental issues.

As Figure 1 show, the electromagnetic fields are not the top health concerns in people's minds, and the responders were most concerned about air pollution than any other environmental health risk. Among the 10 environmental factors presented as potential threats to health, the sources of electromagnetic fields appear in the lowest three positions.

After the study of environmental issues that people perceive as affecting their health, we looked in more detail at the specific area of EMF.

S. Vlad and R.V. Ciupa (Eds.): MEDITECH 2011, IFMBE Proceedings 36, pp. 320–323, 2011.
www.springerlink.com

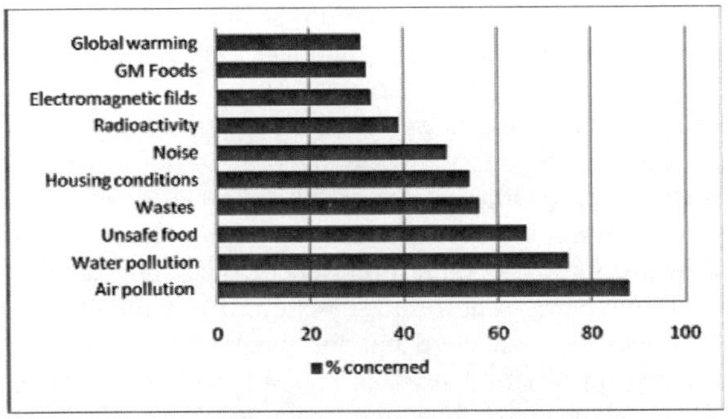

Fig. 1 Main concerns on environmental issues

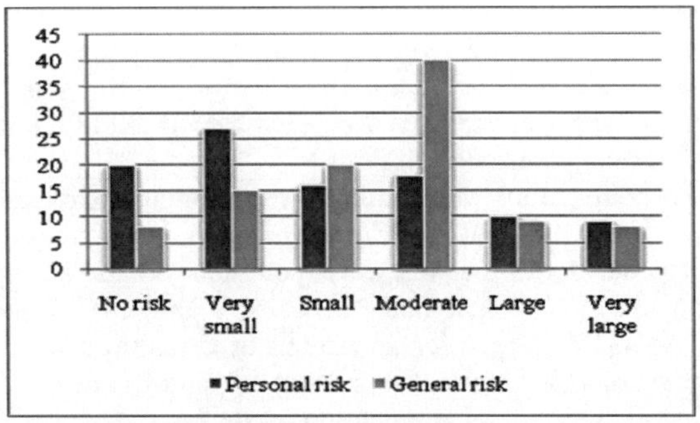

Fig. 2 Distributions of perceived risk according to rating scale

The respondents were presented a list of objects and asked which of them emit EMF in order to assess their knowledge about the possible source of EMF. Broadcasting stations, TV towers and mobile communication masts were cited by a majority of respondents (66%), followed closely by mobile telephones (64%) and power lines or transformer stations (58%). High mentions are also noted for visual display units or TV sets (44%), computers (40%), radar equipment (39%) and electrical household appliances (35%) and much lower mentions are recorded for wireless computer networks (12%) and electric wiring in houses (9%). It is worth noting that only 8% of interviewed say that everything on the list is a source of EMF and 2% of them don't recognize any source of EMF. See Table 1.

Table 1 Public's awareness about the sources of EMF

Possible source of EMF	recorders
Broadcasting stations, TV towers or telecommunication masts	66%
Electrical appliances at home (electric clocks, hairdryers, vacuum cleaners, microwave ovens etc)	35%
Electric wiring in houses	9%
Computers	40%
Wireless computer networks	12%
Light sources (fluorescent tubes or other)	14%
Mobile phones	64%
Power lines or transformer stations	63%
Anti-theft devices (e.g. motion detectors of alarm systems, security gates/barriers)	18%
Radar stations	27%
Railways	24%
Visual display units or TV sets	44%
All of these	8%
None of these	2%

When the subjects were asked to reply to the question "How concerned are you about the potential health risks of electromagnetic fields?" the public is evenly divided on the matter. While 48% of respondents say that they are indeed concerned about the potential health risks of EMF, a slightly larger proportion of the poll (50%) say they are not very concerned or even not at all concerned about this issue and only 2% have no opinion on the matter. A socio-demographic analysis reveals that young people aged 20 to 45 are less likely to be concerned about the adverse health effects of EMF (only 35% of this age group, compared with 61% in the other age group).

Risk perception was studied by means of rating scales from 0=No risk at all to 6=A very large risk. "Don't know" answers were treated as missing data. Because the people tend to believe that others are at greater risk then themselves, the respondents were asked to judge both general risk (the risk to others), and personal risk (risk to themselves). These ratings typically differ as to level. Personal risks are typically judged as smaller than general risks. The differences between personal and general risk suggest that it is probable that people believe they can protect themselves from the hazards. The risks were rated at the level rather small – moderate. The data are illustrated in Fig. 2.

Not to suggest any possible answer, the interviewed were asked to list the five most common symptoms which could be associated with the electromagnetic field exposure. The answers may be classified into the following groups - the first two are further specified:

- Nervous system symptoms; nervousness/restlessness, headache, sleep disorders, stress, anxiety, and neurasthenia.
- Cardiac arrhythmias, tinnitus/ ringing in the ears
- Hormonal disorders, eye symptoms, and digestive problems
- Other responses concerned different types of cancer, reproductive and pregnancy problems and various symptoms attributed to the sick building syndrome.

Overall, the most common symptoms encountered among the responses were nervousness/restlessness, headache and sleep disturbances. Other more specific symptoms such as cancer or pregnancy problems occurred less consistently among these descriptions.

These reactions were analyzed in different situations and have been attributed to different sources. 45% of respondents considered that the most common situations appear “at work”, 31% specified “at home”, and 24% reported “outdoors” as places for apparition of these problems. Excepting mobile phones, the sources of radiofrequency (RF) fields such as telecommunication masts, broadcasting or TV towers and radar stations were not mentioned as being a common source of problem. A quarter (26%) of respondents affirms that mobile phones affect citizens’ health to a major extent. Power lines or transformer stations, visual display units (especially computers and TV screens), and some electrical appliances (especially microwave ovens) were frequently mentioned also. More general suspected sources included "all EMF sources", or "all EMF from household". There were no clear patterns between the reported symptoms and their suspected sources.

One of the most important criteria used when taking a decision or forming an opinion on any particular issue is information. Therefore, the survey looks at the extent to which information on electromagnetic fields has been received by the population. Only 38% of the responders say they have received information on the potential health risks of EMF. The majority of respondents (62%) report that they have not received any information on the matter. Among the majority proportion of the poll who had not received this information, just 36% say they are worried compared to 69% among those who received information. Education and occupation have a significant influence. In terms of education, 39% of those who have high school diploma and above had received information on these potential health risks compared with just 7% of those who left school at age 16 or before. A similar disparity is seen on the basis of occupation. Among students this figure rise to 46%, and approaches one in three for managers and those working in health or education compared with just 9% of house persons, 11% of the retired and 17% of manual workers. The results show that 40% of respondents claim to be not very satisfied with the information received because the material they had received was insufficient or not objective. However, when respondents were asked about the two main ways they received information on potential health risks linked to EMF, television (75%) and newspapers and magazines (54%) were by far the most frequently mentioned. 21% of those questioned say they received information over the Internet, and fewer than 5% of the respondents mentioned specialist publications (science or health journals) and other channels of information, such as official publications, institution manuals, seminars, information in the workplace, books or conferences.

Preference for the Internet is best explained by age, education and occupation. While over a third (38%) of the youngest age group would prefer to receive information regarding EMF via the Internet, only 5% of those aged 46 and above have the same preference. Education is also an important determinant in the selection of this medium: only 4% of the least educated opt for the Internet as a way of receiving information on this subject, and for those educated stands at more than five times this figure (27%). Occupation is responsible for even more striking variations. Combining the factors of education and age, a third (65%) of students opts for this medium. A high figure of 42% is also recorded amongst managers. The figures fall to 16% for manual workers, 10% for house persons and just 4% among the retired.

In the light of the general criticism of the role of public authorities in this aspect of public health protection, it is interesting to look at how respondents feel that the authorities should act in this area. The most opinions were that public authorities should inform the public as to the potential health risks linked to EMF (53.8%). Other recommendations were setting safety standards for products (39%) and developing guidance for public health protection (38%), harmonizing national safety standards and policies with those of EU (36%), and even reviewing of the status of scientific evidence and financing of the research (22%) were proposed as a possible measure.

IV. Discussions

Over the course of the past decades, numerous electromagnetic field sources have become the focus of health concerns. It is recognized that this topic has received different awareness in various European countries: In Sweden, a substantial part of the EMF research and health related efforts is directed towards “electromagnetic hypersensitivity” primarily in relation to office work situations and visual display units. In other countries like Austria and Germany, concerns of people appear to be more concentrated on the exposure at home and focused on power lines and transmitter stations [2].

Studies of the risk perception of EMF have been carried out using various methods, including psychometric (questionnaire) techniques. These studies indicate that people in general do not rate electromagnetic sources as being among the highest sources of risk. The power lines are typically rated as having more severe consequences than other sources being much less controllable and much less equitable then those from electrical appliances at home [3]. According to respondents' assessments, the power lines, mobile telephones, microwave ovens, computer and TV screens are considered the most risky source of EMF, which may affect their health in the fullest extent. Mobile phones, with their widespread adoption across the Romania in the past decade, are not only common place particularly amongst younger age groups but have generated substantial claims and denials as to the possible long-term harm they may cause to users. It is well known that voluntary exposure is an important factor in risk perception. Generally, people who do not use mobile telephones perceive the risk as high from the fields emitted from mobile telephone base stations. However, more a quarter of interviewed consider that mobile phones affect citizens' health to a major extent, but they are less worried by base stations.

In most of the cases (58%), the interviewed suggested their suspected association between an EMF source and some nervous system symptoms (nervousness/restlessness, headache and sleep disturbances). Because scientists cannot exclude that EMF may cause health problems, the application of the precautionary principle is debated heavily. It seems that precautionary measures will increase trust in risk management, which in turn will result in lower risk perceptions [4].

Risk perception depends on several factors, many of which are relevant for electromagnetic fields. They include lack of familiarity with the agent, difficulty in understanding interaction mechanisms, and uncertainty in scientific knowledge [5.] Because of uncertainty about health risks associated with EMF exposure, the public is more likely to experience difficulty in evaluating the available information and rely more on perceptions than facts when drawing conclusions. MacGregor and co-workers provided subjects with a brochure which discussed the 'possible, but not proved' health effects of ELF exposure. After reading the brochure, subjects perceived the health risks of ELF exposure as higher than before reading the brochure [6]. In another study, it was recommended to provide laypeople with information to take away exaggerated fears, thus lowering perceived risks [7]. In our study, 36% of respondents who have not received any information on the potential health effects of electromagnetic fields were worried compared to 69% among those who received information. This suggests that the information may have a backlash effect, increasing worries rather than decreasing them. On the other hand, mass media, which usually tend to scare-mongering, exert a powerful influence on people's perceptions.

V. Conclusions

The sources of EMF appear in the lowest three positions in context of public concerns for environmental issues. Young people tended to be less concerned about the adverse health effects of EMF than older people did. Personal risks are judged as smaller than general risks, probably because the people believe they can protect themselves. Nervousness/restlessness, headache and sleep disturbances were the most often reported symptoms and power lines, and the mobile phones, microwave ovens, computers and TV screens were perceived as the main EMF sources suspected to be associated to symptoms.

Information may have a backlash effect, increasing worries rather than decreasing them. This could be interpreted as an indication that information on these potential health risks has a major impact in increasing levels of concern. It could also be that people who tend to be more worried about the issue are more likely to look for information. Education and occupation have a significant influence, and the Internet is likely to become one of the preferred channels of information in the near future as it is chosen by more young people.

In future work, it would be interesting to relate perceived EMF risks to measures of precautionary attitudes.

References

1. WHO (2005). Electromagnetic fields and public health - Electromagnetic hypersensitivity. Fact Sheet No 296 December 2005 at http://www.who.int/mediacentre/factsheets/fs296/en/
2. Bergqvist U and Vogel E (1997) Possible health implications of subjective symptoms and electromagnetic field. A report prepared by a European group of experts for the European Commission, DGV. Arbete och Hälsa, 1997:19. Swedish National Institute for Working Life, Stockholm, Sweden. ISBN 91-7045-438-8 at http://gupea.ub.gu.se/bitstream/2077/4156/1/ah1997_19.pdf
3. MacGregor DG, Slovic P, Morgan MG (1994). Perception of risks from electromagnetic fields: A psychometric evaluation of risk-communication approach. Risk Anal 14: 815-828.
4. Wiedemann PM, Schütz H (2005). The Precautionary Principle and Risk Perception: Experimental Studies in the EMF Area. Environ Health Perspect. 113(4): 402-405. DOI 10.1289/ehp.7538
5. Siegrist M, Keller C, Kiers HA. (2005) A new look at the psychometric paradigm of perception of hazards. Risk Anal 25: 211-222.
6. MacGregor DG, Fleming R. (1996) Risk perception and symptom reporting. Risk Anal 16: 773-783.
7. Hutter HP, Moshammer H, Wallner P, Kundi M. (2004) Public perception of risk concerning celltowers and mobile phones. Soz Praventivmed 49: 62-66

Author: Curseu Daniela
Institute: University of Medicine and Pharmacy
Street: Iuliu Maniu 3/8
City: Cluj-Napoca
Country: Romania
Email: daniela_curseu@yahoo.com

Consequences of a Stenosed Artery in an Arteriovenous Fistula on the Efficiency of the Hemodialysis Access

I. Decorato, Z. Kharboutly, C. Legallais, and A.V. Salsac

Biomechanics & Bioengineering Laboratory (UMR CNRS 6600), Université de Technologie de Compiègne, Compiègne, France

Abstract— **An arteriovenous fistula (AVF) is a surgical vessel connection between an artery and a vein. It is created in end stage renal disease to provide adequate blood access for hemodialysis. In the present study, the local hemodynamics is investigated in a patient-specific AVF using a computational fluid structure interaction (FSI) simulation. The fluid and solid governing equations are solved using ANSYS (ANSYS, Inc.).**

We focus on an end-to-side AVF between the end of the cephalic vein and the brachial artery. The geometry of the vessel lumen is obtained from CT-scan angiography. The vessel wall is modeled as a monolayer of shell elements of uniform thickness, since the actual wall thickness cannot be obtained from medical images. We investigate the effect of the presence of a severe stenosis upstream of the anastomosis by comparing two different geometries: model 1 consists of the complete patient-specific AVF, presenting a stenosis inside the proximal brachial artery and an enlargement at the cephalic vein; model 2 is obtained from model 1 by substituting the stenosed artery with a straight cylinder. For both models, a physiological time-dependent velocity inlet profile and flow-dependent resistive pressure outlets are imposed as boundary conditions. The hyperelastic, 3rd-order Yeoh model is used as constitutive law to model the vessel wall.

The presence of the stenosis increases the mass flow through the cephalic vein by 9.6%, which could have a positive effect on the hemodialysis access in terms of flow rate. However, the stenosis causes an enlarged area of the vein subjected to low wall shear stresses and high oscillatory shear indices, which are risk factors for atherosclerotic plaque and neointima formation.

Keywords— **arteriovenous fistula, fluid-structure interaction, hemodynamics, wall shear stress, oscillatory shear index.**

I. INTRODUCTION

Most patients with end stage renal disease require hemodialysis. This treatment requires a permanent vascular access easily available and able to provide a blood flow equal to 300-500 ml.min^{-1}. An arteriovenous fistula (AVF) is a surgically-created connection between an artery and a vein. Subjected to arterial pressure, the vein gets arterialized after about three months. When the fistula reaches its maturation state, it acts as a low resistance, high compliance pathway between the high pressure arterial system and the low pressure venous system[1]. The lifespan of an AVF is limited from a few days to about 10 years, the failure being due to insufficient or excessive blood flow inside the cephalic vein. The reason for its failure remains an open question. It has been observed that stenoses are present in about 20% of mature AVF[2]. We would like to investigate whether the stenosis may affect the flow rate distribution in the AVF and therefore the long-term efficiency of the AVF.

Investigating the hemodynamics in the AVF cannot be easily performed *in vivo*: Echo-Doppler measurements can provide the averaged velocity inside vessels but neither local velocity profiles nor parameters such as the wall shear stress (WSS). The hemodynamics inside the AVF has previously been investigated using computational fluid dynamics simulations[3-4], but no previous study has considered the case of a stenosed AVF.

Our purpose is to investigate how the presence of a stenosis at the proximal brachial artery may affect the hemodynamics in a patient-specific geometry with an FSI simulation.

II. METHODS

A. *Geometry, Mesh and Numerical Method*

We analyze a patient-specific end-to-side brachiocephalic fistula at the elbow region. The geometry consists of the AVF lumen reconstructed from CT-scan angiography medical images[3]. All the clinical measurements are conducted on a patient at rest, in supine position and without external constrains on the AVF. We investigate two different geometries. Model 1 consists of the complete patient-specific AVF shown in figure 1. It presents an 80%-stenosis along the proximal brachial artery and an enlargement in the cephalic vein. Model 2 is obtained from model 1 by substituting the stenosed artery with a straight cylinder (box in figure 1). The cylinder is connected to the brachial artery 2.5 cm upstream of the anastomosis.

The vessel wall is modeled as a monolayer of shell elements. The wall thickness is set to be equal to 0.1 times the diameter of the brachial artery, since the actual value cannot be deduced from the medical images.

The mesh of model 1 consists of 89×10^3 shell elements for the vessel wall and 783×10^3 elements for the vessel lumen. In model 2, the vessel wall is meshed with 71×10^3 shell elements and the lumen with 621×10^3 elements.

S. Vlad and R.V. Ciupa (Eds.): MEDITECH 2011, IFMBE Proceedings 36, pp. 324–327, 2011.
www.springerlink.com

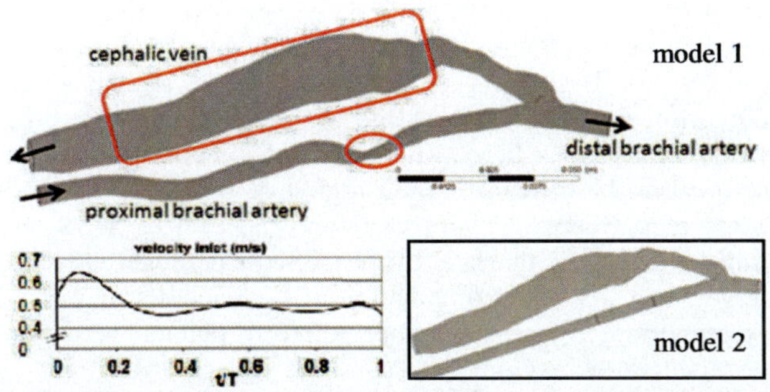

Fig. 1 Geometry of the patient-specific AVF and time-dependent velocity profile imposed as inlet boundary condition. The geometry used for model 2 is shown in the box

The fluid structure interactions are simulated coupling ANSYS-CFX and ANSYS-Mechanical (ANSYS, Inc.) implicitly. A convergence study has been done in order to tune the time-step to guarantee numerical stability. We look for the optimal time-step for which (*i*) the numerical model is stable, (*ii*) the computational time is the smallest, (*iii*) the temporal resolution is sufficient to capture the time-dependent flow features. All the following results are calculated with a time step of 10^{-2} s.

B. *Mechanical and Rheological Properties*

The vessel wall is assumed to be homogeneous, isotropic and hyperelastic in both models. It has to be noted that, to the knowledge of the authors, no AVF wall properties data exists in the literature. We assume both the artery and the vein follow the 3rd-order Yeoh model[5]. The law constants are chosen, so that the diameter of the mid-brachial artery varies by 10% over one cardiac cycle.

Blood is modeled as an isotropic, homogeneous and non-Newtonian fluid; its density is considered equal to 1050 kg.m^{-3}. The apparent viscosity μ follows Casson model:

$$\sqrt{\mu} = \sqrt{\frac{\tau_0}{\dot{\gamma}}} + \sqrt{k}\,, \tag{1}$$

where τ_0 represents the yield stress, $\dot{\gamma}$ the shear rate and k the consistency. We set the model parameters according to experimental data for low shear rates[6]: $\tau_0 = 4\times10^{-3}$ Pa, $k = 3.02\times10^{-3}$ Pa.s.

C. *Boundary Conditions*

A time-dependent velocity profile measured by echo-Doppler on the patient on the same day as the CT-scan angiography is applied at the inlet of the proximal brachial artery (figure 1). The peak Reynolds number is 800, mean Reynolds number 650 and Womersley number 4. At the outlets, we impose purely resistive boundary conditions. The values of the resistances are tuned to obtain a mean flow rate of 350 ml.min^{-1} in the cephalic vein.

D. *Hemodynamic Parameters*

We call wall shear stress (WSS) the modulus of the two-component vector:

$$\boldsymbol{\tau}_{\mathbf{w}} = \mu \frac{\partial \mathbf{v}}{\partial \mathbf{n}}\,, \tag{2}$$

where **v** is the velocity vector and **n** the unit vector normal to the vessel wall. In an healthy brachial artery, the WSS is 1-2 Pa[2], which we will refer to as the healthy physiological range.

The temporal gradients of WSS ($WSSG_t$) are defined as:

$$WSSG_t = \frac{\partial WSS}{\partial t}\,. \tag{3}$$

It has been assessed that the presence at a specific location of instantaneous high $WSSG_t$ and large variations of this index during the cardiac cycle are directly related to endothelial cell proliferation and intimal hyperplasia[7].

The oscillatory shear index (OSI) is defined by:

$$OSI = 0.5\left(1 - \frac{\left|\int_0^T WSS\,dt\right|}{\int_0^T |WSS|\,dt}\right), \tag{4}$$

where T is the cardiac period. By definition, the OSI index varies between 0 and 0.5. It takes into account the oscillations in the flow direction: the larger the index, the more important the wall shear oscillations.

III. Results

A. *Hemodynamics*

The velocity profile analysis in model 1 shows a time-varying velocity profile following the Womersley solution in the proximal brachial artery. A peak velocity of 2.1 m.s^{-1} is observed at the stenosis (location A, figure 2).

We compare the flow rates in the different branches of the AVF in both models to investigate the influence of the stenosis on the flow redistribution. The critical flow rate for hemodialysis treatment is the venous one. We observe that, for the same incoming flow rate, the stenosis in model 1 leads to an increase in the cephalic vein flow rate to 9.6%. A stenosis located on the arterial side may then actually enhance the blood flow entering the vein owing to a flow redirection. Figure 2 illustrates the change in the streamline directions: more streamlines enter the cephalic vein in model 1. It is a consequence of the helical motion induced downstream of the stenosis (Figure 2a).

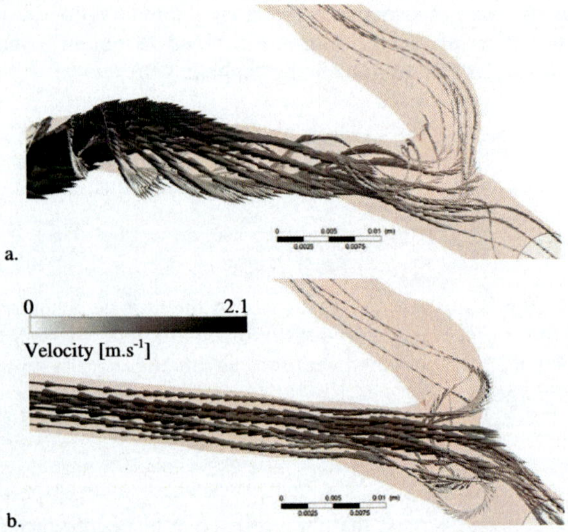

Fig. 2 Velocity vectors along 20 streamlines in the region of the anastomosis calculated at peak systole in models 1 (a) and 2 (b). The grey scale represents the velocity magnitude

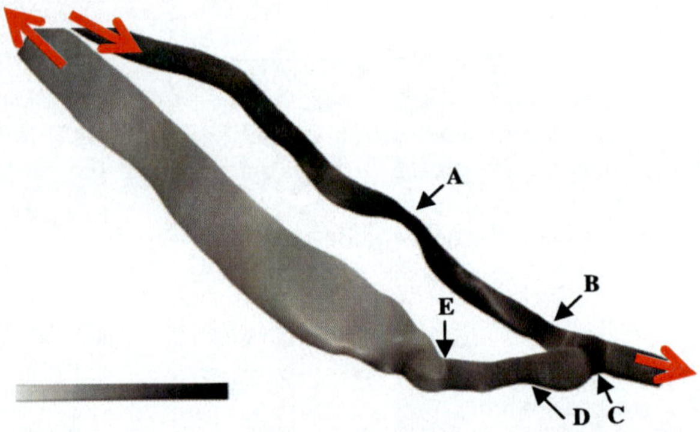

Fig. 3 Spatial distribution of WSS at peak systole. The WSS is the highest at locations A (80%-stenosis), B (15%-stenosis) and C (neck of anastomosis). It is the lowest downstream of location E on the venous side

B. Wall Shear Stresses

The WSS spatial distribution at the peak systole is shown in figure 3 for model 1. The WSS falls within the healthy physiological range in the proximal brachial artery upstream of the stenosis. The vessel tortuosity leads to local WSS values up to 10 Pa. The highest WSS is found at the level of the stenosis (location A), where it approaches 60 Pa. We observed a second region of high local WSS (~15 Pa) at location B, where the lumen cross-sectional area is reduced by 15%. The anastomosis (location C) is the third region subjected to high WSS (~18 Pa): it is induced by the incoming flow impacting onto the bifurcation. On the venous side, the patient endures a venous enlargement over most of the cephalic vein. This entire region is subjected to low WSS (~0.15 Pa). High values of OSI (~0.3-0.4) are observed downstream of the stenosis and the anastomosis. The area with the largest OSI (> 0.4) is observed 1-3 cm downstream of location E, in the enlarged region of the cephalic vein. Most of the cephalic vein therefore experiences WSS ten times lower than the healthy physiological range and OSI larger than 0.3. The combination of both factors may play an important role in the atherosclerotic plaque formation observed in this region[7].

Previous studies highlighted the role of the $WSSG_t$ on the endothelial cell migration processes[8]. We therefore investigated its evolution during the cardiac cycle T in model 1 (figure 4). The maximum absolute values of $WSSG_t$ occur at location C during the systole ($WSSG_t$ = 63.9 Pa.s^{-1} at t/T = 0.1 and $WSSG_t$ = 74 Pa.s^{-1} at t/T = 0.2). The $WSSG_t$ experiences large temporal variations over the cardiac cycle. It continues to oscillate during the diastolic phase of the cardiac cycle, but the oscillations tend to reduce in amplitude (~40 Pa.s^{-1}). The presence of high temporal gradients is typical at the neck of the anastomosis[8]. Temporal gradients remain otherwise small. Their second highest value is found at location D, but figure 4 shows how much lower their amplitude is.

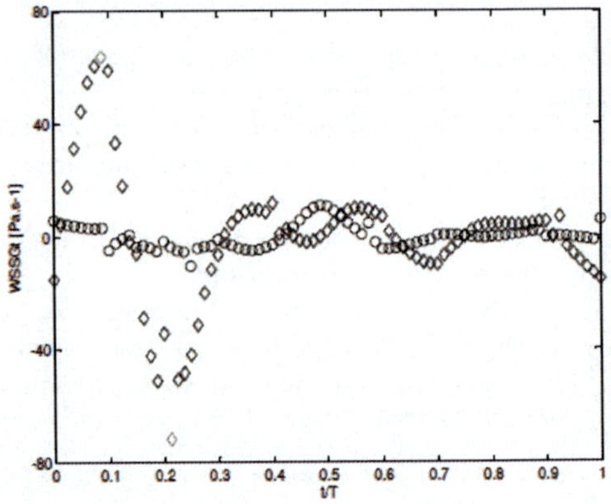

Fig. 4 Temporal variation of $WSSG_t$ at locations C (◊) and D (○) in model 1. The location of point D can be found in figure 3

Hardly any difference is found when we compare the WSS values for both models (model 1: with stenosis, model 2: without stenosis). Regions of non-physiological WSS are found in both models at the same locations. A difference in the WSS values is found at location B, the WSS being 25% higher in the presence of the stenosis: it is a consequence of the higher flow entering the vein. Inside the cephalic vein, the WSS is again similar in both models 1 and 2, the peak WSS being much lower than the physiological WSS.

More significant differences are found between models 1 and 2 for the $WSSG_t$. The presence of the stenosis increases the oscillation amplitude at the anastomosis (location C) by 3.8% and downstream of the anastomosis in the cephalic vein (location D) by 50%. This shows that the presence of the stenosis might accelerate the intima proliferation and wall remodeling, especially downstream of the anastomosis between locations D and E[8]. Another important difference is that downstream of location E the zone subjected to high OSI (> 0.4) is halved in the non-stenosed case (model 2). The presence of the stenosis along the artery might therefore accelerate the atherosclerotic plaque formation inside the cephalic vein.

IV. Conclusions

The originality of this work was to perform an FSI numerical simulation in a patient-specific AVF, modelling the vessel wall with a hyperelastic model and blood with a non-Newtonian model. Such a simulation proves to be highly challenging (large region of interest, non-linear effects of the fluid and solid constitutive laws, instabilities of the FSI coupling, etc). By the work, we prove that it is feasible to use commercial codes to solve the physical problem.

In order to evaluate the importance of taking all these factors into account, we have compared the results with simplified versions of the simulation. Assuming the vascular walls to be rigid, we find that the WSS are overestimated by 10-13%. If instead we assume blood to be Newtonian, the WSS values in the cephalic vein are increased by 17%. It shows that all the physiological non-linearities need to be simulated to get realistic results on the hemodynamics and wall dynamics, and therefore on the AVF long-term efficiency.

The results confirm that AVFs are subjected to complex hemodynamics. We show that the cephalic vein is subjected to pathological values of WSS and OSI for both models with and without stenosis. An unexpected result is encountered with the presence of the stenosis upstream of the anastomosis that increases the mean flow rate in the cephalic vein by 9.6%, due to a redirection of the flow into the vein. Despite this increase the WSS values in the cephalic vein remain lower than the physiological range during the entire cardiac cycle. The size of the area subjected to OSI values higher than 0.4 is 2 times larger in the stenosed case. It is therefore more likely to suffer from the formation of atherosclerotic plaques in this region. The analysis of the $WSSG_t$ shows that the presence of the stenosis could potentially accelerate abnormal neointima proliferation, especially between locations D and E.

Stent placement in the case of stenosed AVFs has been proven to be efficient to avoid early AVF's failure[9]. By considering model 2 as an approximation of the stented AVF, our results show that size of the regions subjected to OSI higher than 0.4 is halved inside the cephalic vein. The $WSSG_t$ peak amplitude is reduced by 3.8% at the anastomosis (location C) and by 50% in the cephalic vein (downstream of location D).

The main limits of this study are the use of purely resistive outlets and of the same wall characteristic law for both the artery and the vein. Further improvements will consist in differentiating the mechanical properties for the different vessels and implementing a Windkessel model for the outlet boundary conditions. Another limit is that the case presented herein is specific to one patient, but more patients will be studied in the future to have points of comparison.

Acknowledgment

This research is funded by the European Commission, through the MeDDiCA ITN (grant agreement PITN-GA-2009-238113). The authors gratefully acknowledge Polyclinique St Côme (Compiègne, FRANCE) for the medical images and J. Penrose from ANSYS-UK (ANSYS, Inc.) for all his help and advice.

References

1. Sivanesan S, How T V, Black R A, Bakran A (1999) Flow patterns in the radiocephalic arteriovenous fistula: an in vitro study. J Biomech 32: 915–925
2. Van Tricht I, De Watcher D, Tordoir J, Verdock M (2005) Hemodynamics and complications encountered with arteriovenous fistulas and grafts as vascular access for hemodialysis: a review. Ann Biomed Eng 33: 1142–1157
3. Kharboutly Z, Fenech M, Treutenaere J M, Claude I, Legallais C (2007) Hemodynamics and vascular alterations in an established arteriovenous fistula. Med Eng Phys 29: 999–1007
4. Ene-Iordache B, Mosconi L, Remuzzi G, Remuzzi A (2001) Computational fluid dynamics of a vascular access case for hemodialysis. J Biomech Eng 123: 284–292
5. Yeoh O H (1993) Some forms of the strain energy function for rubber. Rubber Chem Technol 66: 754-771
6. Merril E W, Pelletier G A (1967) Viscosity of human blood: transition from Newtonian to non-Newtonian. J App Physiol 23: 179–182
7. Markl M, Wegent F, Zech T, Bauer S, Strecker C, Schumacher M, Weiller C, Henning J, Harloff A (2010) In vivo wall shear stress distribution in the carotid artery: effect of bifurcation geometry, internal carotid artery stenosis, and recanalization therapy. Circ Cardiovasc Imaging 3: 647-655
8. Ojha M (1994) Wall shear stress temporal gradient and anastomotic intimal hyperplasia. Circ Res 74: 1227–1231
9. Chan M R, Young H N, Yevzlin A S (2009) The effect of in-stent restenosis on hemodialysis access patency. Hemodial Int 13: 250-256.

Author: Iolanda DECORATO
Institute: Biomechanics & Bioengineering Laboratory (CNRS UMR 6600), Université de Technologie de Compiègne
Street: BP 20529, 60205
City: Compiègne
Country: FRANCE
Email: iolanda.decorato@utc.fr

A Method to Increase the Pulsatility in Hemodynamic Variables in an LVAD Supported Human Circulation System

S. Bozkurt, K.A.M.A. Pennings, S. Schampaert, F. N. van de Vosse, and M.C.M. Rutten

Department of Biomedical Engineering, Eindhoven University of Technology, Eindhoven, The Netherlands

Abstract— **Left Ventricular Assist Devices (LVADs) generally operate at a constant speed in the human body. This causes a decrease in the pulsatility of hemodynamic variables. To increase the pulsatility a stepwise change was applied to the LVAD operating speed over a cardiac cycle. To do this, a numerical cardiovascular system model and a pump model were used. The model was developed by considering the static characteristics of the MicroMed DeBakey LVAD. First, the simulations were performed at constant operating speeds, 8500 rpm, 9500 rpm and 10500 rpm. Pulsatility indexes were calculated for left ventricular (LV) pressure, aortic pressure, LV volume and LVAD flow. Cardiac output (CO) was calculated at constant operating speed and these values used for comparing the pulsatility indexes with stepwise and constant operating speeds. The LVAD was operated at two different constant speeds in the stepwise operating speed simulations. Low and high operating speeds were adjusted so as to obtain the same cardiac output values with the constant operating speed simulations. The operating speeds in the simulations were 7800-11250 rpm, 9300-11250 rpm and 10300-11250 rpm. The same cardiac output values were obtained with an increase in the pulsatility of the hemodynamic variables without significant changes in their shapes except the LVAD flow. The obtained results show that it is possible to obtain more physiological results by applying a stepwise change to LVAD operating speed over a cardiac cycle.**

Keywords— **Heart pumps, LVADs, pulsatility, static characteristics, stepwise speed control.**

I. Introduction

Left Ventricular Assist Devices operate at a constant speed in the human body. This speed causes a decrease in pulsatility in the hemodynamic variables such as LV pressure, LV volume, aortic pressure, flow rate of the LVAD. In full support the aortic valve remains closed over a cardiac cycle and all the blood flows through the LVAD [1]. The remaining pulsatility in an LVAD assisted heart exists because of contractions of the LV. This low pulsatility condition, may lead to aortic insufficiency and other long-term vascular complications. More physiological LVAD operation may alleviate these problems.

Several attempts to improve the assistance of the LVADs in the human cardiovascular system have been described in literature. Moscato et al. developed a control strategy that provides an explicitly definable loading condition for the failing ventricle [2]. The studies show that there is a relation between the motor current and hemodynamic variables in an LVAD assisted heart. This relation was used to develop LVAD control strategies [2-6]. The LVADs show deleterious effects such as suction when the operating speed reaches a relatively high value. There are studies to detect and prevent the suction in the literature [7-11]. These studies show sufficient support to the circulation system and prevent suction.

Pulsatility in the hemodynamic variables decreases in an LVAD supported heart however pulsatility is a desired effect in a VAD assisted heart and can be obtained by using a pulsatile flow VAD. In a continuous flow VAD pulsatility diminishes and afterload increases [12]. In this paper, a different approach is presented to improve the long-term support capability of the LVAD. A stepwise operating speed change was applied to obtain pulsatility in hemodynamic variables over a cardiac cycle.

II. Method

To apply a stepwise operating speed over a cardiac cycle, a numerical human cardiovascular system (CVS) model and a numerical model for Micromed DeBakey LVAD were used. The operating speed was increased instantly when LV pressure reaches its peak value and kept constant until it reaches the lowest value. When LV pressure reaches its lowest value, the LVAD operating speed was adjusted to the low value again and kept at this value until peak LV pressure was observed again. The results were compared with constant speed operation, rendering the same overall cardiac output. For an accurate comparison pulsatility indexes (PI) were defined for the hemodynamic variables.

The numerical CVS simulation model consists of an active left ventricle (LV) and right ventricle (RV), left atrium, right atrium, aortic valve (AV), mitral valve (MV), tricuspid valve (TV), pulmonary valve (PV), aorta, systemic veins, pulmonary arteries and pulmonary veins. The equivalent electric analogue is given in Fig.1. Abbreviations are given in Table1.

S. Vlad and R.V. Ciupa (Eds.): MEDITECH 2011, IFMBE Proceedings 36, pp. 328–331, 2011.
www.springerlink.com

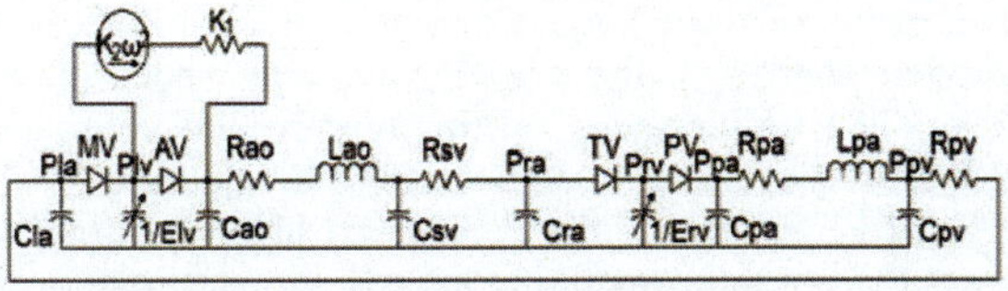

Fig. 1 Equivalent electric analogue of CVS model and LVAD

Table 1 Symbols

Nomenclature		Subscripts	
AV	Aortic valve		
C	Compliance	ao	aorta
e	Activation function	la	left atrium
E	Elastance	lv	left ventricle
K	Pump coefficeint	pa	pulmonary arteries
L	Inertance	pv	pulmonary veins
MV	Mitral valve	ra	right atrium
PV	Pulmonary valve	rv	right ventricle
Q	Flow rate	sv	systemic veins
R	Resistance	d	diastolic
T	Heart beat period,	s	systolic
t	instantaneous time	1	peak systole
TV	Tricuspid valve	1	number of pump coefficient
ω	Rotation speed	2	peak diastole
ΔP	Pressure difference	2	number of pump coefficient

Parameter values and the equations for compartments are taken from the available literature and all the parameter values and equations were used in the numerical model can be found in [13-16]. Elastance and activation functions of the ventricles were taken from [17]. The elastance function (1) of the ventricles is calculated by using the activation function (2). The activation function describes the contraction and relaxation phases in the ventricles.

$$E = E_d + \{(E_s - E_d)/2\}e \quad (1)$$

$$e = \begin{cases} 1 - \cos(t\pi / T_1), 0 \le t < T_1 \\ 1 + \cos\{(t - T_1)\pi / (T_2 - T_1)\}, T_1 \le t < T_2 \\ 0, T_2 \le t < T \end{cases} \quad (2)$$

The symbols are referenced in Table 1. The same equation was used in both ventricle models but the parameter values were different. Es and Ed for the LV were 2.5 mmHg/mL and 0.1 mmHg/mL respectively. Es and Ed for the RV were 1.15 mmHg/mL and 0.1 mmHg/mL respectively [17]. T was adjusted to 0.8 sec and kept constant in all the simulations. T_1 was adjusted 0.3*T and T_2 was adjusted 0.45*T. Dilated cardio-myopathy was induced by reducing the Es value for both ventricles to 0.5 mmHg/mL.

A numerical pump model was developed to simulate the MicroMed DeBakey LVAD by considering the static characteristics of the pump. Static characteristics of this pump are given in Fig. 2. Detailed information about the measurements of pressure across the LVAD can be found in [18]. The model used in the simulations is given below:

$$\Delta P = K_1 Q + K_2 \omega^2 \quad (3)$$

The values of K_1 and K_2 were estimated to be -0.264 mmHg/mL/s and 0.0035 mmHg/s^2 respectively.

The numerical pump model was implemented in the CVS model and simulations were performed for constant operating speed at 8500, 9500 and 10500 rpm. After completing the simulations at constant operating speeds, simulations were performed for stepwise LVAD operating speeds. In the stepwise operating speed simulations the LVAD switched to high speed at the peak LV pressure and kept constant until LV pressure reaches its minimum value. Three different simulations were performed at 7800-11250 rpm, 9300-11250 rpm and 10300-11250 rpm operating speeds to obtain the same cardiac outputs as observed in the simulations for constant operating speeds. The maximum speed was set to 11250 rpm to prevent the suction. The LVAD operating speed and LV pressure are given in Fig 2.

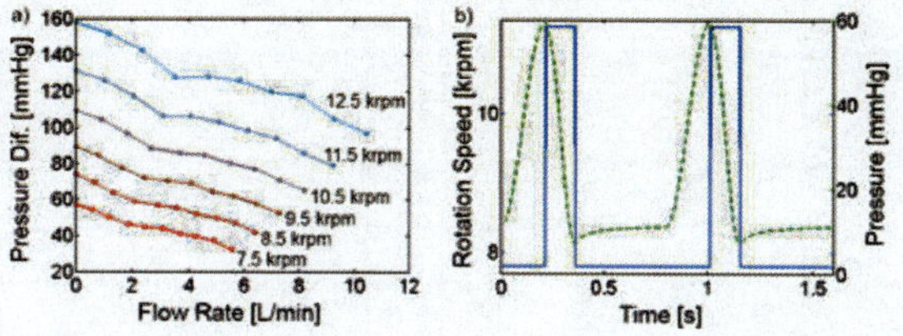

Fig. 2 a) Static characteristics of the MicroMed DeBakey LVAD, b) LVAD speed (-continuous line), LV pressure (--dashed line)

To determine and make a comparison between the pulsatility in the hemodynamic variables, pulsatility indexes were calculated as below.

$$PI = \{X_{max} - X_{min}\}/\{(X_{max} + X_{min})/2\} \quad (4)$$

The parameter X is used for the hemodynamic variables that were considered in the simulations; LV pressure, aortic pressure, LV volume and LVAD flow. The subscripts of max and min in the (4) denote maximum and minimum values of these hemodynamic variables. Simulations were performed using the Matlab Simulink tool. Solver and maximum step size were set to ode15s and 0.0002 s respectively.

III. RESULTS

Simulations were performed for healthy and pathological conditions without LVAD support first. LV pressure, aortic pressure, LV volume and cardiac output (CO) for healthy and pathological conditions are given in Fig.3 and summarized in Table 2.

LV pressure, aortic pressure and LV volume under assisted conditions at constant speeds are given in Fig 4.

For increasing constant LVAD operating speed LV pressure decreases and aortic pressure increases with increasing LVAD operating speed (Fig. 4). End-systolic and end-diastolic volume of the LV decrease with increasing constant operating speed. LV pressure, aortic pressure and LV volume under assisted conditions at stepwise changing speed are given in Fig 5.

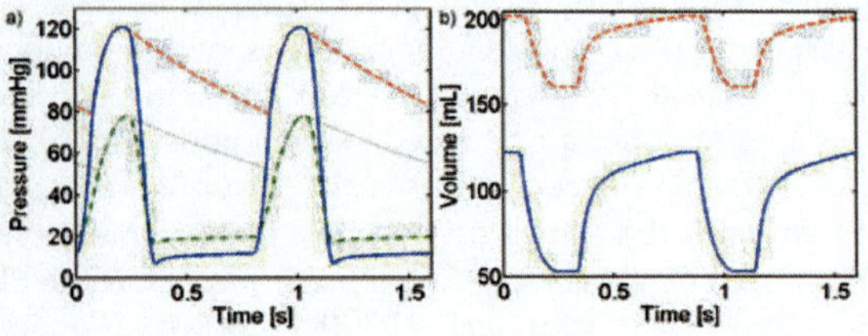

Fig. 3 a) LV pressure (- healthy, -. pathological), aortic pressure (-- healthy, : pathological) b) LV volume (- healthy, -- pathological)

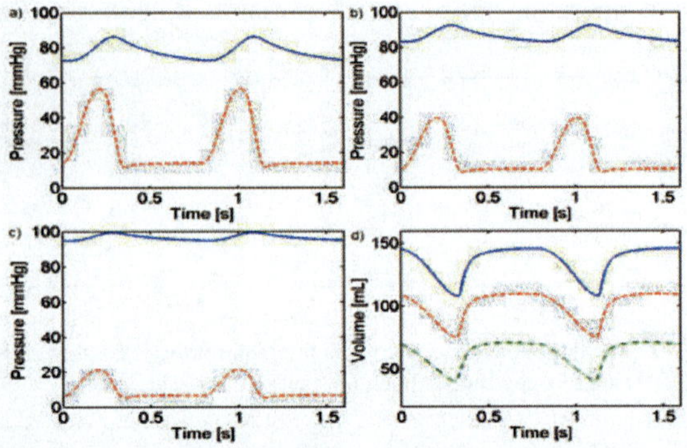

Fig. 4 a) LV pressure and aortic pressure at 8500 rpm, b) LV pressure and aortic pressure at 9500 rpm, c) LV pressure and aortic pressure at 10500 rpm, (-- LV pressure, -aortic pressure), d) LV volume at constant speed (- 8500 rpm, -- 9500 rpm, -. 10500 rpm)

Table 2 Simulation results

	Plv[mmHg]	Pao[mmHg]	Vlv[ml]	CO[mL/s]
Healthy	7-120	79-121	53-122	86.3
Pathological	17-78	58-78	160-202	52.5

The shapes of the hemodynamic signals at the stepwise operating mode are similar to the constant operating speeds over a cardiac cycle. Stepwise change of the LVAD operating speed hardly changes the shape of the hemodynamic signals. However the amplitude of the aortic pressure signal doubles. The LVAD flow for constant and stepwise operating speeds are given in Fig. 6.

The shape of the LVAD flow changes significantly due to sudden change in the LVAD operating speed. The PI values of the LV pressure, aortic pressure, LV volume and LVAD flow were calculated according to equation (4). The change of PI values for the considered hemodynamic variables at the constant operating speeds and stepwise operating speed are given in Fig. 7.

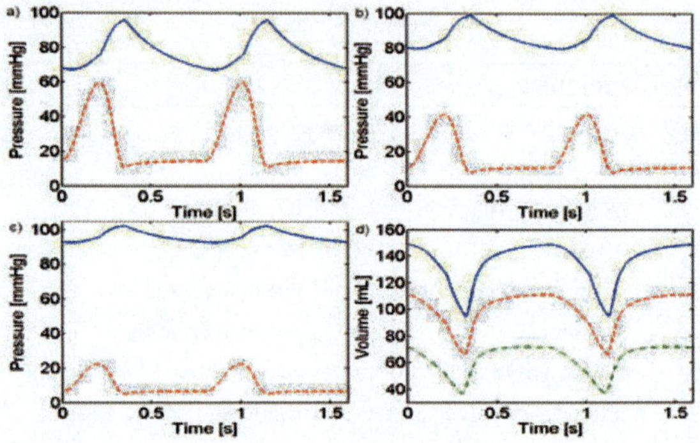

Fig. 5 a) LV pressure and aortic pressure at 7800-11250 rpm, b) LV pressure and aortic pressure at 9300-11250 rpm, c) LV pressure and aortic pressure at 10300-11250 rpm, (--LV pressure, -aortic pressure), d) LV volume (- 7800-11250 rpm, --9300-11250 rpm, -.10300-11250 rpm)

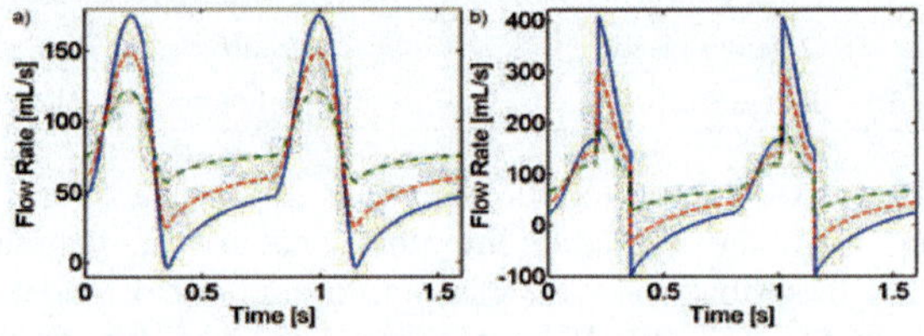

Fig. 6 a) LVAD flow at constant operating speed (- 8500 rpm, -- 9500 rpm, -. 10500 rpm), b) LVAD flow at stepwise operating speed (- 7800-11250 rpm, -- 9300-11250 rpm, -. 10300-11250 rpm)

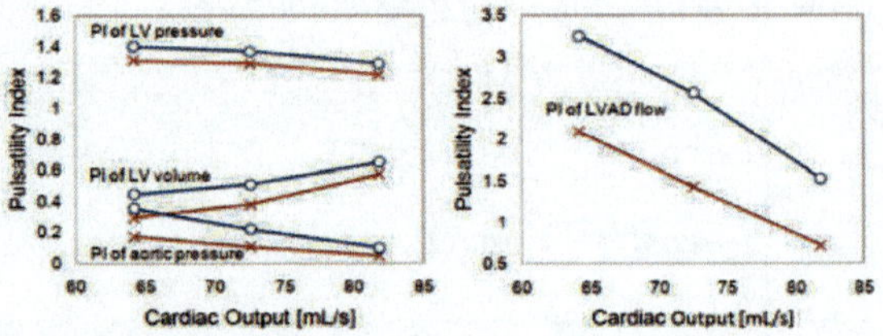

Fig. 7 PI of the LV pressure, aortic pressure, LV volume and LVAD flow for constant operating speed and stepwise operating speed (o: stepwise speed change, x: constant speed)

As shown in Fig 6, maximum value of the LVAD flow increases and minimum value decreases to obtain same mean CO values with the constant operating speed mode at stepwise operating speed. The peak values are not excessively high, the human CVS can handle these short term peak flows because of the large compliance of arterial system.

Stepwise change in the operating speed over a cardiac cycle provides increase in the pulsatility in all the hemodynamic variables considered (Fig. 7). The systolic aortic pressure is increased if the operating speed changes stepwise over a cardiac cycle. PI values decrease for increasing cardiac outputs in both constant and stepwise change of speeds except for the LV volume. The difference between the maximum and minimum values in hemodynamic variables decreases for constant and stepwise operating speeds for increasing cardiac output values.

IV. Discussion

In this paper a method was proposed and applied to increase the pulsatility of the hemodynamic variables in an LVAD assisted circulation system. It is possible to increase pulsatility and systolic aortic pressure if the operating speed changes stepwise without a significant effect on the shape of the hemodynamic variables except the LVAD flow. CO values did not change indicating that support quality of the LVAD is improved. In this study high speed was kept constant at 11250 rpm and the lower speed was changed. The operating speeds could be adjusted by considering heart rate, the inlet and the outlet pressure of the LVAD or any other parameter can be measured or estimated to obtain more physiological results and a better support. Also a pump model was developed by using the static characteristics of MicroMed DeBakey LVAD. A dynamic pump model would give better and more accurate simulation results, because dynamic pump load changes due to contractions of LV. In the simulations operating speed was changed instantaneously which is impossible in a real application. In real pumps, there will be a transition time from one speed level to next. To develop controllers capable of achieving this are subject of ongoing research.

At constant LVAD operating speeds aortic valve incompetence would occur due to change of pulsatility [19]. At stepwise operating speed mode pulsatility increases and it can potentially be a method to solve insufficiency problems in an LVAD supported human CVS.

Acknowledgement

This study is part of the MeDDiCA project and funded under FP7, People Programme, Marie Curie Actions. Grant agreement PITN-GA-2009-238113.

References

1. Karantonis DM, Lovell NH, Ayre PJ (2007) Classification of Physiologically Significant Pumping States in an Implantable Rotary Blood Pump: Effects of Cardiac Rhythm Disturbances. Artificial Organs 31:476-469
2. Moscato F, Arabia M, Colacino FM et al. (2010) Left Ventricular Afterload Impedance Control by an Axial Flow Ventricular Assist Device: A Potential Toll for Ventricular Recovery. Artificial Organs 34:736-744
3. Parnis SM, Conger JL, Fuqua JM, et al. (1997) Progress in the development of a transcutaneously powered axial flow blood pump. ASAIO J 43:M576–580
4. Fu M, Xu L. (2000) Computer simulation of sensorless fuzzy control of a rotary blood pump to assure normal physiology. ASAIO J 46:273–278.
5. Ohuchi KD, Kikugawa K, Takahashi M,et al. (2001) Control Strategy for Rotary Blood Pumps. ArtificialOrgans 25: 366-370
6. Kikugawa D (2000) Evaluation of cardiac Function During Left Ventricular Assist by a Centrifugal Blood Pump. Artificial Organs 24:632-635
7. Arndt A, Nüsser P, Graichen K, Müller J et al. (2008) Physiological control of a rotary blood pump with selectable therapeutic options: Control of pulsatility gradient. Artificial Organs 32:761–771.
8. Giridharan G A, Skliar M (2003) Control Strategy for Maintaining Physiological Perfusion with Rotary Blood Pumps Artificial Organs 27: 639-648
9. Vollkron M, Schima H, Huber L, et al. (2004) Development of a Suction Detection System for Axial Blood Pumps. Artificial Organs 28:709-716
10. Vollkron M, Schima H, Huber Let al. (2005) Development of a Reliable Autonomic Speed Control System for Rotary Blood Pumps. The Journal of Heart and Lung Transplantation 24:1878-1885
11. Vollkron M, Schima H, Huber L et al. (2006) Advanced Suction detection for an Axial Flow Pump. Artificial Organs 30:665-670
12. Travis AR, Giridharan GA, Pantalos GM, et al. (2007) Vascular Pulsatility in Patients with a Pulsatile- or Continuous- Flow Ventricular Assist Device. J Thorac Cardiovasc Surg, 133:517-524
13. Bozkurt S (2009) Dynamic Modeling of Human Cardiovascular System and an Axial heart Pump. Master Thesis, Yeditepe University
14. Heldt T, Shim EB, Kamm RD et al. (2002) Computational Modeling of Cardiovascular Response to Orthostatic Stress, Journal of Applied Physiology, 92:1239-1254
15. Darowski M, De Lazzari C, Ferrari G,et al. (1999) The Influence of Simultaneous Intra-Aortic Balloon Pumping and Mechanical Ventilation on Hemodynamic Parameters Numerical Simulation. Frontiers of Medical and Biological Engineering 9:155-174
16. De Lazzari C, Darowski M, Ferrari G et al. (2000) Computer Simulation of Hemodynamic Parameters Changes with Left Ventricle Assist Device and Mechanical Ventilation. Computer in Biology and Medicine 30:55-69
17. Korakianitis T, Shi Y (2006) A Concentrated Parameter Model for the Human Cardiovascular System Including Heart Valve Dynamics and Atrioventricular Interaction, Medical Engineering & Physics 28:613-628
18. Schampaert S (2009) Evaluation of the Micromed DeBakey Ventricular Assist Device. Internal Report, Eindhoven University of Technology.
19. John R, Mantz K, Eckman P, et al. (2010) Aortic Valve Pathophysiology During Left Ventricular Assist Device Support. Journal of Heart Lung Transplantation 29:1321-1329

Author: Selim Bozkurt
Institute: Eindhoven University of Technology/Dept. of Biomedical Engineering
Street: P.O. Box 513 5600 MB
City: Eindhoven
Country: the Netherlands
Email: s.bozkurt@tue.nl

Blood Flow Analysis in Portal Vein System – Unsteady-State Case Study

C.C. Botar[1], A. Bintintan[3], P.S. Agachi[1], S. Clichici[2], P. Mircea[3], and S. Sfrangeu[4]

[1] Chemical Engineering Department, Faculty of Chemistry and Chemical Engineering, "Babes-Bolyai" University, Cluj Napoca, Romania
[2] Department of Physiology, University of Medicine and Pharmacy "Iuliu Hatieganu", Cluj-Napoca, Romania
[3] Department of Gastroenterology, University of Medicine and Pharmacy "Iuliu Hatieganu", Cluj-Napoca, Romania
[4] Department of Medical Imaging, University of Medicine and Pharmacy "Iuliu Hatieganu", Cluj-Napoca, Romania

Abstract— **Anatomical features and a complex vascular system characterize the liver. The blood flow results as a complex interaction between fluid, vascular system complex geometry and liver functional and structural features. The disease presence produces pathological changes that may induce hemodynamic perturbations not only due to geometry modification but especially due to liver perfusion alteration. The analysis of the blood flow dependence on the geometry variability in physiological condition could emerge in its parameterization and quantification. This may eliminate the confusion between blood flow modification due to geometry variability in physiological conditions and flow alteration due to pathological conditions or congenital anomalies presence.**

In this paper the analysis focuses on portal vein system and consists in blood flow analysis under unsteady-state conditions. The study involves the investigation of 12 patients by MRI techniques followed by 3-D portal vein system geometry acquisition, blood flow simulations based on mathematical models that include constitutive equations describing the hemodymanics and its relations with the deformable vessels wall. The computational technique applied to model the blood flow approaches both the velocity field and the pressure field. The vessels wall was considered elastic, coupling in this way the vessel/wall deformability problem.

The blood flow analysis in physiological conditions enables the improvement of understanding of the complex blood flow behavior in the portal vein system; enables to identify critical information and to parameterize the domain of normal portal vein circulation.

Keywords— **blood flow, CFD, MRI, mathematical modeling, hemodynamics.**

I. Introduction

During the recent years, especially in the field of medicine [1], visualization techniques gain an increased role in the scientific and/or engineering work. In parallel with the increase in computer power the techniques of image acquisition and processing became important investigation tools and data sources for modeling and simulation [2]. The research field employing flow visualization in anatomical features was the first beneficiary of the development of these techniques [3].

Coupling image acquisition and processing techniques with large computational grids originating from computational fluid dynamic (CFD) simulation the quality of feature representation and process analysis is improved.

Through the visualization and preprocessing techniques the real-world models and their associated data (i.e. velocity) are transposed in a virtual one, able to numerically simulate the flow and the fluid/structure interaction, by solving complex mathematical models based on Navier-Stokes equations and gaining scientific data and comprehensive graphical representations.

There are several authors that used this approach for studying the blood flow in the human body [4-6]. The velocity components being acquired either from MRI sequences or form ultrasound Doppler measurements and the data compared qualitatively with the predictions of CFD calculations.

In the present analysis the MRI sequences were used for 3D reconstruction of the portal vein system and the ultrasound Doppler measurements to provide the boundary condition for CFD model implementation.

The accuracy of such approach and the validity of the mathematical model used for blood flow simulation in 3D complex geometries were demonstrated in a previous work of Botar et al. [7]. Based on that, the purpose of this research was tracking physiological features of the blood flow in the portal vein system.

II. Problem statement

The essentials in the present analysis relay on differentiation of physiological characteristics of the portal vein blood flow, its parameterization and quantification. For this reason, after obtaining the approval from the Ethics Committee of the University of Medicine and Farmacy "Iuliu-Hatieganu" and informed consent of patients, there were included in the study a series of 12 patients with physiological liver aspect. The clinical investigation consisted of Magnetic Resonance Imaging (MRI) and Ultrasound Doppler (UD).

The MR images were processed with MatLab software - Image Acquisition and Image Processing Toolboxes

S. Vlad and R.V. Ciupa (Eds.): MEDITECH 2011, IFMBE Proceedings 36, pp. 332–337, 2011.
www.springerlink.com

creating a binary version of the image by using the thresholding approach. The 3D portal vein system geometry reconstruction was performed using the SolidEdge V20 software. Further on, the computational fluid dynamics (CFD) technique was applied to describe the blood flow in the portal vein system.

The elastic wall conditions have been introduced based on mathematical models that include constitutive equations describing the hemodynamic and its relations with the deformable vessels wall. The constitutive model for the vessel wall is considered the hyperelastic one, as this model may cover non-linear stress-strain behaviour at modest strains, or elastic one up to huge strains.

The portal vein hemodynamic analysis considers the contribution of the confluent abdominal veins.

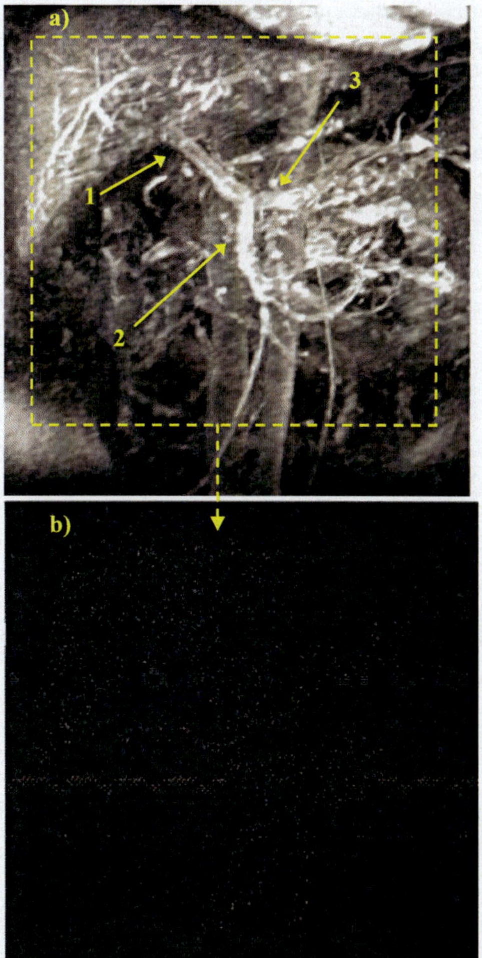

Fig. 1 Portal vein system: a) MRI and b) 3D SolidEdge V20 reconstruction, 1 - portal vein, 2 - Superior mesenteric vein, 3 – Splenic vein

III. Representation of the 3D portal vein geometry

The objective of the 3D reconstruction of the portal vein system was to obtain a 3D realistic volume that reproduces as intimate as possible the complexity of the portal and of the associated veins system.

For each patient the same type of 3D reconstruction has been done in order to incorporate in the simulation the high diversity of the portal vein geometries.

The portal vein geometry acquisition has been done using a 1 T MRI system (Sigma LX, GE Medical Systems) with a 9.1.0723d software, 1270 MHz IP30 processor, main memory size 512 Mbytes, and a phase array TORSOPA receiver. The geometrical information have been extracted from a 3D TOF SPGR vascular sequence acquisition, using the SmartPrep option based on bolus detection (gadolinium). The apparatus settings consisted in: Angio TOF SPGR – 3D acquisition; maximum monitor period 12 s; image acquisition delay 4 sec; imaging options: Fatsat; SmartPrep; TE Min; Prep Time 21; FA 35; Bandwidth 41,67; FOV 40; Sl Th 2.2; Locs per slab 34; Freq/Phase 256/192; NEX 0.50; PhaseFOV 0.90; scanning time 0:23, breath hold.

The flow related data were acquired using in vivo Eco-Doppler technique. The medical investigations have been conducted using an Ultrasound Logiq 7 BT 06 machine, a Convex probe 4C, with the following B mode settings (Freq 4.0 MHz, AO=100%, Gn 78, DR 111), Doppler color settings (Freq 3,3 MHz, PRF 1.2 KHz, Gn 29, WF 175 KHz) and Pulse Doppler settings (SVL 4 mm, GN 23, PRF 3,5 KHz, DR 40, WF 69 Hz).

The 3D geometry reconstruction was done using the SolidEdge V20 software capabilities. An approximation has been made in what it concerns the shape of the geometric section in the distal branches of the geometry; the section shape has been considered circular.

The volume geometry has been imported in GAMBIT. The surfaces mesh was generated, using the face surface Quand/Pave algorithm and smoothed using the length-weighted Laplacian algorithm. The volume mesh was generated using the Tet/Hybrid/Tgrid algoritm. The spacing used was 1. The simulations were carried out by means of the CFD software Fluent. The geometric parameters (Table 1) used in simulations were determined by medical investigations.

The domains of the portal vein system investigated were the main branch and its left and right branches.

An un-steady state model has been used; considering the fluctuations of the blood flow between inspiration and expiration periods. In this way dependency of the blood flow in portal vein system has been considered. The tixotropic properties of the blood have been taken into account. The differential equations have been discretized in a manner of finite element method. Mass flow boundary conditions have been specified at the geometry inlet. The inferior and superior values are provided by Doppler Ultrasound measurements to supply the user defined profile functions. The vessel walls have been treated as elastic. The no-slip condition was imposed to the vessels wall. It is necessary to mention that the geometric variability of the portal vein system from one patient to the other is very high.

Table 1 Values of experimental (clinical) parameters used in simulation

Pacient		1	2	3	4	5	6	7	8	9	10	11	12
Hepatic dimensions	LD [mm]	204	178	167	179	170	180	183	198	175	169	181	171
	LS [mm]	88	80	60	57	75	87	85	86	78	71	59	73
	LC [mm]	48	42	30	38	45	47.9	39.7	44	45	36	40.1	48
Portal vein dimensions	Inspiration [mm]	15	15.5	12.5	11.4	14.3	14.8	13.9	14.7	15.7	13.9	12.3	14.1
	Expiration [mm]	13.9	14.5	11.6	11	13.4	13.7	11.7	13.3	14.1	12.1	11.5	12.9
	length [cm]	6.7	5.4	6	5	7	6.8	6.3	6.5	5.5	6.2	5.2	6.8
Portal vein stream velocity	max. [cm/s]	29	27.2	25	28	26.1	30	29.8	28.2	26.9	25.7	27.1	26.7
	min. [cm/s]	22.9	23.5	19.7	20.1	19.2	25.3	23.6	21.8	21.3	18.9	20.9	19.3
Flow type		L	L	L	L	L	L	L	L	L	L	L	L
Splenic vein	max. velocity [cm/s]	21.8	21.6	22.1	27.2	22.3	26.3	24.1	21.7	21.5	20.1	20.9	26.7
	flow [ml/min]	192	210	177	185	168.5	188.3	181.7	189	206	197	179	147.1
Superior mesenteric vein	max. velocity [cm/s]	15.9	18.2	18	19.4	16.2	17.8	17.3	15.4	17.9	19.2	18.7	17.3

* flow type L = laminar, LD – right lobe, LS – left lobe, LC – caudate lobe.

IV. MODELING APPROACH

The flow model considers the tixotropic characteristics of the blood. The blood is considered as a non-Newtonian fluid, the relation between the shear stress and the strain rate is nonlinear and time-dependent. The blood viscosity was defined according to the non-Newtonian power law:

Table 2 Values of experimental parameters used in simulation [8]

Power law index (n)	0.4851
Consistency index k (kg-s^n-2/m)	0.2073
Reference temperature (K)	310
Minimum viscosity limit ηmin (kg/m-s)	0.001
Maximum viscosity limit ηmax(kg/m-s)	0.003

The hemodynamic was described by the Reynolds stress model (RSM) instead of laminar model as the past experience showed that it gives accurate prediction of the blood flow in complex geometries [4]. The RSM is abandoning the isotropic eddy-viscosity hypothesis, and closes the Reynolds-averaged Navier-Stokes equations by solving transport equations for Reynolds stresses together with an equation for the dissipation rate. Since the RSM accounts for effects of streamline curvature, swirl, rotation, and rapid changes in strain rate, in a more rigorous manner than the one-equation and the two-equation flow models, it has been used in simulation due to its greater potential to give accurate predictions for complex flows [FLUENT 6.3 user guide]. More than that, using the Quadratic Pressure-Strain Model it is possible to obtain superior performance in a range of basic shear flows, including plane strain, rotating plane shear, and axisymmetric expansion/contraction. The non-equilibrium wall functions have been also considered to extend the applicability of the wall function approach by including the effects of pressure gradient and strong non-equilibrium (Fluent 6.3 User Guide, Chapter 12.7.4).

More than that, for low Reynolds numbers past experience showed that RSM model

The model equations are the following ones:

- the Reynolds stresses:

$$\rho\overline{u_i'u_j'} \tag{1}$$

- the transport equations:

$$\underbrace{\frac{\partial}{\partial t}\left(\rho\overline{u_i'u_j'}\right)}_{Local\ time\ derivative} + \underbrace{\frac{\partial}{\partial x_k}\left(\rho u_k\overline{u_i'u_j'}\right)}_{C_{ij}\equiv Convection} =$$
$$\underbrace{-\frac{\partial}{\partial x_k}\left[\rho\overline{u_i'u_j'u_k'} + \overline{p\left(\delta_{kj}u_j' + \delta_{ik}u_j'\right)}\right]}_{D_{T,ij}\equiv Turbulent\ diffusion} +$$
$$\underbrace{\frac{\partial}{\partial x_k}\left[\mu\frac{\partial}{\partial x_k}\left(\overline{u_i'u_j'}\right)\right]}_{D_{L,ij}\equiv Molecular\ diffusion} \underbrace{-\rho\left(\overline{u_i'u_k'}\frac{\partial u_j}{\partial x_k} + \overline{u_j'u_k'}\frac{\partial u_i}{\partial x_k}\right)}_{P_{ij}\equiv Stress\ production} -$$
$$\underbrace{\rho\beta\left(g_i\overline{u_j'\theta} + g_j\overline{u_i'\theta}\right)}_{G_{ij}\equiv Buoyancy\ production} + \overline{p\left(\frac{\partial u_i'}{\partial x_j} + \frac{\partial u_j'}{\partial x_i}\right)} - \overline{2\mu\left(\frac{\partial u_i'}{\partial x_k}\frac{\partial u_j'}{\partial x_k}\right)} -$$
$$\underbrace{2\rho\Omega_k\left(\overline{u_j'u_m'}\epsilon_{ikm} + \overline{u_i'u_m'}\epsilon_{jkm}\right)}_{F_{ij}\equiv Production\ by\ system\ rotation} + \underbrace{S_{user}}_{\text{User-defined source term}} \tag{2}$$

Unsteady state conditions have been used. The differential equations have been discretized in a manner of finite element method. The operation and the boundary conditions have been specified. The vessel was treated as elastic. A dynamic mesh model was used in order to address the movement of the mesh in the unsteady state solver. The no-slip condition has been imposed.

V. SIMULATION RESULTS

The hemodynamic simulations in the portal vein system have been initialized considering the contributions of the splenic and superior mesenteric veins. The resulted data, provided by the computer simulation, supply the values of the blood velocity along the entire portal vein geometry.

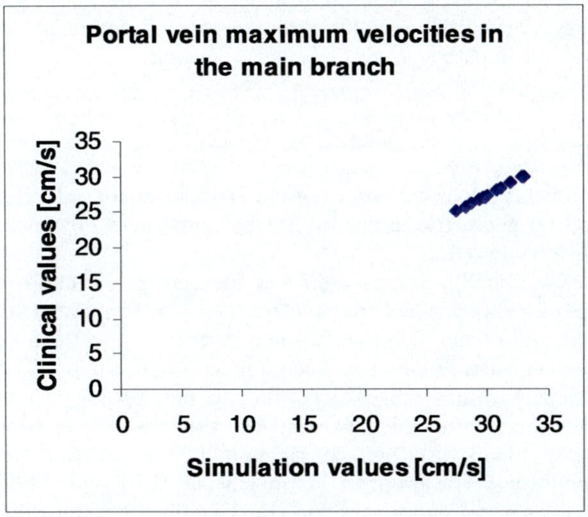

Fig. 2 Correlation between clinical and mathematical modeling data of blood velocity in the portal vein main branches during the inspiration period

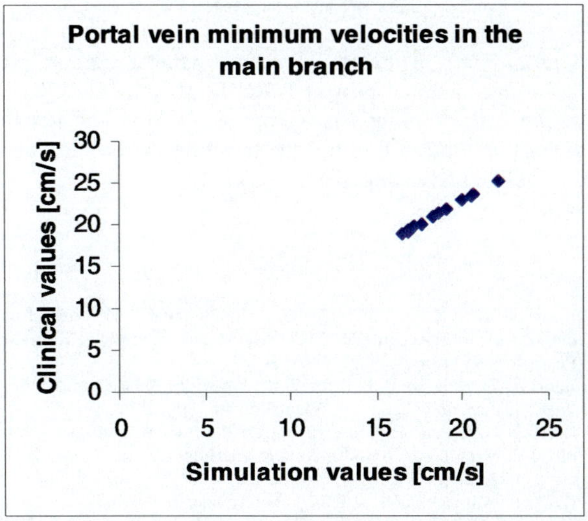

Fig. 3 Correlation between clinical and mathematical modeling data of blood velocity in the portal vein main branches during the expiration period

The analysis of the simulation results has been done according to the clinical investigations (Doppler Ultrasound). In the clinical experiments the velocity was measured in the portal vein main branch at 1 cm ahead from the branches bifurcation. A user profile user defined function was usedin simulations in order to cover the inspiration expiration cycle. The simulation results show a good agreement between clinical and mathematical modeling and simulation data (figures 2 and 3); both in case of inspiration and expiration periods.

To illustrate and to parameterize the behavior of the portal vein blood flow in its main branch, the distribution of the Reynolds number has been computed for all the geometries analyzed. The simulation results (figure 4) demonstrate that there exists a certain laminar domain in which the physiological blood flow takes place.

The range of Reynolds number extends between the values 2600 and 1020 in the region situated at 1 cm ahead the main portal vein bifurcation.

Extending the analysis to the left and to the right branches of the portal vein, the Reynolds number distribution is consistent with the data obtained in the portal vein main branch; even if the domain of Reynolds number differ significantly in the region situated this time at 2 cm after the main portal vein bifurcation.

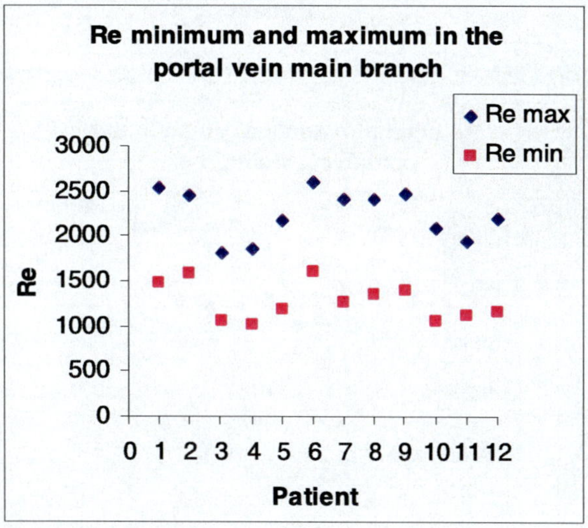

Fig. 4 Flow parameterization in the main branch of the portal vein

For the portal vein left branch the domain of Reynolds number extends between 1255 and 540 (Figure 5a), and for the portal vein right branch the values are comprised between 790 and 460 (Figure 5b).

The flow homogeneity observed in the three regions of the portal vein system, for such a high geometric variety, enables the prediction of the blood flow behavior.

Turbulent blood flow, as anticipated by the clinical investigations, is uncommon in normal circulation, but it could occur when pathological signs are established.

In pathological conditions disturbed flow may result from reflux, outflow obstruction, and/or stasis which leads to venous inflammation and thrombosis, and as a consequence the development of chronic venous diseases.

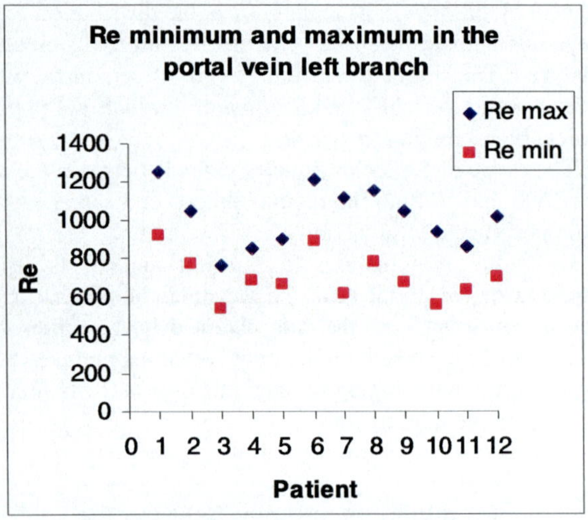

(a)

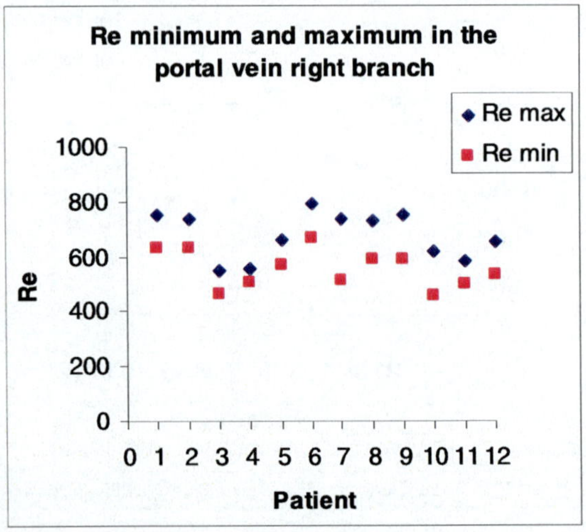

(b)

Fig. 5 Flow parameterization in the left and right portal vein branches

The ability to predict blood flow along the portal vein system can lead to a better understanding of the liver blood perfusion in physiological conditions and to eliminate the confusion between blood flow modification due to geometry variability and flow alteration due to pathological or congenital anomalies presence.

Understanding of the physiological flow may also provide mechanistic insights into the role of complex flow patterns in pathogenesis of vascular diseases.

VI. Conclusions

In this paper a comprehensive framework for exploiting the 3D geometric variability of the portal vein system has been presented.

Specifically, the research was focused on detecting the geometric-dependent parameters capable to predict the behavior of the blood perfusion of the liver. The Reynolds number distribution has been found to be comprised in compact values ranges according to the topology of the investigated domain. This conclusion suggests the possibility of blood flow parameterization in liver physiological conditions. The proposed approach and the results obtain enables a straightforward analytical evaluation of the portal vein flow parameters especially for quick diagnosis purposes.

Acknowledgment

The authors are thankful for the financial support provided by the national project PNCDI2-12131, title "Expert system for the non-invasive prognosis of chronic hepatic pathology evolution through the analysis of the biological and portal hemodynamic parameters."

References

1. Burrowes K S, Tawhai M H (2006) Computational predictions of pulmonary blood flow gradients: Gravity versus structure, Respiratory Physiology & Neurobiology 154:515–523
2. Jafari A, Zamankhan P, Mousavi S M, Kolari P (2009) Commun Nonlinear Science and Numerical Simulation 14:1396–1402
3. Lorthois S, Cassot F, Lauwers F (2010) Simulation study of brain blood flow regulation by intra-cortical arterioles in an anatomically accurate large human vascular network: Part I: Methodology and baseline flow, NeuroImage, DOI 10.1016/j.neuroimage.2010.09.032
4. Marshalla I, Zhao S, Papathanasopoulou P, Hoskins P, Xu Y X (2004) MRI and CFD studies of pulsatile flow in healthy and stenosed carotid bifurcation models, Journal of Biomechanics 37:679–687
5. Long Q, Xu Y X, Ariff B, Thom S A, Hughes A D, Stanton A V, (2000) Reconstruction of blood flow patterns in a human carotid bifurcation: a combined CFD and MRI study, Journal of Magnetic Resonance Imaging 11:299–311.

6. Nanduri J R, Pino-Romainville F A, Celik I (2009) CFD mesh generation for biological flows: Geometry reconstruction using diagnostic images, Computers & Fluids 38:1026–1032
7. Botar C C, Vasile T, Sfrangeu S, Clichici S, Agachi P S, Badea R, Mircea P, Cristea M V (2010) Validation of CFD simulation results in case of portal vein blood flow, Computer Aided Chemical Engineering 28:205-210.
8. Fluent 6.3 Documentation Manual, Chapter 8.4.5.

Author: Botar Claudiu Cristian
Institute: Chemical Engineering Department, Faculty of Chemistry and Chemical Engineering, "Babes-Bolyai" University
Street: Arany Janos 11
City: Cluj-Napoca
Country: Romania
Email: cbotar@chem.ubbcluj.ro

A Review of Atherosclerosis and Mathematical Transport Models

B. Keller[1,2], F. Clubb Jr.[3], and G. Dubini[1]

[1] Laboratory of Biological Structure Mechanics, Structural Engineering Department, Politecnico di Milano, Milan, Italy
[2] Bioengineering Department, Politecnico di Milano, Milan, Italy
[3] Cardiovascular Pathology Laboratory, Texas A&M University, College Station, Texas, USA

Abstract— **The mechanisms and definitive predictive components involved within the clinical initiation and formation of atherosclerosis remain elusive as of yet. Over the years, established fluid and mass transport theories and experimentally agreed upon vasoactive agents have contributed towards a boom of predictive mathematical models concerning atherogenesis. This paper aims to elucidate currently utilized theories available regarding initiation of atherosclerotic proliferation and provide a brief review of available mathematical transport phenomenon models which utilize these theories.**

Keywords— **atherosclerosis, mathematical models, transport phenomena, low density lipoproteins.**

I. Introduction

Atherosclerosis, generally speaking, is related to a thickening and hardening of the elastic and muscular arterial walls due to the infiltration and proliferation of the macrophage white blood cells and smooth muscle cells into the vascular tissue [1, 2]. The event is exacerbated by the insudation of plasma lipoproteins that transport cholesterol and triglycerides in the vascular tissue, and inadequate removal of the lipids and cholesterol, resulting in the formation of a fibrolipid plaque. As atherosclerosis has been identified as a chronic, progressive, disease with an extended asymptomatic phase, the actual cause or reason for initiation of the lesion growth is still conjecture. However, specific cellular elements have been identified during variable stages of the lesion formation, such as endothelial cells, smooth muscle cells, platelets, and leukocytes (also referred to as white blood cells, or more specifically, lymphocytes and monocytes) [3]. As the mechanism behind the interaction of these elements also remains elusive, a review of an established theory has been presented in the following text. This theory has become the hallmark for the starting point of current mathematical transport phenomena models to describe the initiation and progression of the atherosclerotic lesion. The mathematical model provides a means to investigate and predict the lumenal atherosclerotic growth patterns and concentration dispersion within the vascular wall, thereby providing relevant data and predictor statistics within the clinical setting for ongoing and future treatment. It has been suggested that the interaction of the cellular elements with the accumulation of the plaque components, such as the plasma low-density lipoprotein (LDL), within the arterial wall promote the atherosclerotic growth. Thus, the represented transport models, as classified by Prosi et al., depict transport mechanism theories, with focus on interactions of the species LDL, within the arterial wall [4, 5].

II. Research in the Area

A. *Theories on Atherosclerosis Initiation and Formation*

Theories regarding the initiation and proliferation of atherosclerosis stem from experimental and natural animal models and scientific review of the biochemical processes. One accepted theory regarding the initiation process of atherosclerosis includes the response-to-injury hypothesis as proposed by Ross [1, 6]. A drive to understand the proposed theory has led to multiple predictive efforts through mathematical models, thus an understanding of the proposed mechanisms is essential for progression on the topic.

It is known that the diseased vessel state of atherosclerosis is predominantly apparent in elastic and muscular arteries, such as the medium-sized coronaries which crown the heart [7]. As a review, the coronary arterial vascular wall is comprised of three main concentric layers, which are the endothelial cell (EC) lined *intima*, the smooth muscle cell (SMC) laden *media*, and the extracellular matrix (ECM) comprised *adventitia*, as shown in Fig. 1 [7]. The medial layer is enclosed by dense elastic membranes, namely the fenestrated intima-media *internal elastic lamina* (IEL) and a media-adventitia *external elastic lamina* (EEL).

Blood oxygen and nutrients are supplied to the media SMCs by two routes: (1) inner third by way of direct diffusion from the vessel lumen through the IEL and (2) outer two-thirds via an arteriole network, termed *vasa vasorum*, in the adventitia through the EEL [7]. However, prior to diffusion past the IEL, all species must first cross the monolayer of endothelial cells lining the vascular wall. The endothelium

S. Vlad and R.V. Ciupa (Eds.): MEDITECH 2011, IFMBE Proceedings 36, pp. 338–343, 2011.
www.springerlink.com

exercises activities critical for normal vessel homeostasis by metabolizing hormones, regulating inflammation, maintaining a nonthrombogenic blood-tissue interface, and affecting growth of other cell types, particularly SMCs [7]. Yet, tight junctions apparent between ECs can gap under hemodynamic influences (e.g., low wall shear stress or high oscillation zones [8, 9]) and/or vasoactive agents (e.g., histamine, cytokines, vasodilator enzymes [7, 9, 10]) allowing infiltration of proteins (normal conditions) or leukocytes (inflammatory conditions).

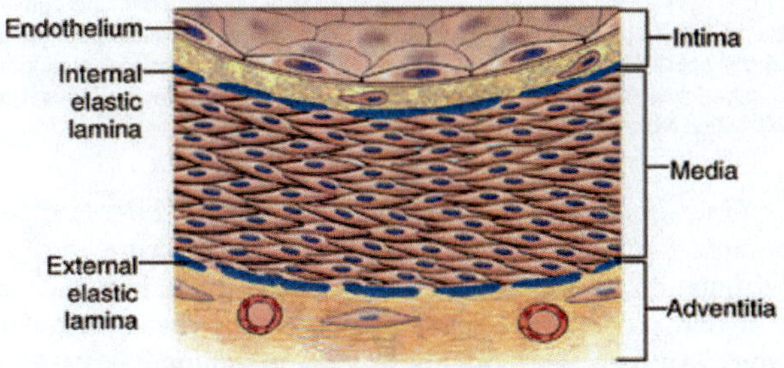

Fig. 1 Representation of distinct layers of normal coronary artery vasculature. Reproduced with permission from [7]

Ross [6, 7] proposed an inflammatory response-to-injury hypothesis which stated that the initial fatty streak formation of atherosclerosis, a condition present in persons of all ages including infants, was due to an endothelial dysfunction as caused by elevated levels of modified or potentially oxidized LDL (ox-LDL), leading to a large infiltration of monocyte-derived macrophages and T-lymphocytes. Furthermore, additional macrophage recruitment occurs via a release of vascoactive molecules, cytokines, and growth factors from the endothelial cells lining the progressively injured site. Over time, an inefficiency in removal of the ox-LDL particles results in a remodeling (intimal dilation) or restructuring (latent fibrous cap formation) via SMC migration, rendering the vessel as a fibrolipid plaque [7]. The fatty plaque is due to the macrophage and SMC metabolism/degradation of the injurious agent, ox-LDL, resulting in the visibly lipid-filled foam cell. The continued intravascular auto-amplification of the inflammatory event (macrophage and T-lymphocyte cell recruitment leading to a release of hydrolytic enzymes, cytokines, chemokines, and growth factors) eventually leads to further interstitial damage and formation of focal necrosis beneath the fibrous cap, at which point the injury site is labeled as an advanced, complicated lesion [7].

Auxiliary theories have been developed regarding the initiation, formation, and progression of the atherosclerotic lesion. However, for the purposes of this paper, only a portion of the theory under discussion has been touched upon in order to explain the biological relevance behind the mathematical model under focus. Further discussion of additional theories and the related models will be discussed in later reviews.

B. Mathematical Models for Transport Phenomena

Mathematical models for transport phenomena are described through structural (vascular wall) and fluid (blood) interactions. The mathematical model provides a means to investigate and predict projected atherosclerotic growth patterns across the arterial surface and LDL concentration dispersion and interaction with surrounding cellular elements within the vascular wall, thereby providing relevant data and predictor statistics within the clinical setting for ongoing and future treatment. The classified fluid transport models for describing LDL transport into the vessel wall include the wall-free model, the fluid-wall model, and the multi-layer model [4]. Within all respective models, blood flow through the arterial lumen is described by the incompressible Navier-Stokes equation for Newtonian fluids (1) and the continuity equation (2);

$$\rho \frac{D}{Dt}\mathbf{u}_l = -\nabla p_l + \mu \nabla^2 \mathbf{u}_l + \rho \boldsymbol{f} \tag{1}$$

$$\nabla \cdot \mathbf{u}_l = 0 \tag{2}$$

where $\mathbf{u}_l$ is the velocity vector field in the artery lumen, p_l is the blood pressure in the lumen, $\boldsymbol{f}$ is the volume force vector (to replace gravitational force), ρ and μ are the density and viscosity of the blood, respectively, and ∇ denotes the del (also referred to as nabla) operator [11, 12].

Table 1 Notations for Transport Equations

Notation	Description
$\mathbf{u}_l$, $\mathbf{u}_w$	Blood velocity in the lumen and filtration (transmural) velocity in wall
p_l	Blood pressure
c_l , c_w	Solute concentration in the lumen and in the wall layer
μ, ν, ρ	Dynamic, kinematic viscosity and blood density
μ_p	Plasma viscosity
μ_e	Effective dynamic viscosity (Brinkman's modified viscosity)
D_w	Solute (macromolecule) diffusivity in the wall
α_w	Wall permeability factor
τ_w	Shear stress vector at the lumen wall
q_w	Diffusive flux at the wall
r	Outward pointing radial vector
β	Constant, proportionality factor
K	Mass transfer coefficient of solute
K_{sl}	Solute lag coefficient
κ	Permeability coefficient of solute
K_D	Darcian permeability of the porous medium
R	Reaction term for mass transport of species concentration (conditional)

For the sake of clarity throughout the remainder of the text, the notations for the following represented equations are indicated in Table 1.

Wall-Free Model: The simplest model is referred to as the wall-free model (WFM) as it incorporates the arterial wall solely by simplified boundary conditions [5, 13]. Through utilization of Fick's law for binary diffusion, the solute transport, e.g. oxygen, albumin, ATP, or LDL, can be measured by a diffusion-advection equation which incorporates the resistance of the arterial wall to transmural transport through use of a constant or shear-dependent permeability term [5, 11, 14]. The transmural diffusive flux q_w at the lumen wall boundary then is described as either a constant where α_w is the constant wall permeability factor and c_w is the wall concentration (3), or with further detail for the permeability factor as a linearly dependent variable on the local wall-shear stress magnitude $|\tau_w|$, and β is a constant (4) [5, 14]:

$$q_w = -D_w \frac{\partial c_w}{\partial r} = \alpha_w c_w \tag{3}$$

$$\alpha_w c_w = f(|\tau_w|)c_w = \beta|\tau_w|c_w \tag{4}$$

An alternative WFM boundary condition proposed by Wada and Karino (1999) and Ethier (2002) incorporates the blood-side solute concentration at the wall surface and an endothelial permeability parameter [15, 16]. In terms of mass conservation, the boundary condition states that the surface concentration (via concentration polarization) is provided through a difference calculation using the amount carried to the arterial by a filtration flow and the amount which returns via diffusion into the mainstream flow:

$$c_w u_w - D_w \frac{\partial c_w}{\partial r} = K c_w \tag{5}$$

where u_w is the filtration velocity at the arterial vessel wall (transmural velocity), illustrated in Fig. 2 as V_w, and K is the net uptake (mass transfer) coefficient [5, 15, 16]. Under different studies, the mass transfer coefficient has been taken as a constant, for example as the permeability coefficient of the solute (κ) [15], or specified to depend on quantities such as wall shear rate, as mentioned in equation [14]. Furthermore, the concentration at the wall can further be improved upon by taking the difference of the bulk concentration in the lumen and the wall concentration [5].

The WFM provides a computationally expedient method to gain qualitative information regarding the mass transfer through the arterial lumen wall. However, this model has gross limitations for computation and understanding of the actual concentration profiles throughout the layer. Thus, in order to provide a more realistic output, the fluid-wall model has been developed.

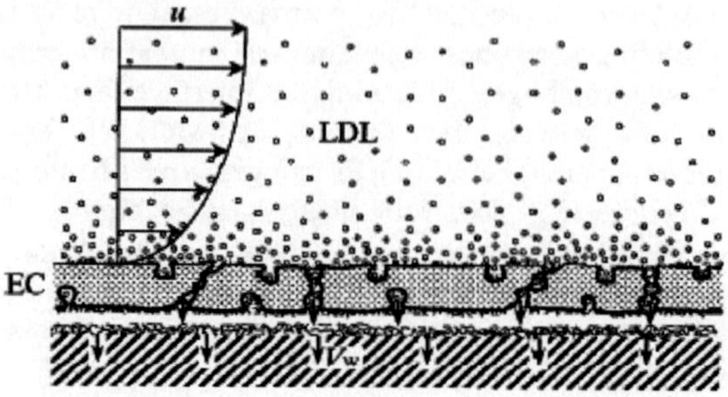

Fig. 2 A schematic of the transmural fluid flux via diffusive and convective transport of low density lipoprotein (LDL) from lumen through the semipermeable endothelial cell wall (EC) [15, 16]. The transmural velocity is denoted as V_w and u is the lumenal velocity vector. Image taken from Wada and Karino (1999) [15].

Fluid-Wall Model: The fluid-wall model (FWM) incorporates the endothelium, intima, IEL and media into the formulation as one single porous layer with homogenous transport properties. The simplification improves the qualitative results of the model at low computational expense. In order to accomplish this task, principle transport equations regarding the velocity, flux, and concentration of the transport species (hence referred to as the solute) and/or the transport carrier fluid (also referred to as the solvent), are applicable.

The velocity in the porous wall region is solved by way of Darcy's law (6) or the Darcy-Brinkman equation (7), which is a generalization of Darcy's law that facilitates matching the lumen/permeable medium interface boundary conditions, that namely being the continuity of the fluid velocity and the shear stress, through use of a volume averaging technique for the transmural velocity,

$$u_w - \nabla \cdot \left(\frac{K_D}{\mu_p} p_w\right) = 0 \tag{6}$$

$$u_w = -\frac{K_D}{\mu}[\nabla p_w - \mu_e \nabla^2 u_w] \tag{7}$$

where K_D is the Darcian permeability of the porous medium, p_w is the pressure in the arterial wall, μ_p is the plasma viscosity, and μ_e is the effective dynamic viscosity (Brinkman term) as related to the porosity and tortuosity of the medium [17] [18, 19]. Assuming constant diffusivity, the concentration field is computed with a mass transport equation (8),

$$\frac{\partial c_w}{\partial t} + u_w \cdot \nabla c_w = D_w \nabla^2 c_w + R \tag{8}$$

and may include a reaction term, R, to incorporate chemical dynamics for advection-diffusion-reaction properties [5, 20]. The boundary conditions for the lumen side mass transfer are satisfied similarly to the WFM by assuming the convection-diffusion difference calculation as shown in equation (5).

Alternatively, the lumen-wall transport can be modeled through use of Darcy's law, in equation (6), and the mass transport in the arterial wall can be coupled with the transmural flow by the following convection-diffusion reaction equation (9):

$$\nabla \cdot (-D_w \nabla c_w + K_{sl} c_w \mathrm{u}_w) = R_w c_w \quad (9)$$

which incorporates the solute lag coefficient K_{sl}, and the consumption rate constant, R_w [5, 21]. In this instance, transport processes in the arterial wall can be coupled with the blood flow through the use of Kedem-Katchalsky (*K-K*) equations thereby incorporating the solvent (10) and solute (11) flux:

$$J_v = L_p(\Delta p - \sigma_d \Delta\pi) \quad (10)$$

$$J_s = \mathcal{P}\Delta c + (1 - \sigma_f) J_v \bar{c} \quad (11)$$

where J_v is the transmural velocity (solvent flux), J_s is the solute flux (mass flux of chemical per unit surface), L_p and $\mathcal{P}$ are the hydraulic conductivity and permeability of the membrane, Δp is the pressure across the membrane, Δc is the solute concentration difference across the membrane, $\bar{c}$ is the mean concentration inside the membrane, and $\Delta\pi$ is the osmotic pressure differential [5, 18]. The selective permeability of the membrane is accounted for with the osmotic reflection coefficient, σ_d (also known as the solvent drag sieving coefficient, where $\sigma_d = 1 - s_d$), and the frictional reflection coefficient, σ_f (also known as the Staverman filtration coefficient, where $\sigma_f = 1 - s_f$). One example of further enhancement through utilization of these parameter values includes the use of a shear-dependent hydraulic conductivity variable, as used by Sun (2006) to estimate LDL and oxygen transport [21].

Provided the computational expense is low, this model is widely proclaimed for its benefits within predictive model schemes. However, the lack of internal transport distinction between the arterial layers, brings potential error into the predictive value through lack of crucial components involved in the formation of atherosclerosis. Therefore, in order to provide the most realistic approach to the transport properties, the multi-layer model was developed.

Multi-Layer Model: The multi-layer model (MLM), developed by [22], provides the most developed and realistic mathematical composition of the arterial wall, taking into account the four layers of the lumen, intima, media, and adventitia. The separating membranes consist of the endothelium, IEL, and EEL. Similar to the FWM, the MLM utilizes the volume averaged stationary convection-diffusion equation with a reaction term, as shown in equation (9), to depict the chemical dynamics of the metabolic processes occurring for mass transport across the intima and media. The filtration velocity within the intima and media are solvable with application of Darcy's law, as shown in equation (6). The transport equations for the lumen, intima, and media can be coupled through use of the *K-K* equations, which incorporate the solute and solvent flux across each membrane, endothelium, and IEL individually, as shown in Fig. 3.

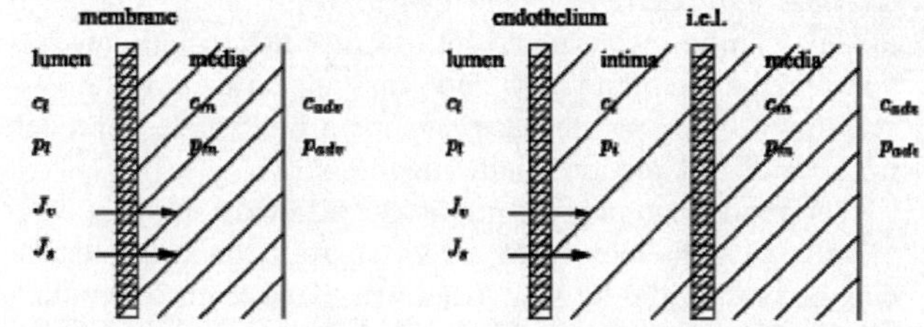

Fig. 3 Figurative approximation of arterial wall layers with separating membranes for the FWM (left) and MLM (left). The flux applied to the membrane couples the corresponding layers. Image taken from Prosi (2005) [4].

Additional Models: Additional models are available which take a single layer of the wall into account, such as the media [23]. Within this context of this FWM, the SMC's are modeled as solid structures within the porous arterial layer. This model shows how the IEL permeability affects the species distribution of ATP and LDL within the arterial layer. Further model development of the individual layers is one move towards a more clear definition of the vascular response, which may provide better definitions when incorporating the layers into MLM's.

III. BIOCHEMICAL MODEL IMPROVEMENTS

The interaction of the fluid and wall structure can be further expanded upon in computational fluid dynamics (CFD) models through combination of the chosen transport equations for boundary conditions with biochemical models that yield the potential of differential calculus equations. The differential calculus model engages ordinary (ODE) and partial differential (PDE) equations for the mathematical model formulation, which enables quantitative estimation of the global behavior of the system and provides a potential to model lipid accrual in and removal from the subendothelial (intimal) layer of the vessel wall (as proposed by Cobbold et al.) or further depict locations of plaque aggregation (as proposed by Calvez et al.) [24-26]. For example, the interactions of high density lipoprotein (HDL) and vitamin C within the oxidation process were described via an ODE model within [24], in order to depict the protective properties of HDL against LDL oxidation by effectively acting as a sponge to free radicals until complete vitamin E removal from the LDL. Additionally, the plaque aggregation model [26] utilized transversal averaged values within PDE's in order to depict the concentration or density evolution of each species of interest (namely immune cells, LDL, ox-LDL, foam cells, cytokines, extracellular matrix).

The biochemical models incorporate chemical and cellular species for description of the coronary event from a microscale approach. The data gained from the biochemical models can be incorporated into further macroscale models that include the blood flow and gross appearance of the vessel, thus leading to the necessity for a multiscale approach for a rational depiction of atherogenesis.

The model can provide predictive information of the vessel plaque vulnerability to increased wall thickness and/or change in the fluid flow environment. Further understanding of the behavior of this disease via numerical validation brings about the potential to improve treatment and outcome for patients at risk or diagnosed with atherosclerosis. The integrative dynamic environment of vascular modeling through incorporation of biochemically relevant mathematical models into CFD models promotes vast potential for expansion on atherosclerosis research.

IV. Conclusions

The substantial amount of evidence that mass transport plays a substantial role in the formation of atherosclerosis has led to the development of a variety of characteristic mathematical models. For the works presented, the transport of specific species such as oxygen, albumin, ATP, and LDL were modeled. The evaluation of these components allows further understanding for description of the formation and progression of the cardiovascular disease atherosclerosis.

Acknowledgment

Financial support provided by the European Commission through MeDDiCA Marie Curie Initial Training Network (www.meddica.eu, EU-FP7/2007-2013 under grant agreement PITN-GA-2009-238113) is gratefully acknowledged.

References

1. Ross, R. (1993) Atherosclerosis - A defense mechanism gone awry. Am. J. Pathol. 143 (4): 987-1002
2. Benditt, E.P. and J.M. Benditt (1973) Evidence for a Monoclonal Origin of Human Atherosclerotic Plaques. Proc. Natl. Acad. Sci. 70 (6): 1753-1756
3. Atherosclerosis: eMedicine Cardiology at http://emedicine.medscape.com/article/150916-overview; Last Updated: Dec. 2, 2010
4. Prosi, M., et al. (2005) Mathematical and numerical models for transfer of low-density lipoproteins through the arterial walls: a new methodology for the model set up with applications to the study of disturbed lumenal flow. J. Biomech. 38 (4): 903-917
5. Khanafer, K. and K. Vafai, Macromolecular Transport in Arterial Walls: Current and Future Directions, in Emerging Topics in Heat and Mass Transfer in Porous Media. P. Vadász, Editor. 2008, Springer Netherlands. p. 219-235.
6. Ross, R. (1982) The Gordon Wilson Lecture: Atherosclerosis--a response to injury gone awry. Trans. Am. Clin. Climatol. Assoc. 93: 78-86
7. Kumar, V., et al. (2007) Robbins Basic Pathology 8ed. Saunders Elsevier,
8. Himburg, H.A., et al. (2004) Spatial comparison between wall shear stress measures and porcine arterial endothelial permeability. American Journal of Physiology - Heart and Circulatory Physiology 286 (5): H1916-H1922
9. Chatzizisis, Y.S., et al. (2007) Role of endothelial shear stress in the natural history of coronary atherosclerosis and vascular remodeling: Molecular, cellular, and vascular behavior. J. Am. Coll. Cardiol. 49 (25): 2379-2393
10. Osterbud, B. and E. Bjorklid (2003) Role of Monocytes in Atherogenesis. Physiol. Rev. 83 (4): 1069-1112
11. Bird, R.B., W.E. Stewart, and E.N. Lightfoot (2007) Transport Phenomena 2ed. John Wiley & Sons, New York
12. Olgac, U., et al. (2009) Patient-specific three-dimensional simulation of LDL accumulation in a human left coronary artery in its healthy and atherosclerotic states. Am. J. Physiol. Heart Circ. Physiol. 296 (6): H1969-H1982
13. Zunino, P. (2002) Mathematical and Numerical Modeling of Mass Transfer in the Vascular System. Thesis, Ecole Polythéchnique Fédérale de Lausanne
14. Rappitsch, G., K. Perktold, and E. Pernkopf (1997) Numerical modelling of shear-dependent mass transfer in large arteries. Int. J. Numer. Methods Fluids 25 (7): 847-857
15. Wada, S. and T. Karino (1999) Theoretical study on flow-dependent concentration polarization of low density lipoproteins at the luminal surface of a straight artery. Biorheol. 36 (3): 207-223
16. Ethier, C.R. (2002) Computational Modeling of mass transfer and links to atherosclerosis. Ann. Biomed. Eng. 30 (4): 461-471
17. Martys, N.S. and J.G. Hagedorn (2002) Multiscale modeling of fluid transport in heterogeneous materials using discrete Boltzmann methods. Mater. Struct. 35: 649-650
18. Formaggia, L., A. Quarteroni, and A. Veneziani (2009) Cardiovascular Mathematics: Modeling and simulation of the circulatory system. Modeling, Simulation & Applications. Vol. 1. Springer-Verlag Italia, Milano
19. Khaled, A.R.A. and K. Vafai (2003) The role of porous media in modeling flow and heat transfer in biological tissues. Int. J. Heat Mass Transfer 46 (26): 4989-5003
20. Yang, N. and K. Vafai (2006) Modeling of low-density lipoprotein (LDL) transport in the artery-effects of hypertension. Int. J. Heat Mass Transfer 49: 850-867
21. Sun, N., et al. (2006) Fluid-Wall Modelling of Mass Transfer in an Axisymmetric Stenosis: Effects of Shear-Dependent Transport Properties. Ann. Biomed. Eng. 34 (7): 1119-1128
22. Karner, G. and K. Perktold (2000) Effect of endothelial injury and increased blood pressure on albumin accumulation in the arterial wall: a numerical study. J. Biomech. 33 (6): 709-715

23. Tada, S. and J.M. Tarbell (2004) Internal elastic lamina affects the distribution of macromolecules in the arterial wall: a computational study. American Journal of Physiology - Heart and Circulatory Physiology 287 (2): H905-H913
24. Cobbold, C., J. Sherratt, and S. Maxwell (2002) Lipoprotein oxidation and its significance for atherosclerosis: A mathematical approach. Bull. Math. Biol. 64 (1): 65-95
25. Pappalardo, F., et al. (2008) Computational simulations of the immune system for personalized medicine: State of the art and challenges. Current Pharmacogenomics and Personalized Medicine 6 (4): 260-271
26. Calvez, V., et al. (2009) Mathematical modelling of the atherosclerotic plaque formation. ESAIM: Proceedings 28: 1-12

Author: Brandis Keller
Institute: Politencio di Milano
Street: Piazza Leonardo da Vinci, 32
City: Milano
Country: Italy
Email: keller@stru.polimi.it

Low-Field Nuclear Magnetic Resonance Relaxometry – A Tool in Monitoring the Melting Transition of Polymeric Capsules with Applications in Drug Delivery

R.E. Nechifor and I. Ardelean

Department of Physics, Technical University from Cluj-Napoca, Romania

Abstract— **Polymeric micro and nanocapsules are often used as drug delivery systems. Their main advantage is that the active substances can be placed inside the capsule and released at the position of the diseased tissue or organ. The release can be performed gradually, owing to a continuous exchange process, eventually influenced by temperature, or suddenly as a consequence of melting the capsule wall. In the last case the determination of the melting transition temperature of the fabricated capsules is an important issue. In the present contribution we are showing that low field nuclear magnetic resonance (NMR) relaxometry can be used as a valuable tool in monitoring the melting transition of polymeric capsules prepared by an interfacial polymerization technique. The NMR relaxation experiments were performed at a proton resonance frequency of 20 MHz. The data were analyzed using a numerical Laplace inversion algorithm that allowed us the determination of the melting point for the fabricated capsules.**

Keywords— **NMR relaxation, polymeric capsules, drug delivery, core – shell colloidal particles, melting transition.**

I. Introduction

Targeted drug delivery is one of the most important issues in nowadays medical therapy. Consequently, to transport the right dose of drug to the diseased tissues suitable carriers are necessary. Such drug carriers may be represented by liposomes [1], cyclodextrins [2] or other micro and nanoparticles [3]. The micro and nanoparticles may be in the form of spheres or capsules and are made of a macromolecular material [3]. The structure of micro and nanospheres is of a matrix-type with the drug adsorbed on the surface or dispersed inside the matrix [3]. The structure of micro and nanocapsules is of a core-shell type with an oily core containing the drug which is surrounded by a polymeric membrane [4-7]. The main advantage is that various active substances such as drugs [3] or even proteins [7] can be transported directly to the diseased tissue, completely bypassing the healthy ones. Thus, using polymeric capsules it is possible to achieve a shielding effect of the physicochemical properties necessary to be delivered.

Nuclear magnetic resonance (NMR) is a completely noninvasive technique and was successfully used to investigate both the structure and the dynamics of molecules in different states of aggregation [8]. The best known application of NMR technique is in medical imaging, where it has proven to be a very important tool in diagnostic. The NMR techniques are also very useful in material science, biology, chemistry and physics. They allow revealing both the structure of molecules and their dynamics. If the structure of molecules is usually investigated in the frame of the so-called high field NMR spectroscopy, a technique which requires expensive superconducting magnets, the dynamics can be studied in low fields using much more inexpensive equipments. Such low field equipments can be even mobile [9] and still able to reveal important information on the dynamics of molecules. The experimental parameter that is mostly measured in a low field experiment is the transverse relaxation time T_2. This parameter characterizes the attenuation of the transverse magnetization component during a specific NMR experiment in the presence of molecular motion. Thus, it provides information on the translational or rotational motion of the tagged molecules, the viscosity of their environment, geometrical restrictions as well as the influence of different parameters (magnetic impurities in the sample, temperature, etc.).

In the present work it will be shown that the low field transverse relaxation measurements can be implemented to monitor the melting transition of biodegradable polymeric capsules prepared by interfacial polymerization of ethyl – 2 – cyanoacrylate (ECA) monomers in an oil-in-water emulsion. The technique allows us both detecting of the melting transition temperature and the observation of the changes that take place during the melting process.

II. Experimental

A. *Sample Preparation*

Two main methods can be used for preparation of biodegradable nanocapsules: interfacial polymerization of dispersed alkylcyanoacrylate (ACA) monomers [10] and interfacial deposition of preformed polymers [3], such as poly (D,L-glycolide), poly (ε - caprolactone), poly (D, L-lactide) and poly (lactide-coglycolide). In the present work the polymeric capsules were produced using an interfacial.

S. Vlad and R.V. Ciupa (Eds.): MEDITECH 2011, IFMBE Proceedings 36, pp. 344–347, 2011.
www.springerlink.com

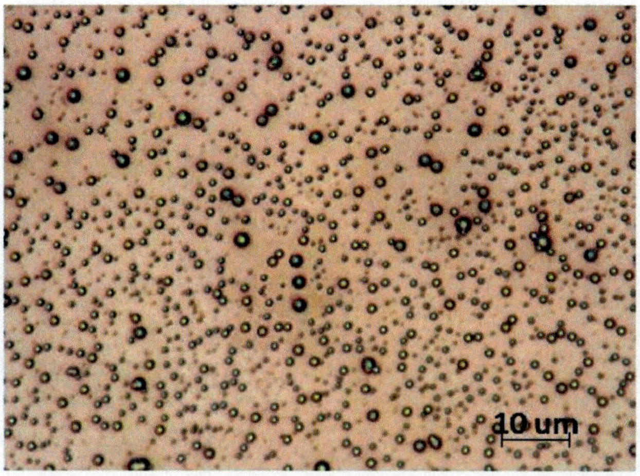

Fig. 1 The optical microscopy image of the produced polymeric capsules

polymerization technique [5-7,10]. The method involves the initial preparation of two phases: an organic phase and an aqueous phase. The organic phase is a mixture of organic solvent (acetone), oil (Miglyol 812 provided by Sasol Gmbh, Germany) and monomer (ethyl – 2 – cyanoacrylate). The aqueous phase contains a mixture of surfactant (Pluronic F 68, provided by Sigma-Aldrich, Germany) and distilled water. The mixture of organic phase was slowly injected through a needle into a magnetically stirred aqueous phase. This mixture immediately became milky and nanocapsules with different diameters were formed. The diluted emulsion was rotary evaporated at ambient temperature to remove the last remainders of solvents and then lyophilized. A dry powder was obtained with no detectable liquid outside. The microscopy image of the produced polymeric capsules was obtained with a Zeiss Axio Lab.A1 MAT optical microscope and is shown in Figure 1. One can observe from this image, a distribution of capsule diameters, some of them bellow optical resolution.

B. *The NMR Relaxation Technique*

A very useful technique for transverse relaxation measurements has been proven to be the Carr – Purcell – Meiboom – Gill (CPMG) technique [11]. The radiofrequency pulse sequence and the echoes train recorded during such an experiment are indicated in Figure 2. The main advantage of the CPMG technique as compared with other techniques which rely on single echo detection is that it allows the elimination of the diffusion effects on echoes decay provided that short enough echo time intervals are implemented. Moreover, such a multiple echo technique is much faster compared to single echo techniques. Consequently, it allows multiple accumulations of the echoes train signal which is an important advantage in low field experiments where the detection sensitivity is strongly reduced relative to the high field experiments.

If the recorded CPMG envelope shown in Figure 2 exhibits mono exponential decay and diffusion effects can be neglected during the time intervals between two radiofrequency pulses, the transverse relaxation time T_2 of the sample can be extracted by fitting the data with the formula:

$$A_n = A_0 e^{-\frac{2n\tau}{T_2}}. \qquad (1)$$

Here A_n is the amplitude of the n-th echo in the echo train and A_0 is a constant that depends on sample magnetization and other parameters of the NMR instrument. The transverse relaxation time T_2 provides information on the translational or rotational motion of the tagged molecules (viscosity, geometrical restrictions) as well as about the content of relaxation centers (paramagnetic impurities) inside the sample.

If the relaxation time varies throughout the sample as result of sample heterogeneity, a multi–exponential attenuation of the CPMG envelope is expected. Assuming a continuous distribution of the relaxation times inside the sample, the amplitude of the n-th echo in the echo train satisfies the formula:

$$A_n(2n\tau) = A_0 \int_0^{\infty} P(T_2) \exp(-2n\tau / T_2)\, dT_2, \qquad (2)$$

where $P(T_2)$ is the relaxation time probability density. The above formula suggests that the analysis of the experimental data using a Laplace inversion algorithm should provide us the relaxation time distribution. It was shown that a reliable numerical algorithm for such an analysis is the CONTIN algorithm [12]. Using such an approach we can estimate the distribution of the relaxation times at different temperatures, and consequently we can monitor the melting transition of the produced capsules.

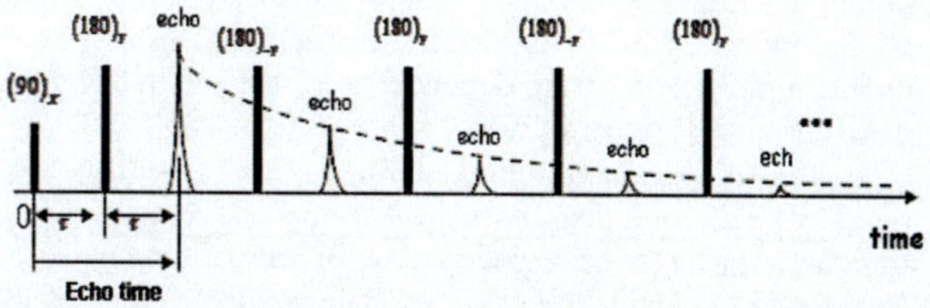

Fig. 2 The CPMG pulse sequence implemented in our relaxation experiments

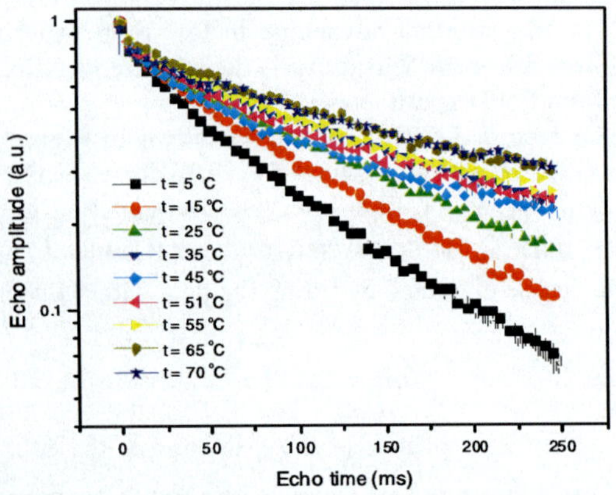

Fig. 3 The echo decay curves obtained at different temperatures in a CMPG experiment

C. Results

The transverse relaxation measurements were performed on a Bruker MINISPEC MQ20 time domain analyzer operating at a proton resonance frequency of 20 MHz. The data were recorded using the standard CPMG [11] pulse sequence as indicated in Figure 2. The length of one hard 90° pulse was set to 2.5 μs and that of a 180° to 5 μs. The echo time interval was chosen as 100 μs in order to prevent the appearance of diffusion effects on the echo decay. A number of 2500 echoes were recorded in each experiment. The repetition delay of one echo train was selected to 5 s, long enough to assure a full recovery of the longitudinal magnetization.

Figure 3 shows the decay curves of the echo train recorded at different temperatures, between 5°C and 70°C. As can be observed, the curves are non exponential indicating a heterogeneous distribution of relaxation times. The non-exponential distribution becomes more obvious at higher temperatures, above the melting transition temperature of 41°C. The shape of the echo decay curves indicates that the experimental data are better described by Eq. (2) with a relaxation time distribution which can be extracted using a numerical Laplace transform. The results of such an analysis based on CONTIN algorithm are depicted in Figure 4. One can observe a strong dependence of the relaxation time distribution on temperature.

In the temperature interval bellow 41°C (melting temperature) only one peak is present in the observed distribution and that can be associated with the presence of oil (Miglyol 812) inside the capsules. The smaller values of the relaxation time as compared with the bulk values are a result of the confinement effect. The relatively broad distribution of relaxation times corresponds to a distribution of capsules diameters as observed in Figure 1. The position of the peak maximum versus temperature is represented in Figure 5 (squares). One can observe an increase in the relaxation time with the temperature. A similar dependence is also observed in the case of bulk Miglyol (triangles) but with higher values of the relaxation time. This increase with the rising temperature can be associated with a decrease of the viscosity of the confined oil.

In the temperature interval above 41°C the relaxation time distribution splits into two components as can be directly observed in Figure 4. The appearance of the second component may be associated with the melting of the polymeric shell (poly ethyl – 2 – cyanoacrylate) of the fabricated capsules. The bimodal distribution can be observed from 41°C to 51°C. Above 51°C only a broad distribution of relaxation times is visible with an average relaxation time increasing with temperature. This behavior is observed since above 41°C the polymer shell starts to melt and as a result a distribution of two relaxation times, one from Miglyol and another from the polymer melt, will be present in the sample. This mixture becomes homogeneous at a temperature above 51°C when all capsule walls are melted and a unimodal distribution of relaxation times is observed.

In the beginning of the melting process the melted mixture contains parts from the partially melted walls (polymers) which act as relaxation centers for the surrounding Miglyol. Consequently the observed relaxation time reduces at the beginning of the melting process as indicated in Figure 5.

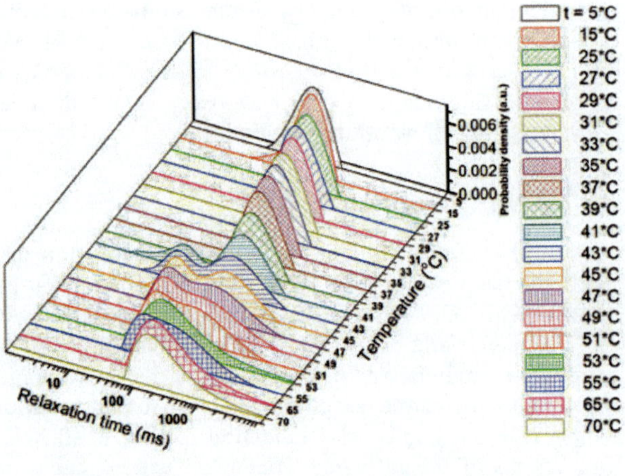

Fig. 4 Relaxation time distribution at different temperatures

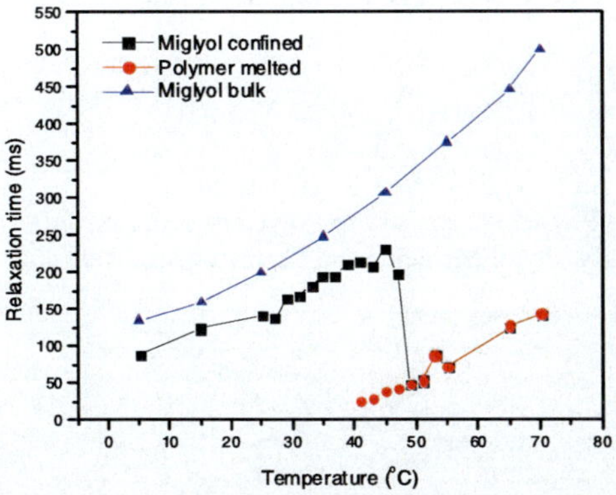

Fig. 5 The evolution of the confined Miglyol (squares), bulk Miglyol (triangles) and the melted mixture (circles) as a function of temperature. The circles correspond to the left peak in relaxation time distribution.

After the polymer shell is completely melted the average relaxation time of the mixture increases as a result of increasing mobility of molecules (circles). Furthermore only one component in the relaxation time distribution starts to be observed by increasing the temperature indicating a better homogeneity of the sample.

III. Conclusions

Delivering drugs as efficient as possible and without any risk is extremely important in any type of therapy, especially in cancer therapy when "the drug" may be a radioactive substance. That is why, numerous preparation approaches were developed and are available nowadays for producing of polymeric micro and nanocapsules with applications in controlled drug delivery. Designing of specific capsules that can melt at a given temperature and thus release the drug at a specific location (obtained by heating) requires new techniques for the characterization of their melting transition. In the present work we have shown that the monitoring of the polymeric capsules with respect to their melting transition can be done using low field nuclear magnetic resonance relaxometry. Thus it was possible to measure the melting temperature of the capsules shell and to monitor the heterogeneity of the sample during the melting process. The results reported here may have implications for the people working in producing and designing of new micro and nanocarriers for drug delivery.

Acknowledgment

Financial support by the European Social Fund (project POSDRU/6/1.5/S/5) is gratefully acknowledged.

References

1. Gregoriadis G. and Florence A. T., (1993) Liposomes in drug delivery: Clinical, diagnostic and ophthalmic potential, Drugs 45:15 – 28.
2. Challa R., Ahuja A., Ali J., and Khar R.K., (2005) Cyclodextrins in Drug Delivery: An Updated Review, AAPS PharmSciTech 6, E329 - E357
3. Kumar M.N.V.R., (2000) Nano and microparticles as controlled drug delivery devices. J. Pharm. Pharmaceut. Sci. 3: 234–258
4. Wende C., Schonhoff M., (2010) Dynamics of water in polyelectrolyte multilayers: restricted diffusion and cross-relaxation. Langmuir. 26: 8352–8357
5. Bogdan M., Nan A., Pop C.V.L., Barbu-Tudoran L., Ardelean I., (2008) Preparation and NMR characterization of polyethyl-2-cyanoacrylate nanocapsules, Appl. Magn. Reson. 34:111-119
6. Nechifor R., Bogdan M., Ardelean I., (2011) The size distribution of core shell polymeric capsules as revealed by low-field NMR diffusometry, Appl. Magn. Reson. 40:205-211.
7. Yan M., Du J., Gu Z., Liang M., Hu Y., et al. (2010) A novel intracellular protein delivery platform based on single-protein nanocapsules, Nature Nanotechnology 5:48 - 53.
8. Hornak J. P., (2008) The basics of NMR, at www.cis.rit.edu/htbooks/nmr/
9. Blümich B., Perlo J., Casanova F., Mobile single-sided NMR. Prog. NMR Spectrosc. 52:197–269
10. Vauthier C., Dubernet C., Fattal E., Pinto-Alphandary H, Couvreur P., (2003) Poly(alkylcyanoacrylate) as biodegradable biomaterial for biomedical applications, Adv. Drug Deliv. Rev. 55:519-548.
11. Meiboom S., Gill D., (1958) Modified spin-echo method for measuring nuclear relaxation times. Rev. Sci. Instr. 29:688-91
12. Provencher S.W., (1982) CONTIN: a general purpose constrained regularization program for inverting noisy linear algebraic and integral equations, Comp. Phys. Comm. 27:229–242

Author: Ioan Ardelean
Institute: Technical University form Cluj-Napoca
Street: 2 Memorandumului St.
City: Cluj-Napoca
Country: Romania
Email: ioan.ardelean@phys.utcluj.ro

Study on Cellulose/Chondroitin Sulfate Hydrogel Used in Drug Release Systems

C.G. Coman[1], M. Al. Macsim[1], A.M. Oprea[2], L. Hurjui[3], T. Petreus[3], and A. Neamtu[3]

[1] Faculty of Medical Bioengineering, University of Medicine and Pharmacy "Gr.T.Popa" Iasi, Romania
[2] "Petru Poni" Institute of Macromolecular Chemistry, Department of Physical Chemistry of Polymers, Iasi, Romania
[3] Center for the Study and Therapy of Pain (CSTD), "Gr. T. Popa" University of Medicine and Pharmacy Iasi, Romania

Abstract— **The aim of this study was to combine the properties of cellulose and chondroitin sulfate in mixed hydrogels in order to obtain new materials for medical and pharmaceutical applications as drug delivery systems. Different swelling profiles and active drug release rates were observed for the proposed formulation compared with pure cellulose hydrogels. Along with this aim, hemotoxicity and subcutaneous implantation experiments had been performed showing a good biocompatibility of the studied hydrogels.**

Keywords— **drug delivery, codeine, cellulose, chondroitin sulfate, hydrogel.**

I. INTRODUCTION

Hydrogels are water-swollen, crosslinked polymeric networks produced by reticulation involving chemical reactions or physical interactions (ionic and hydrogen bonding, hydrophobic interaction, micellar packing). Hydrogels can absorb and retain large amounts of water[1]. Since the pioneering work of Wichterle and Lim[2] of 1960 on hydrogels based on 2-hydroxyethyl methacrylate (HEMA), the field has developed continuously because of multiple applications that hydrogels have especially in medicine and pharmacy[3,5]. The main uses of hydrogels are in the biosensors field, artificially dressing for burns[4], controlled drug release[6] and tissue regeneration due to the remarkable properties that they have[7] in mimicking the extracellular matrix of natural tissues. Their water content and elasticity are similar with natural biological interactions at the molecular level[8]. Natural polymer (e.g. polysaccharide) based hydrogels constitute a very promissing class of substances mainly due to their high biocompatibility.

Chondroitin sulfate (CS) is a glycosaminoglycan (GAG) made by repeating disaccharide units of D-glucuronic (GlcUA) (β1-3) N-acetyl-galatosamine (GalNAc). Both monosaccharides can be sulfated (in position 2 for GlacUA and in positions 4' and 6' for GalNAc) (Figure 1). The impact of this sulfation diversity is very important for molecular biology and not completely understood.

Chondroitin sulphate it is used in medical applications[9], as dietary supplements (capsules and tablets)[10], eye drops[11]. CS can be found in humans[12,13] demonstrating it's important role biological processes[14, 15].

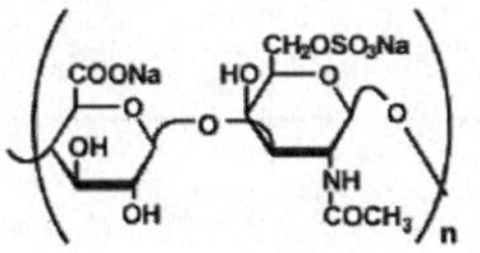

Fig. 1 The chemical structure of Chondroitin 6-sulfate

Chondroitin and Glucosamine[16] are well known for the benefic effects on joint pain, improving joint mobility, increasing and protecting the cartilage. This supplement is recommended for elderly people, athletes and active people maintaining joint health and function.

Cellulose (Figure2) is an organic compound with the formula $(C_6H_{10}O_5)_n$. Cellulose is a β-D-Glucose polymer which, in contrast to starch, is oriented with the $-CH_2OH$ groups alternating above and below the plane of the cellulose molecule thus producing long, unbranched chains.

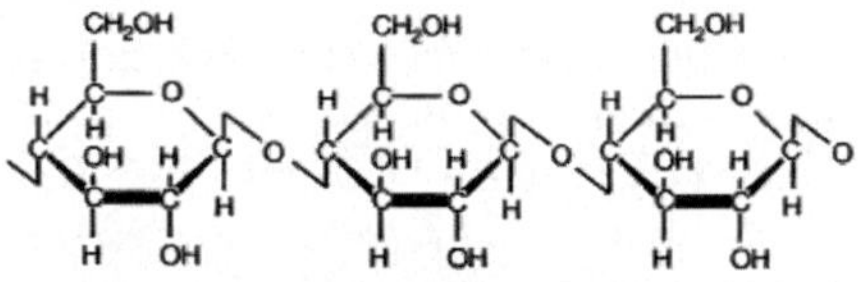

Fig. 2 Cellulose structure

Cellulose is the world's most abundant natural, renewable and biodegradable polymer. Cellulose is hygroscopic, retaining water by hydrogen bonding.

Bacterial cellulose has unique properties. Due to its high water absorbance, biocompatibility, high porosity, mechanical properties, high cristallinity and an ultra-fine and highly pure fibre network structure, bacterial cellulose has become popular in biomedical applications[17,18].

Codeine is a natural alkaloid of opium. Industrial synthesis of codeine is made by O-methylation of morphine. It has much weaker analgesic than morphine. Of all the receptors for opioids, codeine has the highest affinity for μ, subtype μ_2. Linking to subtype μ_1, somewhat weak, is responsible for analgesic effects. Codeine 6-glucuronide is metabolized in the liver by conjugation with glucuronic acid up to 80%. Codeine (methylmorphine) is well-known for the

S. Vlad and R.V. Ciupa (Eds.): MEDITECH 2011, IFMBE Proceedings 36, pp. 348–351, 2011.
www.springerlink.com

antitussive, analgesic and antidiarrheal effects, having a broad safety margin and 8%-12% of the strength of morphin.

The most recent advances in cellulose-based hydrogels aim not only at the sustained release of a bioactive molecule over a long time period, ranging from hours to weeks, but also at a space-controlled delivery, directly at the site of interest. The need to encapsulate bioactive molecules into a hydrogel matrix or other delivery devices (e.g., microspheres) is also related to the short half-life displayed by many biomolecules *in vivo*[19].

Controlled release through oral drug delivery is usually based on the strong pH variations encountered when transitioning from the stomach to the intestine. Cellulose-based polyelectrolyte hydrogels (e.g., hydrogels containing NaCMC) are particularly suitable for this application. For instance, anionic hydrogels based on carboxymethyl cellulose have been investigated recently for colon-targeted drug delivery[20].

The readily water-soluble nature of chondroitin sulfate limits its application as a solid-state drug delivery vehicle. Therefore, it is usual to carry out a crosslinking treatment to tailor the properties of chondroitin sulfate as reported in several works[21], or to combine it with other polymers, such as chitosan[22], gelatin and hyaluronan[23], collagen[24], poly(vinyl alcohol)[25] or poly-(lactic-co-glycolic acid)[26] in order to produce more stable materials.

The present study aims to investigate the properties of a newly developed hydrogel type[27] with a double network made of cellulose and CS as a potential transport matrix for biologically active substances. Along with the release evaluation of the loaded active molecule from the hydrogel matrix, also hemotoxicity tests and *in vivo* implantation studies were performed in order to characterize the biocompatibility of the formulations. The cellulose/ chondroitin sulfate hydrogels combine the biodegradation capacity, the biocompatibility, the transparency and lack of toxicity of cellulose with high water absorption, and biodegradability characteristics of chondroitin sulfate in order to obtain new biomaterials with special applications in processes that do not harm biological environment.

Up to now, after our knowledge, the combination of cellulose and chondroitin sulfate, for obtaining new hydrogels and their potential for biomedical applications, such as drug delivery, has not been yet exploited.

II. Materials and Methods

A. *Material Synthesis and Swelling*

Microcrystalline cellulose (C) Avicel PH-101 with the polymerization degree of 183 was purchased from Sigma-Aldrich. Chondroitin sulfate (CS) was purchased from Roth, Germany and obtained from bovine tracheal cartilage.

The hydrogel samples were prepared as described in Oprea et al (2009)[27]. In short a crosslinking technique was used to prepare pure cellulose (used in this study as reference composition) and mixed hydrogel matrix with a mass ratio of 70/30 C/CS. The obtained composition were purified by washing with water and dried at room temperature (this will be further refered as C/CS). The kinetics of swelling was measured by placing the sample in pH=7.4 Phosphate Buffer Solution (PBS) and measuring the weigths of samples at different time intervals·

B. *Codeine Loading*

The drug loading method was performed by immersing the hydrogels samples in a pH=7.4 PBS-codeine solution with a concentration of $2{,}35{*}10^{-2}$, and left 48 hour for their complete loading.

C. *Codeine Release*

During the drug release study, the release medium (pH=7.4 PBS solution) was maintained at 37 ± 0.5 ^{0}C. At predetermined time intervals, aliquots of the medium of 2 ml were withdrawn from the release medium and analyzed at λ_{max}=285 nm using a RAYLEIGH 1800 UV-VIS spectrophotometer (China).

Codeine concentrations were calculated based on calibration curve determined at specific maximum absorption wavelengths. In order to maintain the solution concentration the sample is reintroduced in the circuit after analysis.

D. *Percent Hemolysis Test*

The used blood was collected from healthy patients. Before testing, the hydrogels were sterilized by ultraviolet light trans-illumination for 4 minutes. Each hydrogels sample was tested with the blood collected from the same patient. Copper powder was used as positive control and low density poly-ethylene (LDPE) as negative control[28]. From each blood sample were withdrawn 0.6mL, placed in Eppendorf containers with copper powder, LDPE, cellulose-based and C/CS hydrogels and incubated at 37°C for 3 h. After the incubation time the samples were centrifuged at 4000 rpm for 10 min. The separated plasmas were diluted 11 fold with Tris (62.5 mmol/L, pH 8.0 adjusted with HCl) prior to spectrophotometrical measurements.

The method used for measuring plasma hemoglobin concentration was the polychromatic method of Noe et al.[29]. Absorbance was measured at 380nm, 415nm and 470nm and the formula used was:

$$C(mg/L) = 1.65 \times m_{A415} - 0.93 \times m_{A380} - 0.73 \times m_{A470} \quad (1)$$

where C - hemoglobin concentration (mg/L); m_{A380}, m_{A415} and m_{A470} - absorbances at 380nm, 415nm and 470nm (miliabsorbance units). The results were expressed as:

$$H(\%) = (C_s - C_n)/(C_p - C_n) \times 100 \quad (2)$$

where H – hemolysis percent, C_S - concentration of hemoglobin in the sample, C_n - concentration of hemoglobin in the negative control and C_p - concentration of hemoglobin in the positive control.

E. Subcutaneous Implantation

The *in vivo* experiments were performed on 4 groups of 6 animals each (Wistar male rats, average weight ~250g) kept under standard conditions, at 20-23 °C, with food and water *ad libitum*. The dimensions of each implant were 6x4mm (after swelling). Prior to the implantation procedure, the samples were swollen into pH=7.4 PBS solution. The implants have been sterilized with UV radiation and packed individually. Each animal was anesthetized with a mixture of ketamin (80-100 mg/kg) and xilazin (5-10mg/kg) injected intramuscular. The samples were implanted paravertebral subcutaneously by making a 1cm incision on disinfected and newly-shaved skin. The incision was sutured in 2 points with nonresorbable thread with a 3.0 needle and cleaned with ethyl alcohol. The implants were collected after 30 days, the examination being performed with an Olympus BX40 microscope and a digital camera. The implants sections were examined for the presence of inflammatory cells, fibrosis, and abnormal cell morphology.

III. Results and discussions

A. Swelling Study

The swollen weights at equilibrium in PBS at 37°C compared with the dryed samples were 183% for C and 317% for C/CS. This fact motivates the choice of the C/CS hydrogel over the C. The difference is due to the negative charges in the CS structure ($-OSO_3^-$ and $-COO^-$, 2 charges/disaccharide repeating unit) which creates affinity for the codeine binding. Codeine molecules link to the CS matrix, but in low concentrations that can be neglected in comparison with the swollen solution.

B. Drug Release Profiles

From the release profiles depicted in Figure 3 it can be observed a decrease of the released amount of Codeine by adding CS in hydrogel composition. Also the rate of the loaded substance release is higher in the Cellulose hydrogel. This can be explained by the fact that the addition of the CS to the cellulose matrix draws it into a polyelectrolyte hydrogel.

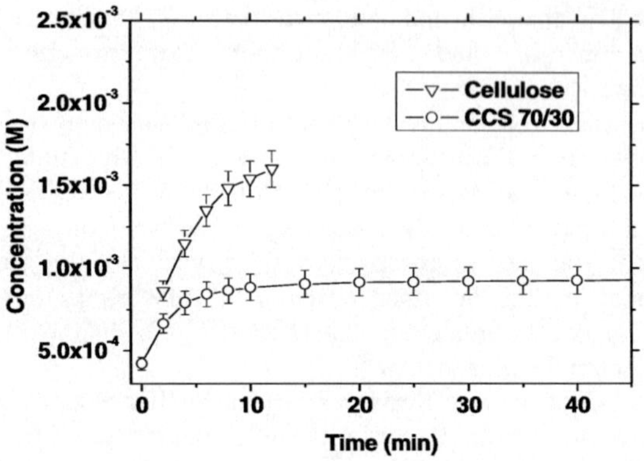

Fig. 3 Codeine release profiles from the Cellulose and C/CS hydrogel matrice

Negative charges are present into the matrix from the carboxylic $-COO^-$ group of the glucuronic acid and from the sulfate moiety $-O-SO_3^-$ of the N-acetylgalactosamine residues. In solution Codeine Phosphate is dissociated, the codeine bearing a net positive charge which triggers the ionic interaction between the drug and the polymer matrix, increasing in this way the retention of the drug inside the matrix.

C. Hemolysis Test

The hemolysis percent for the C/CS hydrogel sample was Hc/cs= -5,79%. The values reflect a high biocompatibility, the negative value obtained suggesting that the variations in the levels of free hemoglobin due to the contact with the material are lower in magnitude than the experimental errors.

D. Subcutaneous Implantation

The implants were easily absorbed without side-effects and proved the theoretical hypothesis of biocompatibility and effectiveness of the C/CS hydrogel.

Figure 4 shows the histological response to the implantation of C/CS. During the first 2 weeks moderate numbers of inflammatory cells were infiltrated into the hydrogel which integrates into the surrounding tissue, this behavior being characteristic to remodeling processes appeared in the implantation sites (figure 4a).

Figures 4b and 4c show the biomaterial-tissue interface, the presence of few inflammatory cells and a reduced infiltration. In this area it can be observed an increased basophilia of fibroblasts to the material interface, which prove the existence of some intense synthesis processes. The presence of inflammatory cells (neutrophils) and a mild chronic inflammatory reaction with multinucleated giant cells in hydrogel mesh (arrow) are presented in figure 4d.

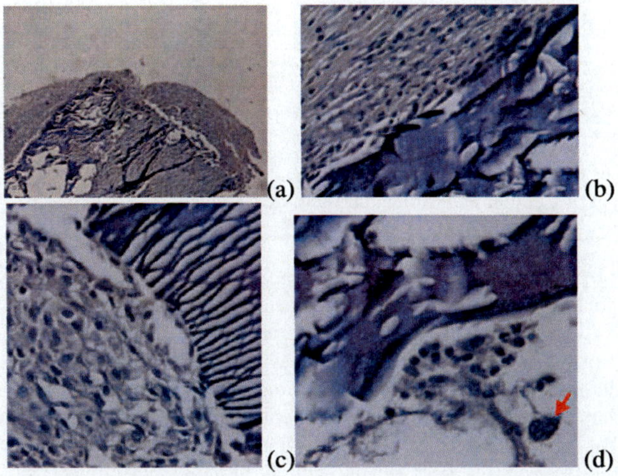

Fig. 4 Microscopic investigations of the inflammatory reactions after subcutaneous implantation of C/CS hydrogel in rats after 4 weeks. Images from 4 (a)(10X), (b)(40X) and (c)(40X) present the detailed zones of biomaterial-tissue interface and image 4 (d)(40X) present a detailed zone with the inflammatory cells (arrow) found at the edge of the hydrogel

The C/CS hydrogel was well tolerated for 1 month after implantation but should also be noted that usually the changes and tissues remodeling occur within the first 4-8 weeks after implantation, which can justify the presence of inflammatory infiltrate into the hydrogel mesh.

IV. Conclusions

The C/CS hydrogel loaded with codeine showed potential as drug delivery system. Its high biocompatibility along with the possibility of controlling the ratio of the released active substance over the retained one, by altering the amount of the negatively charged CS, makes it an attractive target for subsequent studies as a drug carrier.

Acknowledgements

This paper was supported by funding from the research Grant no. 2561/2008 CNCSIS.

References

1. N.A. Peppas, Hydrogels, in Biomaterials Science: An Introduction to Materials in Medicine, 2-nd edition, Editori: B. D. Ratner, A.S. Hoffman, F.J. Schoen, J.E. Lemons, Academic Press, New York, pag. 100, 2004.
2. O. Wichterle, D. Lim, Nature 185, 117, 1960.
3. Y.S Choi, S.R. Hong, Y.M. Lee, K.W., Song, M.H., Park, Y.S., Nam, Biomaterials, 20, 409–417, 1999.
4. A.S. Hoffman, Adv. Drug Deliv. Rev., 43, 3, 2002.
5. R. Langer, N.A. Peppas, AIChE Journal, 49, 2990, 2003.
6. N. Kashyap, N. Kumar, M.N.V. Ravi Kumar, Critical Reviews in Therapeutic Drug Carrier Systems, 22, 107, 2005.
7. R.V. Ulijin, N. Bibi, V. Jayawarna, P.D. Thornton, S.J. Todd, R. Mart, A.M. Smith, J.E. Gough, Materials Today, 10, 40, 2007.
8. I.V. Yannas, E. Lee, D.P. Orgill, Skrabut, E.M. G.F. Murphy, Proceedings of the National Academy of Sciences of the United States of America, 86, 933–937, 1989.
9. H. Ichikawa, Y. Fukumori, Journal of Controlled Release, 63, 107–119, 2000.
10. A. .Matsumoto, S. Ikeda, A. Harada, K. Kataoka, Biomacromolecules, 4, 1410–141, 2003.
11. M. Fried R.M. Lauder, P.E. Duffy, Exp Parasitol., 95, 75—8, 2000.
12. R.M. Lauder, T.N. Huckerby, G.M. Brown, M.T. Bayliss I.A . Nieduszynski, Biochem J., 358, 523—8, 2001.
13. T. Oshika, F. Okamoto, Y. Kaji, T. Hiraoka, T. Kiuchi, M. Sato, Br J Ophthalmol., 90, 485, 2006.
14. K. Sugahara, T. Mikami, T. Uyama, S. Mizuguchi, K. Nomura, H. Kitagawa, Curr Opin Struct Biol., 13, 612—20, 2003.
15. A. Kinoshita-Toyoda, S. Yamada, S.M. Haslam, K.H. Khoo, M. Sugiura, H.R. Morris, Biochemistry, 43, 11063—74, 2004.
16. A. Adebowale, D.S. Cox, Z. Liang, N.D. Eddington, J Am Nutraceut Assoc., 3, 37—44, 2000.
17. E.J. Vandamme, S. De Baets, A. Vanbaelen, K. Joris' P. De Wulf, Polymer Degradation and Stability, 59(1-3), 93-99, 1989.
18. W.K. Czaja, D.J. Young, M. Kawecki, R.M. Brown, Biomacromolecules, 8(1), 1-12, 2007.
19. Sannino, A., Demitri, C. and Madaghiele, M. (2009). Materials, 2:353-373.
20. El-Hag Ali, A., Abd El-Rehim, H., Kamal, H. and Hegazy, D. (2008). J. Macromol. Sci. Pure Appl. Chem., 45 (8):628-634.
21. Sintov, A., Di-Capua, N. and Rubinstein, A. (1995). Biomaterials, 16:473; Wang, S.C., Chen, B.H., Wang, L.F. and Chen, J.S. (2007) Int. J. Pharmacogn, 329:103.
22. Espirito Santo, V., Gomes, M.E., Mano, J.F. and Reis, R.L. The controlled release of platelet lysate from chitosan/chondroithin sulphate nanoparticles controls stem cell proliferation in tissue engineering strategies, Proceeding of ESBP, 5th European Symposium on Biopolymers, Madeira, Portugal, pp. 70-71, 2009; Park, Y.J., Lee, Y.M., Lee, J.Y., Seol, Y.J., Chung, C.P. and Lee, S.J. (2000). J. Control. Release, 67:385.
23. Chang, C.H., Liu, H.C., Lin, C.C., Chou, C.H. and Lin, F.H. (2003). Biomaterials, 24:4853; Fan, H., Hu, Y. Zhang, C., Li, X., Lv, R., Qin, L. and Zhu, R. (2006). Biomaterials, 27:4573.
24. Flanagan, T.C., Wilkins, B., Black, A., Jockenhoevel, S., Smith, T.J. and Pandit, A.S. (2006). Biomaterials, 27:2233; Keskin, D.S., Tezcaner, A., Korkusuz, P., Korkusuz, F. and Hasirci, V. (2005). Biomaterials, 26:4023.
25. Lee, C.T., Kung, P.H. and Lee, Y.D. (2005). Carbohydr. Polym., 61:348.
26. Fan, H., Hu, Y. Zhang, C., Li, X., Lv, R., Qin, L. and Zhu, R. (2006). Biomaterials, 27:4573.
27. A.M. Oprea, A. Neamtu, B. Stoica, C. Vasile, Analele Stiintifice ale Universitatii A. I. Cuza, Sectiunea Genetica si Biologie Moleculara, Tom X, .85-92, 2009.
28. 20.S. Henkelman G. Rakhorst J. Blanton, W. van Oeveren, Material Science and Enginering, C29, 1650-1654, 2009.
29. 21.D.A Noe, V. Weedn, W.R. Beli, Clin. Chem., 30, 627-630, 1984.

Timelapse Monitoring of Cell Behavior as a Tool in Tissue Engineering

C. Niculițe[1] and M. Leabu[1,2]

[1] "Victor Babeş" National Institute of Pathology, Department of Cellular and Molecular Biology, Bucharest, Romania
[2] "Carol Davila" University of Medicine and Pharmacy, Department of Cellular and Molecular Medicine, Bucharest, Romania

Abstract— **Tissue engineering is a research field in biomedicine which has evolved rapidly during the last two decades. It is based on the isolation of cells, their proliferation *in vitro* on a scaffold, followed by the implantation of the construct inside a living organism. Each of the main elements from a tissue engineered construct – cells, scaffold, signaling molecules – has a vital role in assuring the desired characteristics and proper functioning of the whole structure. In order to ensure a normal healing process and no adverse reactions, biocompatibility tests have to be performed in complex experiments needing sophisticated equipments. Two such equipments are presented in this paper: BioStation IM, a microscope that allows capturing images by timelapse videomicroscopy, and xCELLigence, an instrument which determines cell impedance as a measure of cell adhesion, spreading, migration and proliferation. The usefulness of these two equipments in tissue engineering research is discussed**

Keywords— **tissue engineering, timelapse videomicroscopy, impedance, cell adhesion, extracellular matrix.**

I. INTRODUCTION

Tissue engineering is an interdisciplinary field of study using the principles of engineering and life sciences to develop biological substitutes, who will restore, preserve and improve the function of a tissue or entire organ [1]. According to a review on the history of tissue engineering published by Charles A. Vacanti [2], one of the founders of the Tissue Engineering Society, the first use of the term in its current form is as recent as the early 1990s, in a paper published in Surgical Technology International in 1991 [3].

The basic principle of tissue engineering is the generation of a tissue *ex vivo* by seeding cells on a biocompatible scaffold, followed by the implantation of the tissue in the organism. Biocompatibility refers to the ability of a scaffold or matrix to serve as a platform for cell culture and to allow and/or favor cells to perform their activities, in order to facilitate the optimal tissue regeneration without eliciting unwanted local or systemic responses in the recipient [4].

The evaluation of biocompatibility includes assessing cell behavior and cell-matrix interactions *in vitro*. The interactions of cells with biomaterials are influenced by both cell surface features depending on membrane components and physical and chemical properties, such as the chemistry [5], wettability [6] and charge [7] of the materials organizing the extracellular scaffold. Cell-extracellular material interactions are extremely complex, well controlled and modulated, including multiple steps and involving various cellular events. These interactions could also be dependent on the adsorption of proteins from the cell medium onto the scaffold, as a first event (which happens within seconds since they make contact). The next event is represented by the cellular attachment (minutes). Afterwards the cells will adhere more or less to the substrate (hours) and start migrating, proliferating and differentiating (days/weeks). Timelapse monitoring of cell behavior is very useful in observing and understanding these dynamic processes on a long period of time.

In this paper, we will describe two timelapse monitoring equipments, one being an imaging system and the other a device which measures the variation of the cell layer impedance.

II. EQUIPMENTS, MATERIALS AND METHODS

A. *BioStation IM*

The BioStation IM (Nikon Corp., Tokyo) is a mini-incubator with an incorporated imaging system [8]. The temperature, humidity and CO_2 concentration inside the culture chamber are controlled by the operating software. The access to the culture dish placed into the culture chamber is provided by a top gliding lid (Fig. 1), which makes the handling easier. Therefore, changes in the culture medium, by adding supplements, could be performed, even though the appropriate available accessory was not purchased by the lab. Image acquisition is performed automatically in phase contrast (using a red led beam that does not affect cell behavior due to the low energy of the light) and/or fluorescence (with 2 available channels) by the software, according to a schedule defined by the user. The light is automatically turned on only during image collection to minimally influence the cell behavior by factors out of the experimental design. The equipment assures multipoint image collection, at various focal planes (Z-stack), using a 2.0 mega pixel CCD camera. Data analysis is performed with the NIS Elements BR software [9].

S. Vlad and R.V. Ciupa (Eds.): MEDITECH 2011, IFMBE Proceedings 36, pp. 352–357, 2011.
www.springerlink.com

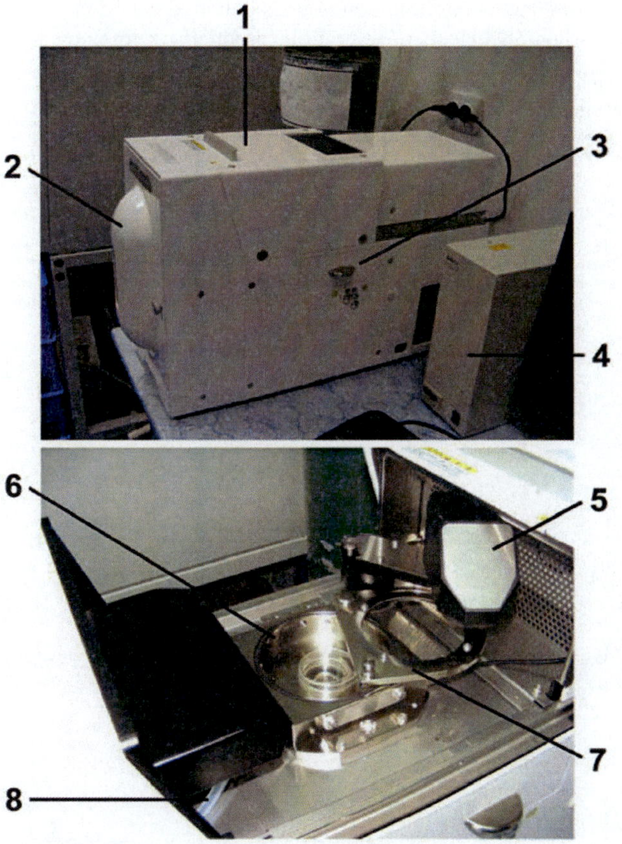

Fig. 1 Image of BioStation IM system (top) and details regarding culture chamber (bottom). 1 – gliding top lid for accessing culture chamber; 2 – front door of the humidity container chamber; 3 – fluorescence filter room; 4 – fluorescence lamp; 5 – phase contrast light holder; 6 – culture chamber; 7 – culture chamber lid; 8 – tube for the access of humidified gases

B. xCELLigence

The xCELLigence system (Roche Diagnostics, Mannheim) measures the electrical impedance of the cell layer across a series of biocompatible microelectrodes placed on the bottom of well plates [10]. The cell analyzer station is properly designed to be used inside an incubator (Fig. 2), while the controller is a laptop provided with RTCA Software 1.2 [11].

Experimental models can be designed to monitor, on a diversity of protein coated surfaces, various cell processes and characteristics: adhesion, spreading, viability, proliferation (by using 16 well E-Plates), or cell motility and invasion. For cell invasion experiments specially designed 16 well CIM-Plates are available. These CIM-Plates and the xCELLigence system replace in an inspired manner the older Boyden chamber or trans-well plates improving the results by recording more objective data, based on a physical parameter, cell layer impedance. The 8µm porosity PET membrane can be coated with various matrix proteins at various concentrations to investigate the cell-matrix interaction effects on cell motility.

C. Data Analysis

Images collected with BioStation IM were analyzed by tracing the contour of the surface area of interest, using NIS Elements BR software [9]. This procedure allows the investigator to determine the area of the object (e.g. cell contour in adherence, spreading and migration experiments, or denuded surface in wound healing experiments) and the centroid (x and y coordinates) of the geometrical figure drawn by the object. These values show the dynamics of cell shape, or decrease of the denuded surface to characterize the cell behavior.

Real time monitoring of the cell layer impedance is reflected in the calculation of a dimensionless value referred to as cell index (CI):

$$CI = (R_{tn} - R_{t0})/F$$

where: R_{tn} represents the cell layer impedance at the experimental time point n, R_{t0} is the background impedance measured before adding cells into the wells, and F is a factor being a constant for the instrument at a specified current frequency (10Ω for 50kHz, the characteristic factor for our system).

CI values depend on cell behavior in terms of adherence strength, spreading extent, cell number (as a cell proliferation or cell death proof).

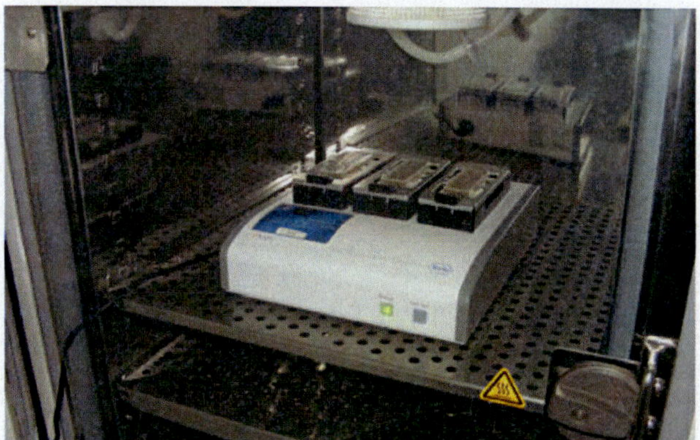

Fig. 2 Image of the cell analyzer of the xCELLigence system, located in a cell culture incubator. Three 16 well E-Plates are placed in the three appropriate holders

III. Results

Both BioStation IM and xCELLigence systems are helpful in the investigation of events like cell adherence and spreading, cell proliferation and viability in various experimental conditions, under different environmental features and treatments. Tissue engineering research depends on such studies that allow the investigators to decipher cellular events and find effective ways to control and modulate the cell behavior *ex vivo*, in order to obtain composite tissues or parts of organs.

A. *Cell Adherence and Spreading Investigation*

Cell adherence and spreading, on different matrix proteins, at various concentrations, can be quantitatively assessed in BioStation experiments with the NIS Elements BR software, by tracing the contour and determining the surface area of individual cells (not shown).

The xCELLigence system can monitor cell adhesion and spreading on different extracellular matrix (ECM) proteins concomitantly, by using different wells in the same plate (Fig. 3). Results in figure 3, prove that the surface chemistry affects the interactions between cells and substrate. The effect of those interactions results in a better spreading of the keratinocytes on collagen in contrast with the spreading on fibronectin.

Both equipments are useful in determining cell adherence and spreading, although the results provided by measuring the impedance are more objective. Moreover, in a BioStation experiment, a relatively large number of cells must be analyzed in order to get statistically relevant data. However, the data acquired by each of the equipments have to confirm the same trend of the investigated biological events. Therefore, the results obtained by one of the equipment have to validate the results recorded by the other.

B. *Monitoring Cell Proliferation*

The degree of cell proliferation can be determined from a BioStation IM experiment by analyzing the number of mitoses during a specific time frame. The facility to collect images from different microscopic fields in a plate favors the statistical analysis of the data collected in fewer experiments than were necessary for older devices.

An easier approach is provided by the measurement of cell impedance, which is proportional to the number of cells (Fig. 4). Cell proliferation is accompanied by the formation of a confluent layer with an increasing thickness, resulting in higher impedance.

In conclusion, both equipments could be useful in the investigation of cell proliferation, producing complementary data.

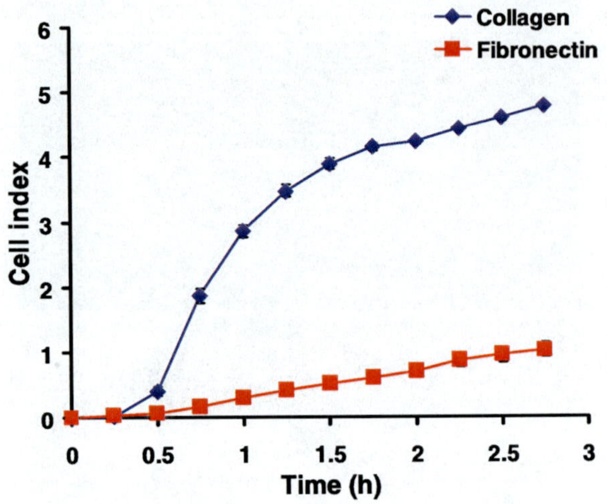

Fig. 3 Keratinocyte adhesion and spreading on different matrix proteins. Experimental conditions: HaCaT cell line, surfaces coated with 5µg/ml collagen/fibronectin, for 18h, at 4°C

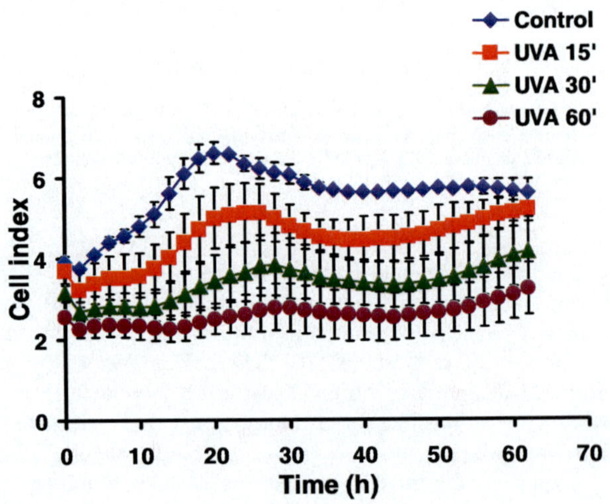

Fig. 4 Keratinocyte proliferation on collagen matrix (5µg/ml). Experimental conditions: HaCaT cell line, UVA irradiated for 15, 30 or 60 minutes or mock-irradiated (control)

C. Cell Migration Studies

Valuable information in regards to cell migration can be obtained using BioStation IM and xCELLigence.

Using BioStation IM both individual cell and collective cell migration can be investigated. A model for collective cell migration is offered by the wound healing experiments, which used in conjunction with timelapse videomicroscopy, represent a productive approach in studying cell movement. Results are analyzed in terms of migration speed, directionality of movement, and cell ability to invade unoccupied culture surfaces. The ability to migrate can be assessed by measuring the surface area of the scratch wound and the time necessary for the cells to cover it (Fig. 5). This is especially useful in the case of cells which migrate in groups (such as keratinocytes), as opposed to those that migrate as individuals.

The directionality of the migration can be estimated by tracking individual cell trajectories using the cell centroid coordinates (Fig. 6). As is suggestively shown in the figure, UVA irradiation of dysplastic oral keratinocytes (DOK) affects the directionality of cell movement.

For the investigation of the invasion ability of different cells on various prepared culture surfaces xCELLigence system is a helpful one due to the specifically designed CIM-Plates [12, 13].

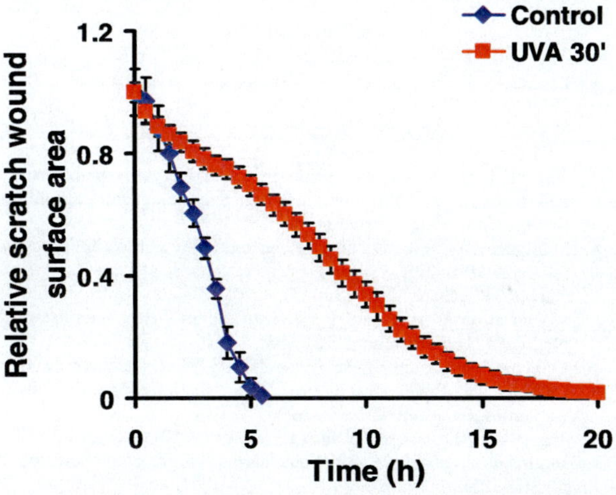

Fig. 5 Keratinocyte migration capacity determined by measuring the scratch wound surface area in a wound healing model. Experimental conditions: DOK cell line, mock-irradiated or UVA irradiated for 30 minutes, collagen matrix 5μg/ml

IV. Discussion

The biocompatibility of the elements in a tissue engineering construct depends – among other factors – on the interactions between cells and their scaffold. During the process of cell adhesion, membrane proteins (integrins) interact with ECM macromolecules, this interaction transducing extracellular signals to cytoskeleton elements (actin, talin, vinculin, etc.) and intracellular signaling proteins (FAK – Focal Adhesion Kinase, MAPK – Mitogen-Activated Protein Kinase) [13, 14]. Binding of the ECM proteins to the extracellular domain of integrins determines the activation of these membrane receptors. Integrin activation mediates the transduction of signals to several effectors through the FAK/MAPK cell signaling pathway, regulating cell motility, survival and proliferation [15, 16, 17]. As we proved above, these processes can be monitored in real time, using the BioStation IM and xCELLigence equipments. This monitoring is helpful in tissue engineering research in order to determine the best conditions for the cells to effectively proliferate, interact with one another, and further organize tissues *ex vivo*.

Studies performed with xCELLigence on cell adhesion, spreading and proliferation can provide useful information regarding the affinity of a cell type for a certain ECM protein, which in turn is revealing for the type of integrin subunits expressed on the cell surface.

In our exemplification, shown in Fig. 2, cell proliferation was monitored after UVA irradiation, but there are many other possible experimental models, more appropriate for tissue engineering, which can be investigated with this device: cytotoxicity assays, the effects of various stimulating or inhibiting factors, etc.

Although information on cell adhesion and spreading can be gathered from BioStation IM experiments, they are more time-consuming and prone to subjective errors due to cell contour measurement that is not always firmly observed. The well plates used with the xCELLigence system provide the possibility of performing several experiments in one, i.e. testing different matrices or compounds, at different concentrations. On the other hand, the latter device doesn't provide any visual information; based on its findings though, the ideal conditions can be chosen and further investigated with timelapse imaging. However, the producer developed plates for BioStation IM with four chambers, allowing simultaneous investigation of the cells in four different experimental conditions. Therefore, the two equipments can be used to produce complementary results once again.

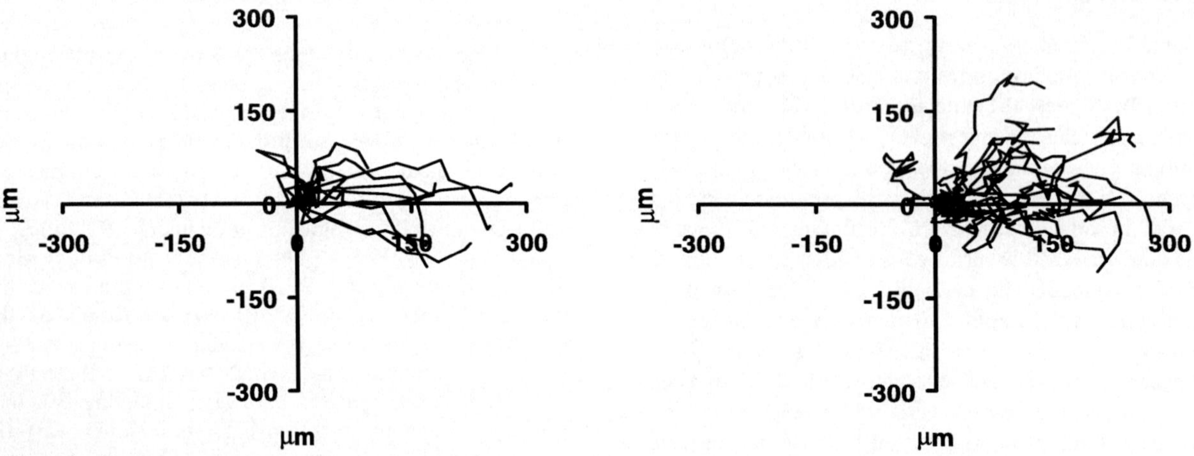

Fig. 6 Keratinocyte directionality of migration in a wound healing model: left – mock-irradiated cells, right – UVA-irradiated cells for 30 minutes. Experimental conditions: DOK cell line, fibronectin matrix 5µg/ml

Cell migration can be examined in videomicroscopy using wound healing models, but several other experimental models too. The examples shown in this paper have focused on the effect of UVA radiation on keratinocyte migration. Although this cell type migrates in groups, changes in the individual movement pattern of cells were observed by determining cell trajectories based on the position of their centroid.

Timelapse experiments indicate the time frames in which various processes occur. All these findings can be used to further investigate the expression level and pattern of different proteins of interest (integrins, FAK, actin filament reorganization) using immunoflourescence, Western Blotting, PCR, etc.

V. Conclusions

This paper considers and discusses the helpfulness of two current equipments designed to investigate cell behavior in real time: BioStation IM allowing cell investigation by timelapse videomicroscopy and xCELLigence, a system destined to monitor cellular actions by the variation of cell layer impedance. The use of these two equipments allows researchers to obtain complementary data and to find solutions in controlling and modulating *ex vivo* the cell survival, proliferation, and complex cell-to-cell or cell-to-matrix interactions, all of which are involved in tissue organizing.

Acknowledgment

Supported by National Authority for Scientific Research of Ministry of Education and Research, Bucharest, Romania (PN II, Grant No. 42-128/2008), and by CNCSIS – UEFISCSU, project number PNII – IDEI code 48/2008 (contract no. 1178/2009, awarded to Dr. Mircea Leabu).

References

1. Langer R, Vacanti JP (1993) Tissue engineering. Science 260:920–926
2. Vacanti CA (2006) The history of tissue engineering. J Cell Mol Med 10:569-576
3. Vacanti CA, Vacanti JP (1991) Functional organ replacement: the new technology of tissue engineering. Surgical Technology International, Century Press, London
4. Williams DF (2008) On the mechanisms of biocompatibility. Biomaterials 29:2941-2953
5. Zhong SP, Zhang YZ, Lim CT (2010) Tissue scaffolds for skin wound healing and dermal reconstruction. Wiley Interdiscip Rev Nanomed Nanobiotechnol 2:510-525
6. Oliver AE, Ngassam V, Dang P et al. (2009) Cell attachment behavior on solid and fluid substrates exhibiting spatial patterns of physical properties. Langmuir 25:6992-6996
7. Vagaska B, Bakacova L, Filova E et al. (2010) Osteogenic cells on bio-inspired materials for bone tissue engineering. Physiol Res 59:309-322
8. http://www.nikoninstruments.com/Vyrobky/Cell-Incubator-Observation/BioStation-IM
9. http://www.lojainterprise.com.br/config/imagens_conteudo/DiscoVirtual/NIS%20%20BROCHURE.pdf
10. http://www.roche-applied-science.com/sis/xcelligence/ezhome.html

11. https://www.roche-applied-science.com/publications/print_mat/RTCA_Software_Manual_SW1.2_Ver3.1.pdf
12. Keogh RJ. (2010) New technology for investigating trophoblast function. Placenta 31:347-350.
13. Ungefroren H, Sebens S, Groth S, Gieseler F, Fändrich F. (2011) Differential roles of Src in transforming growth factor-ß regulation of growth arrest, epithelial-to-mesenchymal transition and cell migration in pancreatic ductal adenocarcinoma cells. Int J Oncol 38:797-805.
14. Michael KE, Dumbauld DW, Burns KL et al. (2009) Focal adhesion kinase modulates cell adhesion strengthening via integrin activation. Mol Biol Cell 20:2508-2519
15. Barczyk M, Carracedo S, Gullberg D (2010) Integrins. Cell Tissue Res 339:269-280
16. Renshaw MW, Price LS, Shwartz MA (1999) Focal adhesion kinase mediates the integrin signaling requirement for growth factor activation of MAP kinase. J Cell Biol 147:611-618
17. Zhao JH, Reiske H, Guan JL (1998) Regulation of the cell cycle by focal adhesion kinase. J Cell Biol 143:1997-2008
18. Sieg DJ, Hauck CR, Schlaepfer DD (1999) Required role of focal adhesion kinase (FAK) for integrin-stimulated cell migration. J Cell Sci 112:2677-2691

Corresponding author:

Author: Dr. Mircea Leabu
Institute: "Victor Babes" National Institute of Pathology
Street: 99-101, Splaiul Independentei, sector 3
City: Bucharest
Country: Romania
Email: mleabu@jcmm.org

Clinical Experience with a Macroporous Synthetic Bone Substitute (Eurocer®) in the Treatment of the Patients with Bone Defects

P.D. Sirbu, T. Petreus, Fl. Munteanu, M. Pertea, S. Lunca, V. Poroch, and P. Botez

"Gr.T.Popa" University of Medicine and Pharmacy, Iasi, Romania

Abstract— **The treatment of bone defects was a major challenge and may still be a problem today. Due to the disadvantages with biologically bone grafts there is a high clinical demand for synthetic bone substitution materials. The aim of this prospective study is to reveal biocompatibility integration and extension of osseous healing for a biphasic synthetic ceramic bone substitute (Eurocer), when used in the treatment of 31patients with 33 bone defects (fractures, nonunions, osteoarthritis). Eurocer® (FH Orthopaedics France) is an osteoconductive ceramic material representing a mixture of 55% hydroxyapatite and 45% tricalcium-phosphate and is available in granular form and in various geometric shapes. The authors used GESTO (Greffes et Substitutes Tissulaires en Orthopedie) protocol for preoperative selection and postoperative follow-up. The mean defect volume for all defects treated with Eurocer was 12cc. According to the size and type of defects the authors used Eurocer® as a single component or mixed with autologous bone graft. Stabilization was achieved by internal fixation in all operation except one (a fracture of humeral head). We have used for osteosynthesis classic plates or plates with angular stability, especially in compression fractures associated with osteoporosis. All patients have been followed-up clinically and radiologic for 2, 3, 4, 6, 9, 12 and 18 months post-operative. The mean time to clinically healing was 3.2 months while the mean time to radiographic healing was 4.5 months. We observed no implant fragmentation and no local inflammation or sepsis. Due to minimally invasive surgery and fast rehabilitation, no joint stiffness or limited joint motion was recorded. This prospective study demonstrates that the biphasic synthetic ceramic material Eurocer® is an effective bone graft substitute for usage in patients with bone defects. To insure a consistent result it is mandatory to strictly follow the three requirements for osteoconduction: proximity, viability and stability.**

Keywords— **macroporous, synthetic bone substitute, osteoconduction, biocompatibility, bone defects.**

I. INTRODUCTION

Bone defects of various etiologies - trauma, osteoporosis, tumors or metabolic diseases - represent an important medical issue with socio-economical implications, due mainly to the lack of spontaneous healing or to the treatment problems and long lasting healing [1]. Approximately 10% of the bone surgery requires bone grafts or bone substitute use [2]. Bone is the most frequently transplanted tissue. Usage of the patient own bone from ilium or other site (autografts) has traditionally been the "gold standard" in treatment of the bone defects [1,3,4]. This is because autogenous bone is osteogenic (viable transplant which contain living cells capable of new bone formation), osteoconductive (it serves as a scaffold in which new bone can deposit) and osteoinductive (it provide growth factors that sustain new bone formation) [1,3,5]. The advantages of autologous bone grafts include long experience in use, availability in most patients, minimal costs and maximal biocompatibility. The disadvantages are represented by the morbidity of the donor site (21% minor complications and 9% major complications - pain, longer than 6 months, infections, dysesthesia, wound drainage, reoperation), limited availability, poor mechanical properties in osteoporotic patients [3,6]. The allografts have osteoconductive and osteoinductive properties (depending on the processing techniques) but their usage involves the infection risk (viral or bacterial) and high costs while it requires a bone bank with all facilities. The disadvantages of the auto- and allografts facilitated the development of bone substitutes and especially the synthetic bone substitution materials that can - theoretically - be prepared in unlimited amounts and with no risk for potential infections [1,4,6]. According to Bauer and Muschler [7] bone substitutes may fall into two categories: osteoinductive and osteoconductive materials [1]. In the first category we distinguish the mineralized bone matrix (DBM) [8] that induces the bone formation when implanted in soft extraskeletal tissues, (compared with conventionally prepared allografts, with a minimal osteoinductive activity). The bone morphogenic protein (BMP) represents a protein (extracted from DBM) that acts as osteoinductive growth factor [1]. The osteoconductive materials are represented by coral hydroxyapatite [9], the calcium sulphate (the oldest osteoconductive bone graft but with very high absorption rate), biovitroceramics [10], ceramic materials or calcium phosphate materials [11] and calcium-phosphate cements [1,11].

II. PURPOSE

The aim of this of this prospective study is to evaluate a biphasic synthetic ceramic bone substitute (Eurocer®)

S. Vlad and R.V. Ciupa (Eds.): MEDITECH 2011, IFMBE Proceedings 36, pp. 358–363, 2011.
www.springerlink.com

regarding intraoperative maneuverability, biocompatibility, integration and extension of osseous healing when used in the treatment of 31 patients with bone defects.

III. MATERIAL AND METHODS

Eurocer® (FH Orthopaedics, France) is a macroporous synthetic bone substitute representing a mixture of 55% hydroxyapatite and 45% tricalcium-phosphate. Eurocer400® is produced in granular form with a granule diameter of 2-3 mm (with a porous structure with 300-500 μm pores) and is recommended for usage in areas not subjected to stress. Eurocer200PLUS® is available in various geometric shapes (cylindrical, disk-shaped, rods, cube and truncated corners) with a porous structure (60% total porosity, partially interconnected by 300-500 μm pores) and must not be used in areas subjected to compression stress greater than 10MPa. These structural properties allow fast osseo-integration followed by gradual resorption.

Between June 2006 - October 2009, 31 patients with 33 bone defects (2 patients with bilateral lesions) were included in a prospective study realized in the Orthopaedic Departments of Emergency Hospital and Rehabilitation Hospital in Iasi, Romania. The mean age for these patients was 57 years, with males averaging of 51 years and females averaging of 65 years. Eurocer was used in various clinical circumstances: Bone defects in proximal tibia fractures - 14 cases with 16 lesions (Fig. 1-4); femur fractures - 2 cases; supracondylar femoral nonunions - 3 cases; delayed union in supracondylar femoral fractures (Fig. 5) - 2 cases; humeral head fractures; - 1 case humeral nonunuions - 2 cases (Fig. 6); malunion of distal radius fractures - 2 cases (Fig. 7); ankle arthrodesis - 2 cases (Fig. 8); subtalar arthrodesis - 3 cases.

The authors used GESTO (Greffes et Substitutes Tissulaires en Orthopedie) protocol for preoperative selection and postoperative follow-up and GESTO classification of loss TOD (type, os, dimensions) for intraoperative quantification of missing bone. In fact, the mean defect volume for all defects treated with Eurocer was 12 cc. Only for metaphyseal defects, the mean defect volume was 8.4 cc. Stabilization was achieved by internal fixation in all operation except one (a fracture of the humeral head). In most cases (24 patients – 77.4%), we have used Eurocer as a single component for bone replacement, mainly for small sized metaphyseal and epiphyseal defects. In limited defects that are conserving bone continuity we have used Eurocer400 (granules) while in defects lacking bone continuity, we have used either Eurocer200 (truncated corners) either Eurocer400 mixed with autologous bone graft harvested from iliac bone. During reconstructive surgery, the surgeons respected the three requirements of the osteoconduction, called the "triad of osteoconduction" [3]: (a) the implant must be in direct contact with the surrounding bone; (b) the surrounding bone must be viable; some factors that decrease this viability are devascularisation, infections and some metabolic bone disease; (c) the interface between the surrounding bone and implant must be stabilized (in most cases, this purpose is reached by internal fixation). The aim of the surgical team was to fill as completely as possible the whole defect area with Eurocer. In some cases (arthrodesis), bone defect was tailored according to the implant contour.

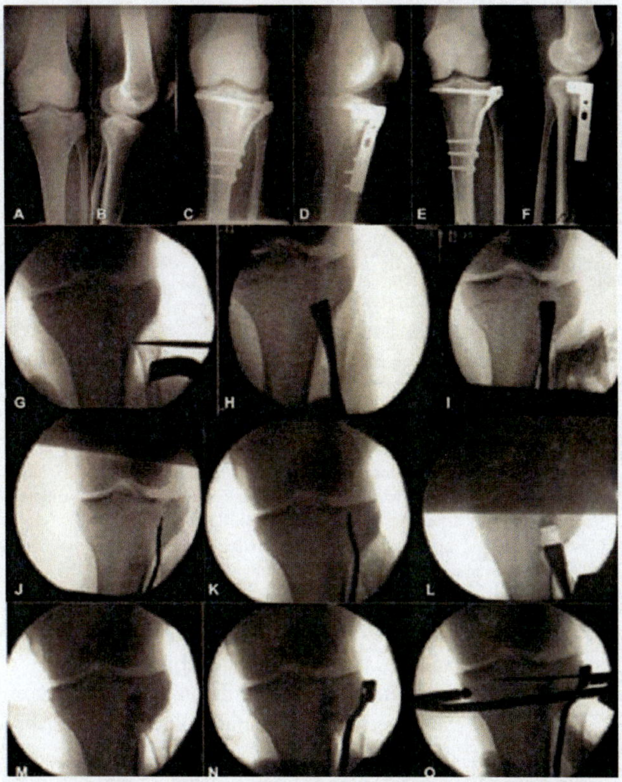

Fig. 1 (A-O) Mixed fracture of the external tibial plateau (type B3/AO). Plate and bone substitute (A,B) preoperative X-rays; (C,D) postoperative; (E,F) X-ray control at 3 months (radiologic evidence of decreasing granular aspect of bone substitute); G - cortical window; (H-K) elevation of the articular surface; (L,M) Eurocer filled bone defect; (N,O) osteosynthesis with L-plate and screws

Thus, 48.4% of the treated defects were located into the proximal tibia, the fracture types being a combination between comminution and compression associated with osteoporosis in many cases.

Our surgical protocol in these circumstances included the following steps: reduction of the lateral articular surface, using a limited cortical window (Fig. 1G) with elevation of the articular surface using a curved instrument (Fig. 1 H-K), filling the epiphyseal-metaphyseal defect with Eurocer, (Fig. 1 L, M) fixation with a lateral plate (Fig. 1 N, O). In some fractures that included both condyles of the tibia we have to reduce first the fracture of the internal plateau followed by fixation with a butress plate placed medially (Fig 2, A–F).

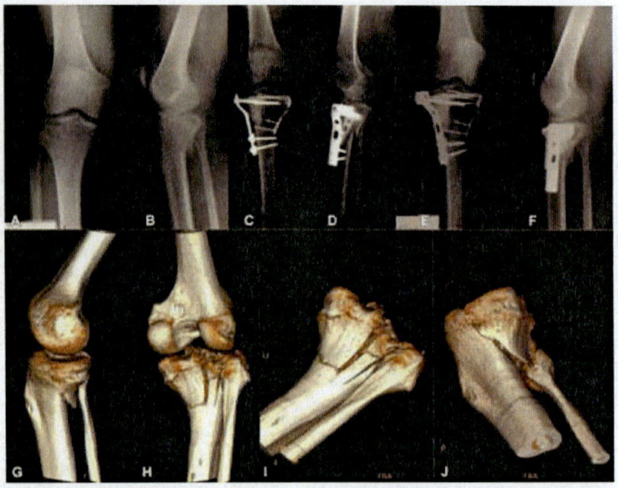

Fig. 2 (A-J) Complex fracture of the proximal tibia (type C3/AO). (A,B) Preoperative radiologic aspect (C,D) postoperative radiologic aspect, medial limited approach, reduction and fixation with small T plate, lateral closed reduction filling the bone defect with Eurocer, L plate and screws (E,F) radiologic aspect at 2 months (vanishing of the radiologic gap at bone-biomaterial interface and homogenisation of the bone substitute structure (G-J) preoperative CT with 3D reconstruction

A CT exam with 3-D reconstruction (Fig. 2, G-J) emphasized the real aspect of the fractures and compressions.

In 2 cases, the complexity of the displacement and compression made us check the articular reduction by arthroscopic surgery (Fig. 3, E-H).

In patients with complex fractures in both tibial condyles, associated with osteoporosis we have used laterally placed plates with angular stability in order to limit the secondary displacement and to allow faster knee rehabilitation.

We have used either plates with monoaxial angular stability type Less Invasive Stabilization System - Proximal Lateral Tibia (LISS-PLT), Locked Compression Plate (LCP-PLT) or plates with polyaxial stability that has the advantages of screw pathway adjustment (their position being adapted to a specific fracture) [12,13].

All patients have been followed-up clinically and radiologic for 2, 3, 4, 6, 9, 12 and 18 months following surgery.

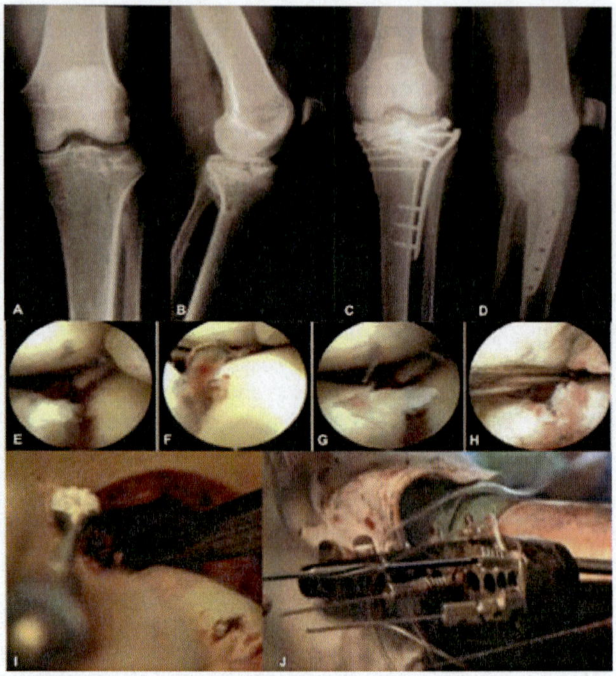

Fig. 3 (A-J) Proximal tibia fracture (type C3/AO) (A,B) preoperative aspect (C,D) medial approach, reduction of the articular surface, small T-plate, lateral approach, close reduction, LCP-PLT, bilateral filling with Eurocer - postoperative aspects (E,H) arthroscopic reduction control; (I) defect filled with Eurocer granules; (J) internal fixation with LCP-PLT on the lateral side.

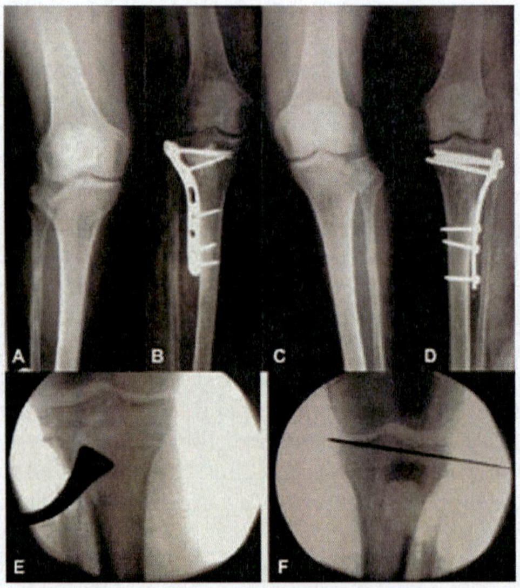

Fig. 4 (A-F) Bilateral complex proximal tibia fractures. (A-D) close reduction, bone defect filled with Eurocer, fixation with plates (polyaxial stability on the right side) (E,F) intraoperative fluoroscopic aspects

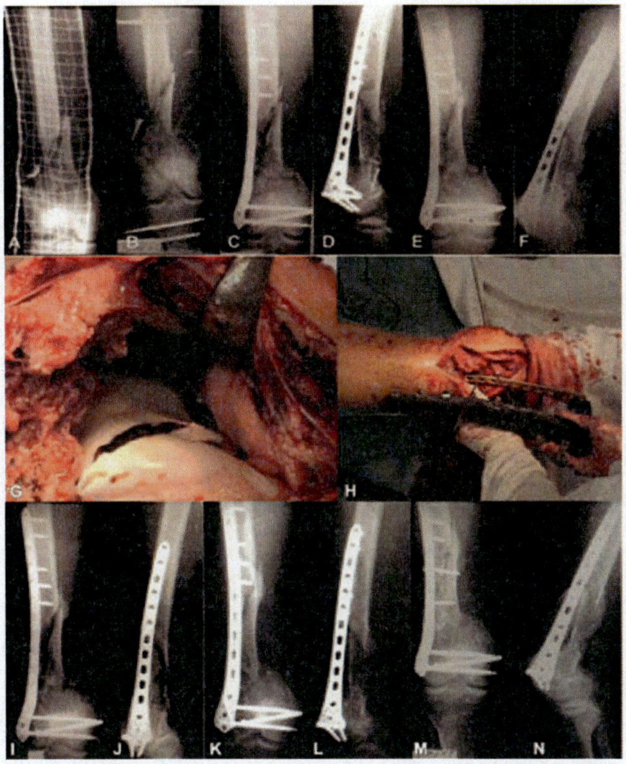

Fig. 5 (A-N) Distal femoral fracture (type C3/AO) with bone loss and open type II Gustilo. (A) preoperative aspect (B) external temporary fixation in damage control period (C,D) Internal fixation with LCP-distal femur, postoperative control (E,F) Radiographic aspect at 1 month postoperatively (G,H) intraoperative aspects (I,J) 3 months postoperative (K,L) 5 months postoperative, delayed union (M,N) defect filled with Eurocer granules mixed with bone graft from iliac crest

IV. Results

Bone substitute osseointegration for all 33 bone defects in 31 patients, treated with Eurocer® was evaluated according to clinical and radiological criteria. In order to determine the effectiveness of clinical healing, all 31 patients were evaluated for the degree of pain at rest, degree of pain during weight bearing or movement and for the degree of movement impairment. The mean time to clinically healing was 3.2 months. None of the patients undergone local inflammation or sepsis. No articular stiffness or limited joint motion was recorded. Due to the fact that performed knee surgery was minimally invasive (mainly in proximal tibia fractures) with indirect reduction using a plate with angular stability, the patients started an immediate rehabilitation and the functional results were excellent.

The radiologic results were interpreted according to three radiographic parameters [5]: interface between biomaterial and the receiving tissue; radiological biomaterial density, eventuality radiological biomaterial fragmentation. The postoperative radiological gap at the bone-material interface was filled at 3-4 months in all bone defects (Fig. 2E-F, 6F, 8G). At 3-6 months, the biomaterial granularity disappeared and became homogenous in 31 bone defects (Fig. 1 E-F), while in two cases with femoral nonunion, this phenomena appeared later, at 9-12 months (due to the large bone defect). The mean time to radiographic healing in all patients was 4.5 months with no implant fragmentation.

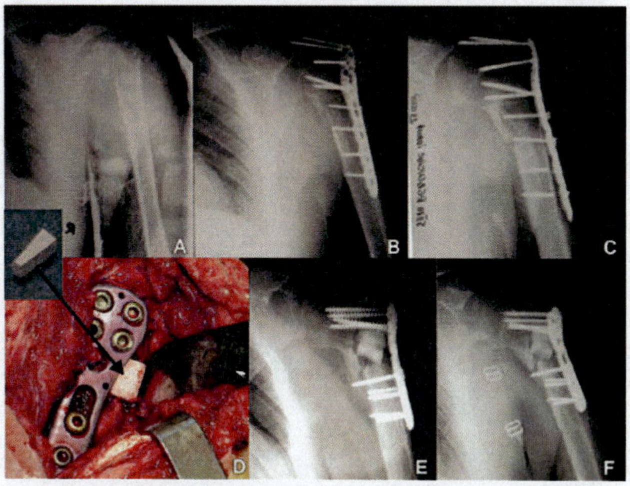

Fig. 6 (A-F) Nonunion of the proximal humerus (A) fracture of the proximal humerus, preoperative aspect (B) internal fixation with plate with monoaxial stability type Phylos (C) construct secondary displacement with broken screws at 3 months (D) reoperation with Phylos plate removal, fixation with plate with polyaxial stability, filling the defect with Eurocer 200 - trapezoidal shape (E) postoperative aspect (F) 2 months postoperative, radiological gap fading with early consolidation

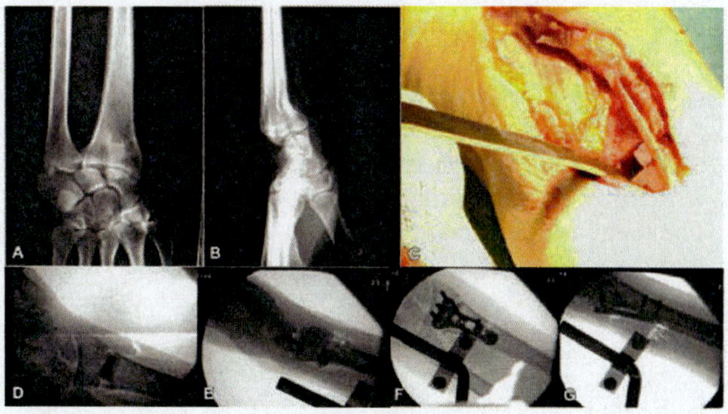

Fig. 7 (A-G) Malunion following a distal radius fracture (A,B) preoperative aspects (C) osteotomy, defect filled with Eurocer 200, intraoperative aspect (D,E) fluoroscopic aspect of the Eurocer block filling the defect (F,G) internal fixation with titanium plate with polyaxial stability

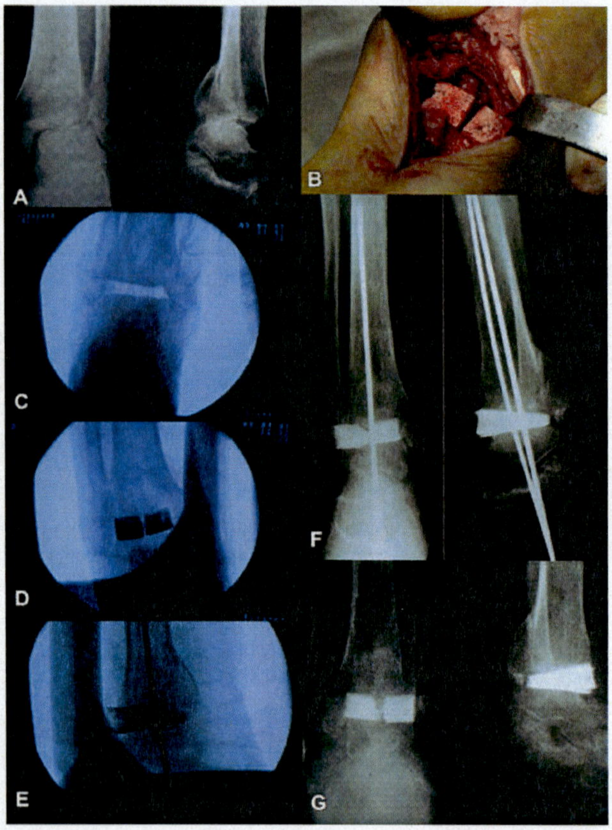

Fig. 8 (A-G) Ankle arthrodesis (A) malunion following a bimaleolar fracture (B) intraoperative aspect of bone defect filling after osteotomy with two blocks of Eurocer 200 (C,D,E) intraoperative fluoroscopic aspects of bone defect following osteotomy, filling with Eurocer 200 and fixation with two Steinmann pins (F) postoperative aspect (G) radiological aspect at 3 months, radiologic gap fading with early consolidation of the arthrodesis site

V. Discussions

Bone defects treatment remains a difficult problem and a challenge for the orthopaedic surgeon. There are various ways to solve these problems but they are usually difficult. The therapeutic means extend from the classical autologous graft to expensive allografts that require a bone bank. The disadvantages of auto and allografts facilitate the development of synthetic bone substitutes. They consist of calcium-based ceramic materials, collagen, non-collagenous proteins, inductive molecules, bioglass and biological degradable polymers. [6]. The ideal bone substitute should be biocompatible, bioresorbable, osteoconductive, osteoinductive, with a structure similar to the bone, with a good intra-operative maneuverability and cost-effective [6]. None of the known bone substitutes fulfill these requirements. Recently, the osteoconductive materials as bi- or triphasic phospho-calcic ceramics played the main roles as bone substitutes. These substances have a composition similar to bone mineral matrix and are biocompatible [1]. The good results obtained by using these materials is due mainly to the physical properties and especially to their macroporosity [5]. The Eurocer was used in experimental study performed on rabbits [12] and we have observed a good radiological osseointegration. Histological findings confirmed that the material is visible and surrounded by lamellar bone (newly formed bone) and the bone sequestration is absent.

Actual research is directed toward the production of macroporous phosphocalcic cements in order to increase the osteoconductivity and biodegradability without altering the biomechanical properties.

VI. Conclusions

This prospective study demonstrated that the biphasic synthetic ceramic material Eurocer® is an effective bone graft substitute for usage in patients with bone defects. The authors appreciated the intraoperative versatility of this product. In all cases, bone consolidation was fast and of good quality, with lack of inflammatory processes. To insure a consistent result it is mandatory to strictly follow the three requirements for osteoconduction: proximity, viability and stability.

References

1. Mihaila R, Redl H., Antonescu D., Schwarz N, Sirbu PD, (2006) Substituenţii de os în tratamentul defectelor osoase, Casa de editură Venus Iasi, Romania.
2. Katsuko SF, Shunsuke M, Daisuke S, Yoshikazu U, Tsuneo S, Takashi U, Miki E, Tetsuya T, (2004) Bone tissue engineering based on bead–cell sheets composed of calcium phosphate beads and bone marrow cells, Materials Science and Engineering C 24, 437–440
3. Shors, E.C., (1999) Coralline bone graft substitutes. Orthop Clin North Am,. 30(4): 599-613.
4. Neumann M, Epple M, (2006) Composites of calcium phosphate and polymers as bone substitution materials, Eur. J. Trauma, 32, 2, 125-131
5. Botez P, Sirbu P, Simion L, Munteanu F, Antoniac I, (2009) Application of a biphasic macroporous synthetic bone substitutes CERAFORM (R): clinical and histological results, Eur J Orthop Surg Traumatol. 19(6): 387-395
6. Larson S., (2007) Bone substitutes in the treatment of fracture, in Lemaire R, Bentley J, European Instructional Course Lectures, 8, 36-41.
7. Bauer TW, Muschler GF. (2000) Bone graft materials. An overview of the basic science. Clin Orthop Relat Res.(371):10-27
8. Urist MR.(1965) Bone: formation by autoinduction. Science; 150:893-9
9. Keating JF, McQueen MM., (2001) Substitutes for autologous bone graft in othopaedic trauma, J. Bone Joint Surg , 83-B, 3-8.

10. Popescu Negreanu T. (2001) Utilizarea biovitroceramicilor şi materialelor compozit pe bază de biovitroceramică şi colagen în ortopedie, Revista de Ortopedie şi Traumatologie (Bucureşti), vol. 11, nr. 1-2, 3-13
11. Bohner M. (2000) Calcium orthophosphates in medicine: from ceramics to calcium phosphate cements, Injury, Int. J. Care Injured 31 S-D37-47
12. Sîrbu PD, Friedl W, Schwarz N, Asaftei R, Bar M, Berea G, Petreus T, Botez P (2010) Polyaxial vs. Monoaxial Angular Stability in Osteosynthesis with Internal Fixators for Complex Periarticular Fractures, AT-EQUAL 2010: 2010 ECSIS SYMPOSIUM ON ADVANCED TECHNOLOGIES FOR ENHANCED QUALITY OF LIFE: LAB-RS AND ARTIPED 2010, 23-26
13. Sîrbu PD, Asaftei R, Danciu M, Mihaila R, Simion L, Cotrutz CE, Petreus T, Botez P (2010), Experimental Study Regarding the Comparative Behavior of Three Osteoconductive Bone Graft Substitutes AT-EQUAL 2010: 2010 ECSIS SYMPOSIUM ON ADVANCED TECHNOLOGIES FOR ENHANCED QUALITY OF LIFE: LABRS AND ARTIPED 2010, 27-29.

Author: Paul Dan Sîrbu
Institute: "Gr.T.Popa"University of Medicine and Pharmacy Iasi
Street: 16, Universitatii
City: Iasi
Country: Romania
Email: pdsirbu@yahoo.com

A Customized Dot Plot Analysis for Alpha Satellite DNA Localization

G.P. Pop[1], A. Voina[1], and E. Onaca[2]

[1] Technical University of Cluj-Napoca/Comm. Dept., Cluj-Napoca, Romania
[2] "Iuliu Hatieganu" Medicine and Pharmacy University/Family Medicine Dept., Cluj-Napoca, Romania

Abstract— **Numerical representation of genomic signals is very important as many of the methods for detecting repeated sequences are part of the digital signal processing field. An original nucleotide sequence representation and a mapping algorithm are used to provide a single numerical sequence for DNA repeats detection which includes information about repeats length. A customized dot plot analysis was used to estimate position of repeats, using two methods of estimating similarity between numerical subsequences. These approaches were combined to isolate position and length of DNA repeats from human alpha satellite DNA.**

Keywords— **DNA Repeats, DNA Representations, Alpha Satellite DNA, Dot Plot Analysis.**

I. INTRODUCTION

The presence of repeated sequences is a fundamental feature of genomes. Detection of DNA repeats can be used for phylogenetic studies and disease diagnosis. A major difficulty in identification of DNA repeats arises from the fact that the repeat units can be either exact or imperfect, in tandem or dispersed, and of unspecified length [1].

Alpha satellite DNA has been identified at every human centromere and consists of tandem repetitions of a 171-bp AT rich sequence motif. Two distinct forms of alpha-satellite are recognized based on their organization and sequence properties: higher-order (HOR) and monomeric. Higher-order alpha satellite is the predominant type in the genome and made up of ~171 bp monomers organized in arrays of multimeric repeat units (ranging in size from 3–5 Mb) that are highly homogeneous. While individual human alpha satellite monomer units show around 30% single-nucleotide variation, the sequence divergence between higher-order repeat units is typically less than 2% [2]. The number of multimeric repeats within any centromere varies between different human individuals and, as such, is a source of considerable chromosome length polymorphism.

Almost all DSP techniques used in repeats detection require two parts: mapping the symbolic data to a numeric form in a nonarbitrary manner and calculating a kind of transform of that numeric sequence. Therefore the numerical representation of genomic signals is very important.

This work presents results obtained using a dedicated numerical representation and a customized dot plot analysis to isolate position and length of DNA repeats in Alpha satellite DNA from human chromosome 19.

II. ASSIGNMENT OF NUMERICAL VALUES

DNA sequences are represented by character strings, in which each element is one of the letters A, T, C and G. Applying a transform technique requires mapping the symbolic domain into the numeric domain in such a way that no additional structure is placed on the symbolic sequence beyond that inherent to it.

One common representation is to map nucleotides to a set of indicator sequences [3] which produces a four dimensional representation yielding an efficient representation for spectral analysis.

Starting from these representations we introduced a novel representation to reduce the dimensionality of representation and generate only one numerical sequence for each DNA subsequence [4].

For a DNA sequence of length L a numerical value is associated in polynomial-like representation:

$$V = \sum_{k=0}^{L-1} V_{\alpha_k} 10^{k}, \quad \alpha \in \{A, G, C, T\} \tag{1}$$

where V_α is the value of a single nucleotide.

One possibility is to use numbers assigned to each nucleotide from the beginning as follows: A=1, G=2, C=3, T=4. For example, consider the sequence TCAGA, then the computed value is: 43121. This approach will be used to represent sequences of certain length.

The following input values are needed:

- a DNA sequence of length N;
- the length of expected repeated sequence, L;
- maximum number of mismatches in the repeated sequences, M_m.

In passing from DNA sequence to numerical values, Hamming distance and consensus value are needed. Hamming distance measure the similarity of two sequences by number of mismatches between sequences: if two sequences are identical the Hamming distance is zero. Given a number of sequences of same length, the consensus sequence is a

S. Vlad and R.V. Ciupa (Eds.): MEDITECH 2011, IFMBE Proceedings 36, pp. 364–367, 2011.
www.springerlink.com

sequence formed by the most frequent nucleotide in the same positions.

The algorithm is summarized bellow:

- Consider all successive subsequences of length L in the DNA sequence;
- Determine all the positions (and the associated subsequences of length L) in original sequence for which the Hamming distance is less or equal the prefixed mismatches number;
- Determine the consensus sequence for all subsequences of length L starting at these positions;
- Compute the numerical value (1) for consensus sequence and assign this value to all these positions.

As output, the algorithm generates a single vector of (N-L) values.

This mapping algorithm has the following properties:

- Even for an L value smaller than the actual length of repeated sequence, the final numerical sequence will highlight a repeat.
- If the L value is a prime factor of repeated sequence length then the entire repeated sequence will be emphasized. This allows a significant reduction of the computational effort.

This algorithm generates only one numerical sequence containing embedded information about the repeated sequences searched for (length and number of mismatches) which can be exploited later. In addition, no additional structures or special memory requirements are needed and if the length of the repeated sequence admits divisors, computing effort can be reduced substantially.

III. DNA DOT PLOT ANALYSIS

Dot plots provide an easy and powerful means of biological sequence analysis. Most commonly they are used for detecting:

- Regions of similarity within a single sequence (i.e. repeats) or between different sequences.
- Sequences that have the potential for forming secondary structure by intramolecular base-pairing.

Dot plots are two-dimensional representations where the x-axis and y-axis each represents a sequence and the plot itself shows a comparison of these two sequences by a calculated score for each position of the sequence. If a window of fixed size on one sequence (one axis) match to the other sequence a dot is drawn at the plot.

The scores that are drawn on the plot are affected by several issues [5]:

- Window size: instead of comparing single residues it compares subsequences of length set as window size; the score is now calculated with respect to aligning the subsequences;
- Threshold: used to filter out noise resulting from random matches; hence you can better recognize the most important similarities.

Some characteristics of patterns appearing in dot plots are:

- A continuous main diagonal shows perfect similarity for symbols with the same indices (Fig. 1-a).
- Parallels to the main diagonal indicate repeated regions in the same reading direction on different parts of the sequences (Fig. 1-b).
- Bold blocks on the main diagonal indicate repetition of the same symbol in both sequences (Fig. 1-c).
- Parallel lines indicate tandem repeats of a larger motif in both sequences. The distance between the diagonals equals the distance of the repeats. (Fig. 1-d).

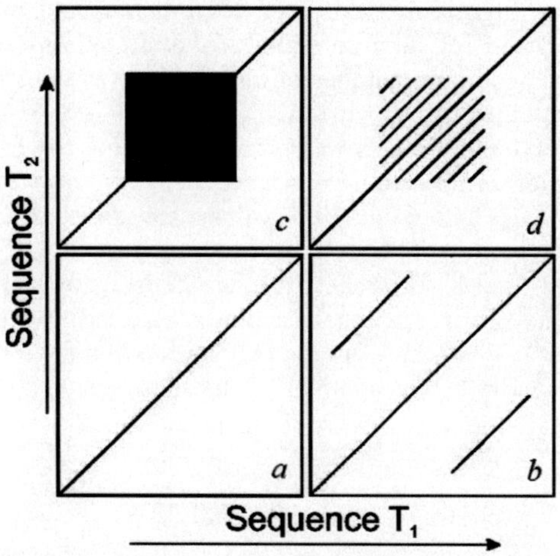

Fig. 1 Characteristic patterns appearing in dot plots

It was designed a customized dot-plot analysis for these considerations:

- It is analyzed a numerical sequence and not a symbolic one.
- Most times the length of analyzed sequence far exceeds the number of points on each axis (in our case, the analyzed sequence is around 40,000 bp length while number of points on one axis is around 1000).
- Due to the large number of nucleotides we need to determine the degree of similarity between subsequences of different lengths to decide if a dot will be plot or not.

To determine the degree of similarity between two numerical subsequences of different lengths, *m* and *n*, we used the following approaches:

- Correlation coefficient (CR): this coefficient is calculated for two sequences of length *n*, (*m-n*) times (using a sliding window), then determine the average coefficient.
- DTW (Dynamic Time Warping): calculate the minimum DTW distance between the two sequences.

Finally a threshold is used to filter the noise arising from the random matches between nucleotides.

IV. Results and discussion

Our case study was the high order repeat in AC010523 Homo sapiens chromosome 19, clone LLNLR-273E6 (GenBank) which contain dispersed alphoid sequences, both higher-order and monomeric alpha-satellite [2].

Numerical representation based on (1) and associated mapping algorithm were used to obtain DNA numerical sequences, using different values for expected repeat length (L) and maximum number of mismatches (M_m). All results were obtained using a custom application written in Delphi.

Several experiments were conducted using the following parameter combinations: L=3, M_m=1; L=9, M_m=2,3,4,5,6; L=19, M_m=3,4,5,6,7. All L values are divisors of alpha satellite length (171 bp).

Next figures shows best results obtained for some combinations of parameters L and M_m, applied to DNA sequence AC010523, using the two approaches for evaluating the similarity between numerical subsequences.

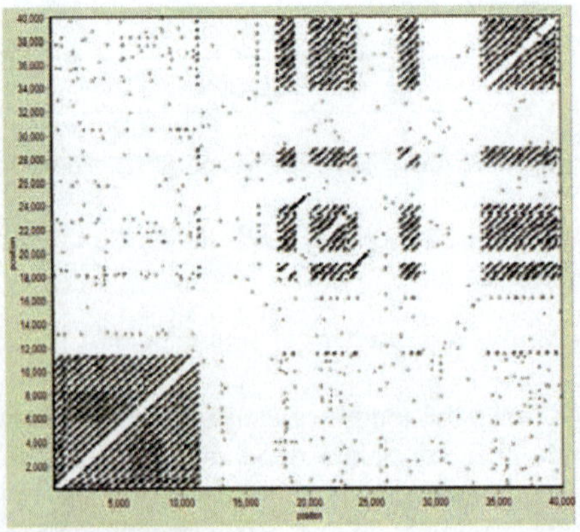

Fig. 2 Dot plot for AC010523 using L=3, M_m=1 and CR

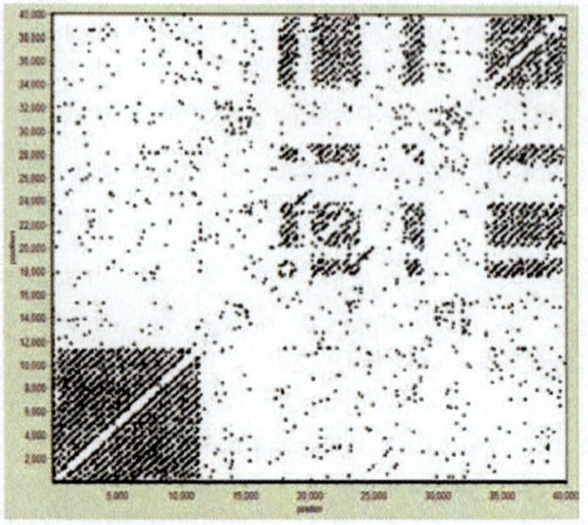

Fig. 3 Dot plot for AC010523 using L=3, M_m=1 and DTW

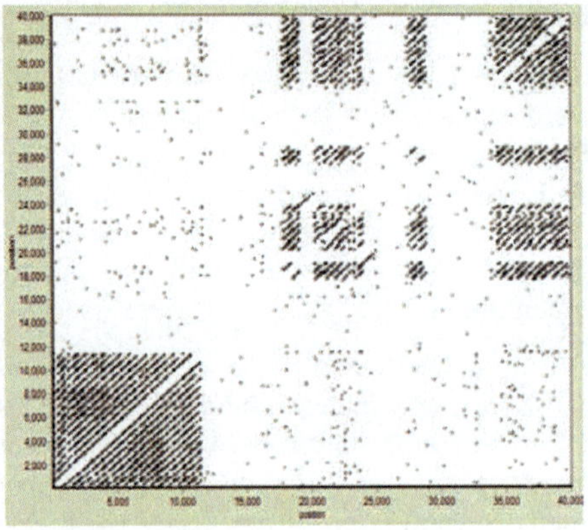

Fig. 4 Dot plot for AC010523 using L=9, M_m=2 and CR

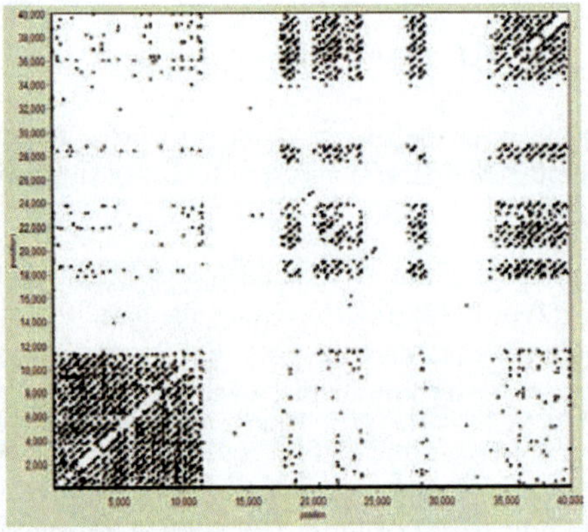

Fig. 5 Dot plot for AC010523 using L=9, M_m=2 and DTW

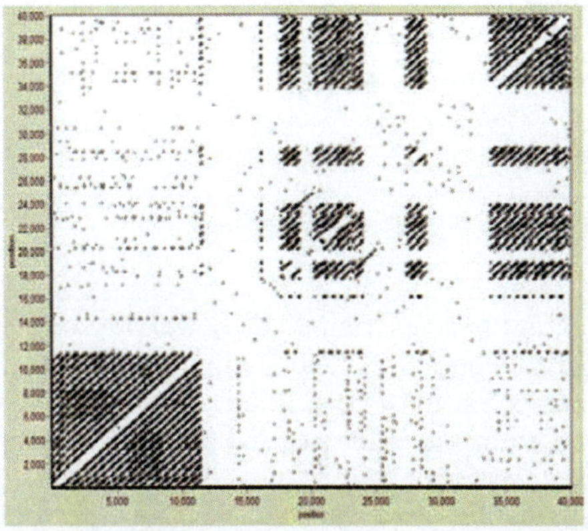

Fig. 6 Dot plot for AC010523 using L=19, M_m=7 and CR

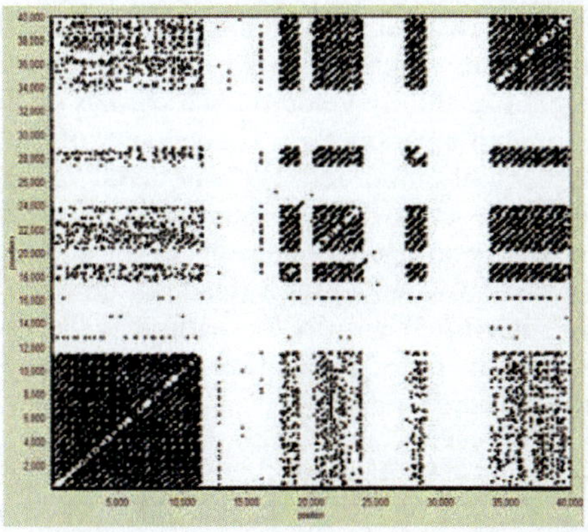

Fig. 7 Dot plot for AC010523 using L=19, M_m=7 and DTW

Analyzing all results including these figures we can say that:

- In all the figures we can easily identify the positions of the patterns associated with repeated sequences.
- Due to the large number of repeated sequences, the results are good even for small values of the parameter L (L=3).
- CR method gives better results, combination L=19, M_m=7 gives the best results.
- In both cases, M_m values affect computational effort and the quality of the results. If the values used for M_m are too small or too large, this may damage results.

Overall, the methods give similar results but computational effort is much higher in DTW method than CR method.

V. Conclusions

An original nucleotide sequence representation and a mapping algorithm are used to provide a single numerical sequence for DNA repeats detection which includes information about repeats length. A customized dot plot analysis was used to estimate position of repeats, using two methods of estimating similarity between numerical subsequences. This approach allows a quick and easy visual localization of repeated sequences within large genomic sequences.

Acknowledgment

This work was partly supported by CNCSIS–UEFISCSU, project number PN2-PARTENERIATE-42127/2008 and partly by the project "Doctoral studies in engineering sciences for developing the knowledge based society-SIDOC" contract no. POSDRU/88/1.5/S/60078, project co-funded from European Social Fund through Sectorial Operational Program Human Resources 2007-2013.

References

1. A. Krishnan and F. Tang (2004), Exhaustive Whole-Genome Tandem Repeats Search, Bioinformatics Advance Access, 20(16):2702-2710.
2. Alkan C, Ventura M, Archidiacono N, Rocchi M, Sahinalp SC, Eichler EE. (2007), Organization and evolution of primate centromeric DNA from whole-genome shotgun sequence data, PLoS Comput Biol. 2007 Sep; 3(9):1807-18.
3. D. Anastassiou (2001), Genomic signal processing, IEEE Signal Process. Mag., 18 (4):8–20.
4. Pop, G.P., Lupu, E. (2008), DNA Repeats Detection Using BW Spectrograms, IEEE-TTTC Int. Conf. on Automation, Quality and Testing, Robotics, AQTR 2008, Cluj-Napoca, Romania, Tome III: 408-412.
5. Yankov, D. Keogh, E. Lonardi, S. (2005), Dot plots for time series analysis, Proc. 17th IEEE International Conference on Tools with Artificial Intelligence: 159 – 168.

Author: Pop G.P.
Institute: Technical University of Cluj-Napoca
Street: G. Baritiu, 26-28, 400027
City: Cluj-Napoca
Country: Romania
Email: petre.pop@com.utcluj.ro

Comparison of Treadmill-Based and Overground Gait Analysis

D.I. Stoia and M. Toth-Tascau

Politehnica University of Timisoara/Faculty of Mechanical Engineering, Timisoara, Romania

Abstract— **This preliminary study was designed to assess the feasibility of gait analysis on a treadmill allowing multiple steps recording at a constant velocity, based on a comparative analysis between treadmill and overground gait. The measurements were realized using Zebris measuring system CMS-HS and a commercially available Hammer Walkrunner Pro Treadmill. One young volunteer (male, 24 years old) with mild instability of left ankle was involved in the study. To evaluate the spatio-temporal and kinematical parameters, the investigated subject performed gait with self-selected velocity both on the treadmill and overground. The main spatio-temporal parameters which have been analyzed are stance and swing phase, stride time, cadence, and velocity. The studied kinematical parameters were the flexion-extension angles of the hip and knee joint, and dorsi- plantar flexion angles of the ankle joint. The comparison of the kinematic parameters was focused on the left limb joints. The walking disorder of the patient has influenced both the kinematic parameters and gait symmetry.**

Keywords— **motion analysis, overground gait, treadmill-based gait, spatial-temporal parameters, joint angles.**

I. INTRODUCTION

Contemporary motion analysis is performed for the purpose of research (measuring system development), diagnosis, rehabilitation, improvement of sport performance and injuries prevention.

Quantitative gait analysis is one of the most used motion analysis due to the alteration of its characteristics when various problems (neurological, skeletal or neuromuscular) occur. Also, it's an objective analysis of the walking ability in healthy persons.

A complex gait analysis consists of three main investigations: kinematic analysis (provide detailed information regarding the spatio-temporal and kinematic parameters of gait), kinetic analysis (pressure distribution and trajectory of mass center and center of pressure are determined during dynamic movements), and EMG (provide data about neuromuscular activity). The gait parameters are grouped to spatio-temporal (swing and stance phases, step length, step width, walking velocity, stride time, and cadence) and kinematic (joint angles of the hip/knee/ankle, and thigh/trunk/foot angles) classes [1], [2].

The relevance of the results is always proportional to the number of valid trials and recorded strides. Usually, gait analysis refers to overground (OG) walking on a walkway located in a laboratory. The ability to obtain significant measurements in these conditions is constrained by the limited length of the walkway and the subjective control of the walking velocity [3].

The evaluation drawbacks of normal and pathological (OG) gait could be overcome by walking on a treadmill [4]. It is often more convenient to perform a treadmill (TM)-based gait analysis since it is possible to acquire more consecutive strides without overcoming the measurement range of the equipment. Another benefit of (TM) gait consists in controlling of walking velocity, which has a significant influence on gait's parameters. The velocity of the treadmill can be sequentially modified. For any given value of velocity, the equipment provides constant belt movement which lead to relatively uniform movements of the subject.

One limitation when using a treadmill consists in adapting of the individual gait to the natural walking which is more difficult in older population and individuals having some disabilities [5].

Some authors have reported no mechanical difference between the two modes (OG and TM gait) while others have documented statistically significant differences regarding the gait parameters [3].

The purpose of the presented study was to perform a comparative analysis of (TM) gait versus (OG) gait in order to understand the relationship between these two modes and to provide a valid foundation about the possibility of treadmill using in clinical gait investigation and rehabilitation. In connection to the clinical target of the investigation, the selected subject presents mild left ankle instability. Thus, the comparison of the kinematic parameters was focused on the left limb joints only.

II. MATERIALS AND METHODS

The measurements were realized in Motion analysis Laboratory of Politehnica University of Timisoara using Zebris measuring system CMS-HS and a commercially available HAMMER Walkrunner Pro Treadmill. The measuring method is based on the determination of spatial coordinates of the miniature ultrasound receptors (markers), by measuring the time delay between the emission of sonic pulses and their reception. The spatial position of the markers is determined by triangulation method.

S. Vlad and R.V. Ciupa (Eds.): MEDITECH 2011, IFMBE Proceedings 36, pp. 368–371, 2011.
www.springerlink.com

The measurement starts with the attachment of two marker triplets on the body in two key points. The first marker triplet is attached on the thigh and the second one on the upper part of the foot. In the next step, the anatomic landmarks are marked with the pointer, and the system software creates the geometrical model of the lower limbs. Signals from the left and right side of the body are measured simultaneously. The spatial positions of the markers and geometrical model are computed and displayed during the subject motion, using the WinGait Software.

Hammer Walkrunner Pro Treadmill has the possibility of adjusting the velocity and inclination of the belt. From the point of view of inclination, all measurements were performed with the treadmill belt in horizontal position.

To evaluate the gait kinematical parameters, one young volunteer (male, 24 years old) having mild instability of left ankle performed gait both on treadmill and overground, with self-selected velocity. In order to adapt his walking to the measurement conditions a trainning session of five minutes was firstly performed. Based on this training session the treadmill velocity was set.

There were performed ten overground valid trials, each consisting of only two strides due to the limitations of the measuring equipment. On the treadmill, the belt velocity was selected according to the self-selected velocity of the subject walking. Because the gait cycles could be endless on treadmill, the duration of trial recording was limited to 50 seconds, which represents around 31 strides.

The sampling rate of the recordings was selected in both cases at 25 Hz, according to the movement velocity. Collection and comparison of spatial-temporal and kinematical parameters of the gait were performed for both limbs executing successive strides, in both modes.

The numerical results of the joint angle measurements were exported for further processing and have been ordered in columns to compute the mean values and standard deviations. The series with large phase difference were eliminated for smoother results achievement. The phase difference series usually corresponds to the velocity changing during walking.

III. Results and discussion

The main spatio-temporal parameters which have been analyzed are: stride time, cadence, velocity, and swing and stance phases of both limbs. The phases of the gait cycle indicate the symmetry between the limbs movement. In case of a healthy person, the left and right limbs have to exhibit a symmetrical behavior with a stance phase of 60 % and 40 % for swing. In our case the average stance percentage in (OG) gait is 65% while average swing is 35%, almost similar for both limbs. In (TM) gait case, the gait phases differ with 1%, for the left limb only. The similar percentages of the gait phases in both walking modes allow a comparison of the results.

The stride time is usually 1.6 seconds in regular walking. The records reveal an average time interval per stride of 1.4 seconds in case of (OG) gait and 1.75 seconds for (TM) walking. In both modes, the gait cycle was computed as a mean of all the recorded strides during the exercise. The difference of 0.35 seconds between stride times in (OG) and (TM) walking is also underlined by the movement cadence, which is 0.7 steps/second in case of (OG) gait and 0.57 steps/second for (TM) gait respectively. Smaller cadence means longer gait cycle which is usually achieved by an extended period of stance phase. This result also indicates that the selected treadmill velocity (0.55 m/s) was smaller than the velocity of the (OG) walking (0.76 m/s).

Three dimensional motion data were obtained for hip, knee and ankle motions. In quantitative approaching of the results, the average variations of the joint angles were represented in time and time-normalized per stride. The average values were computed by manually extracting the data series which represents a stride. The landmark in every series was considered the angle value recorded in knee joint, at the heel contact phase.

Based on the joint angles in different anatomical planes, there were selected the representative movements for each joint: flexion-extension for hip and knee, and dorsi - plantar flexion for ankle. Mild instability in left ankle joint caused correlated effects in joints of both limbs, especially for the left limb. Thus, joint angle variations were determined for both lower limbs, but only for the left limb are presented, the study being focused on the pathology influence on the movements.

Figure 1 presents the averaged joint angle curves in time representation for hip during flexion-extension, in both cases. Both figures illustrate the movement variations and standard deviation of the series. The standard deviation is represented as vertical lines accompanying each data point. The joint motions occur almost entirely within the normal range of motion [1], [6].

Several elements are differencing the two motions. The shape of the curve in (TM) gait is smoother and misses the horizontal complex which characterizes the weight shift from one leg to the other. The standard deviation is smaller for (TM) gait (average of 1.5 compared with 2.6 for overground gait) due to the repetitive motion imposed by the constant movement of the belt. This behavior is manifesting in each joint movement.

Figure 2 depicts the graphs of flexion-extension angles of the hip represented as average, time-normalized stride. As before, the standard deviations of the overground gait recordings are greater than the treadmill recordings. By overlapping the averaged motion curve (figure 3) the differences between the two modes can be analyzed.

Another important parameter in gait analysis is the angular variation of the knee joint, presented in figure 4. The larger standard deviations are occurring in overground gait (average of 6.7 compared with 2.3 for treadmill gait), especially when the curve is changing the slope sign. The inflexions are very sensitive points, thus the measurement could be improved by increasing the sampling rate of the data acquisition.

Figures 5 and 6 depict the graphs of flexion-extension angles of the knee joint as average, time-normalized stride, evidenceing a typical knee behavior in the case of overground gait. Also, the amplitudes of the movements are well shaped for the overground gait.

The last parameter is the angular variation of the ankle joint in dorsi and plantar flexion movement (figures 7, 8 and 9). The plantar flexion is represented by the highest amplitude, while the dorsiflexion is represented by lower peak. The average standard deviation for (OG) gait is 2.2, while for the (TM) gait is 0.57. A notable difference can be observed when overlap the variations of ankle angles for both gait cases. In treadmill gait, the ankle movement has lower amplitudes and lack of smoothness. This can be an effect of the belt movement beneath the feet.

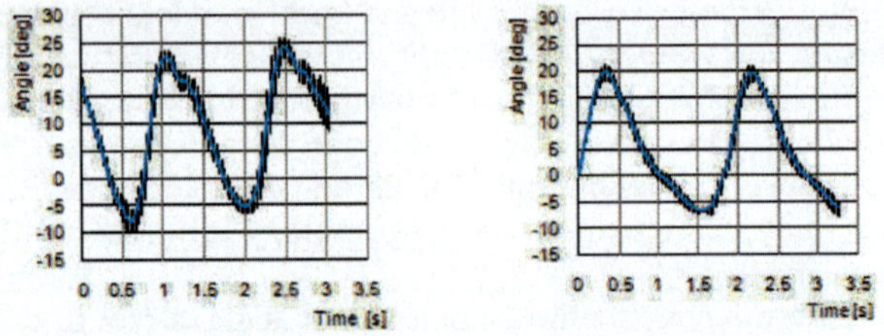

a) Overground gait b) Treadmill gait

Fig. 1 Graphs of flexion-extension angles of the hip joint

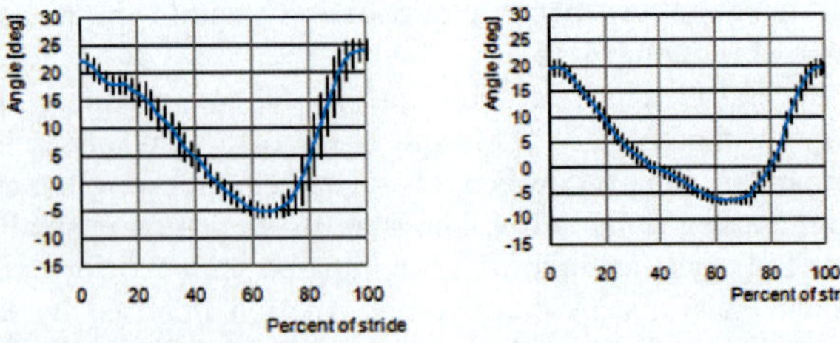

a) Overground gait b) Treadmill gait

Fig. 2 Graphs of flexion-extension angles of the hip joint for one stride

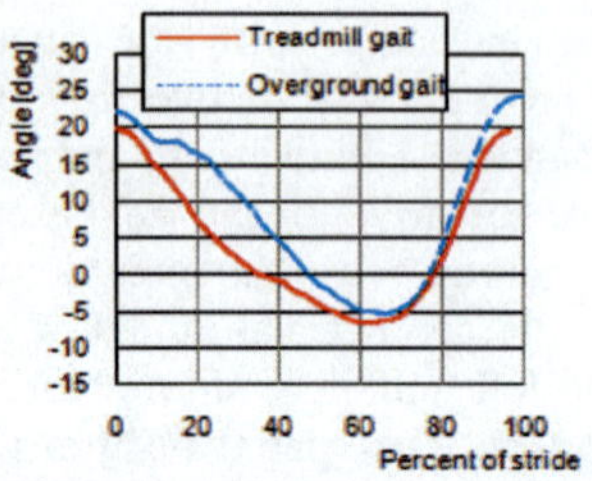

Fig. 3 Comparison of means of flexion-extension graphs of the hip joint for one stride

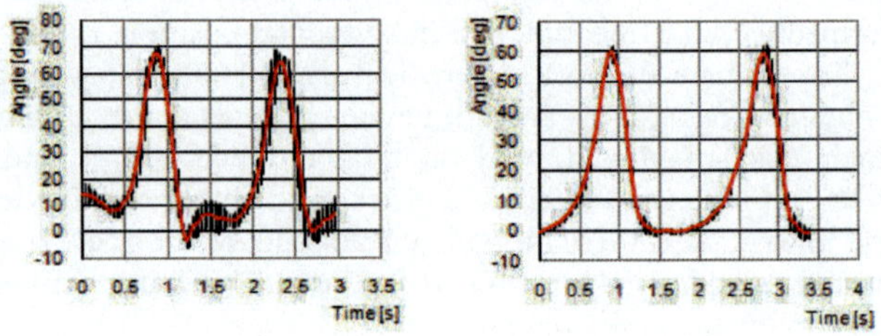

a) Overground gait b) Treadmill gait

Fig. 4 Graphs of flexion-extension angles of the knee joint

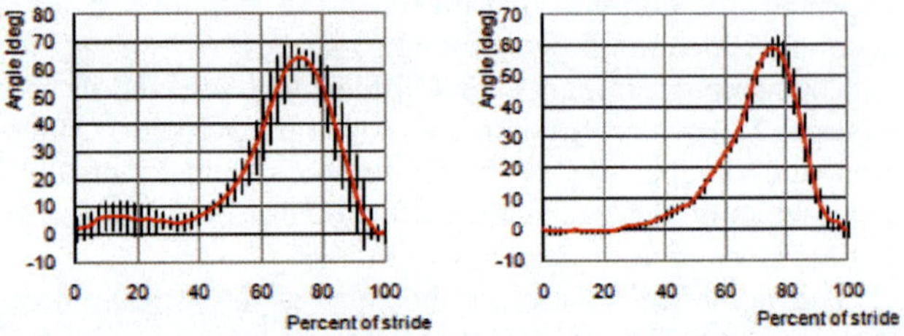

a) Overground gait b) Treadmill gait

Fig. 5 Graphs of flexion-extension angles of the knee joint for one stride

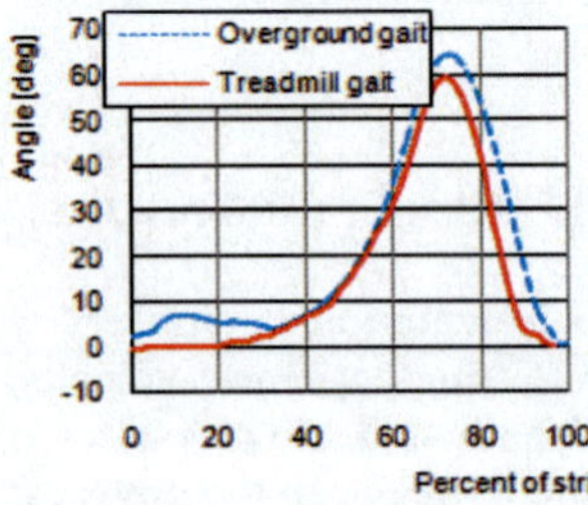

Fig. 6 Comparison of means of flexion-extension graphs of the knee joint for one stride

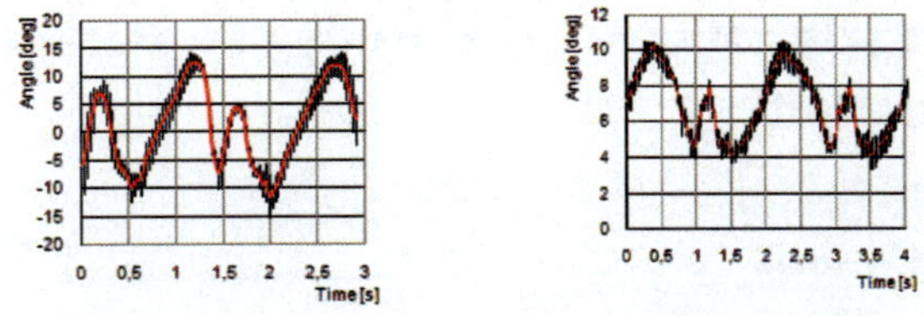

a) Overground gait b) Treadmill gait

Fig. 7 Graphs of dorsi-plantar flexion angles of the ankle joint

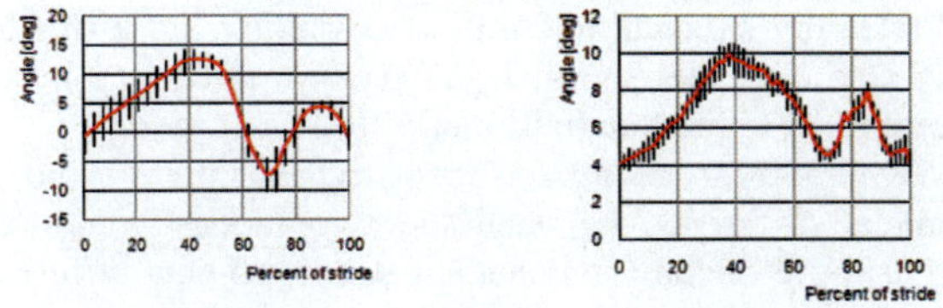

a) Overground gait b) Treadmill gait

Fig. 8 Graphs of dorsi-plantar flexion angles of the ankle for one stride

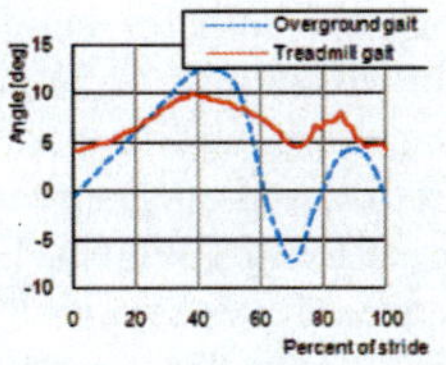

Fig. 9 Comparison of means of dorsi-plantar flexion graphs of the ankle joint for one stride

The means values for discrete angles in (OG) and (TM) gait determined during the performed trials were compared between the two modes by use of an unpaired t-test. Values of $p < 0.05$ were considered statistically significant, while values of $p > 0.05$ indicate no difference between the groups. The flexion-extension angles of the hip ($p = 0.0167$) and dorsi-plantar flexion of the ankle ($p = 0.0086$) were significantly different, while the flexion-extension angles of the knee were not significantly different ($p = 0.112$).

IV. Conclusions

This preliminary study presents a comparative analysis between (TM) and (OG) gait of a patient with a certain walking disorder. The study objective was to underline the gait modifications occurred when moving on the treadmill, and to provide preliminary data about the possibility of using the treadmill in clinical gait investigation and rehabilitation. The walking disorder of the patient has influenced both the kinematic parameters and gait symmetry.

The recorded spatio-temporal parameters present a good similarity. The kinematical parameters reveal that statistical significant differences occur regarding the flexion-extension angles of the hip joint and dorsi-plantar flexion of the ankle joint of the left lower limb. The angular amplitudes in ankle joint are lower when walking on the treadmill than in overground mode and the instability is more obvious. The lack of smoothness of the flexion complex in left ankle in TM-mode is induced by the treadmill movement.

Better results meaning smaller standard deviations are recorded in (TM) case when a certain velocity is imposed. This fact leads to repetitive results, valid for statistical interpretations. On the other hand, accurate recordings and interpretations of the human gait are achieved when the subject is freely waking on the floor. The presence of the moving belt modifies some complexes of the gait cycles, and could lead to inexact interpretations.

The treadmill could be a useful device in routine gait analysis since it allows recording of a large number of successive strides in a short time interval and over a wide range of gait velocities. Still, more practice on (TM) gait analysis should be involved in future studies.

Further research will be performed with a representative lot of subjects, both healthy and having certain disorder to verify the present results and create a data base in order to validate the method for various pathologies.

References

1. Kirtley C (2006) Clinical Gait Analysis, Theory and Practice, Churchill Livingstone Elsevier.
2. DE Gordon Robertson et al. (2004) Research Methods in Biomechanics, Human Kinetics, USA
3. Jung T, Gilgannon M et al. Treadmill-based Gait Analysis for Children with Cerebral Palsy: Biomechanical Comparison of Treadmill and Overground Walking, Kluge Children's Rehabilitation Center at University of Virginia at http://gait.aidi.udel.edu/gaitlab/gcma/info/abstracts/O1.abs20151.pdf
4. Stoquart G, Detrembleur C, Lejeune T (2008) Effect of speed on kinematic, kinetic, electromyographic and energetic reference values during treadmill walking, Neurophysiologie Clinique/Clinical Neurophysiology, 38(2): 105-116 DOI:10.1016/j.neucli.2008.02.002
5. Edginton K A, Güler H C, Ober J J et al. Instrumented Treadmills: Reducing the Need for Gait Labs, Bertec Corporation, Columbus, Ohio, USA at http://bertec.com/uploads/pdfs/papers
6. Perry J (1992) Gait Analysis: Normal and Pathological Function, Slack Incorporated, Thorofare, USA

Author: Mirela Toth-Tascau
Institute: "Politehnica" University of Timisoara
Street: Bd. Mihai Viteazu, No.1
City: Timisoara
Country: Romania
Email: mirela@cmpicsu.upt.ro

Influence of Treadmill Velocity on Gait Characteristics – Case Study of a Patient with Ankle Instability

M. Toth-Tascau and D.I. Stoia

Politehnica University of Timisoara/Faculty of Mechanical Engineering, Timisoara, Romania

Abstract— **The paper proposes a treadmill-based gait study in three particular cases of walking velocity. This study was designed to assess the feasibility of extended gait analysis on a treadmill in relation to the treadmill velocity. The measurements were realized using Zebris measuring system CMS-HS and a commercially available Hammer Walkrunner Pro Treadmill. One young volunteer (male, 24 years old) with mild instability of left ankle was involved in the study. To evaluate the spatiotemporal and kinematical parameters, the investigated subject performed treadmill-based gait with three velocities. The main spatiotemporal parameters which have been analyzed are stride time, cadence, double support time, and velocity. The gait symmetry was also analyzed based on the symmetry of the swing and stance phases between the lower limbs and pelvis obliquity. The studied kinematical parameters were the flexion-extension angles of the hip and knee joint, and dorsi- plantar flexion of the ankle joint. We concluded that higher velocity leads to a lower variability of the gait parameters, while accurate recordings and interpretations of the human gait are achieved when the subject is waking with lower velocity closer to his normal walking speed.**

Keywords— **motion analysis, treadmill-based gait, treadmill velocity, spatiotemporal parameters, joint angles.**

I. Introduction

Contemporary instrumented overground and treadmill-based gait analysis represents one of the most advanced and reliable method both in diagnosis and management of patient rehabilitation.

The use of a treadmill-based gait analysis is increasingly more used due to its major benefits. The main advantage of treadmill-based gait analysis consists in a smaller measurement space, thus being possible to acquire multiple consecutive strides (gait cycles). Another important benefit of treadmill-based gait consists in directly controlling of walking velocity, which has a significant influence on gait parameters. The treadmill provides a relatively uniform walking speed which can be continuously adjusted and monitored. Walking on a treadmill is more safety for elderly people and patients with certain disability because a harness and/or handrails can be used [1].

One limitation of the treadmill consists in adapting of the individual gait to the natural walking, which is more difficult in older population and individuals with disability [2]. Healthy subjects unfamiliar to walking on a treadmill may also alter their normal gait pattern, until they become accustomed to the treadmill inclination and speed.

Some authors have been theorized that there is no mechanical difference between the two modes while some studies have documented statistically significant differences regarding the gait parameters [3].

The influence of the speed on the treadmill-based gait was studied by several authors. There were performed studies on the spatiotemporal parameters, kinematical and kinetic parameters of gait, both for healthy subjects and patients with certain disability, including the benefit of treadmill gait for patient rehabilitation.

Some works were focused on spatiotemporal parameters of gait and certain joint of healthy subjects [4] while other studies reported the influence of systematic increases in treadmill walking speed on gait kinematics after stroke [5], [6].

Study of the effect of speed on kinematic, kinetic, electromyographic and energetic reference values during treadmill walking was performed both to assess the feasibility of extended gait analysis on a treadmill at a constant speed in young healthy subjects and to provide speed-specific kinematic, kinetic, electromyographic and energetic reference values [7].

Some studies reported no relationship between stride time variability and treadmill speed, while other authors reported either a linear or a non-linear relationship. Thus some results highlighted that speed lower than preferred self-selected speed of young healthy adults involves higher stride time variability [8].

Variability of ground reaction forces during treadmill walking has been also investigated and quantified in relation to different constant speeds. Variability of horizontal-anteroposterior force was minimized at the usual walking speed whereas those of the other two components (vertical and horizontal-mediolateral force) increased with increments in walking speed [9].

The purpose of the presented study was to assess information on gait spatiotemporal and kinematical parameters and perform a comparative analysis of treadmill-based gait in relationship with the treadmill velocity. This study is important to establish a valid foundation about the

S. Vlad and R.V. Ciupa (Eds.): MEDITECH 2011, IFMBE Proceedings 36, pp. 372–375, 2011.
www.springerlink.com

possibility of using a treadmill in clinical gait investigation and rehabilitation in our laboratory.

II. Materials and methods

The measurements were realized in Motion analysis Laboratory of Politehnica University of Timisoara using Zebris measuring system CMS-HS and a commercially available Hammer Walkrunner Pro Treadmill. One marker triplet is attached on the thigh and second one on the upper part of the foot. The anatomical landmarks are marked with the pointer, and the system's software creates the geometrical model [10]. To evaluate the spatiotemporal and kinematical gait parameters and evidence the influence of the treadmil velocity, one young volunteer (male, 24 years old) with mild instability of left ankle performed gait on the treadmill at different velocities, having attached the ultrasound recording markers and without using the treadmill handrails.

He gave informed consent before taking part in the experiment. In order to adapt his walking to the measurement conditions a trainning session of three minutes for each selected velocity was firstly performed. After this adequate practice the treadmill velocity was sequentially set at 1, 3, and 5 km/h, corresponding to slow, average, and hurried walk. All the measurements were performed with the treadmill belt in horizontal position. There were performed three trials for each selected velocity, each trial having 60 seconds duration. The trial recording was started after five seconds familiarization period when the treadmill reached full velocity.

A special attention when using treadmill has to be focused on reflection avoidance of the ultrasounds, by the various elements and handrails of the equipment. In order to achieve this goal, the subject was asked to walk along the longitudinal axis of the belt surface, without changing the spatial position of the body during recording period. By changing the spatial position on the belt, the variations of the joint angles could lose the reference.

The sampling rate of the ultrasound system was selected to 30 Hz and maintained for all gait velocities. Using this frequency for the maximum waking velocity, the gait phases are fully described by the recorded values. This is proved by the presence of all gait complexes at maximum velocity, in comparison to other velocities. At much higher speeds, higher sampling rate should be used.

The data acquired by the measuring equipment are the spatial positions of the markers for both limbs during the trial duration. Using these values, the WinGait software computes the characteristic parameters of the gait: swing and stance phases, walking velocity, stride time, cadence, time of double support, and joint angles in all planes.

The numerical results of the joint angle measurements were exported for further processing from the system software and have been computed the mean values, standard deviations, and *P values*. The series with large differences were eliminated for smoother results achievement. The valid results were graphically represented in various combinations and compared in order to express relevant behavior of each spatiotemporal and kinematical parameter of the gait.

III. Results and discussion

The main spatiotemporal parameters which have been analyzed are: symmetry of the swing and stance phases between the lower limbs, stride time, cadence, and double support time. Because the investigated subject presents a mild instability at the left ankle's level, the results are presented in close relation to this issue. Thus, data regarding the symmetry between lower limbs and also the variations of the joint angles of the left limb, at different walking velocity are graphically presented.

The phases of the gait cycles (table 1) indicate a slight difference between the two limbs, which increases with the enhancement of the moving velocity. The average value of stride time was 1.77 seconds (1 km/h), 0.98 seconds (3 km/h), and 0.62 seconds (5 km/h), respectively.

Table 1 Swing and stance percentages for different speeds

Limb/ Phase	1 km/h		3 km/h		5 km/h	
	Swing [%]	Stance [%]	Swing [%]	Stance [%]	Swing [%]	Stance [%]
Left	36	64	37	63	40	60
Right	35	65	40	60	49	51

The average values of cadence and double support time are presented in relation to the walking velocities (figure 1). From this charts an expected tendency can be outlined. When the walking velocity is increased whether by the subject or imposed by mean of the equipment, the cadence of the movement grows, while the double support time significantly decreases. For much higher velocities, the double support time tends to zero.

Three dimensional motion data was obtained for hip, knee and ankle motions. In order to analyze the joint angles tendencies at different velocities, the average gait cycles were computed for each lower limb. There were considered the following joint angles: flexion-extension of the hip and knee joint, dorsi-plantar flexion of the ankle joint, and obliquity of the pelvis. Of these four parameters, the pelvis obliquity is determined by the pelvic girdle while the other three are joint characteristics, so they are represented for each limb joint in relation to the walking velocity.

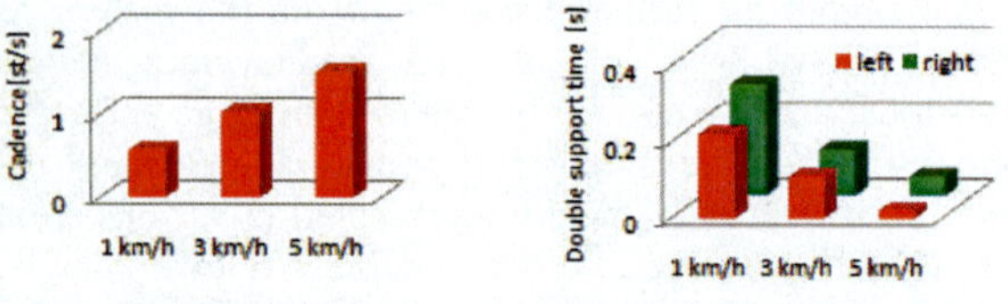

a) Cadence b) Double support time

Fig. 1 Cadence and double support time at different walking velocities

In the paper, the ankle is the only joint presented as a comparison between left and right lower limbs, because the other joint angles indicate no significant statistical difference between limbs. Differences between joint angles in each considered motion during the performed trials were compared between the two lower limbs by use of an unpaired t-test. These values are presented in table 2. Thus, the hip and knee joint angles under different treadmill velocity are presented for the left lower limb only (figure 2).

To evidence the average variability, the standard deviations are represented as vertical lines accompanying each data point. The presence of high values of the standard deviation is usually caused by the phase difference of the movement. The maximum value of the mean standard deviation was 8.7, recorded for angles of the knee joint at lowest movement velocity. The mean standard deviation computed for the gait cycles presents lower values for higher velocities of movement (4 and 3.3 for the knee joint angles at 3 and 5 km/h velocities). Large standard deviations occur when the walking velocity is changed during one exercise. These movements also called non-harmonically, have a higher incidence of occurring at lower movement velocities. The differences of the wave lengths in the graphs are generated by the imposed walking velocities, lower wave lengths being recorded for higher velocities.

A similar behavior regarding the wave length-speed relation was recorded for dorsi-plantar flexion of the ankle joint and pelvis obliquity (figure 3). Larger standard deviations are recorded, due to the evasively movements and lower amplitudes. The dorsi-plantar flexion angles of the ankle joint are presented by comparison between left and right lower limbs. The graphs are represented as average, time-normalized stride for each considered velocity (figure 4).

Table 2 *P values* computed for hip and knee angle series

P value	1 km/h	3 km/h	5 km/h
Hip joints	0.225	0.254	0.424
Knee joints	0.422	0.165	0.457

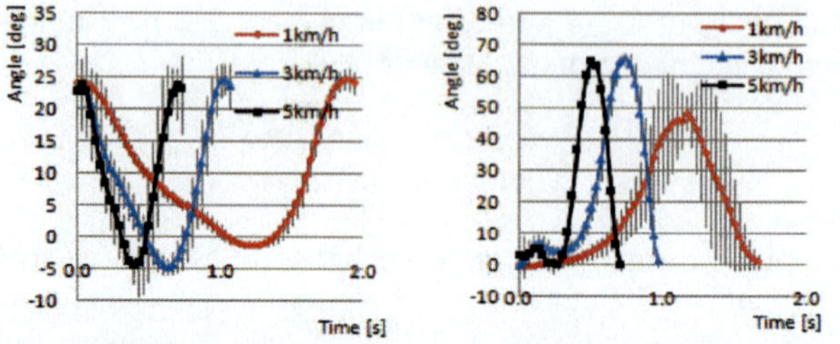

a) Hip joint angles b) Knee joint angles

Fig. 2 Graphs of flexion-extension angles in hip and knee joints, for one stride of the left lower limb

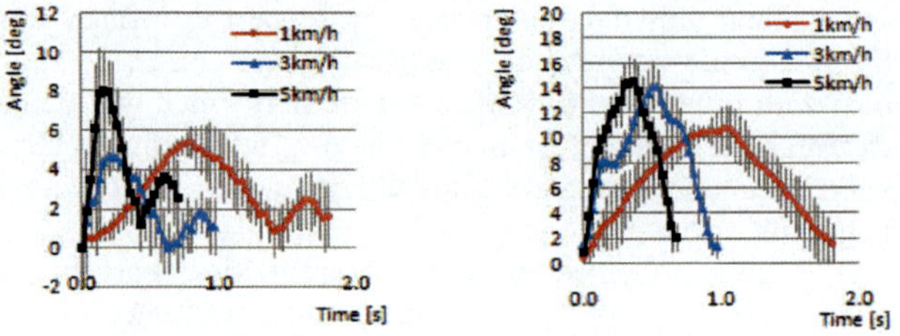

a) Ankle joint angles b) Pelvis obliquity angles

Fig. 3 Graphs of dorsi-plantar flexion of the ankle joint and pelvis obliquity, for one stride of the left lower limb

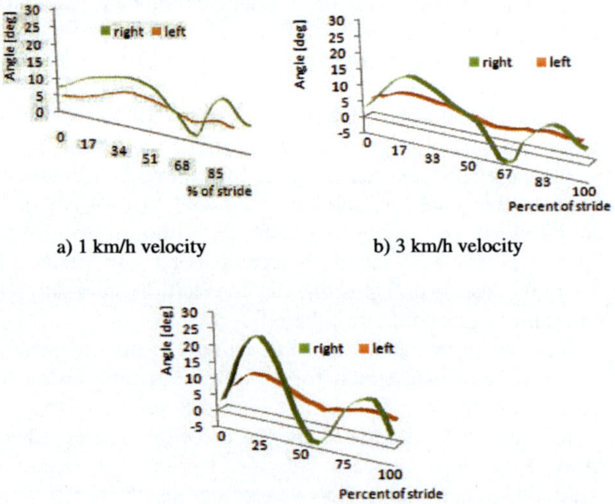

a) 1 km/h velocity b) 3 km/h velocity

c) 5 km/h velocity

Fig. 4 Graphs of dorsi-plantar flexion in left and right ankle movement at different velocities, for one stride

The graphs clearly express the differences between the movement capabilities in the two ankle joints. The healthy ankle exhibits normal amplitudes corresponding to the movement velocity, while the unstable ankle tends to maintain almost a constant angle. This maladjustment of the left ankle mobility can be associated with pain or physical restrictions of the range of motion. Again, higher walking velocities lead to lower wave lengths.

IV. Conclusions

This study offers some particular information regarding the gait characteristics of a selected subject for three current walking velocities imposed by treadmill equipment. The particularities of the results are generated by the subject's mild instability in left ankle.

The influence of the treadmill velocity on the gait spatial-temporal parameters should be always considered when perform treadmill-based gait analysis. The normal (3km/h) and higher waking velocities prove to lead in this case to a lower variability of the gait parameters allowing identifying all the gait complexes. Much higher velocity could modify some complexes of the gait cycles leading to inexact interpretations.

No significant differences were recorded between the hip and knee angles of the left and right lower limbs. But, due to the local instability of the left ankle, significant differences were recorded at this level between the left and right joint angles.

The mild instability of the left ankle produces a reduced effect in knee and hip joints, as we can conclude from the *P* values (table2). In ankle joint, where we expected larger instability of the left lower limb when the velocity is rising, we observed no major changing. On the other hand, at 5km/h waking velocity the gait complexes of the right limb are better evidenced, exhibiting lower standard deviations.

The mean standard deviation computed for the gait cycles presents lower values for higher velocities of movement. This finding indicates repetitive movements during the successive gait cycles at each velocity. Therefore, recordings having lower standard deviations are better representing the characteristics of the gait. The mild instability of the left ankle induces less repetitive results at 1km/h walking velocity, which are evidenced by larger standard deviations.

In the future work, higher velocities will be involved in the study, in order to identify which one offers a better compromise between having low standard deviations and keeping away from the running movement domain.

The treadmill could be a useful device in routine gait analysis since it allows recording of a large number of successive strides in a limited space and over a wide range of steady-state gait velocities. Still, more practice and experience on treadmill gait analysis should be involved in future studies. It will be the task of further studies to determine the optimum treadmill velocity according to the study objective and the subject particularities (healthy or with certain pathology, young or elderly people, freely motion or based on harness/handrails aids).

Further research will be performed with a representative lot of subjects, both healthy and having certain disorder to confirm the present results, collect reference data and validate the method across pathologies.

References

1. Kirtley C (2006) Clinical Gait Analysis, Theory and Practice, Churchill Livingstone Elsevier.
2. Edginton K A, Güler H C, Ober J J et al. Instrumented Treadmills: Reducing the Need for Gait Labs, Bertec Corporation, Columbus, Ohio, USA at http://bertec.com/uploads/pdfs/papers
3. Jung T, Gilgannon M, Munjal R et al. Treadmill-based Gait Analysis for Children with Cerebral Palsy: Biomechanical Comparison of Treadmill and Overground Walking, Kluge Children's Rehabilitation Center at University of Virginia at http://gait.aidi.udel.edu/gaitlab/gcma/info/abstracts/O1.abs20151.pdf
4. Li G, Kozanek M, Hosseini A et al. New fluoroscopic imaging technique for investigation of 6DOF knee kinematics during treadmill gait (2009) Journal of Orthopaedic Surgery and Research, 4:6 DOI10.1186/1749-799X-4-6
5. Tyrell C M, Roos M A, Rudolph K S, Reisman D S (2011) Influence of systematic increases in treadmill walking speed on gait kinematics after stroke, Phys Ther 91(3):392-403
6. Bayat R, Barbeau H, Lamontagne A (2005) Speed and Temporal-Distance Adaptations during Treadmill and Overground Walking Following Stroke Neurorehabil Neural Repair vol. 19 no. 2 115-124 DOI 10.1177/1545968305275286
7. Stoquart G, Detrembleur C, Lejeune T (2008) Effect of speed on kinematic, kinetic, electromyographic and energetic reference values during treadmill walking, Neurophysiologie Clinique/Clinical Neurophysiology, 38(2): 105-116 DOI:10.1016/j.neucli.2008.02.002
8. Beauchet O, Annweiler C, Lecordroch Y et al. (2009) Walking speed-related changes in stride time variability: effects of decreased speed, Journal of NeuroEngineering and Rehabilitation, 6:32 DOI10.1186/1743-0003-6-32
9. Masani K, Kouzaki M, Fukunaga T (2002) Variability of ground reaction forces during treadmill walking, *J Appl Physiol* 92: 1885–1890 DOI10.1152/japplphysiol.00969.2000
10. Stoia D I, Toth-Taşcău M (2011) Comparison of treadmill-based and overground gait analysis, International Conference on Advancements of Medicine and Health Care through Technology MediTech 2011, Cluj-Napoca, submitted

Author: Mirela Toth-Tascau
Institute: "Politehnica" University of Timisoara
Street: Bd. Mihai Viteazu, No.1
City: Timisoara
Country: Romania
Email: mirela@cmpicsu.upt.ro

Assisted Scanning Techniques Optimization with Application in Biomechanics

B.C. Braun, I.C. Rosca, C. N. Druga, and M. Ionescu

TRANSILVANIA University/Advanced Mechatronic Systems, Brasov, Romania

Abstract— **The paper presents a software interface dedicated to indicate the proper method used for the scanning different complex models with application in Biomechanics, like orthopedic orthesis. There is described the way in which certain physical models with application in Biomechanics were scanned using two different methods. The results on the scanning precision were synthesized as statistics into a software application created by us for this reason. Finally a conclusion about the proper scanning method depending on Biomechanics application can be obtained and established for our further research.**

Keywords— **scanning, routine, accuracy, method, programming.**

I. INTRODUCTION

Due to the fact that nowadays the rapid prototyping in Biomechanics is more and more required, the scanning technology is a concept that concern the researches. The most suggestive examples refer to the prosthesis and orthesis prototyping. An efficient and proper prototyping requires previously a suitable scanning process to generate a CAD model faithful to biomechanical element for which the prosthesis or orthesis prototyping is realized.

II. THE RESEARCH OBJECT

Due to the fact that some important parameters, like walking or standing stability affect directly the life's quality, we have choose as study object the orthopedic orthosis.

Our research issue is to find a low cost and efficient method to prototype a family of plantar orthesis which role is to correct progressively any diseases for human subjects.

About foot sole scanning, different dedicated devices having their associated software environment are largely used. *Foto Scan 3D Handheld Scanner* or *Foto Scan 3D for custom insole* (United Kingdom) could be considered as representative examples. The information about plantar pressure can be registered and processed through their associated software systems. As a result, a decision on the plantar orthotic achievement items could be taken.

Until now, different types of plantar orthosis were developed, like: plantar supporters, foot inserts, calcaneal sights orthesis for correction of static foot etc. [1], [2].

Although the actual used methods for the corrective plantar orthosis obtaining are very efficient, their main disadvantage refers to the costs caused to the used complex equipment. For this reason, the main purpose of our studies is to establish an efficient, non expansive and accurate method for orthopedic orthesis prototyping. One of their functions could be to correct progressively the walking or stability parameters for persons having different diseases [1]. Another destination could be the stimulation of certain nervous centers in the plantar area for the persons presenting diseases of internal organs.

Our current researches aim is to develop an efficient method as to ensure a rapid and low cost prototyping process for some elements that could compose such a plantar orthosis.

Our actual research refers to the first stage on prototyping, namely the proper and efficient scanning process for the initial CAD models that will lead to the final design of the prototypes. For this reason, the aspect of interest referred to the scanning precision for some foot sole physical models obtained previously.

The models were provided from 5 persons aged between 25 and 40 years, without disabilities. For each human subject, two physical models were made.

For a more accurate statistical determination in terms of accuracy of scanning was necessary to achieve a larger number of physical models, each of which being scanned. Based on information obtained in this respect, our research found to be sufficient development of physical models for five subjects.

The procedure to obtain a physical model for a human foot sole was the following: the foot was introduced into a plaster mass, the model being obtained as a negative profile of the human foot.

III. THE SCANNING METHODS DESCRIPTION

The applied procedure to obtain the physical model was repeated for both foot of each of the five evaluated persons.

Once the physical models were obtained, we identified two different solutions for their scanning. Each of them relies on the two existing equipment as standard our research department.

The first scanning method invoked the using of the *DEA GLOBAL Performance 05 x 05 x 05* (Italy) series coordinate

S. Vlad and R.V. Ciupa (Eds.): MEDITECH 2011, IFMBE Proceedings 36, pp. 376–379, 2011.
www.springerlink.com

measuring machine permitting the scanning with contact. The main technical characteristics are presented below:

Table 1 The main technical specification of the used coordinate measuring machine [3]

Model	05 x 05 x 05
Measuring range along along the three axes of coordinate system [mm]	500 x 500 x 500
Measuring head ordering	automatic, by software
Software interface environment	PC-DMIS, 4.1.1
Measuring/scanning accuracy [mm]	0.001

The second method was based on the principle of non-contact scanning, for it using the *ExaScan 30144* portable scanner with laser beam, which technical characteristics are presented below:

Table 2 The technical characteristics for the used non contact principle portable scanner [4]

Model	*ExaScan 30144*
Measuring range [mm]	± 22.5
Measuring frequency [no of emission-reception cycles/s]	25000
Measuring resolution [mm]	0.05
Scanning distance [mm]	300

The aim of our research was know which method could be applied successfully for clinical investigations in human subjects for the purpose CAD modeling and orthotic plantar prototyping elements.

When applying the first method, we established the step scan of 3 mm. The reason for choosing this value is related to both the process and the quality of the surface point cloud formed after scanning. Thus, experimentally, after repeating the procedure with different values of step scan, it was observed that for this step value the machine precision scanning was not affected. Under these conditions the process of scanning a physical model lasted approximately 3 hours.

Through the first method, the scanned surfaces were imported as CAD models, in .IGS or .STL format, as cloud of points. For a complete and correct scanning procedure we proceeded to the finite scanning portions, allowing their perfect alignment and CAD model 3D reconstruction [5]. The way to apply the contact scanning is presented in the figure 1.

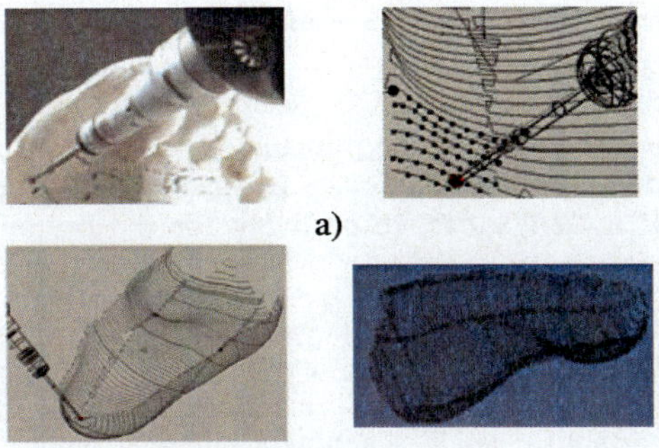

a)

b)

Fig. 1 The contact scanning procedure: a) the scanning process; b) the CAD model obtaining [4]

For the non contact principle scanning method, the scanning parameters on distance and laser beam intensity were established directly by software, using the option adjust automatically. For a better scanning process being used some marks placed at irregular distances [3]. The scanning duration for one model was about 20 minutes. The scanning procedure without contact is illustrated in the figure 2. The 3D CAD model was exported as .STL file, as shape surface. It will be used for further processing to generate its reference surface against which it can be constructed and modeled some orthotic plantar elements.

Once established the best method from the point of view of scanning speed, our research was concentrated on the scanning accuracy.

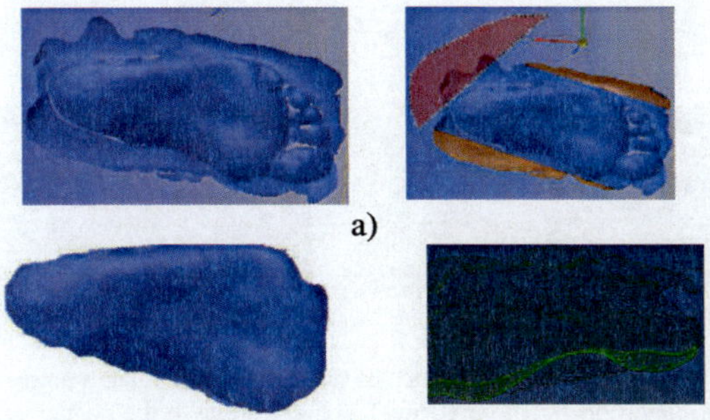

a)

b)

Fig. 2 The scanning procedure without contact: a) the scanning process including the adjusting; b) the CAD model obtaining [6]

The reason that research focused on assessing the two methods is that each of them may involve lower costs because there are not necessary complex equipment for human foot scanning foot.

Furthermore, our research aimed at identifying the scanning applications in Biomechanics, using portable scanner, its principle of operation being close to that used for specific scanning equipment, mentioned in section II.

Specifically, it aims to determine the precision of scanning for applying the principle of non-contact method, to know if this method can be successfully applied in clinical investigations, respectively for the orthotic items inspection.

IV. Scanning accuracy evaluation using PC

To establish which scanning method is better, our research was especially focused on scanning precision. For this reason, we have developed a software interface that allows the automatic determination of the results on the scanning precision for each one of both used methods [7].

To evaluate the best possible the precision in scanning as there it was taken as an example the physical model, that was scanned through the both presented methods. There were taken into account the following aspects:

- the differences between the measured global dimensions on the CAD models (the CAD obtained as cloud of points due to the contact principle scanning and the CAD model obtained as .STL file, due to the non contact scanning principle); the second CAD model was generated as continuous surface, but it could be viewed also as cloud of points;
- the differences regarding the coordinates of some representative points measured on the cloud of points surface reported to the same representative points coordinates measured on the second CAD surface (due to the non contact scanning principle) (figure 3).

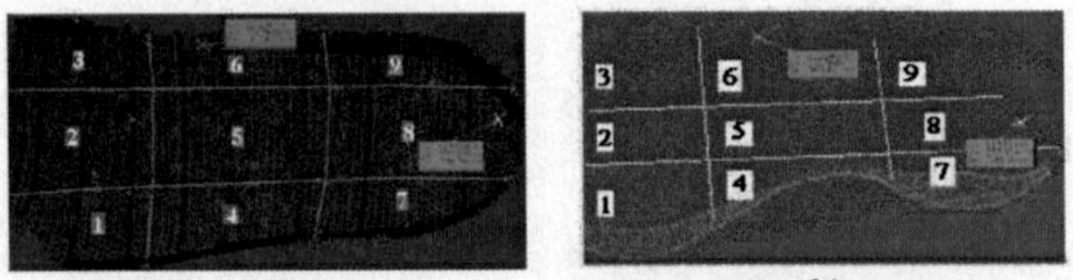

a) b)

Fig. 3 The measured coordinates of the same representative points on the CAD models: a) for the CAD model obtained due to the contact scanning method; b) for the CAD model obtained due to the non-contact scanning method

For this assessment to be made quickly and accurately, from our study on the interpretation of CAD models, it was established that the most effective method is to divide them into 9 areas of evaluation (figure 3). The division on the 9 areas was done as follows: 3 for the metatarsals, 3 for the arch and other 3 for the heel area.

For each area, as points of evaluation, there were taken the points of minimum and maximum share reported to the OZ axis. The differences between OZ coordinates of the same measured points for both CAD models have provided information on the differences between the two scans.

The points coordinates were measured directly on the CAD models and further the information was used for the statistical determination of the scanning deviation.

The entire procedure about the scanning process using both methods, followed by the CAD models evaluation was applied for each set of physical models providing from each of the 5 human subjects. This procedure was necessary for a more precise assessment of the accuracy of scanning physical models (see section II).

Due to our researches on the contact scanning principle leaded to the conclusion that this method could be considered as reference solution on the scanning precision. This conclusion was drawn after the scanning accuracy analyzing for one surface of a parallel-gauge - the measured deviation was about 4 um. Based on this finding it could be evaluated a deviation of about 10 μm in case irregular convex surfaces scanning.

For this reason, our research was focused to calculate the scanning deviation in case of non contact method applying reported to the with contact scanning method.

To obtain a statistical result as accurate and faster as possible, this algorithm has been implemented into a software application created in the *Lab VIEW 7.1* virtual environment. The reason to choose LabVIEW as programming environment was that it is efficient and proper for all statistic determination [8].

The purpose of the application is to determine an overall average deviation for the principle of non-contact scanning in relation to the case of scanning by contact.

For this, in the application software there it were placed the information on the evaluation of each CAD part model. There it was taken into account the scanning and evaluation of all physical models providing from the 5 subjects.To develop the application, a sequential programming structure was used, for each sequence a calculus algorithm being defined. The calculus algorithm is similar for each sequence. For this reason, each sequence includes a programming sequential structure with 2 steps (figure 4).

A program routine of the calculus algorithm is illustrated in the figure 5.

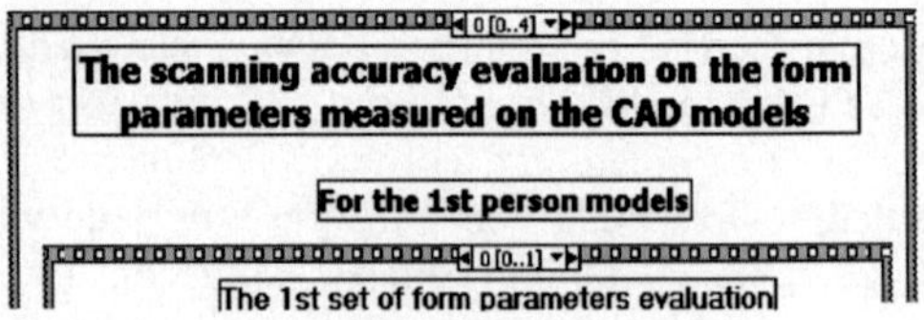

Fig. 4 Sequential structure programming for the data processing providing from the CAD models evaluation

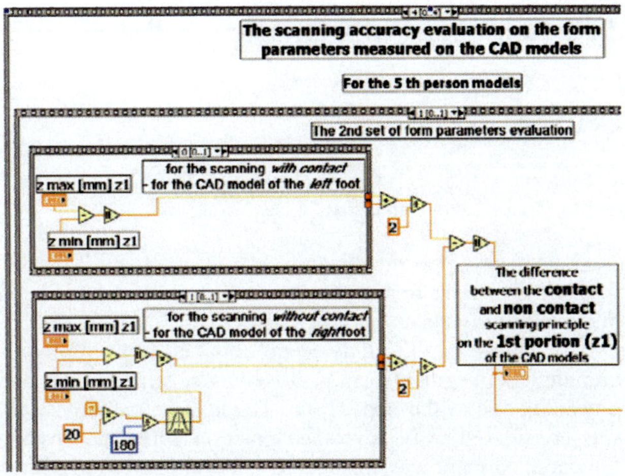

Fig. 5 Programming routine for the calculus algorithm for the scanning accuracy statistic determination on the models form parameters

In figure 5 there is presented a programming routine to calculate the difference between the measured values using both scanning methods, on the 1st portion of the CAD models (figure 3). Similar calculus algorithms were used for all form parameters (both sets). As a result, for the models associated to each tested subjects, the information on the scanning precision when using the portable scanner, non contact principle was synthesized (figure 6).

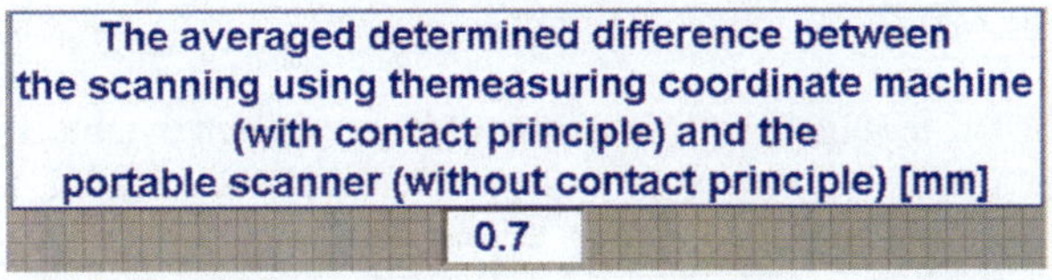

Fig. 6 The information on the scanning precision when applying the non contact scanning method

Taking also into account the determined difference on the scanning accuracy between the two used methods, we found that the scanning accuracy by non contact scanning is between ***0.65 mm and 0.75 mm***.

V. Conclusions

Based on the accuracy of the scan results, it could be concluded that non-contact scanning method using the portable scanner *ExaScan 30144* could be successfully applied to obtain the CAD reference model of the foot sole. It will be used for the some subsequent orthotic elements construction and CAD modeling. Furthermore, this method can be used even for direct scanning of the foot of the subject, if the foot remains fixed while scanning.

The other method being more accurate it could be used for the dimensional inspection and scanning of certain plantar orthotic elements for their design and form adjustment when prototyping.

If the orthotic items should be obtained from deformable material (silicone, shape memory material etc.), if they intended to correct some foot malformations (such as flat feet), could be recommended for non-contact scanning method. In case of progressive prototyping of some elements for the correction of certain deficiencies more difficult to be observed, we propose that for our next research, the non-contact scanning method to be matched by using the *Foot Scan* method. This refers to the subject's plantar pressures determination when walking and standing on a special plate [9]. The plate contains many pressure sensors, piezoelectric principle for a proper plantar pressure variation measuring along the foot sole.

Acknowledgment

This paper is supported by the Sector/Operational Program Human Resources Development (SOP HRD), ID 59323 and ID 59321, financed from the European Social Fund and by the Romanian Government under the project number POSTDRU/89/1.5/S/59323 and POSDRU/88/1.5/S/ 59321.

References

1. Plantar supporters at http://www.sensicare.com
2. Plantar orthosis at http://www.ortoprofil.ro
3. DEA Global 05x05x05, 2007 – The user's manual at http://hexagonmetrology.com
4. Xscan Catalogue - Handy 3D Scanners, 2008, at: http://www.handyscan3d.com
5. Wilcox Associates, Inc. PC-DMIS CMM User guide, for PC-DMIS, 2010
6. Derigent, W.; Chapotot, E.; Ris, G., Centre de Recherche en Automatique de Nancy – Cedex, France Controle des piece par capteur laser – definition de fonctions d'aide a la numerisation, CPI, Casablanca, Morrocco, 2005
7. Rosca, I.; Braun. B.; Marosy, Z. Computer aided optimization of quality inspection methods for complex models used in Biomechanics, DAAAM International Vienna - Austria – EU, pp 0053, 0054
8. Cottet, Fr.; Ciobanu, O. Bazele programării în LabVIEW, Publishing house MATRIX ROM Bucureşti, 2008
9. Papilian, V. Anatomia omului, vol 1 ISBN 973-571-262-8, Publishing house BIC ALL Bucharest, Romania, 1998

Author: Barbu Cristian Braun
Institute: Transilvania University, Advanced Mechatronic Systems Department
Street: 2, Teatrului Place, sc. B, ap. 9
City: Brasov
Country: Romania
Email: braun@unitbv.ro

Statistical Analysis of Anthropometric and Physiologic Performance of the Hand

I. Serban[1], M. Baritz[1], I.C. Rosca[1], and L.D. Cotoros[2]

[1] Fine Mechanics and Mechatronics / Advanced Mechatronics Systems, Transilvania University, Brasov, Romania
[2] Mechanical Engineering / Advanced Mechatronics Systems, Transilvania University, Brasov, Romania

Abstract— **All the experiments are, and should be, based on a strong statistical analysis that give a generalization of the cases experimented. This article is about the statistical analysis of data obtained from a population formed by 200 subjects mean aged of 20 years composed from 116 masculine gender and 91 of feminine gender. These subjects have been analyzed anthropometrically and physiologically. The experiment is looking forward to build up a database consisting of enough data to establish the base for a study regarding the physiology and anthropometry of the upper limb, especially the hand. It would statically evaluate and determine the physiologic limits of the upper limb according to some parameters took into consideration. This study can be useful to guide the production of prosthesis and orthotics, adapted to the upper limb. The study was taken at Transilvania University of Brasov.**

Keywords— **biomechanics, anthropometry, statistical analysis, physiologic performance, upper limb.**

I. Introduction

Anthropometry is a branch of anthropology that studies the physical measurements of the human body to determine differences or similarities in individuals and groups. In the past the major emphasis of these studies has been evolutionary and historical. Nowadays there is a special need coming from technological developments (ergonomically workspace design), as well as from industries that deal with products that need to fit ergonomically to the human body, its movement and its cognitive abilities.

Consideration of the hand anthropometrics in design process is essential [1], as well as the physiologic performance.

The relationship between anthropometry and ergonomics is that in ergonomics, anthropometric data is used as a support in designing equipment, workspaces and devices that need to fit the human body considering the differences between the characteristics, abilities, and physical limits of the human body, to which they interact.

All the studies, starting from Leonardo da Vinci who was the first person that looked at a person from an anthropometry point of view, were conducted to find relationships between the sizes of different parts of the human body and that body's height or weight. The measurements have been made in vitro (cadavers) as well as in vivo.

The values, of this study, were obtained from the measurement of segments and movements, anatomical and physiological, of the upper limb. The anatomical movement was considered to be a relaxed position while the physiological movement was considered to be strained to maxim.

The measurements took into consideration the right hand. We took into account the subjects with pathologies of the upper limb as well as the left handed, in which cases they were tested accordingly.

All the subjects provided informed consent prior to the test. They have been informed about the protocol of the experiment and the devices that were used.

The statistical analysis: the mean, standard deviation, percentile, probability density and cumulative probability for each group of subjects (divided by gender), for each type of measurement in part.

II. Method

Devices used in the measurements are shown in table 1.

Table 1 Devices used

Device/ brand	Range	Precision
Dynamometer for hand/ Jamar (Fig. 1, a)	0-90 (kgs)	0.1
Goniometer/ Jamar (Fig.1, b)	0-180 (degree)	1
Pinch gauge/ Baseline (Fig. 1,c)	0-30 (lbs)	1
Caliper for length/ Lafayette (Fig.1, d)	0-60 (cm)	0.1
Ruler for finger perimeter/ Richardson (Fig.1, e)	0-14 (cm)	0.1

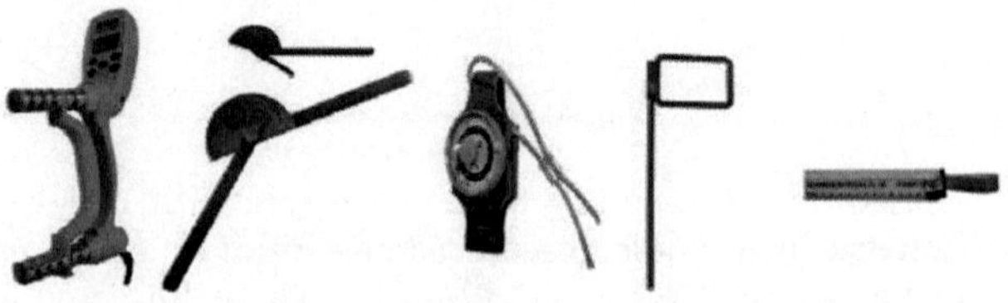

a) b) c) d) e)

Fig. 1 Devices that was used in the experiment

S. Vlad and R.V. Ciupa (Eds.): MEDITECH 2011, IFMBE Proceedings 36, pp. 380–383, 2011.
www.springerlink.com

Mean characteristics of the subjects that have been measured are shown in table 2. The height was measured using a stadiometer, and the weight was measured using a property of the Kistler plate that operates on the piezoelectric principle and has an accuracy of hundredths.

Table 2 Mean characteristics of the measured subjects

All subjects	Male, n=116 Average±Standard Deviation	Female, n=91 Average±Standard Deviation
Mean age (years)	20.9± 3.1	20.5± 1.9
Height (cm)	1.8± 0.1	1.7± 0.1
Weight (kg)	75.9± 13.1	56.3± 6.8

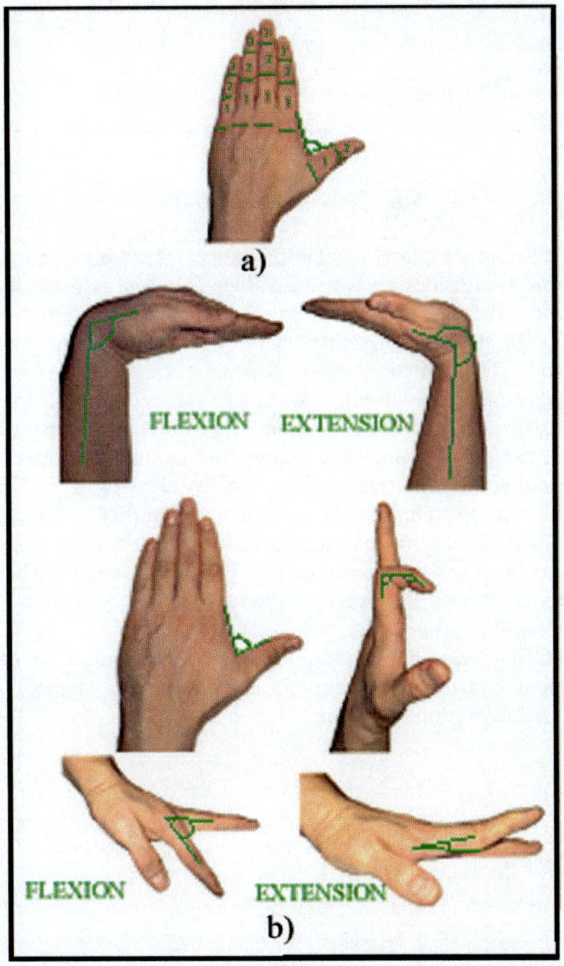

Fig. 2 Illustration of the dimensions that were measured: a) numbering and delimitation of the phalanx, b) position of the segments after the required movement

Figure 2 is an illustration of the segments position from the upper limb. This was also used for a better understanding, from the subjects, of what they are required to do in order to eliminate the human error. This is the reason the ranges of the movements are shown indirectly, normally it is between the neutral position and the final position.

In table 3 there are the values (mean, standard deviation) that correspond to the measurements made from anthropometric point of view. The abbreviations from the table represent, for every finger, the length (L) and perimeter (P) numbered according to the number of phalanx and figure 2.a.

Table 3 Values of length and perimeter of the upper limb segments

Upper limb segments	Male, n=116 Av. ± St. Dev. (cm)	Female, n=91 Av. ± St. Dev. (cm)
Thumb		
L1	41.2±3.1	35.3± 2.6
P1	6.5±0.6	5.7± 0.3
L2	32.5±2.4	28.6± 2.0
P2	5.9±0.5	5.2± 0.3
Ltot	70.2±3.8	62.1± 3.3
Index finger		
L1	54.1±4.2	47.4± 3.4
P1	6.4±0.6	5.8± 0.3
L2	30.8±2.1	27.2± 2.0
P2	5.5±0.5	4.9± 0.3
L3	25.1±2.3	22.0± 1.2
P3	4.7±0.4	4.2± 0.3
Ltot	103.0± 6.0	91.8± 5.0
Middle finger		
L1	58.3± 5.6	52.7± 3.9
P1	6.3± 0.5	5.6± 0.4
L2	36.0± 2.6	31.6± 2.4
P2	5.6± 0.5	4.9± 0.3
L3	26.1±1.7	23.3± 1.4
P3	4.9± 0.4	4.3± 0.3
Ltot	114.5± 6.6	102.6± 5.4
Ring finger		
L1	56.1± 4.3	49.2± 3.3
P1	6.0± 0.5	5.4± 0.3
L2	34.4± 2.4	30.4± 2.2
P2	5.2± 0.5	4.6± 0.3
L3	26.1± 1.6	23.2± 1.2
P3	4.7± 0.4	4.1± 0.2
Ltot	109.4± 9.2	97.3± 5.9
Small finger		
L1	44.8± 3.3	39.4± 3.3
P1	5.4± 0.5	4.8± 0.4
L2	25.9± 2.2	22.9± 2.1
P2	4.8± 0.4	4.2± 0.3
L3	23.1± 1.7	20.5± 1.1
P3	4.3± 0.4	3.6± 0.3
Ltot	87.0± 5.4	77.4± 6.3
Hand length	19.9± 1.1	18.0± 0.9
Hand width	8.3± 0.5	7.4±0.4
Forearm length	27.2±1.6	24.4± 1.2
Arm length	33.2± 1.9	30.8± 1.9

The anatomical reference points for the measurements of the segments were obtained through palpation, visual inspection according to Dempster. [2]

Table 4 includes the angle values obtained according to figure 2.b. The movements that were taken into consideration were mainly the flexion and extension each of them in two different position anatomical (relaxed) and physiological (strained).

Table 4 Angle values

All subjects	Male, n=116 Av. ± St. Dev. (degrees)	Female, n=91 Av. ± St. Dev. (degrees)
Anatomical flexion angle between hand and forearm	85.6± 8.3	82.9± 9.0
Physiological flexion angle between hand and forearm	98.6± 9.5	94.8± 9.8
Anatomical extension angle between hand and forearm	224.3± 57.7	238.6± 43.5
Physiological extension angle between hand and forearm	215.7± 43.8	226.8± 33.5
Anatomical angle between thumb and hand	78.0± 11.6	77.2± 12.6
Physiological angle between thumb and hand	54.4± 10.0	48.0± 8.2
Anatomical flexion angle between index finger and hand	72.3± 7.6	74.1± 7.3
Physiological flexion angle between index finger and hand	37.1± 7.9	40.3± 8.6
Anatomical extension angle between index finger and hand	35.8± 7.5	34.5± 10.0
Physiological extension angle between index finger and hand	19.7± 6.4	15.5± 5.9
Anatomical flexion angle between phalanx 1 and 2 of the index finger	69.4± 6.1	4.6± 69.9
Physiological flexion angle between phalanx 1 and 2 of the index finger	104.8± 11.8	108.5± 11.0
Anatomical flexion angle between phalanx 2 and 3 of the index finger	97.5± 8.5	98.2± 5.5
Physiological flexion angle between phalanx 2 and 3 of the index finger	129.6± 11.1	134.7± 10.6

According to the table included in the user's guide of the Jamar dynamometer for the subjects aged from 20 to 24 years, right handed, the average (Av.) ± standard deviation (St.dev.) is 54.9(kg) ± 9.3 for males and respectively 31.9(kg) ± 6.6 for females. While the values obtained from the test are shown in table 5. There is an obvious difference between other literature references. [3]

Table 5 Force values

All subjects	Male, n=116 Av. ± St. Dev. (KG)	Female, n=91 Av. ± St. Dev. (KG)
Anatomical hand grip strength	48.5± 9.0	28.4± 5.5
Physiological hand grip strength	17.2± 10.1	10.5± 6.6
Anatomical finger force	9.1± 1.9	6.4± 1.2
Physiological finger force	4.0±1.9	3.0± 1.4

III. Statistical Analysis

For the statistical evaluation of the values obtained from the test, considering that the values follow a normal distribution, there has been calculated the probability density and the cumulative probability according to the functions from Excel. The function used, Normdist (x, mean, standard deviation, cumulative), calculates the probability that a measured value, falls at or below a given value. Considering that the abbreviations are known, the cumulative value can be either FALSE returning probability density or TRUE in which case it returns the cumulative probability.

First all the values from a row, determined by gender, have been ordered ascending and assigned a rank. Then it was calculated the average, the standard deviation and the percentile in this order.

Considering the variable x normally distributed, the theoretical equations that express the probability density and cumulative probability are [4]:

$$f_{(x)} = \frac{1}{\sigma\sqrt{2\pi}} \cdot e^{-\frac{(x-m)^2}{2\cdot\sigma^2}}, (x \in R, \sigma > 0) \quad (1)$$

respectively:

$$F_{(x)} = \int_{-\infty}^{x} f_{(x)} dx = \frac{1}{\sigma\sqrt{2\pi}} \cdot \int_{-\infty}^{x} e^{-\frac{(x-m)^2}{2\cdot\sigma^2}} dx \quad (2)$$

where: σ is the standard deviation; m is the average and σ^2 is the variance.

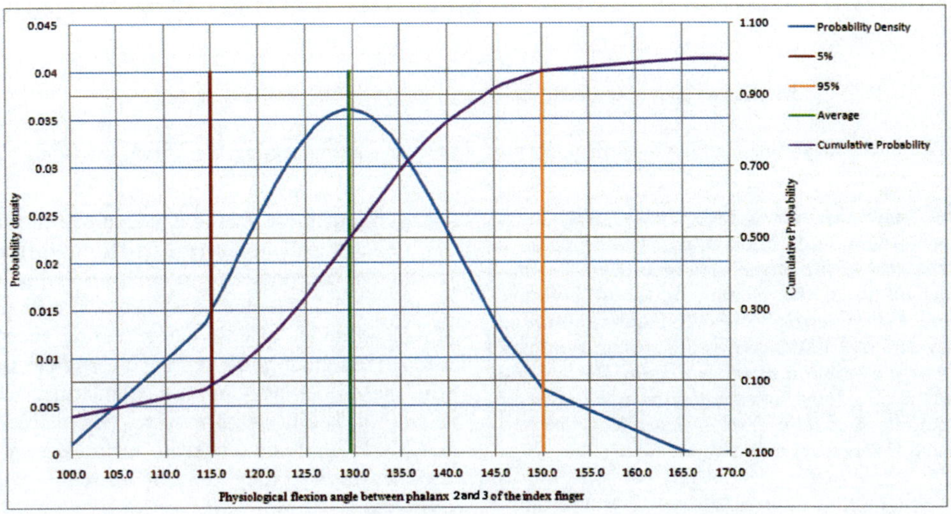

Fig. 3 Distribution of the values of physiological flexion angle between phalanx 1 and 2 of the index finger

In figure 3 is the normal distribution, as an example, of a row of data which reflects physiological flexion angle between phalanx 2 and 3 of the index finger. The graph has three axes (two vertical and one horizontal) for a better integration of the functions used to analyze the data.

This type of graphical analysis can be used for each row in particular. It offers a view of the probability density, cumulative probability, values variation, distribution of values regarding the average, standard deviation and percentile.

IV. Conclusions

This paper aims for an in vivo database of the Romanian subjects with mean age of 20 years. This is intended to be used as a reference for the design and development of anthropometric device for the standardization of sizes of hand orthotics and for recovery devices. The experiment reveals the limits for the performance and personal satisfaction of the subjects.

All the values obtained from the experiment have been analyzed with the statistical tools in order to obtain a view of the density of the measurements according to genre, height, weight, age of the subjects. In the near future there will be made a more closely investigation of the values using other types of statistical tools.

Acknowledgment

This paper is supported by the Sectoral Operational Programme Human Resources Development (SOP HRD), ID 6600 financed from the European Social Fund and by the Romanian Government References.

These researches are part of the Grant PNII-IDEI 722 with CNCSIS Romania and we've developed the investigations with apparatus from Research Project "CAPACITATI" and "IDEI 722" in Mechatronic Researches Department from University Transylvania of Brasov.

References

1. Marras W.S., Karwowski W. (2006) The occupational ergonomics handbook, 2nd ed., Taylor & Francis, London
2. Dempster, W.T. (1955) Space Requirements of the Seated Operator. Aerospace Medical Research Laboratory WADC technical report 55 159
3. Barut C., Demirel P., Kiran S. (2008) Evaluation of hand anthropometric measurements and grip strength in basketball, volleyball and handball players, International Journal of Experimental and Clinical Anatomy, pg. 55- 59
4. Pheasant S. (2003) Bodyspace, anthropometry, ergonomics and the design of work, 2nd ed., Taylor & Francis, London

Author: Ionel SERBAN
Institute: Transilvania University
Street: Vlad Tepes
City: Brasov
Country: Romania
Email: ionel.serban@unitbv.ro
serban_ionel1984@yahoo.com

Determining the Center of Pressure Trajectories during Lumbar Spine Flexion

M. Toth-Tascau, C. Saftescu-Jescu, D. Bugariu, and L. Bereteu

"Politehnica" University of Timisoara, Faculty of Mechanical Engineering, Timisoara, Romania

Abstract— **The paper presents a preliminary study on a healthy female, performing back flexions, while standing on a force platform. The goal of the paper is to quantify how lumbar spine flexions influence the plantar center of pressure (CoP) distribution. Using Zebris CMS-HS ultrasound-based motion analysis system and FDM measuring system for force distribution the range of motion of lumbar spine flexion and CoP displacements during fully forward flexion were assessed simultaneously. Results showed a close symmetrical distribution of ground reaction forces on each subject foot.**

Keywords— **lumbar spine flexion, range of motion, center of pressure, force platform, ultrasound-based motion analysis system.**

I. Introduction

The ability to maintain balance is fundamental in uni- bipedal standing or in performing certain movements, some of them being essential in daily living.

Centre of pressure (CoP) is defined as the application point of the resultant of ground reaction forces acting on the base of support. The analysis of CoP displacements is used as an index of postural stability both in standing and walking. A common technique to determine the postural stability is represented by platform stabilometry. This implies a set of pressure transducers, contained into a force plate, which record the ground reaction forces and determines the centre of pressure (CoP) and body centre of mass (CoM). There are many commercial available instrumented force platforms that can be involved in gait analysis and postural stability studies, allowing accurate recordings of ground reaction force, CoP and CoM displacements.

The studies reviewed by Ruhe et al. [1] show that "bipedal static CoP measures may be used as a reliable tool for investigating general postural stability and balance performance under specific conditions". The influence of the CoP displacements during standing or performing certain exercises has been studied especially for elderly people. Seigle et al. have been investigated whether aging has an influence on the dynamics of the fluctuations of the displacement of the center of pressure (CoP), during quiet standing, taking into account the visual conditions of subjects [2].

The assessment of lumbar spinal disorders requires data about spinal range of motion. There have been proposed and developed several methods to measure the spinal mobility [3]. Some of these methods use an inclinometer, while others depend on measurements from different anatomical landmarks (original and modified Schober method), or use finger to floor distance to asses forward flexion (Viitanen, et al., Heikkilä et al.). In some cases, such measurements, using fixed anatomical points, are subject to errors. Moreover, these methods are not standardized or assessed for reliability, validity, and may not be feasible in clinical practice [3]. Even if some of these methods have good reliability and sensitivity, they do not correlate with radiographic changes.

The presented study proposes a method to simultaneously assess the range of motion of lumbar spine flexion and CoP displacements during fully forward flexion. Zebris CMS-HS ultrasound-based motion analysis system and FDM measuring system for force distribution are accurate and easy to use, allowing unrestricted and comfortable movements of the subject. This preliminary study quantifies the influence of the lumbar spine flexion on the CoP displacement and subject postural stability.

II. Materials and methods

To evaluate the angular amplitudes of lumbar spine flexion, one young healthy volunteer (female, 26 years old) performed exercises with self-selected velocity both overground and on force platform. Prior beginning the experiment, a series of anthropometrical parameters was measured, as shown in Table 1.

The measurements were realized in Motion analysis Laboratory of Politehnica University of Timisoara using two Zebris measuring systems (Zebris Medical GmbH): CMS-HS ultrasound-based motion analysis system and FDM measuring system for force distribution.

Table 1 Anthropometric data

Subject characteristics	
Weight [kg]	76
Height [cm]	167
Body Mass Index [kg/m^2]	26.9
Big toe distance – BTD [mm]	226
Inter-malleolar distance – IMD [mm]	78
Feet opening angle – α [°]	28

S. Vlad and R.V. Ciupa (Eds.): MEDITECH 2011, IFMBE Proceedings 36, pp. 384–387, 2011.
www.springerlink.com

The measuring method used by Zebris CMS-HS system is based on the determination of spatial coordinates of the miniature ultrasound receptors (markers), by measuring the time delay between the emission of sonic pulses by the transmitters and their reception at the microphone capsules. The spatial position of the lumbar spine was calculated by the WinSpine system's software.

The measurement starts with the attachment of two special marker triplets on the subject torso with elastic Velcro strips. The reference marker triplet is attached on the sacrum and triple measurement marker is applied in the area of the lower thoracic spine. The marker spatial positions are calculated and displayed during the subject movements, using the WinSpine Software. The measuring sensor was placed on the right hand side of the patient, according to the WinSpine Operating Instructions [4].

The second measuring system consists of a type 2 FDM platform integrated in a level walking area which contains high-quality capacitive pressure sensors arranged in a matrix form and WinFDM software enabling a simple static and dynamic analysis of the recorded measuring data [5].

The performed experiments consisted in two types of measurements of lumbar spine flexion range of motion (ROM): overground measurements only using the CMS-HS system, and on a force platform, using both the CMS-HS system and FDM platform.

Each experiment consisted in five valid trials, for each trial being recorded five free lumbar flexion movements. The sampling rate of the recordings was selected for both systems at 10 Hz, covering a period of 20 seconds per trial and 140 frames/trial. Each trial was preceded by a calibration, the subject standing in a neutral position. In order to adapt the subject movements to the measurement conditions a training session of five minutes was firstly performed.

Figure 1 illustrates the modules of Zebris CMS-HS system: measuring unit and the two special marker triplets, and neutral and maximum flexion position of the investigated subject.

The subject was free to choose her feet position to perform natural and comfortable movements. In order to obtain data about feet position during the measurements on the FDM platform and maintain the same subject position on the top half of the platform during the measurements, the static footprints were recorded by WinFDM software and drawn on a piece of paper fixed in place using adhesive tape.

The reference frame (RS) attached to the platform has the origin in the left lower corner of the platform, the x axis representing the platform width/medio-lateral (ML) direction, and y axis representing the platform length/antero-posterior (AP) direction, respectively, as shown in figure 2.

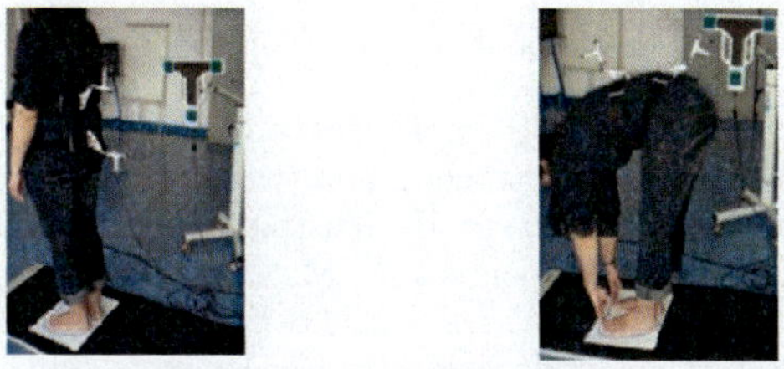

a) Neutral position and the modules of Zebris CMS-HS system

b) Measurement of flexion ROM with subject on the force platform

Fig. 1 Measurements of lumbar flexion ROM

Using the footprints obtained by static measurement of plantar pressure distribution, the parameters of base support were determined (Table 1 and Figure 3): effective foot length, big toe distance, inter-malleolar distance, and feet opening angle was calculated [6]. These parameters influence the postural stability, being of great importance in balance control. They will be considered in future studies in order to establish their influence on overall body balancing.

Using the Zebris CMS-HS system, the angular amplitudes during lumbar flexion were determined. WinSpine software generates a report of each recorded trial. Using the Zebris FDM force platform, the ground reaction forces and time-varying displacements of CoP under each subject' foot during the flexion movement were determined. The displacements of CoP were determined along the antero-posterior (AP) and medio-lateral (ML) directions.

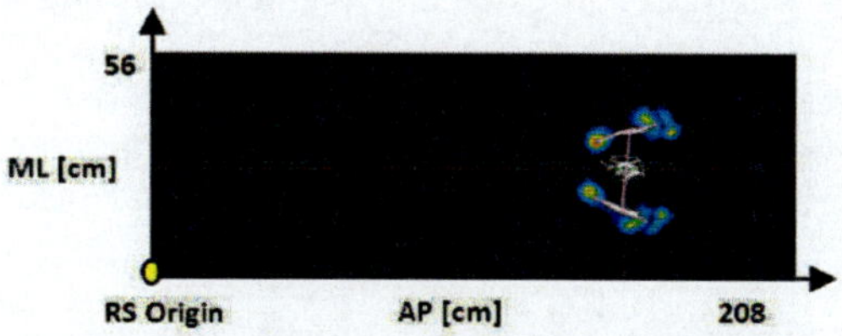

Fig. 2 2D representation of the FDM platform

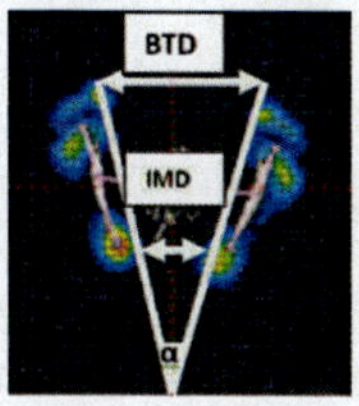

Fig. 3 Base of support measurements from the footprints

III. Results and discussions

Firstly, the angular amplitudes during lumbar flexion performed in overground conditions were obtained from the report generated by the Zebris WinSpine software as ASCII file. All data was processed in MatLab environment and evaluated. The flexion movements occur entirely within the normal range of motion according to normative data [4].

Next following, the same type of exercises were executed on the force platform. The corresponding mean curves of flexion angles with respect to the time for each trial are shown in figure 4 in frame representation, where T1 - T5 represent the trials recorded in overground conditions and on the force platform.

The mean intra-trial maximum flexion value was of 40.30° in overground movement while on the platform reached 39.28°. When comparing inter-trial maximum amplitudes, the mean values were 39.5° in overground movement, and 40.6° for measurements on the force platform. The minor differences obtained from intra- and inter-trial comparison show a good repeatability of the imposed exercise.

The variability of the flexion movements both overground and on the force platform was evaluated in figure 5. To evidence the average variability, the standard deviations are represented as vertical lines accompanying each frame.

In overground movements, the standard deviation of flexion angles ranges from 0.23° to 1.39°, while on the platform the standard deviation ranges from 0.07° to 2.45°, respectively. The mean inter-trial standard deviation of flexion angles was 0.72° in case of overground movements and 1.41° while standing on the platform.

Greater values of standard deviations in case of measuring on the platform are determined by the recording synchronization of the two measuring systems.

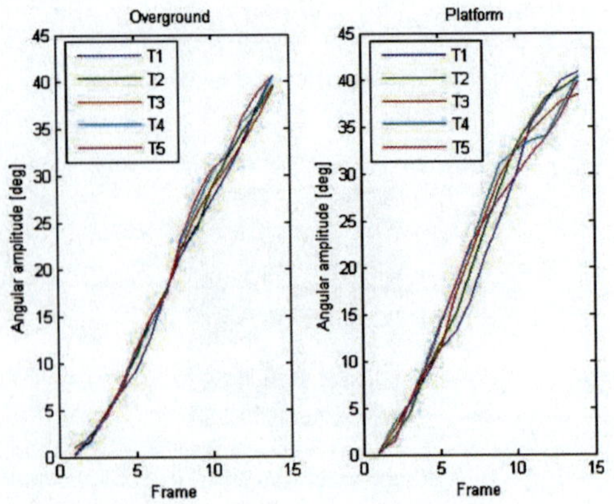

Fig. 4 Lumbar flexion angles measured overground and on the platform

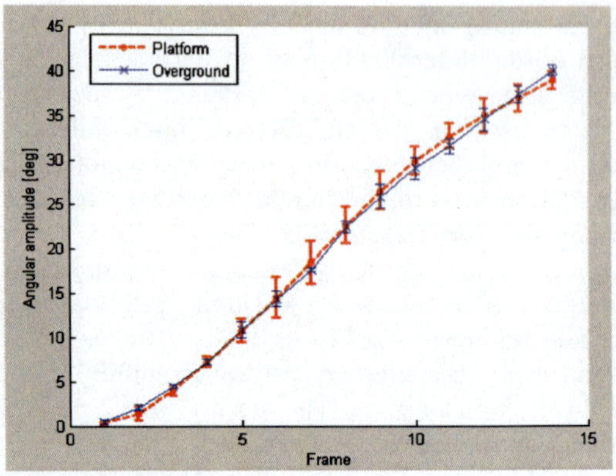

Fig. 5 Mean lumbar flexion angles in both cases

The means of values for discrete flexion angles in overground and force platform measurement conditions determined during the performed trials were compared by use of an unpaired t-test. The flexion angles were not significantly different ($p = 0.9874$) for the two studied cases.

During standing in neutral position, the platform recorded an unevenly distribution of ground reaction forces, the right foot being more loaded (55%) than the left one (45%).

For each flexion frame, both for the left and right foot, the CoP coordinates (x_{CoP}, y_{CoP}) were calculated using the classical mechanics method of determining the centre of parallel forces.

Figure 6 presents both the CoP distribution and mean CoP trajectory for each foot during lumbar flexion movement. The right foot is shown on the left side, with the toe covering the upper left region and the heel covering the lower right area. Analogically one can observe the left foot CoP distribution, as a mirror disposition.

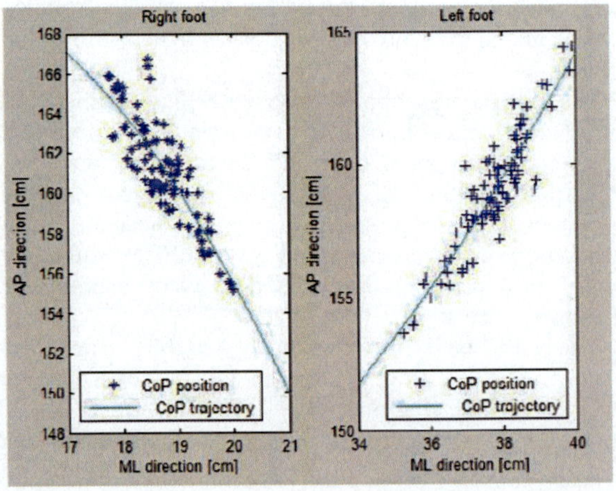

Fig. 6 CoP positions and trajectories overground and on the platform

Analyzing CoP distributions, one may observe the almost linear spread of positions, displayed in a symmetrical layout. The differences could be due to the feet initial pressure distribution (45%-left, 55%-right), or may be caused by a loss of balance during the exercise execution.

Taking into account that the flexion was performed along the AP direction, only these displacements of CoP were compared for the left and right foot. Thus, the means values of CoP positions in AP direction for both feet were compared by use of an unpaired t-test. The CoP positions were not significantly different ($p = 0.3134$) for the two feet. This indicates a strong correlation between both CoP distributions in AP direction during lumbar flexion.

Figure 7 presents the relationship between AP and ML displacements of CoP and flexion angles variation during spinal fully flexion. The CoP displacements along ML direction are almost linear, varying with less than 1 cm. The CoP trajectories in the AP direction show a slight turn, close to the maximum flexion angle. This is explained by the body tendency of maintaining the postural stability when reaching the lumbar flexion amplitude.

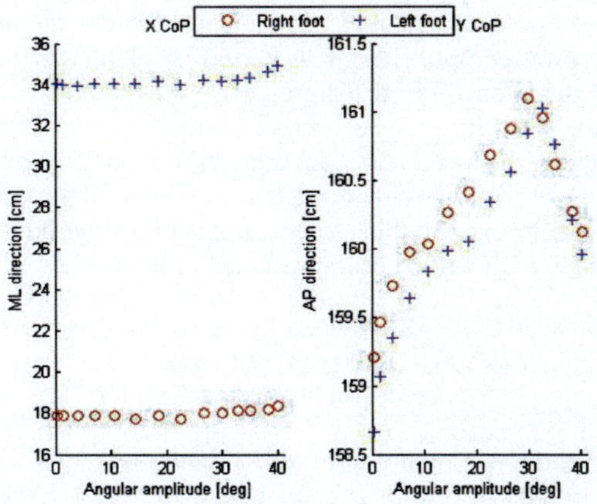

Fig. 7 Relationship between AP and ML displacements of CoP and flexion angles

IV. Conclusions

The presented study investigates a method to simultaneously and independently assess the range of motion of lumbar spine flexion and CoP displacements during the fully forward flexion.

Zebris CMS-HS ultrasound-based motion analysis system and FDM measuring system for force distribution were considered to be sufficiently accurate and easy to use, allowing unrestricted and comfortable movements of the subject.

Statistical analysis demonstrated that the flexion angles were not significantly different for measurements in overground and on platform conditions. The measuring accuracy is however influenced by the recording synchronization of the two measuring systems.

This preliminary study quantifies the influence of the lumbar spine flexion on the CoP displacement and subject postural stability. Results showed a close symmetrical distribution of ground reaction forces on each subject foot.

Further research will be performed with a representative lot of subjects, both healthy and having certain disorder to confirm the present results, collect reference data and validate the method across pathologies. Also, the influence of the subject weight on the CoP displacements will be taken into account. There will be considered different cases such as underweight, normal and overweight according to BMI category. Another index of postural stability both in standing and walking is the center of mass displacement, thus future studies will be focused on the analysis of center of mass dynamics.

Acknowledgment

This work was partially supported by the strategic grant POSDRU 6/1.5/s/13, (2008) of the Ministry of Labor, Family and Social Protection, Romania, co-financed by the European Social Fund – Investing in people.

References

1. Ruhe A, Fejer R, Walker B F (2010). The test–retest reliability of center of pressure measures in bipedal static task conditions – A systematic review of the literature. Gait & Posture, 32 (4). pp. 436-445 DOI:10.1016/j.gaitpost.2010.09.012
2. Seigle B, Ramdani S, Bernard P L, (2009). Dynamical structure of center of pressure fluctuations in elderly people. Gait & Posture 30, 223–226 DOI:10.1016/j.gaitpost.2009.05.005
3. Davis J C Jr, Gladman D D, (2007). Spinal Mobility Measures in Spondyloarthritis: Application of the OMERACT Filte, J Rheumatol Vol 34: No. 4, p. 666-670.
4. Zebris Medical GmbH (2006).WinSpine 2.x for Windows. Operating Instructions. Assessment of the mobility function of the cervical and lumbar spine.
5. Zebris Medical GmbH (2006). Measuring System for Analysis of Force Distribution FDM. Technical data and operating instructions.
6. Chiari L, Rocchi L, Cappello A, (2002) Stabilometric parameters are affected by anthropometry and foot placement, Clinical Biomechanics 17 666–677 doi:10.1016/S0268-0033(02)00107-9.

Author: Mirela Toth-Tascau
Institute: "Politehnica" University of Timisoara
Street: Bd. Mihai Viteazu, No.1
City: Timisoara
Country: Romania
Email: mirela@cmpicsu.upt.ro

CAD Methods for Orthopedic Orthosis Prototyping

B.C. Braun, I.C. Rosca, I. Serban, and C. Coblis

TRANSILVANIA University/Advanced Mechatronic Systems, Brasov, Romania

Abstract— **This paper presents an important stage of our research on finding an efficient and low cost method for the modeling and prototyping on orthotic elements with applications in biomechanics.**

Specifically CAD modeling is presented on the stage of the elements that will compose a family of plantar orthosis for locomotion and posture correction of certain human subjects who exhibit different locomotory diseases.

As a result of our research until now, we could identify the optimal method to scan the subjects' feet, regarding the costs, accuracy and scanning duration. From this it was possible to establish an efficient and proper procedure for the CAD modeling on the plantar orthotic items, the negative surface of the scanned human subject's being considered as reference.

Keywords— **CAD, scanning, orthosis, orthopedic, method.**

I. Current researches on the plantar orthosis obtaining

Stability and locomotion diseases affect largely the life's quality. For this reason, actually many researches are focused on prototyping improving with application in Biomechanics. Plantar orthosis to correct the locomotion and posture deficiencies is one of the main aspects on the research in Biomechanics.

For this reason, actually the modern devices for foot sole scanning, assisted by PC are largely used. Different devices for locomotion and stability parameters are also frequently used, most times associated with the plantar scanning devices. Different materials like silicon, plastic, memory foam are used to obtain the orthosis prototype respecting the subject's foot sole conformation. The information previously obtained due to the equipment used for any locomotion or posture investigation is taken into account for the plantar orthosis prototyping.

Until now, different kind of plantar orthosis were developed, like: foot inserts, correction of static foot orthosis, calcaneal sights, plantar supporters and so on [1], [2].

Although the actual method used for corrective plantar orthosis obtaining are very efficient, their main disadvantage refers to the costs caused to the used complex equipment. For this reason, our research aims is to develop a flexible method which could reduce the costs to obtain some orthosis elements to correct different stability or locomotion diseases.

II. On the CAD modeling for the ortopedic orthosis prototyping

Due to the fact that a lot of products having application in Biomechanics like prosthesis and orthosis have a complex profile, the CAD modeling plays a key role for prototyping in this domain.

The orthopedic orthosis prototyping represents a suggestive example for the CAD modeling with application in Biomechanics. Our research chose as study object the plantar orthosis due to the fact that they could help some persons having problems with their stability or locomotion to correct progressively this kind of diseases [1]. Besides, some parameters like stability when standing, walking or running are very important factors who determine the life quality. For this reason, a proper and efficient method for prototyping components of any kind of plantar orthosis is one of the aims of our research.

Until now, the study was concentrated on identifying and improving the procedure on the CAD modeling of some components which will complete a family of progressive plantar orthosis. In this order, two distinct methods were applied, which will be both presented in the paper.

III. The research on the CAD modeling for a plantar orthosis component

For the CAD modeling, two methods were applied. The first consists in the CAD modeling from a CAD reference model previously obtained by scanning with contact a real primary model. The used equipment to perform the scanning consisted into a 3D measuring coordinate machine.

The second method refers to the CAD modeling based on a CAD reference model previously obtained by scanning without contact the same real primary model. The non contact scanning was performed using a handy portable scanner with laser beam directed to the primary model surface.

The reason for which the two methods were proposed is to demonstrate that both methods could represent a good solution for the CAD modeling.

Due to our researches on the scanning accuracy and process duration, we found that the non contact method could represent the better solution. Thus because this method allows to scan directly the human foot, the

S. Vlad and R.V. Ciupa (Eds.): MEDITECH 2011, IFMBE Proceedings 36, pp. 388–391, 2011.
www.springerlink.com

procedure being similar to existing foot scanning methods for plantar orthosis prototyping.

However, in case of using the portable scanner, an important advantage refers to the relatively low costs of the equipment.

The real model was obtained through fingerprinting of a human subject's foot, into a plaster bed (figure 1).

To obtain the real model to be scanned, the following procedure was performed: the human foot was introduced into a plaster mass, the real model being obtained as a negative profile of the human foot (figure 1).

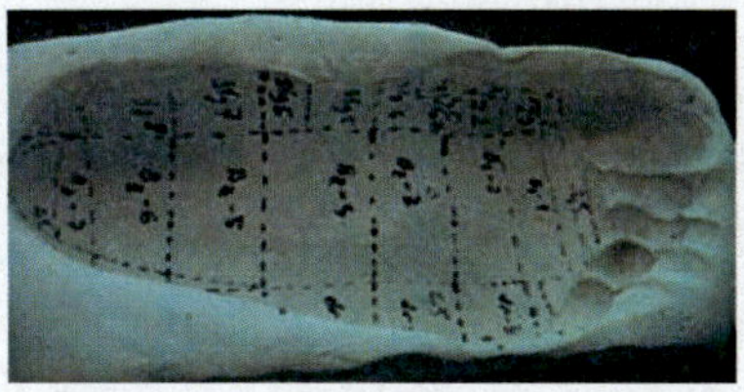

Fig. 1 The real model to be scanned for the CAD reference model

For both methods we identified the following steps on the CAD reconstruction of the foot fingerprinting primary model: the primary model obtaining by fingerprinting, the model scanning, CAD model generating and, finally the CAD modeling of the components that compose the orthopedic orthosis. The final step takes into account the CAD model obtained due to the scanning, as reference.

Our research was concentrated first of all to establish which scanning procedure could be most appropriate from the point of view of CAD modeling. For this reason, before applying both scanning methods, the necessary steps for the reference CAD model obtaining were identified.

When using the first method (scanning by contact), the reference CAD model was exported as .STP or .IGS file, the virtual model being generated as cloud of points (figure 2) [3]. Once exported, the model was imported in CATIA V5 CAD environment in order to be converted into a .STL format. In our research we found that this procedure had to be applied because the model can be converted from cloud of points into a 3D surface. For this reason once imported the model in the *Shape* mode, the conversion from cloud of points into a continuous surface was possible due to a Mesh procedure, then the initial cloud of points was removed (figure 4).

For the second method (involving scanning without contact), the CAD model was generated as .STL file, in this case it consisting into a canvas type surface (figure 3) [4].

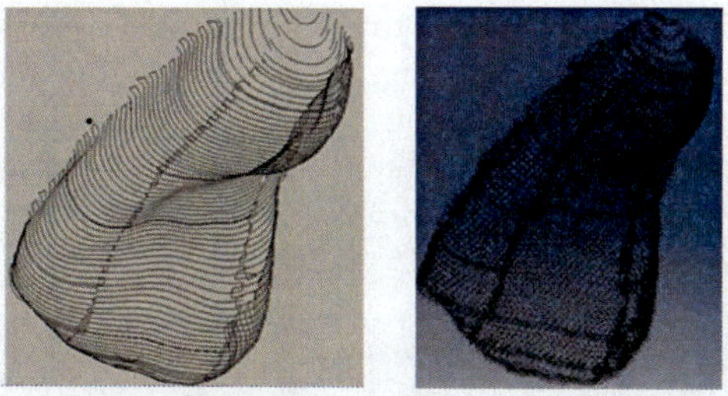

a) b)

Fig. 2 The reference CAD model generated due to the scanning by contact (1st method) obtaining: a) the scanned model viewed in the software environment associated to the coordinate measuring machine; b) the scanned model imported as cloud of points into the CATIA V5 CAD enviromnent

Fig. 3 The reference CAD model due to the scanning without contact (2nd method) obtaining

In this case the initial procedure regarding the cloud of points to surface conversion was no more necessary. Once imported the model into the CATIA V5 CAD environment, the first step was referred to the mesh creation (figure 4).

Once the meshing surface was performed, the established CAD modeling steps were the following: the reference surface generation (figure 5), the reference model for a plantar orthosis construction (figure 7) and, finally, the modeling of some components that could compose a plantar orthosis (figure 7).

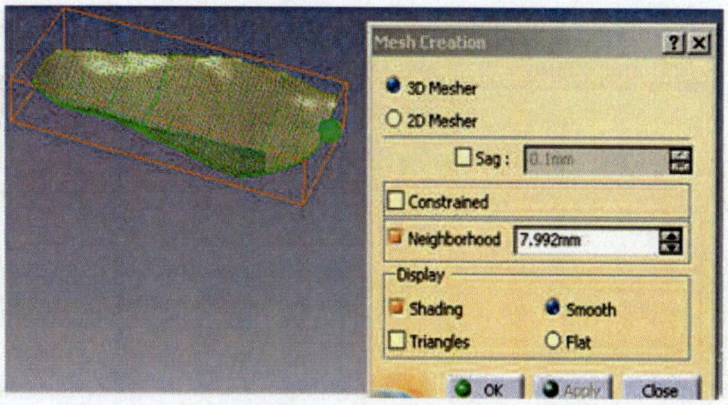

Fig. 4 The Mesh creation for the reference model

After the mesh creation, the following step consisted into the original surface erasing, obtaining just the reference surface to be furtherly processed. After that, the automatic surface was generated (figure 5), the necessary steps being presented [5].

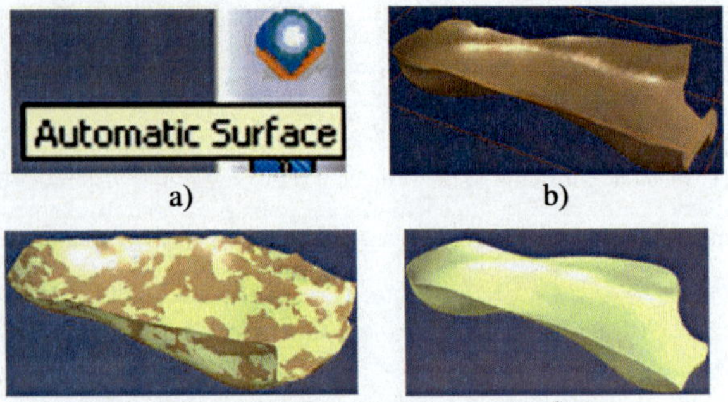

Fig. 5 The automatic surface applying over the reference surface: The automatic surface generation tool; b) The initial meshed surface; c) The meshed surface with the new generated surface supperposing; d) The automatic surface generation

After the automating surface applying, the superposition over the reference surface was obtained, as it can be seen in figure 5, b). This step was followed in order to obtain the reference shape surface for the orthosis construction. For this reason, the initial reference surface was deleted (figure 5, c).

For the global orthosis model construction the following steps were necessary: the selection of the sketch reference plane for the model profile; the profile extrusion, until the reference surface, previously obtained, the global 3D model generation (figure 6) and the insert's final model obtaining, due to some repetitive CAD cutting operation (figure 7).

Fig. 6 The 3D global model obtaining

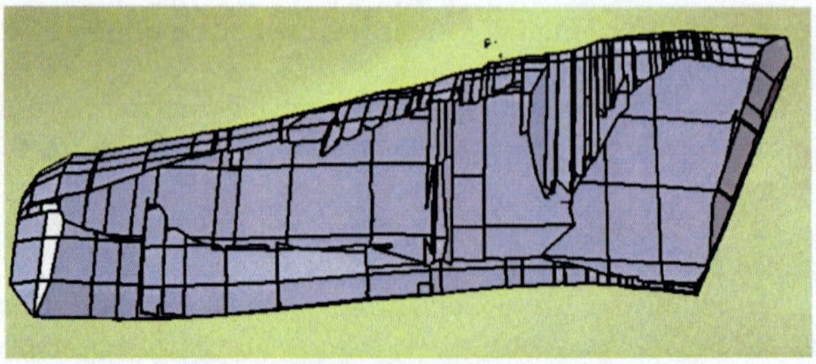

Fig. 7 The 3D reference model for a plantar orthosis respecting the reference CAD model

Once obtained the model the reference model for the plantar orthosis, our research is focused on the prototyping of different specific components that could compose a plantar orthosis.

The aim of our research is to create a family of plantar orthosis, composed by different prototyped components, which role is to correct progressively some diseases that affect the locomotion and stability parameters.

The main prototyping elements refer to different plantar areas, like: heel area, sole arch area, metatarsal or finger zone and so on.

Our actual researches are concentrated on the modeling and prototyping of an orthosis component for sole arch area.

It manifests a special interest because this foot sole area plays a key role on stability and locomotion of the human subject.

At this stage of our research the prototyping elements that will compose a family of orthosis planting for size 43 shoes is experimented, for the flatfoot correction that can affect both flat feet and walking stability. For this reason, until now, we have created the CAD model for the 1st orthosis element (figure 8). It was taken from the 3D reference model for the plantar orthosis, previously created.

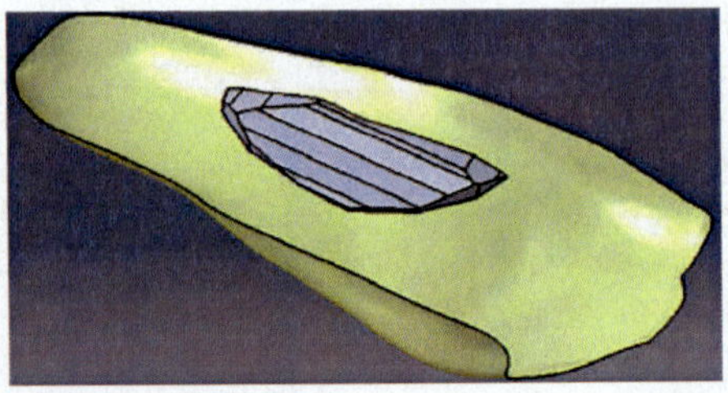

Fig. 8 The 1st element modeling that compose a family of orthosis for the flatfoot correction

On the further our research on prototyping this kind of plantar orthosis components, we have identified the main steps:

- establish which kind of orthosis elements must be prototyped depending by each case of locomotion or stability disease;
- CAD modeling of each one of the established elements in order to define all the necessary elements that will compose the orthosis family, each of which is used to correct a specific progressive disease;
- the rapid prototyping of each of the CAD modeled components; for this purpose each CAD model must be exported as .STL format, which is related software compatible with the existing prototyping machine in our department.

The prototyped components will be slipped insides the shoes and for the foot comfort an insole made of a shape memory material can be used.

IV. Conclusions

Due to the fact that CAD modeling plays an essential role as a step in the process of prototyping of components that compose a plantar orthosis, we find it expedient that in the face of our research to focus primarily on improving the working procedure with the CAD environments.

As a result of our researches on scanning and CAD processing, it could be seen the second CAD method to be more quick and more simple. Thus could be an additional reason for choosing the method of scanning on non-contact principle.

Acknowledgment

This paper is supported by the Sector/Operational Program Human Resources Development (SOP HRD), ID 59323 and ID 59321, financed from the European Social Fund and by the Romanian Government under the project number POSTDRU/89/1.5/S/59323 and POSDRU/88/1.5/S/59321.

References

1. Plantar suporters at http://www.sensicare.com
2. Plantar orthosis at http://www.ortoprofil.ro
3. Wilcox Associates, Inc. (2010) PC-DMIS CMM User guide, for PC-DMIS, 2010
4. Derigent, W., Chapotot, E., Ris, G. (2005) Centre de Recherche en Automatique de Nancy – Cedex, France Controle des piece par capteur laser – definition de fonctions d'aide a la numerisation, CPI, Casablanca, Morrocco, 2005
5. Braun. B, Repanovici, A., Ionescu, M, CAD models obtaining for ECO-TECH and Biomechanics, DAAAM International Vienna - Austria – EU, pp 1281, 1282

Author: Barbu Cristian BRAUN
Institute: Transylvania University, Advanced Mechatronics Systems Department
Street: 2, Teatrului Place, sc. B, ap. 9
City: Brasov
Country: Romania
Email: braun@unitbv.ro

Early Functional Results after Volar Fixed-Angle Plating of Distal Radius Fractures

A. Todor, A. Pojar, C. Arghius, and D. Lucaciu

"Iuliu Hatieganu" University of Medicine and Pharmacy, Orthopedic and Traumatology Clinic, Cluj-Napoca, Romania

Abstract— **Aim of the study was to evaluate the short term results and rehabilitation period after distal radius fractures treated with a volar locking plate and screws. 10 patients with unstable fractures of the distal radius were included in the study. Radiological and functional results were evaluated at 2 and 6 weeks postoperatively which showed no redisplacement of the fracture and faster rehabilitation than with other methods of treatment.**

Keywords— **distal radius fracture, volar plating, functional results.**

I. Introduction

Distal radius fractures are very common injuries and represent about one sixth of all fractures seen in the emergency department [1]. While many of these fractures can be successfully managed nonoperatively, some require surgical stabilization. Debate continues as to the optimal treatment modality of unstable fractures, both intra and extraarticular [2]. Some of the most popular stabilization techniques used to be external fixation [3, 4], pinning [2, 3], dorsal plating or a combination of these [2]. External fixation devices are uncomfortable, need to be kept in place for 6 to 8 weeks and some fracture redisplacement often occurs after the fixation device has been removed [4, 5]. Percutaneous pinning is unsuitable for displaced intraarticular fractures and has less stability in osteoporotic bone [6]. Dorsal plates have the risk of extensor tendons irritation and need to be removed on a regular basis. Volar locking plates are a relatively new concept and are being widely used in the surgical treatment of distal radius fractures [3]. The volar approach is more physiological than dorsal approach and is less disruptive to the tendons because there is more space available on the volar aspect of the radius [7]. Volar approach also maintains dorsal vascularity of the fragments thus allowing early motion of the wrist joint. The plates are well covered by soft tissues in the pronator fossa and implants are not removed routinely. We started using this technique for treatment of distal radius fractures that require surgery and present our early results.

II. Materials and Methods

From August 2010 to January 2011 we included in the study 10 patients with unstable unilateral distal radius fractures. We define an unstable fracture when redisplacement occurs after initial reduction and cast immobilization or if, after closed reduction, displacement still exists and is greater than 15° angulation in any plane, 2 mm of articular step-off, or greater than 2 mm of shortening [9]. . Written informed consent was obtained from each patient prior to enrollment in the study There were 7 women and 3 men with a mean age of 58 years (33 - 71). 7 fractures were extraarticular, Colles type, and 3 had intraarticular involvement. All patients were operated in regional anesthesia (plexus block). We used the extended flexor carpi radialis (FCR) approach as described by Orbay et al. [10]. The skin incision is made directly over the course of the FCR tendon and is about 8 cm long. Over the wrist flexion creases, the incision is made in a zigzag fashion (Fig. 1).

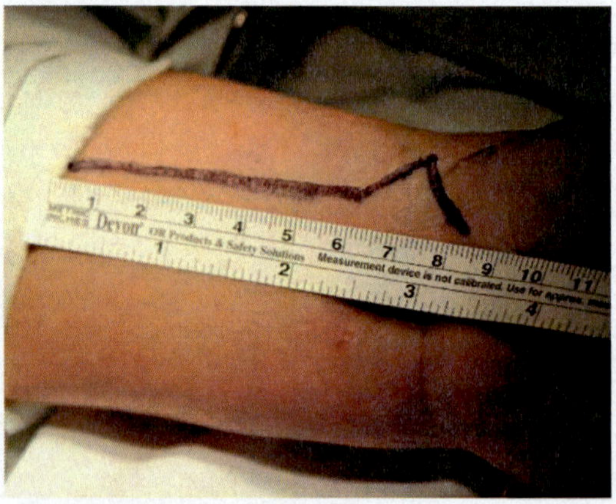

Fig. 1 Skin incision over the FCR tendon and zigzag over the wrist flexion creases

The FCR tendon is retracted medially, protecting the median nerve and the radial artery is retracted laterally. The floor of the tendon sheath is then incised to gain deep access. Distally, the dissection is taken to the level of the scaphoid tuberosity. The virtual space between the flexor

S. Vlad and R.V. Ciupa (Eds.): MEDITECH 2011, IFMBE Proceedings 36, pp. 392–395, 2011.
www.springerlink.com

tendons and the volar surface of the pronator quadratus, the space of Parona, is developed by blunt digital dissection. The pronator quadratus is mobilized with an L-shaped incision along the radial and distal sides and is lifted from its bed by subperiosteal dissection and retracted ulnarly, exposing the fracture site. The extended FCR approach has three steps more than the standard FCR approach. These are represented by the release of the radial septum with opening of the extensor compartment and releasing the insertion of the brachioradialis, pronating the proximal radial fragment out of the way and reduction of the articular fragments by intrafocal manipulation. The release of the brachioradialis is important as this is the primary deforming force of the distal fragment. After fracture site debridement and reduction of the articular fragments to restore normal anatomy the radius is supinated back in place and the plate can be applied. Adequate reduction of the fracture and positioning of the plate and screws is confirmed with fluoroscopy. The screws in the distal fragment must be applied in the subchondral bone. The brachioradialis and the pronator quadratus are sutured back in place covering the plate and separating it from the flexor tendons.

We used two types of fixed angle locking plates specially designed for volar fixation of distal radius fractures (3.5 LCP locking T-plate , Synthes and ChLP System® 4,5 ChM Ltd.).

All patients were immobilized in a below elbow splint for two weeks. Immediate postoperative finger and elbow motion was encouraged. After 14 days patients were called for splint and sutures removal. At that time they were instructed to start progressive wrist motion exercises. Next visit was scheduled at 6 weeks postoperatively when range of motion was assessed and X-rays were taken.

III. Results

No complications were recorded during surgical procedures and early postoperative period. All patients were discharged 2 days after surgery. By 2 weeks finger motion was comparable to preoperative status. At the 6 weeks follow-up wrist range of motion was very good, ranging from 70% to 100% with a mean value of 84% compared to the contralateral side (Fig. 2-4). Comparing the postoperative X-rays with those taken at 6 weeks, there was no loss of reduction in any of the cases (Fig. 5-8).

With regard to pain, 3 patients had none during normal activities, 5 patients reported mild pain at peak motion of flexion and extension and 2 patients complained about moderate pain when also at peak motion of the wrist.

There were no cases of tendon irritation, or other complications like infection, reflex sympathetic dystrophy or hardware failure.

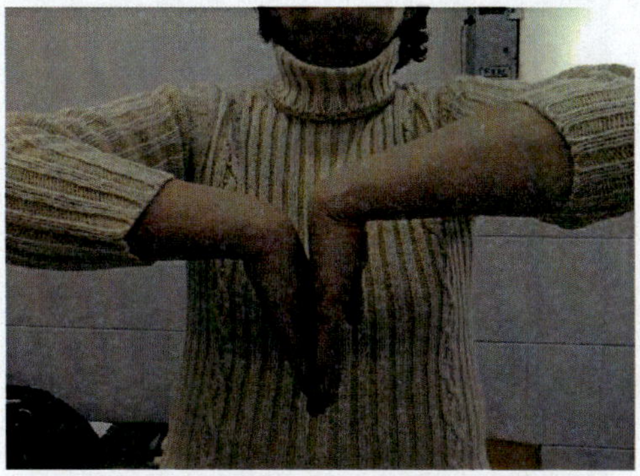

Fig. 2 Range of motion – flexion - 6 weeks after surgery

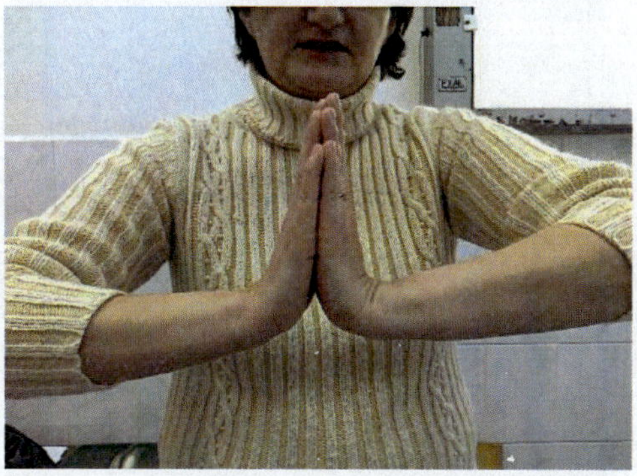

Fig. 3 Range of motion – extension – 6 weeks after surgery

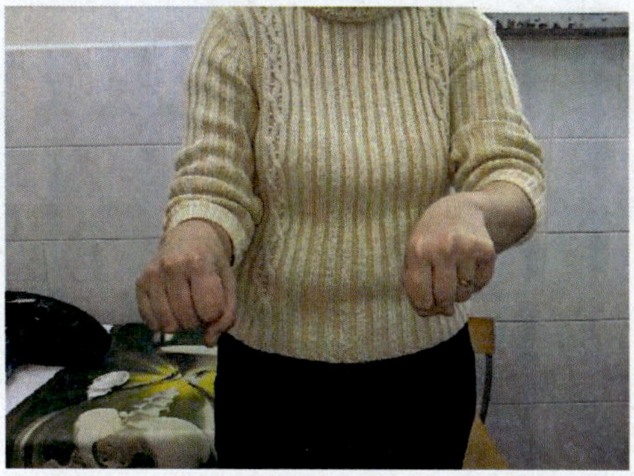

Fig. 4 Range of motion – pronation – 6 weeks after surgery

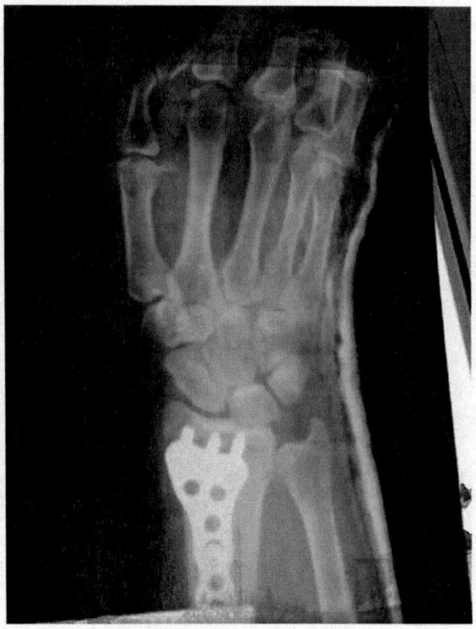

Fig. 5 Postoperative AP view

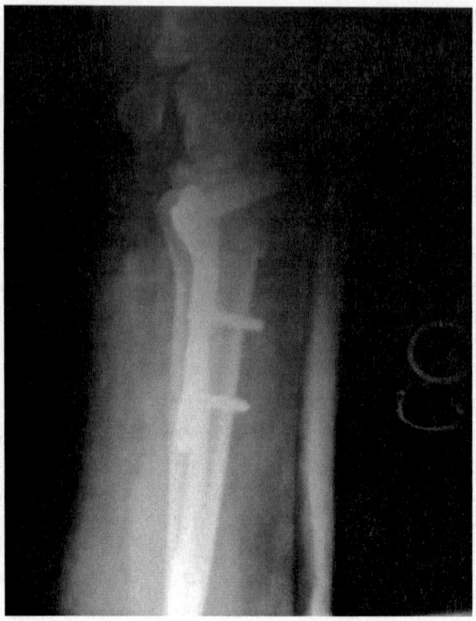

Fig. 6 Postoperative lateral view

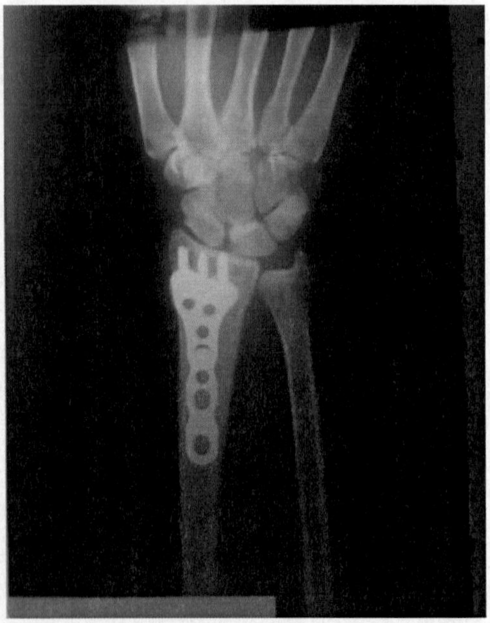

Fig. 7 6 weeks AP view

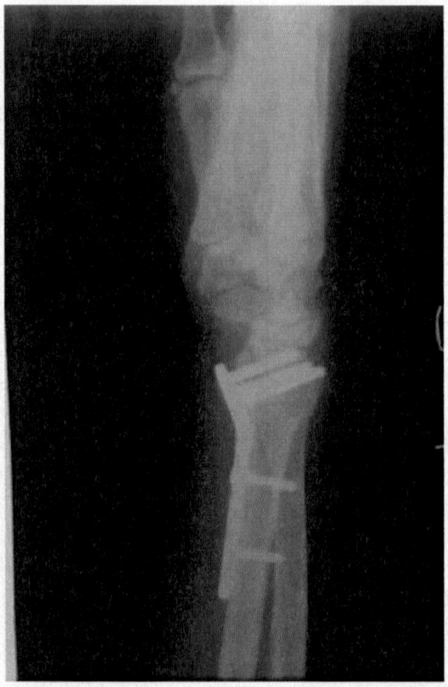

Fig. 8 6 weeks lateral view

IV. Discussions

Treatment of distal radius fractures evolved significantly over the years. Not only the final outcome is important but the time to functional recovery also matters and influences patient satisfaction.

Knox J et al. [11] showed, in a cadaveric model study, that volar plate fixation results in less displacement of intra-articular distal radius fractures with dorsal comminution compared with Kwire fixation. Also, McFadyen I et al. [2] compared volar locked plating and percutaneous pinning for distal radius fractures and found both superior functional and radiological outcomes, 3 and 6 months after surgery in the locked plate group and a significantly less complications than in the pinning group.

Other authors [4] compared functional results after volar plating and after external fixation of distal radius fractures and found that volar locked plating was advantageous in the early rehabilitation period, compared to bridging external fixation.

Other advantages are seen with volar plating in relationship with dorsal plating. Volar plating preserves vascular supply to dorsal metaphyseal fragments and does not cause extensor tendon problems [12]. On the other hand, dorsal approach is particularly beneficial for intrarticular reduction as classical volar approach doesn't permit good intraarticular visualization. This problem was resolved by Orbay JL. et al. [10] who described and popularized the extended volar approach which uses intrafocal reduction and restoration of articular congruency indirectly. Also, locked plates have the advantages of creating a rigid construct that allows the transfer of axial load across the fracture site, as implants possess a mechanical strength similar to normal bone [2]. This is particularly important in osteoporotic bone, many of these fractures involving elderly patients with osteoporosis. Orbay JL et al. [8] showed that open reduction and internal fixation with a volar fixed angle device is effective for the treatment of unstable distal radius fractures in the elderly population.

For all these reasons we started using this technique and we were pleased with the outcome and found the procedure effective. Patients demonstrated early functional use of the hand and were satisfied with the results.

References

1. Browner B, Levine A, Jupiter J, Trafton P. (2003) Skeletal Trauma. Basic Science, Management, and Reconstruction. 3rd. ed. Saunders, USA
2. McFadyen I, Field J, McCann P et al. (2011) Should unstable extra-articular distal radial fractures be treated with fixed-angle volar-locked plates or percutaneous Kirschner wires? A prospective randomised controlled trial. Injury, Int. J. Care Injured 42:162–166
3. Bentley G. (2009) European Instructional Lectures, vol. 9, 10th EFFORT Congress, Springer, Vienna, Austria
4. Wilcke MKT, Abbaszadegan H, Adolphson PY. (2011) Wrist function recovers more rapidly after volar locked plating than after external fixation but the outcomes are similar after 1 year. A randomized study of 63 patients with a dorsally displaced fracture of the distal radius. Acta Orthopaedica 82: 76–81
5. Dicpinigaitis P, Wolinsky P, Hiebert R et al. (2004) Can external fixation maintain reduction after distal radius fractures? J Trauma, 57 4: 845-850
6. Blakeney WG. (2010) Stabilization and treatment of Colles' fractures in elderly patients. Clin Interv Aging. 18:5:337-344
7. Orbay JL, Touhami A. (2006) Current Concepts in Volar Fixed-angle Fixation of Unstable Distal Radius Fractures. Clin Orthop Relat Res 445:58-67
8. Orbay JL, Fernandez DL. (2004) Volar Fixed-Angle Plate Fixation for Unstable Distal Radius Fractures in the Elderly Patient. J Hand Surg 1:96-101
9. Orbay JL, Badia A, Khoury RK, Gonzalez E, Indriago I. (2004) Volar Fixed-Angle Fixation of Distal Radius Fractures: The DVR Plate. Tech Hand Upper Extrem Surg 8(3):142–148
10. Orbay JL, Badia A, Indriago IR et al. (2001) The extended flexor carpi radialis approach: a new perspective for the distal radius fracture. Tech Hand Up Extrem Surg
11. Knox J, Ambrose H, McCallister W, Trumble T. (2007) Percutaneous Pins Versus Volar Plates for Unstable Distal Radius Fractures: A Biomechanic Study Using a Cadaver Model. J Hand Surg 6:813-817
12. Osada D, Kamei S, Masuzaki K, Takai M, Kameda M, Tamai K. (2008) Prospective Study of Distal Radius Fractures Treated With a Volar Locking Plate System. J Hand Surg 33A:691-700Smith J, Jones M Jr, Houghton L et al. (1999) Future of health insurance. N Engl J Med 965:325–329

Author: Adrian Todor
Institute: "Iuliu Haţieganu" University of Medicine and Pharmacy, Orthopedics Clinic
Street: 47 Traian Mosoiu
City: Cluj-Napoca
Country: Romania
Email: adi.todor@yahoo.com

Author Index

Keyword Index